S0-ATH-866

AIR MONITORING FOR TOXIC EXPOSURES

An Integrated Approach

SHIRLEY A. NESS

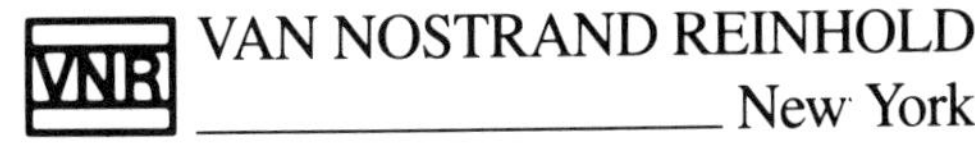
VAN NOSTRAND REINHOLD
New York

Library of Congress Catalog Card Number 90-38159
ISBN 0-442-20639-9

Printed in the United States of America.

Van Nostrand Reinhold
115 Fifth Avenue
New York, New York 10003

Chapman and Hall
2-6 Boundary Row
London, SE1 8HN, England

Thomas Nelson Australia
102 Dodds Street
South Melbourne 3205
Victoria, Australia

Nelson Canada
1120 Birchmount Road
Scarborough, Ontario M1K 5G4, Canada

16 15 14 13 12 11 10 9 8 7 6 5 4 3 2 1

Library of Congress Cataloging-in-Publication Data

Ness, Shirley A.
Air monitoring for toxic exposures/Shirley A. Ness.
p. cm.
Includes index.
ISBN 0-442-20639-9
1. Air—Pollution—Measurement. 2. Threshold limit values (Industrial toxicology) I. Title.
TD890.N47 1991
628.5'3'0287—dc20 90-38159
CIP

Contents

PART IV: SAMPLING WITH DETECTOR TUBES AND MONITORING FOR AGENTS OTHER THAN CHEMICALS

PART V: AIR SAMPLING DECISIONS: GATHERING BACKGROUND INFORMATION

PART VI: GENERAL INFORMATION FOR THE SAMPLER

APPENDIXES

Preface

Sampling to identify and quantify chemical, biological, and physical agents requires specialized techniques and equipment. Sampling can be done to try to quantify the amount of a contaminant present in the air (especially in a worker's breathing zone) or on a surface, or to characterize what might be present in a bulk sample. Urine, blood, or breath samples can also be collected to determine how much of a chemical or other agent an individual actually absorbed. However, the most appropriate characterization of exposure may involve all these areas of sampling.

As concerns increase about environmental contamination and hazards posed to workers, so does the interest in the use of sampling for health hazards. Included are sampling on hazardous waste sites; evaluation of properties for real estate transfer; sampling for asbestos, radon, and bioaerosols; indoor sampling for tight building syndrome cases; and checking worker exposures in industrial jobs. Reliable measurements of airborne contaminants are useful for selecting personal protective equipment, determining whether and which engineering controls can achieve permissible exposure limits, delineating areas where protection is needed, assessing the potential health effects of exposure, and determining the need for specific medical monitoring. The increasing specialization of sampling professionals often leaves them groping when it comes to sampling in new situations. Consequently there is a need for a practical guide on air sampling techniques and equipment that reflects current applications and interests.

This book provides guidance on evaluating potentially harmful exposures to people from hazardous materials, including chemicals, radon, and bioaerosols. It provides practical information on how to perform air sampling, how to collect biological and bulk samples, and how to evaluate dermal exposures, as well as how to make judgments as to the advantages and limitations of a given method. Integrated methods are the best way to monitor exposures to a specific chemical, and a large number of methods are available. Real time or direct reading techniques are increasingly important, since they are the best way to examine the exposure profile and identify sources. Dermal exposure monitoring has predominantly been done for pesticides, but it is a route of entry for many more classes of chemicals and it is not accounted for by air sampling. Biological monitoring that provides the best measurement of the actual dose derived from respiratory, mouth, or skin exposure is the most difficult to validate due to metabolism, timing of collection, invasive methods, and sample stability. The emphasis of sampling should not be on air sampling alone; all types of sampling—air, wipe, bulk, and biological—are needed in today's complex evaluations of hazardous environments. Therefore, the most sensible solution is some combination of these approaches depending on the situation.

This book emphasizes using an integrated approach to air sampling for chemical agents; however, chapters on radon and bioaerosol sampling are included as well, due to their increasing importance in indoor air situations. An example of the integrated approach is a survey including wipe samples to identify surface contamination, air samples to estimate airborne exposures, and urine samples to estimate biological dose. Health physicists should be consulted whenever attempting to do any sophisticated radiation sampling beyond radon, which is discussed in this book.

The equipment and techniques discussed are those representing both new technologies and older ones that are commonly used. In general, the trend in equipment is toward ease of use, even though instruments are becoming more sophisticated. Equipment is becoming smaller as microprocessors are being installed. Computer applications are being developed to provide interfaces with direct reading instruments. For example, data loggers allowing minute-by-minute sampling data to be stored for an 8-hour day are replacing the use of the strip chart as well as expanding the use of many instruments.

Thus, this book contains expanded sections on real time/direct reading methods along with discussions of data logging and computer integration of data. The sampling professional today must be aware of as many monitoring options as possible. Since the overlap between many environmental monitoring and personal sampling situations cannot be ignored, both OSHA (Occupational Safety and Health Administration) and EPA (Environmental Protection Agency) sampling methods are discussed where they apply to exposures. Finally, bulk sampling of chemicals, wastes, and water is discussed in detail, since it can often be a simple and direct method of identifying the hazard but is often ignored in favor of expensive, extensive air monitoring.

The increasing number of applications for air sampling is creating new directions for classical techniques. Until recently, if sampling was required indoors, it was performed by industrial hygienists, and if sampling was required outdoors to document air conditions, it was done by an air pollution specialist. Increasingly, the need for both indoor and outdoor sampling is being felt for both professions, and the distinctions are becoming blurred. The formerly strict dividing line between environmental sampling and occupational sampling is becoming a gray area in certain situations.

As the importance of air monitoring becomes more recognized, many classical techniques are being adapted for it. There are risks when taking methods and instruments developed for one application and using them for another. A key purpose of this book is to minimize these risks by providing an understanding of the original intent, limits, and proper applications of various sampling methods.

There are a number of excellent references on the theories of sampling. The emphasis of this book is on application and use of sampling methods and equipment. Therefore, it is designed to supplement these references and provide practical information in applying sampling techniques in the field. The information on these areas is deliberately kept to a minimum so the field applications can be emphasized. The reader is advised to refer to these references for information on the variety of equipment available under any one category including manufacturers (ACGIH, Annual Equipment Survey from *IH News*) and for theory (ACGIH, *Patty*, NIOSH). Information on additional resources to guide the reader who is interested in specific advanced or in-depth aspects of various areas of sampling and analyses, for example, sampling statistics or setting up a laboratory control program, is at the ends of the chapters.

In summary, this book is organized with the following in mind. First, one needs to determine what types of sampling must be done. What chemicals will be sampled? One usually attempts to identify the exposure first. Then a selection of the sampling methods, equipment, and materials is made. Generally this selection is done to keep sampling options as broad as possible. Although sometimes it is known exactly what types of chemicals must be sampled, in general, even in the most classic type of industrial operation, these chemicals are going to change either as the process changes or as new chemicals are introduced as replacements for older ones. Contaminants previously thought harmless may become a concern and sampling will be necessary. The next step is to plan sampling by selecting methods, calibrating equipment, and making sure that there is adequate quality control so that results are not negated by poor technique or other problems. Finally, evaluation and interpretation of the sampling data are performed to determine how the data can best be applied to assist in minimizing any hazards that might be present.

ACKNOWLEDGMENTS

I wish to thank my publisher, Van Nostrand Reinhold, for patience and assistance; Charles R. P. Keating for his illustrations and editing suggestions; Hank McDermott, CIH, for his advice and encouragement; the manufacturers who freely provided photographs and other artwork; the many industrial hygienists I have known and worked with over the years for their contributions to their profession; and Apple Computer for developing the Macintosh. Most important, I wish to thank Josh Joslin, Bertha Ness, and Vernon Ness for everything.

PART I

Types of Hazards and Sampling Methods

CHAPTER 1

Hazards

Historically the practice of air sampling has been limited to industrial hygiene and air pollution measurements except for a few other very specific applications, such as radiation. A myriad of different instruments, techniques, and methods, operating by means of a variety of chemical and physical principles, are now available to perform proper data collection and monitoring. Although the emphasis of air sampling has always been and continues to be for chemicals due to the frequency of their presence, an increased interest in monitoring for other types of agents is occurring. Radon, a gas producing ionizing radiation, is an example, as are bioaerosols such as bacteria, which are a result of indoor air problems in large office buildings.

Many changes are occurring in the field of environmental and occupational sampling to identify and quantify exposures to agents. Sampling can measure the amount of a contaminant present in the air, especially in a worker's breathing zone, or on a surface. Sampling can characterize what might be present in a bulk sample. Urine, blood, or breath samples might also be collected in order to determine how much of a chemical or other agent an individual actually absorbed. However, the most appropriate characterization of exposure may involve integrating some or all of these areas of sampling, as well as adapting the sampling strategy to reflect the toxic effects of a material. Reliable measurements of airborne contaminants are useful for selecting personal protective equipment, identifying sources of contaminants, determining whether and which engineering controls can reduce exposures, assessing the potential health effects of exposure, and determining the need for specific medical monitoring.

In asbestos, work bulk samples are collected to identify the material and tape may be used to identify surface contamination. Air sampling is conducted before, during, and following abatement work, as well as in conjunction with operations and maintenance (O&M) programs in facilities with asbestos containing materials (ACM) that must remain in place. To accomplish this sampling, large, stationary, high-volume pumps collect area samples with large volumes while smaller, battery-powered pumps are worn by workers for the collection of personal samples.

Hazardous waste work offers a new set of challenges, since frequently the types of agents that may be present on a site are unknown and identifying the exposures usually requires the use of real time instruments (see Part III).

In the indoor air or tight building syndrome, a phenomenon often related to the recirculation of air in many office buildings and schools, inhabitants are often bothered by concentrations of contaminants far below what is commonly encountered in industry. Therefore, special sampling strategies and techniques are required.

Although sampling methods traditionally have focused on air sampling because of the availability of methodologies and emphasis on compliance standards, in order to appropriately assess the degree of exposure an individual has absorbed through all potential routes of exposure including inhalation, oral, and dermal routes, the assessment may require air sampling. Integrated measurements analyzed by a laboratory and real time or direct reading measurements, as well as wipe, bulk, and biological samples may be necessary. Real time (immediate) sensing best describes the immediate environment; however, for many substances an integrated sampling method that requires laboratory analysis may be the only sampling technique available.

Sampling strategies should be organized with the following in mind. First, the sampler must determine what types of exposure hazards are present. Then the sources and degree of exposure to each significant hazard are assessed. Often the sampler knows what chemicals are present, although these may change either as the process changes or as new chemicals are introduced as replacements for older ones. Contaminants previously thought to be harmless may become a concern and sampling will be necessary. Individuals in a building may attribute their illness to an unknown source. The next step is to select methods and equipment, perform calibrations, and assure that adequate quality control exists so that the data are not negated by poor technique or other problems. Last, the sampler evaluates and interprets the sampling data and determines how they can best be applied to assist in minimizing any hazards that might be present.

TYPES

There are two basic types of chemical hazards: health and physical. Health hazards cause diseases and physical hazards refer to the potential for some compounds to be flammable, explosive, and reactive. Effects on the skin are considered health hazards. Flammable gases and vapors have a minimum concentration in air that is necessary for ignition, known as the lower explosive limit (LEL). There is also a maximum concentration of these materials in air above which flames cannot occur, called the upper explosive limit (UEL). Hydrogen, methane, gasoline, and benzene fall into this category. In addition to presenting a health hazard, many dusts can be explosive if they are exposed to a source of heat or spark while sufficient concentrations are present in air. Examples are the famous grain dust explosions that have occurred in silos. Explosive gases and vapors of the greatest concern have flash points below 100°F (37.8°C).

Chemicals are the most common exposures encountered in both the occupational and the community (ambient) environment and consist of various atoms. They may belong to either organic or inorganic categories depending on whether they contain the requisite tetrabonded carbon or not.

Microbials are unique because they are living organisms and thus can multiply and grow. Many have more than one growth stage. In the spore stage the organism is encapsulated, somewhat indifferent to temperature and essentially a particle, whereas in the growth stage it can assume many different appearances but generally must be collected on some type of growth medium. Specific information on sampling for these types of agents is included in the chapter on Sampling for Bioaerosols. Radon, a physical agent, is a gas whose alpha radiation can stick to particles. Although in the strictest terms it is considered a type of ionizing radiation, it must also sometimes be considered as a particulate. Radon is discussed further in the chapter on Radon Measurements.

Organic dusts are derived from vegetable, soil, and animal sources. Examples include grain dusts (oats, wheat, barley), straw, hay, wood dusts (a wide variety of compounds with many diverse effects ranging from asthma to suspect carcinogens), cotton, sewage sludge, and animal dander. In addition to the dust particle itself, organic dusts can contain microorganisms and their toxic products (toxins), enzymes, and other agents capable of eliciting an immunological response. Inorganic dusts include mineral dusts such as silica, asbestos, coal dust, and many less toxic materials such as marble, gypsum, plaster of paris, and emery. Not all chemical hazards are readily identified.

The sampler must not overlook the potential for many products to generate chemicals during use or after storage or standing, even though they in themselves may not appear to present a chemical hazard. For example, many building construction materials, such as fabrics, wall partitions, and particleboard, can offgass formaldehyde. Paper when cut can release paper dust. Carbonless paper may also release formaldehyde from its coating. Oxyacetylene cutting or welding on painted metals or plastics may produce by-products far more toxic than the initial ingredients.

CONTAMINANTS

Physical States

Contaminants are classified in terms of their physical state, which can be a gas, vapor, or type of aerosol. *Gases* are completely airborne at room temperature and can be liquified only by the combined effect of increasing temperature and decreasing pressure such as occurs when creating liquified nitrogen or dry ice from carbon dioxide. A substance is considered a gas if at standard conditions of temperature and pressure (25°C, 1 atm) its normal physical state is gaseous. Examples are hydrogen, helium, oxygen, formaldehyde, ethylene oxide, carbon monoxide, argon, and nitrogen oxides.

A *vapor* is the gaseous state of a substance in equilibrium with the liquid or solid state of the substance at the given environmental conditions.[1] This equilibrium results from the vapor pressure of the substance, which if high enough, causes it to volatilize (evaporate) from a liquid or sublime (evaporate directly from a solid) into the atmosphere. Vapors become airborne as the result of the evaporation of substances that are liquids at room temperature, such as styrene or acetone. The primary reason for differentiating between the two is to be able to assess the potential likelihood of a chemical being airborne in high concentrations. For sampling it is important to note that gases and vapors stay airborne for long periods.

An *aerosol* is a system consisting of airborne solid or liquid particles that is dispersed in a gas stream, usually the atmosphere. Aerosols are generated by fire, erosion, sublimation, condensation and abrasion of minerals, metallurgical materials, organic and other inorganic substances in construction, manufacturing, mining, agriculture, and transportation among other operations. Aerosol classifications depend on the physical nature, particle size, and method of generation. Dusts, smoke, soot, particles, mist, fumes, and fog are all terms used to describe certain types of aerosols. Solid substances are defined as particles to distinguish them from liquid droplets. When there is no need to differentiate between the particle and droplet components of an aerosol, the collective term *particulate* is used.[2] The ability of aerosols to get into the body as well as the rate at which they are absorbed depends on the particle size distribution and solubility characteristics of the aerosol. In general, smaller particles tend to be deposited in the lower regions of the lungs and many occupational diseases are associated with this area and other regions of the respiratory tract.

Types of Aerosols

Dusts are generated from solid materials by mechanical means, such as grinding, crushing, pulverizing, chipping, or other abrasive actions occurring in natural and commercial operations. Dusts can be derived from inorganic minerals, such as asbestos, silica, and limestone, and various metals or organic sources, such as wood dust, flour, and grain dust. Dust can also be derived from animal dander, insects, mites, fungal spores, and pollen. Microbials can often be found mixed with dusts, especially those that are organic-based. Dusts can be quite heterogeneous in composition and often consist of larger particles, although they range from 1 mm to 150 mm in diameter. Most often dusts are somewhat spherical in shape.

Fibers, particles whose length exceeds their width, can be generated from minerals, such as asbestos, and humanmade sources, including fiberglass, if the composition of the material lends itself to disintegration producing such particles. For the purposes of classification some fibers are assigned a minimum size criterion, for example, asbestos particles must be at least three times longer than they are wide to be considered a fiber for occupational sampling purposes. Fibers are thought to behave differently when in the lung than particles in the shape of a sphere.[3] Organic sources, such as hemp and animal fibers, also exist.

Mists are suspended liquid droplets. They are generated by condensation from the gaseous to the liquid state or by mechanically breaking up a liquid into a dispersed state by spraying, splashing, foaming, or atomizing. Oil mists produced from metal-working fluids during parts machining and mists above electroplating tanks are examples. Some mists can have a vapor component as well, such as paint spray mists, which contain volatile solvents. When a mist's droplets evaporate or vapor condensation occurs, the aerosol will contain higher concentrations of very small particles. *Fog* is a term used for a mist that has a particle concentration high enough to obscure visibility.

Fumes are produced by such processes as combustion, distillation, calcination, condensation, sublimation, and chemical reactions. Fumes are solids that are the result of condensation of solids from an evaporated state. When a metal or plastic evaporates, the atoms disperse singly into the air and form a uniform gaseous mixture. In the air they combine rapidly with oxygen and recondense, forming a very fine particulate ranging in size from 1.0 mm to 0.0001 mm. Many fumes have a high chemical reactivity that is thought to explain the high biological activity exhibited by some metal fumes.[4] Examples are welding fumes, ammonium chloride, hot asphalt, and volatilized polynuclear aromatic hydrocarbons from coking operations. The term *fume* is often confusing. When used to refer to exhaust fumes or paint fumes, it is inappropriate, since gases and vapors and airborne mists are not classified as fumes.

Smokes are produced by the incomplete combustion of carbonaceous materials, such as coal, oil, tobacco, and wood. Smoke particles generally range from 0.3 mm to 0.5 mm in diameter. Because of their small size, these particles can remain airborne for long periods of time. Situations where smoke is produced also tend to produce gases.

For the purposes of sampling, several facts are important about aerosols. Particle size is often dependent on the means in which the aerosol was generated. When airborne, particles collide and can stick together. If the purpose of sampling is simply to collect the total (mass of) dust, this factor will not influence the results to a significant degree, since it is likely that these larger particles will still be collected. If size selective sampling is done, the results may not reflect the actual airborne dust composition, because it will appear as if there are fewer particles and they are larger in size. This consideration can be especially important when sampling near combustion sources, because thermal effects can promote coagulation of smaller particles, leading to changes in the particle number and composition of the aerosol. As a result of gravity and diffusion, particles will settle onto surfaces, thus decreasing airborne concentrations.

Chemical Properties

Vapor pressure is a measure of the ability of a compound to become airborne. The higher the vapor pressure (larger the number), the higher the airborne concentration possibility. Vapor pressures should be referenced to the temperature at which they were measured. Compounds that are a gas at room temperature, such as ammonia, will have vapor pressures greater than 760 mm Hg (1 atm). An example of a liquid with a high vapor pressure is acetone: 266 mm Hg at 25°C. For strict comparison of compounds, vapor pressures at the same temperature and units of pressure must be used.

The degree of volatility can affect the sampling method, since it often dictates the physical state of the compound as it is found in the air. Volatile compounds have vapor pressures greater than 1 mm Hg at ambient temperature and exist entirely in the vapor phase. Semivolatile compounds have vapor pressures of 10^{-7} mm Hg to

1 mm Hg and can be present in both the vapor and particle state. Nonvolatile compounds have vapor pressures of less than 10^{-7} mm Hg and are found exclusively in the particle-bound state.[5] Some of the classes of semivolatile organic compounds of particular interest to sampling professionals include polynuclear aromatics (PNAs), organochlorine pesticides, organophosphate pesticides, polychlorinated biphenyls, and chlorinated dibenzodioxins and furans. The challenges of sampling for these compounds are discussed further in the chapter on Special-Case Sampling.

Occasionally a chemical will be described as being "heavier than air," suggesting it will sink to the ground and stay low. Generally in these cases the property of vapor density, defined as the weight of a gas or vapor compared to the weight of an equal volume of air, is the point of reference. Low ambient temperature increases vapor density as does high humidity.[6] Vapor density can be important in situations where high concentrations are present due to an emergency release, buildup to combustible concentrations, or chemical release in a confined space (sludge pits, tanks, and even ditches) where there is a lack of air movement. In reality, when comparing molecular weights, most chemicals are heavier than diatomic nitrogen, the primary component of air with a molecular weight of 28 g/mole, and the most important criterion is the ratio of contaminant to available air currents and air volume for mixing. For example, the molecular weight of methyl isocyanate, the compound released in 1984 in Bhopal, India, is 57 g/mole, twice that of "air," and this compound traveled a significant distance outdoors before dispersing. Table 1-1 shows vapor densities for selected compounds.

In cases where chronic health hazard concentrations are present (generally quite low when compared to the situations discussed above), most of these compounds are going to mix with air and disperse readily throughout an area. A simple calculation can be done to demonstrate this property using 1,1,1-trichloroethane, a common degreasing solvent.[7] Assuming there is ample air for dilution, the following calculations indicate the effective vapor density of a 1,000 ppm 1,1,1-trichloroethane—air mixture:

TABLE 1-1. Vapor Densities of Selected Compounds

Compound	*Vapor Density (air = 1)*
Ammonia	0.59
Carbon dioxide	1.52
Carbon monoxide	0.97
Chlorine	2.45
Gasoline	3.93
Hydrogen	0.07
Hydrogen sulfide	1.17
Methane	0.55
Methyl isocyanate	1.97
Propane	1.52
1,1,1-trichloroethane	4.60
Nitrogen	0.97

vapor density of air = 1

vapor density of 1,1,1-trichloroethane = 4.6
1,000 ppm = 0.1% = 1 part 1,1,1-trichloroethane to 999 parts of air

$0.001\,(4.6) = 0.0046$

$0.999\ \ (1.0) = 0.999$

$0.999 + 0.0046 = 1.0036$, or approximately 1.004, which is the effective vapor density of the mixture

Therefore, the 1,1,1-trichloroethane—air mixture compared to clean air would have a ratio of 1003:1000 and not 4.6:1. Since 1,000 ppm is almost three times the safe occupational exposure limit for 1,1,1-trichloroethane, a mixture normally encountered in these situations would contain much less. Thus, the effects of window ventilation, cross currents, wind, traffic, and heat are often more important than molecular weight and vapor density.

When comparing the densities of liquids, the term *specific gravity* is used. The specific gravity is the relative weight per unit volume of a liquid compared with the weight per unit volume of pure water. Water has an arbitrarily assigned value of 1.000 g/cm^3 and all liquids heavier than water will have larger specific gravities.

For gases and vapors *solubility* in water is the principal characteristic affecting penetration to the deeper areas of the respiratory tract. The more water soluble, the greater likelihood of dissolving in the nasal or oral airways while less water-soluble gases can penetrate the smaller airways and alveoli. Gases that are soluble in water are often upper respiratory irritants, such as hydrogen chloride. The basis for solubility of particles in the lungs of humans may be very different from that in water or other solvents or other species. The lung fluids contain lipids and proteins that modify the solubility in ways that are difficult to evaluate or predict.[8] On the other hand, when referring to the ability of a contaminant to penetrate the skin, the primary concern is solubility in fats because it often influences the ability of a material to penetrate the skin.

Compounds that dissolve in the mucous of the respiratory system and do not produce any immediate local irritation at the site of deposition can reach the bloodstream. Insoluble compounds will deposit in the respiratory system or will be carried on the mucous and cilia in the esophagus to the mouth and will be swallowed. The solubility of most compounds in the gastrointestinal tract is far less than that in the lung.

In occupational exposure standards dusts are classified as either soluble or insoluble according to their solubility in water. Examples of insoluble dusts are minerals, metal oxides, silica, granite, and asbestos. Soluble dusts include some minerals, such as limestone and dolomite, and organic dusts, such as trinitrotoluene, flour, soap, leather, cork, wood, and plastics. One definition that has been used is 5% solubility in water; however this definition may change in the future.[9] Table 1-2 lists soluble and insoluble chemical forms.

A characteristic related to water solubility is hygroscopicity, the ability of a compound to absorb water from the air. Hygroscopic particles increase in diameter as they pass through the high humidity of the respiratory tract. A 1-μm particle may grow by water absorption to 3 μm. When aero-

TABLE 1-2. Examples of Soluble and Insoluble Chemical Forms

Compound	*Soluble Forms*	*Insoluble Forms*
Molybdenum	Ammonium molybdate Sodium molybdate	Calcium molybdate Molybdenum trioxide Molybdenum halides Molybdenum disulfide
Nickel (inorganic salts)	Nickel chloride Nickel nitrate Nickel sulfate	
Platinum	Ammonium chloroplatinate Sodium chloroplatinate Platinic chloride Platinum chloride Sodium tetrachloroplatinate Potassium tetrachloroplatinate Ammonium tetrachloroplatinate Sodium hexachloroplatinate Potassium hexachloroplatinate Ammonium hexachloroplatinate	
Rhodium	Rhodium nitrate Rhodium potassium sulfate Rhodium sulfate Rhodium sulfite	
Silver		
Thallium		

sols of ammonium sulfate or sulfuric acid are inhaled, the individual particles grow and as a result more impact on the upper airways. Generally the effect of hygroscopicity on deposition will be less for particles composed of materials having greater densities and high molecular weights.[3]

Complex Compounds

Some solids, such as naphthalene, found in moth balls, have a vapor pressure high enough for some of this material to exist in an airborne (vapor) state. Therefore, when sampling these types of compounds the collection method must account for the aerosol as well as the vapor component. Compounds that have been found to exist as both an aerosol and a particulate include caprolactam, PCBs, fenamiphos, methomyl, methyl demeton, acrylamide, and diazinon. Ventilation or increases in material use can impact the relative distribution between the amounts of vapor and particle present.

Mixtures, such as paint and pesticide formulations, frequently have both vapor and particulate components. However, in this case the sampler is usually sampling for specific components either the pigment or solvent in the case of the paint and in the case of the pesticide, unless the carrier is a solvent. Usually the pesticide is the primary concern. It has also been demonstrated that the method of use can influence the composition of the material that is airborne. In some operations where spraying of resins is done, the aerosol represents a significant percent of the air concentration. In these situations, particle size is important as well as the potential for changes, such as vaporization and polymerization, which will alter the relative percent of aerosol vapor present. For example, styrene when sprayed has been shown to generate both an aerosol and a vapor with as much as 32 – 33% of the total styrene concentration due to the aerosol component.[10]

Sampling reactive compounds requires special techniques, because through reactions such as polymerization, hydrolysis, and oxidation the sample can be changed or lost. This situation can also occur during sampling, leading to misinterpretation about what types of compounds and concentrations may be present. One group of reactive compounds is the diisocyanates, which react with water to polymerize and release carbon dioxide.

WARNING SIGNS

Many chemicals reveal their presence due to certain physical properties, including, odor, color, or ability to cause irritation. A knowledge of these properties can provide valuable clues to levels of concentration and identification of locations to sample.

Chemicals are considered to have adequate warning properties if the odor threshold is at or below the occupational exposure limit. Exposure limits are discussed in detail in a later section on standards in this chapter. Increases in concentration do not necessarily cause the same incremental increase in odor, and the intensity varies from one substance to another. Some odors, termed *characteristic* because they are very distinctive and tend to be associated with one particular compound, are used to describe other compounds that have a similar smell. Some chemicals, such as hydrogen sulfide, can also cause olfactory fatigue where the senses become "dulled," leading to the inability to detect the material even though it is still present.

Odor thresholds can be defined in more than one way. For example, in one case it might be detection: "I smell something." In another case it may involve recognition: "This smells like acetone." Therefore, comparing odor thresholds can be difficult as the criteria for the test can vary. Some tests determine the odor threshold of a compound mixed in water. The drawback of this data is that they are only valid when the concern is an odor associated with water and not air. There are individual differences in levels of odor perception, including differences in the level of odor detected when panelists come from a "clean air" background versus if they have spent time in an industrial atmosphere, as well as a decrease in the ability to detect odors with increasing age or as the result of allergies. Therefore, most reliable data are given as a range of values. Not surprisingly, compounds with the lowest odor thresholds have the most variability in ranges of detection, whereas others with the highest thresholds have the least.

Several methods have been used to determine odor thresholds. Usually testing procedures involve panels of volunteers. A mask and tubing of chemically inert odorless material are used to expose each subject to the compound. First, a concentration known to be above the odor threshold is used. Once familiar with the odor they are attempting to detect, a zero concentration sample is presented. The concentration in the mask is then raised slowly over a period of several minutes with the subject sniffing the mask at 15-second intervals until it is detected. A concern regarding the estimation of odor thresholds is that laboratory experiments that gauge the capacity to detect and recognize small amounts of the warning agents may not accurately predict detection and recognition in the field.[11] Some test subjects repeatedly challenged by a compound grow more sensitive to the material, whereas others develop a tolerance or have olfactory fatigue.

TABLE 1-3. Examples of Irritation and Odor Thresholds

Compound	*Irritation Threshold (ppm)*	*Odor Threshold (ppm)*	*Odor Description*
Acetic acid	10–15	0.2–24	Vinegar, characteristic
Acetone		100	Nail polish remover, characteristic
Ammonia	55–140	0.32–55	Characteristic
Benzene		4.68	Aromatic
Butyl acetate	300	0.037–20	Fruity
Carbon disulfide		0.21	Characteristic
Chlorine	1–6	0.01–5	Bleach
Chloroform	>4096	50–307	Characteristic
p-Dichlorobenzene	80–160	15–30	Mothballs
Dimethyl amine		0.047	Fishy, ammonia
Epichlorohydrin	100	10–16	Chloroform
Ethyl ether	200	0.33	Characteristic
Formaldehyde	0.25–2	0.1–1.0	Characteristic
Hydrogen sulfide	50–100	0.00001–0.8	Rotten egg, characteristic
Isoamyl acetate	100	0.002–7	Banana oil, characteristic
Isopropyl alcohol	400	7.5–200	Characteristic
Isopropyl amine	10–20	0.71–10	Ammonia
Isopropyl ether	800	0.053–300	Ether-like
Methyl alcohol	7,500–69,000	53.3–5,900	Characteristic
Methyl methacrylate	170–250	0.05–0.34	Characteristic
Naphtha, coal tar	200–300	4.68–100	Aromatic
Phosgene		1	Hay-like
Phosphine		0.021	Oniony, mustardish, fishy
Stoddard solvent	400	1–30	Kerosene
Styrene (uninhibited)	200–400	0.047–200	Characteristic
Sulfur dioxide	6–20	0.47–5	Characteristic
Toluene	300–400	0.17–40	Airplane glue
1,1,1-Trichloroethane	500–1,000	20–400	Chloroform
Turpentine	200	200	Characteristic
Xylene	200	0.05–200	Aromatic

Odor thresholds can be useful for determining when to change respirator cartridges (providing the compounds are suitable) and are the basis for the banana oil fit test. They can also assist in investigating potential engineering control failures. Strange odors are often the basis for indoor air concerns as well as community odor problems, which have a significant impact on real estate values. One area where odor rather than testing has been strongly relied on is in locating liquid propane (LP) and natural gas. Due to its presence in homes and other buildings, odorants such as ethyl mercaptan and thiophane are added to the gas to enhance the likelihood of leak detection. However, unless the sampler knows exactly what compound is in the air and understands its characteristics, it is not a good idea to use odors to identify the presence or attempt to estimate the quantity of a contaminant. In particular, the sampler should never sniff unknown samples of water, soil, or other materials in order to determine if chemicals are present. Prolonged use of this practice could lead to very serious health effects.

Irritation responses are also measured using panels of volunteers similar to the procedures used for odor thresholds. Subjective measurements of irritation can be affected by a variety of psychological and physiological factors, including airflow over the eyes, the presence of dust particles, the amount of sleep the previous night, and anticipation. The time for a response to occur is also important and in general increased concentrations are detected sooner.[12] Table 1-3 lists irritation and odor thresholds.[13,14,15]

Certain chemicals, such as metallic fumes and mercaptans, can cause an odd taste in the mouth upon exposure. And not to be forgotten is the most obvious situation, in which materials are easily identified due to their physical presence. Some gases are colored when present in high concentrations. For example, nitrogen oxides are red-brown and chlorine is yellow-green. A puddle on the floor, a dark stain on the ground, or piles of dusts on rafters or window sills are obvious situations where chemicals may be present. Concentrated vapors of chemicals, such as hydrogen chloride, hydrogen fluoride, and ammonia, form visible clouds in areas where high humidity exists.

TOXIC EXPOSURES

Exposures are characterized by their time period and the concentration of the contaminant present. Acute exposures are the result of exposures to high concentrations over periods of less than 24 hours and can involve either single or repeated exposures within that period. Chronic or long-term exposures involve years of exposures to low levels of a contaminant. Examples of chronic health effects are silicosis and byssinosis.

The responses to a toxic exposure vary from immediate effects to delayed symptoms, such as those associated with carcinogens where there are long latency periods. When identifying chemicals that are of high hazard in an emergency release, acute toxicity is the most important criterion. On the other hand, for long-term, low-level exposures to the community or workers, chronic toxicity is the most important. For a given chemical, its acute and chronic effects can be quite different.

Not all chemicals are capable of causing acute toxicity. An example is asbestos. Even though the hazards of asbestos are well documented, it causes only chronic health effects. Even the worst exposure is unlikely to cause toxic effects for several years. Long-term, high-dose exposures tend to cause asbestosis, which could occur in a shorter period of time than the latency period associated with cancer development, but in both cases the time period is in years.

Routes of Entry

Toxic agents enter the body in the same ways that nontoxic materials do: through the gastrointestinal tract (GI), the respiratory system, or the skin. In many ways the factors affecting the entry of life-sustaining nontoxic compounds also affect the entry of toxic agents. For example, carbon monoxide inhalation is modified by the same parameters that accelerate or inhibit the inhalation of oxygen; similarly, the same liver enzymes responsible for the metabolic breakdown of many toxic compounds are also involved in the metabolism of food constituents.

Contaminants entering the GI tract often produce effects only on the cells that line the tract, although absorption from the tract can also take place. For example, some compounds such as alcohols are absorbed directly through the stomach. Caustic or primary irritants can destroy the mucosal lining of the GI tract, allowing entry of other chemicals into the bloodstream.

Inhalation via the respiratory system is the most common occupational route of entry for chemicals and other agents. It is also the most susceptible site. Substances reach the bloodstream very shortly after inhalation, provided that they have the right properties to pass through the lungs. During studies to determine if there are any differences in the degree of exposure if a person favors breathing through the mouth rather than the nose, it has been noted that the oral airway is far less efficient at removing materials from inspired air than is the nasal airway.[16] Therefore, during mouth breathing, greater doses of toxic materials may reach sensitive sites in the lung than during nasal breathing.

Because larger particles when inhaled have a tendency to deposit in the upper areas of the respiratory system, such as the nasopharynx, and a certain portion of these are likely to be swallowed rather than retained in the respiratory system. The main factors that affect the rate of absorption or deposition of material from the respiratory tract into the bloodstream include solubility, particle size, and concentration.[16] Biological variability between individuals, differences in health, confounding factors such as cigarette smoking, differences that relate to age or breathing styles, as well as inherent differences in airway sizes can also impact exposures. Table 1-4 lists some chemical and physical properties of compounds that influence toxicity.[17]

Dermal exposure depends to a significant degree on the solubility in fat of the compound. In addition, the site of exposure is a factor: Forehead, testicle, abdominal, arm, and back skin all have different permeabilities. Many solvents and pesticides are capable of penetrating the skin. Some chemicals gain entry through the skin very effectively. Table 1-5 compares the results

TABLE 1-4. Factors Influencing the Effects of Inhaled Agents

Physical Properties

Physical state.
Size and density of particles, mist, or aerosol determines the site of deposition.
Shape and penetrability influences propensity for migration.
Solubility
- Particulates: Insoluble agents produce local effects whereas soluble compounds may produce systemic effects.
- Gases and vapors: Insoluble agents, such as nitrogen oxides, are inhaled into small air passages, whereas soluble agents, such as ammonia and sulfur dioxide, seldom pass beyond the nose and nasopharynx.

Hygroscopic particles increase in size as they travel down the respiratory tract.
Electric charge influences the site of deposition.

Chemical Properties

Acidity and alkalinity have a toxic effect on cilia, cells, and enzyme systems.
Agents such as carbon monoxide and hydrogen cyanide have systemic effects, whereas fluorine compounds may have both local and systemic effects.
Fibrogenicity.
Antigenicity stimulates antibodies.

TABLE 1-5. Comparison of Oral and Dermal LD_{50}'s for Selected Pesticides

Compound	*Oral LD_{50} (mg/kg)*	*Dermal LD_{50} (mg/kg)*
TEPP	1.1	2.4
Phorate (Thimet)	1.7	4.3
Phosdrin	4.9	4.45
Disolfoton	4.5	10.5
Demeton (Systox)	4.3	11
Thionazin (Nemaphos)	4.9	14
Ethyl parathion	8.3	14
Chlorfenvinfos	14	30
Bidrin	18	42
Methyl parathion	19	67
Dichlorovos	68	91
Azinphosmethyl	12	220
Malathion	1187	>4440

when lethal doses of pesticides capable of killing 50% of the exposed animals were delivered both by oral and dermal routes of entry.[18] The smaller the value of the LD_{50}, the more toxic the compounds.

While not considered a primary route of entry, due to the concerns about contracting hepatitis and AIDS in the health care field, exposure via injection has become an important matter. Needle sticks are one of the primary means by which health care workers are exposed to infected body fluids.

Elimination of Hazardous Materials

The ability of a compound to accumulate in the body or a target organ also depends on the overall rate at which a chemical gets into the body as well as how fast it is metabolized and eliminated. The primary way that agents are eliminated from the body is in the urine, although other modes include the feces, sweat, tears, hair, nails, and breath. Some chemicals such as ethanol are eliminated from the body very rapidly, within hours, whereas others such as lead and arsenic can persist for days. Biologic half lives describe the length of time it takes for the concentration of a chemical in the body to decrease by 50%. Since the biological half life is one way of estimating the rate of elimination of a compound from the body, it also represents the likelihood of a compound accumulating in the body. Body burdens can be viewed in more than one way, including peak body burden, average body burden, and residual burden once exposure has ended. The shorter the half life, the faster the peak body burden is reached. The chapter on Biological Monitoring describes the collection of bodily fluids to supplement air sampling measurements.

Exposure versus Dose

It is useful to differentiate between concentrations measured outside and inside the body. The term *exposure* refers to those levels measured outside the body and often is reserved for air sampling measurements. In most cases, these do not directly correlate with what actually gets inside the body. *Absorbed dose* refers to the portion of the exposure that actually enters the body through the skin, lungs, or other route and reaches the bloodstream to be transported to the target organs. *Deposited dose* is a term usually reserved for skin exposures and refers to the amount of material that is actually deposited on the skin's surface.[18] Actual body dose differs from the results of air sampling measurements used to approximate inhalation exposure due to many variables: degree of activity of the subject (work versus rest); changes in air temperature (affects body metabolism); fluctuations in airborne concentrations; ability of the contaminant to be eliminated in exhaled air; accumulation in the tissue or ability to be metabolized to more or less toxic metabolites.[19] Individual biovariations, such as age, size, sex, and genetics, can also affect the dose. Exposures are expressed as milligrams per cubic meter (mg/M^3) or parts per million (ppm) and dose is expressed as the mass of contaminant per mass of body weight or mg/kg.

The site of action associated with toxic substances varies widely. Local effects such as a skin burn are those lesions caused at the site of first contact between the agent and the individual. Systemic effects involve the absorption and distribution of the toxic agent from the site of entry, whether via oral, dermal, or inhalation, to a distant site where the toxic response occurs. An example is mercury, which produces toxic effects on the central nervous system.

When two or more substances that act on the same organ systems are present, their combined effect is a consideration. Methylene chloride and carbon monoxide both bind with red blood cells to form carboxyhemoglobin. As most solvents affect the central nervous system for operations where many solvents are used, such as paint and ink manufacturing, additive effects are a concern. Some combinations of exposures, such as smoking and asbestos, can act synergistically, meaning there is a much greater effect than would be predicted from the exposure to levels of the individual contaminants.

Contaminant Exposure Side Effects

Irritation

The eyes, skin, and upper respiratory system can be affected by chemicals that produce irritation. The symptoms of irritation include stinging, itching, and burning of the eyes; tearing; burning sensation in the nasal passages; nasal inflammation; cough; sputum production; chest pain; wheezing and other breathing difficulties. At low levels irritation can be useful as a warning that chemicals are present. Chlorine, ammonia, and formaldehyde as well as many other compounds are irritants.

Sensitization

A sensitization reaction, also known as an allergic reaction, is defined as an adverse response to a chemical following a previous exposure to that substance or to a structurally similar one.[19] A related phenomenon is cross-sensitization. It occurs when exposure to one substance elicits a sensitization reaction, not only upon subsequent exposure to the same substance but also with exposure to a different substance with a similar structure. Common target organs for sensitization are the skin and respiratory system. Sensitivity to a chemical can persist for an individual's lifetime. Examples of sensitizers are isocyanates and poison ivy.[20]

Systemic Effects

Chemicals that have a systemic effect act on tissues other than the site of entry. As the result of exposure, a variety of adverse effects can be manifested at multiple target organ sites. Although the specificity of the systemic effect caused by exposure to the substances may vary, in general these chemicals are capable of interfering significantly with biological processes and impairing normal organ function. Examples include chloroprene, acetonitrile, hydrogen cyanide, and phosphine.[20]

Target Organ Effects

A compound can preferentially act on a given organ to cause a toxic effect. Some compounds such as lead can act on many organs whereas others affect primarily one organ. The ability of a chemical or its metabolite to bind at a given site often determines the target organ affected.

Substances that act on the nervous system can cause either peripheral nervous system (PNS) or central nervous system (CNS) effects or both. Motor function (muscular weakness or unsteadiness of gait), sensory function (alterations in sight or sensations of touch, pain, or temperature), and/or behavior may be affected. PNS toxicants, such as mercury, cause degeneration of the nerve and its structure. CNS toxicants cause seizures, narcosis (dizziness and drowsiness), dementia, cranial neuropathy, and visual disturbances. Chemicals known to cause narcosis include solvents, such as heptane, toluene, and methylene chloride.[20]

Liver effects depend on the dose, duration, and particular chemical agent involved. Acute exposures can cause lipid (fatty) accumulations in the liver cells, cell death (necrosis), and liver dysfunction. Chronic exposures may lead to cirrhotic changes and/or the development of liver cancer. The earliest and most sensitive indicator of liver toxicity is an alteration in biochemical liver functions, such as changes in specific enzyme functions. Chlorinated compounds frequently have adverse effects on the liver as well as many solvents.[20]

Depending on their site of action, chemicals acting on the kidneys can interfere with hydration, excretion of wastes, electrolytic balance, or metabolism. Because a significant degree of kidney cell damage must occur before it is reflected by measurable changes in kidney function, kidney function tests do not necessarily detect early damage. Compounds that affect the kidneys include 1,3-dichloropropene, ethyl silicate, and hexachlorobutadiene.[20]

Effects on the cardiovascular system include cardiac sensitization, which results in disturbances in the heartbeat, and vasodilation when blood vessels expand, resulting in blood pressure and circulation losses. Chemicals that affect the cardiovascular system include carbon disulfide, some freons, and sodium azide.[20]

The bladder can also be a site of toxic effects, most notably cancer. Compounds such as α- and β-naphthylamine cause bladder cancer.

Respiratory Effects

The respiratory system is susceptible to many diseases including cancer. Chronic pulmonary disease can include fibrotic changes as well as other diseases such as emphysema. Fibrotic conditions in the lung result in a loss of elasticity with a decrease in the ability of the individual to utilize oxygen. Fibrosis is often considered the same as pneumoconiosis; however, pneumoconiosis is a more general term indicating the presence of a foreign substance in the lungs. Dusts, such as silica and coal, cause fibrosis.[20]

Biochemical/Metabolic Effects

Many chemicals have the ability to interfere with the normal metabolism or biochemistry of the body. Examples are substances that act on enzymes such as acetylcholinesterase to inhibit their actions, or substances that interfere with the oxygen-carrying capacity of blood. Acetylcholinesterase inhibition results in signs and symptoms such as bronchoconstriction, increased bronchial secretions, salivation, and tearing; nausea; vomiting; cramps; constriction of the pupils; muscular weakness; and cardiac irregularities. If sufficiently severe, this condition can lead to death. Many organophosphate and carbamate pesticides are in this category.[20] Substances that interfere with the oxygen-carrying capacity of the blood often act by tying up the hemoglobin in the red blood cells, which are responsible for oxygen transport. Carbon monoxide and methylene chloride are examples. Chemicals with biochemical and metabolic effects are often selected for biological monitoring. This type of sampling is discussed in the chapter on Biological Monitoring.

Carcinogenic Effects

Cancer is the result of an abnormal tissue composed of cells that have been altered in such a way as to cause unrestricted cell growth invading organ systems. As a result, these cells lose the ability to function normally and the tissue that they comprise interferes with the vital functions of normal organ systems. There is a correlation between substances that are mutagenic (meaning they cause mutations in the DNA) and the ability to cause cancer. While many compounds have caused mutations in various test systems but have not been proven to cause cancer in humans, a positive result in several tests will cause that chemical to be suspect. Chemicals are categorized according to the evidence linking their ability to cause cancer in humans. The International Agency for Research on Cancer (IARC) is one of the primary agencies involved in doing carcinogenic risk assessments. For evidence of possible carcinogenicity in humans the IARC uses two sources: epidemiological studies of groups of people, often workers, and long-term (usually 2 years) animal tests. Some compounds known to cause cancer in humans are asbestos, benzene, and MOCA. Table 1-6 lists IARC categories for carcinogens.[21]

TABLE 1-6. IARC Carcinogen Categories

Group 1	Chemicals carcinogenic to humans; usually determined on the basis of epidemiological studies.
Group 2	Chemicals that are probably carcinogenic; the evidence ranges from "limited" to "inadequate." Group 2A indicates the data are considered more significant than compounds in Group 2B, which is often the subcategory used for compounds for which there are sufficient animal data but inadequate human data.
Group 3	Chemicals that cannot be classified as to their carcinogenicity because the evidence for carcinogenicity to humans and to experimental animals is inadequate to make an evaluation.

Reproductive Effects

Chemicals that cause reproductive effects can act in two different ways. They can cause sterility in either males or females or cause defects in the embryo or fetus, termed *teratogenesis*. When they affect offspring they are called teratogens. Many chlorinated pesticides affect hormone systems and are implicated in sterility effects on females, and dibromochloropropane (DBCP) and kepone are known to cause sterility in males. Lead as well as many other compounds may cause teratogenic effects in offspring when pregnant females are exposed. Lead has also caused decreased sperm counts in males.

Nuisance Effects

High concentrations of some dusts may seriously reduce visibility; cause unpleasant deposits in the eyes, ears, and nasal passages; or cause injury to the skin or mucous membranes either due to their own properties or by the rigorous skin cleansing procedures that may be necessary to remove them.[20] The terms *inert* and *nuisance* dusts have been applied to those dusts that are not known to produce any significant toxic effects when exposures are kept under reasonable control. However, these terms are misnomers because there is no dust that will not evoke some type of response in the lung if inhaled in sufficient amounts. Examples of compounds that have been included in this category include calcium carbonate, cellulose (paper fiber), emery, glycerin mist, graphite, gypsum, kaolin, limestone, magnesite, marble, silicon carbide, starch, and sucrose. Be aware of the fact that since toxicological information on many chemicals including dusts is lacking, the philosophy regarding a particular dust can change. For example, there is currently an increased concern regarding potential toxic effects of fiberglass, which has been considered a nuisance dust for many years.

Toxicity Versus the Hazard

The toxicity of a substance while important should not be the sole criterion used in determining the existence of a health hazard associated with a specific situation. Many factors should be considered, such as the chemical and physical properties of this toxic substance, the ability of other toxic substances to interact with it, and the influence of environmental conditions such as temperature as well as the concentration present. The degree of hazard associated with exposure to a specific substance also depends on the conditions of use. For example, a highly toxic chemical that is processed in a closed, isolated system may be less hazardous in actual use than a low-toxicity compound handled in an open batch process. Therefore, the nature of the process in which the substance is used or generated, the possibility of reaction with other agents, the degree of engineering controls including ventilation, and the amount of personal protective equipment in use all relate to the potential hazard associated with each use of a given agent. Another factor affecting the ability of a chemical to elicit a toxic response is the susceptibility of the biological system or individual. For the relative degree of hazard to be known in a particular instance, knowledge about the chemical agent, the exposure situation, and the exposed subject is required.

STANDARDS AND GUIDELINES FOR AIR SAMPLING

Since the results of sampling usually consist of numbers representing concentrations, proper interpretation requires comparing them to some type of agreed-upon standard. The reasoning and research behind standards are what make these numbers viable. For example, if studies have shown that 500 ppm of a particular substance causes dizziness in workers, 50 ppm might be selected to be a level that will provide a safe margin between exposure and this expected effect.

Many organizations recommend exposure levels; however, unless these levels are incorporated into a regulation, they are simply guidelines whereas regulations are legally binding requirements. Although most commonly the federal and state regulations are referenced, some cities also have their own air emissions regulations. Often these are implemented by local health departments or county environmental agencies.

Chronic Exposures to the General Community

The greatest concern when setting standards to protect the general populace is the wide range of ages and degrees of health that must be considered. The Environmental Protection Agency (EPA), who has the primary responsibility for community standards, has set only a few substance specific air exposure standards for chronic exposures to the general community. Their standards are set for the ambient or general community air. National *primary* ambient air quality standards define levels of air quality that provide an adequate margin of safety to protect the public health while *secondary* standards define levels of air quality that are necessary to protect against specific adverse effects of a pollutant. Allowable concentrations for secondary standards are higher than for primary standards and apply to much shorter periods of time. These air pollution standards are based on a number of different time periods, such as 1, 3, and 24 hours as well as 3-month and 1-year averages, depending on the contaminant. The EPA also publishes guidance documents on air sampling and has developed a number of air sampling methods for outdoor and indoor air.

The American Society of Heating, Refrigeration and Air Conditioning Engineers (ASHRAE) has developed guidelines for acceptable indoor air quality, published as its "Ventilation for Acceptable Indoor Air Quality" standard. These guidelines are used for indoor air situations such as office buildings in which occupants develop illnesses that appear to be of unknown origin but that can actually be due to contamination carried into the building by its HVAC system.

Acute Exposures to the General Community

The potential for a chemical release in or near a community, whether it be from an overturned railroad car or from a factory's exhaust, is very real and so a number of organizations have suggested concentrations that would trigger emergency warning measures. Emergency Response Planning Guidelines (ERPGs), developed by the American Industrial Hygiene Association, are designed to address acute exposures to the general community. Three exposure levels are set for each of a number of chemicals:

ERPG-1: The maximum airborne concentration below which it is believed that nearly all individuals could be exposed for up to 1 hour experiencing only mild, transient adverse health effects or without perceiving a clearly defined objectionable odor.

ERPG-2: The maximum airborne concentration below which it is believed that nearly all individuals could be exposed for up to 1 hour without experiencing or developing irreversible or other serious health effects or symptoms that could impair an individual's ability to take protective action.

ERPG-3: The maximum airborne concentration below which it is believed that nearly all individuals could be exposed for up to 1 hour without experiencing or developing life-threatening health effects.

Chemicals for which these guidelines are available include chloroacetyl chloride, chloropicrin, crotonaldehyde, perfluoroisobutylene, phosphorus pentoxide, ammonia, chlorine, diketene, formaldehyde, and hydrogen fluoride.

Acute community exposure standards will be much higher than those used to protect against long-term, low-level effects. It has been suggested by the EPA that one tenth of the immediately dangerous to life and health (IDLH) level for any given compound might provide a sufficient safety factor to allow its use as public exposure guidelines for emergencies.[22] IDLH values are discussed in greater detail in the section on acute occupational exposures in this chapter. There is much debate as to whether IDLH values would provide sufficient protection of the general public because of difficulties in setting up an effective warning system and the need to have respirators available at these levels.[23]

Other guidelines that have been suggested for use in acute community exposures include emergency exposure guidance levels (EEGLs) developed by the National Academy of Science (NAS) to protect military personnel. These are 1-hour emergency exposure levels intended to prevent irreversible harm for narrowly defined exposure situations. A second set of guidelines called short-term public emergency guidance levels (SPEGLs) have been developed by NAS to apply to the public in the event of a catastrophe.[23]

Chronic Occupational Exposures

Workplace air standards are the most numerous and cover a wide variety of chemicals. In addition, with few exceptions, occupational air monitoring methods have been developed that are chemical specific. These standards apply only to workplace exposures; assume that individuals work an 8-hour day, 40-hour work week; and are designed to be applied to the healthy working individual and do not take into account sex-related or age-related considerations. Some individuals have increased susceptibility and may not be protected by the standards. The most difficult situations are represented by individuals who have heritable (genetic) deficiencies or idiosyncrasies such as sickle cell anemia, preexisting diseases such as asthma, and those who become sensitized to a compound such as an amine. Habits, especially drug use, alcohol intake, and smoking, as well as sex and age, influence the amount of exposure that can be tolerated.[24] See Table 1-7 for the physical characteristics of a model worker used to develop standards and estimate the consequences of doses of various agents.[25] The best way to view these standards is not as fine lines between safe and unsafe conditions but as guidelines applied by professionals to protect workers.[26]

The standards for chemicals that the Federal Occupational Safety and Health Administration (OSHA) promulgates are termed permissible exposure limits (PELs). These include regulated exposure limits for nearly 600 chemicals along with 24 more comprehensive single-substance standards. OSHA standards are considered to be the legal minimum health protection required of employers and are designed to protect against a variety of toxic effects including irritation, target organ toxicity, chronic lung disease, and biochemical/metabolic effects. Table 1-8 contains some examples of the basis upon which OSHA standards for various chemicals have been set.[20]

TABLE 1-7. The Reference Worker

Parameter	*Value*
Weight	70 kg
Height	175 cm
Age	20–30 years
Body surface area	1.8 m^3
Lung weight	1000 g
Lung surface area	80 m^2
Total lung capacity	5.6 L
Vital capacity	4.3 L
Residual volume	1.3 L
Respiratory dead space	160 ml
Breathing rate	15 breaths/min
Tidal volume	1450 ml
Inspiratory flow rate	43.5 L/min
Minute volume	21.75
Inspiratory period	2 sec
Expiratory period	2 sec

The primary mechanism for describing and calculating exposure limits is the 8-hour time weighted average (TWA). Virtually all of the standards designed to be applied to workers are based on this concept. Currently this concept is under examination because it may not be the most accurate method of estimating worker exposures.[27] PEL—TWAs refer to airborne concentrations to which nearly all workers may be exposed for 8 hours per day, 40 hours per week, for a working lifetime without adverse effect. The TWA is calculated as follows:

$$\frac{C_1(T_1) + C_2(T_2) + \ldots + C_n(T_n)}{\text{Length of workday*}}$$

where C is the concentration of a sample and T is the sampling time for that sample. For compounds that cause systemic toxicity increasing

*Usually 8 hours, but where possible it is better to calculate based on the actual length of the workday if it exceeds 8 hours. In this case, the standard of interest must also be adjusted to reflect the new time span.

TABLE 1-8. Examples of the Basis for OSHA Standards

Compound	*Toxic Effect Prevented*
Acrylonitrile	Liver carcinogen
Allyl chloride	Liver and kidney toxicity
Aluminum, soluble salts	By analogy to hydrogen chloride because hydrolysis to hydrogen chloride occurs (pulmonary injury)
sec-Amyl acetate	Irritation to eyes and respiratory tract
Asphalt fumes	Carcinogen
Carbon monoxide	Carboxyhemoglobinemia
Chlorodifluoromethane	No effect level for cardiac sensitization
Chlorodiphenyl, 54% (PCBs)	Systemic effects
Cobalt, metal, fume, and dust	Sensitization
Cobalt carbonyl	Systemic toxicity
Cyclohexanone	Liver toxicity
Dichloromonofluoromethane (Freon 21)	Hepatotoxicity, cardiac sensitization
Diethanolamine	No effect level for impaired vision and skin irritation
Ethyl alcohol	Eye and respiratory tract irritation
Fibrous glass dust	Nuisance dust
Formamide	Testicular toxicity, teratogenicity
Hydrogen bromide	Irritation
Hydrogen chloride	Pulmonary injury
Hydrogen cyanide	Systemic toxicity including cyanide poisoning, weakness, mucosal irritation, colic, nervousness, and enlargement of the thyroid in humans
Hydrogen fluoride	Eye and nose irritation
Hydrogen sulfide	Occular effects
Isoamyl acetate	Irritation of the upper respiratory tract
Isobutyl alcohol	Irritation, narcosis
Malathion	Nuisance dust
Manganese (and compounds)	Manganism
Manganese, fume	Neuropathic
Methyl amyl ketone	Irritation
Methyl butyl ketone	Neuropathic
Methyl isocyanate	Mucous membrane irritation
Methyl parathion	Cholinesterase inhibition
Toluene	Narcosis
Xylene	Irritation

exposure times (> 8 hours) will require decreasing exposure concentrations in order to assure that the total body burden will not exceed that allowed by the 8-hour standard.[28]

When a ceiling is applied to an 8-hour standard, it means that the PEL can never exceed this value. Thus, the ceiling is the maximum allowable concentration over the shift. It is common to have ceiling standards for irritants.

The short-term exposure limit (STEL) is a peak concentration to which workers can be exposed continuously for a short period of time without suffering from irritation, chronic or irreversible tissue damage, or narcosis of sufficient degree to increase the likelihood of accidental injury, impair self-rescue, or materially decrease work efficiency. STELs are valid exposures provided the daily PEL–TWA is not exceeded. The STEL is not intended to be a separate independent exposure limit; rather it supplements the 8-hour TWA limit

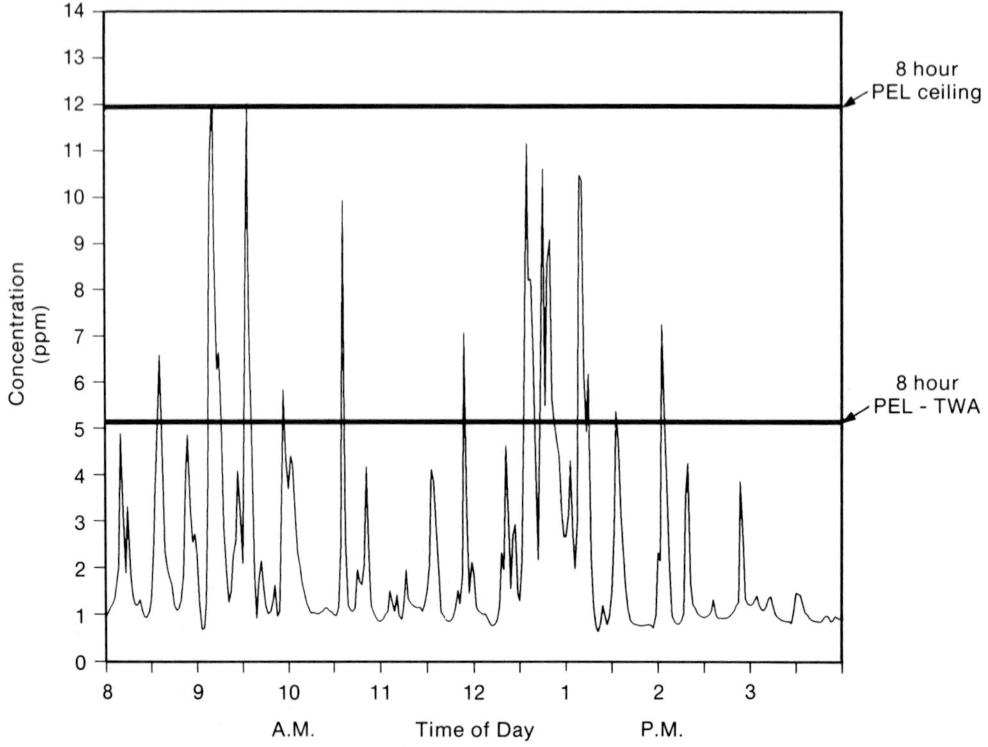

Figure 1-1. Representations of OSHA PELs: The 8-hour TWA showing typical fluctuations, and the 8-hour PEL Ceiling as a maximum value not to be exceeded.

where there are recognized acute effects from a substance whose toxic effects are primarily of a chronic nature.[29]

Exposures at the STEL should not exceed a 15-minute time period and should not be repeated more than four times per day. In addition, there should be at least 60 minutes of unexposed time between successive exposures at the STEL. The incremental difference between the 8-hour TWA and the STEL is not always the same for different chemicals. For example, the 1990 OSHA 8-hour TWA for benzene is 1 ppm, while the STEL is 5 ppm. The 8-hour TWA and STEL for ethyl benzene are 100 ppm and 125 ppm, respectively. Different chemicals can act over different periods of time, including less than 15 minutes, but the use of a generic period simplifies the approach of sampling for peak exposures.

Another approach for identifying peak exposures that has been attempted by the National Institute for Occupational Safety and Health (NIOSH) is to select a short-term sampling period based on the biological activity of a compound. However, this method is time consuming and may not be practical as the result, depending on the acute effect identified for a given chemical, is different short-term sampling periods for different chemicals. Figure 1-1 demonstrates how OSHA's TWA standard is actually an average of the varying concentration levels that usually exist over any given workday, but if the standard is a ceiling the averaging effect is ignored in favor of the maximum level detected. Figure 1-2 shows the STEL, which is also an average concentration, but over a much shorter time period.

An action level is incorporated into many of OSHA's single-substance standards that is one-half of the 8-hour TWA and triggers certain requirements such as medical monitoring, training, and repeat air sampling. The action level applies a safety factor to the 8-hour PEL by taking into account day-to-day variations, statistical error, and

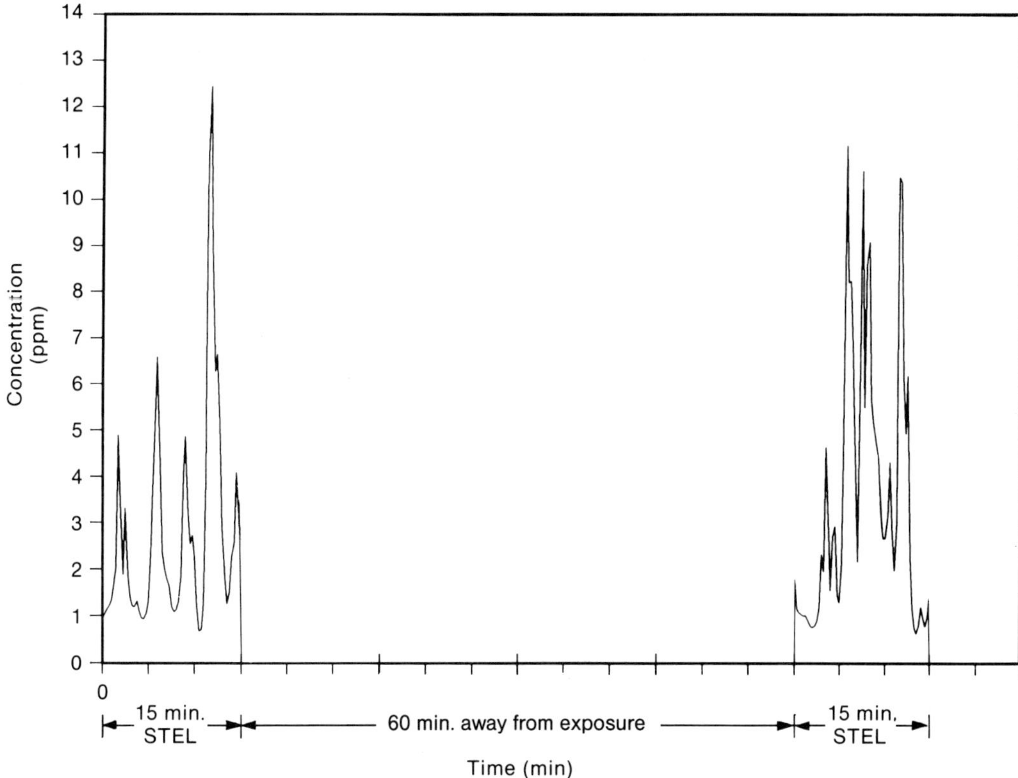

Figure 1-2. The 15-minute STEL.

other aspects that influence the accuracy of sampling results, so although a respirator is not required at this level, repeat sampling increases the likelihood that an overexposure will be detected. The action level for benzene is currently 0.5 ppm calculated as an 8-hour TWA.

While the existence of PELs tends to focus attention on those chemicals with PELs, exposures to other contaminants should also be evaluated. The General Duty Clause, section 5 (a)(1) of the OSHA Act, provides that "each employer shall furnish to each of his employees a place of employment which is free from recognized hazards that are causing or likely to cause death or serious physical harm to these employees." Thus OSHA compliance officers can cite for an overexposure to a chemical for which no PEL exists.

The American Conference of Governmental Industrial Hygienists (ACGIH) was founded in 1938 and is an independent organization comprised of industrial hygienists and other health professionals from academia and government-related institutions. The threshold limit values (TLVs) developed by ACGIH are recommended exposure limits based on a belief that there is a threshold(s) of response, derived from an assessment of the available published scientific information including studies in exposed humans and experimental animals, and at exposures below these levels no adverse health effects will occur to workers. New chemicals are added on a regular basis and exposure levels are often adjusted, usually to new, lower levels, and for each TLV a set of supporting documentation is published. Like the PELs, TLV—TWAs refer to 8-hour exposures and 40-hour weeks and are expressed in parts per million (ppm) or milligrams per cubic meter (mg/M^3) at a standard temperature and pressure of 25°C and 1 atmosphere.

The latest recommendation from ACGIH for evaluating peak exposures is the use of excursion limits. The basis for excursions is more intuitive than that for STELs, but the general concept is that these concentrations should not be allowed to exceed five times the value of the 8-hour TWA for more than 30 minutes within any one workday.[26] ACGIH still has STELs for some chemicals although they are being phased out.

The particle size-selective TLVs take into account not only the inherent toxicity of the particles but also the particle size distribution, their patterns of deposition within the respiratory tract, and the related rate of dissolution and translocation to target tissues. They also take into consideration the diseases associated with the inhaled material, and are based on the physical characteristics of the lung; size, mass, distribution, and dynamics of particles; physical and chemical composition of particles emitted by varying processes; and other factors including dissolution rates in the lung.[30] Figure 1-3 shows the areas of the respiratory system that correspond to the size-selective TLVs.[31] OSHA uses a much simpler approach to regulate dusts based on particle size. However, this may change to reflect ACGIH's particle-size TLVs in the future.

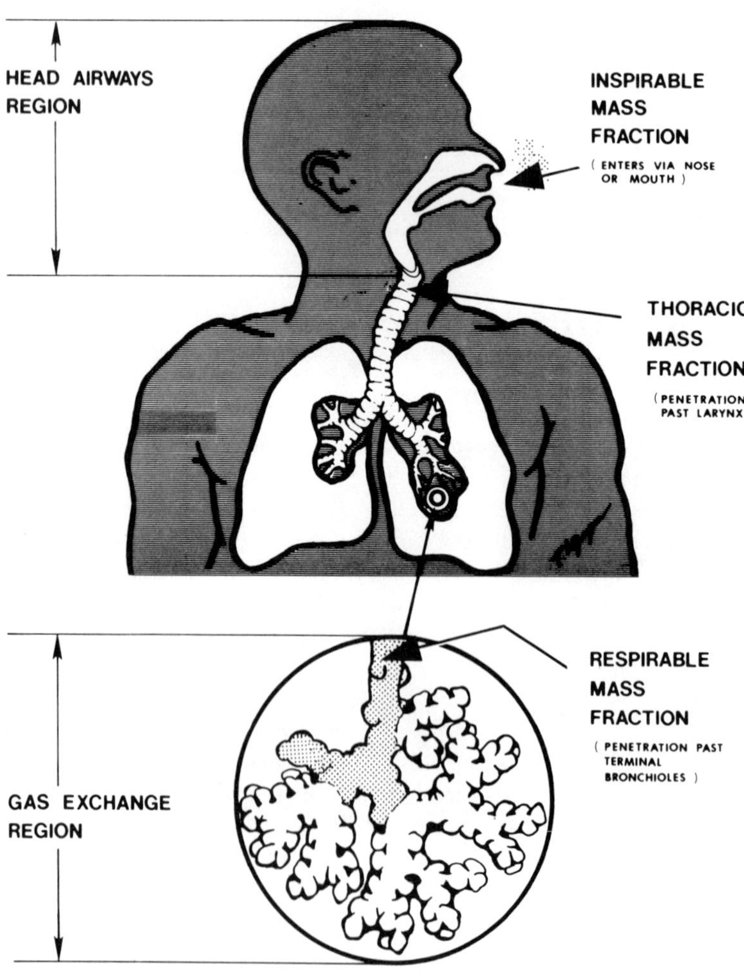

Figure 1-3. Area of the respiratory system corresponding with size-selective TLVs. (From Phalen, R. F. Rationale and recommendations for particle size-selective sampling in the workplace. *Appl. Ind. Hyg.* **1**(1):3–14, 1986; reproduced with permission of *Applied Industrial Hygiene.*)

Inspirable particulate mass (IPM)–TLVs are designed to represent materials that are hazardous when deposited anywhere in the respiratory tract. IPM–TLVs apply to particles ranging in size up to 100 μm.[26]

The thoracic particle mass (TPM)–TLV applies to chemicals that are hazardous when deposited anywhere within the lung airways and the gas exchange region. These particles are capable of penetrating a separator-sampler whose size collection efficiency is based on a median aerodynamic diameter of 10 μm $\pm$ 1.0 μm and a geometric standard deviation of 1.5 ($\pm$0.1).[26]

The respirable particulate mass (RPM) limits are designed to represent contaminants that are hazardous when deposited in the gas exchange region. These particles will penetrate a separator-sampler whose size collection efficiency is based on a median aerodynamic diameter of 3.5 μm $\pm$ 0.3 μm and a geometric standard deviation of 1.5 ($\pm$0.1).[26]

A calculation is also available to tailor a TLV for situations when mixtures of several chemicals are present, each of which is likely to affect the same target organ. A common example is the use of solvents. If a worker was exposed to the full TLV of each compound, it is very likely that an excessive exposure would occur; therefore, in these situations the TLV for the individual compounds is adjusted lower by considering their effects as a group. Although the air sample is analyzed for each component, the TLV for a mixture of exposures to an individual is calculated using the following formula[26]:

$$\frac{C_1}{TLV_1} + \frac{C_2}{TLV_2} + \cdots + \frac{C_n}{TLV_n} \leq 1$$

where

C_1 = concentration of the first compound
C_2 = concentration of the second compound
C_n = concentration of the last compound
TLV_1 = TLV for the first compound
TLV_2 = TLV for the second compound
TLV_n = TLV for the last compound

This formula can also be used for mixtures of biologically active mineral dusts. For example, as there are various types of silica, each with a different TLV, a calculation for additive effects should be considered if more than one type is involved in an exposure.

NIOSH functions as an agency within the Department of Health and Human Services whose principal responsibilities are to conduct research to identify hazards and to develop new techniques that relate to workplace health and safety. This duty includes conducting various types of field investigations, such as health hazard evaluations, in specific workplaces at the request of employees and employers. Health hazard evaluations are generally done when there are worker complaints that cannot be associated with any known chemical or other hazards in an operation. Most importantly, NIOSH also develops and publishes air sampling methods.

NIOSH has recommended a number of standards for chemicals, known as recommended exposure limits (RELs). These often include both 8- and 10-hour TWAs as well as ceiling limits of varying time periods. Once established, RELs are rarely changed. The mechanism for setting an REL is usually through publishing a criteria document. In the process of developing a criteria document an extensive set of research is done to review existing human and animal data. These documents also contain recommendations for exposure limits, warning label wording, exposure and medical monitoring, sampling and analytical methods, and recommended controls for engineering and personal protective equipment. NIOSH standards generally set concentrations that are less than other occupational standards.

The Mine Safety and Health Administration (MSHA) oversees underground and surface mining, associated surface preparation and processing operations, mine construction, and other activities related to mining. MSHA has health standards for air contaminants, such as coal dust, silica, radon, and asbestos, that may be encountered during mining activities.

The American Industrial Hygiene Association (AIHA) is a professional organization that has been in existence since the 1930s. Workplace environmental exposure limits (WEELs) are developed by an AIHA committee whose purpose is to establish workplace environmental limits for chemical substances and physical stresses for which no TLV, PEL, or other limit exists. The committee functions in much the same way as ACGIH's TLV committee in that it utilizes all available information on toxicology, epidemiology, industrial hygiene, and workplace experience information to develop safe exposure guidelines. Like the PELs and TLVs, WEELs are in the form of 8-hour TWAs and 15-minute STELs.

It is important to be aware of the potential limitations of applying standards based on an 8-hour exposure to situations where the shifts are longer. For example, seasonal work often involves 10–12-hour days or work weeks of 6 days, as does project-related work such as remediation on hazardous waste sites and jobs involving concentrated repairs over a period of time common during shutdowns of industrial facilities. An adjusted 8-hour PEL should be developed for substances that are capable of accumulating in the body to prevent the body burden from increasing over the amount the 8-hour TWA standard assumes is safe. Figure 1-4 provides an example of how body burdens resulting from 8-hour and 10-hour workdays can differ for the same chemical.[32] It should be noted that while standards may be lowered to account for increased exposure periods, they are never increased to allow for decreased exposure periods, such as a 7.5-hour day, 37.5-hour work week.

According to OSHA, there are two types of chemicals whose standards can be adjusted: (1) acutely toxic compounds, such as butyl alcohol, carbon monoxide, and hydrogen sulfide; and (2) chronic toxicants, such as lead, carbon disulfide, silica, and chromium. Ceiling standards and those standards designed to prevent sensory irritation or excessive odor are generally not adjusted, nor are standards designed to prevent cancer or compounds that are physical irritants (sometimes termed nuisance particulates).

Several approaches have been recommended for adjusting standards where schedules exceed 8 hours.[33] For acutely toxic chemicals that could accumulate in the body over an 8-hour exposure period, OSHA uses the following equation to reduce PELs and recommends daily adjustment[34]:

$$\text{PEL} = \text{8-hour TWA}\left(\frac{\text{8 hours}}{\text{hours of exposure in 1 day}}\right)$$

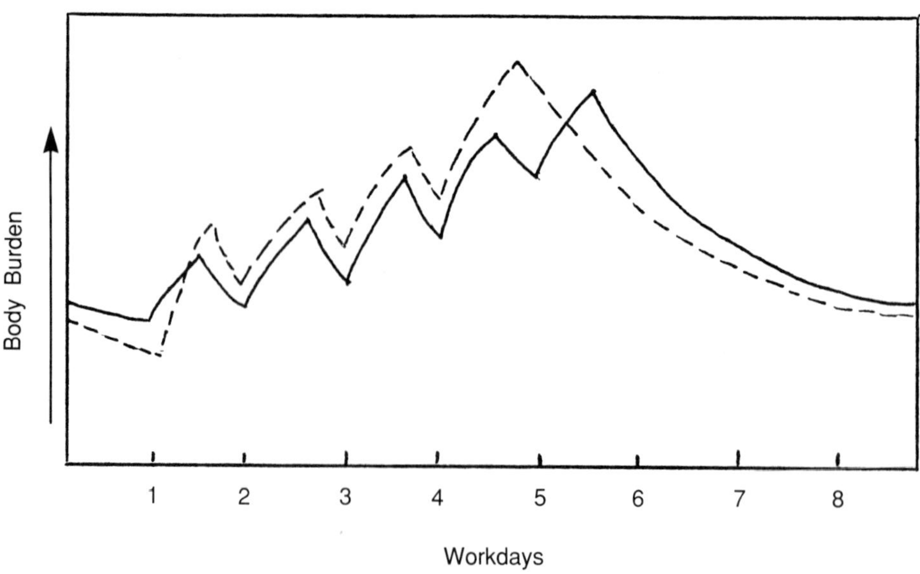

Figure 1-4. Comparisons of differences in body burdens for a chronic toxicant for 8-hour and 10-hour workdays. The solid line represents a 5-day–8-hour workday week; the dashed line, a 4-day–10-hour workday week. (Reprinted with permission by the *American Industrial Hygiene Journal*, Vol. 38, p. A-614, 1977.)

For chronic toxicants that are cumulative over longer periods of time, such as lead or mercury, another equation has been developed by OSHA to address these chronic exposures, which is designed to be adjusted weekly[34]:

$$\text{PEL} = \text{8-hour TWA}\left(\frac{\text{40 hours}}{\text{hours of exposure in 1 week}}\right)$$

One approach the sampling professional can use to estimate the actual dose that exposure to a given concentration represents is to do a calculation based on lung volume and breathing rate. An average-size person takes 15 breaths per minute via the nose. At rest the tidal volume, meaning the amount of air taken in, is 750 ml; during moderate activity, 1450 ml; and for strenuous activity, 2150 ml.[16] If the amount of asbestos that would be inhaled over an 8-hour workday if the concentration were just at the current OSHA PEL of 0.2 fib/cm^3* is estimated, the calculation would be as follows:

tidal volume for moderate work: 1450 ml/breath

average breaths per minute: 15/min

length of day: 480 min

concentration: 0.2 fib/cm^3 = 0.2 fib/ml

$$\frac{\text{15 breaths}}{\text{min}}(\text{480 min})\frac{\text{1450 ml}}{\text{breath}} = \text{10,440,000 ml}$$

$$\text{10,440,000 ml}\frac{(\text{0.2 fib})}{\text{ml}} = \text{2,088,000 fib/8-hour day as a total exposure}$$

Acute Occupational Exposures

The IDLH level is not a standard, but is often used for evaluating exposures. The concept of the IDLH was established by NIOSH and represents the maximum level to which a healthy individual can be exposed for 30 minutes and escape without suffering irreversible health effects or impairing symptoms. The IDLH demarcates the concentration above which only respirators offering maximum protection, such as self-contained breathing apparatus, must be used. In theory if there is a potential for an IDLH atmosphere to occur, a person at risk should be provided with not only air monitoring but also respiratory protection. An existing IDLH atmosphere should never be entered unless the respirator is already donned. Recently it has been recommended that IDLHs not be applied to carcinogens, because respirators should be used when much lower levels of carcinogens are present or suspected.

The lethal concentration low (LC_{LO}) is the lowest concentration to which animals exposed in inhalation studies exhibit lethality. If an IDLH is not available for a chemical, the LC_{LO} can be used to approximate an IDLH.[35] If an LC_{LO} is not available, the lowest LC_{50} value available for a chemical divided by 10 can be used to approximate an IDLH. The LC_{50} is the level at which 50% of the test animals die when exposed to a compound in an inhalation study.

When using the IDLH as a measure of a level of concern, certain factors must be taken into account:

- The IDLH is based on the response of the healthy male working population and does not take into account exposures to more sensitive individuals, such as elderly, children, or people with various health problems.
- The IDLH is based on a 30-minute exposure time that may not be realistic for accidental airborne releases.
- IDLH values do not exist for all acutely toxic chemicals.
- By using the IDLH as the level of concern, this methodology may not identify all quantities of concern that could result in serious but reversible injury.

Biological Monitoring

Biological exposure indices (BEIs) are limits that have been established by ACGIH that apply to concentrations of chemicals inside the body. BEIs represent the levels of the determinants most likely to be observed in specimens collected from a healthy worker who has been exposed to chemicals to the same extent as a worker with inhalation exposure to the TLV—TWA. Therefore, BEIs apply to 8-hour exposures, 5 days a week.[26] Standards have been set for urine metabolites, blood

*OSHA has proposed lowering it to 0.1 fib/cm^3.

measurements, and exhaled air. Biological monitoring is considered complementary to air monitoring. Biological monitoring involves measuring specific chemical substances normally found in the body, but may increase from work experience; substances normally absent from the body; substances resulting from or affected by a biochemical response to exogenous influences.[9]

Biological monitoring can be used to substantiate air monitoring, to test the efficiency of personal protective equipment, to determine the likelihood of dermal and oral exposure, and to detect nonoccupational exposure. This last application is somewhat controversial in that this type of information may or may not be appropriate for release to persons other than the employee.

Dermal Samples

Currently there are no standards to use for comparison of wipe samples or other forms of dermal exposure samples. Therefore, each situation must be evaluated on a case-by-case basis. In some cases, the presence of any material such as a carcinogen like MOCA on work surfaces or solvents found on the inside of personal protective equipment such as gloves is a concern. Often levels in a given area are compared. An alternative is to make a grid showing the relative differences of concentrations over the various areas sampled. If the wipe samples were for a carcinogen, and detectable levels of surface contamination were found, then the detection level of the method would be a likely basis for comparison and all levels greater than this value would be considered significant.

Bulk Samples

There are no OSHA or other occupational standards for bulk samples. Bulk sample results may be expressed in mg/L, μg/L, total μg, or percent. These should be converted to match whatever standard or guideline will be used. For example, a sample of paint analyzed for lead may be reported as mg/L and would be converted to percent because the Housing and Urban Development (HUD) standard for lead in paint is expressed in percent.

A TLV can be calculated for a liquid mixture (usually solvents) when it is assumed that the atmospheric composition of the vapor above the mixture is similar to that of the mixture.[26] The percent composition (by weight) of the liquid mixture must be known:

$$\frac{1}{\dfrac{f_a}{TLV_a} + \dfrac{f_b}{TLV_b} + \cdots + \dfrac{f_n}{TLV_n}}$$

where f_a through f_n = the percent of each material in the mixture, f_a being the percent of component *a*, f_b the percent of *b*, and so forth, and TLV_a through TLV_n = the TLV for each material in the mixture, TLV_a being the TLV for component *a*, TLV_b being the TLV for component *b*, and so forth.

EPA regulations contain tests used to identify materials as hazardous waste that can also be applied to samples of soil and liquid wastes.[36] Many state environmental agencies publish cleanup levels for certain situations, such as underground storage tank removal. The EPA sets standards for chemicals, microbials, and radon in drinking water via the Clean Drinking Water regulations.

A bulk sample is considered asbestos-containing by the EPA if it contains 1% or more, and a product must include mention of asbestos or any other carcinogen on its material safety data sheet (MSDS) under the OSHA Hazard Communication Standard if it contains 0.1% or more. The Hazard Communication Standard also defines hazardous materials for the purpose of including them on MSDS, and noncarcinogens that are considered health hazards must be noted if they are present at a level of 1% or more. Standards applicable to radon and bioaerosols are discussed in their respective chapters.

CONCLUSIONS

Historically chemicals have been selected for sampling based on their degree of toxicity and the amounts in use. Although air sampling is an attempt to best approximate actual exposures, it should be remembered that chemicals can enter the body through many different routes. The dose that results from exposures regardless of the route of entry is what is most important. Therefore, it is becoming increasingly important for sampling professionals to devise sampling schemes that accurately reflect exposures in terms of the body burden, toxicology, and biological half life, and rationale for the exposure limits must be taken into account. Sometimes called biological considerations in sampling, or the pharmacologic approach to sampling, this procedure involves measurements of exposures utilizing the parameters of the biological relationship as guides. The toxicokinetics and pharmacodynamics of the agent and its effects will guide the selection of sample averaging time and total period of interest.[37,38]

A weakness in all air sampling strategies is the dependence on measurements of current exposures. In general, past exposures are not taken into account, although in the case of compounds capable of accumulating in the body it might be possible through biological sampling. However, currently biological methods exist for only a few compounds and the identification is difficult of what levels are significant in exposed air, urine, blood, or other substances.

A result is useless without some comparison or interpretation as to what it means. Generally this involves comparing the result to some legal or recommended standard or recognized guideline. It is a concept that is very important and should be incorporated into every air sampling strategy. The method used will determine in many cases how results are to be interpreted, for example, NIOSH or OSHA methods used to collect personal samples in the workplace should be compared to an OSHA—PEL, ACGIH—TLV, AIHA—WEEL, or other available guideline for workplaces. The basis for a standard must be considered when interpreting sampling results. For example, the current OSHA standard of 0.050 mg/M^3 for lead is designed to prevent reproductive effects (in workers, not the fetus) whereas the previous standard of 0.150 mg/M^3 was designed to prevent systemic effects. A prior OSHA standard of 2 fib/cm^3 for asbestos was designed to protect against developing asbestosis, a pneumoconiosis of the lung, but due to the greater concern of workers developing cancer, the OSHA standard was lowered to 0.2 fib/cm^3 and may soon be decreased to 0.1 fib/cm^3. Yet these levels are not considered safe for the general population and therefore an EPA guideline of 0.01 fib/cm^3 is frequently used for office buildings and schools.

If the standard is an 8-hour TWA, then the sample should be collected for close to 8 hours. If the workday is longer than 8 hours, then the standard may need adjustment. Sometimes the times cannot be exact, such as is the case when the minimum detection level requires a longer sampling time than 15 minutes, the time used for sampling peak exposures. In this case, the industrial hygienist must select a sampling rate and a time close enough to compare to the 15-minute ceiling.

While standards and guidelines are important, the ultimate criterion for a sampling professional such as an industrial hygienist is professional judgment. There may be a situation where the professional chooses to use a guideline that is lower than a standard because there is a hypersensitive worker population or it is necessary to protect individuals other than workers. It is not always possible to use the lowest standard. Therefore, it is essential to be practical as well as judicious. It is the ability to utilize standards and toxicity information, along with the existence of the necessary conditions of use for determining the degree of the hazard, that sets the sampling professional apart from the amateur.

REFERENCES

1. Soule, R. Industrial hygiene sampling and analyses. In *Patty's Industrial Hygiene and Toxicology*, 3rd rev. ed., vol. 1, *General Principles*, G. D. Clayton and F. E. Clayton, eds. New York: Wiley, 1978, pp. 707–769.
2. Dennis, R. *Handbook on Aerosols*. Washington, D.C.: Technical Information Center, Office of Public Affairs, U.S. Energy Research and Development Administration, 1976.
3. Stuart, B. O., P. J. Lioy, and R. F. Phalen. Use of size-selection in estimating TLVs. *ACGIH Ann.* **11**:85–96, 1984.
4. ACGIH Reports: TLVs for soluble and insoluble compounds. *Appl. Ind. Hyg.* **2**(6):R-4–R-6, 1987.

5. Riggins, R. M., and B. A. Petersen. Sampling and analysis methodology for semivolatile and nonvolatile organic compounds in air. In *Indoor Air and Human Health: Major Indoor Air Pollutants and Their Health Implications,* R. B. Gammage and S. V. Kaye, eds. Chelsea, Mich.: Lewis Pubs., Inc., 1985, pp. 351–359.
6. Wray, T. Vapor density. *Haz. Mat. World,* May 1989, pp. 68–69.
7. American Conference of Governmental Industrial Hygienists. *Industrial Ventilation, A Manual of Recommended Practice,* 16th ed. Cincinnati: ACGIH, 1980.
8. Lippmann, M. Dosimetry for chemical agents: An overview. *ACGIH Ann.* **1**:11–21, 1981.
9. Mastromatteo, E. TLVs: Changes in philosophy. *Appl. Ind. Hyg.* **3**(3):F-12–F-16, 1988.
10. Malek, R. F., et al. The effect of aerosol on estimates of inhalation exposures to airborne styrene. *AIHA J.* **47**(8):524–529, 1986.
11. Cain, W. S., and A. Turn. Smell of danger: An analysis of LP-gas odorization. *AIHA J.* **46**(3):115–126, 1985.
12. Bender, J. R., et al. Eye irritation response of humans to formaldehyde. *AIHA J.* **44**(6):463–465, 1983.
13. Environmental Protection Agency. *SOPs for Work on Hazardous Waste Sites.* Washington, D.C.: EPA, 1986.
14. Billings, C. E., and L. C. Jonas. Odor thresholds in air as compared to threshold limit values. *AIHA J.* **42**(6):479–480, 1981.
15. National Institute of Occupational Safety and Health. *Pocket Guide to Chemical Hazards.* Washington, D.C.: U.S. Dept. of Health and Human Services, Sept. 1985.
16. Raabe, O. G. Size-selective sampling criteria for thoracic and respirable mass fractions. *ACGIH Ann.* **11**:55–56, 1984.
17. Morgan W. K. C. The respiratory effects of particles, vapors, and fumes. *AIHA J.* **47**(11):670–673, 1986.
18. Popendorf, W. J., and J. T. Leffingwell. Regulating OP pesticide residues for farmworker protection. *Res. Rev.* **82**:125–201, 1982.
19. Klassen, C. D., M. O. Amdur, and J. Doull (eds.). *Casarett and Doull's Toxicology: The Basic Science of Poisons,* 3rd ed. New York: Macmillan, 1986.
20. Occupational Safety and Health Administration. *Preamble to the OSHA Final Rule Revising Workplace Air Contaminants.* Fed. Reg. vol. 54, no. 12, pp. 2332–2920, January 19, 1989.
21. Vianio, H., and L. Tomatis. Exposure to carcinogens: An overview of scientific and regulatory aspects. *Appl. Ind. Hyg.* **1**(1):42–48, 1986.
22. USEPA, Federal Emergency Management Agency, and U.S. DOT. *Technical Guidance for Hazards Analysis: Emergency Planning for Extremely Hazardous Substances.* Washington, D.C.: Government Printing Office, 1987.
23. Alexeeff, G. V., M. J. Lipsett, and K. W. Kizer. Problems associated with the use of immediately dangerous to life and health (IDLH) values for estimating the hazard of accidental chemical releases *AIHA J.* **50**(11):598–605, 1989.
24. DeSilva, P. TLVs to protect "nearly all workers." *Appl. Ind. Hyg.* **1**(1):49–53, 1986.
25. Phalen, R. F. Airway anatomy and physiology. *ACGIH Trans.* **11**:35–46, 1984.
26. American Conference of Governmental Industrial Hygienists. Threshold Limit Values and Biological Exposure Indices for 1989–1990. Cincinnati: ACGIH, 1989.
27. Atherly, G. A critical review of time weighted averages as an index of exposure and dose and of its key elements. *AIHA J.* **46**(9):481–487, 1985.
28. Anderson, M. E., et al. Adjusting exposure limits for long and short exposure periods using a physiological pharmacokinetic model. *AIHA J.* **48**(4):335–343, 1987.
29. Chemical substances TLV committee study papers. *ACGIH Ann.* **4**:153–157, 1983.
30. Lioy, P. J., M. Lippman, and R. F. Phalen. Rationale for particle size-selective air sampling. *ACGIH Ann.* **11**:27–34, 1984.
31. Phalen, R. F. Introduction and recommendations. Particle size-selective sampling in the workplace. *ACGIH Ann.* **11**:85–96, 1984.
32. Hickey, J. L. S., and P. C. Reist. Application of occupational exposure limits to unusual work schedules. *AIHA J.* **38**(11):613–619, 1977.
33. Panstenbach, D. Occupational exposure limits, pharmacokinetics and unusual work schedules. In *Patty's Industrial Hygiene and Toxicology,* 2nd ed., vol. 3, *Theory and Rationale of Industrial Hygiene Practice,* L. J. Cralley and L. V. Cralley, eds.
34. Occupational Safety and Health Administration. *Compliance Officers' Field Manual.* Washington, D.C.: Dept. of Labor, 1979.
35. Environmental Protection Agency. Chemical Emergency Preparedness Program. Interim Guidance. Washington, D.C. November 1985.
36. 40 Code of Federal Regulations, Part 261. Identification and Listing of Hazardous Waste.
37. Rappaport, S. *Biological Considerations for Designing Sampling Strategies.* Advances in Air Sampling. Chelsea, Mich.: Lewis Pubs., Inc., 1988.
38. Smith, T. J. *Sampling Strategies for Epidemiological Studies.* Advances in Air Sampling. Chelsea, Mich.: Lewis Pubs., Inc., 1988.

CHAPTER 2

Sampling Methods

Air sampling is done to identify occupational or environmental exposures. Occupational samples are usually collected indoors and are used to measure personal exposures to workers over a workday, to identify sources of exposure, or to evaluate the need or effectiveness of engineering controls. Environmental air sampling, often conducted outdoors, is done to determine whether a source of air pollution is present. It is sometimes referred to as ambient air sampling or community air sampling. Sampling might also determine whether volatiles are being released from water or soil. For environmental protection, air sampling is used to monitor emissions potentially damaging to outside surfaces or to plants, animals, or human health outside the emitter's facility. Types of sampling include collection of stack samples, high-volume particulates, volatile organic compounds, and boundary line samples at hazardous waste sites. When air sampling is done for community exposures it often involves very complex stationary monitoring systems usually left to collect massive amounts of data for 24 hours or more. Some situations require that both occupational and environmental samples be collected. For example, in indoor air situations in office buildings a determination must be made as to whether pollution from outdoor sources is a problem.

COLLECTING SAMPLES

Characteristics

There are two primary techniques for collecting air samples: collection on media for laboratory analysis, or real time or direct reading methods that give immediate results on ambient air in the field. Integrated sampling refers to the methods of contaminated air collection onto media (see, for example, Fig. 2-1). Because of the need to collect enough of a contaminant to meet laboratory detection methods, integrated sampling must be conducted over a minimum period of time, usually 15 minutes to several hours, depending on the type and concentration of the contaminant that is present as well as on the sampling situation. The result is a sample representing the average concentration present in the air over the sampling period.

Most sampling is quantitative, that is, a specific concentration of exposure to a specific agent(s) is measured; however, qualitative sampling to identify *what* is present is also done. For screening

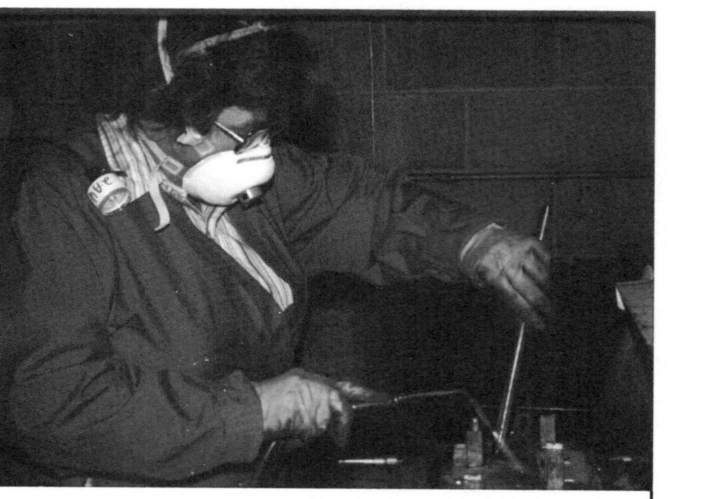

Lead Sampling Results
Acme Battery Works, Inc.

Employee	*Sample No.*	*Sampling Period*	*Result* ($\mu g/m^3$)
Janet Smith	100-10	4 hours	30
Janet Smith	100-11	4 hours	40
8-hour TWA			35
OSHA 8-hour TWA-PEL			50
OSHA Action Level			30

Figure 2-1. Using integrated methods to measure employee exposures to lead under OSHA standards.

samples high accuracy is not important; rather screening is done simply to see if the suspected contaminant is present or not.

Methods are either active or passive. Active methods involve some sort of pump that pulls air into a container, medium, or instrument. Passive methods rely on natural air currents for contact with the medium or on a detector in an instrument exposed to the contaminant (Fig. 2-2). Most passive samplers have been designed for personal monitoring, whether incorporated into a badge or an instrument, although some are used for environmental samples as well, such as the measurement of nitrogen dioxide in residences. There is a growing market for passive monitors due to their ease of use. However, as it is generally accepted that active methods involving the use of a pump provide the best accuracy and precision, most passive monitors are tested against active methods prior to use.

Locations

There are three primary types of samples: area, personnel, and source. Area sampling is most common when doing environmental monitoring. Personnel samples are collected for Occupational Safety and Health Administration (OSHA) sampling. Area samples are collected by putting the sampling apparatus near a source where the sampler wants to get an estimate of whether or not significant contamination may be present. Area sampling can be useful to detect where contaminants are most likely to be generated. A best approximation would involve placing a number of samplers around a given area so all are monitoring at the same time to create a "map" of the levels present. Therefore, the results of area monitoring will have a different interpretation than those of personnel monitoring.

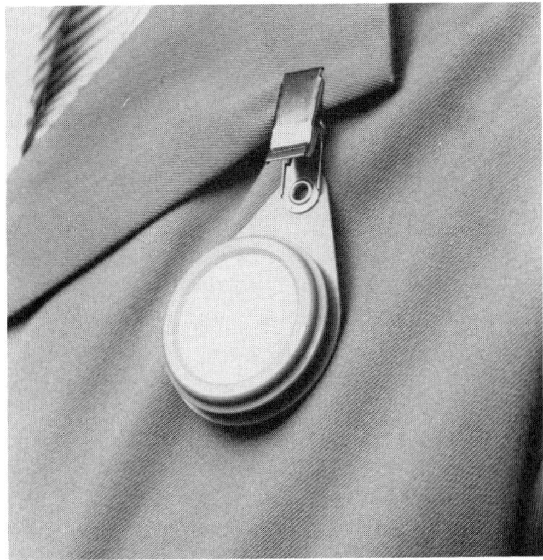

Figure 2-2. Using passive methods to collect integrated samples.

If sampling is being done to determine compliance with an OSHA standard, personnel must be monitored within the breathing zone unless prohibited by the sampling method or the workplace environment. The breathing zone can be defined as the area bordered by the outside of the shoulders and from midchest to the top of the head. Personnel samples are usually collected with a portable sampling apparatus consisting of a pump and media attached to the shirt or jacket collar of an individual, usually a worker. Personnel usually fall into one of two categories: those assigned to a particular station or those who roam throughout an operation in the course of their jobs. An assembly line worker is an example of the first situation and a forklift operator is an example of the second situation. Sampling data on jobs where workers roam is more difficult to interpret than on stationary workers.

Studies comparing personal (or personnel) sampling with area sampling have shown that area sampling may significantly underestimate exposures, especially in environments with moderate to great variability in air concentrations during the workday.[1] Area sampling is more appropriate for the measurement of source emissions than for approximating personal exposures. A significant increase in a worker's exposure compared to the concentration(s) established by area sampling results would suggest that the worker's specific tasks or work practices is increasing his or her exposure, assuming that samples are collected only when the worker is in the area.

Duration

Samples are also categorized by the time over which collection takes place, known as the sampling period. Generally a full period refers to an

8-hour workday, or in the case of environmental sampling, a 24-hour period or longer. One method is to collect a single sample to quantify the exposure over the entire period. A preferable approach is to collect multiple consecutive samples of equal or unequal time duration that equal the total sampling period when combined. An example would be collecting four samples, each 2 hours in length over a workday. There are several advantages to this type of sampling: First, if a single sample is lost during the sampling period due to pump failure, gross contamination, or whatever, at least some data will have been collected to evaluate the exposure. Second, the use of multiple samples will reduce the effects of sampling and analytical errors. And last, collection of several samples provides the sampling professional with more information about the exposure variations throughout the workday including those associated with different tasks.

Partial period air samples are generally an attempt to quantify an exposure or activity that takes place over a period shorter than a full day. This technique is useful when sampling is done to identify what contaminants are present (qualitative rather than quantitative results). Short-term or peak samples are primarily used to determine where highest exposures occur or to evaluate a task that takes a short period of time, such as repairs or batch-type processes.

Grab air sampling can involve collecting a number of short-term samples in one location at various times during the sample period that when combined provide an estimate of exposure over the total period, or it can involve collecting a number of short-term samples in different areas. In some cases, a grab sample may require filling a bag or evacuated canister with air to send to a laboratory for analysis; in other cases, it might involve collecting a bulk sample of a liquid or solid or swabbing a surface to see if contamination is present. For more information on these types of samples, see the chapter on Bulk Sampling Methods.

Screening samples refers to collecting samples to identify what is present (qualitative). These are often grab samples. Screening is often done using real time instruments to identify areas of high concentration so follow-up sampling to quantify specific contaminants will be more successful.

Strategy

For best sampling accuracy many samples should be collected including random sampling over long periods. Although this approach may be the most effective one, limitations in personnel or resources often result in sampling surveys that are limited to a few scheduled days. Costs of analyses and time limit the number of samples that can be collected. Often scheduling is done using worst-case scenarios to reflect typical high production and thus the highest likely exposure concentrations. However, whenever possible it is best to collect a sufficient number of samples to qualify for traditional statistical analyses. For more information on sampling strategies, including statistical techniques for determining the adequate number of samples to collect, see the chapter on Survey Preparations and Performance.

Equipment

Integrated gas and vapor methods for both occupational and environmental sampling use sorbent tubes, impingers, and bags. The primary difference is that environmental sampling trains are often more complex. Sorbent sampling can be done using either passive or active methods, although active are the most common. Sampling sorbents include charcoal, silica gel, and chromosorbs for occupational applications and Tenax and carbon molecular sieves for environmental sampling situations. Since the levels expected to be present are much lower in ambient air, the collection periods are longer, often 24 hours. Sorbents for occupational sampling are often contained in glass tubes while those used for environmental sampling are in stainless steel cartridges to allow for thermal desorption.

Impingers, another integrated method, are usually glass bottles with an inlet tube that allows air to be pumped through a measured volume of absorbing liquid, thus trapping the sample. Fritted bubblers are similar in use and appearance to an impinger, but generally allow a higher degree of air-liquid mixing, which is desirable for certain contaminants. Occupational sampling with impingers is being phased out since it is prone to spills during shipment of solutions and when workers wear the devices during sampling. Most commonly it is replaced with sorbent methods. Environmental methods rely on the same types of impingers; however, most techniques require that an ice bath be used. As is the case with other environmental methods, impinger techniques are designed to detect lower concentrations than occupational methods.

Some methods are designed to simply collect whole air without concentrating the sample. Bags of inert plastic materials are used to collect both grab and full-day samples. Stainless steel canisters, used only for environmental collection, can be operated in either a vacuum or pressurized mode. Canisters are used for collecting trace levels of volatile organic compounds.

Occupational sampling for aerosols is in a state of flux since the criteria for classifying particles into size ranges as they relate to sampling and thus respiratory hazards have been redefined. However, classic methods that for the most part do not reflect these new criteria are still the prescribed techniques. New instruments are becoming available, but testing to identify how the new criteria and classic methods compare is still in progress. Filters in plastic cassettes are the most common media used to measure occupational exposures to aerosols. Filters are also used for environmental sampling, but they are larger in size and in this case the apparatus is much larger, often contained inside shelters to allow for outdoor sampling. An advantage of occupational methodologies over environmental ones is their ability to measure specific types of compounds such as lead, arsenic, and sulfuric acid mists. Specialized techniques for some dusts also exist, such as the use of vertical elutriators for cotton dust. This method is an example of a situation where no personal sampling method exists because the sampling device is very large.

Integrated techniques for gases and aerosols are generally designed to collect single compounds that only exist in one of these phases. This collection is the most simple. In real life, compounds that must be evaluated include those that can exist as both a gas and an aerosol, generally called semivolatiles, or spraying mixtures that can generate both a gas and an aerosol. The effects of processing methods on certain elements can result in different compounds being formed within the same operation, and combustion products of organic materials are varied. Sampling in these situations may involve combining media for both gases and aerosols into the same sampling train or using several integrated methods at once in order to fully evaluate the situation. Sampling strategies are also sometimes adapted in this case, such as a decision to only sample for the most hazardous constituent and to ignore all others.

Real time instruments fall into several categories: instruments engineered to sample for a single chemical; instruments that can be adjusted to sample for several chemicals and produce results specific for each chemical; and general survey instruments that respond to many different compounds at once and cannot differentiate between compounds. The selection will depend on the situation being monitored. There are more integrated methods available for specific chemicals than there are instruments that can monitor for specific chemicals, so in many cases an integrated method is the only way to monitor. In a few cases, real time instrumentation is the primary way of detecting certain chemicals such as carbon dioxide and carbon monoxide. Instruments that monitor gases and vapors cannot be used to sample for aerosols and vice versa. In general, real time aerosol monitors cannot distinguish between contaminants and therefore much additional information is needed in order to properly interpret results.

Real time instruments used for occupational monitoring are battery powered and are not set up to measure very low levels (ppb) as are often needed for environmental measurements. Since few environmental instruments are portable, they are not generally used for survey work unless they are built into fixed monitoring stations such as those used by the Environmental Protection Agency (EPA). The use of instruments offers some advantages over integrated methods by reducing analytical costs and being able to provide instantaneous results, albeit at the expense of some degree of sensitivity or specificity.

Due to the increasing incorporation of microprocessors into many types of real time monitoring instruments, equipment is available that will memorize and format the data for output to a printer or to a computer for permanent data storage and data manipulation. Attributes of micro-

B

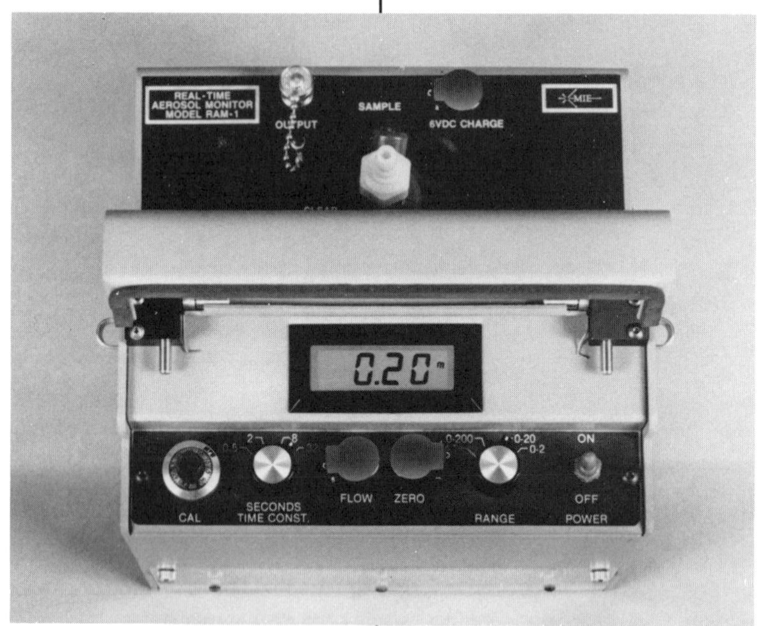
REAL-TIME
AEROSOL MONITOR
MODEL RAM-1
OUTPUT
SAMPLE
6VDC CHARGE
MIE
0.20
2
8
0-200
0-20
0-2
ON
CAL
SECONDS
TIME CONST.
FLOW
ZERO
RANGE
OFF
POWER

A

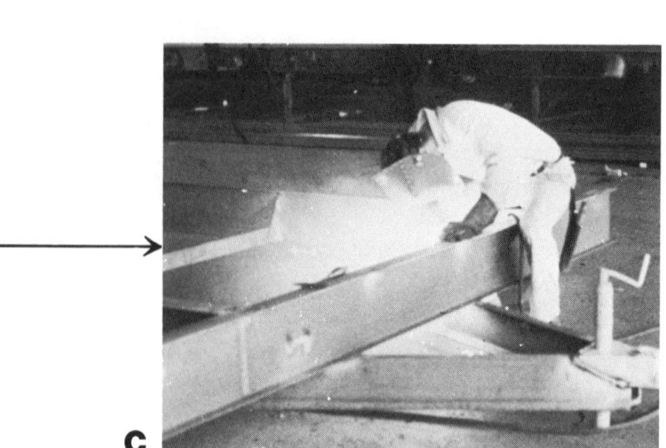

C

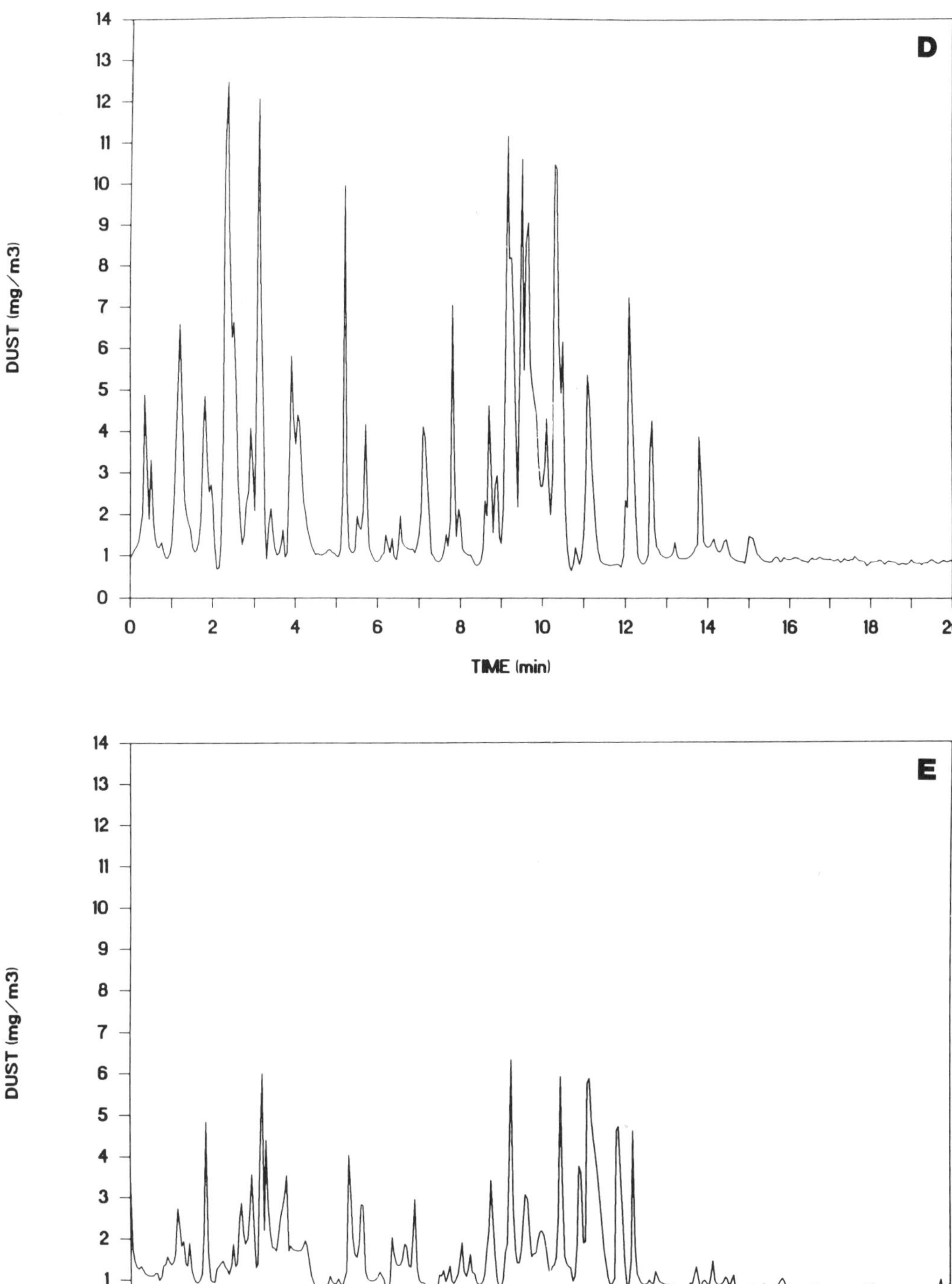

Figure 2-3. Using real time data to view the impact of instituting controls to decrease exposures. *A*: RAM-1 dust monitor (courtesy of MIE, Inc.). *B*: Welder without engineering controls. *C*: Welding with engineering controls. *D*: Real time data for welding without engineering controls. *E*: Real time data for welding with engineering controls. (*D* and *E* from *Appl. Ind. Hyg.* **4**(9):239, 1989.)

processors in sampling are discussed at greater length in the chapter on Reading and Recording Real Time Data. An example of data from real time instrument measurements shows the typical fluctuations to be expected as a result of concentration variations in a facility (Fig. 2-3). Many factors are responsible for this fluctuation, such as the process itself, which may be periodic, random (occasional), or cyclical and may include wind currents and fluctuations in usage of a given chemical. Sampling variables are discussed further in the chapter on Survey Preparations and Performance.

Most instruments designed to evaluate personal exposures are dosimeters, that is, they accumulate an average dose and give an integrated measurement. In most cases, there is an instantaneous readout available as well. Common gases for which these instruments are available include chlorine, ammonia, sulfur dioxide, hydrogen sulfide, and carbon monoxide (Fig. 2-4). Since there are only a few real time instruments that can monitor personal exposures, most real time instruments are used for area measurements.

For continuous area sampling a real time sampler is mounted in a fixed position and left on at all times. One application for this type of approach is leak detection, such as a wastewater treatment plant where incompatible chemicals are present

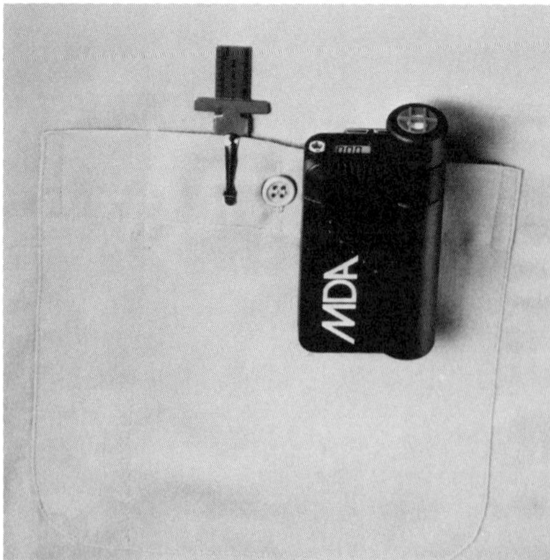

Figure 2-4. Using a real time instrument to measure personal exposures. (Courtesy MDA Scientific, Inc.)

and a hazardous reaction might occur. Often direct reading instruments are useful for identifying situations to repair or correct by installing engineering controls. An example is a leak in an ethylene oxide sterilizer from a poor door seal. A follow-up survey is often done to evaluate the effectiveness of any changes.

Color change media occupy their own sampling category in that they share certain aspects of both integrated and real time techniques. Like integrated sampling methods, there are detector tubes that are specific for a given chemical, but no chemical analysis is needed. Both active and passive techniques exist as well as the ability to pull both grab and long-term (8-hour) samples. Passive detectors can be in the shape of a tube, a badge with a granular spot, or a chemically impregnated tape. The trend has been to develop detector tubes that are increasingly specific for compounds through the use of scrubbing layers and doping the adsorbing layer with materials that promote a reaction with the compound of interest. These methods share the ability to provide on-site results with real time instruments, although a more appropriate description is direct reading. Although their accuracy is not as high as laboratory methods or many real time instruments, an advantage is the ability to provide immediate results for a wide variety of chemicals, particularly in situations where employees have expressed concern over a particular source of exposure.

DETERMINING SPECIFIC NEEDS

Since a large number of screening instruments and methods are available, the sampler must evaluate the sensitivity, selectivity, and potential for interferents in each technique prior to use. The required detection limits will vary considerably depending on the compound of interest, the sampling location relative to the source, meteorological conditions, and available sampling time. In general, the range of detection limits required for sampling will vary. Methods sensitive enough to detect levels in the 1 to 10-ppb range will be needed for ambient air studies where sampling sites are remote from emission sources, whereas 1 to 10 ppm may be sufficient for monitoring in the vicinity of chemical spills, dump sites, and other contaminated areas. The more toxic the compound, the more important it is to have a very low detection limit in an integrated method or low concentration sensitivity in an instrument.[2] However, many of the methods specific for chemicals were initially developed for occupational sampling where relatively high levels might be found; thus, detection levels are much higher than those required for environmental air sampling. Care must be used when applying these methods in environmental situations where levels are likely to be lower.

Alternative Methods

For each sampling situation a number of alternatives to sample collection may have to be evaluated in order to determine the most effective approach. Potential interferences, such as the presence of other similar chemicals, must be considered whenever sampling, especially if they are going to be collected on the same media or are both present in the environment to be sampled. For example, acetone and styrene are commonly found in the same environment because acetone is a good solvent for styrene resin. If styrene is the primary compound of interest, high levels of acetone on the same media may prevent adequate retention of the styrene or may interfere with a real time measurement.

In order to get adequate air sampling data, or to determine what to sample for, other types of sampling are sometimes necessary, such as biological, bulk, and wipe sampling. Biological samples, such as sampling of blood, urine, or exhaled air, are the best ways to determine the actual exposure to the body. However, in most cases the sampler wants to project what the exposure would be before it occurs; therefore, air sampling and other methods are used. Certain standards require routine biological monitoring, that is, blood leads or phenol in urine (a benzene metabolite). Continued high levels of a contaminant in biological fluids when air sampling levels are low can be an indication of skin or oral exposure. A determination of whether biological monitoring could be useful should be done. Table 2-1 lists exposure scenarios.[3]

Since air sampling can only be done once a contaminant gets into the air, bulk sampling is a good way to determine what is there before dusts and other nonvolatile substances are "projected" into the breathing zone. Bulk samples can be collected of air, chemicals, water, and soil. Bulk samples of air are collected in bags, evacuated containers, or sorbent tubes and are then analyzed to identify hazardous constituents. Drums and tanks are the most common types of chemical containers that are sampled.

Bulk soil and water samples are useful for identifying contaminants for which air sampling must be conducted during remediation work. Different collection techniques and apparatus are used depending on whether the material being sampled is a solid or a liquid. When the concern includes the presence of volatile chemicals or

TABLE 2-1. Exposure Scenarios

Exposure Scenario	*Population Characteristics*	*Measurement*
Occupational	Workers, population around the plant	Personal and area integrated and boundary line monitoring; surface wipe and biological samples
Transportation/ storage/spills	Storage, transportation workers, general population in area	General survey real time for releasing ambient levels using integrated techniques
Hazardous waste disposal	Workers at sites; general population around site	General survey real time
Drinking water	General population	Bulk samples of drinking water
Ambient air	General population	Integrated ambient air, bulk samples of surface waters and soil

radon gas in water, procedures are adjusted. Another concern is exterior lead-containing paint on tanks, piping, and other metal surfaces that may be worked on or cut using torches. Bulk sample collection is discussed in the chapter on Bulk Sampling Methods.

Many toxic materials can enter the body through the skin or as a result of oral ingestion of contaminated materials. Sampling for these concerns is primarily done to evaluate occupational exposures. There are two types of methods: those that assess worker exposure by monitoring external surfaces (such as work benches) for contamination, and those that measure contamination on the worker's skin and clothing. The first method requires the use of filters and other media to wipe surfaces or real time instruments to sniff them, while the second technique involves putting patches of gauze or charcoal on the worker or measuring contaminants that are either still on the skin or can be "washed" from it. When used underneath personal protective equipment, these techniques can determine equipment effectiveness in actual usage situations. These techniques are discussed in the chapter on Surface Sampling Methods for Dermal Exposure.

Contaminants Other Than Chemical Agents

While the emphasis of sampling is usually on chemical agents, others such as radon and bioaerosols are also a concern. Radon gas comes from the radioactive decay of uranium and is associated with contamination in residences as a result of soil gas migration. Occupational exposures are also measured, primarily in mines. Radon decay products are also a concern, and techniques for measuring radon decay and its decay products are different. There are both active and passive methods for radon sampling. The most common methods are passive. They are used in buildings and involve the use of charcoal or alpha track detection. Active methods involve the collection of radon on filters using pumps or continuous real time instruments. Grab samples are limited to area sampling and utilize filters. They are the primary means through which occupational samples are collected. Several different sampling periods may be required.

Bioaerosols, meaning airborne particles derived from microbes, viruses, and related agents such as insect parts or microbial toxins, are of concern in many sampling situations. In some cases, sampling is done for both a chemical, say cotton dust, and a microbe, in this case gram-negative bacteria associated with the release of endotoxin, since bract and leaf from cotton plants contain large numbers of these. Since most of these agents are living organisms, their fragility, temperature, moisture, and nutrient needs are concerns when selecting a sampling method. Methods exist that utilize conventional integrated devices, such as filter cassettes and impingers, and others that require very specialized instruments, such as impactors designed to house culturing media. Wipe samples and bulk samples are also used to identify sources of agents suspected of being present in bioaerosols.

Background samples also must be collected if air, bulk, or wipe sampling is being conducted in order to allow for comparison of the natural condition with the potentially contaminated situation. These samples are more critical when conducting ambient air and indoor air pollution monitoring, but they are also useful in other situations.

REFERENCES

1. Rappaport, S. M., S. Selvin, R. C. Spear, and C. Keil. Air sampling in the assessment of continuous exposures to acutely toxic chemicals. Part I—strategy. *AIHA J.* **42**(11):831–838, 1981.
2. Riggin, R. M. Technical assistance document for sampling and analysis of toxic organic compounds in ambient air. EPA-600/4-83-027, Washington, DC, 1983.
3. Preamble to OSHA final rule revising workplace air contaminant limits. *Fed. Reg.* **54**(12):2332–2920, 1989.

PART II

Integrated Air Sampling Measurements for Chemicals

CHAPTER 3

Overview of Integrated Sampling

What distinguishes integrated methods from other sampling techniques is the requirement for a laboratory analysis of the sampling medium. Both occupational and environmental methods exist and the primary difference between the two is the need to collect very low levels when sampling ambient outdoor air for environmental exposures.

There are two basic methods for collecting integrated samples of gases and vapors: The first involves active collection wherein a battery-powered pump pulls air through an appropriate media that collects the contaminant(s) of interest; the second method involves passive collection, thus exposing the sampling media to the contaminated air and allowing the compounds of interest simply to diffuse or permeate it. Collection media for gases and vapors either adsorb contaminants onto a sorbent, allowing them to be easily separated in the laboratory, or absorb them in a reagent as a result of a contaminant's solubility or a reaction that takes place to form another compound. When sorbent tubes are used several different tubes can be hooked up to a single sampling pump, allowing for the collection of several different types of contaminants at once due to variable flow controllers in tube holders that allow for different flows through each tube. Liquid solutions used for sampling are contained within devices called impingers: glass bottles with glass tubes centered inside to force the collected air to bubble through an absorbing solution. The combination of collection media, sampling pump, tubing, and a flow rate measuring device is called a sampling train.

Collection of gases and vapors with sorbents and liquid reagents is done for both occupational and environmental exposures. For occupational sampling a charcoal sorbent is most commonly used while environmental methods rely on Tenax. The collection methods used for impingers are similar but for environmental samples they are usually kept in an ice bath during collection.

When used for integrated samples bags can collect certain types of gases and vapors for which other occupational sampling methods are unsuitable. For environmental collection of bulk air, stainless steel canisters are used. An evacuated canister can be set up to pull a sample over a 24-hour period and the subsequent analytical methods can often detect part per billion (ppb) levels.

There are three categories of aerosol collection: (1) total sampling methods in which the entire mass available to the medium is collected and analyzed; (2) size selective in that only certain size fractions are collected; and (3) size distributions in which various size ranges of the sampled mass are determined. Occupational measurements of total aerosols typically consist of a cassette containing a membrane or fiber filter on which the aerosol is collected. Most filter methods are designed so they do not alter the collected material in the course of sampling. Environmental sampling of aerosols also uses filters; however, the sampling apparatus is much larger. The most commonly used respirable dust sampling device is the 10-mm nylon cyclone that has been used in mines and other types of industries for many years to measure the amounts of certain dusts capable of getting into the lungs. Other size fractions for which there is interest are inspirable and thoracic particles, and equipment is being developed for measuring these. For size selective sampling of environmental aerosols the Environmental Protection Agency (EPA) PM_{10} method is used, which collects particles that measure 10 μm and smaller. Cascade impactors can separate an aerosol into a particle size distribution that allows characterization of the primary size ranges according to their aerodynamic diameters. Cascade impactors have been widely used for occupational measurements including personal and area sampling, environmental air pollution studies, and microbial sampling. In some cases, the aerosol collector is unique to a certain type of collection as the vertical elutriator is for cotton dust.

METHOD SELECTION

Integrated methods used for the air sampling of specific compounds are developed by NIOSH, OSHA, and the EPA. An advantage of integrated methods for occupational sampling of aerosols is that there are methods that can be used to identify and quantify specific contaminants while most integrated environmental methods for aerosols are nonspecific and tend to measure particle counts or mass. For example, gravimetric analyses are done by preweighing a filter, collecting the airborne sample on the filter, and then reweighing the filter, but there is no specific analysis for the type of dust collected. Certain occupational dusts, such as fiberglass, cellulose (paper fiber), oil mists, and wood dust, are also collected this way.

As needs change, integrated methods are developed or discarded; for example, methods for chemicals that are no longer in use. For example, many pesticides have had their registrations revoked by the EPA and new chemicals are being developed and approved for use in manufacturing that do not have methods. Where these are likely to remain in the environment, such as the organochlorines, sampling may still be needed in hazardous waste situations. In addition, exposures to compounds such as drugs, enzymes, hormones, food additives, and other biochemicals have not always been a concern; however, new concerns as to their effects on workers are leading to the development of methods to sample them.

While there are sampling guidelines for compounds that occur in either a vapor or aerosol phase, only integrated methods can address the situation where both states are present. This may be due to a semivolatile compound existing in both states or to a mixture such as paint being sprayed or the by-products of combustion. Integrated methods are the most useful for addressing these situations. These situations are discussed in a separate chapter on Special-Case Sampling.

The selection of an integrated air sampling method depends on the physical and chemical characteristics of the air contaminant. Factors include the presence of high heat, high humidity, interferences, and if sampling is being done for compliance purposes. The availability of laboratory analysis, suitability of the sampling method to the environment to be sampled, complexity of the method and cost of the analysis are also important. In selecting sampling equipment, factors to be considered include the particle size and stability of aerosols involved, and for gases and vapors, solubility, vapor pressure, and chemical stability. Even tubing should be carefully selected, as rubber tubing may absorb certain contaminants and also will readily degrade in the presence of solvents. In general, clear tygon tubing will provide the best resistance to chemicals. While most media can be purchased directly from a supplier, some must be prepared by laboratory either because they are not sold fully assembled for sampling or they must be otherwise processed. Often laboratories will furnish media free or at minimal cost as long as samples are returned for analyses.

Once the sampling media and equipment are assembled, the specific sampling parameters must be determined, such as the number of samples, the length of the collection period, and the flow rate. The next step is to formulate a sampling strategy. It will depend on the standard that the results will be compared to as well as the toxicity of the contaminants and the purpose of the survey. The chapter on Performing Sampling Surveys will provide some guidance. Details on individual types of sampling are provided in the chapters on Integrated Sampling for Gases and Vapors and Integrated Sampling for Aerosols.

CALCULATION OF MINIMUM AND MAXIMUM SAMPLE VOLUMES

Prior to sampling, the minimum and maximum sample volumes allowed by the method must be calculated. The governing standard must also be taken into consideration. Collection of the minimum volume is necessary in order to collect enough sample to meet the detection limit of the method while the maximum volume is to protect the integrity of the compound that has been collected as well as to assure that the sample will not be overloaded. When high sensitivity is desired, longer sampling times should be used, but the sampler must generally stay within the limits of the maximum sampling volume and flow rate of the method. Since final air volumes are used in calculating the concentrations of contaminants, accurate pump flow rates are necessary. This involves doing a flow rate calibration. For more information on the types of pumps, as well as calibration methods, see the appendix on Pumps and Flow Rate Calibrations.

In the sample calculation shown below the permissible exposure limit (PEL) is inserted, but if sampling is being done to evaluate concerns as a result of tight building syndrome the sampler might use an American Society of Heating, Refrigeration and Air Conditioning Engineers (ASHRAE) guideline, which would change the results significantly.[1] For example, if the task being sampled is done for a short period of time, such as 20 minutes, then flow rates may have to be increased in order to collect sufficient sample volume. This situation is common in batch mixing operations that involve tasks such as dumping bags.

$$\text{minimum sample volume (liters)} = \frac{\text{minimum detection limit in } \mu\text{g}}{0.2\,(\text{PEL in } \mu\text{g/L})}$$

$$\text{maximum sample volume (liters)} = \frac{\text{tube capacity for vapors in mg}}{2\,(\text{PEL in mg/M}^3)}$$

Prior to sending air samples to the laboratory, the collection volumes must be calculated. The total minutes a pump ran for each sample are multiplied by the flow rate in liters. If a different flow rate was measured when doing the post-calibration than was originally set, the sampler must make a determination of what flow to use. If working in compliance, the sampler must make a determination of what flow to use. If working in compliance, the sampler would favor the higher flow rate. On the other hand, if the sampler works independently, the lower flow rate would be used, thus calculating a smaller volume, which in turn leads to a higher concentration when the analysis is done by the laboratory. However, in practice most often the two flow rates are averaged together and the average is used. If the pump was calibrated at a different altitude or atmospheric pressure than that at which the samples were collected, further adjustments are needed in this volume.

ADJUSTING AN INTEGRATED METHOD TO COLLECT CEILING AND TWA MEASUREMENTS

There will be situations where users of integrated methods will need to modify them because of interfering compounds, to take advantage of special capabilities of a laboratory or to collect several contaminants on one sample for one analysis. For example, the recommended volume may be reduced if it is suspected that several compounds will be collected on a single sample, and it may be increased if it is suspected that concentrations are very low, such as those frequently encountered in indoor air problems associated with tight building syndromes. In other cases, it may be necessary to collect samples by a number of different methods to determine their concentrations in a mixture of the contaminants.

Methods are meant for sampling over a full 8-hour or 24-hour day. However, some situations require adjustments so shorter periods can be sampled.[2]

The ideal method of measuring compounds with ceiling standards would be continuous monitoring equipment with cumulative display capability as well as an alarm to warn when the ceiling has been exceeded; however, in the absence of this method, standard technique, such as that for styrene, may be adapted to it.[1]

Example: styrene	*NIOSH Method 1501*
Recommended sample volume:	5L
Useful range of the method:	20–600 ppm
OSHA PELs:	200 ppm ceiling 100 ppm TWA
Recommended flow rate:	0.2 Lpm
Breakthrough time:	111 min at 0.2 Lpm and 200 ppm

For the ceiling use 0.2 Lpm for 30 minutes for a total volume of 6 liters. For TWA use 0.02 Lpm for 4 hours to collect two consecutive samples of 4.8 liters each.

When sampling to meet lower concentrations a different adjustment is recommended.

Example: formaldehyde	*NIOSH Method: 2502*
Recommended sample volume:	12 L
Recommended flow rate:	0.05 Lpm
Working range:	0.25–4 ppm
Limit of detection:	1–4 μg per tube, depending on media blank value

In order to collect 4 μg, 33 L of air would have to be collected, which is well above the recommended volume of 12 L. When pushing a method to the limit, such as in this situation, in order to obtain the sensitivity to determine the concentration of interest it is often necessary to sample large volumes while observing the recommended maximum flow rate. If this process is done, there is always the risk that the method may not work outside of the limits for which it was tested in the laboratory. In such cases, the best approach is to consult with the analytical laboratory and then take a sufficient number of samples to determine their useful limits in the particular application.

There will be situations where users of integrated methods will need to modify them because of interfering compounds, to take advantage of special capabilities of a laboratory or to collect several contaminants on one sample for one analysis. For example, the recommended volume may be reduced if the sampler suspects several compounds will be collected on a single sample, and it may be increased if it is suspected that concentrations are very low, such as those frequently encountered in indoor air problems associated with tight building syndromes. In other cases, it may be necessary to collect samples by a number of different methods to determine their concentrations in a mixture of the contaminants.

SAMPLE HANDLING

To prevent mix-ups always label sample holders (tubes, cassettes, impingers) with a unique number before placing them in the sampling train. The last step is to prepare samples for shipment to the laboratory. Sample seals are sometimes necessary. For example, OSHA compliance officers seal their samples prior to shipment. Samples collected for litigation purposes may require seals. Seals, if properly constructed and applied, can provide proof that the samples were not tampered with during transit or storage. The seal should have a glue that will not allow removal of the seal without detection. Another purpose for seals is to keep the caps from coming off of sorbent tubes and the plugs from falling out of the inlet and outlets. Seals minimize the opportunities for contamination of samples. It is also important to seal the caps of any bottles or other containers being shipped to ensure that leakage does not take place. A simple (although not tamperproof) method of sealing in these situations is to use duct tape. Samples should not be left in the work area, should not be easily accessible, and should not be kept at elevated temperatures such as in the trunk or glove compartment of a car.

When submitting samples to a laboratory there should always be an accompanying sampling data sheet. If the samples are incompatible, such as air samples and bulk samples that must be shipped separately, then a copy of the sampling data sheet should accompany each shipment. The sampling data sheet is often accompanied by or integrated into a chain-of-custody record. Chain-of-custody records have been used for environmental samples for some time and are also occasionally used for occupational samples. Generally the following information should be incorporated into an analytical request, also often used as a chain-of-custody form.[3]

Analysis desired
Type of media submitted
Date of sampling
Location of sampling
Description of sampling method
Air volume
Possible interferences
List of samples including client's number for each one
Date shipped
Note of any other samples being shipped separately

The most common way of collecting data in the field is through manually entering it on a Field Sampling Data Sheet. Instruments have been developed to automate this process.

The Buck Cali-Logger calculates flow rates, memorizes the calibration data for each sample, calculates the volumes of air sampled, and then prints out a request for sample analysis. Figures 3-1 and 3-2 show this instrument and a corresponding form to show chain of custody and the sampling report.

QUALITY CONTROL DURING INTEGRATED SAMPLING

The purpose of blanks is to identify sampling and analytical errors. There are two types of sampling blanks: field and media blanks. Field blanks, the most common, are clean sample media taken to the sampling site and handled in every way as the samples, except that no air is drawn through them. Media blanks are unopened and new samplers sent with the samples to the laboratory.

During analyses blanks are used to measure the signal contribution from reagents used by the laboratory in preparing samples for analysis of the collection media plus any contamination that may have occurred during handling, shipping, and storage before analysis.

The specific method being used should be consulted concerning the actual number and type of blanks required. A general rule is at least one blank for each day or shift during which sampling is conducted. The NIOSH recommended practice for the number of field blanks is 2 field blanks for each 10 samples with a maximum of 10 field blanks for each sample set. Sometimes blanks are subtracted from the results of the actual

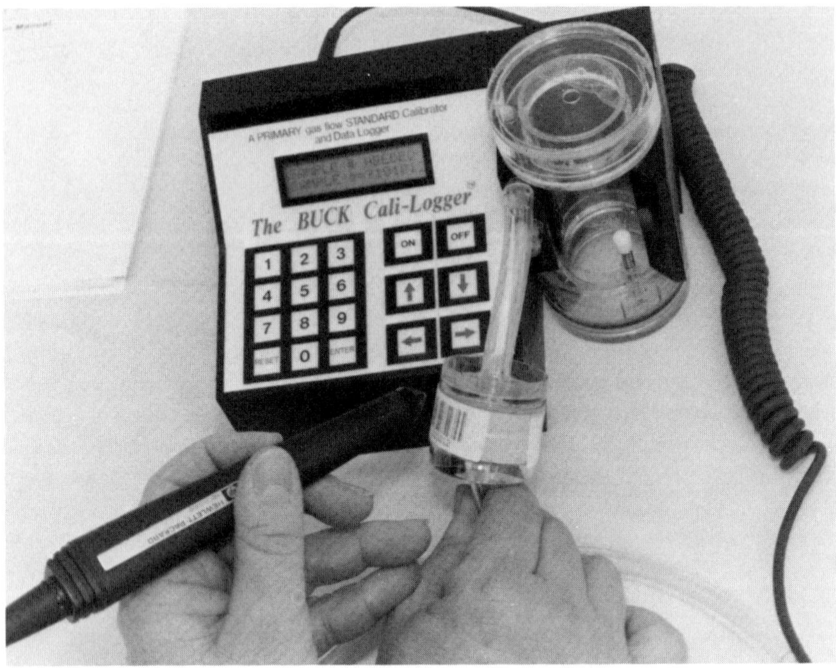

Figure 3-1. The Buck Cali-Logger provides storage of calibration data and generates chain-of-custody forms. (Courtesy A. P. Buck, Inc.)

samples. If this is to be done, it is important to determine whether the levels of contamination are consistent.

There are two types of spiked samples that can be used: laboratory and field exposed spikes. Spiked samples are generally prepared by adding a specific concentration of a chemical to media similar to that expected in the field. A microliter syringe is used to pipet a sample of a compound, such as an organic solvent, onto the collection medium, in this case a charcoal tube, while a pump pulls air through the tube. The most common media to be spiked are passive sorbent badges, sorbent tubes, and impinger solutions. The laboratory spike is taken to the site, but not exposed, and sent back to the laboratory with the actual samples for analysis. Other spikes are prepared in the field and analyzed with the regular samples to provide information on the stability of the sample or the presence of interferences, which will show up on the samples but not on the field spike.

WHAT TO EXPECT IN A LABORATORY REPORT

The detail and nature of the analytical report will depend on the function of the laboratory. As a minimum, the report should include a description or reference to the method used, any deviations or special circumstances encountered with the sample set, estimates of the limits of detection and quantitation, and the results themselves. The report should be dated and signed by the analyst and anyone else who is responsible for approving it. The name of the analyst should be printed as well for easy identification. The name of the client is usually at the top of the report. In some cases, a company will have all laboratory results delivered to one person for consistency; then regardless of who sends in the samples, the results will go to the designated person.

If the client provides a reference number, such as a project number used by most consulting

SAMPLING DATE: 24JAN90

COLLECTOR: MICHAEL

COMPANY: A.P. BUCK, INC.
ADDRESS: 3139 S. ORANGE AVE.
CITY,STATE,ZIP: ORLANDO, FL 32806
TELEPHONE: 1-407-851-8602

SAMPLING SITE: BATTERY PLANT

SAMPLE: Lot Number ______ Cassette Size ________

Batch Number ______ Filter Pore Size ________

SAMPLE NO.	PUMP NO.	START TIME	ELAPSED TIME	START FLOW lpm	STOP FLOW lpm	AVG. BP inHg	AVG. TEMP. degF	TOTAL VOLUME liters	VOLUME (STP) liters	ANALYZE FOR
FZ8900010	106	08:00	08:00	2.002	2.017	30.51	72	964.6	992.9	______
BK890088	444	09:00		1.461		30.31	73	ABORTED SAMPLE		______
123456789	123	06:00	17:00	4.071	4.009	30.31	74	4120.8	4199	______
9589002		08:00	03:00			30.31	73	BADGE SAMPLE		______
B85901	654	08:00	06:00	4.039	4.012	30.31	73	1449	1476	______
89A0134	111	08:00	08:00	4.060	3.727	30.31	72	1869	1912	______
300001	879	08:00	06:30	0.490	0.490	30.26	74	191.3	194.7	______
30BAA02	789	08:00	04:00	1.033	1.009	30.28	74	245.1	249.6	______
098765431						30.41	72	SAMPLE BLANK		______

Sampling Remarks: ____

____

Sample Receiver:

1. ____
name and address of organization receiving sample

2. ____

Chain of Possession:

1. ______________ ______________ ______________
signature title inclusive dates

2. ______________ ______________ ______________
signature title inclusive dates

Figure 3-2. Buck Cali-Logger sampling report and chain-of-custody form. (Courtesy A. P. Buck, Inc.)

firms it should be on the report. Also the sample numbers provided by the client should be stated along with the numbers that the laboratory has assigned to these samples. These will assist the sampling professional in identifying which results correspond to which samples without having to retrieve the sample submission sheet. The laboratory number is a unique number that the laboratory assigns when it receives samples to prevent them from being mixed up. Usually samples are numbered chronologically beginning at the first of the year.

Generally when submitting samples the client will furnish a description of each sample along with the unique sample number. It may include the name of the employee who was sampled or a description of the area that was sampled or the type of bulk material that is sent. If this information is also included on the laboratory report, it assists the client in readily identifying what samples the results are for. Sometimes the sample volume provided by the client is also recorded on the laboratory report.

The type of analysis should be listed according to what type of instrumentation was used. The number and source of the method should be listed, for example, NIOSH 7400 is used for asbestos. Any modification of the method should be stated. For example, there are two different counting rules for asbestos: "a" and "b" rules. OSHA requires the "a" rules. Modifying a method does not mean the results are in doubt. Laboratories often do this because they have learned from experience that the modification works better for them.

The detection limits for the particular analysis should be listed. The limit of detection is defined as the amount of the analyte that can be distinguished from background. The limit of quantitation is the amount of the analyte above which the precision of the reported results is better than a specified level. Since there are numerous methods of determining these quantities, the laboratory should decide on a method for determining them. Some laboratories calculate a new detection limit for each sample based on sample volume only. If very large sample volumes are collected, this procedure can result in very small detection limits, suggesting a sensitivity not originally intended by the method.

Sometimes an individual report is generated for each sample. Usually separate reports are generated for each type of analysis, with all samples for each category listed on each report. Pages should be numbered so that the total number of pages appears on each page of the report as well as the individual page number. Thus, any missing pages will be immediately noted.

The results should not exceed the allowable range for the method unless it has been validated for lower levels prior to sampling. Results for blanks should also be reported.

The appropriate standard depending on the method by which the samples are analyzed may also be listed. In the case of industrial hygiene laboratories, it is usually the OSHA PEL.

Depending on how thorough the laboratory is, additional supporting documentation such as a copy of the chromatogram may be attached to the results. Any problems or oddities that were observed during analysis, such as unexpected discolorations, peaks in the case of gas chromatography, or contaminated blanks, should also be noted. If sampling was conducted for a gas or vapor different from what was actually present, a reputable laboratory should report not only the compound looked for but any extraneous peaks. The sampling professional should be alert for any interferences that may have existed on the sampling site and should try to provide this information to the laboratory. It can account for some of the oddities that result during analysis and alert the laboratory that complications may be present. If breakthrough has occurred it should be noted. Breakthrough is generally defined as the presence of 25% or more of a contaminant in the rear portion of a sorbent tube.[2] In this situation it is very likely that sample has passed through a sampling tube.

If a correction for contamination is necessary, this correction should be performed by the person who submitted the samples. However, sample data should be corrected for recovery or desorption efficiency and for reagent and media blank response by the laboratory.

COMPARING SAMPLES TO STANDARDS

Analytical results must be transformed to represent the standard with which they are to be compared. If the standard is expressed as an 8- or 24-hour standard, and more than one sample was collected, or the sample was collected for a shorter period and there were periods of nonexposure, then these periods must be accounted for.

For particulates, laboratories usually report the results in mg/M^3 of contaminant using the actual volume of air collected at the sampling site. This value can be compared directly to the standard. Sometimes results will be given in μg and the computation will be left up to the client. However, most laboratories have computerized reporting that do these computations automatically.

In order to correctly interpret results of particle sampling, the sampler must be aware that there is much more variability over distance with particle concentrations than with gas and vapor. The results should be supplemented with additional information, such as distance from the source, air flow sources, typical airflow patterns, and obvious buildup patterns if visible.

Laboratories do not measure concentrations of gases and vapors directly in parts per million (ppm). Rather the analytical techniques determine the total weight of contaminant in the collection medium. Using the air volume the laboratory calculates the concentration in mg/M^3 and converts it to ppm at 25°C and 760 mm Hg using a standard equation:

$$\text{ppm}_{(\text{STP})} = \text{mg/M}^3 \frac{(24.45)}{(\text{MW})}$$

where

24.45 = molar volume at 25°C, 760 mm Hg
MW = molecular weight
STP = standard temperature and pressure (25°C, 760 mm Hg)

Thus, for example, samples expressed as ppm are different in Denver than in San Francisco

TABLE 3-1. Formulas for Comparing Results with OSHA PELs

Physical Form of Substance Sampled	*Unit of Air Concentration*	*Formula for Direct Comparison With OSHA PEL Table*
Gas	ppm	$C_v = \frac{m\,(10^3)}{V}\,\frac{(24.46)}{MW}$
Gas	mg/m^3	$C = \frac{m\,(10^3)}{V}$
Aerosol	mg/m^3	$C = \frac{m\,(10^3)}{V}$

Notes:
m = actual mass of substance, in mg, found on the sampling device
V = air volume in liters taken at the sampling site, ambient temperature and pressure
24.46 = the volume in liters of 1 mole of gas at 25°C and 760 mmHg
C_v = air concentration, ppm by volume, at 25°C, 760 mmHg
C = air concentration, mg/M^3
MW = molecular weight in grams/mole

unless referenced to standard temperature and pressure. If the gas or vapor standard is expressed in mg/M^3 it would not have to be adjusted. For situations where it is necessary to know the actual concentration in ppm at the sampling site, it can be derived from laboratory results reported in ppm at STP by using the following equation:

$$ppm_{(\text{sampling site})} = ppm_{(STP)} \frac{(760)}{(P)} \frac{(T)}{(298)}$$

where

P = sampling site pressure in mm Hg
T = sampling site temperature in kelvins
298 = standard temperature in kelvins (25 + 273)

Alternately, the sampling volume can be adjusted prior to sending samples to the laboratory. This approach is described in the appendix on Pumps and Flow Rate Calibrations.

FORMULA FOR DIRECT COMPARISON OF INTEGRATED RESULTS WITH PELS

To summarize the points presented in the previous section, Table 3-1 (on page 49) contains formulas for comparing results with OSHA PELs.

The next step is collecting samples. See the individual chapters as well as the chapters on surveys for more information. An example of a sampling data sheet for collecting field information is in the chapter on Survey Preparation and Performance.

REFERENCES

1. Standard Practice for Sampling Atmospheres to Collect Organic Compound Vapors. ASTM D3686.
2. National Institute of Occupational Safety and Health. *NIOSH Manual of Analytical Methods*, 3d ed., P. M. Eller (ed.). NIOSH, Cincinnati, 1984.
3. AIHA Analytical Chemistry Committee. *Quality Assurance Manual for IH Chemistry*. AIHA, 1988.

CHAPTER 4

Integrated Sampling for Gases and Vapors

Contrary to high dust situations where visible deposits can make it easy to estimate where concentrations are likely to be high, the presence of gases and vapors is often much more complex to detect. Since most gases and vapors cannot be seen, and levels of concern are often below or just at odor thresholds, they are generally detected using sampling methods. Surveys for gases and vapors are rarely ever simple situations involving a single compound; instead most often there is a mixture of contaminants in varying quantities and proportions.

When airborne, gases and vapors are in the form of individual molecules and thus tend to diffuse easily into the air and spread rapidly throughout an area. For air sampling purposes the properties of solubility, vapor pressure, and reactivity can be applied to selecting the proper collection procedure. There are two basic methods for collecting integrated samples of gases and vapors: The first involves a battery-powered pump actively pulling air through an appropriate medium. In the second method the medium is simply exposed to contaminated air, allowing the compounds of interest to passively diffuse or permeate it. The collection medium can be an adsorbent, such as a sorbent in a tube, allowing the gas or vapor to be easily separated from it, or an absorbent, such as a solution, in which the gas or vapor dissolves and/or converts to form another compound. Gases and vapors can also be collected in bags. Most active methods for gases and vapors on sorbents utilize flow rates of 200 ml/min or less and collection into absorbing solutions is most commonly done at 1 Lpm.

Impinger collection of an individual contaminant is also a consideration. When a gas is water soluble, such as chlorine or sulfur dioxide, it can be collected in a water-based solution. The impinger, discussed at length later in this chapter, is used for both occupational and environmental sampling. It is a glass bottle with a glass tube centered inside. Connected to a pump, the contaminated air is pulled through the glass tube into the absorbing solution.

Gases and vapors that are neither soluble in an aqueous solution nor reactive are usually collected on sorbent tubes, although some gases such as formaldehyde can be absorbed into a liquid solution or adsorbed onto a sorbent tube. Aerosols are not collected effectively by most types of sorbent, but when present, they may be collected by a prefilter positioned in front of the sampling medium so as not to interfere with results.

Reagent-treated filters are also used to collect

certain vapors. Nitro-reagent-treated filters are used to collect certain isocyanates, such as hexamethylene diisocyanate (HDI). When glass fiber filters impregnated with mercuric acetate are used to collect methanethiol or butanethiol, the entrapped thiol is regenerated from the relatively stable mercuric mercaptide that has formed on the filter during sampling by treatment with acid, and this solution is then analyzed.[1]

Integrated gas and vapor methods for both occupational and environmental sampling use sorbent tubes, impingers, and bags. The primary difference is that environmental sampling trains are often more complex; for example, impingers are in an ice bath and environmental sampling periods are much longer, often 24 hours rather than the prerequisite 8-hour period for occupational samples. In addition, other methods are used for environmental sampling of gases and vapors: stainless steel canisters and cryogenic trap collection. Canisters are used for very low levels (trace) of volatile organic compounds (VOCs) while cryogenic methods are used to collect compounds that are very volatile or otherwise hard to stabilize on sorbents. As cryogenic sampling requires special techniques that are not widely applicable to most types of survey work, they will not be discussed further.

For a summary of the types of media used for collecting occupational and environmental samples of many gases and vapors, see Table 4-1. This chapter describes the media and methods commonly used to collect integrated samples of gases and vapors (Fig. 4-1; Table 4-1). Other chapters of interest on gas and vapor sampling describe real time instrument methods, color change methods, and collection of bulk (or grab) air samples.

TABLE 4-1. Media Used in NIOSH and EPA Methods for Gases and Vapors

Compound	*Method Number**	*Media*
Acetaldehyde	3507	Girard T Reagent in a bubbler
	TO-5	Dinitrophenyl hydrazine solution in an impinger
Acetic acid	1603	Charcoal tube, 100/50
Acetic anhydride	3506	Alkaline hydroxylamine solution in a bubbler
Acetone	1300	Charcoal tube, 100/50
Acetone cyanohydrin	2506	Porapak QS tube, 100/50
Acetonitrile	1606	Charcoal tube, 400/200
Acrolein	2501	2-(Hydroxymethyl) piperdine on XAD-2 tube, 120/60
	TO-5	Dinitrophenylhydrazine solution in an impinger
Acrylonitrile	1604	Charcoal tube, 100/50
	TO-2	Carbon molecular sieve
Allyl alcohol	1402	Charcoal tube, 100/50
Allyl chloride	1000	Charcoal tube, 100/50
	TO-2	Carbon molecular sieve
Ammonia	6701	Sulfuric acid solution in a passive sorbent badge
Amyl acetate, *n*-, *sec*-	1450	Charcoal tube, 100/50
Aniline	2002	Silica gel tube, 150/75
Anisidine	2514	XAD-2 tube, 150/75
Arsine	6001	Charcoal tube, 100/50
Benzaldehyde	TO-5	Dinitrophenylhydrazine solution in an impinger
Benzene	1500, 1501	Charcoal tube, 100/50
	TO-1	Tenax GC
	TO-2	Carbon molecular sieve

TABLE 4-1. *(continued)*

Compound	*Method Number**	*Media*
Benzyl chloride	1003	Charcoal tube, 100/50
	TO-1	Tenax GC
Biphenyl	2530	Tenax GC tube, 20/10
Bromoform	1501	Charcoal tube, 100/50
Butyl acetate, *n*-, *sec*-, *tert*-	1450	Charcoal tube, 100/50
Bromotrifluoromethane	1017	Two charcoal tubes in series, 400/200 and 100/50
1,3-Butadiene	1024	Charcoal tube, 400/200
Butyl alcohol, *n*-, *sec*-, *iso*-	1401	Charcoal tube, 100/50
tert-Butyl alcohol	1400	Charcoal tube, 100/50
Butyl cellosolve	1403	Charcoal tube, 100/50
p-*tert*-Butyl toluene	1501	Charcoal tube, 100/50
Camphor	1301	Charcoal tube, 100/50
Carbon disulfide	1600	Charcoal tube, 100/50 and sodium sulfate tube, 270 (moisture removal)
Carbon tetrachloride	1501	Charcoal tube, 100/50
	TO-2	Carbon molecular sieve
Chlorobenzene	1501	Charcoal tube, 100/50
Chlorobromomethane	1501	Charcoal tube, 100/50
Chloroform	1501	Charcoal tube, 100/50
	TO-2	Carbon molecular sieve
Chloroprene	1002	Charcoal tube, 100/50
	TO-1	Tenax GC
Chloroacetic acid	2008	Silica gel tube, 100/50
Cresol, *o*-, *m*-, *p*-	2001	Silica gel tube, 150/75
	TO-8	Sodium hydroxide solution in an impinger
Cyclohexanone	1300	Charcoal tube, 100/50
Cumene	1501	Charcoal tube, 100/50
Cyclohexane	1500	Charcoal tube, 100/50
Cyclohexene	1500	Charcoal tube, 100/50
Cyclohexanol	1402	Charcoal tube, 100/50
1,3-Cyclopentadiene	2535	Maleic-anhydride-coated chromosorb 104, 100/50
Diacetone alcohol	1402	Charcoal tube, 100/50
Diazomethane	2515	Octanoic acid-coated XAD-2 tube, 100/50
Dibromodifluoromethane	1012	Two charcoal tubes in series, 150 each
2-Dibutylaminoethanol	2007	Silica gel tube, 300/150
Dichlorobenzene, *o*-, *p*-	1501	Charcoal tube, 100/50
	TO-1	Tenax GC
1,1-Dichloroethane	1501	Charcoal tube, 100/50
1,2-Dichloroethylene	1501	Charcoal tube, 100/50
Dichlorodifluoromethane	1018	Two charcoal tubes in series, 400/200 and 100/50
sym-Dichloroethyl ether	1004	Charcoal tube, 100/50
Dichlorofluoromethane	2516	Two charcoal tubes in series, 400 and 200
1,1-Dichloro-1-nitroethane	1601	Petroleum charcoal tube, 100/50
1,2-Dichloropropane	1013	Petroleum charcoal tube, 100/50
Dichlorotetrafluoroethane	1018	Two charcoal tubes in series, 400/200 and 100/50
2-Diethylaminoethanol	2007	Silica gel tube, 300/150
Diisobutyl ketone	1300	Charcoal tube, 100/50

(continued)

TABLE 4-1. *(continued)*

Compound	*Method Number**	*Media*
Dimethylacetamide	2004	Silica gel tube, 150/75
n,n-Dimethylaniline	2002	Silica gel tube, 150/75
Dimethylformamide	2004	Silica gel tube, 150/75
n,n-Dimethyl-*p*-toluidine	2002	Silica gel tube, 150/75
Dimethylsulfate	2524	Porapak P tube, 100/50
Dioxane	1602	Charcoal tube, 100/50
Epichorohydrin	1010	Charcoal tube, 100/50
Ethanol	1400	Charcoal tube, 100/50
Ethanolamine	2007	Silica gel tube, 300/150
2-Ethoxyethyl acetate	1450	Charcoal tube, 100/50
Ethyl acrylate	1450	Charcoal tube, 100/50
Ethyl benzene	1501	Charcoal tube, 100/50
Ethyl bromide	1011	Charcoal tube, 100/50
Ethyl butyl ketone	1301	Charcoal tube, 100/50
Ethyl chloride	2519	Two charcoal tubes in series, 400 and 200
Ethyl cellosolve	1403	Charcoal tube, 100/50
Ethylene chlorohydrin	2513	Petroleum charcoal tube, 100/50
Ethylene dibromide	1008	Charcoal tube, 100/50
Ethylene dichloride	1501	Charcoal tube, 100/50
	TO-2	Carbon molecular sieve
Ethylene glycol dinitrate	2507	Tenax GC tube, 100/50
Ethylene oxide	1607	Two charcoal tubes in series, 400 and 200
Ethylene oxide	1614	Hydrogen-bromide-coated petroleum charcoal tube, 100/50
Ethyl ether	1610	Charcoal tube, 100/50
Formaldehyde	2502	2-(Benzylamino)-ethanol-coated on Chromosorb 102 or XAD-2, 120/60
	3500	1 μm PTFE membrane filter, 37 mm, followed by two impingers containing sodium bisulfite in series
	TO-5	Dinitrophenylhydrazine solution in an impinger
Furfural	2529	2-(Hydroxymethyl) piperidine-coated XAD-2 tube, 120/60
Glycidol	1608	Charcoal tube, 100/50
n-Heptane	1500	Charcoal tube, 100/50
Hexachloro-1,3-cyclopentadiene	2518	Two Porapak T tubes in series, 75 and 25
Hexachloroethane	1501	Charcoal tube, 100/50
n-Hexane	1500	Charcoal tube, 100/50
2-Hexanone	1300	Charcoal tube, 100/50
Hydrazine	3503	Hydrochloric acid solution in a bubbler
Hydrogen bromide	7903	Washed silica gel tube, 400/200
Hydrogen chloride	7903	Washed silica gel tube, 400/200
Hydrogen fluoride	7903	Washed silica gel, 400/200
Iodine	6005	Alkali-treated charcoal tube, 100/50
Isoamyl acetate	1450	Charcoal tube, 100/50
Isbutyl acetate	1450	Charcoal tube, 100/50
Isoamyl alcohol	1402	Charcoal tube, 100/50

TABLE 4-1. *(continued)*

Compound	*Method Number**	*Media*
Isocyanate	5505	Impinger with solution of 1-(2-methoxyphenyl)-piperazine in toluene
Isophorone	2508	Petroleum charcoal tube, 100/50
Isopropyl alcohol	1400	Charcoal tube, 100/50
Mesityl oxide	1301	Charcoal tube, 100/50
Methanol	2000	Silica gel tube, 100/50
Methylal	1611	Charcoal tube, 100/50
Methyl *n*-amyl ketone	1301	Charcoal tube, 100/50
Methyl bromide	2520	Two petroleum charcoal tubes in series, 400 and 200
Methyl cellosolve	1403	Charcoal tube, 100/50
Methyl chloride	1001	Two charcoal tubes in series, 400/200 and 100/50
Methyl chloroform	1501	Charcoal tube, 100/50
	TO-2	Carbon molecular sieve
Methyl cyclohexane	1500	Charcoal tube, 100/50
Methylcyclohexanone	2521	Porapak Q tube, 150/75
Methylene chloride	1005	Two charcoal tubes in series, 100/50
	TO-2	Carbon molecular sieve
Methyl ethyl ketone peroxide	3508	Dimethyl phthalate solution in an impinger
5-Methyl-3-heptanone	1301	Charcoal tube, 100/50
Methyl iodide	1014	Charcoal tube, 100/50
Methyl isoamyl acetate	1450	Charcoal tube, 100/50
Methyl isobutyl carbinol	1402	Charcoal tube, 100/50
Methyl isobutyl ketone	1300	Charcoal tube, 100/50
α-Methyl styrene	1501	Charcoal tube, 100/50
Mevinphos	2503	Chromosorb 102 tube, 100/50
Naphthalene	1501	Charcoal tube, 100/50
Naphthas	1550	Charcoal tube, 100/50
Nickel carbonyl	6007	Low-nickel charcoal tube, 120/60
Nitric acid	7903	Washed silica gel tube, 400/200
Nitrobenzene	2005	Silica gel tube, 150/75
	TO-1	Tenax GC
Nitroethane	2526	Two XAD-2 tubes in series, 600 and 300
Nitrogen dioxide	6700	Palmes tube with three triethanolamine-treated screens (passive sampler)
Nitroglycerin	2507	Tenax-GC tube, 100/50
Nitromethane	2527	Chromosorb 106 tube, 600/300
2-Nitropropane	2528	Chromosorb 106 tube, 100/50
N-Nitrosodimethylamine	TO-7	Thermosorb/N
n-Octane	1500	Charcoal tube, 100/50
1-Octanethiol	2510	Tenax GC tube, 100/50
Pentachloroethane	2517	Poropak R tube, 70/35
n-Pentane	1500	Charcoal tube, 100/50
2-Pentanone	1300	Charcoal tube, 100/50
Phenol	3502	Sodium hydroxide solution in a bubbler
	TO-8	Sodium hydroxide solution in an impinger

(continued)

TABLE 4-1. *(continued)*

Compound	*Method Number**	*Media*
n-Propyl acetate	1450	Charcoal tube, 100/50
n-Propyl alcohol	1401	Charcoal tube, 100/50
Phosphorus trichloride	6402	Distilled water in a bubbler
Propylene oxide	1612	Charcoal tube, 100/50
Pyridine	1613	Charcoal tube, 100/50
Stibine	6008	Mercuric-chloride-coated silica gel tube, 1000/500
Styrene	1501	Charcoal tube, 100/50
1,1,2,2-Tetrabromoethane	2003	Silica gel tube, 150/75
1,1,1,2-Tetrachloro-1,2-difluoroethane	1016	Charcoal tube, 100/50
1,1,2,2-Tetrachloro-2,2-difluoroethane	1016	Charcoal tube, 100/50
1,1,2,2-Tetrachloroethane	1019	Petroleum charcoal tube, 100/50
Tetrachloroethylene	1501	Charcoal tube, 100/50
	TO-1	Tenax GC
Tetraethyl lead	2533	XAD-2 tube, 100/50
Tetraethyl pyrophosphate	2504	Two Chromosorb 102 tubes in series, 100/50
Tetrahydrofuran	1609	Charcoal tube, 100/50
Tetramethyl lead	2534	XAD-2 resin tube, 400/200
Tetramethyl thiourea	3505	Distilled water in an impinger
Toluene	1500	Charcoal tube, 100/50
	TO-1	Tenax GC
Toluene	1501	Charcoal tube, 100/50
Toluene-2,4-diisocyanate	2535	Tube with N-[(4-nitrophenyl)methyl]-propylamine coated on glass wool
o-Toluidine	2002	Silica gel tube, 150/75
Trichlorofluoromethane	1006	Charcoal tube, 400/200
1,1,2-Trichloroethane	1501	Charcoal tube, 100/50
Trichloroethylene	1022	Charcoal tube, 100/50
	TO-1	Tenax GC
	TO-2	Carbon molecular sieve
1,2,3-Trichloropropane	1501	Charcoal tube, 100/50
1,1,2-Trichloro-1,2,2-trifluoroethane	1020	Charcoal tube, 100/50
Turpentine	1551	Charcoal tube, 100/50
Vinyl bromide	1009	Charcoal tube, 400/200
Vinyl chloride	1007	Two charcoal tubes in series, 150
	TO-2	Carbon molecular sieve
Vinylidene chloride	1015	Charcoal tube, 100/50
	TO-2	Carbon molecular sieve
Vinyl toluene	1501	Charcoal tube, 100/50
Xylene	1501	Charcoal tube, 100/50
	TO-1	Tenax GC
2,4-Xylidine	2002	Silica gel tube, 150/75

Note: All charcoal tubes are coconut charcoal unless otherwise specified.
*TO methods refer to EPA methods.

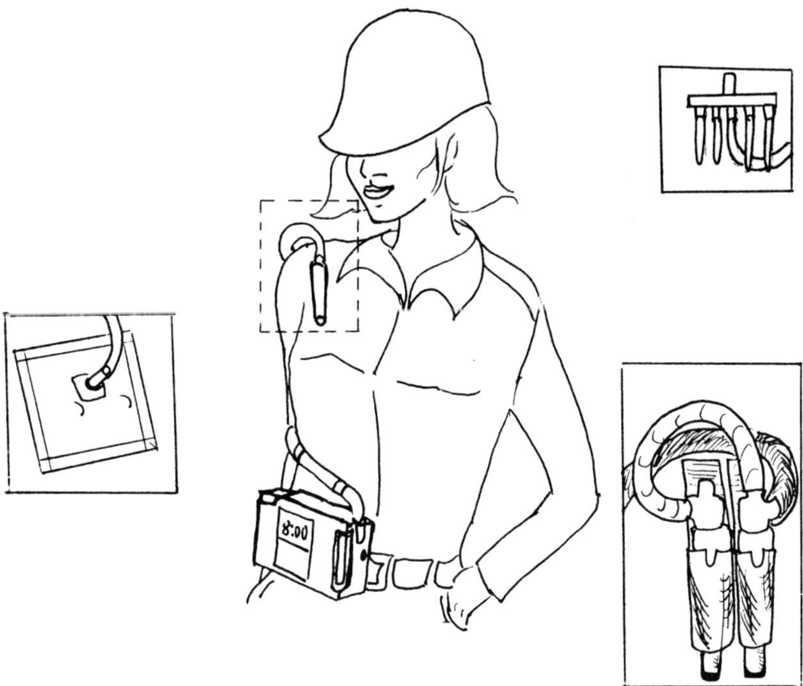

Figure 4-1. Types of personal sampling for gases and vapors. Clockwise from left bag: multiple sorbent tube sampler; impingers; on shoulder, a single sorbent tube.

SOLID SORBENT SAMPLING

Sampling with sorbent tubes is done for both occupational and environmental exposures. In both cases a pump is used to pull air through the sorbent, but here the similarity ends. The sampling media are often different as well as the methods and the calculations. Generally charcoal, silica gel, and chromosorbs are used for occupational air sampling and Tenax and carbon molecular sieves are preferred for environmental samples. Since the levels expected to be present are much lower in environmental samples, the collection period is much longer. Typically environmental samples are collected as a 24-hour average. Sorbents for occupational sampling are often contained inside glass tubes, while those for environmental sorbents are often packed inside of glass or metal cartridges.

Solid sorbents are specific for groups of compounds and one sorbent will not work with all compounds. Most solid sorbents do not differentiate between compounds during collection so unwanted compounds as well as the target compounds might be collected. On the other hand, because of the wide range of vapor pressure and breakthrough volumes for organic gases of interest, there is no one solid sorbent that can collect all gases and vapors simultaneously. Different manufacturers have different names for their sorbent materials and thus XAD and ORBO refer to manufacturers' brands and not a specific material. In the case of the ORBO brand, the tubes are numbered, with the number referring to the type of sorbent used in the tube. For example, ORBO-22 is used for formaldehyde and contains a styrene divinyl benzene polymer coated with N-benzylethanolamine; ORBO-32 contains charcoal granules.

Sometimes sorbents are coated with reagents to enhance collection of a specific compound. For example, acrolein is collected on 2-hydroxymethylpiperidine-coated XAD-2.[2]

When more than one compound is present

that can be collected on a sorbent tube, the amount of any individual component that can be collected is reduced. A reduction in sampling time or volume may be required because of the higher overall concentration being presented to the tube. Other factors that affect the ability of sorbent tubes to be efficient gas and vapor collectors include the size and mass of granules. Most importantly, doubling the mass will proportionately increase the ability of a tube to collect.[2]

The biggest concern in collecting material on a sorbent tube is whether breakthrough can occur. Breakthrough occurs when the front section of a tube is saturated and enough compound accumulates in the backup section that it begins to exit the tube with the airstream. Breakthrough is defined as the presence of 25% or more of a contaminant in the rear portion of a sorbent tube. Some sources consider 20% or more to represent breakthrough.[3] In this situation, it is very likely that sample has been lost through the tube. Some chemicals are prone to breakthrough, such as methylene chloride, which must be taken into consideration when planning sampling strategies. When results indicate breakthrough, the best interpretation is that actual concentrations were higher. Migration is a phenomenon similar to breakthrough, wherein an amount of the contaminant migrates from the front to the rear section of the tube while stored prior to analysis. This situation cannot be differentiated from the other causes of breakthrough.

When selecting sorbents the potential for water absorption must be considered if high humidity conditions are present at the site to be sampled. Competition for the adsorbent's surface will occur between the compound being sampled and the water, with the result that concentrations will appear lower than they actually are.

The flow rate will also impact the ability to retain gases and vapors on a sorbent above the optimum flow rate usually specified by a method. Increasing the flow rate above this level may result in a proportional decrease in the ability of the sorbent to retain contaminants. The variability of the pressure drop on a tube is dependent on the particle size of the sorbent.[4] Sorbents composed of fine adsorbent particles sample most efficiently, but their resistance to airflow is inversely proportional to particle size. Another problem with solid sorbents is that changes in air resistance due to swelling, shrinking, or channeling of the adsorbent are sometimes encountered. Table 4-2 lists factors affecting the collection behavior of solid sorbents.[5]

Many sampling professionals think if they collect an air sample on a charcoal or other sorbent tube and have it analyzed by gas chromatography/mass spectroscopy (GC/MS) that they will find anything that is in the air. It is not true, because although GC/MS will detect many organics, it does not detect inorganics. Depending on the sorbent material, some compounds may have a poor affinity and others too great an affinity. They will either not be collected or once they are collected they cannot be released from the media for detection. Also, libraries of compounds used by these instruments, while large, do not contain all compounds. Although wonderful for many applications, a single technique simply will not provide all the answers.

Charcoal Sorbents

Charcoal is one of the most commonly used sorbents and is useful for sampling a wide variety of organic gases and vapors, including several different compounds at a time. The use of charcoal as an air sampling medium came out of its use in the canisters of gas mask respirators during and after World War I as a defense to chemical warfare. Both the National Institute of Occupational Safety and Health (NIOSH) and Occupational Safety and Health Administration (OSHA) methods specify the use of charcoal tubes for a wide variety of organic compounds.

Charcoal can be derived from a variety of carbon-containing materials, but most of the charcoal used for air sampling is coconut- or petroleum-based. NIOSH recommends coconut charcoal, which has a higher capacity than petroleum charcoal for most compounds, although in some cases such as ethylene oxide collection a different type of charcoal is used. Ordinary charcoal is "activated" by steam at 800–900°C, causing it to form a porous structure. Activated charcoal is not useful for sampling certain types of reactive com-

TABLE 4-2. Factors Affecting the Collection Behavior of Solid Sorbents

Factor	*Effects*
Temperature	Adsorption is reduced at higher temperatures. Increased temperature is proportional to increased breakthrough. Reaction rates are increased at higher temperatures.
Humidity	Water vapor is adsorbed by polar sorbents, increasing the likelihood of breakthrough for chemicals. Increased humidity is proportional to increased breakthrough.
Flow rate	Higher sampling flow rates can lead to breakthrough at lower volumes. These vary with the type of sorbent.
Concentration	Breakthrough capacity increases (but breakthrough volume decreases) with increasing concentrations of contaminant.
Mixtures	When two or more compounds are present, the compound most strongly held will displace the other compounds, in order, down the length of the tube. Polar sorbents hold compounds with the largest dielectric constant and dipole moment most strongly. Nonpolar sorbents prefer compounds of a higher boiling point or larger molecular volume.
Nature of sorbent	Decreases in sorbent particle size are proportional to increases in sampling efficiency and drop in pressure.
Size of tube	A doubling of the sorbent volume doubles the concentration required for breakthrough.

pounds, such as mercaptans and aldehydes, due to its high surface reactivity with these compounds. Inorganic compounds, such as ozone, nitrogen dioxide, chlorine, hydrogen sulfide, and sulfur dioxide, react chemically with activated charcoal. Glass sampling tubes vary in size ranging from 150 mg to greater than 1,000 mg of charcoal. Another type of tube that is also carbon-based is a carbon molecular sieve. This material has a different structure than the conventional charcoal discussed here and is discussed in greater detail in the section on molecular sieves.

The most common charcoal tube in use for occupational sampling was designed according to NIOSH recommended specifications to include fabrication of glass tubing 7 cm long with a 6-mm outside diameter and a 4-mm internal diameter. The charcoal is 20/40 mesh-sized. A 100-mg front section is separated from a 50-mg backup section by a piece of urethane foam. A second piece of foam sits at the outlet to prevent granules from being sucked out of the tube during sampling. A plug of glass wool is in the very front of the tube. An unused tube will have both ends flame sealed. This structure and dimension is common to many tubes, although specialized tubes as well as scaled-up (proportionately larger) versions are also available.

Analysis of charcoal tubes is generally done using the solvent carbon disulfide to collect adsorbed contaminants.

The adsorbing capability of activated charcoal varies from batch to batch in commercial production. Charcoal tube collection efficiency for various hydrocarbons may be affected by such variables as sampling rate, vapor concentrations, and total adsorbed hydrocarbon mass. Nonpolar compounds, preferentially sampled on charcoal, will displace polar compounds in charcoal media. Competitive adsorption also occurs among polar compounds. Factors that may contribute to the affinity of the molecules for charcoal include hydrogen bonding, molecular size, volatility, and the dipole moment.[6]

Breakthrough volumes are variable and a function of the carbon, temperature, humidity, storage times, and pollutant. It has been estimated that for every 10°C increase in temperature, breakthrough is decreased by 1% to 10%.[7]

Charcoal tubes have been shown to be affected by high humidity. Nonpolar compounds, such as toluene and 1,1,1-trichloroethane, are the least affected. However, toluene has been shown to break through at lower sampling volumes if high humidity is present rather than in dry air. Reliable samples can be collected by limiting sample volumes where high humidity exists.[5] Polar compounds, such as ethyl cellosolve and dioxane, are the most susceptible.[8] Water vapor may also carry ethanol through the charcoal, for example. Charcoal has a higher affinity for water vapor than carbon molecular sieves.[9]

The type of compound may affect the flow rate selected for collection. A study of benzene sampling in charcoal determined that benzene was collected more efficiently at flow rates of 2 Lpm and higher. It is the opposite of what would be expected when using traditional charcoal tube sampling methods that specify flow rates of 50–200 ml/min.[10] The desorption efficiency for a particular compound can vary from one batch of charcoal sorbent to another. It may also be affected by the rate of loading, the total loading applied to the sorbent, and the distribution of material within the sorbent.[11]

Even though many compounds can be collected on a single charcoal tube, laboratories often prefer that a separate tube be used for each similar group of compounds sampled. For example, aromatics such as benzene, toluene, and xylene would be collected separately from chlorinated compounds. When sampling air containing mixtures of organic vapors of unknown concentrations, it may be useful to use two tubes in series, so that the backup tube is present in the case of breakthrough. When the mixture contains both polar (especially ethanol and methanol) and nonpolar compounds, the backup tube should contain silica gel as the collection medium.[12]

Silica Gel Sorbents

Silica gel is considered a more selective sorbent than activated charcoal, and gases and vapors are more easily desorbed from it.[13] Silica gel is an amorphous form of silica derived from the interaction of sodium silicate and sulfuric acid. It is the adsorbent recommended for collecting organic amines, both alkyl and aromatic, such as aniline and *o*-toluidine.

Factors that affect the dynamic adsorption of materials onto silica gel include the size range of the gel particles, tube diameter and length, temperature during sampling, concentration of contaminant being sampled, air humidity, and duration of sampling. Since the polarity of the adsorbed compound determines the binding strength on silica gel, compounds of high polarity will displace compounds of lesser polarity. Therefore, when attempting to collect relatively nonpolar compounds, the presence of coexisting polar compounds may interfere with collection on silica gel. The following compounds provide an example of the order of preferential adsorption of polar materials onto silica gel: water, alcohols, aldehydes, ketones, esters, aromatics, olefins, paraffins.[14]

Silica gel will show an increase in breakthrough capacity with increasing humidity. Therefore, under high humidity conditions the sample will be lost due to saturation with water vapor.[14] The ability of silica gel to absorb water vapor and displace collected components is its chief disadvantage.

Some methods specify that silica gel be washed to rid it of any impurities. Washing may be done using distilled water or in some cases inorganic acids. An example is NIOSH Method S138 that specifies sulfuric-acid-washed silica gel for the collection of butylamine.

Molecular Sieves

A carbon molecular sieve adsorbent is the carbon skeletal framework remaining after pyrolysis of the synthetic polymeric or petroleum pitch precursors. The result is a spherical, macroporous structure. The choice of starting polymer or pitch dictates the physical characteristics of the sieve, such as particle size and shape and sieve pore structure. The diameters of the micropores and their number are responsible for the differences in tube retention volume, adsorption coefficient, and equilibrium sorption capacity for a given compound.[9] The limiting factor with molecular sieves is humidity.

Spherocarb, Carboxen, Carbosieve, and Purasieve are examples of molecular sieves. Carbon-based molecular sieves are often called graphitized sieves. Nitrogen dioxide is collected on a triethanolamine (TEA)-impregnated molecular sieve. Carbon molecular sieves are most commonly used to collect environmental samples of highly volatile nonpolar organic compounds.

Porous Polymeric Sorbents

Porous polymers are another class of sorbent used for air sampling. These include Tenax GC, Porapak, Chromosorb, and XAD tubes. Their wide variety offers a high degree of selectivity for specific applications. Most porous polymers are copolymers in which one entity is styrene, ethylvinylbenzene, or divinyl benzene and the other monomer is a polar vinyl compound. Limitations include displacement of less volatile compounds, especially by carbon dioxide; irreversible adsorption of some compounds, such as amines and glycols; oxidation, hydrolysis, and polymerization reactions of the sample; chemical changes of the contaminant in the presence of reactive gases and vapors, such as nitrogen oxides, sulfur dioxide, and inorganic acids; artifacts arising from reaction and thermal desorption; limited retention capacity; thermal instability; and limitations of sampling volume, flow, and time.[14]

The Porapaks are a group of porous polymers that exhibit a wide range of polarity. The least polar member, Porapak P, is used in gas chromatography columns, and the most polar, Porapak T, can separate water and formaldehyde. Porapak QS is used to collect acetone cyanohydrin, Porapak P is used for dimethyl sulfate, and Porapak T is recommended for sampling hexachloro-1,3-cyclopentadiene.

The Chromosorbs are similar to the Porapaks. Chromosorb 101 is the least polar and Chromosorb 104 is the most polar. Chromosorb 106 is a cross-linked polystyrene porous polymer that has been used to sample nitromethane and 2-nitropropane. Chromosorb 102 has been recommended for collecting mevinphos, an organophosphate, and tetraethyl pyrophosphate.

XAD resins are a brand name referring to a number of different porous polymer types. XAD-2, the most commonly used, is equivalent to Chromosorb 102 and is used to collect anisidine and tetraethyl lead. XAD-4 has been used to collect organophosphorus pesticides. Amberlite XAD-2 is a styrene-divinylbenzene polymer.

Tenax

Tenax is one of the most widely used porous polymers, especially for environmental sampling. It has been used for studies of VOC levels in indoor air for comparison with outdoor air levels. Compared to industrial situations, these levels can be expected to be very low. Tenax is a polymer of 2,6-diphenyl-p-phenylene oxide and can be used to collect organic bases, neutral compounds, and high-boiling compounds. Tenax is used mostly for sampling low concentrations of volatile compounds.

Tenax GC has a high thermal stability, being able to withstand temperatures of up to 350°C, which permits it to be used for thermal desorption. Thermal desorption is desirable because the entire sample is introduced into the analytical system, whereas the alternative, solvent extraction, dilutes the sample and allows only a portion of it to be injected. A limitation with thermal desorption is that the sample can be injected only once.

Tenax has other properties besides its temperature stability that make it useful for concentrating compounds of medium volatility. It is relatively inert and has low, but not zero, affinity for water vapor, and due to extensive use, its advantages and limitations have been well characterized. Multicomponent qualitative analysis by GC/MS and quantitative analysis with standards can be done.[2]

However, retention or breakthrough volumes for a variety of highly volatile compounds on Tenax are low.[15] Other limitations of Tenax include a laborious cleanup procedure to control blank problems, a short useful half life after sample collection (approximately one month), and a tendency to decompose during sampling to produce acetophenone and benzaldehyde. Other problems that have been noted are that in the case where Tenax is contaminated blank corrections may not apply equally to all cartridges in a set; very high humidity in the air being sampled may affect sample retention; high concentrations or high numbers of compounds may exceed the retention capacity of the Tenax bed; and artifact formation may occur due to chemical reactions during sampling and/or thermal desorption. Tenax is not effective for low-molecular-weight hydrocarbons (C_4 and below) and mid-range (C_{5-12}) highly polar compounds.[16]

Tenax also reacts with strong oxidizing agents, such as chlorine, ozone, nitrogen oxides, and sulfur oxides, to form benzaldehyde, acetophenone, and phenol.[16] Other documented chemical transformations on Tenax include oxidizers reacting with organics, such as styrene, to produce compounds, such as benzaldehyde and chlorostyrene. One way of removing oxidants before they reach Tenax is to use sodium-thiosulfate-impregnated glass fiber or Teflon filters in front of Tenax in the sampling train.[16]

Other Sorbents

Other sorbents occasionally used for integrated gas and vapor sampling include alumina gel and florisil. Alumina gel, a form of aluminum oxide, is rarely used as an air sampling sorbent except for special applications such as formaldehyde. Alumina gel permits several thousandfold concentration of pollutants. It selectively absorbs polar and higher molecular weight compounds, with the degree of polarity determining binding strength of a compound on this gel.[2] Alumina gel has been used to collect polar compounds, such as alcohols, glycols, and ketones, and aldehydes, such as formaldehyde. It has an affinity for water like all polar sorbents. Florisil, based on silicic

TABLE 4-3. Types of Solid Sorbents

Solid Sorbent	*Characteristics*
Activated charcoal	Very large surface area: weight ratio. Reactive surface. High absorptive capacity. Breakthrough capacity is a function of the source of the charcoal, its particle size, and packing configuration in the sorbent bed.
Silica gel	Less reactive than charcoal. Polar in nature. Very hygroscopic (water adsorbing).
Porous polymers	Less surface area and less reactive surface than charcoal. Absorptive capacity is also lower. Reactivity is lower as well.
Molecular sieves	Zeolites and carbon molecular sieves that retain adsorbed species according to their molecular size. Water may displace organics.
Coated sorbents	A sorbent upon which a layer of reagent has been deposited. Adsorptive capacity determined by the capacity of the reagent to react with the particular analyte.

acid, is used for polychlorinated biphenyl (PCB) collection as well as for some pesticides. It is sometimes acid washed to remove residual magnesium. Table 4-3 lists types and uses of solid sorbents.[14]

USE OF SORBENT TUBES

Occupational Sampling

Collection of gases and vapors on sorbent tubes is the most common technique used in occupational sampling (Fig. 4-2). The most frequently used sorbent is the charcoal tube and analysis is almost always done by gas chromatography. An advantage of sorbent sampling is that samples can be collected for qualitative as well as quantitative analyses. Qualitative samples are discussed further in the chapter on Bulk Sampling Methods in the section on bulk air samples. Most often quantitative sampling is done to identify concentrations to which employees are exposed.

If a single contaminant such as benzene is present at concentrations up to the PEL, a single 50/100-mg charcoal tube is sufficient. On the other hand, if several solvents in the same family are present, such as benzene, toluene, xylene, and ethylbenzene, it may be better to use a larger tube. The NIOSH or OSHA sampling method will dictate what media, flow rate, and time to sample; however, at a typical flow rate of 100 ml/min sampling would be done for 1.5 hours to collect a minimum volume of 10 liters. Passive monitors can be used instead of active techniques. These are described in a later section of this chapter.

Figure 4-2. Personal exposure assessment of an ethylene oxide sterilizer using a sorbent tube.

Procedure

This procedure assumes a pump with a rotameter is being used. Stroke volume pumps can also be used for sampling. Their use is described in the appendix on Pumps and Flow Rate Calibrations.

1. A low flow pump (10–200 ml/min) or a combination flow pump set at the low flow range is used. Hook up tubing to the pump inlet. If the pump is designed to be used with a specific tube holder (e.g., it contains a critical orifice or other flow restricter), that holder must be attached to the end of the tubing.

2. Break the ends off a sorbent tube of the type to be used for sampling and insert it into the holder. If the adsorbent tube is placed directly inside of tygon tubing without a special holder, duct tape should be wrapped around both ends to minimize leaks.

3. If the sampling protocol requires the use of two tubes in series, or a prefilter in front of the tube, then these must be in the calibration train. The airflow to the pump should be in the direction of the arrow on the tube. Usually glass wool is placed at the inlet end of tubes as well.

4. Calibrate the pump using a bubble buret or rotameter as described in the appendix on

Pumps and Flow Rate Calibrations. The flow rate recommended in the procedure for the vapor(s) to be sampled should be used, usually about 100 ml/min with a solid sorbent tube in line, but falling within a range of 50—200 ml/min. Some methods give a range of flow rates. Higher rates are usually used when collecting short-term samples.

5. Determine whether high humidity is present. If so, a drying tube might be added to the front of the sampling train to absorb water. This is more important for silica gel and molecular sieves than for charcoal tubes.

6. Immediately prior to sampling, break off both ends of a new tube to provide an opening at least one half of the internal diameter at each end. There are tube breakers available or small needle-nosed pliers will work. Do not use the charging port or the exhaust outlet of the pump to break the ends of sorbent tubes. Label the tube with a unique number.

7. Connect the tube to the calibrated sampling pump. The smaller section of the tube is used as the backup section and is positioned closest to the pump. Depending on the pump system being used, the tube may have a protective cover screwed over it.

8. Select the employee(s) or area(s) to be sampled. Discuss the purpose of sampling and advise the employee not to remove or tamper with the sampling equipment. Inform the employee when and where the equipment will be removed.

9. Attach the pump to the worker's belt, waist line, or the inside of a shirt pocket. Place the collection device on the shirt collar or otherwise near the employee's breathing zone. (See Fig. 4-1.) Run the tubing either up the worker's back or under the arm. The sorbent tube should be placed in a vertical position, with the inlet either up or down during sampling to avoid channeling and premature breakthrough. Use duct tape to keep the tubing out of the employee's way or else the tubing might be disconnected from the pump.

10. Turn on the pump and record the start time. Get the employee's full name, position or title, or record a description of the area in which the pump is placed. Include the pump number and the number on the sample, beginning sample time, air temperature, relative humidity, and atmospheric pressure if the elevation is above sea level. In some cases, the employee's social security number is also needed.

11. Observe the rotameter on the pump for a short time to make sure it is pulling air. Mark down the setting of the rotameter ball (calibrated mark). As a minimum, check the pump after approximately the first half hour, hour, and every hour thereafter to make sure it is running. Ensure that the tubing is still attached to both the pump and the collection device and that it is not pinched. Make sure the sorbent tube is still in the same position.

12. Collect the sample for a sufficient time to meet the minimum sample volume requirements for the method. For example, organic vapors on a 150-mg charcoal tube require a 10-L sample volume; a 600-mg tube requires 40 L.

13. Periodically monitor the employee throughout the sampling period to ensure that sample integrity is maintained and cyclical activities and work practices are identified.

14. If several tubes will be collected in sequence on the same employee, before removing the pump at the end of the sample period, check to see that the rotameter ball is still at the calibrated mark. If the ball is not at the calibrated mark, the flow rate of the pump must be adjusted prior to inserting another sorbent tube.

15. Record the ending time. Immediately following sampling the tube should be capped off on both ends. Solvent vapors may penetrate (diffuse) through plastic caps on charcoal tubes; therefore, it is important to store collected samples by themselves away from all sources of chemicals.[17] Following sampling at the end of the day the pumps must be recalibrated. Compare the flow rate at the beginning of the sampling period with this postcalibration.

16. Prepare the field blanks about the same time as sampling is begun. These field blanks should consist of unused solid sorbent tubes from the same lot used for sample collection. Handle

and ship the field blanks exactly as the samples (break the ends and seal with plastic caps), but do not draw air through the field blanks. NIOSH recommends 2 field blanks be used for each 10 samples, with a maximum of 10 field blanks per sample set. In addition to the sample tubes and field blanks, it is also recommended that 2 unopened tubes be shipped to be used as media blanks, so that desorption efficiency studies can be performed on the same lot of sorbent used for sampling.

Occupational Sampling with Multiple Tubes

Using the same techniques described for collecting gases and vapors on a single tube, it is possible to sample using several tubes at once. It is often done in indoor air situations where a battery of screening samples are collected to find if any contaminants are present at detectable levels. This technique is also useful for sampling industrial processes where many different solvents are in use. Other applications for this type of system include parallel sampling in situations such as side-by-side OSHA surveys and parallel sampling using tubes for different contaminants, including tubes of different size and composition.

Multiple tube holders are available that allow for sampling with up to four tubes at a time, but only one pump is required. Variable flow controllers attached to each tube allow for regulation of different flow rates for each tube. Tube flows are additive and the total cannot exceed the pump's capacity, which is frequently 500 ml/min for low flows. The sampling rate of the pump is controlled by the characteristics of the limiting orifice assembly built into the tube manifold. The maximum flow rate that can be distributed among the tubes depends on the pump and should be determined ahead of time. This procedure assumes a combination (high-low) constant flow pump is being used.

Procedure

1. Select the correct tube manifold for the maximum number of tubes to be collected, but do not plan to sample with empty slots. Flow rate selection for each tube depends on the minimum sampling volume for each method. It is generally best to set up tubes so that all can be collected for the same time period. This often requires some preparatory calculations in order to match up all the tubes. Since setup can take some time, if repeat sampling is to be done it may be best to dedicate a pump/tube manifold combination for this purpose.

2. Break the ends of each sorbent tube and install the tube into the tube holder housing, with the arrow of the sorbent tube facing toward the manifold. Select the proper size cover for each tube.

3. Follow the procedures discussed in Appendix A for converting combination pumps to low flow. Once this is done, hook up the tube manifold to the pump using Tygon tubing.

4. The end of a tube is hooked up with tubing to a calibration device such as a rotameter or bubble buret. Some tube covers require a fitting at the front end of the housing in order to attach tubing for calibration whereas for others the cover is removed. Tube covers used to hold tubes in place should be screwed down tightly on each end to make sure tubes are correctly positioned. Put a unique number on each tube as it is installed inside the holder.

5. At the base of the manifold there is a screw that is a fine flow adjustment for each tube. This screw is generally visible or underneath a protective cap. Ideally this is the only adjustment that will be necessary, but when as many as four tubes are in use, sometimes the gross flow adjustment on the pump has to be turned up.

6. Ideally, once individual tube holders are set up, changes in the other holders should not affect the system, but as a practical matter they can. Thus, when adjusting the flow rate in one tube, it is best to double-check the flows of other tubes. Make sure the flow is stable for each tube. Shake the tube while the pump is running and watch the rotameter installed on the pump. If the flow through the tube is stable, the pump's flow rate will not change.

7. Once flows are set, if sampling is not to begin right away put caps on the ends of the

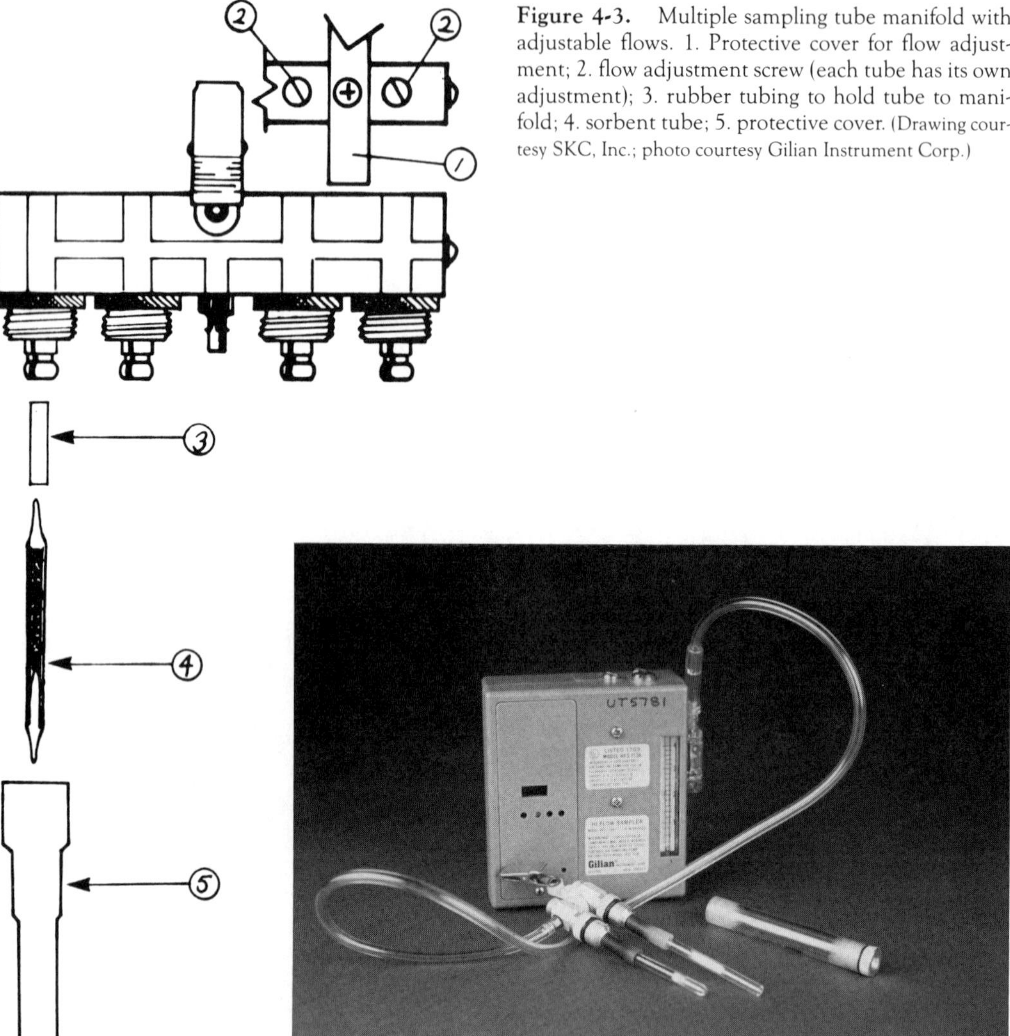

Figure 4-3. Multiple sampling tube manifold with adjustable flows. 1. Protective cover for flow adjustment; 2. flow adjustment screw (each tube has its own adjustment); 3. rubber tubing to hold tube to manifold; 4. sorbent tube; 5. protective cover. (Drawing courtesy SKC, Inc.; photo courtesy Gilian Instrument Corp.)

tubes/tube holders and set the pump manifold system aside while the other pumps are being calibrated.

8. For sampling the manifold is installed with the tubes positioned vertically. The same procedures are followed as with the use of a single sorbent tube (steps 8–16).

Environmental Sampling

As noted, the Environmental Protection Agency's (EPA) air sampling methods for gases and vapors differ significantly from NIOSH/OSHA ones. When monitoring for low levels of contaminant in ambient air, such as a boundary line at a hazardous waste site or outdoor samples to compare with those collected inside a building suspected of having tight building syndrome, EPA techniques that provide enhanced sensitivity over those used for occupational sampling are needed. Most analyses are done using thermal rather than solvent desorption, allowing for an entire sample to be analyzed rather than a small portion.

Two examples of specific EPA methods for gases and vapors are for nonpolar organics.[18] Depending on the volatility of the compounds, different sorbent media are used. For highly volatile compounds having boiling points in the −15°C to 120°C range, carbon molecular sieve (CMS) tubes are used and the method is referred to as TO2. For less volatile compounds in the 80°C to 200°C range, Tenax tubes are used and the method is referred to as TO1. A significant precaution is the need to rigorously avoid contamination of exterior surfaces when cartridges are used to contain the sorbent as the entire surface is subjected to the purge gas stream during the desorption process. Modified versions of these methods allow glass tubes similar to occupational sorbent tubes to be used instead of cartridges.

Sorbent cartridges should be stored in Teflon-capped culture tubes wrapped in foil to limit exposure of sampling cartridges to ultraviolet (UV) light prior to sampling and during shipment. When sampling outdoors, meteorological information is very important. (For more information see the chapter on Survey Preparations and Performance.) Following is a list of volatile organic compounds for which the CMS adsorption method has been evaluated.[18]

acrylonitrile
allyl chloride
benzene
carbon tetrachloride
chloroform
1,2-dichloroethane
methylene chloride
toluene
1,1,1-trichloroethane
vinyl chloride
vinylidene chloride

Ambient air is drawn through the sample cartridges or tube(s) for up to 24-hours. For the 80°C to 200°C range a single Tenax tube is used, and a pair of CMS-filled cartridges are used for the −15°C to 120°C range. Most inorganic atmospheric compounds will pass through the CMS cartridges.

Prior to sampling, perform the following calculations. The maximum total volume (V_{max}) that can be sampled is calculated as

$$V_{max} = \frac{\text{breakthrough volume (weight of Tenax tube in grams)}}{1.5}$$

TABLE 4-4. Retention Volume Estimates for Compounds on Tenax

Compound	*Estimated Retention Volume at 100°F L/g*
Benzene	19
Bromobenzene	300
Bromoform	150
Carbon tetrachloride	8
Chlorobenzene	150
Chloroform	8
Cumene	440
1,2-Dichloroethane	10
1,2-Dichloropropane	30
1,3-Dichloropropane	90
Ethyl benzene	200
Ethylene dibromide	60
n-Heptane	20
l-Heptene	40
Tetrachloroethylene	80
Toluene	97
1,1,1-Trichloroethane	6
Trichloroethylene	20
Xylene	200

The maximum useable flow rate (Q_{max}) is calculated as

$$Q_{max} = \frac{V_{max}\,(1{,}000)}{\text{sampling time}}$$

Table 4-4 lists retention volume estimates for compounds on Tenax.[18]

Procedure

1. The sampling system consists of a pump, with a rotameter, flow regulator, and sampling cartridge. The EPA recommends the use of mass flow controllers instead of needle valves as flow regulators.

2. Calibrate the system for the flow rate to be used before sampling. If the sampling interval exceeds 4 hours, the flow rate should be checked at an intermediate point during sampling as well. The rotameter on the pump allows the flow rate to be checked without disturbing the sampling process. For more information on performing flow rate calibrations, see the appendix on Pumps and Flow Rate Calibrations.

3. Remove fresh cartridges from their sealed container just prior to sampling and hook up using tygon tubing. Use two cartridges in series; the second is a backup in case of breakthrough. If glass tubes are used, handle with polyester gloves and make sure they do not come into contact with other surfaces. Tape the sample number to the pump and to the cartridge container.

4. If high dust levels are present place a filter in a cassette on the inlet to the cartridges. Glass cartridges are connected using Teflon ferrules and Swagelok fittings.

5. Start the pump and record the following parameters: date, sampling location, time, ambient temperature, barometric pressure, relative humidity, flow rate, rotameter reading, and sample number.

6. At the end of the sampling period remove the tubes (cartridges) one at a time and replace in the original container (use the gloves for the glass cartridges). Seal the cartridges or culture tubes in a friction-top can containing a layer of charcoal and store at reduced temperature (20°C) before and during shipment.

7. If the flow rates at the beginning and end of the sampling period differed by more than 10%, mark the cartridges as suspect.

8. Calculate the average flow rate for each set of cartridges:

$$Q_a = \frac{Q_1 + Q_2 + \ldots Q_n}{N}$$

where $Q_1, Q_2, \ldots Q_n$ = flow rates determined by beginning, intermediate, and end points during sampling.

9. Calculate the total volume (V_m) for each cartridge:

$$V_m = \frac{\text{sampling time (average flow rate)}}{1{,}000}$$

10. Adjust the volume (V_s) for differences in temperature and pressure from STP at the sampling site.

$$V_s = \frac{V_m\,(P_a)\,(298)}{(760)\,(273 + t_a)}$$

where P_a = pressure at sampling site and t_a = temperature at sampling site.

11. During each sampling event at least one set of parallel samples (two or more samples collected simultaneously) should be collected, preferably at different flow rates. If agreement between parallel samples is not generally within ±25%, the user should collect parallel samples on a much more frequent basis.

IMPINGERS

The basic design of the all-glass midget impinger was developed in 1944.[19] Originally the impinger was used primarily for collecting aerosols, especially to do particle sizing, but its current application is for gas and vapor sampling. The midget impinger is designed to contain 10–20 ml of liquid, while earlier impingers such as the Greenburg-Smith impinger held much larger volumes. The function of these absorbers is to provide sufficient contact between the sampled air and the liquid surface to provide complete absorption of the gas or vapor.

These devices are usually made of glass with an inlet tube connected to a stopper fitted into a graduated vial such that the inlet tube rests slightly above the vial bottom. A measured volume of absorber liquid is placed into the vial, the stopper inlet is put in place, and the unit is then connected to the pump by flexible tubing. When the pump is turned on, the contaminated air is channeled down through the liquid at a right angle to the bottom of the vial. The airstream then impinges against the vial bottom, mixing the air with the absorber liquid; the necessary air-to-liquid contact is achieved by agitation. Some impingers are coated with a plastic film to hold the glass pieces together in the event of breakage.

Fritted bubblers are similar in use and appearance to an impinger, but generally are used when a higher degree of air-liquid mixing is desired. The fritted glass is at the end of the impinger tube where the air goes into the solution. With these devices, the contaminated air is forced through masses of porous glass, called frits, breaking the airstream into numerous small bubbles. The frits are categorized as fine, coarse, or extra coarse, depending on the number of openings per unit area. The selection of what size frit to use depends on the ease of collection, with coarse frits used for gases and vapors that are soluble and frits of fine porosity used for gases that are difficult to collect. However, the finer the frit, the higher the pressure drop. Airflow through frits must be carefully controlled to avoid the formation of large bubbles. The size of the bubbles depends on the size of these openings as well as the type of absorber solution.

Impingers are suitable for collecting nonreactive gases and vapors that are highly soluble in the absorbing liquid, such as hydrogen chloride, as well as those that react rapidly with a reagent in the solution, such as occurs in the neutralization of strong acids and bases. The reagent is added to enhance solubility or to react with the contaminant once it is in the solution to stabilize it, thus reducing volatility and minimizing losses. Although most absorbing solutions are mixed in water, sometimes solubility in another compound such as a solvent is used. For example, toluene diisocyanate (TDI) is soluble in toluene, but due to the ease with which toluene vaporizes, this solvent would be used only in situations where other TDI sampling methods could not be used. An example is a heated, humid process, such as sampling a vent on a wet scrubber, to determine its efficiency. Fritted bubblers are used for minimally soluble gases and vapors such as chlorine and nitrogen dioxide.

Variations on the basic midget impinger shape have also been developed, including one with a membrane and another with a rounded shape designed to minimize the potential for spills. The liquid media sampler is made of a borosilicate glass tube wrapped with mylar film. At either end of the tube is a hydrophobic Teflon membrane held in place by a polypropylene cap to which tubing and a pump can be hooked. The membrane retains the sampling solution yet allows air to pass through it. The units are designed to operate at 1 Lpm, the same rate used for traditional occupational impinger sampling. The main concern in the use of these units is to measure the pressure drop on a unit to assure there are no significant variations in the filter membrane pore structure. Problems that could result due to these variations include dust entrapped in the sample because of entry through large pores.[20]

The most important factors to consider when sampling with impingers/bubblers are sampling flow rate, solubility of the contaminant in the absorbing solution, rate of diffusion of the contaminant into the solution, contaminant vapor pressure (volatility), volatility of the absorbing solution, and reactivity of the contaminant being collected with the absorbing solution.[21] The size of the bubble determines how much contact there is between the gas being sampled and the absorbing solution. The number of the air bubbles being released into the solution is also important. In general, the lower the sampling rate, the more complete the absorption.

Problems with the use of impingers/bubblers include condensation of material in the sampling lines and losses by adsorption or volatilization from the equipment. Impingers are bulky, and if used for personal sampling, they must be watched carefully as actions such as bending may spill liquid into the sampling lines and the precautions will limit the workers' mobility. If the liquid reaches the pump, the pump will usually break down and require repair. The use of traps in between impingers and pumps will usually minimize this type of damage to pumps, but some sample may still be lost in the lines or the trap.

Impingers are also susceptible to breakage as they are generally made of glass. In some cases, the method requires immediate processing in the field in order to stabilize the collected material and often impinger methods suffer from interferences.

Frits can also create problems, such as retention of the contaminant, excessive frothing and foaming, variable pressure drop, and nonphysical uniformity from unit to unit in the same batch,[21] which precludes interchangeability without recalibration, fragility, and particulate retention not encountered with the use of impinger-type absorbers. Frits can accumulate particles that are not removed by conventional cleaning procedures after repeated use.

Occupational Sampling

Occupational sampling with impingers is being phased out because it is "messy" due to the need to carry reagents and the potential for the impingers to spill when worn by workers. Most commonly it is replaced by the use of sorbent methods; however, there may be a situation where the sorbent tube is not adequate such as high humidity or only an impinger method exists for the contaminant of interest. For certain compounds such as formaldehyde the impinger technique is considered the "tried and true" method and sorbent tube methods are used less often. As analysis is often done using colorimetric techniques, it is especially important to identify any interferences that might be present. The laboratory should report any off-colors that occur during the analysis as this is suggestive of interferences.

The sampling methods for midget impingers and fritted bubblers are the same. Most standard methods rely on one or two impingers to absorb contaminants; however, some may require several connected in series. Two impingers or bubblers connected in series will increase the overall collection efficiency. If the amount collected in the second impinger is greater than 10%, sample breakthrough is a concern. Table 4-5 displays the effect of two bubblers in series on the collection efficiency of various compounds.[21]

There may be cases where there is a significant amount of particules present. Particules may interfere with analysis, clog orifices, and interfere with collection. In this case, a prefilter in a cassette should precede the impinger, which is nonreactive and nonabsorbing to the gases being sampled. For more information on filters, see the Integrated Sampling for Aerosols chapter.

Always pack a lot of bottles to collect samples. If a light-sensitive reagent is being used, brown bottles will be needed. Even when using traps between impingers and pumps it is a good idea to pack extra pumps. For example, if the wrong part of the impinger is hooked directly to the pump, liquid will be sucked into the hose and can saturate the trap. Also pack extra impingers as the bottles and the glass tubes inside break easily. Take care when handling tubes with frits on the end as the fritted portion may snap off if pressure is applied to it. If bubbles do not appear in the impingers/bubblers when the pump is turned on, there is a leak. Often this is due to the tubing not fitting tightly on the impinger. In this case, wrap 1-inch strips of duct tape tightly around these joints. Generally only 10 ml of solution is added to an impinger flask, and absolutely no more than 20 ml should be added. Figure 4-4 indicates how impingers are set up to collect personal samples.

Procedure

1. Fill the impinger/bubbler with 10 ml of the sampling solution. If two impingers in series are to be used, then these must be in place. Care

TABLE 4-5. Effect of Two Bubblers in Series on Collection Efficiency

		Maximum Air Volume in Liters for 95% Recovery	
Vapor	*Solvent*	*1 Bubbler (10 ml)*	*2 Bubblers (10 ml each)*
Acetone	Water	0.8	5.4
Methanol	Water	9.0	62.0
n-Butanol	Water	10.0	68.0
Chloroform	Isopropanol	0.8	5.4
Carbon tetrachloride	Isopropanol	0.5	3.4
Methyl chloroform	Isopropanol	0.5	3.4
Trichloroethylene	Isopropanol	0.9	6.2

should be taken to see that frits or tips are not damaged and that pieces are secured tightly. The impingers should be labeled with the sample number.

2. Place a tygon tube over the outlet stem on the top of the impinger. Correct selection is important. If the wrong stem is selected, liquid could be sucked into the pump. In virtually every case this problem will result in having to have the pump repaired, usually at the factory. One way of identifying the correct stem is to see where the glass tubing goes inside the impinger. The correct one does *not* go into the solution.

3. A second tube is run from the first impinger to the bubble buret or rotameter. This one should be hooked to the stem of the glass tube that goes into the solution of the impinger.

4. A trap containing charcoal granules and glass wool at each end should be placed in between the impinger(s) and the pump using tubing.

5. If no bubbles appear once the pump is turned on, it is likely that there are leaks in the system and the fittings should all be checked. Because the tygon tubing must fit tightly over the glass tubing stems of the impingers, or leaks will result, it is useful to tape over these connections with duct tape.

6. Once the system is working correctly, the same method for calibrating with the bubble buret or rotameter, as described in Appendix A, should be followed. A flow rate of 1 Lpm is generally used with impingers for gas and vapor sampling.

7. If high levels of dust are present, there is a danger that the orifice in the impinger or the interstices in the bubbler frit might get clogged, so a glass fiber filter in a cassette should be put in front of the impinger. This must also be in the sampling train during calibration.

8. The impinger(s) is placed in an employee's pocket or put into an impinger holster and taped to the employee's clothing. It is very important that the employee does not bend over or otherwise tilt the impinger, which can result in reagent flowing into the tube. The purpose of the trap is to catch any liquid that might get into the tubing. Follow steps 8–16 as outlined in occupational sorbent tube sampling for basic procedures during the sampling period.

9. It is important to note that spillproof impingers are available that minimize the likelihood of spillage.

10. In some cases, it will be necessary to add additional solution during the sampling period to make up for losses. The amount of solution should not drop below one half of the original amount.

11. Following sampling, have a labeled (with the sample number) glass bottle ready. Rinse the absorbing solution adhering to the outside and inside of the stem directly into the flask with a small amount (1 ml or 2 ml) of the sampling reagent. Pour the contents of the flask into the sample bottle.

12. Some sampling solutions are light sensitive. In this case, the impingers should be covered with duct tape and the sample bottles should be dark brown glass. The solutions should also be in dark brown glass bottles and should be kept inside a bag, box, or cooler when not in use.

13. If the impingers must be reused, carefully rinse them with the sampling solution using chemist's technique: Pour in about 5 ml of solution and carefully rotate the impinger on its side so that the entire interior is rinsed. If a water-based sampling solution is being used, the impinger can be rinsed using a distilled water and drained thoroughly to preserve sampling solution. If possible, the glass tubing should be rinsed and drained as well.

Environmental Sampling

Environmental samples are collected using the same types of impingers used for occupational samples; however, the units are kept in an ice bath during sampling. While occupational sampling methods using impingers often rely on colorimetric methods for analysis that suffer from interferences, environmental methods use high-pressure liquid chromatography (HPLC), a more sophisticated technique that is capable of detecting many specific compounds with few interfer-

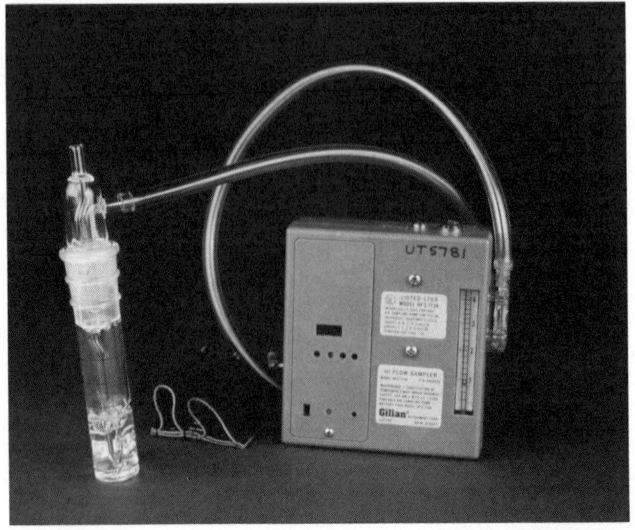

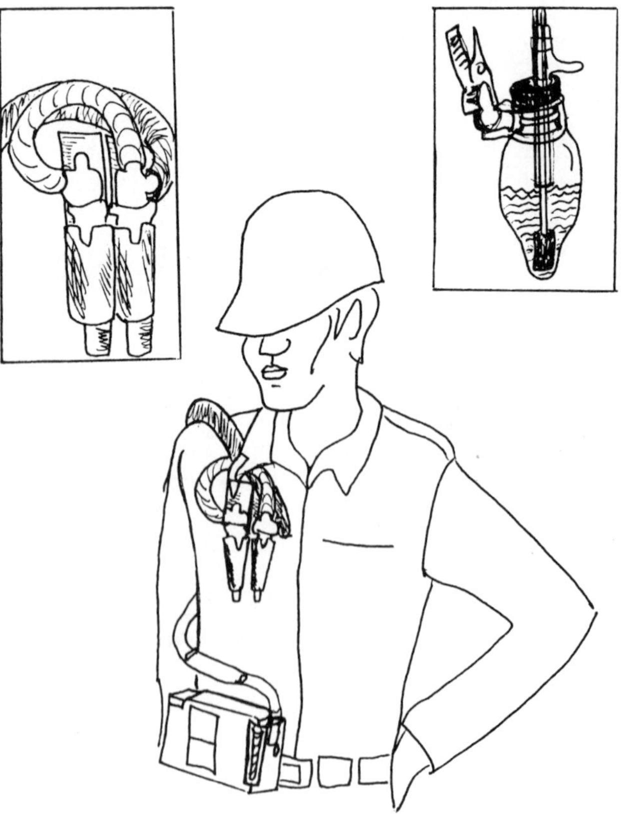

Figure 4-4. Personal sampling impingers/bubblers. (Photo courtesy Gilian Instrument Corp.)

ences. As with other environmental methods, impinger techniques are designed to detect lower concentrations than occupational methodologies.[22]

An example of the sampling technique is contained in EPA Method TO5 for Aldehydes and Ketones (see the following list of those compounds for which the TO5 Method has been evaluated.)[18]

acetaldehyde
acetone
acrolein
benzaldehyde
crotonaldehyde
formaldehyde
hexanal
isobutyraldehyde
methyl ethyl ketone
pentanal
propanal
o-tolualdehyde
m-tolualdehyde
p-tolualdehyde

Ambient air is drawn through a midget impinger containing 2,4-dinitrophenylhydrazine (DNPH reagent) and isooctane over a 2-hour sampling period. To estimate a 24-hour exposure, several samples are collected during this period. Aldehydes and ketones readily form stable 2,4-DNPH derivatives. After sampling, the impinger solution is returned to the laboratory for analysis. Reversed phase HPLC–UV (high-pressure liquid chromatography with an ultraviolet detector) analysis at 370 nm determines the DNPH derivatives.

Procedure

1. The sampling train is assembled (Fig. 4-5). It consists of a pump, needle valve, rotameter, and ice bath with two impingers in series containing DNPH reagent followed by a silica gel tube to act as an adsorber to prevent moisture from reaching the pump. All glassware must be rinsed with methanol and oven dried prior to use. If the pump has a built-in rotameter, a second one is unnecessary.

2. Prior to sample collection the flow rate is calibrated using techniques described in Appendix A. In general, flow rates of 100–1,000 ml/min are useful. Flow rates greater than 1,000 ml/min should not be used because impinger collection efficiency may decrease.

3. Record the flow rate at the beginning of the sampling period. If the sampling period exceeds 2 hours, the flow rate should be measured at intermediate points during the sampling period. If a rotameter is incorporated into the sampling train, the flow rate can be observed periodically.

4. To collect an air sample, two clean midget impingers are each filled with 10 ml of purified

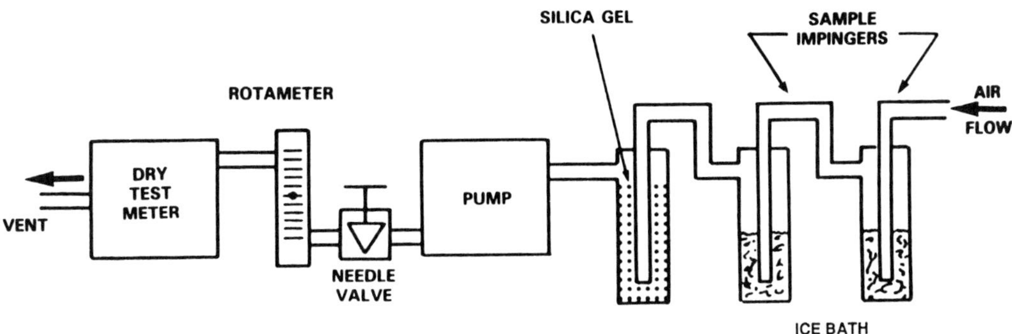

Figure 4-5. Ambient gas and vapor sampling with impingers. (From *Compendium of Methods for the Determination of Toxic Organic Compounds in Ambient Air,* EPA/600/4-87/006 [supplement to EPA/600/4-84-041].)

DNPH reagent and 10 ml of isooctane. The impingers are connected in series to the sampling system and immersed in an ice bath as shown in Figure 4-5. Sample flow is started. The following parameters are recorded: date, sampling location, time, ambient temperature, barometric pressure, relative humidity, flow rate, rotameter setting, DNPH reagent batch number, and pump identification numbers.

5. The sampler is allowed to operate for the desired period with periodic recording of the variables listed above. The total volume should not exceed 80 L. Both the ice bath and, if necessary, the solution in the impingers should be replenished. The purpose of the ice bath is to slow the evaporation of the sampling solution. At least 2–3 ml of isooctane must remain in the first impinger at the end of the sampling interval.

6. At the end of the sampling period the flow rate is again checked. If the flow rate at the beginning and end of the sampling period differ by more than 15%, the sample should be marked as suspect.

7. Immediately after sampling the impingers are removed from the sampling system. The contents of the first impinger are emptied into a clean 50-ml glass vial having a Teflon-lined screw cap. The first impinger is then rinsed with the contents of the second (backup) impinger and the rinse solution is added to the vial. The vial is then capped, sealed with Teflon tape, and placed in a friction top can containing 1–2 inches of granular charcoal and stored in a cooler.

8. Calculate the average flow rate (Q_a) in the same manner as in environmental sampling with sorbents. The total volume is then calculated using the following equation:

$$V_m = \frac{(\text{Time}_2 - \text{Time}_1)\, Q_a}{1{,}000}$$

COLLECTING INTEGRATED BAG SAMPLES

In addition to occupational sampling and calibration applications, bags are often used to collect gases and vapors for high resolution analysis of environmental samples. This method is commonly used for collecting air samples in open fields or from vapor wells. For more information on this technique see the chapter on Bulk Sampling Methods.

When used for integrated sampling, bags can collect (1) certain types of gases and vapors for which other methods are unsuitable due to very low breakthrough volumes on sorbent tubes; (2) gases and vapors for which no alternative method exists; and (3) mixtures where method incompatibilities exist. They can also be used in conjunction with real time instruments for collecting time-weighted average (TWA) samples and for the transport of calibration standards.[23] Although most often bags are used to collect area samples, it is also possible to use a bag to collect a personal integrated sample (Fig. 4-6). This is done when no other suitable method exists, as for carbon monoxide and waste anesthetic gases.

Bags come in different sizes, shapes, and

Figure 4-6. Personal sampling with a gas bag. (Courtesy Calibrated Instruments Inc.)

materials. Most bags have a valve to allow for filling using a pump and a septum for injections using a syringe. Septums can be made of silicone or neoprene rubber and valves commonly are nickel-plated or stainless steel. Stainless steel valves are less likely to corrode than nickel-plate. Fittings on bags are often multipurpose. For example, the fitting may contain both a syringe port with Teflon-coated septum and a hose connection. The fitting also acts as a shutoff valve for the hose connection. A Teflon-coated septum is often used and should be replaced after each use. When replacing the septum, be sure the Teflon side is toward the bag.

Sample bags offer alternatives in the collection of grab and integrated samples of gases and vapors, and they are particularly useful when representative samples are desired. Bags can also be used for grab samples. These bags can be of plastics but more often are of materials, such as Tedlar, Mylar, Scotchpak, and Teflon. Often materials are laminated to aluminum to seal the pores in the plastic and reduce the likelihood of loss of a gas sample by permeation through the walls of the bag. For example, one five-layer material has aluminum foil at its core, but because aluminum reacts with many compounds, this foil is protected on both sides. The exterior is covered by a layer of polyvinylidene chloride followed by a film of polyester. The interior surface is coated with a film of polyamide followed by a layer of polyethylene. Plastic bags are transparent, allowing detection of condensation inside the bag. However, they can stretch, altering their volume size, and they are prone to residual contamination from prior samples.

The sampling period is determined by the purpose for sampling. A personal sample will be collected over 4–8 hours. For a grab sample, the maximum flow rate available on the pump can be used to fill a bag. Precise calibration of the pump is not necessary; however, if a specific sampling period is desired, the size of the bag and the flow rate are important. NIOSH methods recommend Tedlar bags and using new bags for each sample. Although it may prove too expensive to be practical, certainly the sampler must take steps to guard against bags becoming contaminated and being reused.

Factors that can affect the ability of bags to retain samples include the material from which they are constructed, presence or absence of other compounds, humidity, concentration of the sample in the bag, size of the bag, preequilibration, and bag-to-bag variation. Sources of sample losses from bags include leaks either through valves used in filling and sampling from the bag or through poor seam seals during bag manufacture; chemical reactions of the sample with the bag walls or other compounds collected along with it; adsorption of the sample into the bag material; and permeation of the sample through the walls of the bag.

A study examining these factors among a variety of bag materials determined that there were some differences depending on bag composition. For short-term samples ($\leq$4 hours), Tedlar bags are the best choice for minimizing losses. For long-term samples, five-layer aluminized bags are the best choice, because they are the least likely to allow permeation of sample through the bag walls. As most bags appear to develop a memory (retain traces of compounds previously collected), new bags should be used for taking samples of new materials unless there is proof that the bag has been adequately cleaned. This criterion is especially important in the case where air is being collected to analyze for unknowns.[24]

Bags should be leak tested, cleaned, and preconditioned by flushing with the gas to be sampled (if known) or zero air prior to use. If vapor condenses inside of a sampling bag, the bag can be purged by placing it in hot water with the valve open.

Procedure

1. For occupational sampling, the bag should be attached to the exhaust outlet of a low flow pump, with a flow rate of 20–200 ml/min. The sampling period determines the size of the sample bag and the flow rate of the pump. For example, using a 5-liter bag and a sampling rate of 200 ml/min will fill a bag in 25 minutes. Sampling at 20 ml/min will fill the bag in 4 hours.

2. Attach the pump to the employee's belt and place the bag over the shoulder. Open the valve on the bag and turn on the pump.

3. Record start and stop times. After the sample is collected, the bag should be disconnected from the pump.

4. The sample is now ready to be measured using a real time instrument, detector tube, or sent to a laboratory for analysis.

ENVIRONMENTAL SAMPLING FOR VOCs USING STAINLESS STEEL CANISTERS

Collection of ambient air samples in stainless-steel canisters provides convenient integration of ambient samples over longer time periods, such as 24 hours; remote sampling and central analysis; ease of storing and shipping samples; unattended sample collection; analysis of samples from multiple sites with one analytical system; and collection of sufficient sample volume to allow assessment of measurement precision and/or analysis of samples by several analytical systems.

VOCs enter the atmosphere from a variety of sources, including petroleum refineries, synthetic organic chemical plants, natural gas processing plants, and automobile exhaust. VOCs are generally classified as those organics having saturated vapor pressures greater than 10^{-1} mm Hg at 25°C. Conventional methods for VOC determination use solid sorbents such as Tenax, described previously in the section on environmental sampling with sorbents. Canisters are generally not used for the collection of polar compounds, such as methanol, ammonia, and hydrogen chloride, because these adsorb to the walls of the vessel. Following is a list of VOCs that can be detected with EPA canister methods TO-14.

benzene
benzyl chloride
carbon tetrachloride
chlorobenzene
chloroform
1,2-dibromoethane
dichloromethane
m-dichlorobenzene
o-dichlorobenzene
p-dichlorobenzene
1,1-dichloroethane
1,2-dichloroethane
1,2-dichloropropane
cis-1,3-dichloropropane
trans-1,3-dichloropropene
ethyl benzene
ethyl chloride
freon 11
freon 12
freon 113
freon 114
hexachlorobutadiene
methyl bromide
methyl chloride
methyl chloroform
styrene
1,1,2,2-tetrachloroethane
tetrachloroethylene
toluene
1,2,4-trichlorobenzene
1,1,2-trichloroethane
trichloroethylene
1,2,4-trimethylbenzene
1,3,5-trimethylbenzene
vinyl chloride
vinylidene chloride
m-xylene
p-xylene
o-xylene

Canisters can be used both indoors and outdoors. An air sample is drawn from the ambient air into an initially evacuated stainless steel canister that has been treated internally with an inert coating. One type of coating is the SUMMA® process, in which chrome-nickel oxide is formed on the canister's interior. Two sampling modes are used: passive and pressurized. In passive sampling an evacuated canister is opened to the atmosphere and the differential pressure causes the canister to fill. For pressurized sampling, an initially evacuated canister is filled by the action of the flow-controlled pump from near atmospheric vacuum to a positive pressure not to exceed 25 psig (pounds per square inch gauge). Commercial canister units are usually sold with a filter that traps particulate material entering the sampling inlet. A mass flowmeter or valve is used to control airflow into the canisters.

Care must be exercised in cleaning and handling sample canisters and associated sampling apparatus to avoid losses or contamination of the samples. Contamination is a critical issue with canister-based sampling. Cleaning of canisters can be done by a cycle of first evacuating them to a pressure of 5 mm Hg followed by pressurizing them to 40 psig with clean dried air. Cleaning should be verified by injecting a sample of canister air into a GC. Figure 4-7 shows ambient gas and vapor sampling with a stainless steel canister. A field sampling data sheet for use with canisters is shown in Figure 4-8.

Passive Sampling[24]

Procedure

1. This technique may be used to collect grab samples (duration of 10 or 30 seconds) or time-integrated samples (<1 minute) taken

Figure 4-7A. Canister for sampling ambient gas and vapor for VOCs. (Courtesy Andersen Instruments Inc.)

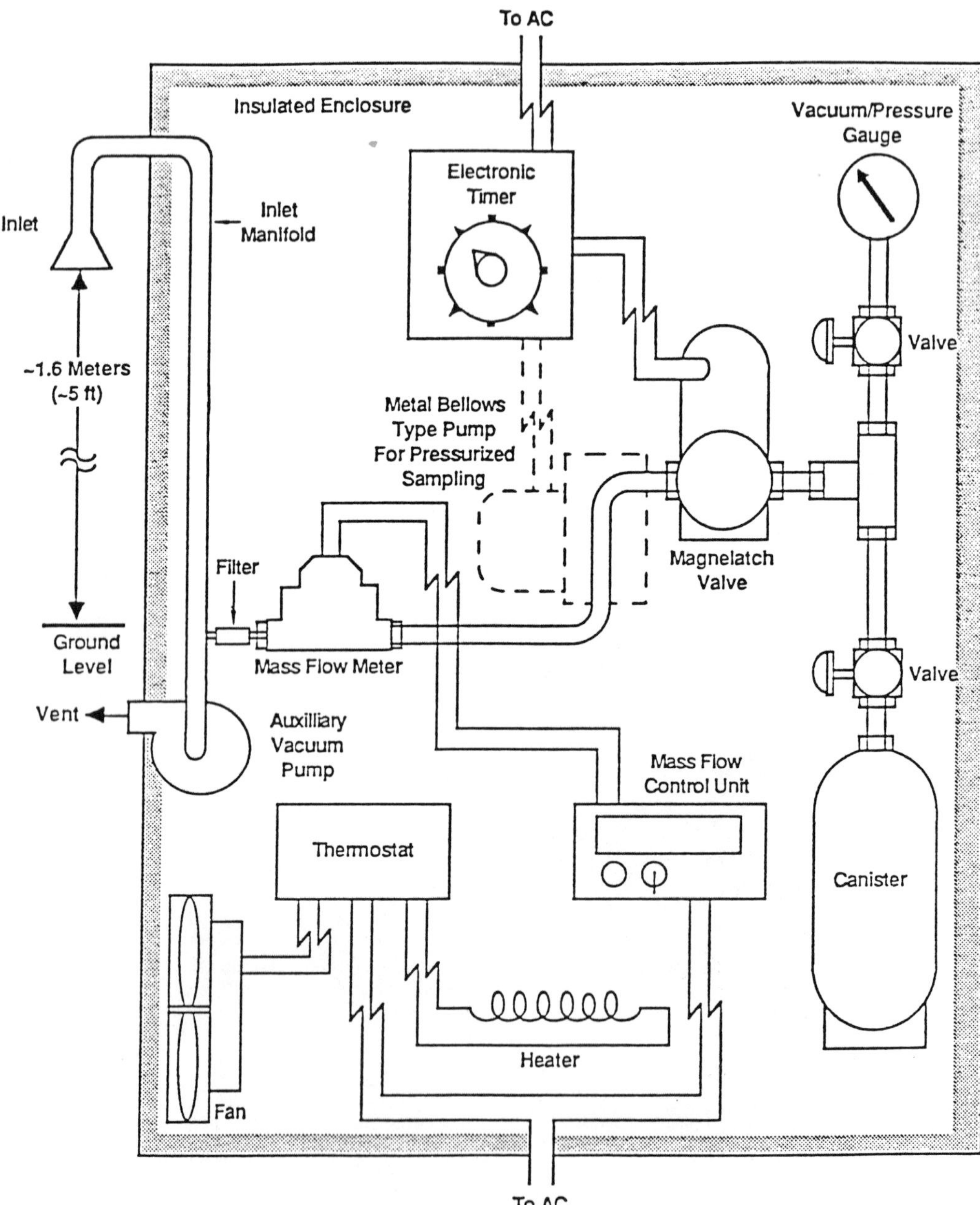

Figure 4-7B. Sampler configuration for pressure or pressurized canister sampling. (From *The Determination of Volatile Organic Compounds (VOCs) in Ambient Air Using Summa® Passivated Canister Sampling and Gas Chromotagraphic Analysis*, EPA, May 1988.)

through a flow-restrictive inlet, such as a mass flow controller or a critical orifice. For passive collection the canister is first evacuated at a laboratory. When opened to the atmosphere containing the VOCs to be sampled, the differential pressure causes the sample to flow into the canister without the need for a pump.

2. In its simplest version, a manual sampling apparatus consisting of the evacuated canister and a valve preset for the desired flow rate is used. The valve is screwed on the canister top for sampling. The canister valve(s) is opened and timers, if in use, are programed for the scheduled sample period and set to the correct time of day and date. The elapsed time meter should be set to zero. When timed sampling is done, the first timer starts the pump about 6 hours prior to the sample period.

3. In a more complex version, a pump draws sample air through the sample inlet and particulate filter to purge and equilibrate these components. The inlet line begins to heat at this time as well to 65–70°C. At the start of the sampling period solenoid valves are activated, stopping the purge flow and allowing sample air to be collected.

4. At the end of the sample period, the second timer stops sample flow and seals off the canister or, alternately, the valve is manually shut.

Pressurized Sampling

Pressurized sampling is used when longer-term integrated samples or higher-volume samples are required. In pressurized canister sampling, a metal bellows-type pump draws in ambient air from the sampling manifold to fill and pressurize the sample canister to a 15–30-psig final pressure. For example, a 6-liter evacuated canister can be filled at 10 ml/min for 24 hours to achieve a final pressure of about 21 psig.

A flow control device is chosen to maintain a constant flow into the canister over the desired sample period. This flow rate is determined so the canister is filled over the desired sample period. The flow rate can be calculated by

$$F = \frac{(P)(V)}{(T)(60)}$$

where

F = flow rate in ml/min
P = final canister pressure, atmospheres
V = volume of the canister in ml
T = sample period in hours

For automatic operation, the timer is wired to start and stop the pump at appropriate times for the desired sample period. The timer must also control the solenoid valve, to open the valve when starting the pump and to close the valve when stopping the pump.

The connecting lines between the sample inlet and the canister should be as short as possible to minimize their volume. The flow rate into the canister should remain relatively constant over the entire sampling period. If a critical orifice is used, some drop in the flow rate may occur near the end of the sample period as the canister pressure approaches the final calculated pressure.

Procedure

1. Insert a "practice" canister into the sampling system. A certified mass flow meter is attached to the inlet line of the manifold, just in front of the filter. The canister is opened, the sampler is turned on, and the reading of the mass flow meter is compared to the sampler mass flow controller. The values should agree within ±10%. If not, the sampler mass flow meter must be recalibrated or there is a leak in the system.

2. Adjust the flow rate to the proper value (e.g., 3.5 ml/min for 24 hours, 7.0 ml/min for 12 hours). Record this final flow rate on the sampling data sheet.

3. Turn off the sampler and reset the elapsed time meter to 000.0 minutes. Any time the sampler is turned off, at least 30 seconds should pass before it is turned on again. Disconnect the "prac-

A. General Information

Site Location: ____________________ Shipping Date: ____________________
Site Address: ____________________ Canister Serial No.: ____________________
____________________ Sampler ID: ____________________
____________________ Operator: ____________________
Sampling Date: ____________________ Canister Leak Check Date: ____________________

B. Sampling Information

	Temperature				Pressure	
	Interior	Ambient	Maximum	Minimum	Canister Pressure	
Start			X	X		
Stop					X	

	Sampling Times	
	Local Time	Elapsed Time Meter Reading
Start		
Stop		

	Flow Rates		
	Manifold Flow Rate	Canister Flow Rate	Flow Controller Readout
Start			
Stop			

Sampling System Certification Date: ____________________
Quarterly Recertification Date: ____________________

C. Laboratory Information

Date Received: ____________________
Received by: ____________________
Initial Pressure: ____________________
Final Pressure: ____________________
Dilution Factor: ____________________
Analysis Results*: ____________________

____
Signature/Title

*Attach Data Sheets

Figure 4-8. Canister sampling field data sheet.

tice" canister and replace it with a certified clean canister.

4. Open the canister valve and vacuum/pressure gauge valve. The pressure/vacuum in the canister is recorded on the canister sampling field data sheet as indicated by the sampler vacuum/pressure gauge. The gauge is then closed and the maximum–minimum thermometer is reset to the current temperature. The time of day and elapsed time meter readings are also recorded. Set the electronic timer for the start and stop time.

5. At the end of the sampling period, the vacuum/pressure gauge valve on the sampler is briefly opened and closed and the pressure/vaccum is recorded on the sampling field data sheet. The pressure should be close to the desired pressure. The time of day and the elapsed time meter readings are also recorded.

6. Following sampling, the maximum, minimum, current interior temperature and current ambient temperature are recorded on the sampling field data sheet. The current reading from the flow controller is recorded.

7. Close the canister valve, disconnect the sampling line from the canister, and remove the canister from the system. The final flow rate is measured and recorded on the canister sampling field data sheet and the sampler is turned off. Attach an identification tag to the canister containing the canister serial number, sample number, location, and date.

PASSIVE COLLECTORS FOR GASES AND VAPORS

Passive monitors are lightweight badge assemblies that rely on natural wind currents rather than pumps to move contaminated air to the collection surface. Most passive samplers or badges as they are called have been designed for personal or occupational monitoring, although some are being used for environmental samples as well, such as the measurement of nitrogen dioxide and VOCs in residences. Most of these units are dosimeters, that is, they accumulate an average dose and give an integrated measurement. Some units are specific for a given gas, such as chlorine, ammonia, sulfur dioxide, hydrogen sulfide, nitrogen dioxide, carbon monoxide, and mercury, whereas others can sample multiple gases, such as organic vapors.

The obvious advantage is the ease and simplicity of use of passive monitors. Generally they must only be slipped out of a package, the cover removed, and clipped onto an individual or a surface to be monitored. When used for personal monitoring, they do not interfere with worker activity and are unlikely to affect the behavior pattern of the wearer, whereas wearing a pump with its tubing and media might. They are also safe for use in flammable atmospheres because there are no moving parts. All of the badges discussed in this chapter require laboratory analysis. There are also passive devices that are designed to change color in proportion to the concentration present. For more information, see the chapter on Sampling of Gases and Vapors Based on Color Change.

The most commonly used passive monitors contain a solid sorbent similar to that used in sorbent tubes. However, devices that use liquid absorption solutions are also on the market. These can take the form of a rectangular badge or a glass cylinder similar in shape and design to an impinger.

Diffusion and Permeation Principles

Passive devices rely on two basic collection principles: diffusion and permeation. With diffusion-controlled monitors, the mass uptake of the monitor is controlled by the length and diameter of the badge cavity(ies) and the physicochemical properties of the contaminant. The badge sampling rate is a function of the diffusion coefficient of the vapor(s) being sampled and the total cross-sectional area of the badge cavity.[25] When permeation is used for sampling, the mass uptake of the monitor is controlled by the physicochemical characteristics of the membrane and the contaminant. The mass uptake is a direct function of the badge permeation sampling rate, the ambient concentration, and the sampling time.[25]

Diffusion devices are the most common. Molecules pass through a barrier or draft shield that minimizes the effect of air currents to a stagnant air layer and then they are collected on an adsorbent material such as charcoal. A concentration gradient is created within the cavity of stagnant air and the amount of gas or vapor transferred is proportional to the ambient vapor concentration. The sorbent surface area, path length from the badge's surface to the sorbent material, and badge sampling rate along with the desorption efficiency are used to calculate the final concentration of material collected.[25]

Over the exposure interval, the sampling rate decreases as the concentration of contaminant that is collected on the badge increases. As the amount of material adsorbed by the badge's surface increases, the molecules must diffuse greater distances, thus increasing the diffusion path length, causing increased resistance to diffusion and a decrease in the effective concentration rate.[26]

Each compound has a unique diffusion coefficient for each type of badge. Diffusion coefficients can be determined experimentally or calculated; however, diffusion coefficients determined experimentally may vary considerably from calculated values. As the diffusion coefficient is necessary to calculate the final concentration after analysis in the laboratory, this limits sampled materials to those for which this value is established.

On permeation dosimeters, the gaseous contaminants dissolve in a polymeric membrane and are then transferred to a collection medium, such as a solution. Permeation across the membrane is controlled by the solubility of the gas or vapor in the membrane material and by the rate of its diffusion across the membrane under a concentration gradient. The permeation constant (ppm-hours/ηg), mass of compound collected, and exposure time determine the concentration collected.[25] These constants vary depending on the design of the monitor and the contaminants involved. Factors influencing permeation include thickness and uniformity of the membrane, affinity of the membrane for the contaminant, swelling or shrinking of the membrane, and possible etching by corrosive chemicals. The efficiency of these devices depends on finding a membrane that is easily permeated by the contaminant of interest and not by all others. Therefore, permeation dosimeters where they exist would be useful to selectively sample a single contaminant from a mixture of possible interfering contaminants due to the selectivity of a properly chosen membrane.

As with diffusion monitors, sampling rates must be determined for each analyte and type of permeation monitor. Difficulties in making such determinations include the need to use thin and fragile membranes to obtain practical sampling rates; the long response times to concentration fluctuations potentially resulting in TWA measurements; and the fact that sampling rate and degree of permeation may be affected by changes in temperature or ambient humidity.[27] In one study it was suggested that variations in concentration and the length of the sampling period might affect diffusion monitors more than permeation-based monitors.[26]

Limitations of Passive Monitors

The accuracy and precision of passive monitors depends on sampling time, air currents, and temperature and humidity effects. In a study on organic vapor badges it was estimated that a temperature change from 25°C to 30°C, if uncorrected, will introduce a measurement error of less than 1%; a change from 5°C to 35°C would introduce an error of 5%.[28] In the same study significant sample losses occurred at 70% relative humidity, whereas only minimal losses were observed at 10% relative humidity. The problem may have been due to competition for adsorption sites between water and solvent molecules. Liquid sorbent badges are also affected by low relative humidity as moisture may evaporate from the badge, reducing the effectiveness of collection.[29] For more information on typical levels of humidity encountered during sampling see the chapter on Survey Preparations and Performance.

In the presence of mixtures passive monitors may be affected by competition between compounds for adsorptive sites, resulting in displacement of one compound by another or preferential adsorption of one compound over another.

All passive monitors require a certain minimum air movement to be present during sampling in order to prevent "starvation" at the sampler surface. In a perfectly stagnant environment, molecules are sampled from the region adjacent to the sampler face, resulting in a reduction of the concentration. If a slight airflow is present, the concentration at the sampler face is continuously renewed and is equal to the ambient concentration. Badge samplers have a high area-to-length ratio, and therefore high uptake rates, requiring a minimum face velocity of 0.05–0.1 m/sec, depending on the type. In some cases, the body's natural movements will provide sufficient air motion for sampling. Perpendicular air currents appear to have the most impact on badge sampling.[30] High face velocities will also affect concentrations. These are usually the result of turbulence, such as an area cooling fan blowing directly at the badge or high wind in the area.

In general, because the effective sampling rate is much lower for passive samplers than for an active method, a smaller quantity of contaminant is collected; therefore, special care must be taken to ensure sampler cleanliness before, during, and after sampling, especially if low concentrations are expected to be present.[31] Sampling in very dusty environments may also be a problem if the gas or vapor being collected is capable of adsorbing onto the dust, as the dust may collect on the sampler surface and cause readings to be higher. In most cases, badge housings are reusable.

Seals on passive monitors are often not as good as those provided with sealed glass sorbent tubes, so there is a potential for contamination both before and after sampling. For this reason many monitors come in evacuated cans or other airproof containers.

Passive monitors must be handled with care as a penetration in the membrane would cause variations in the amount of contaminant collected. Membranes should not be touched with the fingers, due to oil that will clog the pores and decrease the results. A splash of oil or reagent on the surface of the membranes may increase results if it has volatiles in them. Obvious dirt or discoloration on the surface of the membranes will make the sample suspect. If a badge has an obvious odor, it may be suspect because some type of deliberate exposure may have occurred.

Monitor Comparison and Testing

Design characteristics of one monitor versus another can provide advantages. For example, the 3M organic vapor badge is set up so that desorbing solution is added to the monitor body, whereas some others must have the collection matrix transferred to a desorption vial prior to treatment. This transfer can be clumsy, exposing the element to dust and the analyst's hands. Design differences may also affect the amount of a given compound collected when different monitors are compared; just as other sampling devices have a limited collection range over which they are linear, so do passive monitors. Some monitors consistently give low results and others give high results when compared to NIOSH charcoal tube methods.[32]

There is a growing market for passive monitors because of their ease of use, and thus many new devices are being produced. Often these have not been evaluated under a variety of conditions, especially those that might be encountered in the field, including temperature, pressure, and humidity changes and potential interferents. Since it is generally accepted that active methods involving the use of air sampling pumps provide the best accuracy and precision, passive systems should be tested against the active method both in the field and in the laboratory. Therefore, when contemplating the replacement of an established active method with one of these methods, it is best to do side-by-side testing to evaluate the badges for each specific application. When doing this evaluation it is important not to place samplers so close together that they will compete for available contaminant, but they should be close enough to provide confidence that the devices are both measuring the same contaminant(s) and concentrations. Several areas having different concentrations should be evaluated, and the capacity of the passive monitor should be determined. Vendors should also be asked to supply laboratory validation data concerning their passive monitors.

Although the results of existing field studies comparing passive dosimeters with standard monitoring methods are highly varied, in general there seems to be no significant difference between the accuracy and precision of many passive samplers and active monitoring methods. Most potential sources of error, such as interfering contaminants, sorbent capacity, or saturation, and problems associated with analytical determinations are common to both passive and active techniques. Therefore, these approaches can be considered complementary, each one having areas of applicability that may overlap and assist in increasing the flexibility of a monitoring program. Following is a list of important factors in diffusional sampling.[33]

path dimensions
sampling time
concentration range of the contaminant
suitability of the collection medium for the contaminant
accuracy and precision required (NIOSH, OSHA)
type of analytical method to be used
temperature
face velocity
humidity
reactivity/stability of contaminant following collection
presence of interferents
pressure
collection efficiency
storage stability
sample volume

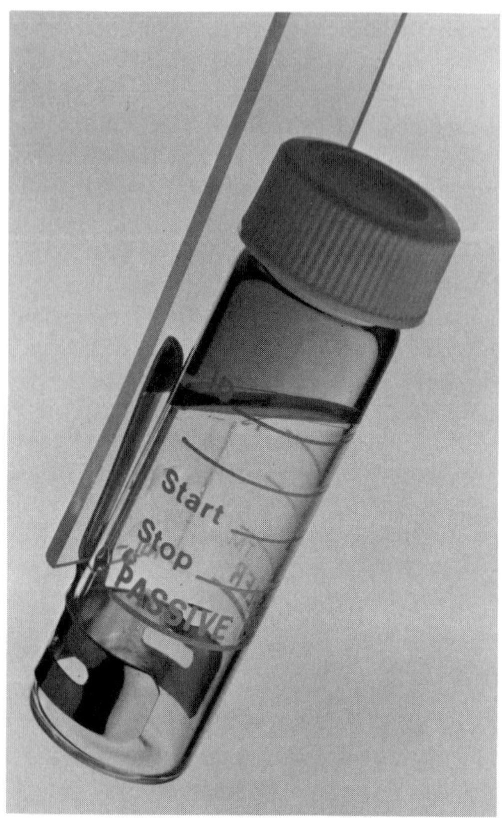

Figure 4-9. Passive bubbler. (Courtesy Air Technology Labs.)

Passive Bubblers

A passive bubbler monitor (Fig. 4-9) consisting of a glass vial with a septum cap and a Knudsen disk to which an absorbing solution is added has been developed for the collection of personal or area samples of gases in ambient air. The Knudsen disk regulates uptake of sample gases by means of Knudsen diffusion, a different process than the molecular diffusion and permeation principles on which passive collectors previously discussed rely.[27] Gas transport by Knudsen diffusion depends only on collisions of molecules with the "walls" of the pores, and is independent of air movements from 25–250 fpm.

Air Technology Labs has designed a monitor using this concept to measure formaldehyde. Use of this monitor requires that water and a tablet be added prior to sampling. The monitor is capable of measuring 0.025–14 ppm of formaldehyde using sampling times between 15 minutes and 8 hours respectively, although a 4-hour sampling period is recommended. The unit requires a mimimum face velocity of 25 fpm. Other aldehydes will interfere and give a positive bias to the analysis. According to the manufacturer, the unit has an accuracy of $\leq$25%. It has been recommended that disks should not be reused more than twice as results will reflect an increasing negative bias.[27]

Specific Monitors

Organic Vapor Badges

Collection of organic vapors is one of the most common uses for passive monitors and different styles of monitors are available (Fig. 4-10). Compared to a typical charcoal tube, organic vapor

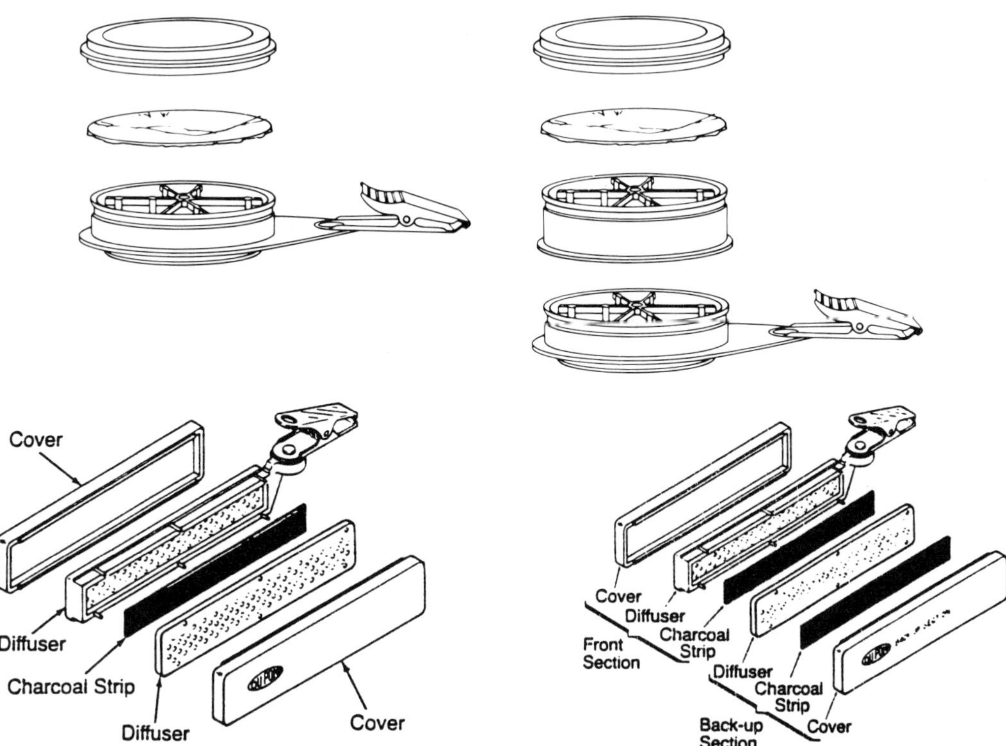

Figure 4-10. Organic vapor monitors, including layout showing badges with and without backup sections. (Top drawing courtesy 3M Corp.; bottom drawing courtesy Protek Systems.)

badges contain twice as much charcoal: 300 mg. Some badges have petroleum-based charcoal whereas others use coconut-based charcoal. The charcoal is usually impregnated in some sort of a matrix, such as a sheet, which can lead to quality control problems. Poor correlation between charcoal tubes and organic vapor badges or between different brands of badges would not be surprising if the units contain different types of carbons or a different arrangement in the carbon matrix. There have been problems in the past with the quality of charcoal used in some organic vapor badges, but in most cases, this problem exists only when sampling for contaminants at low levels such as benzene. Many organic vapor badges contain two different layers of charcoal. The purpose of the backup sections is to trap contaminant that may break through the top sorbent layer. These should be used when high concentrations, highly volatile compounds, or high humidity are present. Reduced sampling periods should be also considered in these circumstances to avoid saturation of the sorbent and loss of sample resulting from breakthrough. Regardless of the design of the organic vapor monitor, the analytical procedure is generally the same: desorption with carbon disulfide and analysis by gas chromatography. Thermal desorption of badges may be possible in the future as it is with tubes.

The affinity of various compounds for the activated carbon in a badge will vary, and because diffusion coefficients exist only for certain gases and vapors, the use of badges is limited to these compounds regardless of their wide applicability for sampling organic vapors. A concern is the need to rely on manufacturers' data for these coefficients. The sampling rate of a badge is specific for a given chemical.

The accuracy of organic vapor monitors can vary depending on the concentrations being sampled as well as on the exposure period. In one study, badges exposed to continuous concentrations for 8 hours were found to undersample, while results for exposure to the same compound at an eightfold concentration for 1 hour were much closer.[28] It was also determined that loss of organic vapors from badges may occur when a period of exposure to high concentrations is followed by a period of exposure to very low concentrations of highly volatile solvents.

A passive monitor using a charcoal sorbent will also experience the same collection limitations, such as temperature and humidity effects and breakthrough, as a charcoal tube sampler. Another variable is the presence of co-adsorbed organic vapors that can limit the capacity of activated charcoal and alter the concentration gradient, as well as interfere during analysis. The presence of several compounds of dissimilar characteristics at varied concentrations in the sampling area may affect individual monitor capacities and therefore cause displacement of specific compounds.[26] However, sampling of multiple vapors using diffusional samplers can be done accurately as long as the adsorbing capacity of the charcoal layers is not exceeded and the ability of that particular monitor to collect those compounds has been characterized prior to use in the field.

A number of studies have been done comparing various organic vapor badges with active methods using charcoal tubes. In general, most organic vapor badges compared well with the active charcoal tube method, although in some cases there were significant differences between badges. It has been suggested that for passive organic vapor samplers to perform as well as the active sampling method a high degree of quality control in determining diffusion rates and for minimizing background contamination is needed.[34]

In summary, the differences in the design of the individual monitors, type of charcoal utilized, quality control and type of adsorption pad, particle size of the charcoal, amount of charcoal present, surface area of the adsorption material, and membrane or draft shield selection create each monitor's unique performance with respect to the compounds collected.[26]

Formaldehyde

Formaldehyde can be measured by a variety of passive techniques. While difficulties in the development of reliable formaldehyde monitors have resulted in withdrawal of some samplers from the market, due to the widespread potential for exposure to formaldehyde it is likely that badges will continue to be available. Before selecting a badge for formaldehyde, it is especially important to determine if any field studies have been performed. Otherwise it may be necessary to perform field studies prior to using these badges. In one case, it was determined that a humid environment was necessary during sampling for a monitor to perform adequately.

Many monitors utilize sodium bisulfite as the collection medium either in solutions or impregnated into filters. High-moisture environments in storage areas can deplete the sodium bisulfite in dry badges. Phenol is the most common interferent in bisulfite sampling methods for formaldehyde. Figure 4-11 shows types of passive formaldehyde badges.

In one study comparing methods for monitoring formaldehyde, a visible dust accumulated during collection on the exterior of the passive monitors.[35] It was concluded that the dust in these plants may have contained free formaldehyde, either as unreacted formaldehyde from the manufacturing process or as airborne formaldehyde gas that had adsorbed onto the particulate. The particulate, when deposited on the pouch of the passive dosimeter, may then have offgassed formaldehyde. The longer the dust remained, the more formaldehyde would have offgassed.

Nitrogen Dioxide

A well-known diffusion nitrogen dioxide sampler, the Palmes, uses a tube with three stainless steel grids coated with triethanolamine (TEA) (Fig. 4-12). The grids are analyzed colorimetrically for absorbance. A study on the Palmes sampler found it to be accurate for use in atmospheres with fluctuating concentrations over 8-hour sampling periods, but recommended caution when using it for short-term exposure limit (STEL) samples.[36] A laboratory chamber study performed by NIOSH in 1983 indicated the Palmes tube successfully sampled nitrogen dioxide at levels of 0.5, 5, and 10 ppm.[37] Another study found that reduced humidity led to reduced sampling rates at low pressure and it was concluded that the

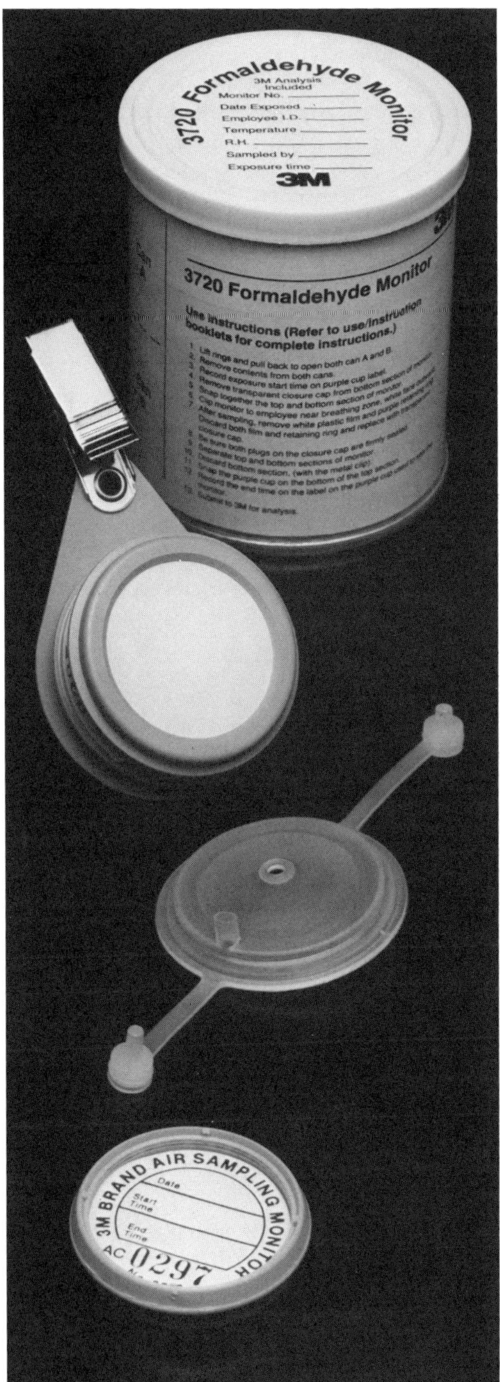

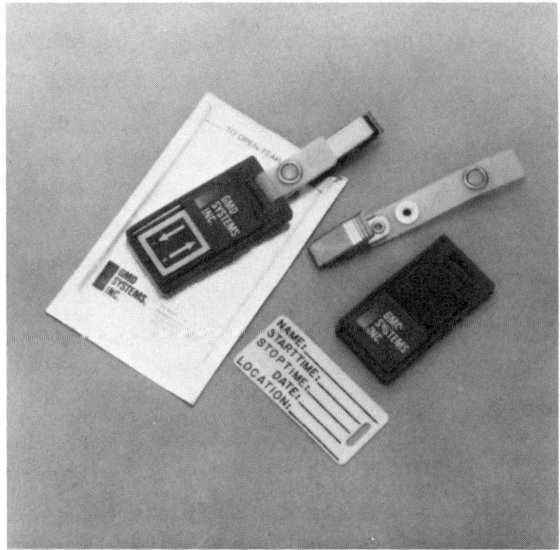

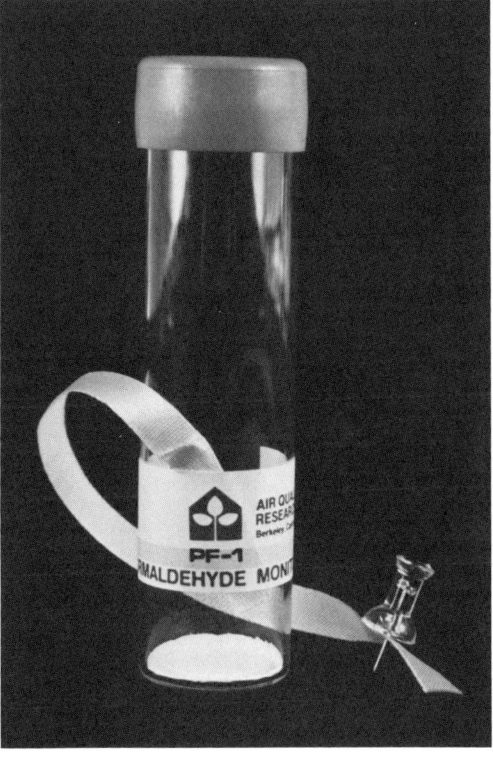

Figure 4-11. Passive formaldehyde monitors. (Left photo courtesy 3M Corp.; top right photo courtesy GMD Systems, Inc.; bottom right photo courtesy Air Quality Research.)

reaction products of TEA and nitrogen dioxide appeared to be different under wet and dry conditions.[38]

Mercury

Collection of inorganic mercury vapor is commonly done using diffusion through a membrane onto gold foil. The gold foil is quite specific for the collection of mercury, so very few interferences are possible. The collected vapor is quantitatively measured through the change in electrical conductivity across the gold foil, the change in conductivity being related to the amount of mercury amalgamated on the gold foil. The diffusion rate of mercury into these types of dosimeters, especially the sampling rate, is affected by large temperature changes, but this can be accounted for with temperature correction calculations. Thus, it is important that the sampling professional record the temperature at which the monitors were exposed. The 3M badge has shown good precision over the range of 0.05–0.2 mg/M^3 (Fig. 4-13). A similar monitor from Advanced Chemical Sensors has an accuracy of ±10% at 24°C and ±20% at 10°–30°C.

Another mercury monitor from SKC, also based on molecular diffusion, uses HYDRAR, which is a material developed from manganese dioxide catalyst materials.

Chlorine has been known to be a negative interferent in mercury collection. In the case of gold foil it reacts with the mercury already collected to form a volatile chloride compound, the net result being a reduced estimate of actual mercury concentration. Some badges specify that chlorine is not an interferent. In other cases, a manufacturer may offer two mercury badges—one for use when chlorine is present.

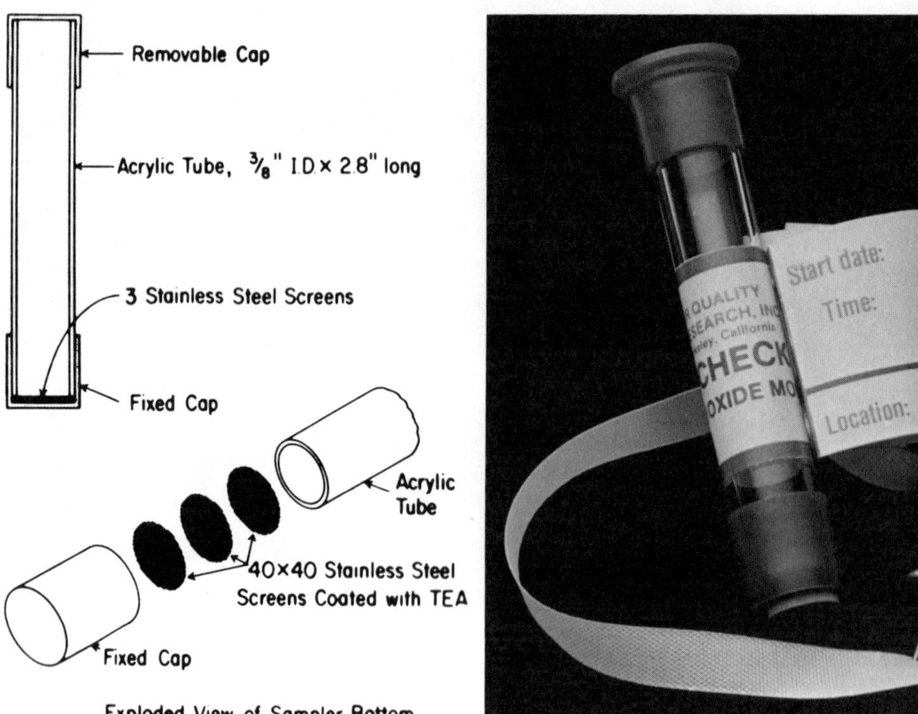

Figure 4-12. Passive monitor for nitrogen dioxide. (Drawing from E. D. Palmes NO_2 and NO_x diffusion techniques, *Am. Conf. Govt. Ind. Hyg. Ann.* **1**:263–266, 1981, reproduced with permission of the American Conference on Governmental Industrial Hygienists; photo courtesy Air Quality Research.)

USE OF PASSIVE MONITORS FOR OCCUPATIONAL SAMPLING

As noted, their ease of use has led to passive monitors replacing active methods in many cases. However, if a sampling period other than 8 hours is to be used, the minimum detection level for the period must be calculated. It is possible to use badges for 15-minute STEL measurements; however, it is very likely that the detection level will exceed the 8-hour PEL while being less than the STEL. In this case a series of 15-minute samples could not be used to calculate an 8-hour PEL.

The use of passive monitors does not eliminate the need to remain in the work area to observe work practices or make sure that the monitor does not get covered up by an employee's collar or fall off.

For OSHA compliance use, it has been recommended that only badges with backup sections be used. Review the manufacturer's instructions prior to use. Because of the ease with which they can be removed it is critical that badges be labeled before they are placed for sampling and not after sampling is done. When marking badges, use an indelible marking pen—water might smear the sample.

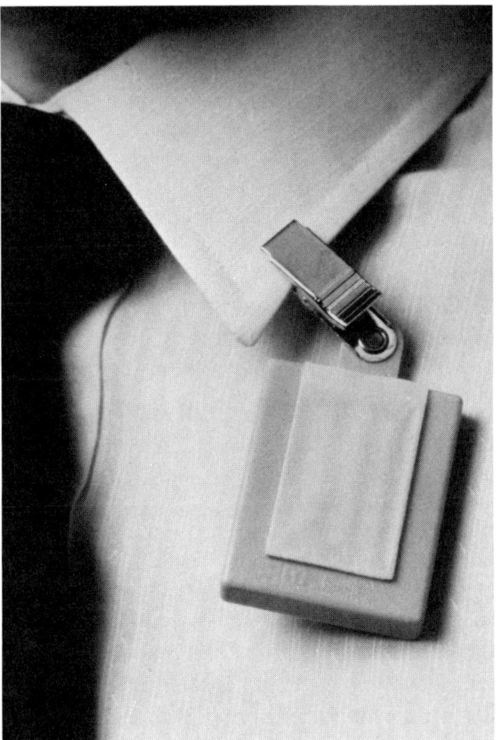

Figure 4-13. Passive mercury vapor monitor. (Courtesy 3M Corp.)

Procedure

1. Before monitoring record the start time, sampling date, employee name, and other information, such as temperature and relative humidity, on the container. Open the sample container when sampling is to start.

2. Some badges have covers that must be removed and replaced with a membrane. The covers are used for shipment later. Other badges come in more than one piece and must be assembled for sampling.

3. Hook the unit to the collar or shirt neck if a worker is to be sampled. Ensure that the open face of the sampler is facing toward the environment and exposed for the entire sampling period.

For area samples, attach it at least 1 meter above the floor in the area to be sampled. Ensure that the ambient air velocity at the sampler position is above the minimum velocity recommended by the manufacturer. Avoid sampling stagnant areas, such as against walls or in corners of rooms.

4. The minimum sampling time is governed by the sampling rate and the sensitivity of the analytical method. The maximum sampling time is determined by the sampling rate and by the adsorptive capacity of the charcoal adsorbent. When the calculated maximum sampling time is less than the desired sampling period, two or more samplers should be used in sequence to accommodate the desired exposure period.

$$\text{minimum sampling time (min)} = \frac{\text{minimum detection limit, } \eta\text{g}}{(0.2)(\text{PEL, } \eta\text{g/cm}^3)(\text{sampling rate, cm}^3\text{/min})}$$

$$\text{maximum sampling time (min)} = \frac{\text{sampler capacity, } \eta\text{g}}{(2)(\text{PEL, } \eta\text{g/cm}^3)(\text{sampling rate, cm}^3\text{/min})}$$

5. At the end of the sampling period, remove the sampler, separate sampling and backup layers if necessary, and record the end time on the container. If a backup section is incorporated into the sampler, separate it immediately after sampling and put into its own container. If a disposable section is present, remove and discard it and cover both sides of the sampling section. Seal the sampler with the cover and refrigerate. It is critical that covers on badges be replaced immediately and securely following sampling, or the collected material may begin to diffuse out of the badge.

6. Send to the laboratory for analysis as soon as possible. Prepare a blank for each set of monitors at the monitoring site by removing a monitor from its sealed container, taking off the cap or seals, and replacing it with the shipping cover. Label the monitor as a blank and submit the blank along with the exposed monitors to the laboratory.

7. The following calculations can be used to determine badge concentrations. In most cases, at least some of these calculations are done by the laboratory.

$$\frac{\text{mg}}{\text{M}^3} = \frac{(W)(10^6\ \text{cm}^3/\text{m}^3)}{(r)(K_o)(t)}$$

where

W = corrected weight in milligrams
r = recovery (desorption) efficiency
K_o = monitor sampling rate, cm^3/min
t = sampling time, minutes

Converting to ppm at STP of 298 kelvins and 760 mm Hg:

$$\text{ppm} = \frac{\text{mg}}{\text{M}^3} \times \frac{24.45}{MW}$$

where MW = molecular weight.

Temperature correction if sampling temperature was significantly different than 298 kelvins:

$$C_0 = \frac{\text{mg}}{\text{M}^3} \times \frac{(298\ \text{K})^{1/2}}{(T_s)}$$

$$C_0 = \text{ppm} \times \frac{(298\ \text{K})^{1/2}}{(T_s)}$$

where T_s = temperature at sample site in kelvins. This correction eliminates an error of approximately 1% for every 3°C increment above or below 24°C.

USE OF PASSIVE MONITORS FOR RESIDENTIAL AND AMBIENT MONITORING

Badges are potentially useful for ambient air monitoring where much lower levels are expected, although this usefulness should be evaluated for each compound, because some compounds such as 1,2-dichlorethane and chlorobenzene are not effectively sampled at very low levels. A criteria for determining whether a badge is suitable is to assume that the lowest useful airborne concentration for sampling is 10 times the median blank or 10 times the badge's detection limit, whichever is greater. Sometimes extending the length of the sampling period will increase the detection limit. For example, organic vapor badges have been used with good results for residential monitoring using sampling periods of 2 weeks or longer.[39]

Procedure

1. Select at least two rooms in the home for monitoring, preferably high-activity areas such as the living room and bedroom. Each floor of the home should have at least one monitor. Because of significant humidity fluctuations, do not select the bathroom as a monitoring site.

2. Place a monitor in each room so that it is well positioned at least 0.3 meter from any wall. The monitor should not be placed in any nonrepresentative ventilation area, such as directly in front of an air conditioner or heating vent.

3. Due to room condition variations, measurements should be taken for at least 24 hours for each monitoring point. In some cases, monitors are designed to be left for up to 7 days. During the monitoring period, the room temperature should be maintained above 21°C and 50% relative humidity. In addition, no smoking, cooking, or other combustion sources should be allowed in the room during the monitoring period.

REFERENCES

1. Knarr, R. D., and S. M. Rappaport. Impregnated filters for the collection of ethanethiol and butanethiol in air. *AIHA J.* **42**(11):839–841, 1981.
2. National Institute of Occupational Safety and Health. *NIOSH Manual of Analytical Methods*, P. M. Eller (ed.). NIOSH, Cincinnati, February, 1984.
3. Pagnotto, L. *Gas and Vapor Sample Collectors. Air Sampling Instruments*, 6th ed. ACGIH, Cincinnati, 1983.
4. Burnett, R. D. Evaluation of charcoal sampling tubes. *Fundamentals of Analytical Procedures in IH.* AIHA, Akron, pp. 64–72, 1987.
5. Saalwaechter, A. T., et al. Performance testing of the NIOSH charcoal tube technique for the determination of air concentrations of organic vapors. *Fundamentals of Analytical Procedures in IH.* AIHA, Akron, pp. 363–373, 1987.
6. Fraust, C. L., and W. R. Hermann. Charcoal sampling tubes for vapor analysis by gas chromatography. *AIHA J.* **27**:68, 1966.
7. Nelson, G. O., A. N. Correia, and C. A. Harder. Respirator cartridge efficiency studies. VII. Effect of relative humidity and temperature. *AIHA J.* **37**:280, 1976.
8. Rudling, J., and E. Bjorkholm. Effect of adsorbed water on solvent desorption of organic vapors collected on activated carbon. *AIHA J.* **47**(10):615–620, 1986.
9. Betz, W. R., et al. Characterization of carbon molecular sieves and activated charcoal for use in airborne contaminant sampling. *AIHA J.* **50**(4):181–187, 1989.
10. Levine, M. S., and M. Schneider. Flowrate associated variation in air sampling of low concentrations of benzene in charcoal tubes. *AIHA J.* **43**(6):423–426, 1982.
11. Feigley, C. E., and J. B. Chastain. An experimental comparison of three diffusion samplers exposed to concentration profiles of organic vapors. *AIHA J.* **43**:(4):227–234, 1982.
12. Goller, J. W. Displacement of polar by non-polar organic vapors in sampling systems. *AIHA J.* **46**:170–173, 1985.
13. Standard Recommended Practices for Sampling Atmospheres for the Analysis of Gases and Vapors. ASTM D-1605. 1979.
14. Crisp, S. Solid sorbent gas samplers. *Ann. Occup. Hyg.* **23**:47–76, 1980.
15. Brown, R. H., and C. J. Purnell. Collection and analysis of trace organic vapor pollutants in ambient atmospheres. *J. Chromatogr.* **178**:79, 1979.
16. Gordon, S. M. *Tenax Sampling of Volatile Organic Compounds in Ambient Air. Advances in Air Sampling.* ACGIH, Lewis Pubs., Chelsea, MA, pp. 133–142, 1988.
17. Dharmarajan, V., and R. N. Smith. Permeation of toluene through plastic caps on charcoal tubes. *AIHA J.* **42**:(9):691–693, 1981.
18. Environmental Protection Agency. Compendium of Methods for the Determination of Toxic Organic Compounds in Ambient Air. EPA-600/4-84-041, 1984.
19. Linch, A. *Evaluation of Ambient Air Quality by Personnel Monitoring.* CRC Press, Cleveland, 1974.
20. Young, R., J. R. Stetter, and G. E. McCarty. Evaluation of a novel impinger. *ACGIH Ann.* **8**:101–105, 1984.
21. National Institute of Occupational Safety and Health. *The Industrial Environment, Its Evaluation and Control.* NIOSH, Cincinnati, 1973.
22. Riggin, R. M. Technical Assistance Document for Sampling and Analysis of Toxic Organic Compounds in Ambient Air. EPA-600/4-83-027, 1983.
23. Possner, J. C., and W. J. Woodfin. Sampling with gas bags. I: Losses of analyte with time. *Appl. Ind. Hyg.* **1**(4):163–168, 1986.
24. Environmental Protection Agency. *Compendium Method TO-14: The Determination of Volatile Organic Compounds (VOCs) in Ambient Air Using Summa Passivate Cannister Sampling and Gas Chromatographic Analysis.* Quality Assurance Division, Environmental Monitoring Systems Laboratory, U.S. EPA, Research Triangle Park, NC, May 1988.
25. Lautenberger, W. J. Theory of passive monitors. *ACGIH Ann.* **1**:91–99, 1981.
26. Perkins, J. Laboratory evaluations of passive organic vapor monitors. *ACGIH Ann.* **1**:125–166, 1981.
27. Miksch, R. R. A "Passive Bubbler" Personal Monitor Employing Knudsen Diffusion: Development and Application to the Measurement of Formaldehyde in Indoor Air. Air Technology Labs, Inc., 548 E. Mallard Circle, Fresno, CA 93710.
28. Gregory, E. D., and V. J. Elia. Sample retentivity properties of passive organic vapor samplers and charcoal tubes under various conditions of sample loading, relative humidity, zero exposure level periods and a competitive solvent. *AIHA J.* **44**(2):88–96, 1983.
29. Standard Practice for Sampling Workplace Atmospheres to Collect Gases or Vapors with Liquid Sorbent Diffusional Samplers. ASTM D-4598-87.
30. Samimi, B., and L. Falbo. Comparison of standard charcoal tubes with Abcor (NMS) GasBadges within controlled atmosphere. *AIHA J.* **46**(2):49–52, 1985.
31. Berlin, A., et al. Diffusive sampling—An alternative approach. *Appl. Ind. Hyg.* **3**(2):R-2–R-6, 1988.
32. Einfield, W. Diffusional sampler performance under transient exposure conditions. *AIHA J.* **44**(1):29–35, 1983.
33. Coulson, D. M., et al. Diffusional sampling for toxic substances. In *Sampling Techniques.* Wiley, New York, 1977.
34. Spielman, C. R., K. D. Blehm, R. M. Buchan, and R. N. Hagar. An evaluation of benzene in the presence of gasoline by active and passive sampling methods. *Appl. Ind. Hyg.* **2**(2):66–70, 1987.
35. Stewart, P. A., D. Cubit, and A. Blair. Formalde-

hyde levels in seven industries. *Appl. Ind. Hyg.* **2**(6):231–236, 1987.

36. Bartley, D. L. Passive monitoring of fluctuating concentrations using weak sorbents. *AIHA J.* **44**(2):879–885, 1983.
37. Douglas, K. E., and H. J. Beaulieu. Field validation study of NO_2 personal passive samplers in a "diesel" haulage underground mine. *AIHA J.* **44**(10):774-778, 1983.
38. Palmes, E. D., and E. R. Johnson. Explanation of pressure effects on a nitrogen dioxide (NO_2) sampler. *AIHA J.* **48**(1):73–76, 1987.
39. Shields, H. C., and C. J. Weschler. Analysis of ambient concentrations of organic vapors with a passive sampler. *JAPCA* **37**:1039–1045, 1987.

CHAPTER 5

Integrated Sampling for Aerosols

The dynamics of sampling for aerosol exposures are becoming more complex as the behavior of the many different types of aerosols is becoming better understood. As described previously, an aerosol is a system consisting of airborne solid or liquid particles that are dispersed in a gas stream, usually the atmosphere. Dusts, smoke, soot, particles, mist, fumes, and fog are all terms used to describe certain types of aerosols. Solid substances are defined as particles to distinguish them from liquid droplets. A unique aspect of aerosol sampling is the concern regarding particle size, which to an extent determines where the aerosol will interact with the respiratory system and the type of inhalation exposure, as mouth and nose breathing affect the composition of the aerosol inhaled as well as the relative amounts swallowed or absorbed into the lungs.

Occupational sampling for aerosols is in a state of flux since the criteria for particle size as it relates to sampling and thus respiratory hazards has been redefined, but classic integrated methods for measuring aerosols are still being relied on and do not reflect this redefinition. The problem is few instruments and methods are available and tested that reliably measure new criteria, and current dust standards are based on the old criteria. This matter is further discussed in the sections on inspirable, thoracic, and respirable sampling in this chapter. For the time being, the sampling professional must be willing to remain flexible and know that in the near future methodologies must change. When this change occurs, comparisons should be made between methods so that old results can be compared with new ones and interpreted in terms of new standards.

An advantage of integrated methods for occupational sampling of aerosols is that there are methods that can be used to identify and quantify and differentiate between specific contaminants whereas real time and most environmental methods are nonspecific and tend to measure particle counts or mass. Environmental exposures are the result of ambient air pollution from sources such as transportation and industry. The primary difference between occupational and environmental samples for aerosol is the flow rate. For the most part, environmental methods use high (flow rate) volume pumps to collect large volumes of samples because the levels of concern in ambient air are much lower. Sampling periods for environmental samples are also correspondingly longer than occupational sampling periods. In its simplest form, aerosol sampling is the use of a prefilter to prevent contamination or clogging of gas and vapor samples.

Results of air sampling measurements for aerosols are expressed in mg/M^3. Recent studies indicate that long accepted conversions from data expressed in millions of particles per cubic foot (mppcf) to mg/M^3 for mineral dusts may not be accurate enough for current data needs.[1] This finding is important because if the sampling professional wants to compare sampling results from measurements made with the dust-counting impinger method to results made with filter collection, a correlation between the two different units is necessary; otherwise, every time a new method is used years of previous sampling data are invalidated.

AEROSOL DESCRIPTIONS

Particle size is most important from a physiological consideration of where various-sized particles might end up in the respiratory system. Particle size is the most important single parameter useful in predicting or explaining the behavior of airborne particles; however, if the sampler wishes to compare the behavior of two aerosols having approximately the same-size particles, other factors such as density and shape are also important since they would alter the predicted behavior of the aerosols. Denser particles have more momentum and more gravitational sedimentation. The shape of a particle influences the resistance of its motion through the air. If it is not a sphere, it has more drag, which slows its motion.[1]

There are a number of different definitions used for particle size. Because a sample of an occupational or environmental aerosol will always contain particles of many different sizes (polydisperse) and shapes, the description of the term *particle diameter* is not as simple as it would be if applied to a circle or sphere. Particle sizes can be determined either through physical or dynamic measurements with each measurement method measuring somewhat different quantities in the case of irregular-shaped particles. The resulting diameters are often called statistical diameters since large numbers of particles must be measured and the results must be averaged if the values are to have much significance.

Physical size measurements are associated with particle geometry, such as the diameter of a sphere or the length and width of a fiber, and are generally done under the microscope. As measurements that identify a single dimension, such as are described by Martin's or Feret's diameters, are difficult and time consuming to do and subject to much difference in definition, most commonly projected area diameters are used for optical size measurements. With this method the observer tries to visualize the size of an irregular particle in terms of the diameter of an equivalent area circle on a graticule mounted in the eyepiece. Using this comparison, an estimate is made of the particle's size.

Dynamic size measurements refer to an observed aerosol property that can be related to particle size and are usually done with sampling equipment such as cascade impactors and cyclones. The type of instrument will determine what property is to be measured. In impactor collection the range of sizes collected on each stage depends on the density and shape of the particles as well as the diameter, and therefore represents the aerodynamic behavior of the particles. Cyclones separate particle sizes based on the velocity at which the air is being collected. For example, the 10-mm nylon cyclone when operated at 1.7 Lpm matches the ACGIH curve used to describe the inhalation of particles less than 10 μm in size. Another method that can size particles is light scattering, which uses real time instruments for measurements. These instruments are discussed in the chapter on Real Time Sampling Methods for Aerosols.

One way of choosing the appropriate diameter definition is to select a method for determining particle size that is closely related to the application of the results. It is important for the sampling professional to select the definition of particle size appropriate to his/her interests, and a method sensitive enough to measure the size range of interest. Results from different techniques may not correspond.

The most widely used particle-size definition for air samples is aerodynamic equivalent diameter (AED) based on the way a particle behaves when airborne in a force field such as air rather than on its geometry under a microscope. Aerodynamic diameter is commonly used rather than geometric diameter when discussing collection efficiency as a function of particle size for specific devices. For example, inertial impaction and gravitational forces are proportional to the particle mass, and in these cases, aerodynamic diameter is the particle size that governs particle motion. The cut diameter is the particle diameter for which collection efficiency and penetration equals 50%. This is useful for comparing efficiencies of various devices and efficiencies of the same device under various operating conditions even when penetration does not reach 50%.[2]

After aerosol size data have been obtained for a specific system, statistical methods are required to impart quantitative meaning and utility to the measurements. Usually the first step is to obtain a frequency (number-size) distribution or at least classification according to size range. Three types of distribution useful for examining aerosols are number, mass, and surface area distributions.[3] As an example, the mass distribution tells how the total airborne mass is distributed among the various particle sizes. When plotted on log-probability paper, the mass-median diameter (MMD) or the size below or above which half of the mass of the particles would occur can be obtained. The MMD identifies the middle of the distribution of mass; that is, the size for which half the total mass is contributed by smaller particles and half by larger particles. When using aerodynamic diameter methods, sometimes the mass median aerodynamic diameter (MMAD) is identified for an aerosol. The size mass distribution is used for predicting the actual dose to the lung resulting from inhalation of a given amount of dust or the weight of material collected by a filter or other collection device that is efficient only for particles larger than a given size.

For a practical sampling system to simulate the collection by the respiratory tract, the sampler must make approximations and assumptions in order to be able to deal with the reality, which is that there is a very complex interaction of variables governing respiratory tract deposition. In measurement, particles are collected by an instrument that has the same efficiency (collects the same aerosol size fraction) as a certain portion of the respiratory tract, and thus the justification exists for expressing aerosol standards in terms of concentrations. In practice, breathing rate varies widely so that the concentration is actually not a perfect prediction; however, the average biological effect in a population is related to the concentration, so the breathing rates of individuals are averaged out. In standards for particles, an average breathing rate is assumed for the entire population of interest, so that the mass deposed is proportional to the concentration measured with an instrument that collects the same fraction as the respiratory tract.[4]

AEROSOL COLLECTION

When selecting samplers for aerosols, the physical, chemical, and biological properties and toxic effects, as well as the site of action of the aerosol, must be taken into account. If a material is absorbed by the body wherever it deposits, then it is important to collect all the material that may enter the body in order to have a sample representative of the hazardous exposure. On the other hand, if the contaminant is slow to dissolve in bodily fluids or is essentially insoluble, then the site of action will determine what type of sampler (inspirable, thoracic, or respirable) should be used; thus, for aerosols known to cause aggravation of chronic bronchitis, the traecheobronchial size fraction should be collected.[5]

Aerosol samplers for occupational sampling typically consist of a cassette containing a membrane or fiber filter on which the aerosol is collected. Cassettes and filters come in many sizes, and some cassettes have an extension cowl such as is used in asbestos sampling to decrease electrostatic effects. In some cases, the cassette is very small, such as when it is used as a prefilter to screen unwanted material from gas and vapor samples. In other cases, the cassette is contained within another piece of sampling equipment, such as a cyclone, to provide size-selection characteristics. The filter can also be contained within the equipment, such as the NR701 personal inspirable particulate sampler or, in the case of impactor stages, the aerosol is collected directly on the greased surface of the metal plate. Figure 5-1 shows various types of collectors for personal sampling for aerosols.

Environmental samplers also use filters, but the apparatus is very large and typically designed to be used outdoors. Some size selective sampling is also done and all techniques utilize high-volume pumps usually collecting for a 24-hour period.

Filters

Filters in plastic cassettes are the most common media used to collect occupational aerosols. Filters come in a variety of materials and are prepared through different techniques, although there are two basic types used for aerosol sampling: compacted or felted fibers and membranes. Collection characteristics are different for each type of filter medium. For example, because fiber filters offer comparatively low resistance to airflow, their use is recommended when a large volume of aerosol must be sampled. The choice of filter medium depends on the physical and chemical properties of the aerosol to be sampled, the sampler, and the analysis to be performed.

For a given aerosol and a given filter, the collection efficiency varies with the face velocity and the particle size. With an appropriately chosen filter, the air will pass through, leaving the particles behind. If the filter is too porous, some particles penetrate and others become embedded within the filter, making recovery and later analysis difficult.[3] If the filter is too impervious, airflow resistance (pressure drop) may be too great. Ideally, particles should be trapped at the front surface of the filter if size properties are to be maintained.

The type of filter medium is specified by the method, so in most cases the sampling professional does not have to understand how to select a given filter medium. However, it is important to understand the limitations of each type of filter and why particular types are preferred for various uses.

All filters may contain major, minor, and trace contaminants that should be considered during filter selection. Important filter characteristics are collection efficiency, pressure drop, mechanical strength, hygroscopicity, chemical purity, and possible artifact production (gas to particle conversion).[5]

Membrane filters are made from cellulose, PVC, or polytetrafluoroethylene (PTFE). Since most particle collection takes place at or near the surface, membrane filters are useful for applications where the collected particles will be examined under a microscope. Since the mixed cellulose ester filters (MCE) dissolve easily with acid, they are also used for collection of metals for atomic absorption analysis. Pore sizes of 0.45 μm and 0.8 μm are the most commonly used. With respect to cellulose, the main concern is their susceptibility to water. PVC membrane filters are used for sampling silica, nuisance dusts, and zinc oxide. PTFE filters come in a variety of pore sizes, 1 μm, 2 μm, and 5 μm, and are used for pesticides, alkaline dusts, and many other compounds.

The primary fiber filter used is made of fiberglass. Filters are available with and without binders. For occupational sampling, glass fiber filters are used to collect pesticides; 2,4-D is sampled with binderless glass fiber filters. They are also used in high-volume environmental sampling for atmospheric aerosols and lead. Environmental filter media are much larger, often 8-by-10-inch sheets. Quartz fiber filters are used for high-volume environmental particulate sampling. They have minimal artifact generation. The biggest limitation is their fragility.

Another type of filter medium, polycarbonate filters, also known as Nucleopore, have superior strength and chemical and thermal stability. They are essentially transparent with a slight green tinge, and they average 5 cm to 10 cm in thickness. Polycarbonate filters are recommended for asbestos sampling for transmission electron microscopy (TEM) analysis.

When impregnated with special reagents, filters can be used to collect certain vapors, such as isocyanates, as well as particles. Table 5-1 provides a description of media used in NIOSH methods for aerosols.

As discussed, filters are most often contained

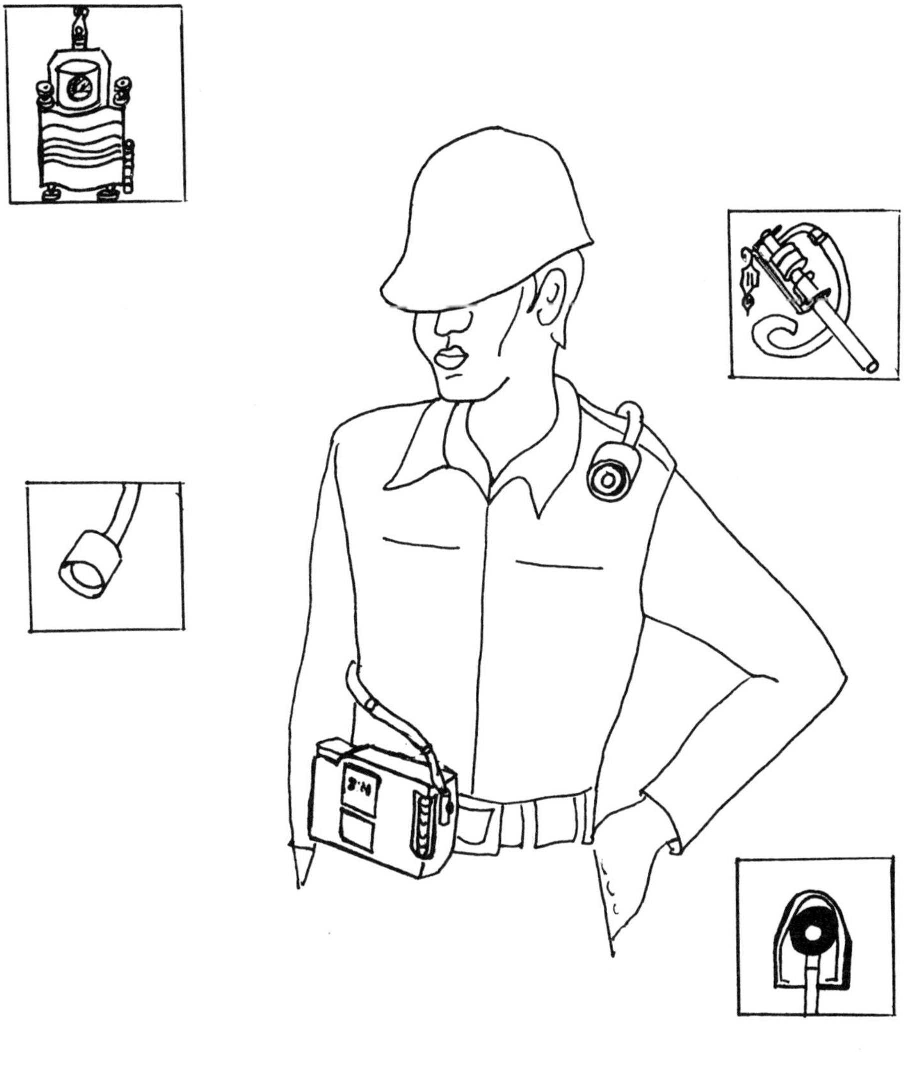

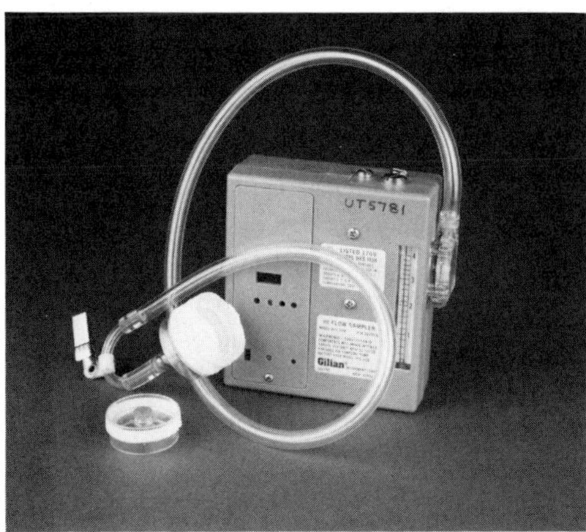

Figure 5-1. Personal sampling with filter cassettes. Clockwise from left: filter cassette, open-face; cascade impactor; cyclone; inspirable particulate sampler; on shoulder, filter cassette, closed-face. The photo shows a pump setup for sampling with an open-face filter cassette. (Photo courtesy Gilian Instrument Corp.)

TABLE 5-1. Media Used in NIOSH Methods for Aerosols

Compound	*Method Number*	*Media*
Alkaline dusts	7401	1-μm PTFE membrane filter, 37 mm
Aluminum and compounds	7013	0.8-μm MCE filter, 37 mm
p-Aminophenylarsonic acid	5022	1-μm PTFE filter, 37 mm
Asbestos	7400	0.8 – 1.2-μm MCE filter, 25 mm with cowl
Azelaic acid	5019	5-μm PVC membrane filter, 37 mm
Barium	7056	0.8-μm MCE filter, 37 mm
Benzoyl peroxide	5009	0.8-μm MCE filter, 37 mm
Beryllium and compounds	7102	0.8-μm MCE filter, 37 mm
Boron carbide (respirable and total)	7506	5-μm PVC membrane filter, 37 mm
Bromoxynil	5010	2-μm PTFE membrane filter, 37 mm
Bromoxynil octanoate	5010	2-μm PTFE membrane filter, 37 mm
Cadmium and compounds	7048	0.8-μm MCE filter, 37 mm
Calcium and compounds	7020	0.8-μm MCE filter, 37 mm
Carbaryl	5006	Glass fiber filter (type A), 37 mm
Carbon black	5000	5-μm PVC membrane filter, 37 mm
Chlorinated diphenyl ether	5025	0.8-μm MCE filter, 37 mm
Chlorinated terphenyl	5014	Glass fiber filter (Gelman no. 64877), 37 mm
Chromium and compounds	7024	0.8-μm MCE filter, 37 mm
Chromium (V)	7600	5.0-μm PVC membrane filter, 37 mm
Coal tar pitch volatiles	5023	2-μm PTFE membrane filter (Zeflour), 37 mm
Cobalt and compounds	7027	0.8-μm MCE filter, 37 mm
Copper (dust/fume)	7029	0.8-μm MCE filter, 37 mm
2,4-D	5001	Binderless glass fiber filter (type AE), 37 mm
o-Dianisidine	5013	5-μm PTFE membrane filter, 37 mm
Dibutylphosphate	5017	1-μm PTFE filter, 37 mm
Dibutyl phthalate	5020	0.8-μm MCE filter, 37 mm
Di(2-ethylhexyl) phthalate	5020	0.8-μm MCE filter, 37 mm
Dimethylarsenic acid	5022	1-μm PTFE filter, 37 mm
EPN	5012	Glass fiber filter (type AE), 37 mm
Ethylene thiourea	5011	5-μm PVC or 0.8-μm MCE filter

in plastic cassettes that have two or three stages and a pad on which the filter can rest. Many analyses involve just weighing filters, assuming they only contain the desired contaminant, and dividing the weight by the sampling volume to obtain mg/M^3. Examples are total (nuisance) dust, respirable dust, welding fume, and oil mist. Filters must be weighed in advance of sampling and inserted in the cassettes. Matched weight filters in cassettes are sometimes used instead. In this case, there are two filters placed inside each cassette whose difference in weight does not exceed 0.1 mg. The analysis involves weighing both filters and comparing the weights. The advantage is that preassembled cassettes can be purchased and stored. Preweighed cassettes require more advance planning. Only the laboratory that assembled them can do the analysis, since it has the prerecorded weights. If 0.1 mg is sufficient sensitivity for the contaminant to be sampled then the

TABLE 5-1. *(continued)*

Compound	*Method Number*	*Media*
Hydroquinone	5004	0.8-μm MCE filter, 37 mm
Lead	7082	0.8-μm MCE filter, 37 mm
Lead sulfide (respirable)	7505	10-mm nylon cyclone and 5-μm PVC membrane filter, 37 mm
Malathion	5012	Glass fiber filter (type AE), 37 mm
Mineral oil mist	5026	Membrane filter, 37 mm, 0.8-μm MCE or 5-μm PVC, or glass fiber filter
Methylarsonic acid	5022	1-μm PTFE filter, 37 mm
Total dust	0500	5-μm PVC filter, 37 mm
Respirable dust	0600	10-mm nylon cyclone and 5-μm PVC
Paraquat	5003	1-μm PTFE membrane filter, 37 mm
Pyrethrum	5008	Glass fiber filter, 37 mm
Rotenone	5007	1-μm PTFE membrane filter, 37 mm
Silica, amorphous (respirable)	7501	10-mm nylon cyclone and 5-μm PVC filter, 37 mm
Silica, crystalline (respirable)	7500, 7601	10-mm nylon cyclone and 5-μm PVC filter, 37 mm
Strychnine	5016	Glass fiber filter, 37 mm
Parathion	5012	Glass fiber filter (type AE), 37 mm
2,4,5-T	5001	Binderless glass fiber filter (type AE), 37 mm
o-Terphenyl	5021	2-μm PTFE filter, 37 mm
Thiram	5005	1-μm membrane filter, 37 mm
o-Toluidine	5013	5-μm PTFE membrane filter, 37 mm
Tungsten and compounds	7074	0.8-μm MCE filter, 37 mm
1,4,7-Trinitrofluorene-9-one	5018	0.5-μm PTFE membrane filter, 37 mm
Vanadium oxides (respirable)	7504	10-mm nylon cyclone and 5-μm PVC membrane filter, 37 mm
Warfarin	5002	1-μm PTFE membrane filter, 37 mm
Welding and brazing fume	7200	0.8-μm MCE filter, 37 mm
Zinc and compounds	7030	0.8-μm MCE filter, 37 mm
Zinc oxide	7502	0.8-μm PVC membrane filter, 25 mm

use of matched weight filters in cassettes can provide more flexibility in the field.

There are two major sources of filter cassette leakage: (1) external, when air enters the cassette through the tapered joints of the stages, for example, two-stage and three-stage cassettes; (2) internal, when air enters the cassette through the inlet, traveling around the edge of the filter, bypassing the surface, and causing uneven distribution of aerosols.[6] The filter holder must seal securely to prevent air leakage around the filter or to the outside.[5]

Problems

Many changes can occur to alter the composition of aerosols during the sampling period. Therefore, aerosols are a challenge to sample. Wind currents, cross drafts, sources, and condition of the material are important and can affect particle concentration and size distributions. Even point sources outdoors do not distribute their particles evenly in all directions; rather the particles travel with the wind, which is much more variable. Generally particle concentrations remain closer to the ground and their point of origin rather than spreading great distances and remaining airborne. Bombardment of airborne particles by gas molecules in air produces random zigzag motions that cause very small particles ($<0.5\ \mu m$) to move and mix even under tranquil conditions. All of these variables impact the sampling strategy.

Where there is a steep concentration gradient, the location of the sampler can be critical. Artifacts and changes can occur to aerosols during collection on filters due to the potential for agglomeration (smaller particles sticking together) and shattering (breaking larger particles into smaller particles).[7] Potential contributors to bias during air sampling are inlet effects, filter efficiency, self-dilution, electrostatic effects, resuspended dust, and variability of concentrations within the breathing zone.[8]

Some particles may fail to enter the inlet of the sampling device because the inertia of these particles prevents them from accelerating, decelerating, or changing their direction fast enough to move with the air. In addition, when particles are far away from the sampling inlet, they will not be collected using conventional flow rates. The three primary variables affecting the sampling efficiency of an inlet are the inlet's diameter, the suction velocity created by the flow rate of the pump, and the diameters of the particles being sampled.[9] In one experiment, larger inlets were found to be capable of sampling larger particles with less sampling bias. For inlets with long sampling tubes, some particles may be lost within the sampling tube and not reach the collector.

Air currents moving across a closed-face filter cassette's sampling surface will decrease sampler efficiency as well.[10] Overloading the sampling capacity of a filter can cause problems. For example, when sampling for asbestos a heavy dust loading will mask fibers, thus making an accurate count impossible. Another source of inaccuracy is filter weighing.[11]

While most dusts, mists, and fumes are relatively inert and remain stable after collection, they can also contain volatile or chemically reactive particles that are very susceptible to undergoing changes during sampling. Sulfuric acid droplets can increase in size due to water absorption and can react with ammonia, so accurate measurement of these droplets involves monitoring humidity during sampling and taking precautions to preserve the chemical identity on the sampling media.[8]

Total Aerosol Samplers

The total aerosol that can be collected by a filter in a cassette is generally used for occupational air samples if the air contaminant is capable of passing through the lungs to the bloodstream. For example, when a lead pigment is inhaled, some particles might be carried out of the lungs by pulmonary clearance mechanisms, and subsequently swallowed. Another application of total dust sampling is for nuisance dusts, which are certain types of biologically inert dusts that in high concentrations may seriously reduce visibility; cause unpleasant deposits in the eyes, ears, and nasal passages; or cause injury to the skin or mucous membranes. OSHA also terms these dusts physical hazards. Nuisance dust standards apply to both organic and inorganic dusts, but they cannot be applied to chemicals that have specific toxic effects.

The total sample mass collected in a two- or three-stage filter cassette has in the past been used exclusively to characterize occupational exposures to many mineral and metallic compounds. However, this type of sample is not representative of compounds that are collected at various sites in the upper respiratory tract or tracheobronchial tree.[12] Therefore total dust samples collected in the traditional manner are not related to any collection curve.[13] In the future, inspirable particulate mass sampling may replace total dust methods because it will eliminate the current concerns about the variability of sampling characteristics of these total dust samples and the accuracy of predicting aerosol health hazards.

Until then, however, "total dust" sampling methods are important for correctly comparing sample results to OSHA standards based on those methods.

Environmental methods for sampling total atmospheric and lead aerosols involve using high-volume pumps inside enclosures. Wind and moisture have a significant impact on this type of sampling. Samples are generally collected for 24 hours or more. As opposed to occupational dust sampling where a specific process and source of dust are usually involved, atmospheric particles have a much broader composition, and may include carbon, metal, and mineral dusts.

Occupational Exposures to Total Aerosols

For occupational exposures to total aerosols the sampling is very simple: a portable pump with a built-in rotameter hooked up via tubing to a filter in a cassette. Collection of a representative sample of the aerosol through the inlet of a filter cassette depends on physical factors, such as particle size, inlet size, sampling velocity, sampler shape and orientation, and ambient air velocity.[3] Generally closed-face cassette sampling techniques where only a small colored plug is removed from the top and bottom of the cassette prior to sampling have been used for the majority of aerosols in total dust sampling, since the closed-face technique lends protection to the filter and its accumulated contaminant.[14]

When sampling with filter cassettes is done closed-face, the collected material is concentrated in the center of the filter near the inlet port. However, it has been found that collection accuracy decreases as particle size increases and as the angle of the cassette increases away from the direction of the wind.[15] One of the primary concerns raised regarding the use of the closed-face cassette is that the basis for its design is to facilitate ease of handling and analysis of the collected dust samples rather than to represent any particular set of collection characteristics. When facing into the wind, oversampling has been reported if the flow rate is different than the wind velocity. It has also been reported to occur in calm air. Undersampling occurs when the inlet is at an angle to the wind.[14]

The alternative to closed-face cassettes for total dust sampling is open-face cassettes. The open-face mode uses three-stage cassettes so the entire top can be removed, exposing the entire filter, thus allowing even distribution of the contaminant during sampling. This situation is vitally important for analysis such as fiber counts for asbestos that assume an analysis of a portion of the filter is representative of the entire filter. Three-stage cassettes must be used for open-face sampling because the second stage keeps the filter from falling out of the cassette when the pump is turned off, cutting off the suction.

A comparison of the two sampling methods indicates a significant difference in their ability to collect certain types of particles. Differences in particle sizes, particle densities, sampling flow rates, inlet radii, and ambient airstream velocities may be responsible for the differences between open- and closed-face cassette sampling techniques. It has been determined that closed-face filter cassette sampling techniques are significantly less efficient than open-face filter cassette sampling techniques for paint spray mist, chromic acid mist, portland cement dust, grain dust, and wood dust.[14]

There are four documented sources of sampling error in total dust sampling using filter cassettes: sedimentation, geometric orientation of cassettes, airflow rate variation associated with personal pumps, and ambient wind speed and direction.[14]

Traditionally samplers for personal exposures are placed on the collar or lapel. Some studies have suggested that samplers placed on the lapel may result in higher results than if they were placed on the forehead or nose; however, for practical purposes the lapel is the best position to simulate the breathing zone.[16] Dust deposited on a worker's clothing may be released and collected by a lapel-mounted sampler, although this resuspended dust is not necessarily carried into the breathing zone of the worker. The job being performed and individual work practices will also influence the results. For additional information on conducting sampling surveys for personal exposures to particulates, see the chapter on Survey Preparation and Performance.

The orientation of the cassette will affect results. The best position for the cassette during sampling is attached to the collar area with the inlet pointing horizontally and slightly downward to prevent very large particles from settling into the inlet. The least desirable angle is 90°, which could easily occur during area sampling when the cassette is suspended from a length of tubing propped over the pump.[15]

The worker's breathing zone consists of a hemisphere of 300-mm radius extending in front of the face, and measured from a line bisecting the ears.[17] The sampling head should be placed to prevent dust from falling into it and to avoid restricting the inlet. The pump is usually placed on the worker's belt.

Procedure

1. Calibrate a sampling pump with a representative filter in line at a flow rate between 1 and 3 Lpm (usually 2 Lpm) according to the procedures listed in Appendix A.

2. Take the plugs out and connect a preweighed, labeled, two-stage filter cassette containing the appropriate filter to the pump by attaching the tubing to the outlet end. If the outlet end is sufficiently long, the tubing can be put directly over it. If not leur slip adapters can be used. If leur slips are used, the calibration should be done with them in the sampling train to account for any leakage. The inlet end of the filter cassette is the end farthest away from the filter. Make sure the plug is out of the inlet prior to turning on the pump. Preweighed filters should have a sample number attached to the cassette that has been assigned by the laboratory.

3. Select the employee or area to be sampled. Discuss the purpose of sampling and advise the employee not to remove or tamper with the sampling equipment. Inform the employee when and where the equipment will be removed. Clip the filter holder onto the worker's collar, T-shirt neck, or area where sampling is to be done and hook the pump onto the worker's belt or waistband. The inlet orifice should be in a downward vertical position to avoid contamination. Use duct tape to keep the tubing out of the employee's way. Collect at least the minimum sample volume. Do not put any tubing in front of the filter.

4. Turn on the pump and record the start time. Record pertinent field data including area, the employee's full name, job, position or title, or put a description of the area in which the pump is placed. Record start and end times, initial and final air temperatures, relative humidity and atmospheric pressure, or elevation above sea level and pump rotameter setting. Include the pump number and the number on the collection device. In some cases, the employee's social security number must also be recorded.

5. Check the pump flow rate by visually observing the rotameter for any changes at least hourly. If there is a change, check the flow rate using a rotameter.

6. As a minimum, check the pump rotameter after the first half hour, hour, and thereafter every 2 hours. Ensure that the tubing is still attached to both the pump and the collection device and that it is not pinched. Make sure the cassette is still in the same position.

7. Periodically monitor the pumps throughout the workday to ensure that sample integrity is maintained, and cyclical activities and work practices are identified. Change the filter if excessive loading is noted or if a significant change in the flow rate has occurred.

8. Before removing the pump at the end of the sample period, check the flow rate to ensure that the rotameter ball is still at the calibrated mark. If the ball is not at the calibrated mark, the flow rate of the pump must be checked prior to inserting another cassette.

9. Disconnect the filter after sampling and immediately cap the inlet and outlet using plugs. Turn off the pump and record the ending time. Remove the collection device and seal off the ends immediately.

10. At the end of the day the pumps must be recalibrated.

11. Prepare field blanks at about the same time sampling is begun. These field blanks should consist of unused filters and filter holders from the same lot used for sample collection. Handle and ship these field blanks exactly as the samples, but do not draw air through them. Two field blanks are recommended for each ten samples with a maximum of ten field blanks per sample set. In addition to these, two unopened filter cassettes from the same lot are often included as media blanks.

Environmental Collection of Total Aerosols

There are two different types of high-volume sampling: one for general ambient air samples, such as might be compared to the Environmental Protection Agency (EPA) primary and secondary particulate standards, and the other is the type associated with asbestos clearance sampling. The sampling set up for ambient air is much more complicated than that used for clearance. For more information on clearance sampling, see the section on asbestos in the chapter on Specific Sampling Situations. For the balance of this section, high-volume filter sampling will refer to those techniques used for ambient air collection rather than clearance sampling.

In high-volume sampling (Fig. 5-2) air is drawn through an inlet to an 8-in. by 10-in. filter installed within a large sampling enclosure at 40 cfm (cubic feet per minute). The filter is weighed prior to and after sampling. The peaked roof is the most common type of inlet. The primary purpose of this inlet is to protect the filter from dust fallout. The sampling effectiveness of this inlet varies depending on its orientation with respect to wind direction and wind speed. Although the peaked roof does not serve to specifically address a sampling curve, in general over 50% of all particles less than 30 μm to 50 μm penetrate this inlet. The addition of various inlets allows for size-selective sampling, such as PM_{10}, an environmental measurement that corresponds to the thoracic sampling criteria. For survey work the most likely use of this technique will be indoor air concerns in buildings where an exterior source of contaminants is suspected or for boundary line monitoring of hazardous waste sites.

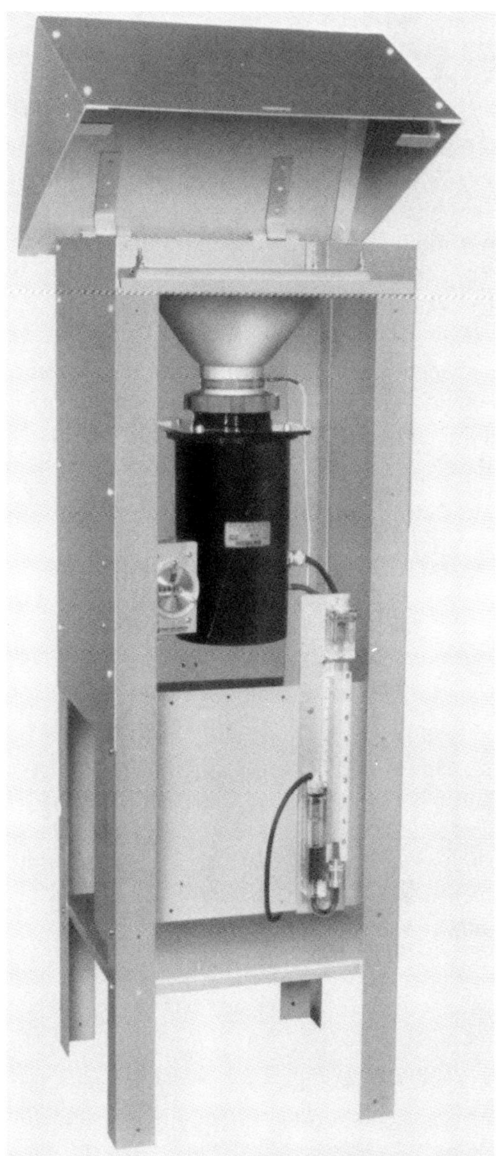

Figure 5-2. High-volume environmental sampler and enclosure. (Courtesy General Metal Works, Inc.)

Generally high-volume samples are collected over a 24-hour period. Filter media can be cellulose fiber, glass fiber, quartz fiber, Teflon-coated glass fiber, and Teflon membrane. The selection of filter type depends on the sampling professional's purpose. Quartz fiber filters are often used, although they are very fragile. Filters must be conditioned before and after the sampling period for 24 hours; therefore, measurements are not available for a minimum of three days. For ambient air monitoring, inlet probes are usually 1–5 meters aboveground. Following is an overview of the basic procedure. The manual provided with the sampler should be reviewed for specifics.

High-Volume Sampler Procedure

1. Tilt back the inlet of the enclosure and secure it. Calibrate using the orifice calibrator available for this unit. Record the flow rate.

2. Place the sampler and filter holder in the servicing position by raising up both the sampler motor/blower unit and filter holder until the filter holder is above the top level of the shelter. Then rotate the unit one quarter turn so that the filter holder hangs in the rectangular hole in the sampler support pan.

3. Remove the faceplate by loosening the 4 wing nuts. Allow the swing bolts to swing down out of the way.

4. Carefully center a new filter, rougher side up, on the supporting screen. Properly align the filter on the screen so that when the faceplate is in position the gasket will form an airtight seal on the outer edges of the filter.

5. Secure the filter with the faceplate and four brass swing bolts with sufficient pressure to avoid air leakage at the edges.

6. Rotate and lower the filter holder and blower/mover assembly to its normal operating position.

7. Wipe any dirt accumulation from around the filter holder with a clean cloth.

8. Close the lid carefully and secure with the aluminum strip, and plug all cords into their appropriate receptacles to start sampling of this type.

9. When sampling is complete, the motor is turned off and the time is recorded.

10. Reversing the procedure, the filter is removed and carefully placed in a container to return to the laboratory for weighing.

11. The final volume is calculated by multiplying the flow rate by the sampling time.

12. The weight in mg is divided by the volume in M^3 to get the results.

PARTICLE SIZE-SELECTIVE SAMPLING

The basis for particle size-selective sampling is the assumption that the likelihood of the particle's capture by a dust capturing system and its extent of penetration into the respiratory system can still be determined even though the geometries of lost particles are irregular in shape.[18]

The criteria for occupational exposures to respirable dust in use by OSHA since 1970 as defined by ACGIH were based on a curve developed by the Atomic Energy Commission (AEC) and applied to particles less than 10 μm in size. A slightly different criterion is used by the Mine Safety and Health Administration (MSHA) for respirable dust measurements in mines. However, as noted in the chapter on Hazards, a new set of definitions for particle size-selective sampling has been proposed by ACGIH, namely, inspirable, thoracic, and respirable. As shown in Figure 5-3, the inspirable mass fraction is the total mass of aerosol that can be expected to enter through the nose and mouth. The thoracic mass is the amount of this mass that can penetrate the respiratory system past the larynx, and the respirable mass is the amount capable of depositing within the lungs, particularly the gas exchange region. The collection curves that aerosol samplers need to emulate to describe these fractions are also shown.

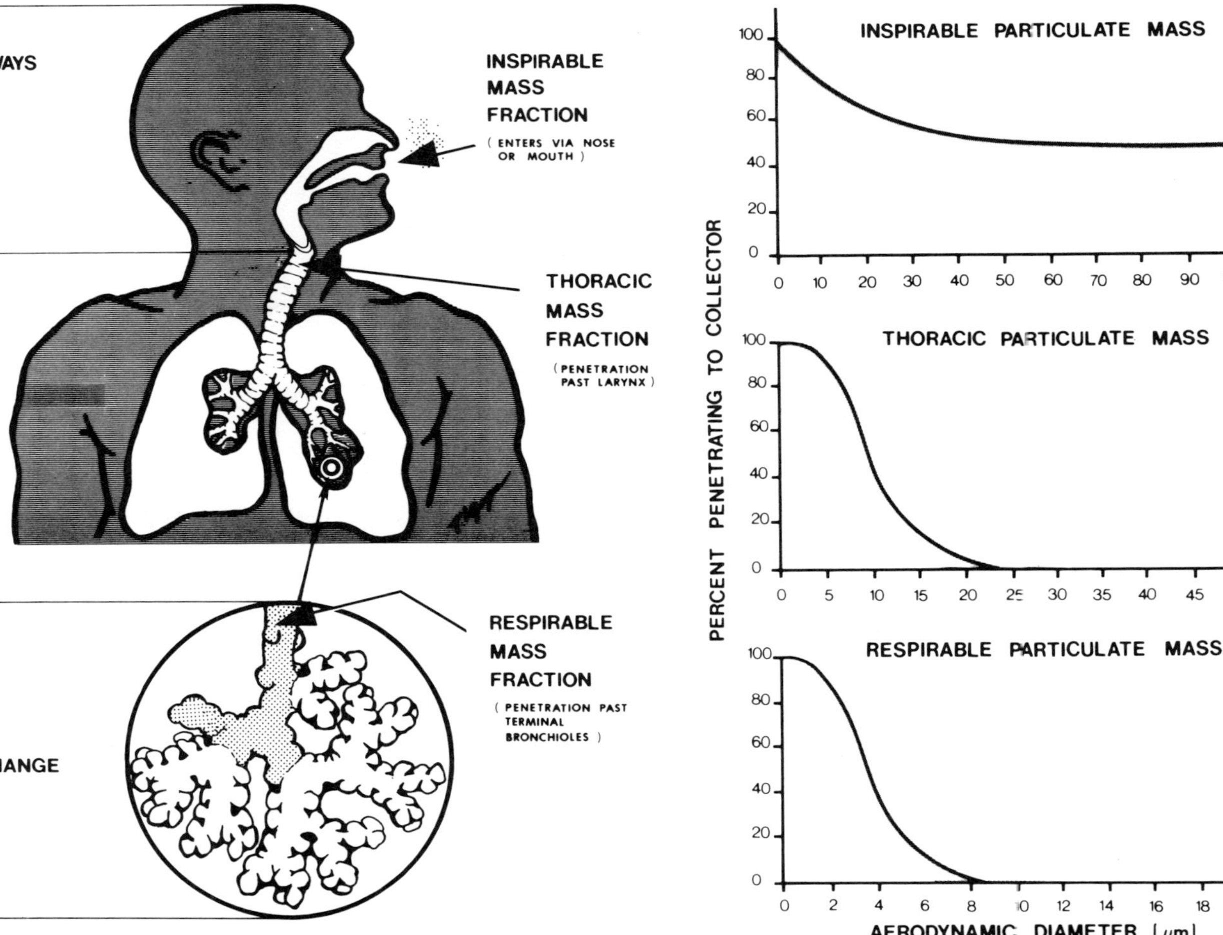

Figure 5-3. Characteristics of aerosol mass fractions. (From Phalen, R. F. Rationale and recommendations for particle size-selective sampling in the workplace. *Appl. Ind. Hyg.* **1**(1):3–14, 1986. Reproduced with permission of Applied Industrial Hygiene, Inc.)

TABLE 5-2. Inspirable Mass Fraction Characteristics

Aerodynamic Diameter (μm)	*Percent*
0	100
1	97
2	94
5	87
10	77
20	65
30	58
40	55
50	52
100	50
185	0

Inspirable Samplers

The inspirable particulate mass (IPM) fraction of an aerosol is the fraction of the ambient airborne particles that can enter the uppermost respiratory system compartment, the head. Airborne material that deposits in the head may be absorbed and/or swallowed, although some may be expelled directly from the body by bulk cleaning mechanisms, such as sneezing, spitting, or nose blowing.[13] Inspirability depends on wind speed and direction, breathing rate, and whether breathing is by nose or mouth.[4] Table 5-2 gives mass fraction characteristics of inspirable samplers.[13]

It has been recommended that IPM sampling replace the present methods of total dust sampling using closed-face filter holders. The primary concern is that there are several methods of collecting total dust samples whose collection characteristics vary. Therefore, comparison as well as making an interpretation of the biological consequence of a sample are difficult. The traditional open-face filter cassette does not measure total dust due to several collection problems, such as blunt face design, wind speed sensitivity, changes related to orientation, and particle size collection characteristics.[19] However, at least two samplers have been developed that are capable of meeting the criteria for IPM collection: a personal and an area sampler.

There are at least three general classes of chemical compounds for which sampling should be done for all inspirable particles: (1) highly soluble materials that can quickly enter the blood becoming available to pass through membranes in many regions and exhibit their toxicity, such as nicotine and soluble metal salts; (2) materials that can be toxic after oral ingestion, such as many metals including lead; (3) compounds that can exert toxic effects at their deposition site, such as hardwood dusts and acids.[13] Important categories of aerosols that are capable of exerting toxic effects following inhalation and deposition anywhere in the body include pesticides, many metals, and acids. An example of an IPM sample of soluble material will include the amount of a substance available to the systemic circulation, as well as the amounts that may deposit in the tracheobronchial region, which would be transported to the systemic circulation.[20]

In many work environments particle sizes are within the thoracic particulate mass (TPM) fraction rather than the IPM because larger particles are often removed via filtration or deposition on surfaces; however, in situations where there is a source of aerosol, such as sprayers, cutting or abrasive machinery, or easily resuspendable material, there may be a significant IPM concentration.[13]

An example of an exposure that would be sampled using an inspirable particulate sampler is a determination of the potential for nasal carcinoma from wood dust. In the case of large-sized particles or resin-impregnated sawdust, the materials would be expected to deposit and remain in the nasal passage.[20]

Types of IPM Samplers

The NR-701 is an inspirable dust sampler for personal sampling. It has a streamlined inlet that protrudes from the body of the device to minimize problems of particle impaction and bounce-off. The inlet is also an integral part of a sealed cassette that contains the filter. After use, the cassette is removed in total from the sampling head, sealed, and returned to a laboratory for weighing or chemical analysis. Several preweighed cassettes can be used with a single sampling head (if it is designed to collect a series of air samples), and the cassettes can be cleaned, reloaded with

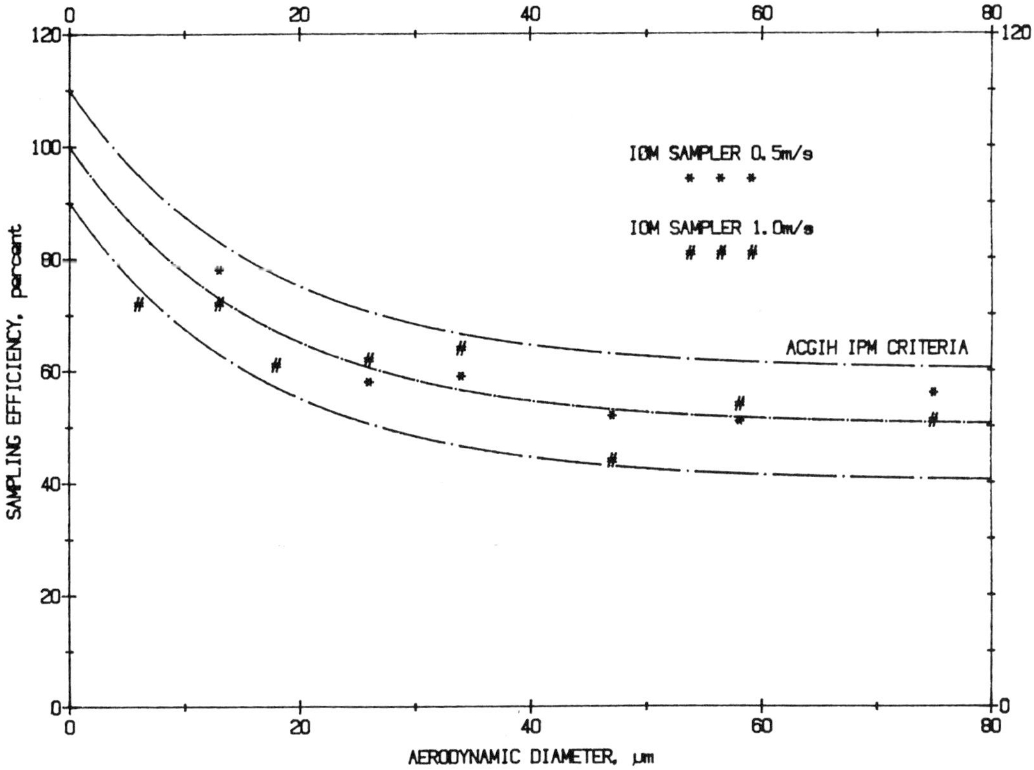

Figure 5-4*A*. IPM personal sampler performance. (From Hinds, W. C. Basis for particle size-selective sampling for wood dust. *Appl. Ind. Hyg.* 3(3):67–72, 1988. Reproduced with permission of Applied Industrial Hygiene, Inc.)

new filters, and reused (Fig. 5-4).[21] The NR-701 has been tested and found to be independent of wind speed over a representative range of wind conditions.[22]

The IOM static inspirable-area sampler operates at 3 Lpm and is designed for area sampling. It has a single entry slot to a dust-collecting cassette located inside a 50-mm-diameter cylindrical sampling head that rotates slowly about a vertical axis while sampling. The dust-collecting cassette incorporates a 37-mm filter and makes a leakproof seal with the body of the sampling head aided by the spring-loaded cap that presses the capsule down onto an O-ring seal. The entire capsule is weighed before and after each sampling run so that the dust evaluation is based on all material that enters through the plane of the sampling slot. According to the developers of this unit, any problems of internal wall losses or particle blowoff have been eliminated. It is impor-

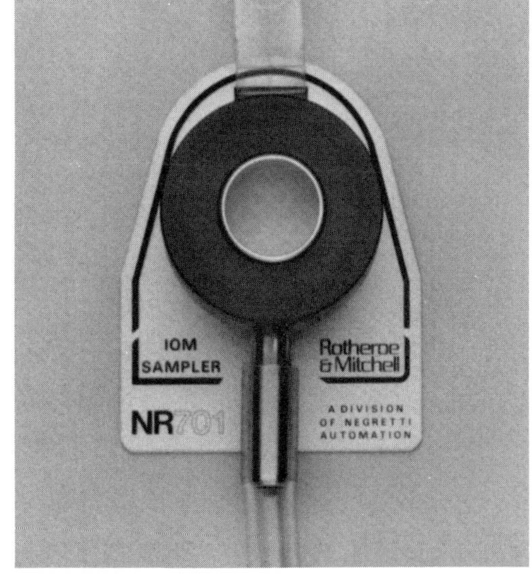

Figure 5-4*B*. IPM personal sampler. (Courtesy Air Quality Research.)

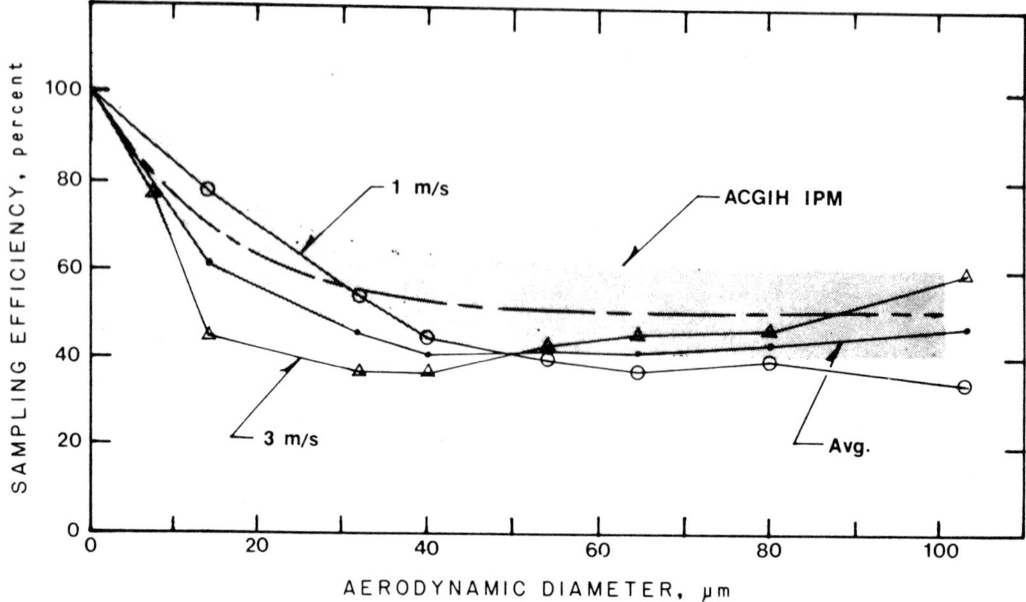

Figure 5-5A. Area sampling, sampling performance. (From Hinds, W. C. Basis for particle size-selective sampling for wood dust. *Appl. Ind. Hyg.* 3(3):67–72, 1988. Reproduced with permission of Applied Industrial Hygiene, Inc.)

tant to stabilize the capsule overnight prior to weighing in a desiccator.[23]

The sampling head is mounted on a combined pump and drive system. The unit uses an electrically driven pump and therefore can only be operated where electricity is available.[23] Its sampling efficiency has compared well with the ACGIH curve for a range of wind speeds and can collect particles up to 100 μm. A similar unit that operates at 10 Lpm is also available.[24] Figure 5-5 shows the IOM area sampler for collecting IPM samples and its performance curve.[21]

Collection of Personal IPM Samples

As shown in Figure 5-4B, the NR-701 personal IPM sampler consists of a filter cassette and outer container. The geometry of the air inlet is such that at a flow rate of 2 Lpm the inspirable fraction of airborne dust is collected over a range of environmental conditions. The filter paper is shielded against particle blowoff. Weighing is done before and after sampling. If only the filter is weighed, it minimizes the need for several extra cassettes for

Figure 5-5B. IOM area sampler. (Reprinted with permission of *American Industrial Hygiene Association Journal*, vol. 46, p. 131, 1985.)

each unit. On the other hand, undersampling may result because dust deposited on the walls of the sampler inlet will not be included. Undersampling also becomes more significant as the local wind speed increases. The degree of undersampling may also vary as the orientation of the sampling head changes with respect to the wind. In any case, either just a filter or the cassette (filter assembly) are weighed. To ensure that dust deposited on the walls of the inlet system is included in the measured dust, the whole cassette can be weighed before and after sampling. Thus, handling of the filter is minimized. If weighing is done in-house, a micro balance capable of 0.00001-gram sensitivity should be used. Since the filter cassette is a part of the sampler, additional filter assemblies are necessary for sampler reuse.

Procedure

1. Unscrew the outer cover and remove the cassette.

2. Turn the nozzle gently until the recess is opposite the indent in the outer shell. Remove the nozzle and the PTFE washer. Using tweezers, insert a suitable 25-mm filter. Replace the PTFE washer and the nozzle, rotating the latter to lock it into position.

3. Before weighing, the cassette should be allowed to stabilize in the balance room atmosphere overnight. It must be labeled for identification after use by being transported in a numbered bag or container. Weigh the cassette. If only the paper is to be weighed, allow the filter to stabilize overnight in the balance room and then weigh.

4. To collect a sample, first insert the preweighed cassette or a cassette loaded with a preweighed filter paper into the base of the sampler. Place the PTFE ring over the cassette and screw on the cover until it is finger-tight.

5. Place a length of plastic tubing on the sampler stub outlet and connect the other end to an air pump. The air pump should be calibrated to pull 2 Lpm through this system.

6. Before sampling, check the sampling train for leaks by running the pump and blocking the nozzle inlet with a finger. The flow rate should fall to zero and the pump will shut off.

7. The sampling head is worn in the lapel of the worker within the breathing zone position with the pump fastened onto the wearer's waist belt. The tubing should be arranged so that it is comfortable for the wearer and does not get in the way.

8. When sampling is complete, place a protective plastic cover over the sampler body and handle the sampler assembly carefully. Transport with the air inlet upward, preferably in a suitable container.

Thoracic Samplers

As noted, the TPM fraction represents those airborne particles that are capable of entering the upper respiratory area and trachea during mouth breathing. It has been described as representing "the worst case potential exposure of the whole lung to particles." The TPM size-selective sampling criterion has been established as a tolerance band consisting of those particles that can penetrate a separator whose size collection has 50% of its particles (50% cutoff) 10 μm in size and a geometric standard deviation of 1.5 ± 0.1.[25]

A substance for which it would be useful to measure the TPM fraction is asbestos. Fiber levels are related to their potential to cause bronchogenic cancer since the types of fibers causing this effect deposit in the tracheobronchial and gas exchange regions.[20]

Another compound for which TPM sampling has been recommended is sulfuric acid aerosol.[26] The human health effects of major concern for this compound are bronchospasm in asthmatics and bronchitis, both of which affect the TPM region. The bronchospasm is the result of acute exposure, and the bronchitis is the result of chronic exposure. Due to the potential for hygroscopic growth of sulfuric acid droplets in the airways, the particle size favors deposition taking place within the upper respiratory tract, trachea, and larger bronchi. However, it should be noted that most compounds that absorb water may not be good candidates for collection using TPM or respirable particulate mass (RPM) methods because of the unpredictability of where these compounds will actually penetrate in the airway as they grow in size after encountering the humidity in the respiratory tract.[25]

Due to a concern regarding a higher than normal incidence of bronchitic symptoms in miners, a study was conducted to determine whether compliance based on measurements using the criteria for the respirable standard for dusts $<10\,\mu m$ in size equally protected against thoracic dust exposures that would affect the bronchial region. It was found that thoracic dust levels were five to seven times higher than respirable levels depending on the area of the mine being sampled. As a result, it was determined that the separate TPM samples were useful and additional dust control techniques were required.[27]

Currently there are no TPM samplers available for personal monitoring, although one has been proposed. There are, however, a number of TPM samplers available for ambient air monitoring. In general, most are designed for use in monitoring ambient outdoor air for the EPA PM_{10} standard and can be described as stationary devices rather than portable. TPM samplers used for PM_{10} sampling can be classified by low volume (<20 Lpm), medium volume (20–150 Lpm), and high volume (>150 Lpm).[5]

PM_{10} sampling (Fig. 5-6) involves the use of the same apparatus used for other ambient air particle samples; however, a size-selective inlet and an impactor are fitted into this large, stationary unit. Suspended particles in the air are sampled at 40 cfm through this inlet accelerating the particles through multiple impactor nozzles. By virtue of their larger momentum, particles greater than the 10-μm impactor cutpoint impact onto the greased impaction surface. PM_{10} particles smaller than $10\,\mu m$ are carried vertically upward by the air flow and down multiple vent tubes to an 8-in. by 10-in. quartz fiber filter, where they are collected. The large particles settle in the impaction chamber on the collection shim and are removed/cleaned during prescribed maintenance periods. The filter is weighed before and after sampling, and the increase represents the mass of particles smaller than $10\,\mu m$. Sampling is done for 24 hours. Impactors are described in more detail in a later section. Actual use of these monitors is described in detail in the manufacturers' operating manuals.

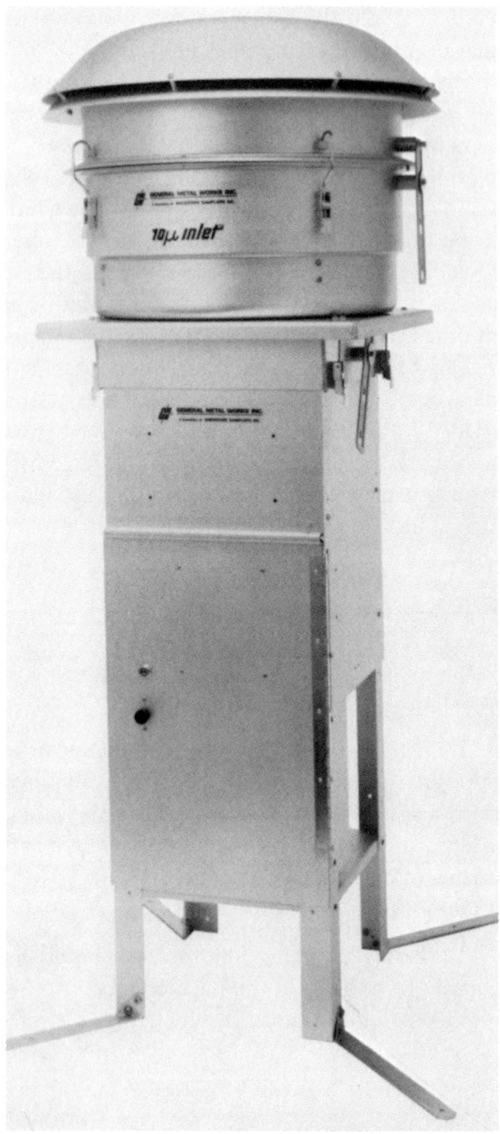

Figure 5-6. PM_{10} environmental sampler and enclosure. (Courtesy General Metal Works, Inc.)

Respirable Samplers

The respirable particulate mass (RPM) fraction is considered the portion of an aerosol available to the gas exchange region during nose breathing. It is based on ACGIH's previous recommendations with a tolerance band added. The 50% cutoff is the same as it was previously, 3.5 μm, and the geometric standard deviation is the same as for the other fractions, 1.5 ± 0.1.[25]

Iron oxide is an example of an insoluble compound that causes disease in the gas exchange regions of the lungs including massive pulmonary fibrosis. Therefore, the appropriate sampler for this exposure is an RPM sampler because iron oxide particles that might deposit in the nose or tracheobronchial region are most likely to be swallowed.[20]

Historically the most commonly used personal respirable dust sampling device in the United States is the 10-mm nylon cyclone. The cyclone is a centrifugal separator. Cyclones are commonly conical or cylindrical in shape, with an opening through which particle-laden air is drawn along a concentrically curved channel. Larger particles impact against the interior walls of the unit due to inertial forcess and drop into a grit chamber in the base. The lighter particles continue through and are drawn up through the center of the cyclone, where they are collected on a filter cassette. Cyclone samplers are based on fluid mechanics (centrifugal force), not actual diameters of the particles. Table 5-3 describes collection characteristics of cyclones.

Cyclones (Fig. 5-7) were developed to sample according to the size distribution curve developed by the ACGIH. Therefore, at a flow rate of 1.7 Lpm, this cyclone passes 50% of 3.5-μm aerosol particles. The selection of a flow rate of 1.7 Lpm for occupational sampling using NIOSH methods was made on the basis of its use in silica sampling. However, MSHA standards recommend 2 Lpm.

As noted, the dynamics of collection have been well tested for using the cyclone to collect respirable samples for silica. However, due to the designation as a respirable dust sampler, and ready availability, its use has been extended to other types of samples, often without careful evaluation of the changes that may occur in the system. When using cyclones for organic dusts, such as wood dust, there are problems due to particle size and density. While the most commonly used flow rate is 1.7 Lpm, other rates might be used if the purpose were to make the sampler fit a different collection curve or the characteristics of an aerosol were such that at a different flow rate collection of this dust would simulate the ACGIH curve. It has been noted that a modification of the results obtained at 1.7 Lpm may be necessary in some situations to provide equivalent concentrations for 1.4 Lpm and 2 Lpm may be necessary for some situations.[28]

Cyclones have minimal particle bounce and reentrainment, a large capacity for loading, and an insensitivity to orientation.[29] One drawback of the nylon cyclone is that it can accumulate a

Figure 5-7. Personal sampling with the cyclone.

TABLE 5-3. Characteristics of Cyclone Respirable Dust Collectors

Particle Size (μm)	*% Passing Selector*
<2	90
2.5	75
3.5	50
5.0	25
>10	0

static charge, and when highly charged aerosols are being sampled it can lead to variability in results.[30]

In general, single orifice collectors such as the cyclone are considered less effective at collecting particles than multiple orifice collectors such as the cascade impactor. The most critical locations in the cyclone that may leak, therefore limiting the cyclone's efficiency, are those associated with the adapter collar.[31] Sources of error include differences between cyclones, pump performance, concentration nonhomogeneity, electrostatic effects, and filter weighing procedures. The overall precision of the cyclone is about 2%. The following maintenance procedures should be done before sampling.[32]

1. Remove grit pot from the vortex tube taking care not to lose the O ring. Empty contents of the grit pot.
2. Use a small brush to clean the inner bores and mating surfaces of the cyclone's vortex and finder.
3. Wipe all parts externally with a damp cloth.
4. Allow sufficient time for the parts to dry, then reassemble them, making sure that the grit pot seals on the O ring.

Cyclone samplers are the most commonly used sampling devices for respirable dust. While they have been validated for use with silica, respirable samples of many other dusts can be collected.

Frequently, in order to get a sufficient sample, a full shift sample is collected on a single cassette. Shorter sampling periods may be required if the filter becomes overloaded or the operation being sampled does not last the full shift.

For silica, sample weights of 0.1 mg to 5.0 mg are acceptable. Sample weights of 0.5 mg to 3.0 mg are preferred. If heavy sample loading is noted during the sampling period, it is recommended that the cassette be changed to avoid collecting a sample with a weight greater than 5.0 mg.

Procedure

1. Remove the cyclone's grit cap and vortex finder prior to use and inspect the cyclone interior. If the inside is visibly scored, discard this cyclone since the dust separation characteristics of this cyclone might be altered. Clean the interior of the cyclone to prevent reentrainment of large particles.

2. Insert a labeled two-stage 37-mm cassette with a preweighed 5-μm PVC filter inside of the cyclone by taking out both plugs and inserting the cassette upside down. Check and adjust the alignment of the cassette and cyclone inside the cyclone to prevent leakage. It should be tight. Connect the outlet of the cyclone to the pump via tubing as in Figure 5-8.

3. Calibrate the pump using the method designed for cyclones to 1.7 Lpm just prior to use. Note the point on the pump's rotameter that corresponds to this flow rate. A piece of cellophane tape can be wrapped at this point and marked. For more information on calibration see Appendix A.

4. Clip the cyclone assembly to the worker's collar or T-shirt neckline and hang the sampling pump on the worker's belt or waistband. Ensure that the cyclone hangs vertically. Make sure the sampling inlet is not blocked. Duct tape may be useful in securing it in place.

5. Turn on the pump, recording the flow rate and time. At the end of the sampling period, record the final flow rate and the time. Replace the filter cassette caps. Calibrate the pump again following sampling.

Laboratory results for respirable silica samples are usually reported under one of four categories.

1. *Percent quartz (or cristobalite)*: Applicable for a respirable sample in which the amount of quartz in the sample was confirmed.
2. *"Less than or equal to" ($\leq$) in units of percent*: Less or equal to values are used when the adjusted 8-hour exposure is found to be less than the permissible exposure limit (PEL), based on the sample's primary diffraction peak. The value reported represents the maximum amount of quartz that could be present. However, the presence of quartz was not confirmed using secondary and/or tertiary peaks in the sample since the sample could not be in violation of the PEL.

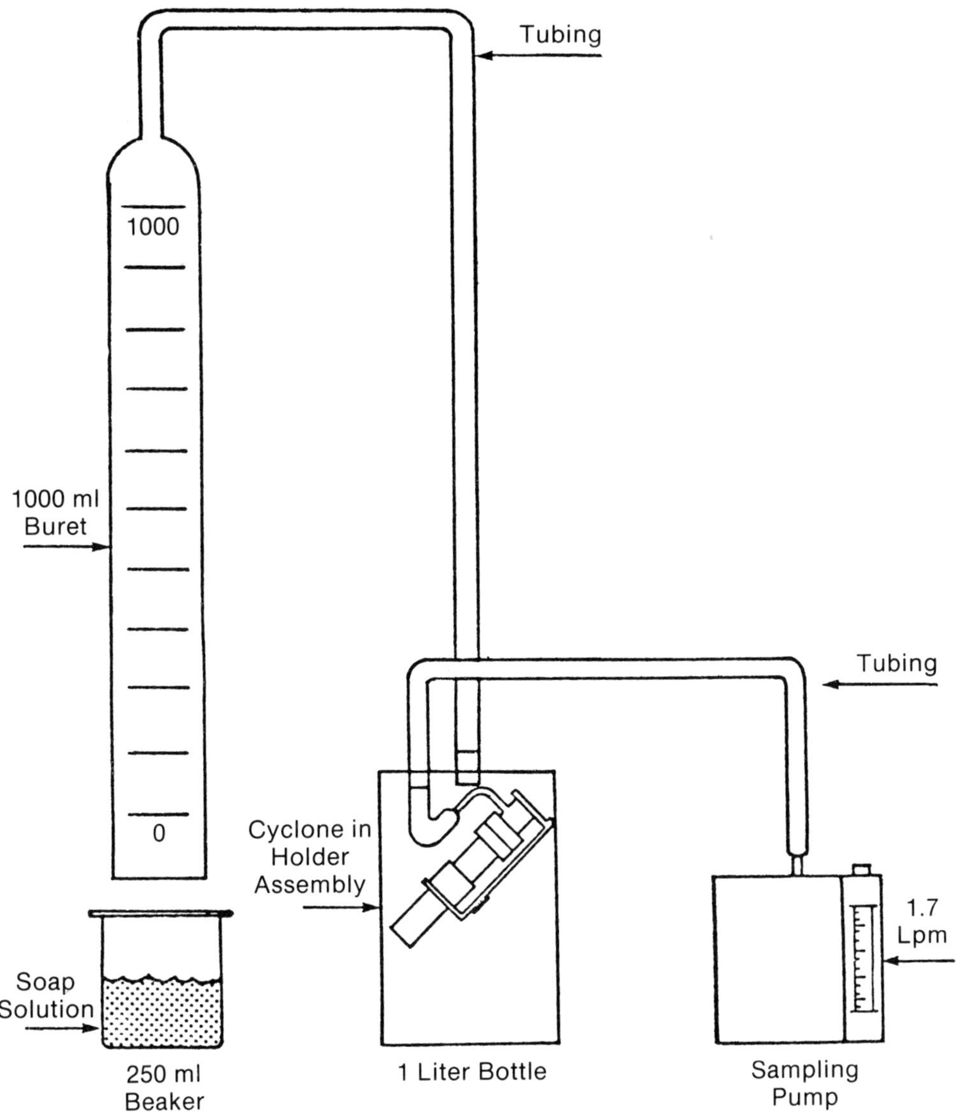

Figure 5-8. Calibration of the 10-mm nylon cyclone using a bubble buret. (From *OSHA Industrial Hygiene Technical Manual.*)

3. *Approximate values in units of percent:* The particle size distribution in a total dust sample is unknown and error in the x-ray diffraction (XRD) analysis may be greater than for respirable samples. Therefore, for total dust samples an approximate result is given.
4. *Nondetected:* A sample reported as nondetected indicates that the quantity of quartz present in the sample is not greater than the detection limit of the instrument. The detection limit is usually 10 μg for quartz and 50 μg for cristobalite.

Particle Size Distribution Sampling: Cascade Impactors

Cascade impactors can separate an aerosol into a particle size distribution that allows the sampler to characterize the primary size ranges according to their aerodynamic diameters. Cascade impactors have been widely used for occupational measurements, including personal and area sampling, environmental air pollution studies, and microbial sampling (with modified instruments). A major use of impactor data is determining the complete particle size distribution of a sampled aerosol, which allows the particle mass concentration in any size range to be calculated, including the inhalable, thoracic, and respirable fractions. Chemical characteristics of the aerosol must be considered when selecting aerosols to sample with impactors. For example, corrosive and combustible aerosols require special techniques and should not be sampled with impactors without special precautions. For more information on microbial sampling techniques, see the chapter on Sampling for Bioaerosols.

Impactors separate particulates in an airstream by directing them toward a coated flat surface. Particles enter the inlet jet and pass through a series of progressively smaller jets with which there is an associated collection surface (plate) usually at right angles to it. The aerosol stream passes through the first jet, flows around the impaction surface that is obstructing its flow, and then through the next jet and its associated impaction surface, and so on. Progressively smaller particles are collected on each plate. As the aerosol moves through the plates, larger particles are deposited on the top stages and smaller ones are deposited near the bottom. Particles larger than the cutpoint of the first stage impact on the precut collection substrate. As the airstream flows through the narrower slots in the second impactor stage, smaller particles impact on the second collection substrate, and so on. The widths of the radial slots are constant for each stage but are smaller for each succeeding stage; thus, the jet velocity is higher for each succeeding stage. After the last impactor stage, remaining fine particles are collected on a filter or plate.

When the impactor has been properly calibrated to define the aerodynamic media (cut) size characteristic of each stage, a size analysis may be made by calculating the percent by weight on each stage using weighings, radioactivity, or chemical analysis to determine the amount of deposited material. The cutpoint refers to the particle size for 50% collection. Cutpoints can be calculated or found in manufacturer's literature.

These devices come in different sizes. The larger ones are used for area samples and the mini-impactor can be used for personal samples. Individual impactor stages may be of the single jet or multijet variety. The latter are often preferred because resuspension of deposited particles is minimized and collection of larger samples is possible. Particles deposited on each stage may be examined microscopically. Impactors can be designed over a wide range of flow rates and can be operated in any orientation. Modified impactors are also used to sample bioaerosols. For more information on these techniques, see the chapter on Sampling for Bioaerosols.

The range of sizes collected by each stage of the cascade impactor depends on the density and shape of the particles as well as the diameter, and therefore represents the aerodynamic behavior of the particles. The flow rate determines the relative distribution. For example, when the Sierra Model 260 cascade impactor, which contains six impaction stages followed by a built-in 47-mm diameter filter, is operated at 3.0 Lpm, the cutpoints for the respective stages are 17, 11, 6.1, 3, 1.3, and 0.68 μm, and when it is operated at 2.0 Lpm, the cutpoints for the same stages change to 20.9, 13.6, 7.5, 3.6, 1.6, and 0.9 μm.[33]

The Sierra Model 260 cascade impactor for area sampling contains six impaction stages followed by a built-in 47-mm diameter filter. This unit can be used for either occupational or environmental sampling. The flow rate can vary from 0.3 Lpm to 20 Lpm. Both circular and rectangular nozzles are available. With the circular nozzles the cutpoints range from 20 μm to 0.5 μm. Particles are impacted on preweighed 18-mm-diameter glass microscope slides coated with grease to reduce particle bounce or reentrainment. Adjacent impactor stages can be rotated relative to each other to obtain a series of particulate deposits, which allow more sample to be collected than most impactors can tolerate without developing reentrainment effects.[33]

Generally cascade impactors are very good at collecting large particles as these particles tend to impact on the plates because it is harder for them to make the turn necessary to go around the edge of the plate. Under standard conditions, this includes particles with aerodynamic diameters greater than 0.2 μm. The range can be extended to 0.05 μm with micro-orifice impactors. In certain situations, even when particles are in the correct size range, another type of sampler may be more appropriate. This is the case when precise sizing of monodisperse particles is desired, or for distributions where one stage will be overloaded before sufficient material for analysis can be collected on another.

Typical problems that result in differences between impactors are particles lost to bounce, reentrainment due to overloading of the sample on a particular stage, inaccurate calibration, and lack of a sharp cutoff for each stage. Solid particles are capable of bouncing from the collection surfaces of impactors and being carried to subsequent stages or the backup filter, the result being that the size distribution is distorted toward the smaller sizes.[34] Another source of error when comparing impactors with total dust samplers is interstage losses. Adhesive may lose its effectiveness during sampling, leading to the nonuniformity of the deposit. Sample collection can also be biased by inlet configuration and sampling flow rate.[33]

The best prevention for particle bounce is the use of silicone grease, oil, or a similar adhesive on the surfaces of the collection stages. Grease-coated surfaces have the limitation of being good for particle retention only when there is less than one monolayer of impacted particles. As particle loading increases on grease-coated impaction surfaces, the incoming particles no longer strike the grease layer but impact on the already collected particulate matter, and may bounce.[34]

The Personal Cascade Impactor

A line of impactors designed for personal sampling has been developed: the Series 290 Impactors. These impactors consist of a baffled inlet and eight, six, or four impactor stages followed by a built-in filter holder. After the last impactor stage, remaining fine particles are collected by the built-in 34-mm filter. Each stage has six radial slots with beveled inlets. The widths of these slots are constant for each stage, but are smaller for each succeeding stage. Thus, the jet velocity is higher for each succeeding stage, and smaller particles eventually acquire sufficient momentum to impact on one of the mylar or stainless steel collection substrates. A cowl on the inlet eliminates ashes and debris from the sampler.

The Model 298 cascade impactor contains eight impaction stages followed by a built-in 34-mm diameter filter holder (Fig. 5-9). The device is designed to be a personal sampler and to operate at a flow rate of 2.0 Lpm, although it can be reliably operated at any flow rate between 1 Lpm

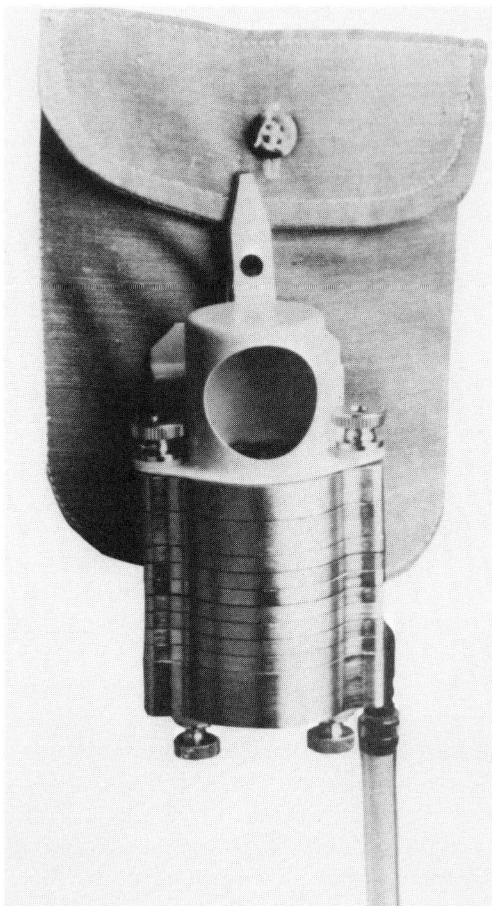

Figure 5-9. Size-selective sampling with a personal impactor. (Courtesy Andersen Instruments, Inc.)

TABLE 5-4. Cutpoints for the Series 290 Impactors when Operated at 2 Lpm

Stage:	*1*	*2*	*3*	*4*	*5*	*6*	*7*	*8*	*Backup Filter (μm)*
Model									
298	21.3	14.8	9.8	6	3.5	1.55	0.93	0.52	0
296			9.8	6	3.5	1.55	0.93	0.52	0
294	21.3	14.8	9.8		3.5				0

and 3 Lpm. Although it has been suggested that the range of flows can vary from 0.5 Lpm to 5 Lpm, at flow rates above 3 Lpm, significant particle bounce and internal losses may occcur.[33] Table 5-4 lists cutpoints for the personal impactor when operated at 2 Lpm. Particles are impacted on preweighed substrates that are coated with grease. The four-stage unit has been recommended for wood dust sampling.

An example of an application for the personal impactor is a study that was done to determine air-lead particle sizes during battery manufacturing. As a result, it was determined that where lead particles are predominantly <5 mm, a greater hazard would exist than is normally assumed using total lead collection techniques, and more stringent exposure controls are needed. Also, to minimize contamination at the site, the impactors were prepared off site and transported inside Ziplock plastic bags. New bags were used for each day of sampling. A daily field control impactor was prepared, transported, and unloaded in the same manner as the samplers. Following sample collection, each substrate and backup filter was placed in a screw-capped vial, previously cleaned with nitric acid, for transport to the laboratory.[35]

During a comparison of the personal impactor and the 10-mm nylon cyclone, an advantage of using impactors over cyclones that was noted is the ability to use multistage impactors to develop size mass distribution data, which can then be used to estimate particle deposition at all levels of the respiratory tract. However, the cyclone is easier to use, less expensive, and provides adequate data for the respiratory fraction.

Occupational Sampling with the Personal Cascade Impactor

The flow rate is critical with these devices (Fig. 5-10). If it is not constant and calibrated correctly, the collecting efficiency will be hampered. New impactors should be examined for imperfections, for example, small deviations in the jet diameters such as burrs, and the jet diameter should be verified before use.[36]

As the flow rate is usually fixed for a given instrument in order to maintain the desired size distribution, the orientation of the inlet probe is the primary variable. The best approach is to align the probe parallel to the air currents in the area. Critical orifices can be used to maintain a constant flow rate.

Calculate the mass to be collected ahead of time, since collection of an insufficient or excessive mass can cause erroneous results. Too much deposit will result in overloading the stages and can cause particle bounce, particle reentrainment, and changes in particle collection characteristics. Insufficient mass will result in errors in the analysis of deposited material on each stage and will lead to large uncertainties in interpreting the particle size distribution. A general rule is that material depositing on each stage should be <0.5% of the total mass collected on all stages.[36]

Impaction grease is applied as a suspension or solution of 10–20% grease in a solvent such as toluene. The mixture is applied to the substrate with a brush, eyedropper, or sprayer. Place the substrate on the bottom plate of the template with the two locating pins through opposite perforations. Place the top plate on top and apply

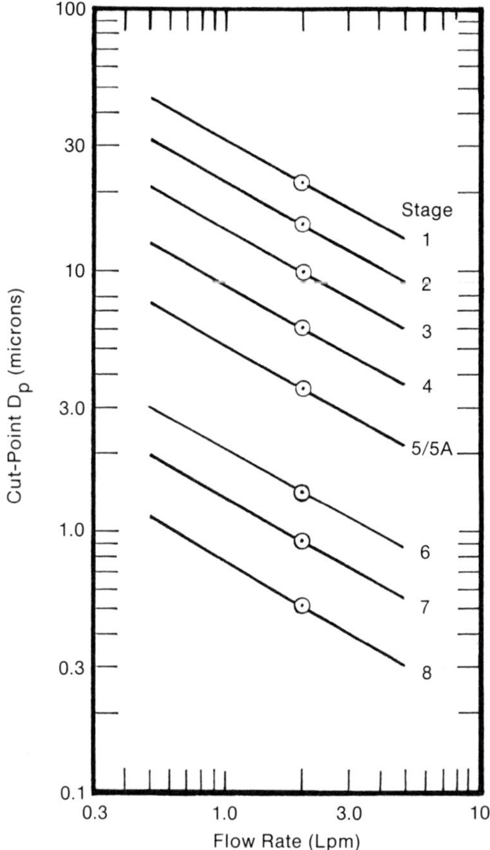

Figure 5-10. Series 290 personal impactor cutpoints versus flow rate for air at 25°C and 1 atm. (Courtesy Andersen Instruments, Inc.)

the solution within the slots in the template. Note that the outer edge and center of the substrates must be devoid of grease or else the substrate will stick to the upstream impaction stage. Allow the solvent to evaporate. After drying, a cloudy white film is visible. The final greased substrate should be tacky but not slippery, with a film thickness about equal to the diameter of the particles to be captured, 1–10μm thick. Figure 5-11 is an example of an impactor sampling data sheet used during field monitoring.

After using the cascade impactor, it is critical that the head be opened properly and not inverted. If it does become inverted, the sample will be lost. The plates that do not use filters are more difficult to weigh than samplers that do use them, because of the potential for loss of sample.

Procedure

1. Select the sampling flow rate based on the particle cutpoints desired and the vacuum capability of the sampling pump.

2. If the collection substrates are to be greased, use the procedure described above. Preweigh all the substrates and 34-mm, 5-μm PVC backup filters. Record the weights. The substrates and filter should be equilibrated with the laboratory environment for 24 hours at a relative humidity of 50% or less before being weighed and passed over a static eliminator if they have a static electricity charge.

3. Assemble the cascade impactor. Each impactor stage is numbered along its edge. In assembly, all numbers should line up along one of the two threaded studs in ascending order from inlet to exit. No number should be upside down. The two threaded studs orient the impactor stages so that the slots in each stage are staggered from the slots in adjacent stages. The substrates are placed using a tweezer on the top surface of each impactor stage so that the perforations in the substrates match the slots in the impactor stages.

4. The unit is attached to a personal sampling pump and calibrated in a jar similar to a cyclone to 2 Lpm.

5. Connect tubing to the sampling pump and set the flow rate on the pump.

6. Attach the impactor to the lapel or pocket of the worker, the pump to the belt, and the interconnecting tubing to the impactor.

7. Turn on the pump. Record the start time and other data on the sampling data sheet.

8. Turn off the pump at the end of the sampling period. Record the time.

9. During sampling the unit is attached to the worker's collar. After sampling the impactor

Test No. ____________________ Date ____________________

Name of Subject ____________________

Plant ____________________

Location ____________________

Sampling Flow Rate Q (Lpm) ____________________ Impactor Model No. ____________________

Starting Time T_1 (Hrs.) ____________________ Impactor Serial No. ____________________

Ending Time T_2 (Hrs.) ____________________

Total Sampling Time $T_2 - T_1$ (Hrs.) ____________________

Sampling Volume V $(M^3)^{(1)} =$ ____________________

Total Mass Concentration C_{tot} $(mg/M^3)^{(2)}$ ____________________

Stage Number	Stage Cut-Point $D_p^{(3)}$ (microns)	Initial Weight W_1 (mg)	Final Weight W_2 (mg)	Particulate Weight $W^{(4)}$ (mg)	Concentration $\Delta C^{(5)}$ (mg/m^3)	$\log_{10}D_p$	$\Delta\log_{10}D_p^{(6)}$	$\Delta C/\Delta\log_{10}D_p$ $(mg/m^3/\log micron)$	$GMD^{(8)}$ (microns)	$\frac{W}{W_{tot}}$ (%)
1										
2										
3										
4										
5										
6										
7										
8										
Back-Up Filter										
$W_{tot}^{(7)}$	–	–	–			–	–	–	–	100

NOTES:
(1) $V = \frac{1}{16.7} Q (T_2 - T_1)$
(2) $C_{tot} = W_{tot}/V$
(3) Cut points are defined for each stage of an impactor.
(4) $W = W_2 - W_1$
(5) $\Delta C = W/V$
(6) $\Delta_i\log_{10}D_p = \log_{10}D_{p_{i-1}} - \log_{10}D_{p_i}$
(7) W_{tot} = sum of particle masses for all stages including back-up filter.
(8) $GMD_i = \sqrt{D_{p_i}(D_{p_{i-1}})}$

Figure 5-11. Impactor data worksheet. (Courtesy Andersen Instruments, Inc.)

is disassembled and the substrates and filter are removed and weighed. An advantage is that the unit does not have to be inverted for sampling. The weight change on each substrate represents the mass of particles in the size range of that impactor stage. The total weight of particles on all stages and the filter are added together and the percent particle mass in each size range is calculated. A graph is developed for the cumulative size distribution showing the percent of particle mass smaller than the aerodynamic article diameter. The respirable particle mass fraction is determined from the particle size distribution.

10. For best results, return the impactor to the laboratory fully assembled in the upright position and with the inlet sealed to prevent sample contamination.

Impactor Data Analysis

A major use of impactor data is determining the complete particle size distribution of sampled aerosol. Given the complete particle size distribution, the particle mass concentration in any size range can be calculated, including the inhalable, thoracic, and respirable fractions.

The most common data analysis involves developing a cumulative mass distribution, graphing this distribution, and using the graph to identify the mass median diameter and other parameters of interest. However, the sampling professional should be aware that correct interpretation of cascade impactor data requires experience and an understanding of what the calculations mean. The following is an overview of some typical uses for impactor data; however, there are other references that can provide more comprehensive information and the sampling professional is encouraged to consult them.[36]

The differential particle size distribution is a plot on log-log graph paper of the particle mass concentration in each particle size band (stage) versus the geometric mean diameter. The differential particle size distribution gives the details, or "fine" structure, of the particle size distribution.

The cumulative particle size distribution is a plot on log-normal (or log-probability) graph paper of the total particle mass smaller than a given particle size. This cumulative distribution gives an overall picture of the size of the particles.

The mass mean diameter of the log-normal distribution is the particle size where 50% of the particle mass is contained in particles larger than the already-defined MMD, and 50% of the particle mass is contained in particles smaller than the MMD.

The geometric standard deviation is the ratio of the MMD to the particle size for which 16% of the mass is borne by particles smaller than this particle size. It is a measure of the "spread" in the particle size distribution. If the geometric standard deviation equals one, all of the particles are of the same size, otherwise known as monodispersed. Table 5-5 is an analysis of impactor data.[37]

TABLE 5-5. Analyzing Impactor Data

Distribution	*Formula*
Differential particle size (plot on log-log graph paper)	$\dfrac{\Delta C_i}{\Delta_i \log_{10} D_{p_i}}$ (log) vs. GMD_i (log)
	$GMD_i = \sqrt{D_{p_i}(D_{p_{i-1}})}$
Cumulative particle size (plot on log normal paper)	$\dfrac{\sum_{i=j-1}^{N} W_j}{W_{tot}}$ % (normal) vs. D_{p_i} (log)
Mass mean diameter	$\dfrac{\sum_{i=j-1}^{N} W_j}{W_{tot}} = 50\%$
Geometric standard deviation	$\dfrac{MMD}{D_{p16\%}} = \dfrac{D_{p84\%}}{MMD}$

ΔC_i = Particle mass concentration in each particle size range
D_{p_i} = Particle size or cutpoint of each individual stage
GMD_i = Geometric mean diameter of each size interval
W_j = Particle mass on impactor stage i, where i = size interval for that stage
W_{tot} = Sum of particle masses on all stages plus that on the backup filter
MMD = Mass mean diameter

SAMPLING FOR SPECIFIC AEROSOLS

Coal Mine Dust

Coal dust exposures are primarily related to mining, either underground, strip mining, or augering, with the highest concentrations related to underground mining. There are four types of coal mined: lignite, subbituminous, bituminous, and anthracite.

Coal dust causes coal workers pneumoconiosis (CWP), which can range in severity but is characterized by fibrosis and is found predominantly in the upper lobes of the lung. Symptoms of CWP are indistinguishable from those typical in other chronic obstructive lung diseases, especially chronic bronchitis. Most often there is an exposure to other mineral compounds, such as silica and talc, during coal mining.[38]

Currently, MSHA regulations require that respirable dust samples collected in coal mines should be sampled using a 10-mm nylon cyclone containing a 5-μm PVC filter operated at a flow rate of 2 Lpm rather than 1.7 Lpm, with the results multiplied by a factor of 1.38 to convert the results to the equivalent concentration that would be measured by an instrument meeting the criteria of the Mining Research Establishment of the National Coal Board of England.[39] The ability of this method to be a good estimator of the respirable dust concentration has been questioned and in the future may change, but for the present for compliance sampling in coal mines this method must be used.[40] It has been suggested that the use of a 1.2-Lpm flow rate and a multiplier of 0.91 would be better when using the 10-mm nylon cyclone to estimate BMRC respirable dust.[41]

Prior to use, the rotameter on the pump should be calibrated at 1.6, 1.8, and 2.0 Lpm. The point at which the pump with the cyclone hooked up is pulling 2 Lpm should be marked on this rotameter so that the flow rate of the pump can be monitored during the survey. Otherwise, sampling methods are similar to those previously described for use of the cyclone.

The MSHA standard calls for designated occupational and area samples to be collected in mines and submitted to MSHA for analysis of a schedule that requires a minimum of six sample collection periods over the year (Table 5-6).[39] The standard specifies the specific work location where samples need to be collected for each of 10 different mining sections. A minimum of 5 samples must be collected in each location. Occupational samples can be collected on the worker (personal) or by placing the sampling apparatus near the normal working position of the miner. The production must be at normal levels and the sample must be collected for either the full shift or a minimum of 8 hours. The purpose of the area samples is to identify sources of respirable dust generation. The above description is a somewhat simplified view of the sampling strategy, and if results exceed the standard, sampling must be repeated with more frequency and can involve more positions and more attempts to randomize the samples.

Variation in coal dust results can be due to changes in the ventilation used to control dust, the cutting speed, and the amount of rock being cut above or below the coal seam.[33] In mines large concentration gradients have been known to occur in many prospective sampling locations. At such locations, moving a sampler a foot one way or the other may affect the dust-level readings more than any other factor. These concentration gradients are impacted greatly by airflow patterns. An example of an area with a steep concentration gradient is a long wall, depending on the distance between the source and sampling point. When sampling is conducted downwind of mining machines, the measured concentration is not always a reliable indicator of the amount of dust produced by that machine. Ideally, sampling should be conducted only where gradients are low; however, since steep dust gradients are often unavoidable, the only recourse is to use multiple sampling points.[42]

The presence of rock bands—an irregularly occurring band of harder, noncarbonaceous material such as shale—layered within the coal seam will cause a wide variation in dust levels. When equal amounts of material are cut, the dust from rock is an order of magnitude greater than coal, so even a relatively minor rock band will cause

TABLE 5-6. Locations for Respirable Coal Dust Sampling Under MSHA Standards

Mining Section	*Sampler Location*
Conventional section using cutting machine	Cutting machine operator On cutting machine within 36 inches of operator
Conventional section—shooting off the solid	Loading machine operator On loading machine within 36 inches of operator
Continuous mining (other than thin auger)	Continuous mining machine operator On machine within 36 inches of operator
Continuous mining—auger type	Jacksetter nearest the work face on the return air side of the continuous mining machine A location representing the maximum concentration to which the miner is exposed
Scoop section—using cutting machines	Cutting machine operator On cutting machine within 36 inches of operator
Scoop section—shooting off the solid	Coal drill operator On coal drill within 36 inches of operator
Longwall section	Miner nearest the return air side of the longwall working face On the working face on the return air side within 48 inches of the corner
Handloading section using a cutting machine	Cutting machine operator On cutting machine within 36 inches of operator
Handloading section—shooting off the solid	Hand loader exposed to highest concentration A location representing the maximum concentration of dust to which miner is exposed
Anthracite mining section	Hand loader exposed to the greatest dust concentration A location representing the maximum concentration to which miner is exposed

dust levels to double. Variations in production will also cause average dust-level changes due to such events as equipment breakdowns and low production due to hard cutting caused by rock intermingled with coal, or individual work habits on the part of machine operators. One recommendation for such situations is to use both integrated and real time sampling methods to provide a comprehensive set of data along with close monitoring of variables such as airflow, water flow, and differences in work practices.[42] In order to collect samples of respirable coal dust to satisfy MSHA standards, a sampling professional must be certified by MSHA.

Cotton Dust

OSHA's cotton dust standards were developed from health studies that frequently used a sampler that has become known as a vertical elutriator, and in some circles, a "bazooka." This sampler also became part of the cotton dust standard. Although alternate samplers are also allowed, equivalency to the vertical elutriator must be established and currently none have been approved by OSHA. This sampling method is an example of a situation where no personal sampling method exists, and area samples are collected using a sampling strategy designed to approximate the employee's exposure. Another sampling approach used to evaluate exposures to cotton dust involves the collection of gram-negative bacteria and associated endotoxin since bract and leaf from cotton plants contain large numbers of these. These microbial materials have been implicated as possible causative agents of byssinosis, the lung disease attributed to cotton dust exposure.[43] Fungal and actinomycete spores as well as dust have been identified during cotton dust sampling.[44] For more information on techniques for sampling bioaerosols, see the chapter on Sampling for Bioaerosols.

Prior to the survey, the plant is surveyed in order to determine where to set up the instruments. Questions are asked about the employees' work activities. For example, in cotton ginning, the ginner and helper spend most of their time in the gin stand area and only a little time in the seed cotton-cleaning and lint-cleaning areas. The head pressman and crew spend most of their time operating the bale press. Some press crews may trade off duties with the suction-pipe operator and the yardman.[45] Approximately four to five elutriator measurements are used to determine an employee's exposure in most mill operations. In some areas, if an employee's work area is limited, two elutriators will be sufficient. Generally this technique is limited to sampling employees who work in specific areas rather than those who roam throughout the plant, such as maintenance or supervision people. In general, placement of vertical elutriators depends on the purpose of the study, the number of elutriators available, areas where employees spend most of their time, access to electric power, and best locations for instrument survival during emergencies such as during cotton fires.[45]

The instruments should be placed as close to the employees' work area as possible. However, the elutriators must not be located in areas with strong air currents. Cotton ginning requires a significant amount of processing air. Much of this processing air is drawn from within the building and must be replaced by air from outside through vents or doorways into the building. As a result, it is not uncommon for many areas to have air drafts whose velocity exceeds 33 m/min. One way of minimizing the effect of these drafts on the sampler is to baffle the elutriator inlet using 6-inch diameter metal funnels suspended by masking tape to prevent turbulence within the elutriator.[45] Elutriators separate out particulates of varying sizes by gravitational effects under low velocities, and are commonly classified as horizontal or vertical based on their design. The vertical elutriator was selected as the appropriate device to sample cotton dust by OSHA.

The elutriator by definition is a separator or purifier. This sampler uses a cylinder standing in a vertical position as a separator; hence, the term *vertical elutriator*. The vertical elutriator was designed to collect dust particles that are less than 15 μm in aerodynamic diameter. The particle size cutpoint ranges from 9.7 μm to 10.5 μm AED.[46] The unit has an inverted cone as a bottom section. A pump and flow control device is used to draw air into the small end of the cone at 7.4 Lpm. Samplers use a critical orifice as a flow controller.

The velocity of the air is highest at the small cone entrance, and gets progressively slower as it moves onto the wide cone area and into the cylinder. Heavier pieces of material will slow down and fall back while the smaller-size particles will be carried to the collection filter. In this process, the elutriator separates cotton dust by size. As long as filter loading does not exceed 0.5 to 1 inch Hg, this device will operate as needed; however, higher loading is a problem because the orifice is located downstream of the sampler filter, which makes it susceptible to upstream pressure changes.

Another type of flow controller available consists of two valves: a manual one and a pneumatically operated one. This unit is capable of handling filter loading up to 8 inches of Hg, so it is better in situations where heavy loading is expected.

One study reported that the coefficient of variation when vertical elutriators sampled side-by-side in a cotton mill was 34–43%.[47] Factors contributing to precision problems in vertical elutriator sampling include low sampling rate, a penetration curve with a shallow slope, and inverted horizontal filter, and variability of performance with environmental factors. For example, there can be a loss of sample during removal of the inverted horizontal filters. The low flow rate either causes problems with reproducibility in weighing the filters or it forces the operator to sample for inordinately long periods of time.[46]

Vertical elutriators (Fig. 5-12) have been found to be durable under field conditions if proper precautions are taken. Vertical elutriators tend to be top heavy and prone to falling due to their construction and operating height. Therefore, they should be provided with bases that are as compact and stable as possible. The critical orifice calibrations were stable unless the orifices suffered mechanical damage or became partially plugged with lint or dirt. A vertical elutriator should not be operated under very dirty conditions without a filter cassette attached to avoid plugging the critical orifice. Routine checking and cleaning of orifices before each sampling period is recommended.

The pressure-relief valves on the vertical elutriators usually require no readjustment to their initial setting during a normal 6- to 8-hour running period. Filter loadings under normal ginning conditions are not heavy enough to cause airflow restrictions that require relief valve adjustment.[45] Ginning variables that affect vertical elutriator results arise from three primary sources: input material, layout of the gin, and management practices. Although gin layouts are standardized to some extent, dust produced by each machine varies from gin to gin. The location of machinery within the gin and the structure of the building affects the velocities, quantities, and direction of cotton flow and of ventilating air into the building. The quantity and type of foreign matter, such as bract, stems, leaf particles, and dirt in the input material being processed, have an effect on the overall dust level within the gin. However, the inherent dust content of the raw cotton may have less effect on actual dust concentrations than dust leakage from individual machines.[45]

Procedure

1. Obtain three-piece filter cassettes 37 mm diameter with 5-μm PVC filters inside. Plan to have enough so that 10% can be field blanks.

2. Calibrate the elutriator to a flow rate of 7.4 ± 0.2 Lpm. Secure the rubber stopper made for calibration checks into the elutriator body inlet. Wet the inside of a large bubble buret and set up. Allow the elutriator to run 1 minute prior to taking readings.

3. Electrical power is required for operation of this instrument. The outlets must provide 60-cycle alternating current and 110–120 voltage. It is preferred that outlets with a ground plug connection be used. It is critical if there are moist floors, metal pipes, or other sources of electrical shock within 8 feet of any of the selected sampling locations that could serve as grounding return paths from the sampler. It is possible to provide auxiliary conductive grounding paths by means of two-slot to three-slot adapters that have a grounding connection wire. The adapters are plugged into the two-slot outlets and the grounding connection wires are attached to grounded surfaces to form three-slot outlets. The grounded surfaces must be determined using a voltmeter or ohmmeter. Since the retaining screw for the outlet's cover plate is the most convenient place for attaching the grounding connection wire, it is the logical place to test first. Locate the circuit breaker associated with the outlet and determine if it is appropriate. The fuses or circuit breakers must be no more than 15-amp capacity if protecting outlets for two-prong plugs, or not more than 20-amp capacity if protecting outlets for three-prong plugs. Be sure that the amp load on the circuits that will power the samplers will not overload them. These samplers typically draw 4 amps of current.

4. Because of space requirements, elutriators are generally transported disassembled. While

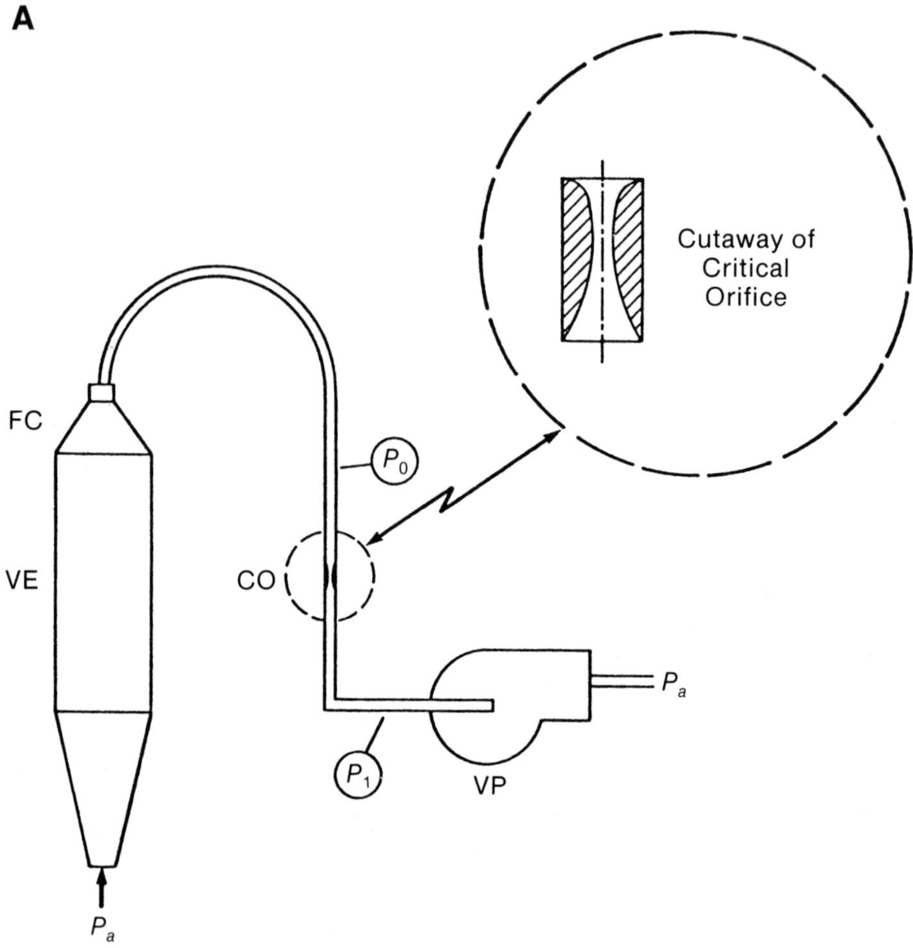

Figure 5-12. Vertical elutriators. *A*: Using a critical orifice as a flow controller. *B*: Using a modified pressure regulator as a flow controller. VE, vertical elutriator; FC, filter cassette; CO, critical orifice; RG, regulator; VP, vacuum pump; RM, rotameter; MV, manual valve; P_a, atmospheric pressure; P_o, pressure at input of regulator; P_1, pressure at output of regulator; P_2, pressure at input of manual valve. (From OSHA Field Service Memo No. 8: *Vertical Elutriator Cotton Dust Samples.*)

carrying the assembled elutriator to the sampling location, the filter should not be in it.

5. At the sampling location, run a clean, lint-free cloth through the elutriator body to remove any dust that may have accumulated inside.

6. Insert the cassette after removing its plugs into the opening at the top of the elutriator as is done for collecting a sample. The cassette is used while the sampler is run 2 to 3 minutes, to assure that any extraneous dust is flushed out that may have been left in the elutriator after the cleaning.

7. Stop the sampler. Remove this cassette and insert another cassette for collecting the sample. Since the cassettes can be loosened by vibration, the cassette must be inserted securely and taped to the vertical elutriator body using electrical tape to prevent leakage during sampling.

8. Secure tubing to the pipe with tape, especially at the top where the weight of the tubing may cause a pinch at the cassette connection.

9. Changing filters during the sampling shift depends on the dust concentration (see Table 5-7).

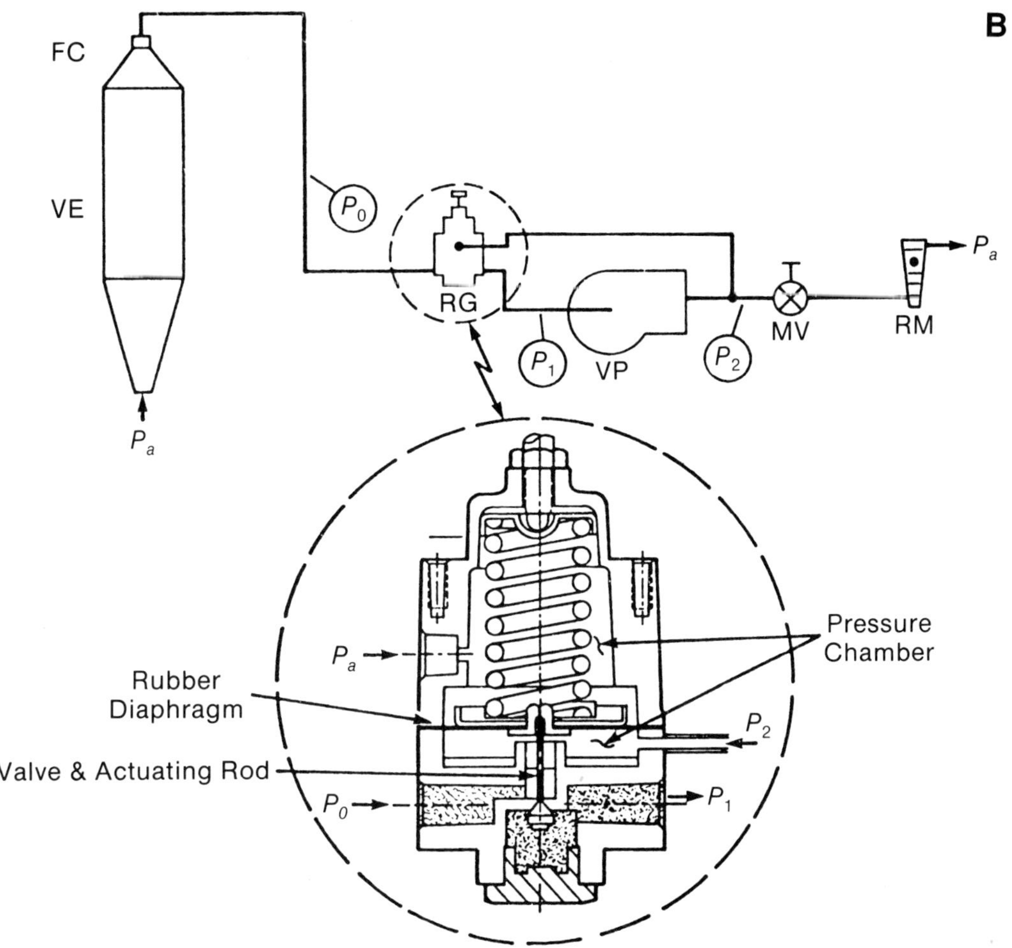

Figure 5-12. *Continued*

TABLE 5-7. Filter Changes for the Vertical Elutriator During Cotton Dust Sampling

Dust Concentration	*Changes*	*Number of Filters/Shift*
<300	0	0
300–600	1	1
>600	2	3

10. When changing the filter, the vertical elutriator pump must be turned off because the filter may be ruptured while working under negative pressure.

11. The filter can be changed by tilting the elutriator body. For convenience, one end of the tape for sealing the cassette can be stuck to the body of the vertical elutriator so that it will be handy to reach while holding the instrument in the tilted position.

12. Make checks on the sampling operation every hour. Look to see if the cassette is taped down tightly, if all tubing is connected, and if the rotameter designates the proper flow rate.

13. Following sampling clean the inside of the elutriator body with a damp cloth. This should be done after each day of sampling. Use a cloth or a brush to clean the dust from the pump motor and body also.

The 6-hour sampling period is representative of the full-shift exposure. Therefore, there is no adjustment for unsampled time unless the worker was not present in the locations covered by the elutriator measurements (such as in the lunchroom). The individual elutriator measurements collected in each employee work area are averaged arithmetically. This average value is then converted to a time-weighted average (TWA) to determine the employee's exposure. In most cases, the worker will move about the work area tending to machines.

The average of the elutriator measurements will represent the worker's exposure. For example, if five samples are collected in one area throughout a day (6 hours), the results are added up and divided by 5. The average result is then multiplied by the *actual* workday, which is the time the workers were in the area, and divided by 8.

Observations during sampling are also important for interpreting results. For example, differences in fungal spore concentrations were found to be related to the quality of cotton being processed and to the cleanliness of the mill in one study. In the same study, it was also determined that the distribution of cotton fibers and dust particles varied with the stage of processing, with the carding and speed frame stages having the highest levels of cotton fibers. Dust particles were most numerous during carding and opening.[44]

Wood Dust

In general, except for some sanding operations, studies indicate that more than 80% of wood dust is contained in particles greater than 10 μm.[48] Recent studies have determined that the ACGIH IPM criteria may be more appropriate than classical methods to use for assessing general wood dust concentrations in furniture, cabinet making, and other industries where wood dust is present. To do otherwise would exclude from the sample the major source of particles depositing in the nose. IPM sampling collects all particles that would enter the nose or mouth in a typical work situation. To a large degree, this sampling reflects the particles that deposit in the nose; consequently, IPM sampling will be the environmental measurement that is most closely related to the risk of developing nasal cancer from hardwood dust exposure. The furniture woods associated with nasal cancer are oak, beech, mahogany, maple, walnut, teak, and birch.

Until recently, different types of wood dust have been treated as a uniform entity; however, studies indicate that there are many differences. For example, recent data suggest that the use of hardwoods may lead to more dust of respirable size (<15 μm) than the use of softwoods, and that very little dust from any given sample of wood dust is in the inhalable ranges. Therefore, it can be postulated that the primary sites for toxic effects are the upper, larger airways.[49] Different types of woods have been known to cause different types of toxic effects.[16] Some woods, particularly those considered tropical woods, have insecticidal and antibacterial compounds in them. It has been recommended that wood dusts be categorized as toxic, biologically active types and those with low biological action.[46] Table 5-8 gives toxic effects of different wood types.[21, 50]

Since wood is an organic medium, it can provide a conducive environment for the growth of microbials. Certain types of diseases associated with wood are actually due to microorganisms that tend to grow on the wood. Therefore, if sampling is being done to evaluate exposures to this wood, microbial sampling may also be necessary in certain situations. For more information on these methods, see the chapter on Sampling for Bioaerosols.

For furniture operations, sanding departments have been shown to have the highest maximum and average dust levels. Assembly areas and finish mills where detailed wood machining is done have relatively high dust levels, but in general they are likely to be substantially lower than sanding areas. In some areas, such as assembly, the only source of exposure is reentrainment of often heavy dust deposits on furniture parts. In plywood and plywood products operations, less dust is generated than in many furniture manufacturing operations.[48]

It should be noted that all of the aforementioned operations use dry wood and generate dry wood dust. In saw mills the wood dust is wet because the wood has not been dried. Wet wood

TABLE 5-8. Toxic Effects of Different Types of Wood

Type of Wood	*Toxic Effect*
Redwood	Allergic alveolitis in the lung
Western red cedar	Allergic alveolitis in the lung
Oak	Nasal cancer, skin and eye irritation, allergic skin response, asthma
Beech	Nasal cancer, skin and eye irritation, allergic skin response, asthma
Mahogany	Nasal cancer, asthma
Maple	Nasal cancer
Walnut	Nasal cancer
Teak	Nasal cancer
Birch	Nasal cancer
White cedar	Skin and eye irritation, allergic skin response
Douglas fir	Allergic skin response
Tropical woods	Skin and eye irritation, allergic skin response
Exotic woods	Asthma
Hardwood (general)	Rhinitis, nasal dryness, pulmonary function changes
Softwood (general)	Pulmonary function changes
Fungal spores associated with wood	Skin and eye irritation, allergic skin response, asthma

affects the density and the likely size of the particles to be generated. The nature of woodworking tends to generate high dust levels on floors and surfaces. In general, many of the processes do not have local exhaust ventilation; rather the dust, considered inert, is simply swept up at the end of the day.

The four-stage personal impactor, as discussed previously, has been evaluated for personal sampling for wood dust and found to provide a good description of the size ranges associated with wood dust sampling.

Metal-working Fluid Mists

Traditionally mineral oils have been used as metalworking fluids but they are gradually being replaced by aqueous, synthetic machining fluids and soluble oils due to cost advantages and industrial waste disposal requirements. Metal working fluids are most often collected if they are mineral oil-based as the OSHA standard for oil mist applies to only mineral oils. Also called cutting oils, lubricants, machining fluid, or coolants, they are used in metal cutting to cool the cutting tool and the work; they reduce friction between tools, chip, and work to produce a smooth finish on the work, and to flush out the cutting area and wash away chips. There are several types of cutting oils: straight oils that are mixtures of alkylated aromatics, alkylated napthalene, and paraffin hydrocarbons; soluble oils that contain emulsifiers, such as soaps; lubricating oils that are highly refined oils and additives, such as rust inhibitors; and synthetic water-soluble coolants.[51]

Many operations now use water-based coolants for grinding and other operations. When evaluating the toxic effects of these compounds, the potential for other exposures must be considered. These may include additives with acute and/or chronic toxic effects, such as 2-chloroacetamide and N-hydroxy methyl-2-chloroacetamide. In the past, nitrite has been used as an additive in cutting fluids. Since diethanolamine has often been present as an impurity in synthetic fluids, it has reacted with the nitrite to form the carcinogen N-nitrosodiethanolamine.[52] Occasionally dermatitis problems attributed to metalworking fluids are the result of hydraulic fluid leakage into the coolant reservoir.

A study done to evaluate size characteristics of metal-working fluids found that both large droplets >8 mm in aerodynamic mass median diameter and fine particles in the 0.1 – 1-mm size range existed. The large aerosols were formed from the direct spraying of fluids or from emissions from centrifuges in salvage operations to separate metallic chips from oil, whereas the fine aerosols were produced from the actual machining operations at high shear forces and elevated temperatures. It was also found that differences in the type of fluid can affect the size distribution of the aerosol as occurred when a synthetic fluid was replaced with a semisynthetic compound resulting in a shift in the size distribution toward larger particles.[53]

An evaluation of exposures to metal-working fluids might include collection of bulk samples for analysis for contaminants such as heavy metals or microbials as well as air samples. Sampling for mists is usually done with the top off of a three-stage filter cassette, thus exposing the entire filter. It is known as open-face sampling. The same procedure is used for calibrating a pump to be used with an open-face filter as for cyclone calibrations except the top of a three-stage filter is removed and the filter is attached to the fitting inside the jar.

CONCLUSIONS

In the actual sampling environment, atmospheres of aerosols are seldom simple. For example, some industrial processes generate either dusts or fumes but not both, whereas in other cases mixed exposures to both physical forms are common. For example, the air in a melting or pouring area of a foundry will contain fume from the molten metal and dust reentrained from the floor or generated in the processing of castings. Similarly, welding fumes in a busy metal-working shop will be mixed with dust from other operations.

Since in some cases the toxic effect of an aerosol depends on its form, zinc dust is considered to act as a nuisance dust and zinc oxide fume is known to cause metal fume fever. Dust particles generally average 5 μm or more in diameter and fume particles are usually less than 1 μm. When a total dust sample is analyzed chemically it will not distinguish between dust and fume; therefore, in order to fully evaluate the potential exposure hazard the sampling professional needs to use other methods.

In this situation, one approach to sampling mixed exposures is to use a personal cascade impactor or other sampler with a sharp size cutoff, allowing the fraction with particle diameters less than 1 μm to be weighed. Although this approach would effectively collect all fume and dust particles with the same size range, it would make the actual fraction of fume hard to determine. A 10-mm nylon cyclone could also be used; however, it also passes a significant fraction of particles larger than 1 μm, thus overstating the fume concentration to a greater extent than the cascade impactor. However, in spite of these limitations, a better understanding of the hazard will be gained.[54]

Finally, in order to correctly interpret the results of particle sampling, the sampling professional must be aware that there is much more variability over distance with particle concentrations than with gases and vapor. The results should be supplemented with additional information, such as distance of the sampler from the source, sources of airflow, typical patterns of airflow, and patterns of obvious buildup if visible.

REFERENCES

1. Tomb, T. F., and R. A. Haney. Comparison of number and respirable mass concentration determinations. In *Advances in Air Sampling*. ACGIH, Cincinnati, OH, 1988.
2. Hosketh, H. E. *Fine Particles in Gaseous Media*, 2nd ed. Lewis Pubs., Inc., Chelsea, MI, 1986.
3. National Institute of Occupational Safety and Health. *The Industrial Environment: Its Evaluation and Control*. NIOSH, Cincinnati, OH, 1974.
4. Ad Hoc Working Group to Technical Committee 146-Air Quality, International Standards Organization. Recommendations on size definitions for particle sampling. *AIHA J.* **42**:A-64 – A-68, 1981.

5. John, W. Sampler efficiencies: Thoracic mass fraction. *ACGIH Ann.* **11**:75–79, 1984.
6. Frazie, P. R., and G. Tivoni. A filter cassette assembly method for preventing bypass leakage. *AIHA J.* **48**(2):176–180, 1987.
7. Furst, M. W. Air sampling. *Appl. Ind. Hyg.* **12**(3):F-21, 1988.
8. Knutson, E. O., and P. J. Lioy. Measurement and presentation of aerosol size distributions. In *Air Sampling Instruments,* 6th ed. ACGIH, Cincinnati, OH, 1983.
9. Agarwal, J. K., and B. H. Liu. A criterion for accurate aerosol sampling in calm air. *AIHA J.* **41**:191–197, 1980.
10. Cohen, B., et al. Bias in air sampling techniques used to measure inhalation exposure. *AIHA J.* **45**(3):187–192, 1984.
11. McCawley, M. A. Performance considerations for size-selective samplers. *ACGIH Ann.* **11**:97–100, 1984.
12. Lioy, P. J., et al. Rationale for particle size-selective air sampling. *ACGIH. Ann.* **11**:27–34, 1984.
13. Soderholm, S. Size-selective sampling criteria for inspirable mass fraction. *ACGIH Ann.* **11**:47–52, 1984.
14. Beaulieu, H. J., et al. A comparison of aerosol sampling techniques: "Open" versus "closed-face" filter cassettes. *AIHA J.* **41**:758–765, 1980.
15. Buchan, R. M., et al. Aerosol sampling efficiency of 37 mm filter cassettes. *AIHA J.* **47**:825–831, 1986.
16. Martinelli, C. A., et al. Monitoring real time aerosol distribution in the breathing zone. *AIHA J.* **44**:280–285, 1983.
17. ASTM. Standard test method for respirable dust in workplace atmospheres. ASTM D4532-85, 1985.
18. Dennis, R. *Handbook on Aerosols.* Technical Information Center, Office of Public Affairs, U.S. Energy Research and Development Administration, Washington, DC, 1976.
19. Hinds, W. M. Sampler efficiencies: Inspirable mass fraction. *ACGIH Ann.* **11**:67–74, 1984.
20. Stuart, B. O., et al. Use of size-selection in establishing TLVs. *ACGIH Ann.* **1**:85–96, 1984.
21. Hinds, W. C. Basis for particle size-selective sampling for wood dust. *Appl. Ind. Hyg.* **3**(3):67–72, 1988.
22. Mark, D., and J. H. Vincent. A new personal sampler for airborne total dust in workplaces. *Ann. Occup. Hyg.* **30**:89–102, 1986.
23. Mark, D., et al. A new static sampler for airborne total dust in workplaces. *AIHA J.* **46**(3):127–133, 1985.
24. Vincent, J. H., and D. Mark. Comparison of criteria for defining inspirable aerosol and the development of appropriate samplers. *AIHA J.* **48**(5):454–457, 1987.
25. Raabe, O. G. Size selective sampling criteria for thoracic and respirable mass fractions. *ACGIH Ann.* **11**:53–65, 1984.
26. Lippmann, M., et al. Basis for a particle size-selective TLV for sulfuric acid aerosols. *Appl. Ind. Hyg.* **2**(5):188–199, 1987.
27. Potts, J. D., M. A. McCawley, and R. A. Jankowski. Thoracic dust exposures on longwall and continuous mining sections. *Appl. Occup. Env. Hyg.* **5**(7):440–447, 1990.
28. Aerosol Technology Committee, AIHA. Guide for respirable sampling *AIHA J.* **3–4**:133–137, 1970.
29. John, W. Sampler efficiencies: Respirable mass fraction. *ACGIH Ann.* **11**:81–83, 1984.
30. Lippman, M. Size-selective health hazard sampling. In *Air Sampling Instruments,* 6th ed. ACGIH, Cincinnati, OH, 1983.
31. Carsey, T. P. An investigation of the performance of the 10 mm nylon cyclone. *Appl. Ind. Hyg.* **2**(2):47–52, 1987.
32. Mine Safety Appliance, Inc. *Cyclone Reference Manual.* Pittsburgh, PA.
33. Treaftis, H. N., et al. Comparison of particle size distribution data obtained with cascade impaction samplers and from Coulter counter analysis of total dust samples. *AIHA J.* **47**(2):87–93, 1986.
34. Hinds, W. C., et al. Particle bounce in a personal cascade impactor: A field evaluation. *AIHA J.* **46**(9):517–523, 1985.
35. Hodgkins, D. G., et al. Air-lead particle sizes in battery manufacturing: Potential effects on the OSHA compliance mode. *Appl. Occup. Env. Hyg.* **5**(8):518–526, 1990.
36. Lodge, J. P., and T. L. Chan. *Cascade Impactor: Sampling and Data Analysis.* AIHA Monograph Series, AIHA, Akron, OH, 1986.
37. *Series 290: Instruction Manual, Marple Personal Cascade Impactors.* Bull. No. 290I.M.-3-82, Andersen Samplers, Atlanta, GA.
38. Levy, S. A. Occupational pulmonary diseases. In *Occupational Medicine, Principles and Practical Applications,* C. Zenz, ed. Year Book Pub., Chicago, IL, 1975.
39. 30 CFR Mine Safety and Health Administration, Subchapter O.
40. Corn, M., et al. A critique of MSHA procedures for determination of permissible respirable coal mine dust containing free silica. *AIHA J.* **46**(1):4–8, 1985.
41. Bartley, D. L., and G. M. Breuer. Analysis and optimization of the performance of the 10 mm cyclone. *AIHA J.* **43**(7):520–528, 1982.
42. Kissell, F. N., et al. How to improve the accuracy of coal mine dust sampling. *AIHA J.* **47**(10):602–606, 1986.
43. Fischer, J. J., et al. Environment influences levels of gram negative bacteria and endotoxin on cotton bracts. *AIHA J.* **43**:290–292, 1982.

44. Lacey, J., and M. E. Lacey. Micro-organisms in the air of cotton mills. *Ann. Occup. Hyg.* **31**(1):1–19, 1987.
45. Hughs, S. E., et al. Methodology for cotton gin dust sampling. *AIHA J.* **42**(6):407–410, 1981.
46. McFarland, A. R., et al. A new cotton dust sampler for PM-10 aerosol. *AIHA J.* **48**(3):293–297, 1987.
47. Suh, M. W., et al. Statistical analysis of CAM/LVE data in the major cotton manufacturing process areas. *4th Special Session on Cotton Dust Research Proc.* Technical Research Service/National Cotton Council, Memphis, TN, pp. 103–111, 1980.
48. Whitehead, L. W., et al. Suspended dust concentrations and size distributions, and qualitative analysis of inorganic particles, from woodworking operations. *AIHA J.* **42**(6):461–467, 1981.
49. Whitehead, L. W. Health effects of wood dust—relevance for an occupational standard. *AIHA J.* **43**(9):674–678, 1982.
50. Hausen, B. *Woods Injurious to Human Health: A Manual.* Walter deGruyter, New York, 1981.
51. National Safety Council. *Cutting Oils, Emulsions, and Drawing Compounds.* Data Sheet 501, Chicago, IL, 1977.
52. Wu, W. S., et al. Determination of nitrites in metal cuttings fluids by ion chromatography. *AIHA J.* **43**:942–945, 1982.
53. Chan, T. L., J. B. D'Arcy, and J. Slak. Size characteristics of machining fluid aerosols in an industrial metalworking environment. *Appl. Occup. Env. Hyg.* **5**(3):162–170, 1990.
54. ACGIH Committee Reports (Inorganic Dust Subcommittee). TLVs for chemical substances. *Appl. Ind. Hyg.* **2**(6):R-4–R-6, 1987.

CHAPTER 6

Special-Case Sampling

Vapors and aerosols may be present when (1) semivolatile organic compounds have vapor pressures that allow them to exist in the atmosphere as a vapor and/or aerosol under certain conditions; (2) different compounds generated from the same element as a result of different processing methods occur within the same operation; (3) different compounds are generated as a result of combustion of a single material; (4) mixtures contain both volatile and nonvolatile compounds that are sprayed; (5) certain gases and vapors are capable of existing in the gaseous phase and adsorbing to particles.

Some solids, termed *semivolatile,* have a high enough vapor pressure that evaporation losses occur during the sampling period when traditional filtration methods are used. During sampling, air is continuously passing through the collected aerosol; thus, organic aerosols with significant vapor pressures could be partially or totally lost during the sampling process unless special precautions are taken. Semivolatile organic air pollutants include polynuclear aromatic hydrocarbons (PAHs) with four or fewer fused rings and their nitro derivatives, chlorobenzenes, chlorotoluenes, polychlorobiphenyls, organochlorine, and organophosphate pesticides, and the various polychlorodibenzo-*p*-dioxins. These are the most common groups found in both the vapor and particulate phases. Some semivolatile organic compounds have vapor pressures that allow them to exist in the atmosphere as vapors, in the condensed aerosol phase, or in both phases depending on the ambient conditions at the time of sampling[1]; for example, certain solids such as arsenic trioxide and benzidine sublime, meaning they go directly from the solid phase to the vapor phase without going through a liquid phase.

It has been suggested that different states of a semivolatile compound, particles or vapors, will have different effects on the body.[2] For example, if a particle is surrounded by a thin shell of essentially saturated vapor, and it gains entry to the respiratory passages, it may have a more potent effect on the mucosal surface than the same substance in a pure vapor state, which is more likely to be diluted by room air. Therefore, a particle cloud will have a greater effect than the same weight of a vapor. The effect of these particles can be distinguished from the vapor only if they can be collected and analyzed separately.

Multiple species of the same element may exist in the same industrial environment. Inorganic antimony compounds are all possible workplace contaminants during the lead-acid battery manufacturing process and may coexist. Antimony fume can be generated during torching or welding on aluminum and other scrap; stibine (antimony hydride) gas can be generated during the battery boosting and forming process as a result of the antimony-contaminated lead plates coming into contact with sulfuric acid and nascent hydrogen gas. A concern in this case is the varying degrees of toxicity each compound might possess. Dusts of antimony compounds cause irritation of the mucous membranes and a variety of systemic effects, and stibine gas causes hemolysis of red blood cells.

Another situation of interest in which sampling involves both vapors and mists is when a solvent-based mixture is sprayed. The toxic effects of mixtures, such as paints and pesticide formulations, will depend on the components of the mixture. When polyurethane paint containing lead pigment is sprayed on automobiles there are several toxic concerns: the isocyanate, sometimes toluene diisocyanate; the pigment, often lead chromate; and the solvent, usually a mixture of several.

Toluene diisocyanate is a lung sensitizer, and based on animal data, is suspected of being a carcinogen. Lead acts on many different organs including the kidney and liver. Solvents affect the central nervous system. Depending on the conditions of exposure (type of respiratory protection, effectiveness of engineering controls such as a paint spray booth, concentrations, length of exposure), a variety of acute or chronic effects could result.

Some gases and vapors are capable of adsorbing to particles. Many dusts are highly adsorbent and often can contain a significant amount of a gaseous component, such as sulfur dioxide or formaldehyde. The adsorption of gases and vapors onto particulate matter can also create a potential for reactions as a result of catalytic effects on the surface of particulates.[3]

COLLECTION METHODS FOR SEMIVOLATILE COMPOUNDS

Occupational Sampling

For occupational sampling of semivolatile compounds, most often a filter cassette is used in front of a sorbent tube. The filter collects the aerosol; the tube collects the vapor.[4] When filters precede sorbent tubes to collect aerosol, they are often very small in diameter—13 mm—compared to those used strictly for aerosol collection. Once the sampling train is assembled, the procedures are the same as described in the chapters on Integrated Sampling for Gases and Vapors and on Integrated Sampling for Aerosols. The flow rate is determined by the method and should be acknowledged. Calibration must be done with the entire train assembled. Table 6-1 describes media used in NIOSH methods for semivolatile compounds.

Examples of compounds that would be collected using occupational methods are PAHs. These compounds may consist of both an aerosol and gaseous portion. They can be a concern in an active industrial process, for example, the manufacturing of coke, as well as on hazardous waste sites, such as coal gassification plants. Sampling would most likely be done outdoors in these instances. Vapor pressure data for a number of PAHs (Table 6-2) indicate that compounds less volatile than chrysene would exist primarily in the particle state.[5]

The following conclusions reached during a study that evaluated factors affecting the sampling of airborne PAHs show how important it is to identify the characteristics and physical behavior of a semivolatile contaminant prior to sampling: (1) A fraction of the more volatile PAHs (three- and four-ring) may exist in the vapor phase at ambient temperatures, with the percentage increasing as temperatures elevate; (2) even though it is assumed that higher molecular weight PAHs will be collected and retained on a filter at ambient temperatures, a vapor trap is needed as a backup to the filter; (3) care must be taken when conducting particle-size analyses on PAH-containing aerosols using cascade impactors, since PAHs may be stripped from the material collected on one stage and be readsorbed by particles on subsequent stages, and compounds in the vapor phase may condense on certain stages of the sampler as a result of cooling during sampling; (4) care should be taken during sampling, since PAHs may react with ozone or may be degraded by sunlight and ultraviolet light; (5) losses of PAHs can occur from filters during storage.[6]

TABLE 6-1. Media Used in NIOSH Methods for Combination Aerosol and Vapor

Compound	*Method Number*	*Medium*
Acenaphthene	5506, 5515	2-μm PTFE filter, 37 mm, and washed XAD-2 tube, 100/50
Acenaphthylene	5506, 5515	2-μm PTFE filter, 37 mm, and washed XAD-2 tube, 100/50
Aldrin	5502	Glass fiber filter (Gelman type AE) in combination with isooctane in a bubbler
Anthracene	5506, 5515	2-μm PTFE filter, 37 mm, and washed XAD-2 tube, 100/50
Arsenic trioxide	7901	Sodium carbonate-impregnated 0.8-μm MCEF, 37 mm
Benz(*a*)anthracene	5506	2-μm PTFE filter, 37 mm, and washed XAD-2 tube, 100/50
Benzidine	5509	Glass fiber filter (type AE), 13 mm, and silica gel, 50
Benzo(*b*)fluoranthene	5506, 5515	2-μm PTFE filter, 37 mm, and washed XAD-2 tube, 100/50
Benzo(*k*)fluoranthene	5506, 5515	2-μm PTFE filter, 37 mm, and washed XAD-2 tube, 100/50
Benzo(*ghi*)perylene	5506, 5515	2-μm PTFE filter, 37 mm, and washed XAD-2 tube, 100/50
Benzo(*a*)pyrene	5506, 5515	2-μm PTFE filter, 37 mm, and washed XAD-2 tube, 100/50
Benzo(*e*)pyrene	5506, 5515	2-μm PTFE filter, 37 mm, and washed XAD-2 tube, 100/50
Chrysene	5506, 5515	2-μm PTFE filter, 37 mm, and washed XAD-2 tube, 100/50
Cyanides	7904	0.8-μm MCE filter, 37 mm, and potassium hydroxide in a bubbler (Filter collects cyanide salts; bubbler collects HCN.)
Demeton (Demeton is a liquid with a low vapor pressure—0.001 mm Hg.)	5514	2-μm MCE filter and XAD-2 tube, 150/75 (Filter collects aerosol; tube collects vapor.)
Dibenz(*a,h*)anthracene	5506,5515	2-μm PTFE filter, 37 mm, and washed XAD-2 tube, 100/50
Diborane	6006	PTFE filter and oxidizer-impregnated charcoal tube, 100/50 (Filter removes boron-containing particulates.)
Dibutyl tin *bis*(isooctyl mercaptoacetate)	5504	Glass fiber filter, 37 mm, and XAD-2 tube, 80/40
Ethylene glycol (Ethylene glycol is a liquid with a low vapor pressure—0.1 mm Hg.)	5500	Binderless glass fiber filter, 13 mm, and silica gel tube, 520/260
Fluoranthene	5506, 5515	2-μm PTFE filter, 37 mm, and washed XAD-2 tube, 100/50
Fluorene	5506, 5515	2-μm PTFE filter, 37 mm, and washed XAD-2 tube, 100/50
Fluorides	7902	0.8-μm MCE filter, 37 mm, and sodium carbonate-treated cellulose pad, 37 m in series (Fluoride salts liberate HF in the presence of acids; the method collects both aerosol and gaseous fluorides.)
Indeno(1,2,3-*cd*)pyrene	5506, 5515	2-μm PTFE filter, 37 mm, and washed XAD-2 tube, 100/50

(continued)

TABLE 6-1. *(continued)*

Compound	*Method Number*	*Medium*
Kepone	5508	0.8-μm MCE filter, 37 mm, and impinger with sodium hydroxide (May have something to do with the conversion of mirex to kepone, since mirex is converted to kepone in the presence of sodium hydroxide.)
Lindane	5502	Glass fiber filter (Gelman type AE) in combination with isooctane in a bubbler
Mercury	6000	Glass fiber prefilter, 13 mm, and silvered Chromosorb P, 30 (Filter collects particulate mercury; tube collects mercury vapor.)
Naphthalene	5506, 5515	2-μm PTFE filter, 37 mm, and washed XAD-2 tube, 100/50
Naphthylamines	5518	Glass fiber filter, 13 mm, and silica gel tube, 100/50
PCBs	5503	Glass fiber filter, 13 mm, and florisil tube, 100/50
Pentachlorobenzene	5517	PTFE fiber mat filter, 13 mm, and amberlite XAD-2 tube, 100/50
Phenanthrene	5506, 5515	2-μm PTFE filter, 37 mm, and washed XAD-2 tube, 100/50
Phosphorus	7905	0.8-μm MCE filter, 37 mm, and Tenax GC tube, 100/50
Pyrene	5506, 5515	2-μm PTFE filter, 37 mm, and washed XAD-2 tube, 100/50
Sulfur dioxide	6004	0.8-μm MCE filter, 37 mm, and cellulose filter impregnated with potassium hydroxide, 37 mm, in series (Sulfur dioxide is collected on the treated filter and sulfuric acid, sulfate salts, and sulfite salts are collected on the front filter as particulates.)
1,2,4,5-Tetrachlorobenzene	5517	PTFE fiber mat filter, 13 mm, and amberlite XAD-2 tube, 100/50
Tributyl tin	5504	Glass fiber filter, 37 mm, and XAD-2 tube, 80/40
Tributyl tin chloride	5504	Glass fiber filter, 37 mm, and XAD-2 tube, 80/40
1,2,4-Trichlorobenzene	5517	PTFE fiber mat filter, 13 mm, and amberlite XAD-2 tube, 100/50
Tricyclohexyltin hydroxide	5504	Glass fiber filter, 37 mm, and XAD-2 tube, 80/40

When sampling PAHs, NIOSH recommends a combination of 2-μm-pore-size polytetrafluoroethylene (PTFE) filters in 37-mm cassettes in series with a 100/50 XAD-2 resin sorbent tube at a flow rate of 1.7 Lpm.[4]

Environmental Measurement Methods

For environmental collection of semivolatile compounds, the most common type of medium is polyurethane foam (PUF). It has low resistance to airflow; thus, large volumes of air can be collected on it. It is easy to purify and handle; however, it has been shown to have a poorer retention for the more volatile pesticides and polychlorinated biphenyls (PCBs) compared to XAD-2 or Tenax resins, and it does not differentiate between the vapor and particulate phases. The addition of a Tenax cartridge or tube to the sampling train (following the PUF) may be useful in collecting the vapor phase. There is also a concern that mutagenic artifacts may form during sampling.[5]

The primary semivolatiles of concern in environmental (outdoor) measurements are PCBs,

TABLE 6-2. Vapor Pressures of Selected PAHs

Compound	Vapor Pressure (at 25°C, torr)
Phenanthrene	1.7×10^{-4}
Anthracene	2.5×10^{-5}
Pyrene	6.8×10^{-6}
Fluoranthrene	4.9×10^{-6}
Benzo(*a*)anthracene	1.1×10^{-7}
Chrysene	9.0×10^{-9}
Benzo(*a*)pyrene	5.5×10^{-9}
Benzo(*e*)pyrene	5.5×10^{-9}
Perylene	4.3×10^{-9}
Benzo(*g,h,i*)perylene	1.0×10^{-10}
Coronene	1.5×10^{-12}

PAHs, and pesticides. In EPA Methods TO4, TO9, and TO13, an ambient air sample is drawn over a 24-hour period through a glass fiber filter with a PUF backup absorbent. The PUF adsorbent specified by the EPA method is a polyether-type polyurethane foam (density no. 3014, or 0.0225 g/cm^3). This type of foam is used for furniture upholstery. It is white and yellows upon exposure to light. The PUF inserts are 6-cm cylindrical plugs cut from 3-inch sheet stock, and should fit with slight compression in the glass cartridge, supported by the wire screen.[7] The filter and backup foam are returned to the laboratory for analysis. PCBs and pesticides are recovered for Soxhlet extraction and then subjected to GC/ECD (gas chromatography/electron capture detector) or other analysis. The method is capable of detecting >1 ηg/M^3 using a 24-hour sampling period. Figures 6-1 and 6-2 show semivolatile sampling equipment.

Procedure

1. The airflow through the sampling system is monitored by a venturi/Magnehelic assembly. A multipoint calibration of this assembly must be conducted every 6 months using an audit calibration orifice.

2. Prior to calibration, a blank PUF cartridge and filter are placed in the sampling head and the high-volume sampling pump is turned on (Table 6-3). The flow control valve is fully opened and the voltage is adjusted so that a sample flow rate corresponding to approximately 110% of the desired flow rate is indicated on the magnehelic (based on the 6-month multipoint calibration curve). The motor is allowed to warm up for approximately 10 minutes, and then the flow control valve is adjusted to achieve the desired flow rate. The ambient temperature and barometric pressure should be recorded.

3. The calibration orifice is then placed on the sampling head and a manometer is attached to the tap on the calibration orifice. The sampler is momentarily turned off to set the zero level of the manometer. The sampler is then switched on and the manometer reading is recorded. Once a stable reading is achieved, the sampler is then shut off.

4. The calibration curve for the orifice is used to calculate sample flow from the data obtained in the previous step, and the calibration curve for the venturi/magnehelic assembly is used to calculate sample flow from the data obtained during the 6-month calibration. If the two values do not agree within 10%, the sampler should be inspected for damaged and flow blockages. If no obvious problems are found, the sampler should be recalibrated (multipoint).

5. A multipoint calibration of the calibration orifice, against a primary standard, should be obtained annually.

6. The samples should be located in an unobstructed area, at least 2 meters from any obstacle to airflow. The exhaust hose should be stretched in the downwind direction to prevent recycling of air.

7. A clean sampling cartridge and quartz fiber filter are removed from sealed transport containers and placed in the sampling head using forceps and gloved hands. The head is tightly sealed into the sampling system. The aluminum foil wrapping is placed back in the sealed container for later use.

8. The zero reading of the magnehelic is checked. Ambient temperature, barometric pressure, elapsed time meter setting, sampler serial number, filter number, and PUF cartridge number are recorded.

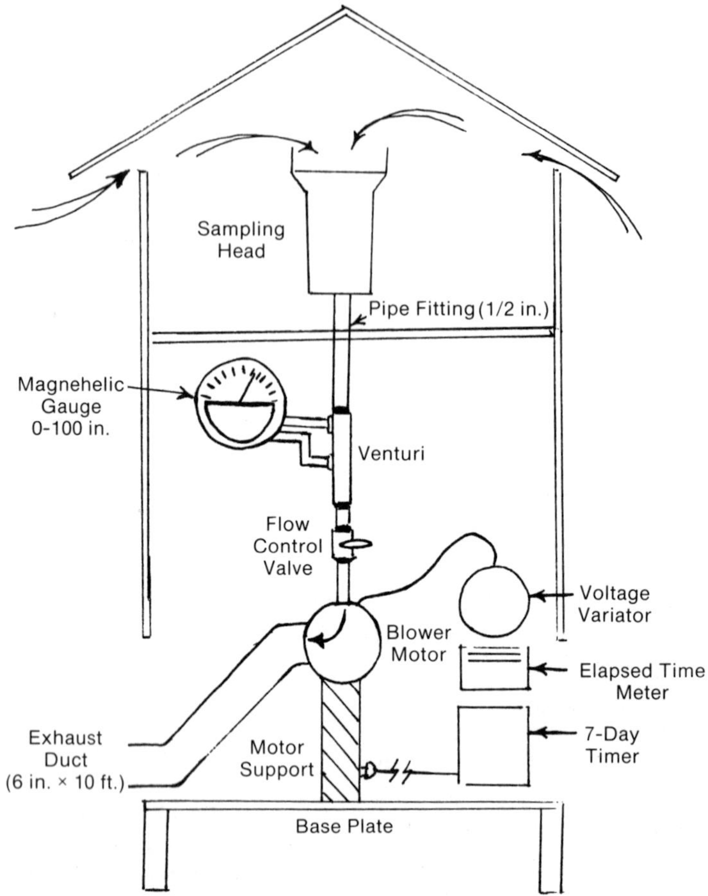

Figure 6-1. High-volume ambient air samplers set up for semivolatile sampling. (Courtesy General MetalWorks, Inc.)

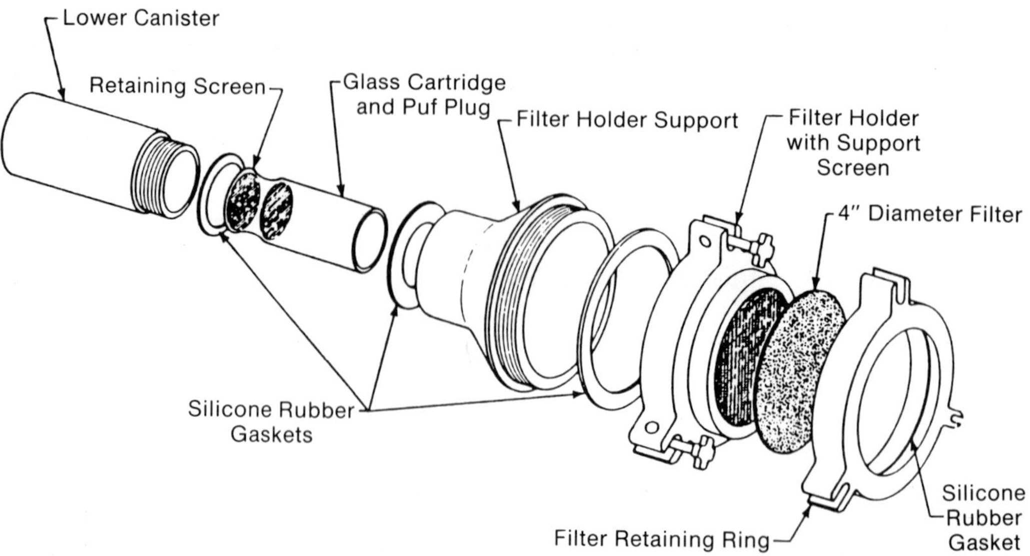

Figure 6-2. Sampling head for semivolatile sampling. (Courtesy General MetalWorks, Inc.)

TABLE 6-3. Selected Components Determined Using High-Volume PUF Sampling Procedure

Compound	*Air Concentration ($\eta g/m^3$)*	*% Recovery*
Aldrin	0.3–3.0	28
4,4′-DDE	0.6–6.0	89
4,4′-DDT	1.8–18	83
Chlordane	15–150	73
Chlorobiphenyls		
4,4′-Di-	2.0–20	62
2,4,5-Tri-	0.2–2.0	36
2,4′,5-Tri-	0.2–2.0	86
2,2′,5,5′-Tetra-	0.2–2.0	94
2,2′,4,5,5′-Penta-	0.2–2.0	92
2,2′,4,4′,5,5′-Hexa-	0.2–2.0	86

9. The voltage and flow control valve are placed at the settings used with the calibration orifice and the power switch is turned on. The elapsed time meter is activated and the start time is recorded. The flow (magnehelic setting) is adjusted, if necessary using the flow control valve.

10. The magnehelic reading is recorded every 6 hours during the sampling period. The calibration curve is used to calculate the flow rate. Ambient temperature and barometric pressure are recorded at the beginning and end of the sampling period.

11. At the end of the desired sampling period, the power is turned off and the filter and PUF cartridges are wrapped with the original aluminum foil and placed in sealed, labeled containers for transport to the laboratory.

12. The magnehelic calibration is checked using the calibration orifice. If the calibration deviates by more than 10% from the initial reading, the flow data for that sample must be marked as suspect and the sampler should be inspected and/or removed from service.

13. At least one field blank will be returned to the laboratory with each group of samples.

14. The calculation of total sample volume is

$$V_m = \frac{Q_1 + Q_2 \ldots Q_n (T)}{N(1000)}$$

where $Q_1, Q_2 \ldots Q_n$ = flow rates determined at the beginning, intermediate, and end points during sampling (Lpm).

An alternate, more portable, sampling setup for semivolatiles involves the use of a PUF cartridge and a personal air sampling pump (Fig. 6-3).

COLLECTION OF MULTIPLE SPECIES: ARSENIC

Industrial operations where multiple species can be generated either during the same process or in processes within the same operation require an understanding of the chemistry of each phase of the process, as well as the properties of each compound that might be formed, or sampling will not fully evaluate the situation. Therefore, the methods for evaluation involve the use of occupational sampling techniques.

Multiple species of inorganic arsenic can coexist in the lead-acid battery manufacturing process.[8] Arsenic fume can be generated during torching or welding on aluminum and other scrap; arsenic trioxide vapor may also form around welding torches, and arsine gas might be released during the battery forming and boosting processes as a result of arsenic-contaminated lead plates coming into contact with sulfuric acid and nascent hydrogen. Arsenic trioxide can exist in both particulate and vapor phases at ordinary room temperatures. The equilibrium vapor pressure of this compound is influenced by the temperature. Concerns in this situation are the varying degrees of toxicity associated with each compound and the need to use different collection methods to identify different compounds or the degree of the hazard may be underestimated.

As an illustration of the importance of selecting the correct sampling method, the following problems were identified in the course of selecting sampling methods to use in a lead-acid battery manufacturing operation. Arsenic trioxide vapor generated from torching or welding operations could condense, forming a fume that coexists with the vapor phase. When collected on a filter, the arsenic trioxide fume would ordinarily

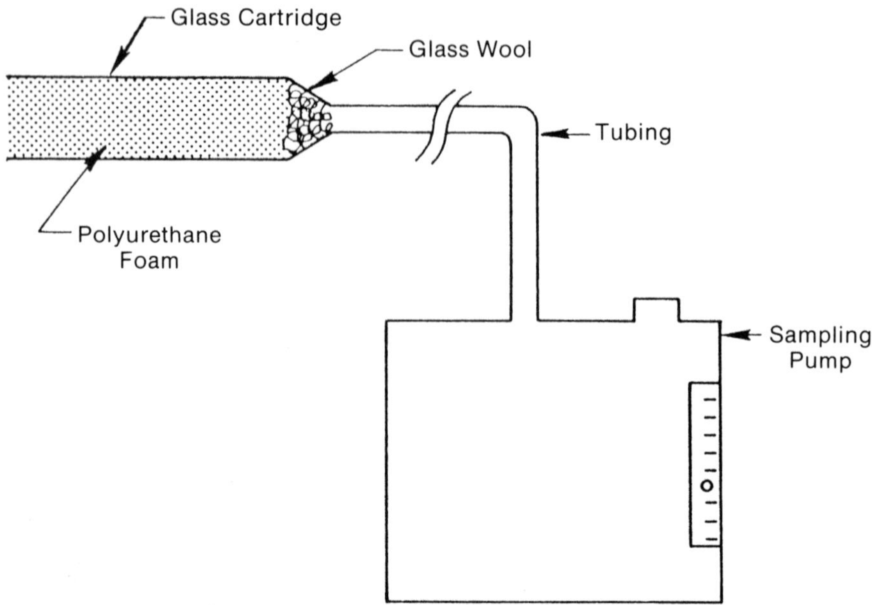

Figure 6-3. Portable ambient sampling apparatus using PUF cartridges. (From EPA: *Characterization of Hazardous Waste Sites—A Methods Manual,* vol. 2: *Available Sampling Methods.* EPA-600/4-83-040. Sept. 1983.)

be interpreted as particulate arsenic due to the need to use atomic absorption for analysis of this medium. If the arsenic trioxide migrated through the filter to the backup pad, it would not be identified at all, since backup pads are not routinely analyzed. When collected on charcoal tubes, arsenic trioxide would be identified as arsine due to the limitations of this analytical method. In this case, a separate method must be used to collect the arsenic trioxide. The arsenic solids can be trapped on a 13-mm filter cassette and the arsine vapor can be adsorbed on a charcoal tube. Following analysis, these results are adjusted to deduct the amount of arsenic trioxide present.[8]

Other inorganic arsenic-containing compounds, such as arsenic tribromide, arsenic triiodide, and arsenic monophosphide, also have equilibrium vapor concentrations (as does arsenic) in excess of 1 $\mu g/m^3$ at the existing air temperature. Therefore, in these cases a sampling approach that collects both aerosol and vapor must also be used.

COMBUSTION PROCESSES: CIGARETTE SMOKE COLLECTION

Combustion processes can occur both outdoors and indoors. Due to their complexity, sometimes sampling is done only for the compounds expected to exist in the highest or most toxic concentrations. For example, welding produces many gases and fumes and would involve occupational methods. Cigarette smoke would be of concern in an indoor air situation, and the methods would be a combination of environmental and occupational sampling techniques, depending on what was to be sampled.

Cigarette smoke is a complex mixture of liquid droplets containing a large variety of organic and inorganic chemicals that are dispersed in a gaseous medium. The smoke aerosol is a continuously changing entity, and aging results in changes in its physical and chemical properties. The composition of the smoke from a given product will depend on the chemical composition and physi-

TABLE 6-4. Percent Distribution of Cigarette Smoke

Material (% by Vol.)	*Weight (mg/cigarette)*	*Weight of Total Effluent (%)*
Particulate matter	40.6	8.2
Nitrogen (67.2)	295.4	59.0
Oxygen (13.3)	66.8	13.4
Carbon dioxide (9.8)	68.1	13.6
Carbon monoxide (3.7)	16.2	3.2
Hydrogen (2.2)	0.7	0.1
Argon (0.8)	5.0	1.0
Methane (0.5)	1.3	0.3
Water vapor	5.8	1.2
C_2-C_4 hydrocarbons	2.5	0.5
Carbonyls	1.9	0.4
Hydrogen cyanide	0.3	0.1
Other known gaseous materials	1.0	0.2
Total	500	100

TABLE 6-5. Major Toxic Agents in the Gas Phase of Cigarette Smoke (Unaged)

Agent	*Concentration in One U.S. Cigarette*
Dimethylnitrosamine	13.0 ηg
Ethylmethylnitrosamine	1.8 ηg
Diethylnitrosamine	1.5 ηg
Nitrosopyrrolidine	11.0 ηg
Hydrazine	32.0 ηg
Vinyl chloride	12.0 ηg
Urethane	30.0 ηg
Formaldehyde	30.0 μg
Hydrogen cyanide	110.0 μg
Acrolein	70.0 μg
Acetaldehyde	800.0 μg
Nitrogen oxides	350.0 μg
Ammonia	60.0 μg
Pyridine	10.0 μg
Carbon monoxide	17.0 μg
Acrylonitrile	10.0 μg
2-Nitropropane	0.92 μg

cal properties of the tobacco leaf. These in turn depend on the genetic makeup and environmental factors, such as mineral nutrition, soil properties, moisture supply, temperature, and light intensity during the growth cycle of the leaf. Finally, during the manufacturing process many modifications of cigarettes are possible, including the use of different parts of the tobacco plant (leaves, stems, vines), all of which vary in composition, as well as additives and papers. Table 6-4 shows distribution of cigarette smoke[9] and Table 6-5 lists toxic agents in the gas phase of cigarette smoke.[9]

Reasons for monitoring cigarette smoke are listed below.[10]

Certain toxic agents in tobacco products and/or their smoke may also occur in the workplace, thus increasing exposure to the agent. These include carbon monoxide, acetone, acrolein, aldehydes, arsenic, cadmium, hydrogen cyanide, hydrogen sulfide, ketones, lead, methyl nitrite, nicotine, nitrogen dioxide, phenol, and polycyclic aromatic compounds.

Workplace chemicals may be transformed into more harmful agents by smoking due to the heat generated by burning tobacco. Polymer fume fever from degradation of heated Teflon has been reported as a result of cigarette smoking.

Tobacco products can become contaminated by chemicals used in the workplace, thus increasing the amount of toxic chemicals entering the body.

Smoking may contribute to an effect comparable to what can result from exposure to toxic agents found in the workplace, thus causing an additive biological effect. For example, combined exposure to cigarette smoke and chlorine can enhance the damage done by chlorine alone. Other agents that can act additively with tobacco smoke include cotton dust, coal dust, and beta-radiation.

Smoking can act synergistically with toxic agents found in the workplace to cause a much more profound effect than that anticipated simply from the separate influences of the occupational exposure and smoking. The synergistic effect of smoking and asbestos exposure that enhances the likelihood of an individual developing cancer is well known. Other chemicals and physical agents that ap-

> pear to act synergistically with tobacco smoke include radon daughters, gold mine exposures, and exposures in the rubber industry.

Due to the complexity of cigarette smoke, usually monitoring involves sampling for a specific constituent. Often carbon monoxide is selected, due to the ease in monitoring for it and the low levels expected to be present without smoking. However, for comparison, samples should be taken outside at the same time to account for background levels from exhausts that have been entrained into the building's heating, ventilating, and air conditioning (HVAC) system. Carbon monoxide can be sampled using real time instrumentation and long-term detector tubes. For more information, see the chapters on Monitoring Instruments Dedicated to a Single Chemical and on Sampling of Gases and Vapors Based on Color Change. Another approach is to measure nicotine as it is considered a carcinogen.

A passive monitor is also available—the Smoke-Check—that measures nitrogenous organic bases including nicotine. It reads out in cigarette equivalents. It is designed to be used for a 1- to 4-hour test period.

COLLECTION OF MIXTURES

In the case in which a mixture is being sprayed, the type of collection methods will vary significantly and depend on the compounds present. When liquid paint is sprayed from an atomizing nozzle, the generated aerosol conditions are dependent on the liquid properties, spray-nozzle flow conditions, and droplet evaporation. Initial droplet size directly affects droplet evaporation rate. Once the nozzle hardware and paint have been chosen, the aerosol concentration and particle-size distribution are primarily dependent on the flow rates of the liquid paint, nozzle air, and diluent air. As the volatiles evaporate, the aerosol is changing from "wet" to dry," and a steady particle-size distribution can be attained only after the evaporation is complete. Paint spray nozzles generate droplets typically in the range of 20 μm to 100 μm in diameter. The droplets form as a result of the shearing action of the air jet in the nozzle on the liquid film. The diameter of the droplets decreases as evaporation progresses, causing an unstable particle-size distribution that changes over time.[11]

An example of the complexity of paint spray sampling is illustrated by a study to measure the epoxy content of paint spray aerosol.[12] The purpose for sampling the epoxy was to determine the concentration of the reactive epoxide during spraying. Compounds containing the epoxide group have demonstrated a wide number of toxic effects. Aerosol epoxide content was collected using midget impingers containing dimethyl formamide, which stops the epoxide reaction with the amine from the curing agent to preserve the epoxide for analysis. Nonvolatile aerosol mass samples were collected on 1-μm PTFE filters and a cascade impactor was used to characterize the size distribution of the generated aerosol. The sampling location was selected to keep the instruments out of the direct spray from the spray guns and to remain clear of the paint hoses as the painters moved across the work area. In this case, a wide range of aerosol epoxy concentrations was observed and the aerosol generated during spray finishing was found to be unstable, changing in physical and chemical properties while it was airborne.

The primary sources of variation that can affect the results of paint spray sampling are the type and effectiveness of the ventilation systems, and where the spraying takes place relative to the positioning of the ventilation system; the size of the item being painted; and the amount of pigment in the paint. Paint pigment formulation could have an effect as well, because the percentage of pigment in paints varies. One study found a variation of 0.2% to 8.9% lead by weight alone.[13]

GASES ADSORBED TO PARTICLES

Situations in which gases may be adsorbed to dusts may be sampled using either environmental or occupational methods depending on the situation and the levels of concern. Sulfur dioxide can occur as a common outdoor pollutant and can be generated by many industrial processes, such as bleaching pulps from the kraft process and combustion of coal. In a typical occupational sampling situation using an impinger, the primary goal would be to prevent particles from entering the solution. If they did, the results would be higher than the actual exposure.

REFERENCES

1. Clements, J. B., and R. G. Lewis. Sampling for organic compounds. In *Principles of Environmental Sampling*. American Chemical Society, Washington, D.C., pp. 287–296, 1987.
2. Chemical Substances TLV Committee Study Paper. *ACGIH Ann.* **4:**153–157, 1983.
3. Soderholm, S. C. Aerosol instabilities. *Appl. Ind. Hyg.* 3(2):35–40, 1988.
4. National Institute for Occupational Safety and Health. *NIOSH Manual of Analytical Methods*, 3rd ed. NIOSH, Cincinnati, 1984.
5. Riggin, R. M., and B. A. Petersen. Sampling and analysis methodology for semivolatile and nonvolatile organic compounds in air. In *Indoor Air Quality*, H. Kasuga, ed. Springer-Verlag, pp. 351–359, 1989.
6. Leinster, P., and M. J. Evans. Factors affecting the sampling of airborne polycyclic aromatic hydrocarbons—a review. *Ann. Occup. Hyg.* **30**(4):481–495, 1986.
7. Environmental Protection Agency. Compendium of Methods for the Determination of Toxic Organic Compounds in Ambient Air. EPA-600/4-84-041, 1984.
8. Costello, R. J., P. M. Eller, and R. D. Hull. Measurement of multiple inorganic arsenic species. *AIHA J.* **44**(1):21–28, 1983.
9. Surgeon General's Report on Cigarette Smoking. 1975.
10. National Institute for Occupational Safety and Health. *Current Intelligence Bulletin 31*. Adverse Health Effects of Smoking and the Occupational Environment. NIOSH, Cincinnati, February 5, 1979.
11. Beaulieu, H. J., et al. A comparison of aerosol sampling techniques: "Open" versus "closed-face" filter cassettes. *AIHA J.* **41**(10):758–765, 1980.
12. Herrick, R. F., M. J. Ellenbecker, and T. J. Smith. Measurement of the epoxy content of paint spray aerosol: Three case studies. *Appl. Ind. Hyg.* 3(4):123–128, 1988.
13. Ackley, M. W. Paint spray tests for respirators: Aerosol characteristics. *AIHA J.* **41**(5):309–316, 1980.

PART III

Real Time Measurements

CHAPTER 7

Reading and Recording Real Time Data

Instruments capable of making "real time" or instantaneous measurements may be defined as those devices in which the sampling and analysis are carried out within the instrument, and the results read out on a meter or digital indicator within a few seconds or minutes. Another term often used for real time monitors is direct reading monitors. Real time monitors can be set up to continuously monitor operations or take periodic samples.

TECHNOLOGICAL DEVELOPMENTS

The trend toward development of more sophisticated electronics is present in air sampling equipment just as it is with most other electronic-based equipment. Digital readouts are replacing meters with needles. Some microprocessor-based instruments have data-logging capabilities built into them whereas others can be hooked up to independent data-logging units.

Real time instruments can be used for personal monitoring, for conducting surveys in work sites to identify sources such as leaks and potential levels of contaminants, and for continuous fixed station monitoring. Documentation of the effects of work practices on exposures can be done with real time instruments, and these results can in turn be used as an educational tool with workers to get them to modify their practices in order to reduce exposures. The documentation can also be an evaluation of the effectiveness of control systems. One leak-testing application is fugitive emissions. Another common application is testing specific equipment such as ethylene oxide sterilizers for leaks.

In some cases, real time instrumentation is the best available method for monitoring a chemical. While there are many specific methods available, some chemicals are very difficult to collect on media. In addition, the trend is to identify immediate chemical levels over short time periods rather than using an integrated measurement period of 2, 4, or 8 hours to provide an average value. A problem has been that field instruments lack the specificity and detection limits of laboratory methods. Although commercial technology is not perfected, progress is being made. The biggest problem for the sampling professional in the future will not be the availability of equipment but the cost: As instruments become more tuneable or specific, their reliability increases, and their detection capability becomes more sensitive to lower levels, their cost increases. Many high-grade field instruments are in the price range of a luxury car.

Real time instruments fall into several categories: those that are specific for a single gas or vapor; general survey instruments, which respond to many different gases and vapors at once, but cannot differentiate between them; more sophisticated analyzers such as gas chromatographs (GC) that can either sample for several gases and vapors at once, or infrared (IR) instruments that can be tuned in the field to provide a specific measurement for a number of different gases and vapors, one at a time; and instruments that monitor for aerosols. In each case, one or more different principles of detection are involved. Most monitors dedicated to a single specific chemical are electrochemical or IR-based.

General survey instruments can be classified as catalytic combustion, photoionization (PID), flame ionization (FID), and metallic oxide semiconductor (MOS) and are used to measure both combustible and health hazard levels. GCs and IRs are examples of laboratory analytical instruments that have been modified for field use. Real time aerosol instruments do not differentiate between types of aerosols, but are useful for developing sampling strategies and represent an area of monitoring that is expanding.

There is a variety of technology available. Since the catalytic combustion-type sensor was developed in 1923, there are instruments based on this principle that were developed more than 30 years ago and are still sold. In some cases, these are as reliable as newer, more automated instruments. Other instruments have been around for some time, but only recently came into popular use. An example is the PID general survey instrument. Prior to the concerns of health and safety in hazardous waste work, this instrument was rarely used due to its extreme nonspecificity. With the advent of hazardous waste work, nonspecificity became a virtue, making these units highly desirable for detecting a wide range of contaminants. In other cases, changes in sampling needs have also dictated changes in equipment needs. For years, IR was the method of choice for detection of ethylene oxide, but when the Occupational Safety and Health Administration (OSHA) standard was lowered to 1 ppm, the portable GC with a PID became the preferred sampling mode because of its increased sensitivity.

Microprocessor capability has allowed miniaturization of real time monitors so that they can be worn for personal monitoring. Many are dosimeters, meaning an 8-hour time-weighted average (TWA) can be integrated. Others can be attached to a data logger while being worn by the individual. Some have their own data-logging capability and can provide a comprehensive printout of the day's exposure. Some instruments for gases and vapors have alarms for high levels to provide warning capabilities that are essential for work in confined spaces, repair of tanks, process vessels, and pipelines. In some cases, these units do not have a readout, only an alarm.

As more instruments become microprocessor-based and data-logger compatible, sampling professionals will need to become familiar with the basics of communicating with printers and computers if they want to be able to utilize all the features and advantages of these units. The following list shows the history of the development of chemical sensors.

1923	Catalytic combustion-type sensor
1952	Galvanic cell-type gas sensor (O_2)
1961	Solid electrolyte-type sensor
1962	Metal oxide semiconductor-type gas sensor
1964	Piezo quartz crystal sensor for aerosols
1965	Practical use of catalytic combustion-type sensor
1967	Practical use of metal oxide semiconductor-type gas sensor
1974	Practical use of electrochemical gas sensor

THE BLACK BOX

It is convenient to think of a real time instrument as a black box. If the box is sophisticated enough, it can be mounted in a stationary place and lots of "bells and whistles" can be attached, such as an immediate printout, high-sensitivity detection, more than one sensor at different locations for feeding information, and audible and visual alarms. These systems can be very useful as emergency leak detectors or as area monitoring stations to develop data concerning concentration variation over time to assess exposures, evaluate controls, or identify problem areas in the workplace.[1]

If the same box was made small enough to carry around so that surveys could be done, for example, for spot leak detection, it might be necessary to sacrifice a few of the options. Certainly only one site at a time could be monitored; however, the box could be used in many different places. If the instrument was compatible, a data recorder could be attached to it to store its output or alternately integrated into it. If a long-term battery was put on it, such as a nickel cadmium battery inside of it, it could be used indoors or outdoors; however, the basic operating principle, be it chemical, IR, combustion, or chemical sensitive tape, will not change. If the box is hooked up to a strip chart recorder, it is not as portable and may require a cart, but it could be used to measure carbon monoxide emissions in a warehouse or carbon dioxide increases in an office building at several different points.

If the black box is even smaller—small enough to hang on the belt or put into the shirt pocket of a worker—more bells and whistles will be sacrificed. Now there may or may not be a readout; instead there may be just an audible alarm to indicate that a certain level has been reached, and data storage may be minimal. Diffusion rather than a pump will probably be used to get the sample to the sensor. However, the minimal demands of the electronics will allow small lightweight batteries to be used, and now the unit can go into confined spaces and workers may be monitored continuously in difficult places, because of the unit's portability and long-lasting battery.

There are three different levels of real time monitoring (Fig. 7-1): (1) stationary fixed sensing systems that can involve several sensors and cover a number of different areas of a plant; (2) small personal monitors, many of which store data for later output to a computer; and (3) portable survey instruments that are used for taking a variety of different types of samples.

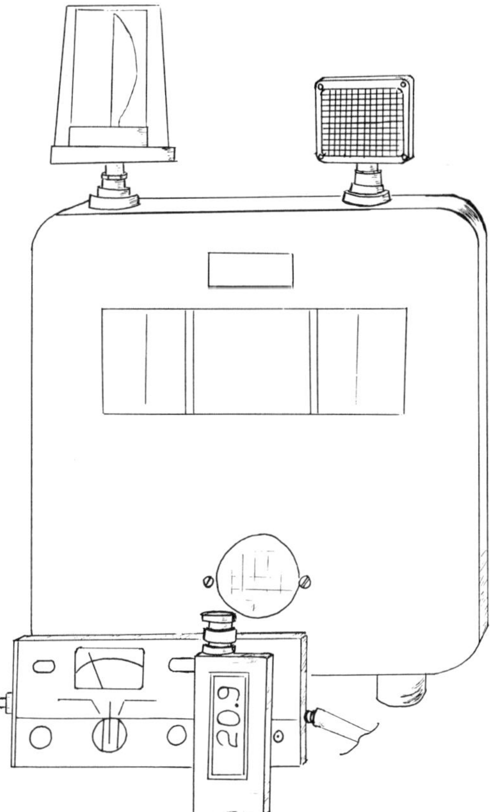

Figure 7-1. Types of real time instruments: stationary, survey, and personal.

STATIONARY MONITORING SYSTEMS

Although the focus of this book is personal and portable monitoring survey equipment, it is useful for the sampling professional to have an understanding of the purpose of stationary monitoring systems. Often due to the system's perceived complexity, sampling professionals will be overwhelmed at the prospect of selecting, evaluating, or using data from a stationary monitoring system, especially one that is multipoint. The biggest problem with stationary monitoring systems is a lack of regular scheduled maintenance and calibration on the part of many facilities. Once installed, the units tend to be forgotten and taken for granted.

Stationary monitoring systems in actual use vary in complexity from simple single-point systems, used for warning only, to complex multipoint systems with data acquisition, data averaging, and exposure estimation. The complexity of the system depends on the use to which the data will be put. Data recording allows the data to be used to identify problem areas and assist in the implementation of controls. If the system is to be used for warning only, then the outputs from the system can be very simple and no data storage is necessary. In choosing a system, it is important to know how the data will be used so that the proper system components can be chosen.

Continuous multipoint monitoring of workplace air has demonstrated a potential for solving a wide variety of workplace exposure problems, and is valuable in both identifying problems and by keeping workers and management aware of pollutant concentrations in the workplace. These types of monitoring systems are an important part of workplace controls in terms of measuring exposure concentrations over time in many areas of a plant, thus developing a historical information base about concentrations and the sources contributing to them in these areas. Since personal monitoring and portable direct reading surveys are performed much less often, they often provide much less information than these types of systems.

Use

Stationary monitoring systems (Fig. 7-2) can be used for a variety of purposes:

- Real time alarm warnings to workers that high pollutant levels are present
- Direct feed as to the operation of control devices
- Definition of exposure problems
- Identification of troublesome areas in order to aid in the development of controls
- Historical record of hazardous material concentrations
- Monitoring of performance of controls and work practices

The most important purpose of any stationary monitoring system is to detect emergencies and alarm workers rather than to measure worker exposure. However, estimates of worker exposures can often be obtained from the data provided by the system. Other examples of uses for these systems include monitoring of combustible levels of gases and vapors in operations such as LPG loading racks and dust levels in clean rooms. Examples of industries where stationary monitoring units have been installed include plastics and resins, textiles, parking garages, steel production, ethylene oxide sterilizers, wastewater treatment in industrial facilities, processes using isocyanates, and semiconductor manufacturing.[2]

Design and Selection

There are three main components of a stationary monitoring system: the sensor, the sampling system, and the data acquisition and display box. The main configurations for multipoint monitoring systems are: a single instrument with multiple sensors, a single analyzer and sensor that can sequentially analyze from different points via the sampling line, and multiple instruments. In a single-instrument multisensor system, there are several sensors, each continuously monitoring a certain point where sample lines are used. A sample switching system is used to time-share the analyzer among the various sampling points. In the multiinstrument system an analyzer is used for each sampling point.[2]

The number of sensors required, their detection principle and specificity in the environment to be monitored, as well as the components of the sampling system must be considered when selecting a stationary monitoring system. The sample collection system must be able to take data on samples from sensors at each sampling point or pull contaminant through sampling lines to a centralized analyzer. In the latter case, the multipoint system must be able to draw in the samples reliably with minimum degradation, and to switch from one sample to the other without mixing samples. The most important aspect of collection, regardless of the type of system, is that it be done in a timely manner. If the process being monitored is a constant operation with little variation, readings can be taken with the gas monitor every half hour, and these readings and detector tube readings should be recorded if the sampler is taking compliance measurements. If the process is variable, then it may be necessary to take the readings as often as every 5 minutes or less.[2]

Some systems are designed to provide an alarm function only. It is important that the alarm be heard and seen in both the areas where the sam-

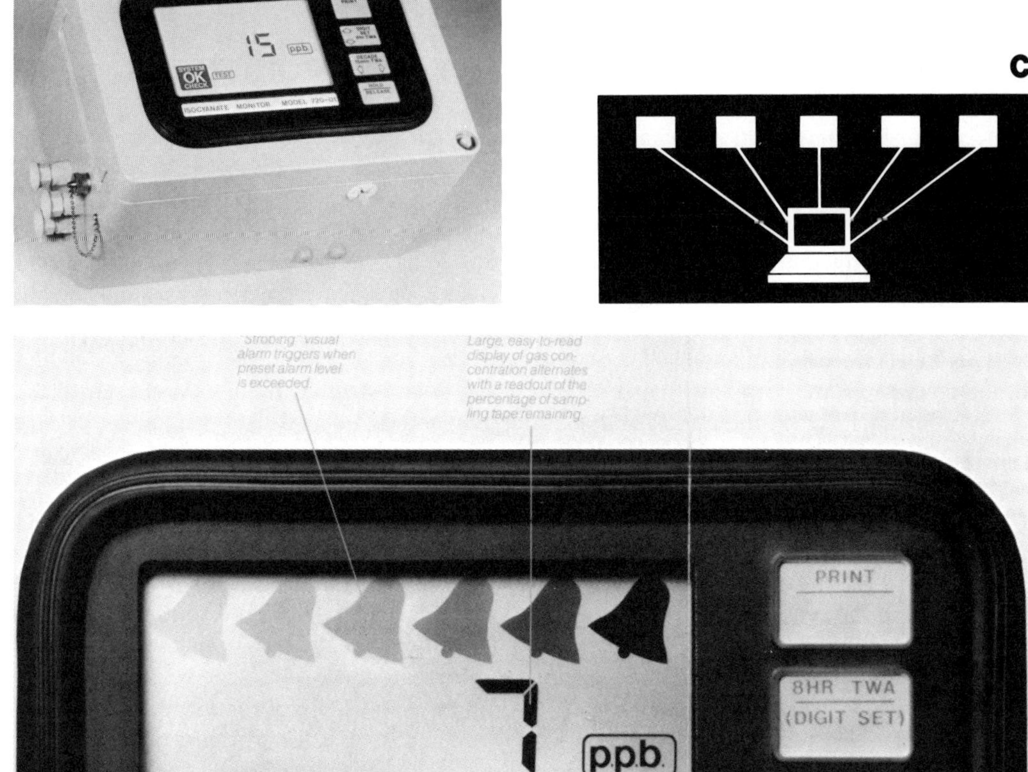

Figure 7-2. Stationary monitoring systems. *A*: Remote sensor for isocyanates. *B*: Closeup of sensor readout. *C*: Computer linkup of several instruments. (Courtesy GMD Systems, Inc.)

ples are collected and in some central location, such as the control room, where supervisory personnel can take appropriate actions. Data storage and analysis are also desirable. Computer control of the entire monitoring system is probably the best available approach as opposed to just recording information on strip chart recorders that will require significant analysis time later.[2]

Often stationary monitoring systems are perceived as complicated and mysterious because they look very complicated, and many purchasers leave the selection of a system or a unit up to a salesperson who is frequently not well versed in the idiosyncrasies of the contaminants generated by their operation. Once thousands of dollars have been spent purchasing and installing such a system, the tendency is to stick with it even though it may not be capable of doing the necessary monitoring and of providing adequate warnings. Factors to consider include[2]:

- Sensitivity, specificity, and accuracy along with being rugged enough for use in the workplace.
- Placement of the sampling probes.
- Selection of a detection method.
- For single-instrument systems, a reliable purging and sample switching mechanism. If the sample lines are not purged quickly enough, a long time delay between sampling and analysis will result.
- Compatibility of the material from which the sample lines and valves of the system are made with the substance to be measured. Adsorption, absorption, or condensation of the components of the sample onto the surfaces of tubing will cause loss of sample.

System expandability is a desirable feature. Some systems may allow the addition of more sample points at a lesser expense, and may allow several substances to be monitored without major system modifications. Other systems may be very specific, but cannot be expanded to monitor more than one substance without major modification. A system with a great deal of computer data acquisition and control may allow more flexible formats for data acquisition and also allow data to be analyzed in a number of different ways. Vendor support, supplies, and spare parts to keep the system running must also be considered.

PERSONAL MONITORING

Although personal real time monitors with datalogging capabilities have been available since 1980, only within the last two years has the market really started to expand to offer a reasonable variety of units. A compelling factor is the increased amount of work on hazardous waste sites and the increasing interest in the ability to monitor the air for specific chemicals at all times, due to the many unknowns regarding exposures on these types of sites.

While still somewhat limited in terms of the types of contaminants for which personal real time monitors are available, dedicated units are now available for carbon monoxide, chlorine, hydrogen, hydrogen cyanide, hydrogen sulfide, hydrazine, oxygen, phosgene, sulfur dioxide, and nitrogen dioxide, all using electrochemical sensors for detection as well as for oxygen deficiency monitoring. Because of the low power requirements and small size, the electrochemical sensor is often used for alarm and dosimeter systems designed to monitor for specific chemicals and for measuring oxygen levels. Some units claim as much as ±5% or ±3 ppm accuracy, whichever is greater (Ecolyzer 241 hydrogen sulfide personal alarm). One ppm is often the minimum detection level. The principle of detection for these systems is discussed in the chapter on Monitoring Instruments Dedicated to a Single Chemical.

Small personal monitors are also being used to detect combustible gases. These units can be worn in a confined space and monitor continuously, drastically improving the protection to the worker over the established method, which is to monitor briefly prior to entering the space. Because hydrogen sulfide is such a hazard in confined spaces, several continuous monitoring personal-sized units have become available for this contaminant. The need arises from the fact that hydrogen sulfide causes olfactory fatigue, that is, the odor-detecting cells in the nasal passages become tired after a period of exposure. As a result, the worker may think that the gas has

Figure 7-3. Personal real time monitoring. (Courtesy Interscan Corporation.)

gone away or may not be aware of an increase in concentration. Figure 7-3 shows a worker wearing a real time personal monitor.

Personal Dosimeters for Gases and Vapors

The trend is to incorporate the ability to store data into personal-sized monitors. As a result, real time dosimeters are capable of collecting and averaging a worker's dose, making the measurement comparable to the OSHA permissible exposure limit (PEL). Dosimeters frequently read out in ppm-hours; therefore, to obtain the reading in ppm the dose must be divided by the number of hours in the measurement period. The alternatives to using instrumental dosimetry are active or passive integrated sampling methods. These are discussed in the chapter on Integrated Sampling for Gases and Vapors. In general, all of these methods require laboratory analysis.

Monitors used as dosimeters may have electronic circuitry that provides a TWA, but not necessarily a continuous readout of the concentration. Personal units are available that read out directly, allowing them to provide instantaneous results as well as cumulative average. For some monitors with storage capabilities to download data, a separate unit is required to collect and print out the data that is specific for that manufacturer and application. Some units provide a display of the current TWA that can be updated periodically or at the user's command. An example of a personal data-logging instrument is the Dräger Model 190 Datalogger (Fig. 7-4), which has units available for carbon monoxide, hydrogen sulfide, nitrogen dioxide, and sulfur dioxide. A pocket-sized device with an electrochemical sensor, it can record 1-minute average concentrations for up to 12 hours, and then through an RS 232 interface it can download both data and a graphical analysis to a printer. Graphics software is available that allows these data to be stored and manipulated in a computer. While monitoring, it will provide a digital ambient concentration display, a total average display, and a minute peak exposure.

This area of monitoring is likely to continue to grow. The data-logging capabilities of many of these units are opening up new monitoring possibilities, such as monitoring inside vehicles and in other situations where the ability to make a continuous record of measurements was once limited by the need for alternating current (AC) voltage for a strip chart recorder or by the bulk of the equipment needed.

Personal Alarm Monitors for Gases and Vapors

The term *go/no go* is often applied to units that have only alarms and no concentration display. The alarm circuit is designed to activate whenever the preselected value is reached or exceeded for a predetermined time. Some units can be set for two alarm points. Alarms are often preset at the factory and cannot be changed by the user. Some units are more flexible and can be adjusted.

Alarms can be audible or visual. For example, the MDA MSTox8600 makes a steady beep, accompanied by a flashing red light-emitting diode (LED) for a low-level alarm and a series of three quick beeps with a synchronized red LED for a high alarm. Most audible alarms are 80 dbA or louder at a distance of 12 inches. Some units have a jack for earphones for use in high-noise situations.

Calibration of these monitors with the gas or vapor being measured is necessary. Also impor-

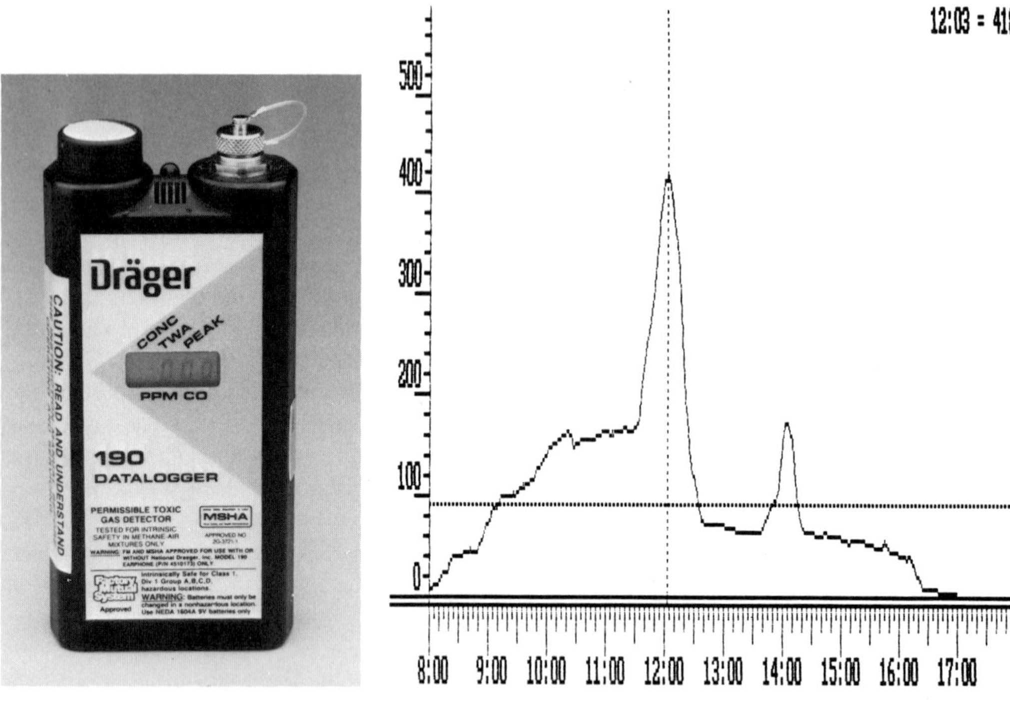

Figure 7-4. Personal real time monitor and data logger (photo) and printout of data. (Courtesy National Draeger, Inc.)

tant is the determination of field conditions, such as temperature, humidity, and air velocity, and potential interferences encountered during sampling as well as interpretation of the effect of these variables on the measurements. Therefore the sampling professional should not simply place the devices and leave for the duration of the sampling period.

Advantages, Disadvantages, and Other Important Considerations

The use of personal real time monitors offers a number of advantages if done correctly. Below are advantages of these devices.[3,4]

- They are portable, can be conveniently worn by workers, and can be used in confined areas.
- They can collect data under a wide variety of changing environmental conditions at high and low levels with typically better than ±15% accuracy.
- In most cases, the measurement is specific for the chemical being measured.
- They can collect rapidly fluctuating contaminant concentrations with precision and accuracy.
- In operation they are independent of vibration and orientation, and thus follow the workers about their normal duties.
- They offer the ability to obtain a detailed contaminant exposure profile through printouts of the monitoring data.

Disadvantages include the limited number of chemicals for which personal monitors are available and increased maintenance.

When using personal real time monitoring instruments, it is essential that not only the sampling professional understand the use of the instrument but the worker as well. The worker should be informed of the concentration at which the alarm goes off as well as what actions to take as a result. The purpose for the monitor as well as the toxic symptoms of the gas or vapor being monitored should also be reviewed.

The location of the readout is important. If it is on the front or bottom, a worker may find it difficult to view while wearing the monitor. Some units are RF (radio frequency)-shielded; others must have shielding added. Some monitors have extension cords that allow the sensor to be placed near the breathing zone while the electronics remain in the shirt pocket. Since these units are often placed near the collar, the user must take care not to block the sensor inlet. A disadvantage of the miniaturization of sensors is the increase in drift and lower accuracy that can result from lower signal–background ratios when compared to conventional survey-sized units, which is due to the fact that sensor output is a direct function of the sensing electrode area.

In some cases, several single-point stationary full-time area monitors are more useful than personal alarm units. For example, if there are many workers in an area, the area monitors would warn many individuals rather than one at a time. Also, there would be fewer monitors to maintain as it is not practical to fit large numbers of workers with these units. On the other hand, in the case of a confined space entry, the personal units are the best choice.

INSTRUMENT CHARACTERISTICS

Direct reading equipment can be specific for chemicals or very nonspecific, and the selection of an instrument will depend on the sampler's needs. In some cases, such as sampling on hazardous waste sites with unknown contaminants and concentrations, a nonspecific instrument is needed. In other cases, such as using an instrument to measure employee exposure to a known contaminant, a very specific instrument is needed.

Regardless of the type or brand, each instrument has certain characteristics which make that device unique. Understanding these characteristics allows the prospective user to compare instruments and to select the one most appropriate for the intended use. If the user already owns a device, understanding its operating characteristics can be very helpful in determining the types of applications for which the device is best suited. For example, some instruments have incorporated speakers that feature clicking noises similar to a Geiger counter. As the concentration increases, so does the rapidity of the clicking, which may be useful in situations where the meter is difficult to read—not unusual for liquid crystal screens in dimly lit areas. Most speakers, actually most optional features including alarms, can be deactivated.

The design of an instrument can make it easier to use in the field. For example, a probe with the readout integrated into it allows the sampler to make measurements and observe the readout without being distracted from where the probe is being pointed, which is especially important when conducting leak testing. The advantage of probes attached via tubing to an instrument is that they can be raised to collect a sample or extended to get closer to a source while the readout is still at eye level.

The minimum detectable limit must also be taken into consideration. If the instrument is to be used to provide a warning for occupational health purposes, then its minimum detectable limit must be below OSHA's PEL. If the instrument cannot respond to the highest observed concentrations, then a total exposure estimation will not be possible and the instrument can only be used for warning purposes.[5]

When a compound can be measured by greater than one real time method, these methods will not necessarily give the same information. The application, expected concentration, whether the measurement is for compliance, if any interferences are expected, and if the measurement is qualitative rather than quantitative can impact the selection of an instrument. The basic elements found in direct reading instruments are similar. These include a sensor or detector, electronics, readout, function switches, sampling probe, pump, power source, and sometimes an alarm. There are various types of sensors and detectors. Instrument characteristics follow.

Ability to store or transfer data: The ability to store or transfer data is extremely important since microprocessors are being incorporated into more and more instruments. It also eliminates the need for such bulky items as a strip chart recorder.

Alarm: It can be a desirable feature depending on how reliable it is and how easy it is to change the setting. The alarm is usually audible, but it can also be a visible signal that is activated at a predetermined level of contaminant or reading of the meter. Some devices have preset alarms, which can be useful too. Generally, audible alarms are more useful than visual ones.

Ease of use: Features such as portability, ergonomic design, nonbulky shape, location of the readout, and ability to carry while leaving the hands free are important.

Electronics: This section consists of the amplifier and the associated electronic circuitry necessary for the instrument to function. It receives the signal from the sensor, amplifies it, and displays it visually. This signal can be stored for subsequent readout using data loggers.

Filters: Filters are often incorporated into an instrument to remove particulate and interfering gases before the sample reaches the sensor. Filters can affect the response time of an instrument. Sometimes the choice of a filter requires some trade-offs between specificity and response time.

Fuel source: Some instruments also require a source of fuel, for example, those with flame ionization detectors use hydrogen gas.

Function switches: These are switches with associated dials used to adjust the electronics to properly operate the sensor and to process the signal. Common switches are span adjust, scale select, and zero adjust.

Lag time: It is the period of time between when the meter is exposed to the contaminated atmosphere and a step change is seen in the meter response.

Latching: It refers to alarm instruments. If a high concentration triggers the alarm, it will not turn off until the concentration goes below the alarm level. The instrument must often be reset.

Linearity: It is the deviation between actual meter readings and the readings predicted by a specified straight line. Some instruments are not linear over their entire scale, meaning that actual concentrations are not represented by the readout on certain portions of the meter scale. If a meter has multiple scales, it should be linear of each scale or at least calibrated on each scale. Instruments are often linear over only the portion of the concentration range they are designed to monitor. The actual range of linearity will vary from chemical to chemical.

Locking out: Locking out is the same as latching.

Peak hold: A peak hold function is useful for situations when an instrument is lowered into a confined space or when it is difficult to see the readout. In this case, the maximum value is memorized and can be read when returning the unit to better light.

Power source: It can be batteries or alternating current. The source provides the electric power for the electronics, pump, sensor, detector, and other accessories. The power going to the instrument may have to be conditioned if the instrument manufacturer has not made provisions for line voltage fluctuations of the magnitude of those that can be encountered in the workplace.

Probe: Many instruments have probes that are extensions to enable sampling of air at a distance away from the instrument. The distance varies from a few feet to several depending on the instrument.

Pump: A pump allows the instrument to pull air into it rather than having to passively wait for a sample, which is particularly important for those instruments that have sampling lines. Devices that utilize the diffusion principle for sample collection do not require a pump.

Range of concentration: Some instruments have several ranges, which is an advantage when sampling for a variety of concentrations, but also requires the user to be aware of what scale is on. Otherwise, erroneous results will be recorded and dangerous exposures may occur. The monitoring range of an instrument will determine what situations it can be used to sample. For example an instrument set up for ambient air monitoring will be far too sensitive and have too low a range to use for industrial monitoring. The range on a scale can determine the accuracy of a reading. For example a scale of 0 to 500 marked at 5 ppm intervals will make individual readings more difficult than a scale of 0 to 10 or 0 to 50 with marks to indicate one-half or one-tenth of a ppm.

Range of detection: The lower limit of detection or sensitivity is the lowest concentration that an instrument is capable of accurately reading. The detection range is important when selecting an instrument, because if the lowest level the instrument is capable of measuring is greater than the concentration of interest, it will not be useful. An example is using a CGI to measure health hazard levels of hexane. Since the GCI measures in percent, it will not detect low ppm levels (1% = 10,000 ppm).

Readout: The readout is the display of the chemical's concentration in units appropriate to the measurement. For example, the combustible gas indicator (CGI) displays percent of lower explosive limit (LEL). Most gas and vapor instruments measure in ppm, and dust monitors display fibers/cc *or* mg/M^3.

Recovery time: It is used to describe the time it takes the needle to return to zero after the instrument has been removed from the contaminated atmosphere. Recovery time is normally longer than response time, and depends on the concentration the instrument has been exposed to, especially if saturation has occurred.

Response factor: General survey instruments can usually only be calibrated with one chemical at a time; therefore, the response of the instrument to any other chemical that is detected will be relative to the calibration chemical.

Response time: Real time instruments making continuous measurements do not perfectly identify every concentration fluctuation since every instrument requires a certain amount of time in order to collect enough sample to respond. This period is known as the response time. The result is a certain amount of averaging, even in real time instruments; however, since response times are on the average of seconds, averaging is minimal compared to the length of time required for integrated measurements. Response time is defined as the time it takes for the instrument to give a 90% (of the actual concentration) reading after the sample collection has begun. Some samplers break down this time into lag time, the time interval between introduction of the contaminant to the instrument and the first observable response, and rise time, the time period between the initial response and 90% of the final instrument reading.

Rise time: It is the period after a lag time until full response is reached.

Sample outlet: This feature allows samples to be collected from an exhaust port in the instrument for further analysis. This is only possible with nondestructive detection methods.

Saturation: It occurs if the instrument is exposed to a higher concentration than it is designed to measure and purge under normal circumstances. The result can be sustained deflection of the needle or readout. Usually the instrument must be removed from the contaminated atmosphere and purged with clean air in order to get it to return to zero, unless this high exposure has contaminated the inside of the instrument.

Scrolling mode: It has become common for the displays on instruments with multiple sensors to "scroll" or sequence the display every few seconds to display the measurement of a different sensor.

Sensor: The sensor is the component that responds to a chemical or physical property of the contaminant. The terms *detector* and *sensor* are used interchangeably. In a sensor, an electrical signal proportional to the chemical concentration is generated and transmitted to the electronics section, which in turn displays it to the observer. Sensors can be either destructive or nondestructive, meaning the sample may or may not be "consumed" or destroyed over the course of the measurement. With nondestructive instruments, often the sample can be collected following measurement for additional laboratory analyses.

Sensor life: It is highly variable and differs from instrument to instrument. It is dependent on the type, frequency of instrument use, presence of contaminants (to the sensor), concentrations of chemicals measured, and storage conditions.

Stability: An instrument is considered to be stable if it maintains its electronic balance over the time required for sampling. *Zero drift* or *span drift* are terms also used to refer to stability. Drift refers to a gradual change in the needle setting that is unrelated to the contaminant being measured, the result being that the reading becomes inaccurate. An unstable instrument may require frequent zeroing and calibration, which is an important consideration for extended periods of monitoring.

Time constant: The shorter the time constant of a direct reading instrument the more readings it makes in any given period giving a closer approximation of real time concentrations. An example is for an instrument to continuously indicate a reading for a short time after it has been removed from the area where the contaminant(s) are present. This is a function of the principle of detection of the instrument as well as its electronic circuits. Some instruments have the capability of varying the time constant for their measurements.

ACTIVE AND PASSIVE COLLECTION

Active and passive collection are the two basic methods for introducing a test atmosphere to a sensor. The active method uses a battery- or line-powered vacuum pump that draws the air through the external sampling system to the instrument sensor. Diffusion head sensors are characterized by relatively low cost, ease of installation, and simple maintenance. Sensor heads can be removed from the electronics housing and replaced by the user in most cases. In general, these sensors are the best choice for sampling in remote locations because they can be linked electronically to a remote monitoring station. Passive sampling depends on air currents and convection to push the gas, vapor, or aerosol through the detection region.

The advantages of actively drawing the air to the sensor include[6]:

- Hard-to-reach sample locations are more easily sampled. For example, a probe can be inserted into a duct or other restricted space stack, and the air can be drawn to the sensor.
- The temperature and humidity of the test atmosphere can be controlled more readily by drawing the air over a water reservoir or through a water bath. This feature is important when the air being sampled is outside the operating temperature or humidity range of the sensor.

Instruments using the passive sampling method rely on the principle of gaseous diffusion to provide contact of the test atmosphere with the sensor. The advantages of the diffusion method are[6]:

- Instruments are lighter and more compact.
- The vacuum pump, its maintenance, and the increased power needs of associated airflow system are eliminated.
- It is possible to sample remotely in the case of confined space entry. This sampling is accomplished by placing the sensor in the test atmosphere and connecting it to the remotely located readout. When the proper electronics and cable are used, a rapid readout of the contaminant levels is possible without the delay of pumping the gases to the sensor, thus reducing the response time.

Temperature changes can result in exaggerated error in diffusion instruments. A lower signal output can occur with this mode, because in the pump-drawn mode, where a directed stream of gas flows across the electrode, the sensor output is much higher, since the gas molecules are replaced quicker than they can be consumed. For a diffusion mode sensor, the number of gas molecules available per unit time is much smaller. The lower current occurs because the base is consumed faster than it can be replenished. The depletion effect gives an even lower signal—background ratio.[6]

HAZARDOUS ATMOSPHERES

Never take a direct reading instrument or any other instrument into a potentially explosive atmosphere without checking with the manufacturer to ensure that a specific instrument meets the appropriate instrinsic safety requirements of Underwriters Laboratory (UL), Factory Mutual (FM), or the Mine Safety and Health Administration (MSHA). UL is an independent testing and approval agency that examines electrically operated equipment and accessories, primarily from the standpoints of safety and freedom from hazard. Approved equipment carries the UL label. FM is a combination of insurance companies who have formed a testing and approval agency for fire protection and other industrial safety equipment. Products submitted to them are tested for safety in performance, and if acceptable, receive the FM stamp. Hazardous atmospheres may contain volatile flammable liquids and gases, combustible dusts and easily ignitable fibers, and other materials as defined by various classes within the National Electrical Code. Some instrument manufacturers, rather than having a UL or FM rating, have chosen to get approval by the Canadian standard for hazardous atmospheres. The term *intrinsically safe* is often applied to instruments that have been certified for use in hazardous atmospheres.

CALIBRATION AND MAINTENANCE OF REAL TIME INSTRUMENTS

The primary concerns for quality control when using direct reading instruments are understanding and minimizing sources of error through proper calibration, maintenance, and use of the instrument. Initial checks and adjustments consist of battery check and zero and calibration procedures, all of which should be performed at least once a day. Of these, the most critical first step is proper calibration. Two different types of calibrations are often required for real time instruments: flow rate and chemical concentration. There are different levels of chemical calibration, ranging from factory calibration with several concentration ranges of different gases, to a specified concentration of a single gas done by the user, to a field check with an unknown gas concentration.

Units are sometimes calibrated electronically rather than with a gas or vapor. This is often the case with very small instruments, such as those used for personal monitoring and for instruments for which the calibration gas is difficult to generate by the typical user, such as mercury. The electronic calibration for the MSTox8600, a personal electrochemical-based unit, is done only on new sensors, and a monthly gas calibration is recommended by the manufacturer. For the electronic calibration, generally a voltmeter is used or a button is pushed to generate a readout adjusted to match a number on the cell via turning a screw. A variation of an electrical calibration is used in tape-based detection instruments that often utilize cards with colored spots to serve as calibrations. This is because the instrument interprets the degree of discoloration as concentration.

Most instruments using vacuum pumps are designed to sample at a specified flow rate. Know that rate and what rate would represent a significant shift higher or lower than that flow. The instrument's manual may indicate what effect these changes could have on the instrument's response. Flow rates of portable instruments should be checked every 40 to 50 hours of operation with at least a secondary standard (rotameter), which in turn is compared frequently with a primary standard, such as a bubble buret. For a detailed description of techniques for using the bubble buret or rotameter, see the Appendix on Pumps and Flow Rate Calibrations. For instruments with pumps, a leak test will confirm that the sample is being drawn into the sensor.

The preferable way to calibrate concentrations is with the specific contaminant that is going to be present during the survey. The precision of a direct reading measurement is a function of the concentration of the contaminant present and the range of its linear measurements. The concentration(s) of calibration gas should be similar to the level(s) of contaminant expected to be found in the field. Calibration before and after each use should be performed for personal monitoring instruments designed to measure specific chemicals using a gas standard that provides a concentration at or near the OSHA PEL of the compounds to be measured. If calibrating a general survey meter, then a gas or vapor should be selected that best fits the user's need. This selection depends on the application. It is often recommended that the calibration of most survey instruments be carried out using two or more concentrations of the reference gas in order to verify the response over the full range of concentrations anticipated in use.

If a single concentration is used, calibration is

commonly performed using a span gas having a concentration approximately equal to the midscale value of the meter. Never connect a gas cylinder directly to the intake port of most instruments, as doing so will pressurize the instrument and cause serious damage to the sensor. Instead, fill a sample bag with the calibrating gas and attach it to the intake port of the unit. For a detailed discussion of the various techniques involved in generating gas and vapor concentrations for calibrating real time instruments, see the Appendix on Gas and Vapor Calibrations. Also see the discussions included with each type of detection method for more specific information on calibrating a particular type of instrument.

Although calibration with known concentrations of the chemicals of interest is the best method for calibrating real time instruments, preparations of these materials are not always available or they are problematic. For example, formaldehyde is not stable as a gas preparation and instead the gas must be derived from crystals of paraformaldehyde. A similar problem exists with reactive gases such as isocyanates. In some situations, the best calibration gas is toxic, so another gas that causes a similar response is used. For example, most PIDs are calibrated with isobutylene and adjusted to be equivalent in response to benzene, a carcinogen.

Aerosols of appropriate concentrations for calibrating instruments are even more difficult to generate than gas and vapor atmospheres; therefore, unless the sampler has extremely good skills, knowledge, and equipment, calibrations of this type are best achieved by sending the instrument to the factory. Aerosol monitoring instruments based on light scattering must be calibrated against a dust sample of the specific dust compound that has been collected using integrated methods. Changes in aerosol size distribution or optical properties of the dust being measured will necessitate a new calibration.[7] The user can change the calibration constant of an aerosol monitoring device for a specific type of aerosol by performing integrated sampling on a filter concurrently with taking measurements with the instrument. The ratio of the two concentration values can then be used to adjust the instrument's response.

A zero calibration gas may or may not be required. Zero gas is a high-purity air. Most manufacturers do not require calibration with zero gas to make the zero setting adjustment. Instead an adjustment is used to electronically balance the circuitry to zero an instrument. Some instruments have an electronic zero as well as a zero potentiometer that can be adjusted with a screwdriver. The use of an electronic zero provides the capability to accurately set an instrument to zero in the presence of atmospheres containing background contaminant levels.

Another technique used in the field is to check the response of the meter with a known substance without necessarily knowing its exact concentration. It is known as a field check, and although not useful for calibration, it will show if there are any serious malfunctions in an instrument.

Instrument manufacturers often offer their own calibration kits. Many of these kits are specially set up for a given instrument, and using different equipment may cause problems. For example, a regulator on a gas cylinder may be set up to deliver a very low pressure, which is important if the cylinder is attached directly to an instrument. Use of the wrong cylinder—regulator combination could result in pressurizing an instrument. Always heed the manufacturer's instructions when selecting calibration methods for any given instrument. The user should determine prior to purchasing from another vendor if the combination will be incompatible or result in less accuracy. Generally kits contain a gas cylinder with a regulator, tubing, and a bag.

Prior to calibrating any instrument it is a good practice to review the manufacturer's manual. Most manuals give step-by-step instructions for use with their specific instrument. In addition, any idiosyncrasies or problems often encountered during calibration are often noted. While the instrument's manual will provide guidance in calibration frequency, the user should make a determination based on experience. The following guidelines are representative of instruments in general.[8]

- Calibrate before and after each use to provide assurance of measurement accuracy.
- Always calibrate an instrument when it is first purchased regardless of what was done in the factory.
- Calibrate whenever the instrument has been

"bounced around," such as can occur during shipping, heavy usage, climbing ladders, and so forth and after an instrument has been exposed to potential sensor poison and severe weather conditions.

- Calibrate following repairs or replacement of any of the instrument's parts.
- The frequency of calibration should be increased when a sensor nears the end of its service life or when abnormally high variations are observed. The frequency of tests should also be increased when a new instrument or sensor is installed. Sensors often exhibit higher shift during the first month of operation.
- If an instrument cannot be calibrated, the span potentiometer may be at such a low setting that the instrument cannot respond properly. Turn the span potentiometer several turns in the direction of increase and recalibrate.

Generally most instruments are a complete system and parts should not be interchanged with other instruments or substituted for those provided by the manufacturer. Very few malfunctions can be easily corrected in the field; thus, users should check out an instrument prior to a survey. A preventive maintenance program will also prevent or minimize problems in the field. The troubleshooting section may be helpful; if not, call the manufacturer. When filters on inlets are discolored or heavily clogged with dust they should be changed. The exact schedule will depend on the instrument.

Consult the manufacturer's instructions for when to use interference filter, organic filters, and humidity adapters. For continuous monitoring applications in excess of 3 hours, electrochemical and solid-state devices often require an external source of humidity. This source usually consists of a bottle with distilled water installed in-line with the intake port. If a filter is incorporated into the inlet of the instrument, it should be removed because it will trap the water vapor intended for the sensor. The air filter should be reconnected to the front of the humidity control bottle.

Most chemical sensors depend on a specific chemical reaction to generate a signal. This reaction can occur on a solid or liquid surface, light, or flame. The reactions involved in these sensing mechanisms may or may not be reversible, and exposure to other contaminants that can degrade the sensor surface also occur, with the result that the sensor material is used up. The result can be sensor drift or a high readout even in a clean atmosphere. Therefore, sensors or certain parts of them will have to be replaced periodically. Sensor replacement is simple in most personal units; the sensor just plugs into the top of the unit. Sensors generally must be replaced at least once a year. The sensor should be removed from the monitor whenever it will be out of service any length of time. To preserve the sensor, some units, such as the MSTox8600, must have their shorting socket replaced. For proper readings batteries must have a sufficient level of charge. Usually it is not a good idea to operate an instrument while it is being charged.

READING DATA

Most instruments generate a voltage proportional to the concentration of compounds detected by the sensor. Initially almost all instruments used an analog scale where readings were made by viewing the movement of a needle on a scale reading ppm or compared to a calibration graph to get the concentration. Analog output is defined as a continuously variable voltage or current that is proportional to the instrument's measurement. It is suitable for use with a recorder and alarm system, but analog to digital conversion would be required for use with a computer system. If a monitor has an internal microprocessor, the data can be transmitted directly by the monitor to another computer or printer.[9] This is time consuming and requires caution in that errors in reading, interpreting, and copying the data can occur. The incorporation of digital readouts and storage capabilities into instruments has streamlined this. A digital readout is the result of an analog to digital converter and requires less interpretation.

Most readouts are linear, that is, there is an equivalent increment in the deflection of the needle for equal increments in concentration. For example, the needle will move the same distance on a 0 to 200 scale if the concentration of a compound increases from 10 ppm to 40 ppm as it will during an increase of 130 ppm to 160 ppm. Linear scales usually have two to four concentration ranges, because a scale that would provide sufficient space to adequately distinguish between measurements by 1 ppm would require a large space on the front of an instrument. Also, usually at the lower ranges it is more important to differentiate between fractions of a ppm whereas at higher ranges it is not. On a few monitors the readout is presented as a range rather than as a specific value. Some instruments have a logarithmic scale that, like log paper, compresses the range at higher concentrations, the result being that the sampler does not need to switch between ranges. These are most common in instruments designed to measure leaks where high concentrations are expected.

A comparison of logarithmic readouts to linear readouts follows[10]:

1. A single meter setting is required for a logarithmic readout whereas different scales (ranges) are required for linear readouts.
2. A logarithmic meter may be more compatible with measurement of air contaminants in typical workroom situations, because their concentrations are considered to fit a lognormal statistical distribution.[3]
3. The zero can be set on the linear scale at the background level, allowing the instrument to automatically subtract this value during measurements.
4. If a device produces a logarithmic output current, logarithmic scales do not suffer from a lack of linearity at low (approximately 10%) scale readings because there is no conversion necessary, whereas linear scales have this problem. If the readout on an instrument is indirect, that is, in units such as percent absorbance or percent transmission, it must be converted to concentration and a calibration curve must be consulted in order to determine the actual concentration.

Often when making measurements needles or digital values will jump around because of fluctuations in airborne concentrations and air currents. In these situations, deciding where to make a reading can be difficult. Even in the case of digital instruments, it is not uncommon for the readings to change rapidly unless the instrument is set to average over a certain period. Some instruments allow the user to control the frequency of changes in the readout, for example, displaying a 10-second average instead of a 1-minute average measurement, but most have their own individual response times.

In some instruments more than one averaging period is available and in this situation a slower period is desirable. A conservative rule of thumb would be to select the maximum value over a given time period. Another option would be to record the range for the values, for example, 56 to 72, or 64 to 74, and so on. The sampler could use the midpoint each time a needle moves. In general, the upper and lower 10% of any instrument's range should be avoided, and if possible the sampling professional should operate it only in the middle 50% of the overall range.

Parallex effects that cause readings on scales to appear higher or lower unless the user is looking directly down on the needle can occur with certain readout designs. The use of a mirror in the background of the meter can eliminate this problem, so that the user is assured of looking straight down at the needle by lining up the needle with its mirror image.

In strong sunlight a digital display on a liquid crystal panel can turn black, making it unreadable. Although after a period of time the panel returns to normal, it can be a nuisance when trying to make measurements. Therefore, precautions should be taken when doing measurements outdoors, for example, on roofs and balconies or in areas brightly illuminated by skylights or windowed walls.

RECORDING EQUIPMENT

In the past, the common method for recording data has been to manually write down readings on a form and manually perform any needed statistics or graphs. Data recording can be incorporated into an instrument or an external device can be used that either serves as a memory (data logger) or immediately makes a tracing (strip chart recorder) or printout (printer) of the instrument's response. In either case, the result is the same: Measurements made by the instrument are preserved for analysis later rather than having just a few random measurements recorded as the sampling professional watches the readout. Microprocessors are being increasingly utilized in instruments to provide them with memory, but also many other capabilities as well. Most instruments can be equipped with continuous recording devices, such as strip chart recorders that show a visible curve plotted by a pen on a moving piece of paper or data loggers that can discharge data into a computer or print them out later, giving concentrations at various times.

Strip Chart Recorders

Strip chart recorders are the most traditional data recording devices and plot measurements on chart paper after a voltage scale has been selected that corresponds to the expected output of an instrument. These units range from very basic, just providing the ability to record an output on a single channel, to having several channels, to having programmable functions such as alarms and periodic printouts. The most common method for displaying the output is to calibrate the strip chart recorder and compare the peaks measured to the calibration peak to determine the concentrations that occurred during the measurement period. Often in a situation where concentrations build up gradually over time, rather than in a series of distinct peaks, a very broad curve with only a few peaks is the result. This situation is typical of carbon monoxide measurements in parking garages. The peaks represent the heavy auto movement periods, morning and evening, and in the event of an inadequate ventilation system the curve gradually rises. The speed of the chart is selected so the time of day can be identified on the chart following the measurements.

In other cases, such as with a GC, a few distinct peaks are the result of the measurement. In this case, the entire area represents the concentration of interest rather than using the maximum needle deflection to estimate concentrations at any given time. In some cases, distinct peaks also result during the measurement period, usually a day or longer.

When a manual integration of peaks is determined to be necessary, the area of each block may be estimated and the readings may be averaged to determine compliance with an 8-hour TWA or ceiling standard. However, the peaks are often not nice and neat; instead, the user can get lots of narrow peaks of varying heights or broad peaks that are difficult to separate from baseline drift, making manual integration tedious and difficult.

Many recorders can be used in vertical or horizontal operating positions with the roll chart installed. Here are some factors to consider in the selection of strip chart recorders.

Chart paper: Most paper is specific for the instrument and other charts may jam and damage the recorder. Some strip chart recorders use pressure-sensitive paper and it can be purchased from only a few sources. Paper usually comes in roles or folded.

Chart speeds: Most instruments have more than one chart speed; however, some have more than others. Very slow speeds are desirable for long monitoring periods.

Chart speed accuracy: All AC-powered recorders use synchronous motors to drive the chart, so a properly functioning recorder will not have timing errors. Recorders with unregulated motors can be expected to gain or lose several hours a day in the worst case. Recorders with inverter motors will keep time to within 7 minutes a day.

Inputs: Some recorders can support a wide range of inputs: DC (direct current) voltage, different types of thermocouples, resistance temperature detection, and converters for pressure, differential pressure, flow rate, dew point, humidity, and pH.

Measurement accuracy: It ranges from ±0.5% to ±2% for different recorders and usually does not include errors due to chart paper expansion or shrinkage and errors due to source resistance.

Measurement range: Recorders are available that will accept voltage inputs from 10 mV to 500 V. However, field instruments generally do not produce voltages greater than 5 V full scale, which would be a high concentration in most cases.

Number of channels: Some strip chart recorders with more than one channel are capable of recording DC voltage measurement differences on one channel in relation to a reference value set in another channel. A different color ink is often used for each channel on multichannel recorders.

Operating conditions: Allowable operating temperatures are usually specified: 0°C to 50°C is common, although this may be lower for NiCad batteries. Humidity is also important as it affects chart paper. Most manufacturers test and set up their recorders under standard operating conditions, often 25°C and 55% RH. This is especially important when using recorders outside or in extreme environments such as hot processes.

Power requirements: One of three types of power sources—AC voltage, NiCad batteries, or +12 DC—is used to power strip chart recorders.

Printing capability: Models are now available that can combine the pen tracing with digital printout information, such as chart speed, time, date, and scale markings.

Programmability: With programmable strip chart recorders, settings are entered by a key pad. These devices require a substantial amount of manual study in order to set them up properly. Settings vary depending on the input being received as well as on how the user desires the recording to reflect this input. The most common setting used with sampling instruments is measurement of DC voltage with a linear scale recording. Alarm points can also be programed into some units. If the level is reached during sampling, the instrument will identify it with a printout after the peak. Tags are used to identify the object being measured and recorded. The tag number can be used to identify what channel measurements are being made and will appear during a period printout. Microprocessor-based strip chart recorders can often do some self-diagnostics. Usually they are done when the unit is initially powered up.

Scale: Linear scaling values are selected so that a given voltage input is proportional to a given concentration. The voltage range must be specified. Most sampling instrument output voltages are in the range of 10 V.

Writing system: Some have a striker bar rather than a pen; thus, if examined closely, the tracing actually consists of a series of dots.

All strip chart recorders require a certain warm-up period. It can often be longer than that specified by the manufacturer. Thirty minutes in the environment to be used is not atypical. If not warmed up, measured concentrations will appear lower than they are as the baseline stabilizes (lowers) after warm-up.

Strip chart recorders have their problems. Pens go dry and the paper runs out; therefore, they must be monitored. Many can use only special sizes and types (pressure-sensitive) of paper, making it important to keep a stock on hand because of waits when ordering more paper from the manufacturer. Unless information is collected while peaks are being recorded, it is hard to correlate what caused them to occur. Baseline drift is a problem, and periodic calibration during operation is required. A drawback of very small-size strip chart recorders is that some data, such as GC peaks, can be hard to read. An advantage of strip chart recorders is that the user can make notes on the paper next to the peaks of interest as to events that occur during the sampling period.

Calibration and Maintenance

If strip chart recorders are used for compliance measurements, they must be calibrated before and after each use. They should also be calibrated periodically regardless of use. Calibration is done using a known voltage or current standard that is equivalent to the full scale on the recorder. Adjustments are generally done with small screws. Span should be done while calibrating the instrument with a gas or voltage input representing 100% of scale. Then zero is adjusted with or without gas, or if necessary using a cylinder of zero gas. A nice feature is a digital readout of the recorder at the same time it is running, which allows the user to select an absolute baseline rather than an arbitrary one. All recorders without signal conditioners are calibrated by zeroing the galvanometer. The following methods of zeroing are useful:

Zero on scale: With the signal disconnected, rotate the zero adjustment to give a recording on the appropriate line of the paper. To check accuracy, apply a known voltage or current equivalent to full scale.

Zero off scale: Apply a known voltage or current equivalent to the low-end scale and adjust the mechanical zero adjustment to give a recording on the most left-hand line of the chart paper. Check the full scale by applying a known full-scale signal.

Maintenance needs include chart paper replacement, pen replacement, and battery replacement. Most chart paper has marks to indicate it is running out. The marks usually start when only 40 cm of paper are left. Do not use partial rolls of paper on units that can reroll the paper. Do not pull paper backward through the recorder because of the danger of snagging the printer.

Pens pop in and out. They are sometimes directional, and when installed backward they do not stay in or work properly. When the recorder is not going to be used for more than an hour, replace the pen caps to prevent the ink from drying out. Since the pen tips are often made of felt, do not crush by applying strong pressure. When felt pens are new, sometimes the ink does not flow from the pen tip. In this case, rub the pen lightly against the paper until the flow starts.

If chart speed is fast, the brake spring may be too weak or the clutch may be too tight. If the chart speed is slow, the brake spring may be too tight or the clutch may be too loose. A slow chart can also be caused by a very soft or spongy roll of chart paper that is sometimes indicated by tears through the top side of the sprocket holes. Other causes of slow chart speed are the gear train slipping out of engagement because of a weak gear spring or an intermittent or low torque chart motor.[11]

Do not store strip chart recorders where they will be exposed to direct sunlight or high temperatures, high humidity, dust, dirt, salt, corrosive gases, vibration, strong magnetic fields, or electrical noise. Pens can be stored for a year. Longer storage may dry out the ink. If a recorder is to be unused for a long period, remove the chart paper from the instrument; otherwise the paper may distort and not fold properly after recording. Chart paper expands and contracts as ambient humidity varies.

Before shipping or transporting strip chart recorders, remove the pen cartridges and pack them in a separate box. Most manufacturers recommend that their units be shipped in the original packing box.

Use

1. The unit should initially have power off, pen lifted up, chart drive off, input at zero, and the correct power source connected. Connect the instrument to be monitored to the recorder.
2. Depress the power button to turn on. The power light will come on. Allow the recorder to warm up for 15 to 45 minutes, depending on how long it takes to stabilize.
3. With the input switches at zero, adjust the position knob on each channel to zero the pens. Turn the right-most gear (chart manual, advance roller) backward to minimize the backlash; otherwise there is a time lag before the chart starts moving in low chart speed mode.
4. Set the voltage sensitivity with the range selector knob in accordance with the input voltage. The maximum allowable input voltages must be taken into consideration. Be careful when the input voltage may be high.
5. Set the chart feed speed with the chart speed push buttons. Press the chart drive start button to start the chart moving.
6. Set the pen lift levers to down to begin recording with all pens.

Data Loggers

A data logger is a modified form of a computer with limited functions that can collect, store, and receive information. Data loggers have made it more practical to use real time instruments to do personal sampling and continuous monitoring. Previously, stationary instruments with strip chart recorders were used for contaminants for which there was no good integrating method, for example, carbon monoxide. Some instruments have data-logging capabilities incorporated into them whereas in other cases a separate unit is attached to an instrument that has no data-logging capabilities (Fig. 7-5). Some can also do limited processing, such as averaging and singling out certain peaks. Data loggers log the voltage response input from an instrument's measurements. Since the data are stored digitally, analysis can be performed on computers at a later time.

In the case where data-logging capability has been incorporated into an instrument, the capabilities, storage, and other aspects of the system vary significantly. Some instruments can record only one event—usually either the maximum or average value for each measurement— whereas others can record the maximum, minimum, and average concentrations and log concentration data every few seconds. Some instruments must be turned off in order to erase data; if simply left on and reused, they will store up to the maximum number of data points and then cease storage whereas others will continue and write over the early measurements.

Several different data loggers are now available. Some instruments can provide just a tabular time/event reading printout and others can do graphs directly to a printer. Some can be hooked up to a computer so that information can be integrated into a data base program and stored; others will dump to a printer only and cannot store information once it is printed out. Benefits of utilizing a computer rather than just dumping to a printer are the ability to do quick and effective data manipulations and analysis, and having a graphic capabilities available as well as the capability to create data files for use with other commercially available software packages. Time and money spent on manual calculations and graphs can be saved. Results can be evaluated immediately after each day of testing, thus increasing the flexibility, capability, and accuracy of the field evaluation process.[12] Converting stored data to a printout or computer is discussed later in the chapter.

Uses

In order to use a data logger with an instrument, the instrument must have a connection that provides an analog output as well as a linear voltage

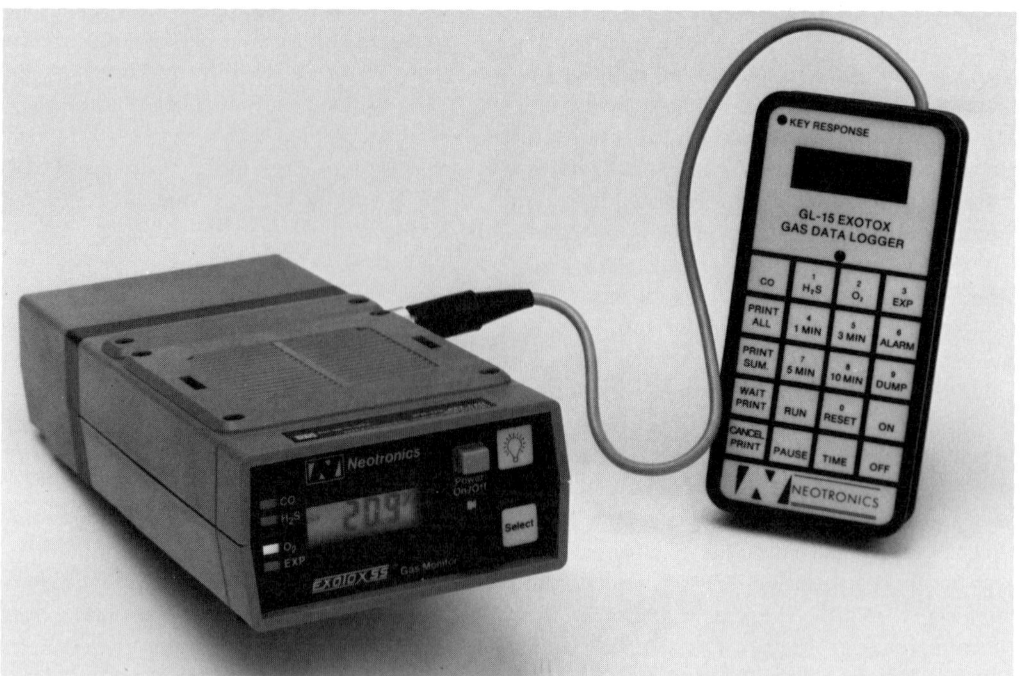

Figure 7-5. Survey instrument with attached data logger. (Courtesy of Neotronics of North America.)

output. In the case of instruments that perform more than one measurement at a time, there might have to be an output for each sensor. The output connection varies, for example, the Metrosonics dl-332 has a spade lug connection.

Data loggers can be used to record manual tests, scheduled runs, and alarm exceedance logging. Manual logging is when the user manually turns the unit on and off. A scheduled run is done by programming start/stop dates/times, and the unit then functions automatically. Alarm exceedance logging is a mode in which the data logger only records concentrations that exceed a preset value.

Data loggers can be used to update older instruments that do not have storage capability or when readouts must be translated. An example is the MIRAN 1A IR analyzer, until recently the best available IR analyzer for general field use; however, it reads out in absorbance, which must then be translated to ppm using a graph. This instrument can be updated with a data logger. The calibration curve for a particular compound is entered in the data logger, and during measurements with the IR the data logger will display the ppm concentration directly as well as store the sampling data.

Sometimes a data logger will provide a more accurate readout than the instrument to which it is connected, because the instrument may only sense increments of 5 ppm, for example, whereas most data loggers show increments of 1 ppm or less.

Generally a data logger must be set up for each instrument with which it will be used. The unit must be programed to read out in the correct measurement units as long as they do not exceed four characters. Examples are ppm, degrees Fahrenheit, degrees Centigrade, volts, and mg/M^3. For critical monitoring operations the unit can be used to trigger an alarm system. When using data loggers that are compatible with more than one instrument, the output parameters of the instrument in use must be known. Also, if the instrument is programed to alarm at certain points, these should correspond with the data logger's alarm points. If the output parameters are not known, the data logger can be calibrated. It is done by defining zero versus the span. For example, the user sets a certain millivolt (mV) level to equal 10 ppm; therefore, measurement data are defined to the data logger in terms of voltage.

When switching a single-channel data logger to a new instrument, the user will need to reprogram the data logger for the parameters, calibration, and other needs of this instrument. Multichannel data loggers dedicated to a single instrument will have only one plug that is able to accept data from all the sensors of this instrument. Universal data loggers with multiple channels will have separate input plugs for each channel. Therefore, dedicated multichannel data loggers generally cannot be used with any other instruments.

Most data loggers will store data once a battery pack is removed or once it has lost its charge for a given period; however, this period may range from a few hours to a few days. For safety, periodically change battery packs on data loggers that are regularly used and always keep a spare battery pack available.

Calibration

Data loggers must be calibrated like any other instrument. Usually calibration is done at two data points. Generally it can be done in two different ways. The first way is to define the points by selecting and programming certain points into the data logger. Usually this procedure involves setting a certain concentration value equal to a given voltage value. The second way is to calibrate the instrument with a known concentration while it is hooked up to the data logger.

Some data loggers offer the option of choosing either defined or sensed calibration modes. In the defined mode, the dl-332 is calibrated to the specified output of the interfaced sensor. It is accomplished by noting the specified sensitivity of the sensor, and then setting the dl-332 accordingly. In the sensed mode, the dl-332 is calibrated to the actual sensor output while it is operating and is connected to the logger. This procedure ensures that the dl-332 is properly calibrated to an individual sensor rather than to a general specification, which may involve greater tolerances. For each mode, two calibration points are established to allow for zero offset of suppression.

Data loggers are complex to use. The more universal their compatibility, the more setup procedures will be required. The individual not familiar with using computers may find setup difficult. Generally programming data loggers requires study of the user manual and practicing time. A single key often has multiple modes, which can be confusing if the user does not understand how to get both in and out of these modes.

Metrosonics dl-332

The Metrosonics dl-332 Industrial Hygiene Data Logger (Figs. 7-6, 7-7) is an example of a unit that is compatible with a variety of direct reading instruments. A different cable is required to hook up the data logger to each instrument. A second cable is necessary to hook up the data logger to a computer in order to store the data, or to a printer if just a printout is desired. Communications software is necessary in order to transmit information to the computer. The unit is adaptable to instruments that measure toxic gases and vapors, combustible gas indicators, aerosol monitors, as well as other types of instruments that measure airflow, temperature, and heat stress.

Figure 7-6. A versatile, portable data logger. (Courtesy Metrosonics, Inc.)

The dl-332 has a built-in microprocessor that produces a time history profile based on user-selected time intervals. In this mode the unit simulates a strip chart recorder. For example, users select a sample rate of either one or four samples per second and time history segments from as short as 1 second to as long as 4 hours. The dl-332 then samples the incoming signal and computes any or all of the following for each successive interval: minimum sample value, average value, and maximum sample value. It also provides cumulative statistics for the overall test, including amplitude distribution, minimum, average (TWA), and maximum readings and short-term exposure limits (STELs). The dl-332 can log the maximum or minimum values in 1-second intervals for almost an hour, or in more typical applications it can save all three statistics in 5-minute intervals for tests longer than 4 days.

A scheduled run option is built into the unit. The dl-332 can monitor its internal time clock until the current time matches the programmed start time and then it will turn on. The storage capacity allows a certain number of test intervals to be saved depending on how many test statistics, meaning minimum, maximum, or average sample values are being measured at a time.

Three statistics saved	1185 Test intervals
Two statistics saved	1777 Test intervals
One statistic saved	3555 Test intervals

The dl-332 has three different output modes: report generation directly to a printer, analog-to-digital conversion, and direct computer interface. The computer interface allows transmission of all stored data without formatting (ASCII) to a computer. The dl-338 is a variation of the dl-332 in that it can collect data from four instruments at the same time, and thus it is called a four-channel instrument.

For use with an instrument like a portable PID or FID, where the user might expect the levels to change rapidly, the dl-332 would be set for a high sample rate and short time interval. This setup will ensure that all measurements are recorded after they hit the detector. On the other hand, for heat stress monitoring or in a production operation where the levels are relatively constant or slowly build up over the shift, a slower sample rate and longer time interval may be selected. A log of time versus event would be useful for providing worker information to interpret data-logger measurements.

Connecting Data Loggers to Printers and Computers

Most data loggers can send data in three different modes: formatted or compressed reports or a raw data dump. Formatted reports are designed for direct printing on a printer in an understandable, readable format. Compressed reports are designed to be sent to a computer for further processing. A compressed output report contains no descriptive headers or graphs, and is significantly shorter than a formatted report. It is necessary that the computer receiving a compressed report have knowledge of the data format so that it can be properly interpreted, which usually involves sending the data to a software program.

Data signals are transmitted using either parallel or serial connections. Serial transfer is much slower than parallel, but more practical for sending data over long cables and long distances due to inherent problems in parallel data transfer. Parallel transfer is used for hooking up devices such as printers and computers that are likely to be close together. Serial transfer is used when long cables are required and for transferring over phone lines via modems.

In addition to the analog ports for the instrument's connection, some loggers have serial ports and others have both serial and parallel ports. In the case of a logger that is designed to dump to a printer only, in order to utilize a computer a program may have to be written by the user. Generally, when downloading to a printer or computer, a logger will use an RS-232 serial port. Most data loggers are capable of sending a formatted output to a printer. If the output were not formatted, it would be just a string of numbers with no indication of what each number represented or when it was collected.

Each instrument has certain expectations as to how it needs to send and receive data, including the following: parity, start bits, data bits, stop bits, and baud rate. Baud rate is the speed at which characters are sent and received by the data logger and other devices that use serial communications. If the baud rates of each device do not match, the transmission will not work. The baud rate can vary from 150 to 9,600 depending on the instrument. Some instruments are flexible and any of several rates can be selected. Others will operate only at one specified baud rate, often 1,200.

Often a separate software package is required in order to transfer data and store it in a usable and manipulatable format in a computer. Sometimes these are software programs available from the data-logger manufacturer that allow the data to be saved, viewed, plotted onscreen, or exported in an ASCII format for use with other programs. The type of graphics capability of these programs often varies. Sometimes loggers can also be dumped into a commercial data base program. Since most loggers can send information in ASCII code, most any data base program can be used. This operation will require a certain amount of computer skill and knowledge on the part of the sampling professional to generate a report format and set up the data base to properly store and manipulate the data.

In some cases, when information is downloaded to a computer it is erased from an instrument's memory; in other cases, the instrument will continue to store data until the user purges them, or the number of allowable data storage events is exceeded and the instrument starts to record over the earlier data. In still other cases, switching to the standby mode prior to dumping data may be necessary to prevent data loss.

Many loggers have the capability of being operated off-site via a modem connection to a computer. Long-term monitoring is possible, and via the

```
"TEST START DATA:   3/09/90"
"TEST START TIME:  09:01:17"
"  TEST LOCATION:  WEST WING"
"  EMPLOYEE NAME:  SMITH, MARY"
"EMPLOYEE NUMBER:  123-56-3838"
"     DEPARTMENT:  HYGIENE"
"COMMENT FIELD 1:  ----"
"COMMENT FIELD 2:  ----"
"  NUMERIC CODES:  ----   ----   ----   ----   ----"


METROSONICS dl-332 SN 1109 V2.6  12/87

CURRENT DATE:  3/09/90
CURRENT TIME: 17:45:24

CALIBRATION
  0.0000  mV =    0.0  ppm
 20.0000  mV =   20.0  ppm

LOWER ALARM:     0.0  ppm
UPPER ALARM:    22.0  ppm

UNITS: ppm


 TIME HISTORY
PERIOD LENGTH:  0:15:00
NO. OF PERIODS  COMBINED:   1

    MIN        AVG        MAX

DATE:  3/09/90  TIME:  9:01:17
   11.259     13.294     15.842                  -     *      +
   11.643     13.891     16.563                   -      *      +
   11.881     14.135     16.357                    -      *    +
   12.120     14.559     17.041                     -      *     +
   11.497     13.592     15.902                   -     *     +
   11.341     13.986     16.748                  -       *      +
   11.104     13.178     15.242                 -      *    +
   10.831     13.205     16.688                 -      *        +
   10.770     13.533     16.746                 -       *       +
   11.048     14.075     17.347                 -         *         +
   11.280     13.834     16.269                  -       *     +
   11.585     14.078     17.348                   -      *        +
   11.409     13.929     16.867                  -       *       +
   11.581     13.831     15.993                   -      *    +
   11.643     13.747     16.233                   -     *      +
   11.525     13.449     17.282                   -    *          +
   11.856     14.256     17.223                  -       *        +
   11.076     13.412     18.126                 -      *            +
   10.988     13.175     16.230                 -      *       +
   11.101     13.923     16.624                 -        *      +
   12.451     14.766     17.131                      -      *      +
   12.540     14.640     16.539                      -      *     +
   12.128     14.491     16.806                     -       *     +
   11.826     14.490     18.331                    -        *          +
   11.345     13.951     16.537                  -        *       +
   11.250     13.237     15.990                  -     *      +
   10.957     13.386     16.052                 -      *      +
   11.133     14.285     17.223                  -       *       +
   11.614     13.680     16.297                   -       *   +
   12.360     14.557     16.801                    -         * +
   12.157     15.000     17.919                    -          *    +
@  12.035     14.228     16.956                   -         *  +
```

```
"TEST START DATA:    3/09/90"
"TEST START TIME:   09:01:17"
"  TEST LOCATION:   WEST WING"
"  EMPLOYEE NAME:   SMITH, MARY"
"EMPLOYEE NUMBER:   123-56-3838"
"     DEPARTMENT:   HYGIENE"
"COMMENT FIELD 1:   ----"
"COMMENT FIELD 2:   ----"
"  NUMERIC CODES:   ----   ----   ----   ----   ----"


METROSONICS dl-332 SN 1109 V2.6  12/87

CURRENT DATE:  3/09/90
CURRENT TIME: 17:45:24

CALIBRATION
  0.0000  mV =    0.0  ppm
 20.0000  mV =   20.0  ppm

LOWER ALARM:     0.0  ppm
UPPER ALARM:    22.0  ppm

UNITS: ppm


AMP DIST
SAMPLES LOGGED:    20079

   ppm        SAMPLES                                                    %

  10.800        186  *                                              000.92
  11.200        410  **                                             002.04
  11.600        642  ***                                            003.20
  12.000        693  ***                                            003.45
  12.400       1068  *****                                          005.32
  12.800       2128  ************                                   010.60
  13.200       3596  ******************                             017.90
  13.600       3094  ***************                                015.40
  14.000       1445  *******                                        007.20
  14.400       1110  *****                                          005.52
  14.800        930  ****                                           004.63
  15.200        666  ***                                            003.32
  15.600        941  ****                                           004.69
  16.000        833  ****                                           004.15
  16.400        756  ****                                           003.77
  16.800        555  ***                                            002.76
  17.200        599  ***                                            002.98
  17.600        216  *                                              001.57
  18.000        107  +                                              000.53
  18.400          4  .                                              000.02
```

Figure 7-7. Printout formats: time history distribution *(left)* and amplitude distribution *(above)*. (Courtesy Metrosonics, Inc.)

modem the data logging unit can be told to stop, dump data, and clear. There will be certain specific commands to use to access the unit. Usually the commands involve telling the unit to start/stop logging and send data in a given report format. These remote request commands can be different for each data logger.

Sending stored information from an instrument to a printer involves several variables: Some require that a printer have a buffering capability (memory storage capability), and since data can be lost without substantial buffering capability, most require that printing be done on draft mode rather than letter-quality mode. Each instrument/printer setup may require a different type of cable because the connectors on printers vary by manufacturer and type. The selection of the connector to attach the data logger to the output device can vary. The pins attached to the lines designed to send and receive data can be set in two different ways depending on the device. The maximum cable length depends on the instrument. Low-voltage instruments are limited to shorter cable lengths. A resister may be required along with a specific cable and pin connection depending on the voltage output of the instrument. If it exceeds that of the data logger, a resister is needed so the user does not "blow out" the data logger.

In general, when in the printer mode instruments respond to commands from their own keypad to send information (characters) out of the RS-232 port to a printer (peripheral device). When in a computer mode the instrument responds to commands from the computer to send and receive data and perform other tasks.

Important Facts

The addition of data-logging capability does not enhance the detection sensitivity of the instrument, but it does allow the user to study the data in more detail later, and frees the user from the need for extensive manual recording. The detectors in most cases have not changed. Following are some important considerations about data loggers.

Alarm-only mode: If an alarm-only mode is available, many loggers will only record values that exceed alarm setpoints while in standby. This function allows the user to store only high concentrations. Standby can be used if data are to be stored only periodically wherein the user will select certain test periods. It saves on memory space and minimizes having to sort through unnecessary data.

Battery time: Nine-volt batteries are common in data loggers and often provide up to 100 hours of operation. Most data loggers will store data once a battery is removed or has lost its charge for a given period; however, this period may range from a few hours to a few days. For safety, change battery packs on data loggers that are used regularly and always keep a spare battery pack available.

Cable: Usually a different cable is required to hook up a logger to each brand, model of instrument, computer, and printer. In the case where an external battery is used to power an instrument to provide longer sampling time in the field, a single cable with a "T" connection is used to connect the three units together, because many instruments have only one connector used for battery charging and output. Cable compatibility includes number of pins, pin definitions, and gender.

Cable length: The maximum cable length depends on the instrument. Low-voltage instruments are limited to shorter cable lengths. A resister may be required along with a specific cable and pin connection depending on the voltage output of the instrument. If it exceeds that of the data logger, a resister is needed so the user does not "blow out" the data logger.

Clock capability to record time intervals: Some manufacturers have incorporated the ability to pause the unit for breaks and lunch periods, which can save memory space. The timer function should continue so that future measurements are attributed to the correct time periods.

Communication protocols: In order for data loggers to send data to computers, the specific requirements of each instrument must be identified and set on the computer. These include baud rate, number of data bits, and parity.

Ease of programing: In most cases, as part of the initial setup the user must input the date and time and then choose the desired input voltage range, TWA storage period, scale, range, calibration, and units of the stored data. Also programed are what data to store. They can range from all measurements, to the maximum value detected in each measurement period (or average value), to the minimum value, or all three. Some data loggers are tutorial and have self-prompting and error messages whereas others require extensive study of the user manual. An example of a self-prompting message is: Connect Span Gas Then Press Enter. Another technique is to have flashing letters symbolizing the opportunity to enter various data.

Due to the need to keep instruments small, many keypads rely on the use of +/− for a variety of functions, including changing numbers in dates and times and entering other numerical parameters. Units that use arrow (↑, ↓) keys to toggle among digits use a cursor to indicate which digit is ready to be changed. These designs can take a lot longer to program than keypads incorporating the alphabet and numbers. However, keypads with the alphabet and numbers take up more room on the instrument.

Format of printout: The format of data-logger printouts varies widely. The simplest is a numerical listing of values. Some plot data as minimums, averages, and maxima for each sampling interval. Other formats include graph and histogram construction. In some cases, there is only a single format; in others there is a choice of different formats.

Amplitude distribution: An amplitude distribution report contains a graph showing what percent of the data is distributed within certain ranges. Often this graph indicates concentration, the number of samples collected at that concentration, and the percent of the total number of samples represented. Therefore, the most frequently occurring concentration will have the highest plot and percentage.

Cumulative average: Some units can also provide a cumulative average; thus, an 8-hour cumulative average would represent an 8-hour TWA.

Time history report: These reports often show a minimum, maximum, and average value stored for each sampling period. The average will depend on the number of samples made over a given period.

Instrument compatibility: Data loggers are not universally compatible. In some cases, a unit will be compatible with several instruments; in other cases, it will be dedicated to a specific instrument.

Modes: Most data loggers have three modes: display, program, and output. The display mode allows the user to observe the input reading and may include date/time, alarm settings, and other data. The program mode is for programing parameters. The output mode is used to output the input readings, any statistics the unit is capable of computing, and other data such as time history, amplitude distributions, and graphs.

Number of channels: Most current data loggers designed to be attached to portable survey instruments have just one channel, that is, they can log from only one instrument or one sensor at a time. However, multichannel data loggers are now becoming available. Some log for instruments that have more than one sensor; in such cases, they are generally dedicated to these instruments. Another has the capability of being hooked up to four different instruments at a time, which of course limits the portability of the system. For stationary monitoring, more sophisticated data-logging units have been available for some time.

RS232 connection: This is a standard interface through which devices can communicate. Handshaking is the process by which one device monitors the status of another and responds accordingly.

Sampling periods: Sampling periods represent the periods over which the data logger averages or integrates the sample points it receives. They also determine how fast storage memory will be filled. As an example, if the user selects a sample period of 1 minute, and the unit is sampling at 1 sample/second every minute, the unit will average the sixty 1-second readings it has received and store them as one sampling period. The shorter the sampling period, the closer the unit comes to storing real time data, or instaneous measurements.

Separate readout device: Most current data loggers and data-logging instruments have enough processing capability that they can be independently hooked up to a printer/computer without having to go through a secondary processor. However, some units still require the use of an intermediary device in order to transmit data to a printer or computer, because such data loggers only compute and store averages. Therefore, in order to do further calculations and analyses of the data, the data must be transferred to a computer with a data base program. In some cases, companies have developed specific software programs for inputting computer data.

Software to interface with computers: When converting information to a computer where it can be stored on a disk and analyzed further, two types of software are needed. The first is a communications program, allowing the instrument to "talk" to the computer. The other is a data base or other program to allow the user to put the data into some sort of usable report format, develop graphs, and so forth. Otherwise, the information must be stored in ASCII (American Standard Code for Information Interchange), which is a universal text that most computers can understand. It may look like gibberish until the data are processed by a program that can manipulate and organize the data into a usable format.

Storage memory: When comparing data loggers for storage capability, the amount of room required (in bytes) for each measurement as well as other bytes necessary to store other information, such as parameters, should be considered. Data loggers will either record continuously, and after the maximum number of readings are collected begin to overwrite the existing data, or stop recording once memory is full. Therefore, the user must plan to dump data at regular intervals and be aware of the maximum sampling period available.

Stored points: The total number of data readings that can be stored along with the length of required sampling time will often determine the sampling intervals. Typical sampling rates for data loggers are one sample per second or four samples per second. The rate represents the number of sampling points received from an instrument. For example, a unit with a maximum of 700 storage points collected at 1-second intervals will use up its storage in 11.7 minutes; if the collection interval is increased to 1 minute, the unit will store data for 700 minutes, or approximately 11.7 hours; and if the storage interval is further increased to 5 minutes, the length of data logging can be extended to 58.3 hours. When selecting data to log, some instruments can log either points (instantaneous values) or average values at period intervals. Often preselection is necessary.

Tag numbers: Tag numbers can be used to label data with a unique code for a given sensor. They are also required to separate various events. In this way, a data logger can be moved from instrument to instrument in a single survey if tag numbers for sensors are used, which allow distinguishing data during the test.

DEDICATED PRINTERS

Many instruments have as accessories dedicated printers that allow printing out of the data that have been collected over the sampling period. These instruments have a limited amount of built-in data-logging capability, and must be used exclusively with a given instrument. They are generally hooked up with cables specific for that instrument—printer combination. These printers commonly use thermal paper that may have to be obtained from the manufacturer of the instrument. Thermal paper contains the ink that forms the printout when the printer head comes in contact with it, thereby eliminating the need for purchasing ribbons. However, thermal paper is delicate and can discolor if left in the sun or near heat sources. In some cases, it may also be pressure-sensitive.

Since these printers are designed to do only a few types of printouts, they are usually very small, making it possible to use them in the field. Often these printers can be powered off the batteries contained in the instrument.

ADVANTAGES AND DISADVANTAGES OF DIRECT READING EQUIPMENT

Although microprocessors and other innovations have enhanced the intelligence of many instruments, they are not black boxes that can replace the skill and judgment of the user. Different instruments offer different advantages and present their own set of limitations. The biggest problems occur when an instrument is used for the wrong application and the data are then incorrectly interpreted. An example would be the use of an infrared (IR) monitor for soil gas monitoring or any other testing where there is likely to be high humidity and multiple unknown contaminants.

The primary advantage of a tunable IR analyzer is its ability to quantitatively sample relatively specifically and continuously for a variety of compounds, one at a time, with a minimum of waiting time, and thus it can be used to do leak testing in situations where several contaminants might be present. To do this type of testing with an instrument like a GC would only provide snapshots of concentrations at different periods of time given the length of time between sample runs, and it would take a long time to leak test all the sample points in the system. In another situation where a single contaminant or related group of contaminants is being leak tested, such as benzene or petroleum emissions, a general survey instrument that will continuously monitor and provide levels of total hydrocarbons is the best choice and specificity for a given compound is not that important. Some very important advantages of real time instrumentation include[13]:

- The ability to measure concentrations of contaminants on-site with almost instantaneous results
- Continuous fixed station monitoring to alert personnel to emergency situations via audible and visual alarms
- The ability to collect very short samples and therefore break down a task or process and analyze emissions at each step
- Identification sources of problems, such as leaks, and the opportunity to try out the effectiveness of solutions as soon as they are in effect
- Ongoing records of contaminant levels (or nonlevels) to protect companies from false allegations
- Less work on the part of the sampling professional once the instrument is set up when compared to conventional integrated methods
- Reduced laboratory costs
- Immediate samples without concerns over loss of samples during storage and shipment
- Sample results without having to wait out the standard 10-day laboratory turnaround time or pay a surcharge for quicker turnarounds

While real time instruments can be very good for quantitating exposures to known contaminants or estimating exposures to nonspecific compounds, they are very limited for identification of unknowns. For this purpose integrated sampling with laboratory analyses is better. Also, the number of instruments capable of monitoring for specific chemicals is limited. Another problem is that users often do not understand how to correctly interpret real time monitoring results.

Real time instruments, in particular stationary monitoring systems, can be valuable tools in identifying potential overexposures in the workplace and protecting workers and others. However, often audible alarms are disabled from devices designed to act as warning systems because they have been going off. If there does not appear to be a connection between alarm occurrences and the presence of high concentrations such as employee complaints, odors, or known releases, often the assumption is that it is the alarm that is malfunctioning and the nuisance is turned off. However, the instrument is left to run on the assumption that it is still somehow performing its function; but what is more likely occurring is the presence of an interferent, one that will activate the instrument at either lower or higher levels than the compound for which it was calibrated. If the instrument is left running with the alarms disabled, employees may not become aware of dangerous situations. In the case of multipoint sampling monitors, it may be only one or two sites that cause the alarm to go off. By avoiding these areas, often basements or other rarely entered areas, it is thought everyone will be safe.

Microprocessors

The trend to add microprocessors to real time instrumentation has been a significant improvement, allowing the user access to a wider variety of sampling and data collection modes. However, in most cases the detection mechanisms (sensors) have not been significantly changed from the original devices. Overconfidence in the reliability of these units can lead to lack of calibration and maintenance. Sometimes users assume that as long as the lights are on an instrument is working properly. The reality is that frequent calibrations and maintenance are even more important in these cases because often the instruments are depended on to act as warning systems, so that people will exit an area in the event of a hazardous release and don proper respiratory protection prior to reentry. Microprocessor/data-logging capability can be very useful. However, if a user is not computer fluent or does not own the proper equipment, it is better to spend less money and buy a manual instrument. As more manufacturers recognize the lack of computer literacy among sampling professionals, it can be expected that more devices will incorporate what has been termed user interactive intelligence or self-guiding instructions to make downloading of data simpler.

Purpose

A common misconception among sampling professionals seeking to buy real time instruments is that there is one instrument that will do everything. As more instruments become microprocessor-controlled, and appear to be more sophisticated, it becomes even more confusing. The reality is that different instruments are based on different operating principles and have different applications. For example, one monitor may detect lower levels of a compound, but it may be too sensitive to use for measuring high concentrations, such as those involved in explosive situations. The reverse is also true in that CGIs that measure concentrations of explosive gases in the percent range are not appropriate for measuring health hazard levels.

Although most instruments are relatively straightforward to operate, some are more complicated than others, either due to poor design or as a result of poorly written and incomplete manuals. Many microprocessor-based units require several steps to be memorized just to program the units, or the user must carry the manual around at all times. Manuals are getting thicker, especially for programmable instruments, often being an inch thick or greater. The instruments are getting smaller as a result of the increased number of functions on a single chip. Sampling professionals are often tempted to buy the higher-priced models with the most functions programed into them, but then are confused by the profusion of buttons and need to open the case to adjust various poorly marked tiny screws and switches, the result being that they use the instrument for only the most basic of functions.

Factory Calibrations

Some vendors imply in their literature that a factory calibration is all that is ever needed. The reality is that frequent calibrations are needed that involve making up or purchasing calibration standards. Not all instruments are capable of being calibrated with a gas; some utilize indirect methods of calibration, such as using a spot card to calibrate a sensitized tape monitor.

Increased maintenance and repairs are required for real time instruments over what is typical for personal sampling pumps used for integrated sampling methods. For example, electrochemical sensors, which are one of the most common types of detection, wear out and must be replaced regularly. They can also leak and damage an instrument.

Portability and Power

While there are a number of instruments available that can be taken into the field to do direct reading measurements, portability is questionable due to the weights of some systems (up to 25 lb), bulkiness because they are frequently square boxes or can be quite long (up to 3 ft), and often they have poor ergonomic design that makes their use clumsy. In order to be truly portable an instrument should be capable of being operated while being carried by one person. In some cases, if the sampler wants to store data, a separate data recorder is required and can result in having to use the instrument on a cart. Generally, for reliability for long-term or continuous measurements, the worker must plug the units in and use strip chart or other data recorders, which limits portability and often requires long extension cords and carts.

The National Institute of Occupational Safety and Health (NIOSH) had an instrument testing program during the seventies that generally amounted to testing instruments in a group and comparing their performance. Tests were done on combustible gas indicators and hydrogen sulfide, sulfur dioxide, and carbon monoxide monitors. However, this program was suspended, and except for the testing programs that certify instruments for use in explosive atmospheres, the only available means for independent review of instruments is through journal articles.

Interferences

Most sensors have at least one known compound that will interfere with their ability to accurately measure the compounds they are designed to monitor, and there may be many others for which sensors have not been tested. Therefore, interferences must be considered when selecting an instrument, including interferences by other chemicals on measurements of specific chemicals and ones such as humidity or the lack of it. Often these data are available from manufacturers. Sometimes these data are presented as ratios, for example, 300:1. The larger the ratio, the less significant the interference. Calculation of these ratios is done by dividing the amount of interferent used in the test by the change in the meter reading, for example, a challenge of 150 ppm of carbon monoxide to an instrument designed to measure hydrogen sulfide caused an increase of 2 ppm in the response of the meter to 10 ppm hydrogen sulfide:

$$\frac{150 \text{ ppm CO}}{2 \text{ ppm change in reading}} = 75:1$$

Interference charts are most often done by exposing instruments to single gases and then determining their response. Manufacturer's data on interferences may be misleading as they may not contain information on all potential interferents. Only those compounds for which the manufacturer has tested are listed, usually 6 to 10 compounds. The manufacturer's literature should note whether measurements were made with or without a filter in place.

Radio frequency interference (RFI) from sources such as high-voltage lines and radio transmitters can be a problem for many instruments. Some instruments have cases specifically designed to protect against RFI. In order not to lose this protection, often the case must not be opened by the user.

If there are interferences present, it must be determined whether the levels are high enough to give unacceptable measurements or require increased maintenance. There may be trade-offs in this area: Nonspecific instrumentation may be inexpensive and have fast response time, but if instruments have an unacceptably high number of false alarms, production costs could be unacceptably high. Therefore, before purchasing instrumentation the sampler should know what other chemicals can be released into the plant atmosphere in addition to the compound of interest.

CONCLUSIONS AND NEW DEVELOPMENTS

The user can approach a discussion of real time instruments by either categorizing them by principle of operation or by use. There are many instruments available, and a few are considered classics. Since the purpose of this book is to assist with practical field measurements, only a few instruments have been selected for discussion based on their being "tried and true," versatile, state-of-the-art or representing a new trend, or a preferable monitoring method to the integrating method for a particular chemical. This discussion is intended for use as guidance in operating these types of instruments in the field, but does not represent a specific endorsement by the author of any particular manufacturer or instrument.

Manuals provided by instrument manufacturers usually provide step-by-step instructions regarding the function and use of their instrument. The information and procedures in this book are intended to supplement that of the manufacturer and provide the user with background information about field use of the instrument before attempting its use. The use of any instrument without proper training of the user can result in measurements that are either meaningless or under certain circumstances could present a hazardous situation. Therefore, regardless of how simple an instrument appears to be to operate, the user should be thoroughly knowledgeable in its limitations, calibration and maintenance needs, and interpretation of the results.

Manufacture of direct reading equipment is a fast-moving field, and when contemplating selection, potential purchasers should ask many questions of vendors and makers regarding sensor life (and replacement cost), frequency of calibration, temperature limits, cross sensitivities, and detection limits.

Purchasing Tips

Some instruments have been in use for so long that they are referred to by a specific manufacturer's name, such as the HNU PID and the Miran series of IR analyzers. This reference, however, does not represent an endorsement of any specific manufacturer's product. Technology is improving; therefore, the sampler should evaluate a number of instruments in a category for merit prior to purchase. Be aware that some manufacturers have a tendency to stick to a tried and true technology when a better one may have been recently developed. So it may be advantageous to evaluate instruments by less well-known manufacturers. Although custom-made instruments are also available, the primary drawbacks of this type of purchase can be the quality of documentation as well as assistance, including maintenance if the manufacturer goes out of business. One-person operations can be especially susceptible to this type of problem.

A drawback of purchasing equipment from manufacturers that are not well established is that if they go out of business there is often no support for repairs. It is especially a problem with a very specialized piece of equipment. On the other hand, established companies may adhere to the same product line for years and rarely add new models or types of instruments. Poorly written documentation can also be a problem. Sometimes it may contain omissions or inaccuracies and the only way to determine this is to call the manufacturer. To find out how much support is likely to be available, it is a good idea to call a manufacturer when planning a purchase and ask for technical assistance. If you are forced to wait several days for an answer or you are assisted by a salesperson with little technical knowledge, this consideration may be primary when making the purchase decision. The selection criteria for real time instruments are listed below.

Need for portability
Amount of time and expertise available for training
Need for high sensitivity (low detection levels) and accuracy
Conditions of use—heat, humidity, use of a cart versus strap
Appropriateness for the application
Purpose for sampling
Instrument availability and complexity of use
Specificity
Personal choice based on past experience
Measurement period required
Preparation time available prior to use of an instrument
Minimum detectable limit
Required dynamic range
Time constant
Presence and classification of hazardous (explosive) locations
Type of data acquisition and display required
Interferences
Personnel for operation and maintenance
Security

Security codes and keys are being incorporated into instruments so that only authorized persons can alter calibrations or programming. In many cases, the microprocessor has replaced the need to make adjustments with potentiometers during calibration. Once a microprocessor-based instrument is set up and programmed, if it is provided to a minimally trained person —someone who can perform measurements but does not understand how to make changes in the programming and may, indeed, not understand what these changes do in terms of the instrument

—it is a good idea to take advantage of the security option on the instrument. The user should be provided with very specific instructions as to where and what the instrument can be used to measure, and should be warned that the interpretation of results from any other measurements may lead to a dangerous situation.

Price

Now that prices of the more sophisticated field equipment, such as gas chromatographs and programmable infrared units, are in the $17,000 range and beyond, with an average instrument in the $6,000-plus range, more rental equipment is becoming available. Methods are now being written that specify the use of direct reading instrumentation, for example, there are a number of NIOSH methods that specify either an infrared or gas chromatographic unit. Atmospheric sampling has relied on these types of instruments for years, although they are slightly different than the ones used to survey in the field, because generally they are fixed station monitors. While most of the survey instruments are available in "stripped-down" incarnations, the amount of work, such as having to manually integrate GC peaks, increases drastically over the computerized models and this is a trade-off with price.

Trends

A trend, due to miniaturization, is to combine many different sensors into a single unit. An example is the Neotronics Exotox 70/75 Ambient Atmosphere Monitor/Logger. This system can be configured with up to five different continuously monitoring gas sensors along with temperature and relative humidity (RH) monitoring and data logging. Sensors include oxygen, combustible gases, and any three of five available toxic gas sensors (carbon monoxide, hydrogen sulfide, sulfur dioxide, chlorine, and nitrogen dioxide). A separate unit into which the monitor can be plugged allows the user to interface with computers and through both RS232 serial and parallel ports.

A variation of this concept is a single instrument with interchangeable sensors for a number of different gases such as carbon monoxide, hydrogen sulfide, sulfur dioxide, and chlorine. The PM-7700 Toxic Gas Monitor from Metrosonics is designed to provide personal monitoring for any of these gases. Sensors, all electrochemical, are encased in a polycarbonate housing at the end of a 0.9-meter cable. The microprocessor-based instrument automatically identifies the sensor in use, which is important as different monitoring ranges are assigned to each sensor. For chlorine, the range is 0 ppm to 20 ppm, while for sulfur dioxide it is 0 ppm to 200 ppm.

A future trend in the area of stationary monitoring will be to collect data from analyzers using different detection principles and send them to a central computer where the data can be interfaced with other sampling data, such as integrated measurements, as well as biological monitoring data. The installation of local area networks (LANs) will provide for this interconnection via several small computers.

Standards for protocols on data exchange are being developed and implemented that should allow diverse types of equipment to communicate in both the factory and office environment. This will allow the development of software based on these standards to use in networks. This includes a concept being developed by the Instrument Society of America called a Field Bus Standard as a means for allowing data to be collected from different sensors and control information to be sent to them.[9]

This type of system has been attempted. In this demonstration three light-scattering aerosol monitors, two MOS-based hydrogen sulfide monitors, and a thermocouple for temperature measurements were linked. The voltage outputs from the air monitors were obtained by data acquisition modules and transferred to remote computers where the voltages were converted to a concentration using predetermined calibration constants.[14]

The benefit of such systems would be an integrated data communication system to help in the management of emergencies identified by continuous monitoring systems. Also, data from continuous monitoring systems could be combined with data from sources such as process control, ventilation, and energy management to decrease emissions and to run the process and ventilation systems more efficiently.[9]

Real time instruments used in conjunction with data loggers, computers, and video taping allow an evaluation of activities affecting exposure patterns beyond that of traditional integrated sampling. Such a system has been used successfully in several NIOSH studies. In one, a study of manual weigh-out and transfer of powders, the data acquisition system identified the scooping operation as the component of the work cycle that most affected the worker dust exposures. Because of the short work cycle, typically 30 seconds or less, filter sampling could not isolate this work element as the major cause of worker dust exposure. In another study, an evaluation of various control methods used during automotive brake servicing, several control devices were reviewed and the work routines associated with each device were observed. The data acquisition system enabled the investigators to identify the use of an air-driven impact wrench as a significant contributor to dust exposure levels.

In order to put such a system together, it is necessary for both the data logger and the video to have internal clocks. The video clock should be recorded on-screen during filming. The video tape of the work cycle or process can be reviewed while tracking the worker's exposure from a printout or plot of the real time data. The activities can be compared with the exposure results. In some cases, where the job is stationary, a computer can be used to store the data directly through an analog-to-digital hookup from the real time instrument.

A computer is used when the data loggers will not provide sufficient data resolution or the data must be displayed while collection is taking place. When using the computer with an analog-to-digital converter, the data can be sent from the worker-mounted instrument to the computer via an umbilical cord or a telemetry system. The telemetry system should be used if eliminating the umbilical cord is necessary due to the work process or cycle.[13]

It has been suggested that the time constants of real time instruments could be adjusted to match the cumulative behavior of a substance in the body. Using one guideline, the time constant could be set to equal 1.44 of the biological half life of a chemical. Therefore, the rise and fall of the instrument response would be similar to the buildup in the body.[15]

REFERENCES

1. Roach, S.A. A most rational basis for air sampling programmes. *Ann. Occup. Hyg.* **20:**65–84, 1977.
2. Smith, J. Uses and selection of equipment for engineering control monitoring. *AIHA J.* **44**(6): 466–472, 1983.
3. Stetter, J. R., and D. R. Rutt. Instrumental carbon monoxide dosimetry. *AIHA J.* **41**(10):704–712, 1980.
4. Langhorst, M. L. Comparative laboratory evaluation of six chlorine monitoring devices. *AIHA J.* **48**(5):347–361, 1982.
5. Smith, J. Uses and selection of equipment for engineering control monitoring. *AIHA J.* **44**(6): 466–472, 1983.
6. Shaw, M. More Straight Talk About Toxic Gas Monitors. Interscan Corp., P.O. Box 2496, Chatsworth, CA 91313-2496.
7. Willeke, K., and S. J. Degarmo. Passive versus active aerosol monitoring. *AIHA J.* **3**(9):263–266, 1988.
8. Long, S. E., and R. W. Lawrence. Understanding Combustible Gas Sensors—Types, Installation Considerations, Costs and Maintenance Costs. Mine Safety Appliance, Inc., Pittsburgh, PA.
9. Smith, J. P. Use of data communications with air monitoring data. *Appl. Ind. Hyg.* **5**(4):213–221, 1990.
10. *Organic Vapor Analyzer (OVA) 108 Operating Manual.* Foxboro Instrument Co., Inc., Foxboro, MA.
11. *Rustrak DC Signal Recorder Instruction Manual.* Graphic Instruments Division, Manchester, NH.
12. Cecala, A. B., J. J. McClelland, and R. A. Jankowski. Substantial time savings achieved through computer dust analysis. *Appl. Ind. Hyg.* **3**(7):203–205, 1988.
13. Gressel, M. G., et al. Advantages of real-time data acquisition for exposure assessment. *Appl. Ind. Hyg.* **3**(11):316–320, 1988.
14. Smith, J. P., et al. Demonstration of the use of a LAN to collect and analyze data from continuous air monitors. *Appl. Occup. Environ.* **5**(12):870–878, 1990.
15. Rappaport, S. Biological considerations for designing sampling strategies. In *Advances in Air Sampling.* American Conference of Governmental Industrial Hygienists, Lewis Pubs., Chelsea, MA, 1988.

CHAPTER 8

Monitoring Instruments Dedicated to a Single Chemical

There are many instruments designed to specifically monitor for a single chemical. Interferences are minimized and concentrations in the health ranges (permissible exposure limit [PEL], threshold limit value [TLV]) can be reliably and accurately measured. When used correctly and appropriately, often measurements made with these "dedicated" instruments can be used in lieu of traditional integrated monitoring methods that require laboratory analysis (Fig. 8-1).

Dedicated units are available for several chemicals. Whether by interest or technology, availability is most commonly limited to the following mostly inorganic chemicals: ammonia, sulfur dioxide, hydrogen sulfide, mercury, chlorine, carbon dioxide, carbon monoxide, isocyanates, nitrogen oxides, oxygen, phosgene, formaldehyde, ozone, and hydrogen cyanide. For organic compounds most often either a general survey instrument is used or an instrument such as an infrared (IR) monitor is used that can be adapted for many different compounds.

Generally instruments dedicated to a single chemical are less expensive and easier to use than instruments that can be used to detect many different chemicals, such as gas chromatographs (GC) and infrared (IR) units. Instruments with broader measurement capabilities can also be dedicated to the measurement of a single chemical, but this is generally not cost-effective. In the case of IR, instruments which are dedicated to the measurement of a single chemical are constructed more simply than those with broader monitoring capabilities. An available variation is a single-instrument housing that can be adapted to monitor different chemicals by replacing the sensors.

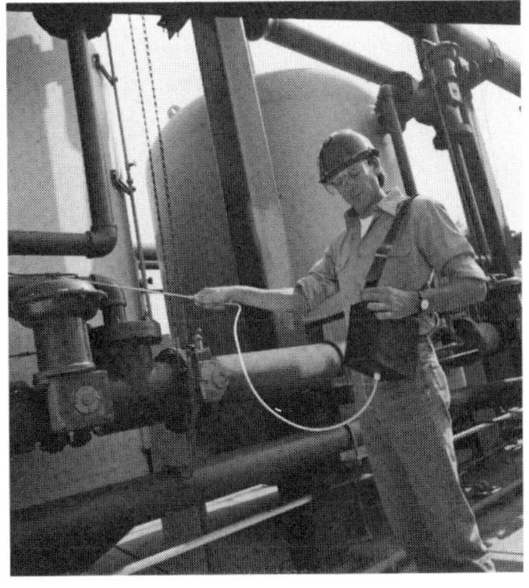

Figure 8-1. Sampling professional monitoring with a survey instrument. (Courtesy Interscan Corp.)

The greatest number of instruments specific for a single chemical are based on electrochemical detection. These are available as both survey and personal-sized monitors. Monitors using taped-based detection are available for certain chemicals, most importantly isocyanates, for which no other reliable real time system exists. Other detection principles are chemiluminescence, ultraviolet (UV), and gold film, each of which is only used for a few compounds. Colorimetry is another principle relying on wet chemistry that can be used to develop instruments specific for a chemical, although only one instrument using different interchangeable reagent systems now exists.

For a given chemical there may be several available types of real time monitoring instruments available or very limited technology. For example, carbon monoxide can be detected using electrochemical, solid-state, and IR methods whereas isocyanate detection is available using chemically impregnated tapes only. The selection of a given type of sensor is often dictated by the type of circumstances in which an instrument must be used. Different measurement principles vary in many characteristics, including sensitivity to low concentrations of a chemical as well as their usefulness under certain conditions. Therefore, which instrument is selected will be determined by cost, minimum detection level, reliability, ease of use, and other factors such as those discussed in the chapter on Reading and Recording Real Time Data.

Advantages of purchasing instruments dedicated to a single chemical are lower cost than systems designed for monitoring many different gases and vapors, specificity, in many cases simplicity of operation, and the large number of personal monitoring instruments available. In some cases, such as mercury, the only real time method available for monitoring is dedicated instruments. It is best to purchase a dedicated instrument when there is a high frequency of monitoring for a single contaminant such as often occurs with carbon monoxide, ubiquitous because of its presence as a by-product of combustion.

Selecting a monitor that is specific for a given chemical can be a case of buyer beware. Instruments are sometimes marketed for a specific chemical when they are based on detection principles that allow only semispecificity or are actually capable of only general survey monitoring. An example is a unit marketed as a leak detector for ethylene oxide that is based on solid-state detection. It is not totally specific for ethylene oxide. The only time it will be is when ethylene oxide is the only chemical present. Therefore, it is important to be aware of what methods are actually specific for a given chemical. Table 8-1 describes different detection principles that can be used to monitor for a number of different compounds. These are described here and in subsequent chapters. Often a strip chart recorder or data logger is used in conjunction with these types of instruments.

DETECTION PRINCIPLES: ELECTROCHEMICAL SENSORS*

Electrochemical sensors are commonly used for the detection of specific chemicals such as chlorine, sulfur dioxide, hydrogen sulfide, nitrogen dioxide or nitric oxide, phosgene, carbon monoxide, hydrogen cyanide, and oxygen among other gases. Instruments are available for survey and personal measurements (Figs. 8-2, 8-3). Electrochemical detectors use a number of different types of operating principles. However, these methods have not arisen logically over the years from a single branch of electrochemistry, and therefore there has been no systematic approach for naming them, particularly since this science is relatively "old." Many names date back to the early nineteenth century.

Many different commercial products use similar detection principles, but because no single nomenclature is used consistently for the description of these devices, the commercial product literature can be confusing. A clarification of the classification and description of the operating principles of the electrochemical devices used in toxic gas monitoring is needed.

It is as unhelpful to generalize a device as an electrochemical sensor as it is to lump chemiluminescence, IR, and UV absorption into the single category optical devices. Accordingly, electrochemical-type gas sensors can be differentiated according to which parameter of Ohm's Law (voltage = current × resistance) is the depen-

*Source: Dr. Manny Shaw, Interscan Corp.

TABLE 8-1. Specific Compounds and Real Time Detection Principles

Compound	*Detection Principle*
Ammonia	Colorimetric, chemical impregnated tape, electrochemical, solid-state
Arsine	Colorimetric, IR, chemical impregnated tape, PID, electrochemical
Bromine	Colorimetric, electrochemical, chemical impregnated tape
Carbon dioxide	IR, electrochemical
Carbon monoxide	Electrochemical, IR, solid-state
Carbon tetrachloride	Gas chromatography, IR
Chlorine	Electrochemical, solid-state, colorimetric, chemical impregnated tape
Combustible gases	Catalytic combustion, solid-state, thermal conductivity
Diborane	Chemical impregnated tape
Disilane	Chemical impregnated tape
Ethylene oxide	IR, electrochemical
Fluorine	Colorimetric, electrochemical
Formaldehyde	Colorimetric, IR, electrochemical
Formic acid	Electrochemical
Germane	Chemical impregnated tape
Hydrazine	PID, colorimetric, chemical impregnated tape, electrochemical
Hydrochloric acid	Electrochemical
Hydrogen	Electrochemical
Hydrogen bromide	Chemical impregnated tape
Hydrogen chloride	Colorimetric, chemical impregnated tape, IR, solid-state, electrochemical
Hydrogen cyanide	Colorimetric, chemical impregnated tape, electrochemical
Hydrogen fluoride	Colorimetric, chemical impregnated tape, electrochemical
Hydrogen selenide	Chemical impregnated tape
Hydrogen sulfide	Electrochemical, gold film, PID, solid-state, chemical impregnated tape, colorimetric
Isocyanates	Chemical impregnated tape
Mercury	Gold film, UV
Methyl chloride	Solid-state
Nitric acid	Electrochemical, chemical impregnated tape
Nitric oxide	Electrochemical
Nitrogen dioxide	Chemical impregnated tape
Nitrogen oxides	Chemiluminescent, colorimetric, electrochemical, solid-state
Nitrous oxide	IR
Oxygen	Electrochemical (galvanic cell)
Ozone	Chemiluminescent, coulometric, UV, chemical impregnated tape
Phosgene	Colorimetric, IR, chemical impregnated tape, electrochemical
Phosphine	Chemical impregnated tape, electrochemical
Silane	Electrochemical, chemical impregnated tape
Stibine	Chemical impregnated tape
Sulfur dioxide	Electrochemical, IR, solid-state, colorimetric, chemical impregnated tape
Sulfuric acid	Chemical impregnated tape
TDI	Chemical impregnated tape

Figure 8-2. Personal electrochemical monitor. (Courtesy MDA Scientific, Inc.)

dent variable of a system's operating principle, and thus categorized as either potentiometric, conductimetric, or amperometric. Amperometric devices are further divided into the subclasses of coulometric and voltammetric, these two having totally different operating principles.

Electrochemical sensors can also be categorized as galvanic or electrolytic. A galvanic cell is any arrangement consisting of two electrodes and an electrolyte that is capable of undergoing a spontaneous chemical reaction to produce an electric current. The term *galvanic* really refers to

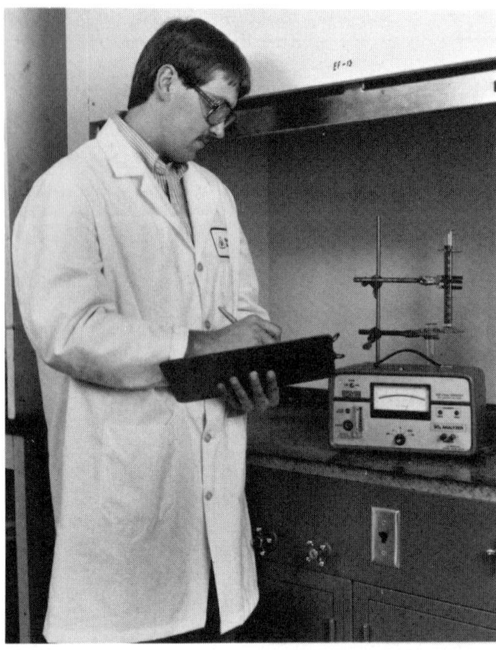

Figure 8-3. Carbon monoxide electrochemical monitor for surveys. (Courtesy National Draeger, Inc.)

any device that converts chemical energy to electrical energy, be it in a gas sensor, battery, fuel cell, or rusty nail. An internally biased sensor has been called a galvanic type to distinguish it from the type requiring an external bias. This is a misunderstanding since both types are galvanic. Bias is discussed further in the section on voltammetric cells.

Electrolytic is a term applied to the conversion of electrical into chemical energy as in electroplating or charging a battery. Some gas sensors, particularly oxygen sensors, are categorized as either polarographic or galvanic, depending on whether an external bias voltage is applied or not. It is a meaningless division as far as operating principles are concerned. In both cases, the sensing electrode is kept at the fixed potential necessary to keep this electrode in the range of the limited diffusion current, where the generated current is a linear function of the gas concentration. Since the output current levels are lower in polarographic cells than in the galvanic cells, the associated electronics are generally more expensive and complex in these types of instru-

Figure 8-4. Electrochemical sensor (photo) and a schematic of an electrochemical cell. (Photo courtesy MDA Scientific, Inc.; schematic from J. Stetter, Instrumentation to monitor chemical exposure in the synfuel industry, *Ann. Am. Conf. Govt. Ind. Hyg.* **11**:225–247, 1984 and reproduced by permission of the American Conference of Governmental Industrial Hygienists.)

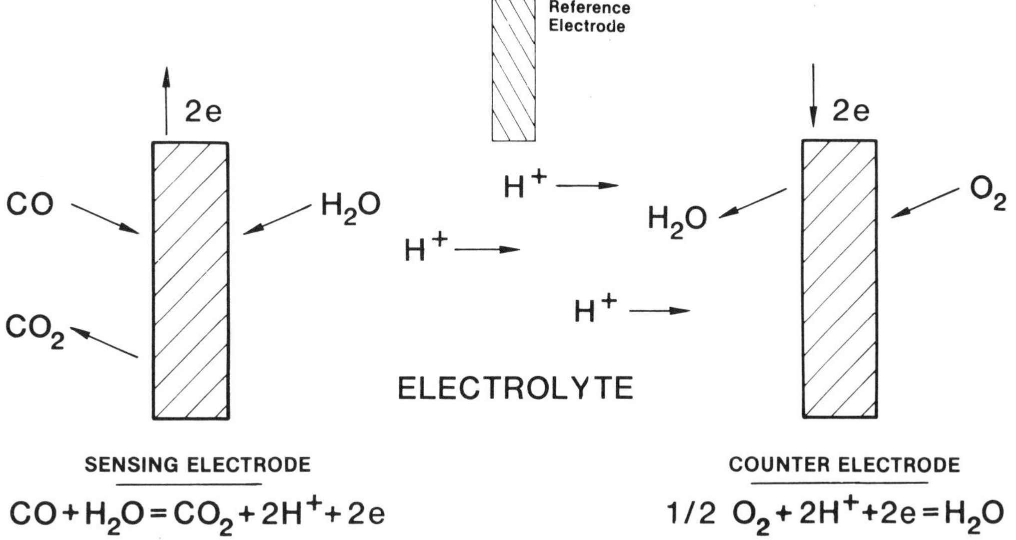

ments. Whether sensors are called galvanic or polarographic, the less confusing and more accurate term is voltammetric.

In most electrochemical gas sensors the gas diffuses through a membrane and a thin film of an internal electrolyte separating the membrane and the working electrode (Fig. 8-4). In some cases, the electrolyte is a liquid; in other cases, it is a gel or it is immobilized in a solid matrix. In order to make instruments specific, the electrolytes in each electrochemical cell must be unique, as well as the electrodes.

There are three types of electrodes: working (sensing), reference, and counter, although sensors can be constructed of either two or three electrodes. In a three-electrode sensor all three electrodes are constructed of the same material. It is only the way they are connected to the circuit that determines which are sensing, counter, and reference electrodes. The sensing electrode interacts with the gas to be monitored. The purpose of the counterelectrode is to act as an electron sink for anodic sensing reactions, and as an electron source for cathodic sensing reactions, and simultaneously to complete the circuit for the sensor. The third electrode is a floating electrode not seeing any current that acts as a reference. The function of the reference electrode is to be a bias reference for either an applied or internal potential so that the sensing electrode is at a potential within the region of the limited diffusion current. By using a potentiostatic feedback between it and the sensing electrode, the latter is maintained at a constant potential. The three-electrode sensor is less expensive to make than the two-electrode sensor.

Various modifications of sensors are possible among manufacturers. For example, one manufacturer may choose to slightly acidify the hydrogen peroxide electrolyte in a sulfur dioxide conductimetric sensor to decrease the effects of certain interferents. The gas exposure path can be modified so that the differing amounts of gas reach the electrode, thus affecting the magnitude of the signal. In units with active sampling, high flow rates are used to pull gases to the sensing electrode in order to obtain the highest sensitivity. Control of the electrode exposure area is important and in practical designs as much catalyst as possible is packed into this area (by using materials with high surface areas) to ensure adequate reacting capability and high sensitivity. Each chemical species has its own unique interaction at an electrode. But for almost all the electrochemical cells, the primary variables are current, working electrode potential, analyte concentration, and time.

The construction materials of the sensor will also influence its operating characteristics. Choosing construction materials and sensor geometry is critical and has a profound influence on the accuracy, precision, response time, sensitivity, background, noise, stability, lifetime, and selectivity of the resulting sensor.[1] For example, selection of a gold rather than a platinum electrocatalyst for the sensing electrode allows for selective determination of hydrogen sulfide in the presence of carbon monoxide.

The main purpose of a membrane is to eliminate the need for a flowing electrolyte solution across the electrode. Membranes are usually chosen for their ability to protect the sensing electrode. However, if a membrane has a low permeability to air, the sensor will have a slower response time. Materials used for membrane construction are typically Teflon and high-density plastics like polypropylene, because such materials must be compatible with reactive gases and corrosive electrolytes.

It is extremely important to select the proper electrolyte for each sensor as the electrolyte composition can affect the solubility and the rate of diffusion of the reactant gas to the electrode (catalyst) surface. Electrolyte composition can also alter the chemical being monitored before it reaches the electrode surface. For example, the use of an acidic electrolyte to detect ammonia causes the formation of ammonium ion, which may not be as electrochemically active as ammonia under the conditions of the cell. The electrolyte profoundly influences the response characteristics observed for sensors with strongly acidic and basic electrolytes.

Another means of chemically controlling the sensor properties is by altering the composition of the electrocatalyst. Each catalyst formulation will have unique properties. The activity of platinum for carbon monoxide oxidation has been found to be 10^3 to 10^6 times better than that of gold. Although the reactions occur on both metals, one is orders of magnitude faster than the other.

The selectivity of the sensor can also be improved by controlling the electrochemical potential of the working electrode. For example, proper selection of a gold electrode potential will allow the determination of nitrogen dioxide in the presence of nitric oxide.[1] Response time, linearity, zero drift, repeatability, sensor stability, and even sensor life are dependent on sensor design and methodology.

As noted, gas sensors are often covered with a membrane that is selectively permeable to a given contaminant. This membrane minimizes the likelihood of poisoning of the electrodes by an electroactive or surface-active species. The resolution of these systems is also enhanced if other contaminants (interferents) that would also undergo electron transfer at the electrode can be excluded by tuning the circuitry. The appropriate choice of electrocatalyst can help in achieving selectivity in some applications. However, while the choice of membrane or electrolyte to achieve selectivity is a nice idea, it is not going to eliminate all possible interferents, since virtually any compound for which a similar type of detection method is used is a potential interferent.

Conductimetric Sensors

Conductimetric sensors are those in which the gas being measured reacts chemically with an electrolyte, changing its conductivity. These older-type gas sensors require frequent maintenance to replace reagents and parts, and in general this principle is rarely used. An instrument based on conductimetric detection is described in the section on sulfur dioxide.

Potentiometric Sensors

If a cell allows a change in the potential of the sensor, it is classified as potentiometric. Potentiometric devices have been well studied and can serve as sensors for a variety of gases, such as ammonia, hydrogen cyanide, carbon dioxide, hydrogen sulfide, and sulfur dioxide. Potentiometric devices include ion-selective electrodes (of which the pH electrode is a special example), and at least one commercial toxic gas monitor. These operate on the basis of the Nernst equation, in which electrode potential is a function of concentration:

$$E = E^* - \frac{RT}{z} \log C$$

where

- E = potential difference, in volts
- E^* = electromotive force for the cell, in volts
- R = a constant
- T = temperature
- C = concentration, a product of the activities of each species that can be considered as effective concentrations
- z = charge number for the cell reaction

Potentiometric sensors for gases utilize the direct chemical reaction of the gas with the electrolyte, thereby changing the potential of the sensing electrode from the initial potential to give an electrochemical couple with one of the electrolyte ions. A change in potential produces a change in current and thus a change in voltage. Thus, the potentiometric sensor observes the potential difference between the sensing electrode and the counter electrode that occurs when the chemical species of choice is detected. The change in potential is related to concentration. The output is logarithmically dependent on the concentration of the species being detected. When such a sensor sees a relatively high ppm, it takes time for the electrolyte to recuperate. With very high ppm the electrolyte may never recuperate.

Amperometric Sensors

As noted, amperometric sensors can be subdivided into two classifications: coulometric and voltammetric. However, often the term *amperometric* is used to describe a voltammetric instrument, while it is never used to describe a coulometric monitor. The indicating electrode is polarized against a reference electrode to a constant potential. The sample flows through the cell and the electric current proportional to the concentration of the species to be measured is recorded. A voltammetric sensor uses a constant voltage circuit, that is, the sensing electrode is held at a fixed voltage. Commercial coulometric monitors are constant current devices. Voltammetric amperometry involves galvanic devices while coulometric amperometry involves electrolytic devices. The process that occurs in these sensors is electrooxidation or electroreduction of the species to be analyzed, with the result generating a current. Some require oxygen to function and others do not.

Amperometric sensors are important in portable instrument design because they are relatively small, inexpensive, and lightweight, and use very little power to generate significant signals. The parameters most frequently observed to influence the characteristics of amperometric sensors include sample flow rate, working electrode composition, electrolyte, membrane type, and electrochemical potential of the sensing electrode.[1] Accuracy is higher than other methods such as potentiometry. These sensors exhibit fast responses, can detect ppm levels of electrochemically active gases and vapors, can be engineered to have significant selectivity, and can be operated over a wide range of temperatures.

Coulometric Sensors

Coulometric detectors measure the quantity of electricity (in coulombs) that passes through a solution during the occurrence of an electrochemical reaction. Therefore, measurement of the amount of electrical energy transferred across an electrode-solution interface (number of electrons) in terms of the coulombs required to carry the reaction of a specific substance to completion is called coulometry. During coulometric sensing the gas being monitored is consumed by electrolysis during passage through the sensor. The current is controlled by the feed rate of the sample and corresponds to the charge passed in a given time unit:

$$\text{Current} = \frac{\text{Charge}}{\text{Time}}$$

It is generally a wet chemical titration method where one of the reactants is generated in the test cell by electrolysis of a solution. These detectors can be made specific by adjusting concentration, pH, and composition of the electrolyte. An important feature of the coulometric sensor is that the current is, within certain limits, independent of the change in the working electrode's activity; thus the electrode characteristics are unimportant. Electrodes may be constructed from platinum foil, wire, or mesh. The accuracy of the sample flow rate controls the accuracy of measurement and must be kept as constant as possible; the signal is, however, independent of the temperature. Coulometric cells are reusable when periodically cleaned and recharged with a fresh electrolyte solution. Since this technique is used almost exclusively for ozone, it is discussed further in that section in this chapter.

Voltammetric Sensors

A typical voltammetric sensor consists of six major parts: filter, membrane, working or sensing electrode, electrolyte, counter electrode, and reference electrode. The gaseous species of interest is transported (by pump or diffusion) across the membrane to the sensing electrode of the cell; it then migrates to the electrolyte boundary, dissolves in the electrolyte, diffuses to the electrode surface, and reacts electrochemically. The products of the reaction diffuse away from the electrocatalyst surface.[2]

The signal from the sensor is sent to a main current amplifier, and is read out as a voltage across the feedback resistor of the amplifier. Different monitor ranges use different feedbacks. Only three-electrode voltammetric sensors in which the counter electrode acts as an oxygen electrode require oxygen. The Interscan two-electrode sensor can operate in totally anerobic conditions because it uses a nonpolarizable second electrode that acts as both a counter and reference electrode. In voltammetric sensors designed for gas monitoring, electrode characteristics are very important, including the size of a sensing electrode. Voltammetric sensors are constructed in such a way that very little depletion of the gas being monitored occurs.

As a way of understanding this method of detection, a typical voltammetric curve is shown in Figure 8-5. Oxidation and reduction current is represented on the $+y$ and $-y$ axis, respectively, while voltage is plotted along the x axis. Below the reaction potential of a given electrooxidizable gas there is no reaction, and therefore the current is zero (except for a small charging effect). As the reaction potential is approached and exceeded, the current rises sharply until it reaches a maximum limited only by the diffusion of the gas to the reaction site. The magnitude of this current is directly proportional to the number of electrons per mole, the Faraday constant, the diffusion coefficient, the surface area of the reaction site, and the concentration of the gas, and is inversely proportional to the diffusion gradient resulting from the varying concentration of reacting gas that ranges from the highest concentration that is found in the electrolyte to zero concentration at the electrode surface. There is a linear relationship between the current generated and the ppm of the gas.

If the voltage is increased beyond the limited diffusion current, which is represented by the *a* to *b* section of the curve in Figure 8-5, the current will increase again due to electrooxidation of the

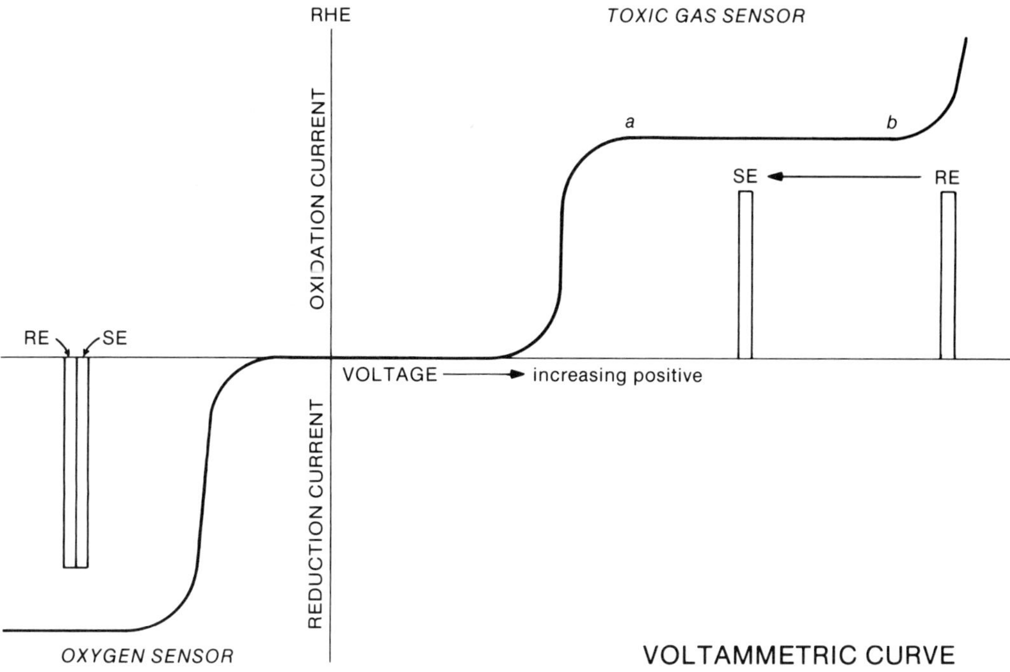

Figure 8-5. Basic principles of operation of a voltammetric sensor. (Courtesy Interscan Corp.)

electrocatalyst comprising the electrode, and a further increase of voltage results in a very high current due to electrooxidation of water to oxygen gas, and no detection takes place.

To have a working device, the sensing (working) electrode must be held at a potential within the region of the limited diffusion current. This is represented by SE in Figure 8-5. A reference electrode (RE) that has a lower or higher oxidation potential than that of SE is used to maintain the sensing electrode in the region of the limited diffusion current. The potential of the reference electrode must not change with the passage of current. Depending on whether the potential of RE is higher or lower than SE, it will have a voltage that is biased in either the negative or positive direction, respectively. The height of the diffusion plateau from the voltage axis is linear with the ppm of the gas. This is an example of an externally biased sensor.

It is also possible to have an internally biased sensor, an example being the oxygen sensor containing potassium hydroxide and a lead anode. The lead–lead hydroxide reference electrode potential has a value coinciding with the limited diffusion current for oxygen reduction. It is therefore unnecessary to apply an external bias. The sensing electrode in this case is immediately polarized from its rest potential to assume the potential of the reference electrode. This is shown to the left of the y axis in Figure 8-5.

Typical features of voltammetric sensors are fast response, linear response in a broad concentration range, nonlinear dependence on the flow rate, and the temperature dependence of the signal.

LIMITATIONS OF ELECTROCHEMICAL SENSORS

Most manufacturers of electrochemical sensors specify the lower temperature limits for operation, usually 0°C to 10°C, and upper limits, typically 50°C to 60°C. These do not present a problem for most applications, but the user should be aware that lower temperatures tend to result in longer response times. When used outdoors in cold climates, ambient temperatures should be monitored to assure that the lower limits are not exceeded.

Electrochemical sensors designed to measure toxic gases may have cross sensitivities to other compounds. Response specificity is determined by the semipermeable membrane selected, the electrode material, and the retarding potential (the potential used to retard the reaction of gases other than that being sampled). Major interferences are those gases for which other electrochemical sensors are designed to monitor, because they are easily electrooxidizable or electroreducible. Gases that cannot be oxidized or reduced, such as methane, do not interfere.

Another way of understanding the phenomenon of interference is to return to Figure 8-5. On the voltammetric curve, if any other electroactive gases are present during monitoring that have a lower oxidation potential than the SE potential (i.e., to the left of the SE bar), they will act as an interference. The extent of this interference will depend on the reaction kinetics of the interfering gas. An example of adjusting the SE potential to achieve selectivity is the measurement of sulfur dioxide in the presence of nitric oxide. This can be done because nitric oxide has a higher oxidation potential than sulfur dioxide.

Methods of eliminating interferences are to use selective chemical filters or to prescrub the sampled atmosphere. Both of these methods have been used by manufacturers for some applications. However, not all interferences can be eliminated and the sampling professional needs to be familiar with those that may pose a problem.

Poor stability of a voltammetric sensor is primarily due to the effect of temperature on the inherent background current present in every sensor. This current is due to internal electrooxidation or electroreduction. This accounts for the zero drift found in such sensors. Most passive electrochemical sensors should not be exposed to the air when not in use as the water loss by evaporation through the porous membrane will shorten sensor life. Most commercial diffusion mode sensors have a limited volume to hold water.

Sensor longevity is also a problem. Since most chemical sensors depend on a specific chemical reaction to generate a signal, the reactions involved in these sensing mechanisms are often not sufficiently reversible, so that the sensitive material (electrolyte) is used up during the life of the sensor.

CALIBRATION AND MAINTENANCE OF ELECTROCHEMICAL UNITS

For electrochemical sensors used to monitor chemicals, such as hydrogen sulfide, carbon monoxide, nitrogen dioxide, and sulfur dioxide, stable cylinders of calibration gases in the concentration range of interest are best to use for calibration. These sensors may also be zeroed in fresh air or zero air from a compressed gas cylinder or clean air prepared by filtration. The frequency of calibration cannot be prescribed exactly, but a good rule is to calibrate at least once a day at the start of a shift. The manufacturer's instructions or user experience may adjust this frequency.

A special cup designed to fit over a sensor is available from at least one manufacturer. It allows a direct hookup of a gas bag to instruments that have a protruding sensor.

Calibration is best performed using a span gas with a concentration about equal to the mid-scale value of the meter. Never connect a gas cylinder directly to the intake port of an electrochemical instrument, as doing so will pressurize the instrument and cause serious damage to the sensor. Instead, fill a sample bag with the calibrating gas and attach it to the intake port of the unit.

For continuous monitoring applications in excess of 3 hours, electrochemical devices (Fig. 8-6) often require an external source of humidity. This setup often consists of a bottle with distilled water installed in-line with the intake port. If a filter is incorporated into the inlet of the instrument, it should be removed because it will trap the water vapor intended for the sensor. The air filter should be reconnected to the front of the humidity control bottle.

The accuracy of most electrochemical monitors is dependent on the age of the sensor. Electrochemical sensors often need to be replaced approximately every 6 months. Sensors should be replaced when they can no longer be calibrated or zeroed easily. Sensor life is highly vari-

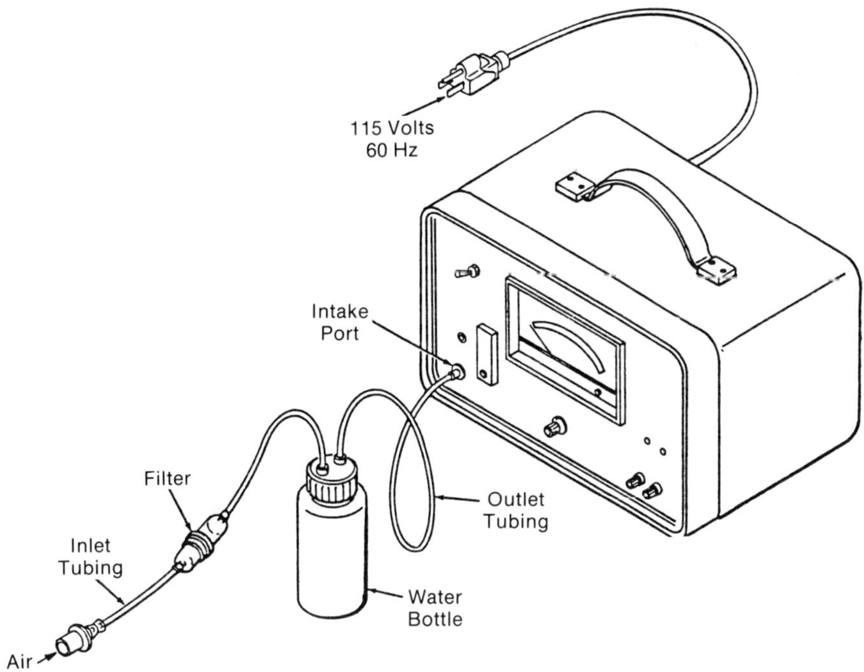

Figure 8-6. Setup of an electrochemical analyzer for continuous monitoring. (Courtesy National Draeger, Inc.)

able and differs among instruments. Most sensors are given a rated "use life" in hours by the manufacturer. The actual life will depend on the type of sensor, frequency and duration of use, concentration levels measured, presence of compounds that also react with or contaminate the sensor, temperature, humidity, maintenance effort, and storage conditions for the instrument. For example, a sulfur dioxide sensor exposed to 5 ppm continuously may last 10,000 hours, or more than a year. If the concentration is lower or higher than 5 ppm, a longer or shorter life will result. It is important to remember that estimates of sensor life are for normal usage, and misuse can greatly shorten the life of the cell. Adequate warmup time is necessary following sensor replacement; from 5 to 24 hours may be necessary to allow the sensor to stabilize.[3] Some cells can be reactivated by the manufacturer.

Signs indicating a need to replace a cell include an inability to zero or calibrate an instrument or readings that become erratic or are strongly affected by moving the instrument. If a cell starts to leak, it should be immediately removed from the instrument and discarded while wearing protective gloves.

For some units, proper readings can only be obtained when the battery has a sufficient level of charge. Removal and separate packaging of the electrochemical sensor prior to shipment is often necessary, especially if it contains a wet electrolyte. This packaging is also needed prior to long-term storage. When shipping electrochemical instruments, remove the cell from the instrument, wrap it in a plastic bag, and pack it with the instrument. The instrument should be in an upright position when shipped. Handle electrochemical sensors carefully; they are sealed units that contain electrolytes that are often acidic or caustic. If electrolyte is leaking from a sensor, exercise caution so that the electrolyte does not contact skin, eyes, or clothing, thus avoiding burns. If contact occurs, rinse the area immediately with a large quantity of water. In case of contact with the eyes, immediately flush eyes with plenty of clean water for at least 15 minutes and contact a physician.

Figure 8-7. A unit for personal monitoring for TDI *(at left)* and the setup for dumping data *(below)*. (Courtesy GMD Systems, Inc.)

TAPE-BASED MONITORS

Chemically impregnated tapes that change color upon exposure to a given chemical have been available for some time. The most common compounds for which the tape-based monitors are used are isocyanates, because this procedure is the only continuous monitoring method available for isocyanates. Hydrogen sulfide was one of the first contaminants for which a tape-based system was developed. Other compounds for which monitors are offered include phosgene, hydrazine, ammonia, bromine, chlorine, hydrides (arsine, diborane, disilane, germane, hydrogen selenide, phosphine, silane, stibine), hydrogen cyanide, hydrogen sulfide, nitrogen dioxide, ozone, sulfur dioxide, and acid gases (hydrogen bromide, hydrogen chloride, hydrogen fluoride, nitric acid, sulfuric acid).

Tape-based monitors are available as small passive badges, stationary monitors, survey instruments, and personal monitors (Fig. 8-7).

Models are available that have alarms, both audible and visible, continuous monitoring, and battery-operated personal units as well as fixed station monitors. Tape-based instruments can be purchased as dedicated systems or as instruments in which different cassettes can be inserted. In either case, the instrument will sample for only one chemical at a time.

An advantage is this technique's ability to directly monitor for reactive compounds, such as isocyanates. A disadvantage is that tape-based monitoring suffers from interferences, for example, anything that can stain the tape, including humidity or grease, will cause a response. Hydrogen sulfide is an interferent in hydride monitoring because most instruments use a generic tape. Tape-based monitors are discussed in more detail in the section on isocyanates.

OTHER DETECTION PRINCIPLES

UV detectors are used for compounds that absorb UV light. These units are currently available for ozone and mercury and in each case a specific wavelength is chosen. Monitors that measure ozone convert it to oxygen first, which in turn absorbs UV. Since mercury absorbs strongly in the UV region, no chemical conversion is necessary. UV-based monitors often have a number of interferences, because many compounds including some organic materials absorb strongly in the UV spectrum. These include benzene and related compounds, alcohols, acetals, ketones, and pyridine.[4] Iodine absorbs strongly at 352 ηm and 440 ηm. Nitrates and nitrites absorb at 270 ηm and 225 ηm, respectively. UV instruments are discussed further in the section on mercury.

Chemiluminescence is used to detect ozone and nitrogen oxides. Chemiluminescence, the emission of visible light, is the result of a specific case of chemical excitation wherein the excited state reaction products are luminescent, thereby producing a reaction intermediate or product in an electronically excited state. The radioactive decay of the excited state is the source of luminescence. The amount of energy emitted at a specific wavelength during such a reaction is proportional to the number of molecules reacting.

In a typical instrument, a known volume of ambient air is introduced into a sample chamber where it reacts with a specific reagent that will then cause radiant energy to be released. The light emitted may then be detected by an instrument calibrated for the specific contaminant. The use of narrow-band optical filters enhances the selectivity of these instruments because interferences are decreased. The instruments are frequently more expensive than other instruments specific for a given chemical.

In colorimetric instruments the sample interacts with a reagent solution that changes color in proportion to the concentration. The degree of color change of the chemically developed solution is measured at a specific wavelength in the visible region of the spectrum in an optical spectrometer and referenced to a "blank" of the colorimetric reagent. An instrument of this type is described in the section on formaldehyde.

In gold film detection a thin gold film that is part of a piezoelectric sensor undergoes an increase in electrical resistance proportional to the concentration of a contaminant. The most common application of this principle is mercury monitoring, although it has also been applied to hydrogen sulfide.

SPECIFIC CHEMICALS

Oxygen

Oxygen measurements are commonly made in conjunction with combustible gas measurements usually for confined space entry. For a discussion of combustible gas measurement techniques, see the chapter on General Survey Instruments for Gases and Vapors.

Oxygen monitors generally use galvanic electrochemical cells. This cell could be described as a small battery with two electrodes, a noble metal and a base metal, in an electrolyte, covered by a permeable Teflon membrane. Typically, the noble electrode is gold or platinum, and the base metal is lead or zinc. The reactions within the cell lead to oxidation of the base metal, and generate a corresponding current.

The galvanic cell generates a current at the

two-electrode terminals of the cell directly proportional to the rate of oxygen diffusion into the cell. At the end of the useful life of the galvanic cell, its output current drops dramatically, which is one way of determining the need for cell replacement. No refurbishing of the sealed fuel cell is possible. Therefore, oxygen cells are inherently self-consuming and must be replaced or reactivated periodically. The life of the cell is shortened by exposure to heat or dry gases. Operating hours do not influence cell life, since the cell is constantly exposed to ambient air and constantly generating its output current, regardless of whether the instrument switch is on or off. Oxygen cells usually last 6 months to 1 year.

The role of water vapor as an air diluent often goes unrecognized as a cause for error when testing warm spaces for oxygen content using a diffusion-type instrument. An oxygen indicator is usually set to read 21% on atmospheric air, which contains a certain amount of water depending on the temperature and relative humidity (RH). A closed vessel or highly confined space, however, can have close to 100% RH. If the temperature in this space is substantially above the ambient, the oxygen will be noticeably diluted; therefore, at 43°C, for example, the highest reading to expect on a tank at 100% RH is 19%.

Oxygen cells are temperature-dependent as well. This dependence is compensated for by design to some degree, but when first exposed to a change in temperature, the thin membrane will respond to the change before the thermistor compensator has time to do so, and the reading will temporarily overshoot. The sensors are generally temperature-compensated from 0°C to 40°C. The indicator's response time is frequently increased in temperatures beyond the compensated range, particularly below 0°C. However, it is a good idea to avoid big temperature changes between the calibration and the operating conditions.

Electrochemical sensors for oxygen provide either a continuous or on-demand display of the present percent oxygen in the atmosphere, and the alarm circuitry is designed to activate at the moment the concentration drops to 19.5% oxygen. Some models, designed to be used in hospitals or as area monitors, have both upper and lower alarm levels so that oxygen-enriched atmospheres may also be monitored. Many of these models have output suitable for a strip chart recorder, so that a permanent record of the oxygen in the atmosphere may be maintained. The output of these devices is in percent oxygen.

Strong oxidants such as fluorine, chlorine, bromide, and oxone may lead to erroneously high oxygen readings when these substances are present in significant concentrations. Sulfur dioxide and nitrogen oxides can also interfere. Acid mists or other corrosive atmospheres can poison oxygen probes, decreasing their sensitivity. Examples are mercaptans and hydrogen sulfide in high concentrations (>1%), which can occur in confined spaces. Acid gases, such as carbon dioxide, will shorten the service life of an oxygen sensor. In at least one instrument the sensor life will be reduced to 2 days in 100% carbon dioxide, 50 days in 5%, and 100 days in 1%.

Calibration and Maintenance of Oxygen Monitors

Due to the temperature dependency described earlier, it is important to avoid large temperature changes between the calibration and the read conditions. Try to calibrate the instrument immediately before testing, and at nearly the same temperature as the testing conditions.

The most simple example of field calibration of an electrochemical sensor is the oxygen monitor. Calibration can be done by placing it in fresh outdoor air and adjusting the calibration potentiometer to make the readout at the specific oxygen percent indicated in the manufacturer's literature, usually 20.8% or 20.9%. To determine if it responds to oxygen deficiency, a person should hold his or her breath for a few seconds, then slowly exhale, directing the exhaled breath to the sensor. Air that has been in contact with the lungs for 5 seconds or so will be depleted of oxygen down to about 16%. If it is functioning properly, the meter will deflect downscale and the alarm circuit will be activated.[5] A need for more frequent and larger concentration indicates the need for sensor replacement. This calibration is not precise, but it is a quick and reliable test.

It is important to verify that the instrument goes to zero, or close to zero, when oxygen is removed. This test can be made with nitrogen, argon, helium, methane, propane, or any gas that will exclude oxygen.

Because oxygen monitors are dependent on the partial pressure of oxygen, readings may not be valid unless the total pressure is less than one atmosphere. For example, the MSA Model 245 oxygen indicator, when calibrated at sea level, will indicate 20% in fresh air at 1,000 feet, 19.3% at 2,000 feet, 18.6% at 3,000 feet, and so forth. This problem can be avoided if the instrument is calibrated in an atmosphere containing at least 20.9% oxygen.

Routine instrument maintenance is important. Check the sensor cell diffusion barrier for dirt and moisture. Clean with a soft cloth or tissue if needed. Replace the sensor cell every 6 months to 9 months with average use. Store the cell in nitrogen when not in use to prolong its life. Some models have nondisposable, user-rechargeable sensors that require charging every 2 to 4 weeks. Check batteries and replace when needed.

Use of Oxygen Monitors

Always check the atmosphere for explosives prior to activating the oxygen indicator. In sample areas where the temperature is not constant (it changes by more than 17°C), or in sampling atmospheres that differ in temperature from that of calibration air (by more than 17°C), the fresh-air reading (calibration) should be rechecked every hour to obtain the greatest accuracy possible. The operating range for RH is 10% to 90%. Avoid touching the sensor with the hand or sharp objects, since the membrane is easily damaged.

Sampling conditions that lead to condensation of moisture on the sensor face will cause erroneously low oxygen readings. These conditions, such as taking a cool sensor into a warm, moist atmosphere, produce a film of water on the sensor face that decreases the transport of oxygen from the atmosphere to the inside of the sensor. To minimize this problem, the sensor should be kept as warm or warmer than the sample area—before and in the intervals between sampling.[5]

Procedure

1. Start the monitor as the manufacturer recommmends. Allow sufficient warm-up/equilibration time if the monitor has been in a different environment. If a remote sensor is located in a different area than the monitor's electronics, allow extra time to reach thermal equilibrium. Generally warm-up is 10 minutes or less, but may be up to 1 hour if the sampled atmosphere varies 5°C or more from the present ambient temperature.

2. For ambient air monitoring, place the monitor in the atmosphere to be analyzed. Allow it to equilibrate. Record the initial percent oxygen readout. Allow the instrument to continue monitoring. Observe any downward trends in percent oxygen levels. If a concentration of 19.5% oxygen is approached, ventilate with fresh air if possible. If the level falls below this, evacuate the area.

3. For oxygen-enriched air measurements, place the sensor in an air line adapter. Allow it to equilibrate. If the line is pressurized, bleed off a stream to the sensor at atmospheric pressure in order to not pressurize the sensor. Record the percent oxygen.

4. Periodically throughout the shift perform an on-site calibration at 20.9% oxygen by exposing the sensor to fresh air or to standard tank air. Replace the sensor cell when the instrument can no longer be adjusted to read 20.9% oxygen in fresh air. Take care not to pressurize the sensor by a direct hookup to a cylinder, as doing so will cause high readings and could do irreparable damage to the sensor cell.

5. Perform a downscale calibration by exposing the sensor to 16% to 19% oxygen in nitrogen at least once per day.

6. Check for alarm function by exposing the unit to 16% to 19% oxygen in nitrogen from a cylinder or by blowing self-exhaled breath directly onto the sensor. Following the manufacturer's instructions for this procedure, adjust the readout on the meter to agree with the calibration standard.

Interpretation of Oxygen Measurements

If the oxygen content is below 19.5%, according to the Occupational Safety and Health Administration (OSHA) definition, the air is oxygen-deficient and air-supplying respirators are required to work in the area. If the oxygen content is greater than 25%, an oxygen-rich situation exists and there is a potential for explosions. If toxic contaminants are present, a pressure demand self-contained breathing apparatus (SCBA) is required for less than 19.5%, and a cartridge respirator may be worn for greater than 19.5% as long as the concentrations do not exceed the cartridge respirator's capability to protect the user. A low oxygen reading in an enclosed atmosphere indicates that some other gas has displaced much of the air or some process has consumed much of the available oxygen.

Changes in barometric pressure due to altitude will have an effect on the meter reading. If the instrument is calibrated at sea level, it will indicate a lower percentage of oxygen by volume at higher altitudes. However, adequate oxygen to sustain life is dependent on partial pressure rather than percentage by volume, and a lower reading at a higher altitude is acceptable. It is best to compensate for this situation ahead of time.

Carbon Monoxide

Carbon monoxide (CO) most commonly occurs as a by-product of combustion. The most common detection methods for CO are electrochemical cells, solid-state sensors, and IR methods. With the proper mixture of metal oxides and the appropriate operating temperature, solid-state sensors can be made relatively specific for CO and can have a severely reduced response to other gases and vapors. These sensors are further discussed in the chapter on General Survey Instruments for Gases and Vapors. IR-based instruments when dedicated to the measurement of CO are frequently stationary devices. For a discussion of these devices see the chapter on Monitoring Instruments for Many Specific Gases and Vapors: GC and IR.

The most common specific operating principle for detecting CO is electrochemical oxidation at a potential-controlled electrode. The current generated by this electrochemical reaction is then directly proportional to the CO concentration. The basic components of most instruments are a three-electrode electrochemical sensor, a sampling system (pump), a sample preconditioning unit (water bottle, filter), an electronic control circuit (potentiostat), and a measuring circuit.[6] The potential of the sensing electrode is controlled by the potentiostat. The sensor consists of a sensing electrode, a counter electrode, a reference electrode, a housing containing sulfuric acid solution, and two face plates. The overall cell reaction is

$$2CO + O_2 \rightarrow 2CO_2$$

Common measurement ranges are 0–100 ppm and 0–500 ppm. Fast response and recovery are a property of these electrochemical sensors as well as minimal drift. Interference filters must be used in the intake port to remove oxides of nitrogen and other compounds. Other compounds known to significantly interfere with CO measurements by electrochemical detection include hydrogen, hydrogen sulfide, sulfur dioxide, ethane, methane, ammonia, and propane. The degree of interference depends on each manufacturer's instrument.

Hydrogen Sulfide

Hydrogen sulfide is commonly monitored using real time instruments rather than integrated measurement techniques due to its prevalence, especially in confined spaces, and its high toxicity. Hydrogen sulfide also causes olfactory fatigue, that is, the odor-detecting cells in the nasal passages become tired after a period of exposure to hydrogen sulfide. As a result, the worker may think that the gas has gone away or may not be aware of an increase in concentration. Measurements are often made in conjunction with measurements for combustible gases and oxygen deficiency. Hydrogen sulfide is the primary toxic agent responsible for deaths in confined spaces. As has already been noted, hydrogen sulfide is frequently incorporated into monitors that can measure for combustible gases and oxygen deficiency. Other situations where hydrogen sulfide is encountered include kraft pulp mills and sewers. In both cases, there is the likelihood of encountering mercaptans, another type of sulfur-based compound.

Instruments that measure hydrogen sulfide are based on several different principles, the most common being solid-state and wet electrochemical sensors. Most solid-state sensors are used in general survey instruments since they have low specificity. However, a special case is presented for hydrogen sulfide solid-state sensors where improvements in technology have resulted in a sensor that is somewhat specific for hydrogen sulfide and carbon monoxide only. Hydrogen sulfide can also be detected by PID, but currently there is no dedicated instrument available based on this principle. There is also an instrument available based on gold film; however, this instrument is too sensitive to be useful for health hazard levels. For more information on solid-state and PID sensors, see the chapter on General Survey Instruments for Gases and Vapors.

Most instruments designed to measure health hazard levels are capable of detecting hydrogen sulfide over a range of 0.1 ppm to 50 ppm. All hydrogen sulfide meters should have an audible alarm that can be preset at a desired level due to the high toxicity of this compound.

The primary types of monitors available for hydrogen sulfide are small personal units capable of continuous monitoring usually set to alarm at 10 ppm and combination units where a hydrogen sulfide sensor is included along with a combustible gas and oxygen sensor. Personal monitors are generally based on wet electrochemical sensors. For more information on the use of real time personal monitors, see the chapter on Reading and Recording Real Time Data.

A similar electrochemical cell to that used to detect carbon monoxide but of a different material and different voltage is used for hydrogen sulfide. A common hydrogen sulfide electrochemical cell utilizes a reaction in which hydrogen sulfide is electrooxidized to sulfuric acid in an aqueous electrolyte at a catalytically active electrode according to the following equation:

$$H_2S + 4H_2O \rightarrow H_2SO_4 + 8H^+ + 8e^-$$

The potential at which the electrode is maintained is such that neither the electrooxidation of water nor the electroreduction of oxygen occurs at a measurable rate. Therefore, the current measured in the detector is a result of the electrooxidation of hydrogen sulfide, and is proportional to the partial pressure of hydrogen sulfide in the gas sample. The current generated in the sensor is amplified and displayed in units of ppm.

Hydrogen sulfide is a very reactive gas and tends to absorb in or react with many materials; therefore, the best material for sampling hoses is Teflon. The most common interferents for hydrogen sulfide electrochemical sensors are mercaptans, although some cells are sensitive to low levels of sulfur dioxide, propane, nitric oxide, ethylene, acetylene, or ethanol.

There are three calibration methods available for hydrogen sulfide: permeation tubes, cylinders of span gas, and ampules. For more information on the use of permeation tubes, see the appendix on Gas and Vapor Calibrations. A test kit is also available that allows the sampler to perform an operational test check prior to use. It consists of a plastic bottle and H_2S test ampoule. The ampoule is broken by vigorous shaking of the bottle and the sensing element is inserted. For this kit, an instrument is considered operational if a reading of greater than 10 ppm is obtained.

It has been recommended that a daily (or prior to each use) calibration check be performed on instruments that measure hydrogen sulfide due to the highly toxic nature of this gas. Instruments should not be zeroed in the field wherever there is the possibility that hydrogen sulfide could be present, due to the problems that even small amounts can cause.[7]

Sulfur Dioxide

Sulfur dioxide can be detected using conductimetric, coulometric, voltammetric, and infrared methods. An example of a voltammetric cell for sulfur dixode functions as follows: A current is generated by charge transfer of sulfur dioxide at the surface of an externally biased electrode. This current is proportional to the concentration of sulfur dioxide in the region of the electrode. The linear relationship between cell current and sulfur dioxide requires a diffusion limiting condition, which is a charge transfer whose rate is limited by the concentration gradient in the electrolyte enveloping the electrode (the arrival rate of sulfur dioxide molecules at the electrode surface).

Conductimetric sensors detect sulfur dioxide by trapping the sulfur dioxide in a dilute solution of hydrogen peroxide. The air sample is bubbled through the solution in the electrolytic cell, and sulfur dioxide in the air reacts with the peroxide solution to form sulfuric acid. The conductivity change of the cell is proportional to the sulfuric acid concentration. Thus, the sulfur dioxide in the sampled air is determined by measuring the conductivity before and after sampling and the total sample volume. Conductivity instruments often require the meter response (in conductivity units) to be converted to concentration using a calculation involving initial and final conductivity, sampling time, and a calibration constant, or on some units, a given percent of the scale is proportional to ppm. Sulfur dioxide calibration can be done using permeation tubes or a mixture in gas cylinders. Ozone is a common interferent in the electrochemical and conductimetric techniques.

Nitrogen Oxides

Real time measurement of nitrogen oxides (NO_x) can be done using chemiluminescence, electrochemical, or tape-based detection methods. Chemiluminescent analyzers can monitor for either nitric oxide or nitrogen dioxide. To measure nitric oxide concentrations, the gas sample being analyzed is blended with ozone in a flow reactor. The resulting light emissions are monitored by a photomultiplier tube. To measure total oxides of nitrogen, the gas sample is diverted through an NO_2 to NO converter before being admitted to the flow reactor. To measure nitrogen dioxide, the gas sample is intermittently diverted through the converter, and the NO signal is subtracted from the NO_x signal. Some instruments utilize a dual-stream principle with two reaction chambers.

$$NO + O_3 \rightarrow NO_2^* + O_2$$

$$NO_2^* \rightarrow NO_2 + h\nu$$

The chemiluminescent detection of NO_x with ozone is not subject to interference from any of the common air pollutants, such as ozone, nitrogen dioxide, carbon monoxide, ammonia, and sulfur dioxide. Possible interference from hydrocarbons is eliminated by means of a red sharp-cut optical filter.

When these instruments are operated in the nitrogen dioxide or NO_x modes, any compounds converted to nitrogen oxide in the thermal converter are potential interferences. The principal compound of concern is ammonia; however, it is not an interferent for converters that are operated at less than 300°C. Other nitrogen compounds, such as peroxyacetyl nitrate (PAN) and organic nitrates, decompose thermally in the converter to NO and may represent interferences in some polluted atmospheres or in smog chambers.[8] Chemiluminescent analyzers often have very low detection limits of 0.002 ppm to 0.01 ppm, with maximum ranges of 0 – 5 ppm to 0 – 10,000 ppm and accuracies of $\pm 1-2\%$.

Most electrochemical monitors are usually designed to detect nitrogen dioxide only and work similarly to those discussed under carbon monoxide. Sulfur dioxide and hydrogen sulfide have been known to interfere with measurements of nitrogen dioxide by these instruments. Calibration of nitrogen dioxide analyzers requires the use of permeation tubes.

Ozone

Due to the difficulties in the use of integrated sampling methods to measure ozone, the sampling professional must rely on real time instruments to measure this compound. Ozone can be monitored using chemiluminescence, UV, and coulometric methods. Ozone can be generated by electric sources such as plate makers, x-ray machines or UV generators, arc welding equipment, electric arcs, mercury vapor lamps, linear accelerators, electrical discharges, and photocopy machines. It is also used in the organic chemical industry, in cold storage rooms as a disinfectant for food, in water purification, in textile and paper pulp bleaching, in aging liquor and wood, in the perfumery industry, in treating industrial wastes, in the rapid drying of varnishes and printing inks, and in the deodorizing of feathers. It is a common outdoor pollutant and occurs naturally in the higher atmosphere where it absorbs solar UV radiation.

In this regard it is often lumped into a group called total oxidants.[9] In a situation where oxidation is suspected, such as when metals are rusting, measurement of total oxidants may be preferable to the use of an ozone-specific monitor. Other known oxidants commonly included in this term are chlorine, bromine, fluorine, iodine, nitrogen dioxide, nitric oxide, hydrogen peroxide, peracetic acid, peroxyacetyl nitrate (PAN), and chlorine dioxide.

In a typical chemiluminescent monitor for ozone the system provides a flow of ethylene concentric with a continuous air sample into a reaction chamber. The two streams merge and react, and the light emitted by the reaction is sensed by a photomultiplier tube. The output of the tube is amplified and displayed on a meter. Some of these monitors can detect ozone down to 0.001 ppm.

It is generally advisable to expose a new chemiluminescence analyzer or one that has been in disuse to a relatively high concentration of ozone (1 ppm or greater) for a short period of time before use or calibration. This procedure will passivate the surfaces to ozone. The period takes from 5 to 30 minutes, depending on how long it has been since the monitor was used. Chemiluminescence monitors have cylinders of ethylene that must be kept fully charged. Portable instruments have their own internal supply of ethylene.

A typical coulometric oxidant meter is one that detects ozone through the oxidation reduction of potassium oxide (KI). The sampled air is brought into contact with a chemical sensing solution containing the proper amount of KI as it is metered into the sensor at the cathode of a unique cathode-anode electrode support structure by way of a solution supply tube. The solution flows in a fine film down the electrode support, upon which are wound many turns of a fine wire cathode and a single turn of a wire anode. An air sample is pumped through the sensor, where it comes into contact with the solution contained on the electrode support, and exits through a precision vacuum pump. This reaction takes place on the cathode portion of the electrode support and is as follows:

$$O_3 + KI + H_2O \rightarrow O_2 + I_2 + 2KOH$$

In this region, any ozone in the air sample reacts with the sensing solution. At the cathode, a thin layer of hydrogen gas is produced by a polarization current. When a voltage is applied to the electrodes, the hydrogen layer builds up to its maximum and the polarization current ceases to flow. As free iodine is produced by the reaction with ozone, it immediately reacts with the hydrogen, reducing it to produce hydrogen iodide. The removal of the hydrogen from the cathode causes a repolarization current to flow in the external circuit, reestablishing equilibrium. Thus, for each ozone molecule reacting in the sensor, two electrons flow through the external circuit. The rate of electron flow, or current, is directly proportional to the mass per unit time of ozone entering the sensor. This reaction is based on ozone chemistry, but the halogens react similarly. Therefore, the detection of other oxidants such as chlorine, bromine, fluorine, and iodine is also possible.

In a coulometric sensor for oxidants, reducing compounds such as hydrogen sulfide and sulfur dioxide are negative interferents while oxidizing compounds are positive interferents. Most coulometric instruments require a source of current other than a battery and can also require long warm-up periods if the meter is turned off for several hours. Replenishment of the consumable sensing solution must be done in a laboratory.

Calibration of ozone monitors is done using an ozone generator, which is usually a mercury arc emitting 190 ηm radiation. This calibration can be done either indirectly by a colorimetric technique utilizing the liberation of triiodide ion when ozone reacts with a buffered potassium iodide solution, or directly by a UV spectrophotometric method using a Dasibi meter. If an instrument is linear, it must be calibrated only with one ozone concentration.

Formaldehyde

Only a few real time methods exist that can be used to monitor for formaldehyde: IR, voltammetric methods, and colorimetric methods. The IR is generally not used, since the levels of concern for formaldehyde, 1 ppm and less, are below the reliable detection limits of most instruments. Only a single electrochemical instrument is available, possibly due to the same problem encountered with IR—the need to monitor at very low levels. The colorimetric instrument has been available for some time.

The CEA TGM 555 is a portable wet chemical analyzer that is capable of continuously monitoring for a number of compounds utilizing colorimetry. Sample air is continuously drawn through an absorbing reagent that removes the compound of interest from the airstream and transfers it into a liquid reagent system that undergoes a color change. The intensity of the color is read at a specific visible wavelength. The unit can monitor for other compounds besides formaldehyde, mostly inorganic, but this monitoring requires a separate set of reagents for each type of contaminant. In order to change the unit over to monitor for a different contaminant, the analytical module tray that contains the necessary glassware and reagents is replaced. While this can be done in the field, it may be preferable to convert the unit in a laboratory so it can be calibrated beforehand. Like tape-based units, this instrument can monitor for only one compound at a time.

For formaldehyde detection a two-step process first reacts sample air with a sodium sulfite/sodium tetrachloromercurate solution and then with acid-bleached pararosaniline to create a measurable color change. In an evaluation of the instrument's ability to monitor for formaldehyde it was determined that manufacturer detection levels can be difficult to achieve and modifications to improve sensitivity during formaldehyde sampling have included an increase in pararosaniline concentration, a longer period for the reaction to develop following the sample collection, and an increase in the sampling flow rate.[10] Other modifications to decrease electronic noise reduction included smoothing of all liquid flow throughout the instrument; modified cleaning, storage, and warm-up protocols; and the substitution of glass tubing for several of the liquid flow lines.

Like other instruments colorimetric monitors suffer from interferences; in the case of formaldehyde these are acetaldehyde and propionaldehyde. The CEA TGM 555 has been found to be insensitive to humidity between 15% and 90% RH at 26°C, but temperature-dependent between 16°C and 38°C.[10]

Formaldehyde gas is not available in cylinders. Calibration mixtures have been prepared from *p*-formaldehyde crystals. Formalin, often assumed to be liquid formaldehyde, is actually a mixture of methanol and formaldehyde in water.

Isocyanates

Isocyanates are monitored using tape-based instruments (Fig. 8-8). In these instruments, the tapes either move continuously or sequentially past a sampling window at a fixed rate. Air is drawn through the tape at a known rate. As the contaminant-laden air passes through the tape, it reacts with the chemicals on the tape to produce a continuous, well-defined, characteristic stain. The tape then passes through a photoelectric detector. The reflected light/optical density of the stain, which is proportional to the concen-

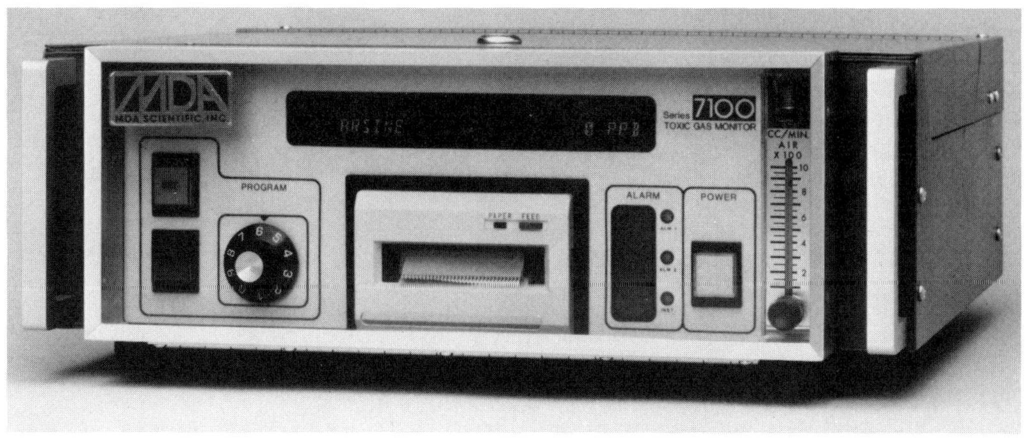

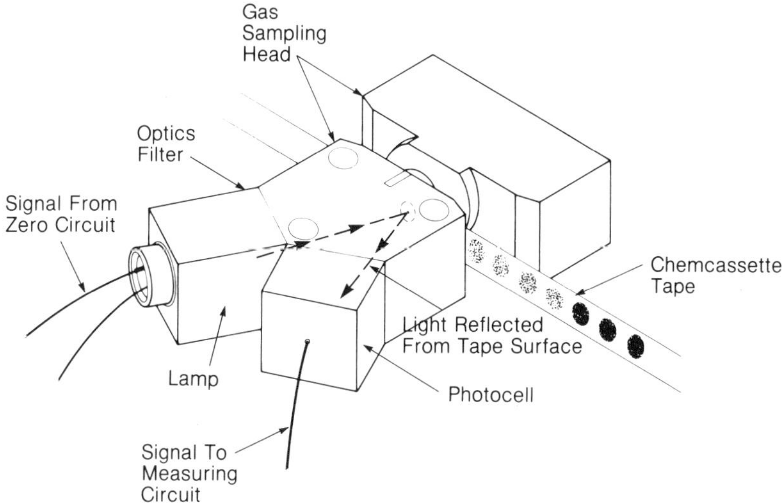

Figure 8-8. Tape-based detection using a stationary monitor (photo). The diagram illustrates the principle of operation. (Photo and diagram courtesy MDA Scientific, Inc.)

tration of the contaminant in the air, is compared electronically to the reference reflectance, or stain, and displayed on a meter and recorded on a chart paper. Tapes can be reel-to-reel, but most instruments use pop-in cassettes due to their ease of use. The response, resolution, and sensitivity of the measurement is dependent on the rate of movement of the tape, the size of the sampling window, and the window size of the optics.[11]

Even in monitors in which the tape moves continuously there is a lag time between when the gas is first sampled and when the tape is scanned by the detector. It depends on the tape speed and the size of the sample window. A simple calculation can be used to determine what the time required for detection is:

$$\text{Time for detection} = \frac{\text{Window width (in mm)}}{\text{Tape speed (in mm/min)}}$$

One instrument, the MDA AUTOSTEP, uses changes in tape speed to vary the monitoring approach. In its "search" mode, designed for leak detection, the paper tape advances rapidly; in the "survey" mode, designed for general area monitoring, the paper tape is advanced sequentially at preset intervals; in the "monitor" mode, designed for stationary (fixed) monitoring, the unit takes a sample every 4 minutes, and then waits another 4 minutes before taking the next sample.

The tension with which the paper tape is held in the sampling head is critical. If the tension is too great, it tends to impede the motion of the paper, sometimes tearing it. When the tension is less than critical, there is a risk of leaking air.[11] Tapes must be replaced often, every 2 weeks, or more frequently in some instruments. Similarly the optics must be verified approximately every 2 to 4 weeks. This procedure is performed by inserting a testing card that has a spot whose discoloration can be related to a specific concentration on that instrument. Each unit should be gas calibrated at the time of manufacture to verify stain concentrations. Under normal use, the optical systems are stable and the instrument does not require periodic recalibration.

A number of evaluations have been done on survey and personal tape-based monitors for isocyanates.[11,12] In general, the problems determined during these evaluations are representative of those that might be encountered by any of these types of instruments. Flow rates should be checked to assure that they conform to manufacturer's specifications. The response has been found to be quite variable among individual units; therefore, calibrations are important for accurate use. It has been found that some monitors have a tendency to "flatten" peak exposure values, meaning brief high levels may be underestimated due to the time required for detection.

It is often assumed that 2,4-TDI is the only isomer of toluene diisocyanate (TDI) present, when in fact TDI rarely occurs as a single isomer in many operations. For example, in the polyurethane foam industry the isocyanate used in the process is an 80% 2,4-TDI and 20% 2,6-TDI mixture; however, 2,4-TDI and 2,6-TDI will react differently on the same tape.[12]

Toluene diiamine may be an interferent in TDI monitoring.[12] In some cases, it may be readily apparent that an interferent was present because the color of the stain will be different. For example, whereas TDI commonly stains a tape red-purple, nitrogen dioxide and chlorine will stain the same tape dark brown. Some compounds, such as sulfur dioxide, can bleach tape, causing a lower response. The only way to determine such interferents is to review the process or sample for the presence of sulfur dioxide. Nonuniformities in the paper tape have been a problem in the past. Therefore, it is a good practice to routinely inspect tape following monitoring.

Mercury

The primary use of mercury vapor monitors is to detect mercury spills. Many devices still contain mercury, such as thermometers, sphygmomanometers, barometers, and manometers. Many older switches also contain mercury. The problem with any mercury spill is that mercury easily vaporizes at room temperature, where it can be breathed or absorbed through the skin. Some of the most dangerous mercury leaks occur in ovens or when mercury is dropped on a floor and gets into a heating vent, because when the material is heated the concentration of airborne mercury increases as the spilled material volatilizes. Dental offices, laboratories, and hospitals are particularly prone to mercury spills.

During a spill, fugitive mercury droplets can roll everywhere. It is not uncommon in these situations for mercury to be tracked all over the room on shoes and the wheels of carts, or for mercury to absorb into porous surfaces such as concrete. In the haste that can follow a mercury spill, inappropriate cleanup procedures are often used, such as sweeping or using a vacuum cleaner that does not have a charcoal filter at its exhaust port. It takes very little mercury to create an unsafe environment. Quantities as low as 1 ml can evaporate over a period of time and contaminate millions of cubic feet of air to levels in excess of allowable limits.[13]

In addition to being used for spill detection, mercury monitors are also used for determining if contamination has occurred to shoes, cart wheels, and other mobile items. It is also possible to monitor smaller articles for contamination. For a discussion of this technique, see the section on mercury monitoring in the chapter on Bulk Sampling Methods.

Industrial operations for which mercury-specific instruments (Fig. 8-9) are useful include

- Manufacture and repair of electrical meters, mercury arc rectifiers, and dry-cell batteries
- Mercury cells in chlor-alkali plant production of chlorine
- Manufacture of neon signs, mercury arc lamps, and electronic tubes
- Mining and refining of cinnabar and gold and silver ores

There are two principles of detection used for mercury: UV light and gold film. Mercury absorbs heavily in the UV region of the spectrum, thus allowing this principle to be used to detect it. The Bachrach M-2 mercury vapor sniffer is an example of an instrument that utilizes the UV method of detection to monitor levels as low as 0.1 mg/M^3. Air to be sampled is drawn into an absorption chamber where a selective 253.7-μm UV light is absorbed by the sample. At the other end of the chamber, a photo-resistive element measures the intensity of radiation passing through the intervening space. The presence of mercury vapor will reduce the radiation energy reaching the photoresistive element in proportion to the vapor concentration. The change affects a photoresistive element, which is connected as one arm of a Wheatstone bridge, creating an unbalanced condition that is detected and displayed on the meter as the mercury vapor concentration in mg/M^3.

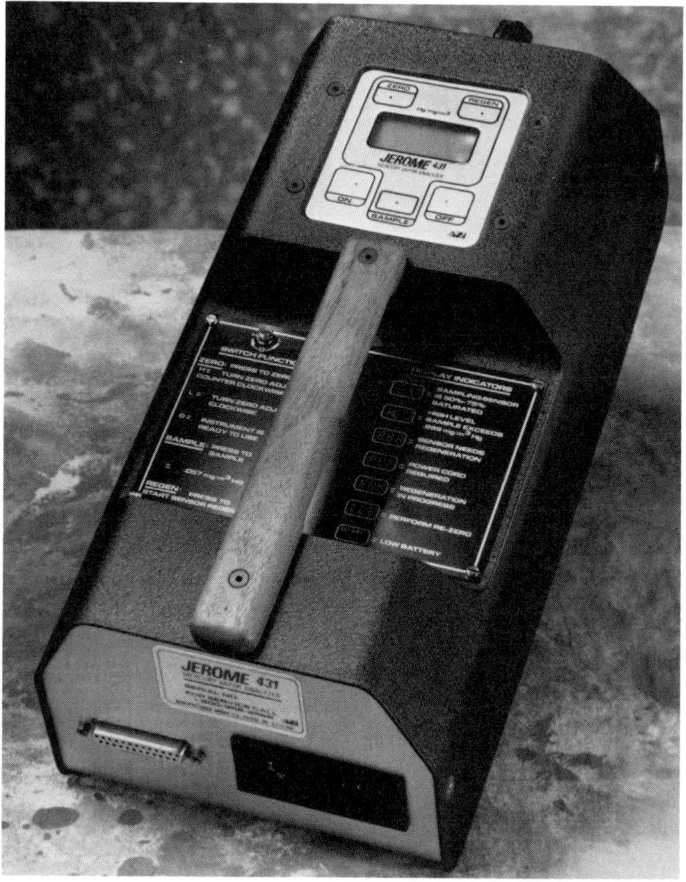

Figure 8-9. Gold foil mercury detection. (Courtesy Arizona Instrument Corp.)

With the low exposure levels currently of concern for mercury, UV-based instruments are used for spill and leak detection and not for verification that PELs are being met.

Often when a UV-based instrument has been subject to relatively cold or hot environments, it will fail to start as quickly as usual. This condition is temporary, and the mercury vapor lamp will ignite once the mercury has sufficiently vaporized. It may require as long as 5 minutes with the power on; time can be shortened if the condensed mercury is agitated. Once the lamp has been ignited and the instrument has been used routinely, it should continue to start within a few seconds after power is applied.

For UV-based detection units it is recommended that users perform a voltage measurement as a calibration check every three months if the instrument is regularly used. This generally involves removing the case of the instrument, covering it with a dark cloth to protect the user from exposure to the UV lamp, and making voltage measurements and trim pot adjustments.

It should be emphasized that any organic compound that absorbs at the same UV wavelength that can be pulled along with the mercury vapor into the gas cell will be a positive interferent, including particulates such as cigarette smoke. Various hydrocarbons such as acetone, benzene, acetylene, and gasoline, and hexane, trichloroethylene, water vapor, and sulfur compounds also interfere. Exposure to chlorine and mercury together or sequentially may produce a negative bias in a UV-based instrument's readings.[14] Due to the potential for water vapor interference, it is best to zero the instrument at the same humidity as the atmosphere to be tested. Magnetic fields have also been reported to interfere with UV-based instruments. One of the chronic problems with UV-based mercury detection instruments is the meter needle's tendency to drift, making readings difficult.

In the gold film technique for direct mercury detection a thin gold film that is part of a piezoelectric sensor undergoes an increase in electrical resistance proportional to the mass of mercury in the sample. The Jerome 411 mercury monitor has an internal pump that draws a precise volume of air over the gold film sensor at a flow rate of 750 ml/min. Mercury in the sample is adsorbed and integrated by the sensor, registering it as a proportional change in electrical resistance. It can collect a sample every 10 seconds. The measurements are displayed on a digital readout. Its sensitivity is reported as 0.003 mg/M^3.

There are two measurement modes: sample and survey. In the sample mode samples are collected in 12-second intervals while in the survey mode the instrument collects samples every 3 seconds. The sample mode is used when the greatest precision is necessary while the survey mode is useful for locating spills or to assess areas of potentially high mercury concentration. During the sample cycle, bars on the digital readout represent the percent of sensor saturation. Approximately 65 samples containing 0.1 mg of Hg/M^3 may be taken before saturation occurs. A 10-minute heating cycle is used to desorb contaminants off the gold film in order to increase sensitivity.

Generation of specific mercury vapor concentrations requires a mercury vapor generator setup and considerable experience. Usually it is done by laboratories with specific expertise in this area, such as those operated by equipment manufacturers. Therefore, for most users, it will be best to send the unit back to the factory for calibration.

Prior to each use, the sampling professional should check an instrument's general response to mercury and the condition of the charcoal in the zero filter. The procedures below are followed or a kit can be purchased from the manufacturer for the response check.

1. Place one drop of mercury in a small glass container or thermos and cover while working inside a laboratory hood.
2. Turn on the instrument and allow it to warm up for 15 to 25 minutes.
3. Zero the instrument with the zero adjust knob, using the filtered air setting if one exists. In this setting the air is pulled through a mercury absorbent often made of iodine-treated charcoal.
4. Set the knob on the sampling position, and lifting the cover insert the probe into the glass container so that mercury vapor is drawn into the intake port. Observe that the meter

pointer deflects upscale and then drops toward zero as the mercury vapor concentration is reduced in the container over this sampling period.

5. Check the condition of the absorbent in the filter by using the "filtered air" position and exposing the probe to the mercury in the glass container. Note whether the meter rests at zero; if not, the absorbent is saturated with mercury and must be replaced.

Hydrogen Cyanide

Hydrogen cyanide can be found in electroplating plants and in reclamation areas of coke oven batteries. Since this compound is acutely toxic, the best type of monitoring is to use a small personal real time monitor mounted on a worker and running at all times during work. These instruments are generally electrochemical based and are similar to those described for hydrogen sulfide. For more information on real time personal instruments, see the chapter on Reading and Recording Real Time Data.

Carbon Dioxide

Carbon dioxide is a by-product of human respiration. It is often measured in indoor air to determine the effectiveness of ventilation systems. The most common instrument used for measuring this compound is the IR, often a dedicated instrument rather than the more versatile one described in the chapter on Monitoring Instruments for Many Different Gases and Vapors: GC and IR. In any case, the detection principle is the same. Usually monitoring is done in several areas of a building, each time over the full course of a day, with a constant recording of the results. If levels are seen to gradually increase, a lack of air change may be the problem. An example of a situation where carbon dioxide monitoring is performed is described in the section on Indoor Air in the chapter on Specific Sampling Situations.

REFERENCES

1. Stetter, J. R. Electrochemical sensors, sensor arrays, and computer algorithms for the detection and identification of airborne chemicals. In *Detection of Airborne Chemicals*. American Chemical Society, Washington, D.C., 1986.
2. Stetter, J. R. Instrumentation to monitor chemical exposure in the synfuel industry. *Am. Conf. Ind. Hyg. Ann.* **11**:225–248, 1984.
3. Shaw, M. More Straight Talk About Toxic Gas Monitors. Interscan Corp., P.O. Box 2496, Chatsworth, CA 91313-2496.
4. American Industrial Hygiene Association. *A Manual of Recommended Practices for Combustible Gas Indicators and Portable, Direct Reading Hydrocarbon Detectors*. AIHA, Akron, OH, 1982.
5. *Miniguard II Manual*. Mine Safety Appliances, Inc., Pittsburgh, PA.
6. *Ecolyzer Carbon Monoxide Analyzer Manual*. National Draeger, Inc., Pittsburgh, PA.
7. Thompkins, F. C., and J. H. Becker. *An Evaluation of Portable, Direct-Reading Hydrogen Sulfide Meters*. NIOSH, Cincinnati, OH, July 1976.
8. Standard Test Methods for Continuous Measurement of Oxides of Nitrogen in the Ambient or Workplace Atmosphere by the Chemiluminescent Method. ASTM D 3824-79.
9. Que Hee, S. S. Formaldehyde evolution from an ethylene-using ozone chemiluminescence meter: A new method of calibration. *AIHA J.* **42**(7): 511–514, 1981.
10. Mathews, T. G. Evaluation of Modified CEA Instruments, Inc. Model 555 Analyzer for the monitoring of formaldehyde vapor in domestic environments. *AIHA J.* **43**(8):547–552, 1982.
11. Dharmarajan, V., and R. J. Rando. Critical evaluation of continuous monitors for toluene diisocyanate. *AIHA J.* **41**(12):869–878, 1980.
12. Walker, R. F., and M. A. Pinches. Chemical interference effects in the measurement of atmospheric toluene diisocyanate concentrations when sampling with an impregnated paper tape. *AIHA J.* **42**(5):392–397, 1981.
13. Easton, D. N. *Management and Control of Hg Exposure*. American Laboratory, July 1988.
14. McCammon, C. S., and J. W. Woodfin. An evaluation of a passive monitor for mercury vapor. *AIHA J.* **38**:378–386, 1977.

CHAPTER 9

General Survey Instruments for Gases and Vapors

Traditional real time monitoring for chemicals has emphasized the use of monitors that are specific for a given compound, therefore minimizing interferences and making measurements as accurate as possible, particularly in situations in which multiple contaminants are present. On the other hand, the major difficulty in monitoring only for specific compounds, especially in unknown atmospheres, is that the presence of other hazardous compounds may be missed and the atmosphere is determined to be safe when in fact it may not be. Another problem is the limited number of chemicals for which specific monitoring instruments are available and the few instruments that can be used to identify chemicals in the field. General survey instruments are capable of detecting a large number of contaminants but generally cannot distinguish between them. In addition, the response will be different for each chemical measured and the result will be difficult to interpret.

General survey instruments fall into three categories: (1) combustible gas measurements higher than health hazard levels and immediately dangerous to life and health (IDLH) values; (2) health hazard measurements in the permissible exposure level (PEL) or parts per million (ppm) ranges; (3) and instruments capable of measuring both levels. Of these the last category is only beginning to be developed but is important since when combustible levels are checked low levels are often ignored because they do not present an immediate hazard. Sampling professionals often balk at the idea of carrying two or even three instruments around, especially when measurements are short and must be taken often.

General survey instruments are used almost exclusively to collect area samples, although occasionally a measurement may be taken by putting a probe in the breathing zone such as is done on hazardous waste sites. Combustible gas indicators (CGIs) are used primarily for occupational measurements for confined space entry and performance of "hot work" near sources of combustible materials, although they have also been used to screen basements of residences for the presence of sewer gas. Measurements may be done indoors or outdoors. Health hazard measurements with general survey instruments are done almost exclusively for occupational purposes, although these instruments have also been used to screen buildings and homes where indoor contamination is suspected. An important environmental application is the identification of fugitive emissions from industrial processes to comply with Environmental Protection Agency (EPA) standards. For occupational purposes only combustible gas measurements would satisfy Occupational Safety and Health Administration (OSHA) standards while other measurements would be used to screen atmospheres to determine if follow-up sampling with a more specific method was needed.

Screening instruments used to measure volatile organic compounds (VOCs) have wide applications, including measurements for fugitive emissions and hazardous waste site work. The term VOCs is used more for environmental sampling than occupational sampling, but in both cases refers to the same types of compounds. These constitute a wide variety of organic compounds that can exist in the gaseous phase at significant concentrations at ambient temperatures, including aliphatic and aromatic hydrocarbons, chlorinated hydrocarbons, amines, and other well-known groups such as alcohols, ethers, ketones, aldehydes, and esters. The most common detection methods used in this type of work are flame ionization detection (FID) and photoionization detection (PID). FIDs as well as PIDs are often incorporated into gas chromotographs. In this situation, the instrument serves a different purpose than when used in the general survey mode. For more information on this application, see the chapter on Monitoring Instruments for Many Specific Gases and Vapors. When used as general survey instruments, these instruments are not designed to be used in lieu of CGIs, since they measure ppm levels in most cases and are not always sensitive to combustible gases.

The usefulness of a general survey instrument depends on the application. In situations where the sampling professional can be relatively certain that only a single contaminant is present, these instruments can be calibrated for that compound and used as selective samplers for quantitative area monitoring and leak detection. In other situations where very little is known about what contaminants are likely to be present, these instruments can be used for qualitative information and a warning that something is present. Applications for general survey instruments are likely to increase, taking advantage of the inability of these instruments to differentiate between compounds. However, the user should be mindful of the limitations of these instruments in order to make sure the use is appropriate and the results are meaningful. Even, when engineered to be somewhat selective, such as is the case with certain semiconductor-based instruments, the selectivity is limited and the units are still susceptible to a number of interferents. Also, these instruments are generally not useful for accurately detecting low ppm levels of hazardous compounds, such as benzene, due to limitations in technology and due to the likelihood that the response to any other compounds present is likely to "mask" the low-level response.

Although the concept of a general survey instrument is that it can monitor for many different chemicals at once, in principle more than one type of general survey or other instrument may be needed in most situations. For example, the use of a combustible gas monitor alone in a sewer will not detect dangerous levels of hydrogen sulfide. The choice of instrument will depend on the situation to be monitored; in general, for atmospheres where the types of contaminants are well characterized, the more specific the monitor the better; on the other hand, with a number of unknowns the more appropriate instrument will be a nonspecific general survey instrument.

Response factors can provide a convenient interpretation of the actual levels most likely to be present. The identification of response factors for specific chemicals in specific monitoring environments, including where known mixtures are present, provides a mechanism for using general survey instruments to develop useful data. A response factor depends on the chemical used to calibrate the instrument, the actual compounds being monitored, and the sensitivity of the detector to each of these compounds. For example, an instrument intended for measurement of gasoline vapors would be calibrated with benzene and its response would be compared to a variety of concentrations of gasoline to develop a response curve:

$$\text{response factor} = \frac{\text{actual concentration}}{\text{instrument observed concentration}}$$

Some instruments have a response factor setting used to equate the response of one gas or vapor with another, usually that of the calibration gas. If this option is selected, the current reading is always multiplied by the response factor in order to display the accurate concentration.

A response factor of 1.0 means that the instrument readout is identical to the actual concentration of the chemical in the gas sample. As the response factor increases, the instrument readout is proportionally less than the actual concentration. If the meter readout is 100 ppm and the sampling instrument is set up so that it has a response factor of 0.1 to the chemical that is present, the actual concentration would be only 10 ppm. It is desirable to set up an instrument so that its response is as close as possible to 1.0 for the specific compounds of interest. When published response factors for the organic compounds being monitored are much greater than 1 (approaching 10) or much smaller than 1 (approaching 0.1), it is prudent to measure the response factor for these specific compounds.

Integrating an automatic response factor into the instrument's readout is a useful feature, but it should be used with care if the sampler is unsure of what contaminants might be present. Response factors are discussed in greater detail in a later section.

Most of these instruments are used for measuring a wide range of levels of gases and vapors. Even though they are generally calibrated with a single concentration of one specific compound, yet they can respond to many different compounds simultaneously. Calibrating an instrument within the concentration range likely to be encountered will provide some assurance, since most general survey instruments are linear over only certain ranges, meaning they can make accurate measurements over these concentrations. However, most general survey instruments when used with the calibration provided and recommended by the manufacturer calibrate for only one point. If it is 50 ppm or 50% of the lower explosive limit (LEL), as is commonly used, it is unlikely that an instrument is going to be accurate at levels of 10 ppm or 10% of the LEL, regardless of how many divisions the scale has. The best way to ensure accuracy is to calibrate an instrument for the range of interest. For example, if measuring for leak detection, 500 ppm may be in the correct range. On the other hand, if the purpose is to do breathing zone measurements, the calibration should be in the 1-ppm to 20-ppm range. Calibration procedures for specific types of general survey instruments are discussed in each section.

In order to interpret general survey measurements, there are certain data the sampling professional should have. For combustibles the LEL of any chemicals suspected to be present should be identified. If any of these compounds is also toxic the IDLH and PEL should also be identified. Sometimes there is a wide range between these values, as is the case for toluene where the PEL is 100 ppm, the IDLH is 2000 ppm, and the LEL is 13,000 ppm (1.3%), and in other cases the range is not great, such as is demonstrated by ethyl ether where the IDLH and LEL are the same, 19,000 ppm (1.9%), and the PEL is much lower at 400 ppm. In another example, ethylene chlorohydrin, the PEL is 1 ppm, the IDLH is 10 ppm, and the LEL is 49,000 ppm (4.9%). Another value of interest when conducting measurements with PIDs is the ionization potential (IP) of a chemical. This must be known in order to determine if a PID-based instrument is properly set up to detect a chemical. Table 9-1 lists IPs, LELs, and IDLHs of selected chemicals.

MEASUREMENT OF EXPLOSIVE ATMOSPHERES: COMBUSTIBLE GAS INDICATORS

Combustible gas indicators are used for measuring explosive levels of gases, often in confined spaces. Other terms for the combustible gas indicator are explosimeter and heat of combustion analyzer. In the past, CGIs have also been called methanometers. Some sources also call them hydrocarbon analyzers, although this term is somewhat of an overgeneralization since there are other groups of instruments that also measure hydrocarbons. For the purposes of this discussion, only monitors designed to detect explosive levels of combustible gases are considered. CGIs were the first direct reading instruments to be developed. Their initial use, which is still very common, was detection of explosive atmospheres of methane in mines. The first meter was a squeeze-bulb,

TABLE 9-1. Ionization Potentials (IP), LELs, and IDLHs of Selected Chemicals

Compound	*IP (eV)*	*LEL (%)*	*IDLH (ppm)*
Acetaldehyde	10.21	4	10,000
Acetic acid	10.37	5.4	1,000
Acetic anhydride	9.88	2.9	1,000
Acetone	9.69	2.6	20,000
Acetonitrile	12.22	4.4	4,000
Acrolein	10.10	2.8	5
Allyl alcohol	9.67	2.5	150
Allyl chloride	10.20	3.3	300
Ammonia	10.15	16	500
Aniline	7.70	1.3	100
Benzene	9.25	1.3	2,000
Benzyl chloride	10.16	1.1	10
1,3-Butadiene	9.07	2	20,000
n-Butyl amine	8.71	1.7	2,000
Carbon disulfide	10.13 (10.06)	1.3	500
Carbon monoxide	13.98	12.5	1,500
Chlorine	11.48	—	30
Chlorine dioxide	10.7	10	10
Chlorobenzene	9.07	1.3	2,400
Crotonaldehyde	9.73	2.1	400
Cyclohexane	9.98	1.3	10,000
Cyclohexanol	10.00	2.4	3,500
Cyclohexanone	9.14	1.1	5,000
Cyclohexene	8.95	1	10,000
Diborane	11.4	0.8	40
1,1-Dichloroethane	11.06	6	4,000
1,2-Dichloroethylene	9.65	9.7	4,000
Dimethyl amine	8.24	2.8	2,000
Dimethyl aniline	7.14	1	100
Ethyl acetate	10.11	2.2	10,000
Ethyl amine	8.86	3.5	4,000
Ethyl benzene	8.76	1.0	2,000
Ethyl bromide	10.29	6.7	3,500
Ethyl butyl ketone	9.02	1.4	3,000
Ethyl chloride	10.98	3.8	20,000
Ethylene chlorohydrin	10.90	4.9	10
Ethyl formate	10.61	2.8	8,000
Ethyl mercaptan	9.29	2.8	2,500
Formic acid	11.05	18	100
Heptane	10.07 (9.90)	1.1	5,000
Hydrogen	15.43	4.1	—
Hydrogen cyanide	13.91 (13.69)	5.6	50
Hydrogen fluoride	16.01	—	30
Hydrogen sulfide	10.46	4.3	300
Isoamyl acetate	9.90	1	3,000
Isoamyl alcohol	10.16	1.2	8,000
Isopropyl acetate	9.99	1.8	16,000
Isopropyl alcohol	10.16	2	8,000
Isopropyl amine	8.72	2	4,000
Isopropyl ether	9.20	1.4	10,000
Liquified petroleum gas (LPG)	10.95	1.9	19,000
Methanol	10.85	6.7	25,000

(continued)

TABLE 9-1. *(continued)*

Compound	*IP (eV)*	*LEL (%)*	*IDLH (ppm)*
Methyl acetate	10.27	3.1	10,000
Methyl acetylene	10.36	1.7	17,000
Methyl acrylate	10.72 (9.19)	2.8	1,000
Methylal	10.00	1.6	10,000
Methyl amine	8.97	5	100
Methyl ethyl ketone	9.53	2	3,000
Methyl formate	10.81	5	5,000
Methyl mercaptan	9.44	3.9	400
Morpholine	8.88	1.8	8,000
Nitrobenzene	9.92	1.8	200
Nitroethane	10.88	3.4	1,000
Nitrogen dioxide	9.75	–	50
Nitromethane	11.08	7.3	1,000
1-Nitropropane	10.81	2.2	2,300
m-Nitrotoluene	9.79	1.6	200
Octane	9.9	1	5,000
Pentane	10.35	1.5	15,000
2-Pentanone	9.39	1.5	5,000
Phosphine	9.96	–	200
Propane	11.07	2.2	20,000
n-Propyl acetate	10.04	2	8,000
n-Propyl alcohol	10.20	2	4,000
Propylene dichloride	10.87	3.4	2,000
Propylene oxide	10.22	2.1	2,000
Styrene	8.47	1.1	5,000
Toluene	8.82	1.3	2,000
o-Toluidine	7.6	1.5	100
Trichloroethylene	9.47	11	–
Triethylamine	7.50	1.2	–
Vinyl chloride	10.00	3.6	–
Water	12.61	–	–
m-Xylene	8.56	1.1	1,000

sample-drawing model with a purely resistive, nonelectronic circuit and a meter graduated in explosibility units to show concentration. Instruments of this type are still available today (Fig. 9-1).

Although CGIs are the easiest instruments to operate, they are also the most misunderstood and taken-for-granted of all real time sampling instruments, and incorrect measurement techniques can lead to dilemmas.

Utility companies use them for leak detection of underground storage tanks (USTs) and associated pipelines, even though the levels of concern for soil and groundwater contamination can be in the ppb level.

Confined space testing is often left to laypersons, for example, shift supervisors or the employees designated to do the work who may not understand the limitations of the instruments or any of the potential problems in the interpretation of their measurements. A layperson may be unaware that combustible gas readings may be inaccurate in a space where there is an oxygen deficiency.

A homeowner calls the fire department because of a funny odor in the sewer or basement and the fire department brings a CGI, even though the odor may be due to mercaptans or hydrogen sulfide or other gases.

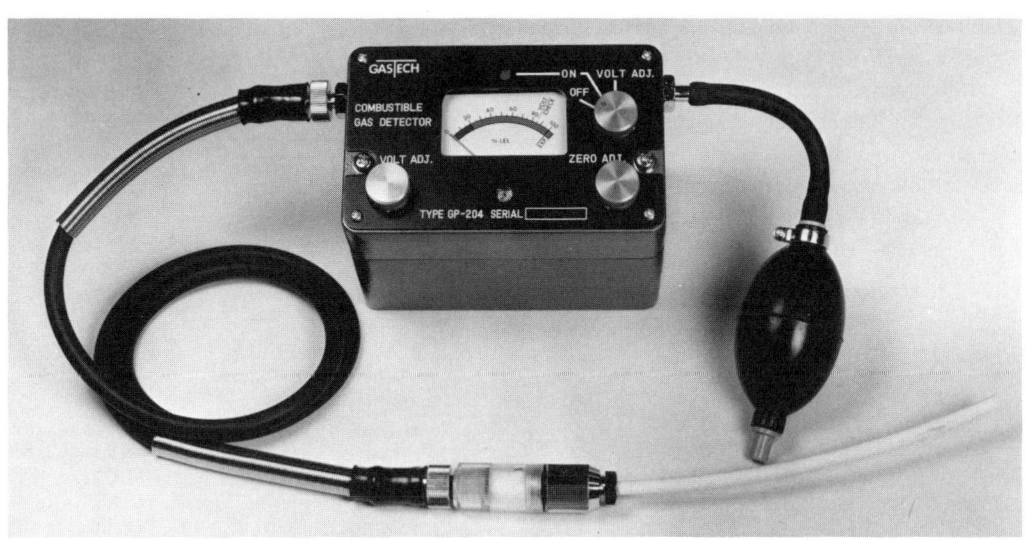

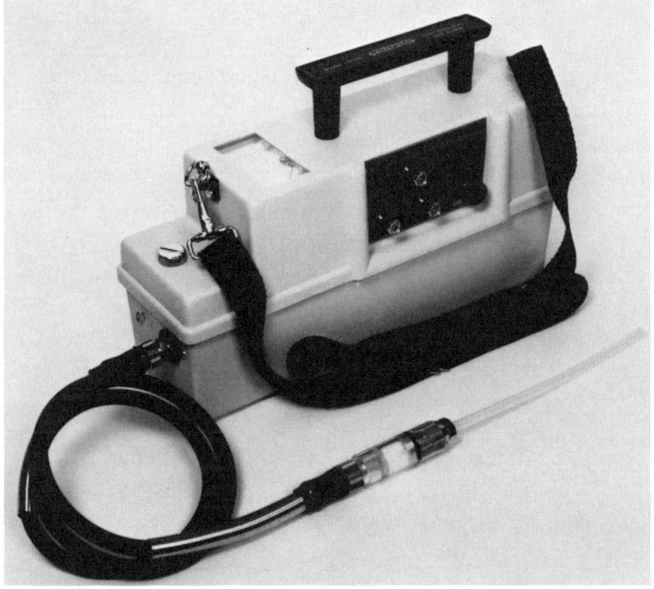

Figure 9-1. Combustible gas indicators. (Top and center, courtesy GasTech, Inc.; bottom, courtesy Neotonics.)

Methane has been known to occur in sanitary landfills and seep into the basements of homes in nearby residential areas, but in its natural state it has no odor and not all CGIs are designed to measure methane.

Basis for Combustible Gas Detection

There are many variations available in CGIs. Some instruments measure both LEL and explosive percentages while others are designed to measure LEL percentages and also lower ppm levels. Some instruments utilize selector switches that change the temperature of the sensor. In one position an instrument operates with the detector hot enough to burn methane along with any other combustibles present, and when the switch is changed to the petroleum vapor setting, the sensor temperature is reduced below the point at which methane is burned, allowing for detection of only those combustibles having lower combustion temperatures. Some instruments can detect hydrogen whereas others cannot, yet hydrogen is a very explosive gas. Therefore, it is important to be familiar with the principle of detection for any instrument along with the type of atmospheres it can measure as well at its limitations.

The detector can be inside of an instrument if it has a vacuum pump to pull air inside or outside if the instrument operates on the diffusion principle. The diffusion sensing elements are often capable of being taken off the instrument and attached to the end of a cable to be lowered into a tank. This is an advantage over pump methods that must pull a sample to the sensor as the air to be sampled surrounds the sensor.

Heat of combustion refers to the heat released by the complete combustion of a unit mass of combustible material. It is a measure of the maximum amount of heat that can be released by a certain mass of a combustible chemical. Heats of combustion for various chemicals are available (Table 9-2).[1] Some sources give two different values: one for the heat of combustion for a material in its liquid state and another for the gross heat of combustion. The heat of combustion for the liquid will always be less than the gross heat; however, in some cases the values are the same. As a rule of thumb, the higher the heat of combustion of a compound, the higher the heat required in the detector.

TABLE 9-2. Heats of Combustion for Some Representative Compounds

Compound	*Gross Heat of Combustion (mJ/kg)*
Acetone	30.76
Benzene	41.83
n-Butane	49.5
Carbon disulfide	10.32
Carbon monoxide	10.10
Gasoline	43.0*
Hydrogen	141.79
Methane	55.50
Methanol	22.70
n-Octane	47.89
n-Pentane	48.64
Phosgene	1.72
Propane	50.35
Toluene	42.43

*Heat of combustion is for the liquid.

Catalytic Sensors

Virtually all heat of combustion instruments in use today are based on catalytic combustion. The use of a catalyst in conjunction with a basic Wheatstone bridge allows combustible gases to combine with oxygen (oxidize) at much lower temperatures than would be required for normal combustion. The sensor consists of a pair of opposing platinum filaments, placed to form two legs of a Wheatstone bridge, with one element exposed to the gas being sampled in a combustion chamber and the second filament sealed to prevent contact with the atmosphere.

In this design one filament is coated with a catalyst (usually platinum or palladium) that initiates combustion (oxidizes the gas mixture) on its surface, thereby increasing its temperature and consequently increasing its resistance. The catalytic filament is connected in series with a second, uncoated reference filament that operates at the same voltage but does not cause oxidation and therefore no temperature increase occurs on it. This inactive compensator filament acts to offset any electrical changes caused by fluctuations in flow conditions, sample temperature, pressure, and/or humidity. The voltage applied to the two filaments in series is divided so that the greater voltage drop across the exposed filament unbalances the Wheatstone bridge. Coiling the platinum filament to form a "bead," a design many instruments have, protects it from contact with materials that could damage it. Older designs, a few of which are still manufactured, utilize hot wire filaments in a Wheatstone bridge without the catalytic coating.

The catalytic bead requires oxygen levels usually above 15% to sustain combustion; therefore, in an oxygen-deficient space erroneous readings may result. Heating causes a slow deterioration of the catalyst with the result that the detector must be periodically replaced. It has been claimed that since catalytic bead sensors operate at lower temperatures, they need less power than other sensors, they have a longer, more stable sensor life, and they have more reduced zero drift than simple hot wire sensors.

Some heat of combustion instruments are compensated and others are uncompensated. In an uncompensated instrument, the active and reference elements are similar or identical, but gas is exposed only to the active element while the reference rests in an isolated cavity. In compensated designs, both the active and the reference elements are exposed to the gas, but the reference element is noncatalytic. This design gives better stability under conditions of pressure, temperature, and background inert gas variations. For example, the uncompensated element, when exposed to 100% gas, responds to the cooling effect of that gas, which produces a substantial downscale reading. The compensated detector, on the other hand, experiences the same cooling effect, acting on both elements, and they cancel out. Thus, the catalytic activity continues to the point where all of the oxygen is gone, and the reading stays at or above zero.

Thermal Conductivity Detectors

Thermal conductivity is another detection method used for explosive concentrations that uses the specific heat of combustion of a gas or vapor as a measure of its concentration in air. It is used in instruments designed to measure very high concentrations, namely percent of gas as opposed to percent of the LEL of a gas, which most CGIs indicate on their readouts. When an instrument has dual scales for both percent gas and percent LEL, both a catalytic combustion and thermal conductivity sensor are incorporated into the instrument.

A thermal conductivity filament is substituted via a selector switch into the Wheatstone bridge. Combustibles in the sample then cool this filament, decreasing its resistance and unbalancing the bridge, which is just the opposite of the catalytic combustion techniques. This filament is not as susceptible to "poisoning" or oxygen levels as the catalytic filament. This method is nonspecific and not sensitive to low levels; consequently, it is used in CGIs that measure in total percent of combustibles. Because of the extremely high concentrations this instrument is capable of measuring, it must be used with extreme caution.

Semiconductor Sensors

Semiconductor sensors are also used for detecting combustible gas levels. These types of instruments are discussed in greater detail in the section on combination measurements of explosive and health hazard levels in this chapter.

Instrument Safety and Trends

In order to be used in a potentially combustible atmosphere, instruments must be qualified as intrinsically safe. They must pass the testing requirements of the Underwriters Laboratory (UL) or Factory Mutual (FM). Following this testing, the codes for the type of atmospheres for which the instrument can be safely used according to the class division and group classification system of hazardous atmospheres of the National Fire Protection Association (NFPA) are incorporated into the label on the instrument. The more stringent the classification given to an instrument, such as class 1 (usage for gases and vapors) and division 1 (usage in areas of ignitable concentrations), the broader its use. Therefore, prior to selecting an instrument, the types of explosive atmospheres most likely to be encountered must be considered.

The current trend is for combustible gas-detecting instruments to have additional monitoring capabilities in addition to combustible gases and oxygen. Because OSHA requires that oxygen deficiency be measured at the same time as combustible gases, when entering a confined space most monitors generally provide both of these measurements. Examples are carbon monoxide and hydrogen sulfide gases, because these are commonly found in confined spaces. Some units have individual sensors for each measurement whereas others have "toxic" combustible and oxygen gas sensors.

Many instruments combine catalytic combustion sensors to detect combustible gases with individual electrochemical sensors for oxygen, carbon monoxide, and hydrogen sulfide. For more information on oxygen, carbon monoxide, and hydrogen sulfide monitors, see the chapter on Monitoring Instruments Dedicated to a Single Chemical.

Limitations

All CGIs, regardless of their detection principle, have limitations associated with their conditions of use. Before making measurements with any CGI, sampling professionals should be aware of the limitations associated with the specific instrument they are planning to use. Some of these limitations are discussed here. Others may be described in the manual provided by the manufacturer. If the manufacturer's manual does not list at least some of these, a different instrument should be considered.

CGIs should not be used for health hazard determinations or to measure the head space of soil or water samples, since the difference is immense between percent as measured by CGIs and ppm levels generally of concern for health hazards and ppb (parts per billion) levels of concern in soil and water contamination. This difference becomes more apparent when it is realized that 1% is equivalent to 10,000 ppm.

$$
\begin{aligned}
1\% &= \frac{1}{100} \\
&= \frac{1{,}000}{100{,}000} \\
&= \frac{10{,}000}{1{,}000{,}000} \\
&= 10{,}000 \text{ ppm}
\end{aligned}
$$

Using a similar calculation it can be shown that 1% is equivalent to 10,000,000 ppb.

Many instruments require oxygen to function properly. For example, catalytic bead systems require oxygen levels usually above 15% to sustain combustion. Oxygen-enriched atmospheres (22% or greater) will also change the response of an instrument that was calibrated in the presence of normal oxygen-laden air. Some sources have indicated that flashback arrestors may not function in this type of environment. Since the level of oxygen can effect readings, only instruments that have both oxygen and combustible gas detection should be used whenever measurements are being done in confined spaces, and it is important that the oxygen measurement be done first.

Since oxidation of the combustible gas is used to obtain a signal from an instrument, if the concentration of oxygen in the air being sampled is too low for complete oxidation of the combustible gas to occur, then the signal obtained will be low. One situation that can cause oxygen deficiency is high concentrations of a gas being present, generally above the UEL, displacing the available oxygen otherwise known as being too rich to burn. As an example, units based on the hot wire and catalytic combustion principle have been known to "peg" and then return to zero when used in an atmosphere that is greater than the upper explosive limit (UEL). Another indication that this condition might be occurring if it was not initially noted is the meter's response rising above 100% of the LEL before returning to zero as the instrument is withdrawn in spite of the levels of gas that are present. Therefore, an instrument's readout should be observed whenever making measurements of potentially explosive atmospheres. Once the UEL is reached, the curve drops off almost immediately, the result being a return to zero on the meter in spite of the hazardous levels of gas that are present (Fig. 9-2).

Thermal conductivity detectors are designed for high concentrations and do not have this limitation. However, thermal conductivity detectors are less sensitive than catalytic combustion detectors. The most caution must be taken when making measurements with dual scale units, since a failure to note that measurements were being made on a less sensitive scale (higher range) could lead to interpreting results as being lower than are correct. On units that can measure both percent LEL and percent gas, caution is the most critical.

Catalytic combustion instruments may be affected by high humidity. When testing heated spaces or atmospheres that contain high levels of alcohols, high humidity may be present. Therefore, the use of a moisture trap to keep water from getting inside the instrument may be necessary.[2]

Although CGIs can measure a wide variety of flammable gases and vapors, they cannot measure all of them. Since often they are used to measure unknown atmospheres this is a concern. Not all instruments can measure methane or hydrogen, which are also explosive gases. Catalytic combustion units must have sensor temperatures sufficient to exceed the ignition temperature of methane, which is relatively high. High levels of some inorganic compounds can also be explosive, for example, hydrogen sulfide has an LEL of 4.3%. Therefore, atmospheres that may contain inorganics should be evaluated for their explosive potential as well as for the ability of the particular instrument to detect them. CGIs designed to measure explosive gases should not be used to indicate the presence of explosive or combustible mists or sprays, such as lubricating (mineral) oils, or for measuring the percent of flammable vapors in steam or explosive (grain) dust levels.

Many compounds, such as silanes, silicones, and silicates, have been known to "poison" catalytic combustion instruments by coating the catalyst so it does not function properly, the result being a decreased response of the instrument. In atmospheres with high concentrations of halogenated hydrocarbons, the thermal decomposition products generated by these compounds will corrode the sensor, causing readings to be low. Corrosive

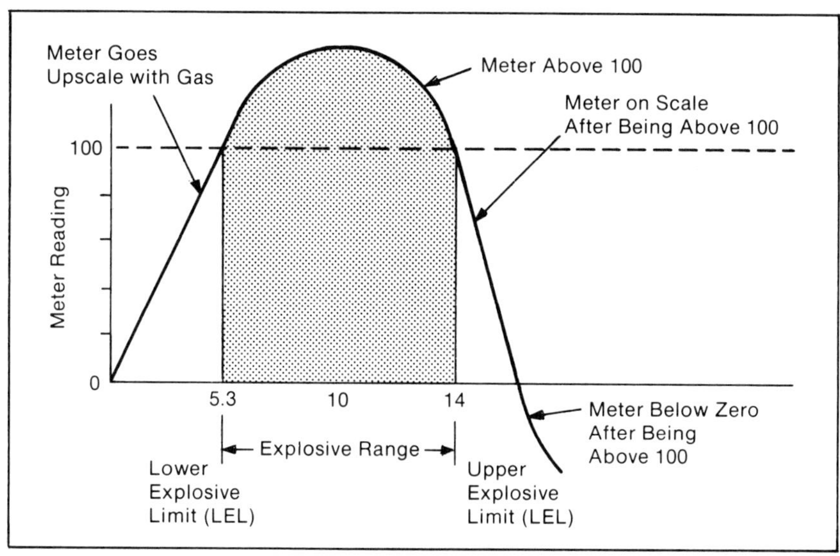

Figure 9-2. Sensor response range from LEL to UEL. (Courtesy MSA.)

gases will deteriorate sensing elements, causing circuitry resistance changes and instability. Hydrogen sulfide can be corrosive, as well as phosphate esters, reduced sulfur compounds, high levels of organic acids such as acetic acid, and acid gases. Nitro compounds, such as nitromethane, nitroethane, and nitropropane, and also be corrosive to instrument interiors.[2] If lead gasoline is suspected of being present, tetraethyl lead will reduce the ability of the filament to detect other compounds, especially those like methane, that have high ignition temperatures. Oxidation of tetraethyl lead can reduce condensation to a solid lead combustion product on the filament.

Arsenic and other heavy metals, as well as dusty atmospheres, can interfere with sensors. This poisoning is usually only partial, with the first loss being shown on methane because of its high heat of combustion. Therefore, if the element still has a normal methane response, it will have a normal response to other gases. An alternative approach to testing in the presence of leaded gasoline vapors is to use an inhibitor filter. This filter promotes a chemical reaction with tetraethyl lead vapors that yields a more volatile combustion product, thereby preventing contamination of the filament. One way to prevent "poisoning" is to have a high enough filament temperature to prevent condensation of products such as lead.

If the material being sampled is hot, then its flash point is a consideration. If propane is being sampled it will pass through a cool hose without changing from its gaseous state. A vapor having a high flash point may condense readily as it passes through the hose, and the instrument's reading will not be representative of what is in the tank's air. As an example, a tank containing isophorone (flash point of 184°F [84°C]) that is heated to 250°F (121°C) could have an explosive atmosphere inside; however, if the sample is withdrawn and cooled to room temperature, the maximum airborne reading would be around 5%, with most of the sample having condensed in the hose. Heated materials with high flash points are best handled with a diffusion-type detector on a cable.

Atmospheres with pressures that differ significantly from normal atmospheric pressure may also affect the accuracy of the reading, for example, in tunneling work. Stray magnetic fields may be encountered in manholes and electric utility vaults. These magnetic fields can affect meter movement unless they are compensated for with a core magnet. These situations are specialized ones for which outside assistance should be obtained before proceeding with work.

High air velocities tend to cool sensing elements and adversely affect their performance. The use of protective baffles or deflectors permits sensors to be positioned in fast airstreams as well as in locations where water streams may be encountered during cleanup operations. Wind currents will affect the readings of diffusion detectors, but they will give a more accurate reading if lowered into a confined space where there is little air movement.

Although real time sampling instrumentation is rapidly changing, many older models are still in use. Companies purchase one or several instruments and continue to use them even though far more accurate models may now be available. For example, until recently one of the most common CGIs employed a rubber squeeze bulb for a pump. This accuracy of a squeeze bulb sample can be far more variable than an electronic pump, and the number of samples that can be collected is limited. When sampling lines are used with squeeze bulb CGIs, the number of squeezes necessary to get accurate readings increases. These instruments should be used with caution and a program to upgrade to more modern instruments should be instituted.

Since CGIs are used to detect situations where an explosive atmosphere could be present, these measurements have life-and-death potential, yet because these instruments have been available for so many years and are sold in so many versions, often requiring very few skills to operate, measurements are often done in a haphazard or cursory manner. Individuals should never be assigned to sampling duties without training in the use and limitations of the instrument as well as the correct way to interpret the results. Some companies have instituted certification programs that require individuals to attend a course and pass a test prior to being able to perform CGI measurements.

Calibration and Maintenance

There is no best gas for calibration of CGIs; instead selection depends on the intended use of the instrument. This should be a tip to the sampling professional that a single CGI calibrated to one gas cannot necessarily be used for all situations where flammable and explosive compounds are present.

Since the heat of combustion varies from one compound to another, the CGI should be calibrated in terms of the compound for which survey information is required if this information is known. Alternatively, if any of several gases and vapor may be present in the space being monitored, the sensor should be calibrated on the least sensitive of the gases to provide a wide margin of safety. In other words, the gas requiring the highest concentration to get a 20% LEL response should be selected. Thus, other gases will cause the same or a greater response at lower concentrations.

Manufacturers calibrate CGIs with a variety of gases, although the most common are methane, pentane, propane, and hexane. Each manufacturer selects the gas with which an instrument is calibrated according to preference. However, this will affect how the instrument responds to other gases. Some gases, such as pentane and hexane, make an instrument read "high" when measuring methane and certain other gases with higher heats of combustion, whereas the reverse is true when calibrating to methane or propane. The fact that a compound has a high heat of combustion means higher concentrations are required in air for a potentially combustible concentration to exist, whereas others such as toluene have both low heats of combustion and low LELs. Pentane is often selected as a calibration gas where petroleum vapors are of concern, since it has a relatively low LEL and its heat of combustion puts it right in between methane and most aromatics.

A problem when calibrating at higher levels is that the sampler is potentially dealing with a flammable atmosphere. This type of calibration can only be done in a laboratory that is properly equipped. A similar situation exists when trying to verify the functioning of a flashback arrestor. The purpose of the flashback arrestor is to act as a barrier to assure that the flame in the sensor chamber during measurements of explosive range concentrations does not travel upstream through the sample tube to the source of explosive gas.[2] The instrument must be exposed to an atmosphere between the LEL and the UEL. Do not introduce 100% methane gas to an instrument located near a source of ignition; otherwise an explosion may occur.

There are private laboratories available that can do these calibrations for a fee. In some cases, a calibration requires a voltage adjustment. This necessity is common for methane. Voltage adjustments may also be required for zeroing.

For field use, instruments should be calibrated weekly, if used regularly, or prior to use if used infrequently. Calibration should be done under the same circumstances as field use. For example, if a long sampling line will be used, then the same length of line should be used during the calibration check. The response time should be noted. Check the flow rate of the instrument if it has a vacuum pump, and compare it to the manufacturer's specifications using a precision rotameter or bubble buret. Variations in the flow rate will result in an inaccurate calibration. For instructions in flow rate calibrations, see Appendix A.

Some instruments have replaceable batteries and others require charging. Generally the only maintenance permitted is replacement of batteries, flashback arrestors, particulate filters, and filaments in select meters. The two filaments in a Wheatstone bridge are usually matched, one filament being a reference and the other being the operational detector, and thus should be replaced as a pair to maintain calibration. Note that if filaments are replaced a recalibration is necessary.

Detectors also must be replaced periodically, the frequency depending on the type of compounds to which they have been exposed. Because catalytic combustion requires that the catalyst be heated, this heating causes a slow deterioration of the catalyst. Also, when detector life is described it is usually in the absence of any effects, such as "poisoning." It has been recommended that a replacement detector be taken into the field. This depends on its ease of replacement for a given instrument. Some units require soldering. Also a longer warmup period than normal and a calibration prior to use is required when the sensor is changed.

Never interchange parts of different makes or models of CGIs. In general, it is better for repairs to be done at the factory so that the instrument can be calibrated and inspected. If flashback arrestors become contaminated, they will have to be replaced. Inspect the instrument for any possible damage marks, bent parts, and so forth after each use. If the instrument is not used regularly, removal of the batteries may be necessary, and some types of sensor cells and other parts may require special handling prior to storage.

Field Use

A quick field check to be sure an instrument is responding should also be done prior to doing measurements, but away from the atmosphere to be tested. One way is to expose the instrument to an unlit butane lighter. Press the top lever down, but do not spin the sparking device.

Response times can vary from 15 seconds to 2 minutes, depending on the instrument. It is critical that the user know the response time prior to taking an instrument into the field. The use of a remote monitoring line also increases this time. Never directly insert a probe into a liquid. If liquid is sucked into the inside of an instrument, it will require repair and maybe even replacement. If sampling under dusty conditions, use a filter in front of the probe. Instruments with aspirator bulbs should be purged by squeezing the bulb 8 to 10 times in clean air. The bulb should inflate completely between squeezes. These instruments usually have a "ready" indicator. The "ready" indicator must be on prior to making measurements. To use an instrument with an aspirator bulb, squeeze the bulb 7 to 8 times to draw in the sample. When the needle stabilizes, the meter indicates the concentration of gas in air in percent volume.

Probes permit samples to be taken in areas that cannot be reached with a sampling line. Bar holes, man holes, sewers, and spaces behind obstructions or areas accessible only through narrow openings can be examined by connecting the probe to a sampling line. Probes can be made of steel, brass, or plastic (dielectric nonconducting). Do not use a brass probe where shock hazards exist. Instead the dielectric plastic probe should be used in these situations. A solid probe is useful to prevent liquids from entering an instrument by keeping the sampling line above the liquid in a tank.

Some manufacturers provide dilution tubes (1:1, 10:1, 20:1) with their instruments to allow for measurements above the LEL. When using these, it is necessary to calculate the actual gas concentration. Generally, additional scales on the meter are not provided for use with dilution tubes. Therefore, an instrument should never be stored or left sitting around with the dilution tube installed if there is any possibility that an inexperienced user might pick it up.

When monitoring in an open area, such as a tank farm, or when monitoring for leaks, use slow sweeping motions to assure "hot pockets" are not bypassed. Some instruments are designed to be dropped into a tank. In this case, if the instrument has a "peak hold" function it will store the highest concentration it encounters. Other methods for collecting a sample from inside a confined space include using an extension probe to reach farther into it or dropping a piece of extension tubing. In both cases, an instrument should have a pump so that it will pull the contaminated air up to the detector. Some passive instruments have detachable detectors that attach to the instrument via a long extension cord, allowing the sampling professional to observe the instrument's output while standing on the outside of the tank. The most difficult situation to monitor is that in which the manway is at the side of a large tank.

Prior to making measurements make sure instruments have time to stabilize. Most instruments are capable of measurements over a limited temperature range, so if a measurement is made outside of this temperature range, it may not be accurate. As a result, some instruments are temperature-compensated. Condensation of vapors can occur in sampling tubes if used at cold temperatures, resulting in a lower readout than the actual concentration present. Sampling heated atmospheres can also be a problem, because if the probe, the sampling line, or the instrument is at a lower temperature than the gas being sampled, a portion of the sample may condense within the sampling line, thus reducing the concentration of combustibles reaching the sensor.

Some instruments have a safety that is activated by exposure to high concentrations of combustible gases. When the readout indicates greater than 100% of the LEL, the readout "latches." When latched, the LEL readout blanks. This feature is a warning that the gas concentration has exceeded the LEL and that all personnel must be evacuated from the area.

Procedure

1. Before starting and after finishing the measurements, the instrument settings should be as follows: All on/off switches should be in the off position. The scale select switch should be in the least sensitive position, meaning on the highest range, e.g., 1 to 10,000. All accessory tubing and cables, including the battery charger, should be properly connected or disconnected as appropriate.

2. Inspect the scale on the meter and determine whether it reads % LEL, % gas concentration, or both. Select the scale of interest.

3. Turn on the instrument and pull air into it from a clean area. This procedure is called zeroing the instrument. If the instrument has standby and battery check settings, make sure the dial is not on one of these. In some instruments the circuit zero is automatically adjusted during the warm-up, but the span setting (meter reading during calibration) is adjusted on a potentiometer (variable resister) that may be located inside the instrument. If no clean air is available, a plastic bag should be filled beforehand with uncontaminated air and it can be used to zero the instrument. A charcoal tube can be fitted to a piece of tubing at the instrument's inlet, provided it utilizes a vacuum pump to screen out larger hydrocarbons (pentane and up), but not methane and its relatives or carbon monoxide. Compressed air can also be used. It should be similar in humidity content to the air at the site.

4. Do a field check using a butane lighter to be sure the instrument is responding. Allow the instrument to return to zero in order to make sure all connections are tight if using a diffusion sensor on a remote sampling line.

5. Put the probe into the atmosphere to be tested. Attention should be kept on the meter readout at all times during the measurement. If this is not possible, repeat the measurement after allowing the instrument to purge. If for any reason the needle pegs and returns to zero, do not allow anyone to go into the space; have detector tubes as a backup and check the space with these.

INTERPRETATION OF MEASUREMENTS OF EXPLOSIVE ATMOSPHERES

CGIs cannot differentiate between individual compounds; thus, the readout for a mixture is unlikely to be specific for any of the individual components including the gas for which it was calibrated if more than one combustible gas is present. Most units are set to alarm at 20% to 25% of the LEL for this reason. Depending on the calibration and the compounds to which the instrument is exposed, this may or may not be safe. It should be noted that any time an instrument detects a percent of the LEL, the potential exists for an immediately dangerous to life and health (IDLH) atmosphere to be present.

Some instruments read in percent, others in decimals. Digital instruments as well as those with a needle and scale for the readout have minimum and maximum detectable quantities. However, unlike a meter whose scale is readily apparent, it is necessary to read the manual to know the digital monitor's maximum. In some instruments, when the concentration reaches a certain point, for example, 100% of the LEL, the meter simply reads "over."

Since these instruments read in percentages, they cannot be used to determine health hazards expressed in ppm levels. (Table 9-3 explains CGI measurements.) The LEL for methane is 5.3%, and therefore 25% is 1.3%, or 13,000 ppm. This level is much greater than any threshold limit value (TLV) or permissible exposure limit (PEL).

TABLE 9-3. Understanding Combustible Gas Indicator Measurements

Meter Readout	*% Methane*	*% LEL*	*% Explosive*
20	20% Methane, 200,000 ppm; for most other gases concentration will read lower; this instrument is always calibrated to methane	20% of the LEL, the compound the instrument was calibrated for; if methane = 1% or 10,000 ppm; if pentane = 0.3% or 3,000 ppm; for other gases depends on the response factor	a concentration of 20% is present depending on what gas the instrument was calibrated on; = 200,000 ppm of this gas; for other gases, it depends on the response factor

Meters that read percent methane are also available. When these instruments read 5% methane, the actual concentration is near the LEL and not 5% of the LEL. For OSHA compliance, a reading in excess of 25% of the LEL of the combustible gas meter indicates an explosive hazard.

When measuring contaminants in known atmospheres, the manufacturer's response curves can assist in determining the actual concentration. As shown in Figure 9-3, a typical catalytic combustion device calibrated with a mixture of 50% of the LEL for pentane (curve 3) will read 85% when exposed to a 50% mixture of methane (curve 1). If the same meter were exposed to a 50% LEL of xylene (curve 6), it will read approximately 26%. Adding the fact that the typical error for these instruments is 10% due to temperature and humidity constraints, the reading for a 50% xylene LEL concentration on an instrument calibrated with pentane would be 23–29%, while the reading for a 50% LEL concentration of methane on the same instrument would be 76–94%. Correspondingly, if the instrument were calibrated to methane, the readout for the other compounds would appear less than the actual concentrations. For this reason, unless methane is actually likely to be present, it is better to calibrate with a compound that has a lower heat of combustion, such as pentane. This is commonly done in petrochemical operations.

Some manufacturers provide conversion factors rather than response curves. In this case, a calculation must be done. For example, if a meter calibrated on pentane is used to test for propane, the conversion factor for a particular unit might be 0.9. If the readout indicates 50% LEL, the true propane concentration is

$$(50\%)(0.9) = 45\% \text{ of the LEL}$$

When response curves or conversion factors are used, it is a good practice to use an accuracy tolerance of ±25% in the interpretation of any meter response.

An oxygen deficiency generally indicates there is a high concentration of one or more other compounds. These may or may not be combustible. One practice used in unloading tank trucks containing flammable and explosive compounds is to "inert" the tank by deliberately displacing the oxygen-containing air with an inert gas, such as nitrogen. Thus, the risk of an explosive atmosphere while moving the truck to a cleaning station is decreased. A similar practice is used when removing underground storage tanks; in this case, dry ice is used to create an atmosphere of carbon dioxide, which also displaces oxygen, thus allowing the tank to be pulled while minimizing the risk of explosion.

If the meter response initially reads off scale and then goes downscale rapidly, it is a sign that the concentration of gas is very high, possibly

No.	Compound	Formula	LEL
1	Methane	CH_4	5.0
2	Acetylene	CH_1CH	2.5
3	Pentane	C_5H_{12}	1.5
4	Ethyl Chloride	C_2H_5C1	3.8
5	1,4-Dioxane	$OCH_2CH_2OCH_2CH_2$	2.0
6	Xylene	$C_4H_4(CH_2)_2$	1.1

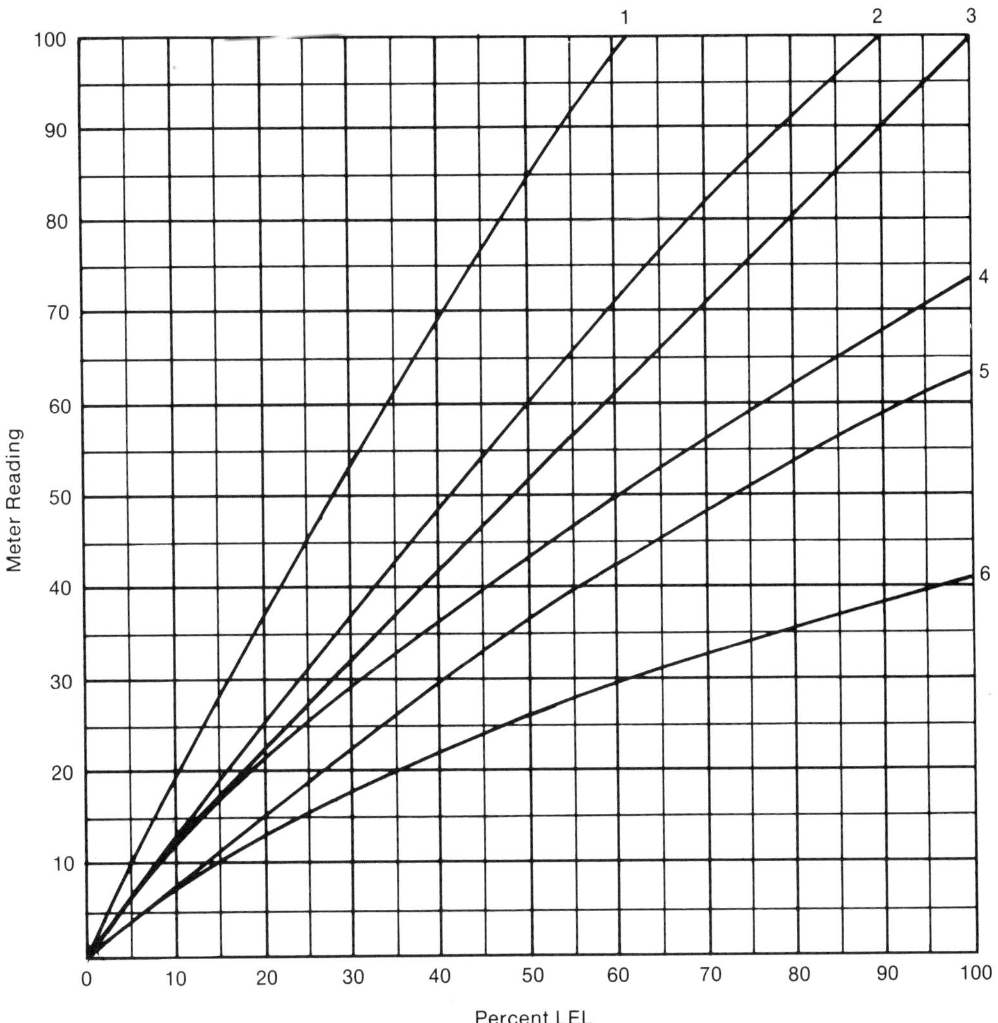

Figure 9-3. Response curves. (Courtesy MSA.)

exceeding the UEL, because the gas absorbs heat from the filament, increasing its resistance. If the gas concentration appears to be above the UEL, a potential explosion hazard still exists as the addition of air to the gas/air mixture will create a concentration in the explosive range.

MONITORING FOR HEALTH HAZARD LEVELS OF VOLATILE ORGANIC COMPOUNDS: FIDs and PIDs

General survey instruments are also used to monitor health hazard levels in situations where immediate results are needed even though the actual compounds present may be unknown or a mixture is present. The primary detectors used are PIDs and FIDs. These instruments can be used for general surveying on hazardous waste sites to determine compliance with EPA action levels and to establish site work zones. They are used for screening of air, soil, water, and drum bulk samples to establish priorities for laboratory analysis, usually for VOCs, due to the high cost of this type of analysis. They can be used to determine if decontamination procedures are effective. They are often used in a qualitative manner just to see if any volatiles are present, for example, when an odor is present in a basement. Leak detection is another common use. Another application involves boundary line sampling for total hydrocarbons to detect airborne releases from an industrial operation or hazardous waste site to the surrounding areas, especially in the case where residences, schools, or other sensitive populations are nearby. Perimeter monitoring is often done by placing analyzers in north, south, east, and west locations along with a roving unit. Continuous monitoring utilizing a data logger or periodic checks at the stationary points by the sampling professional are possible.

Photoionization Detectors

Another term used for PIDs is total ionizables present monitor. Most organic compounds and some inorganic compounds can be ionized when they are subjected to ultraviolet (UV) light. The PID converts the concentration of ionizable chemicals in a sample to an electric signal. In essence these instruments consist of a pump or a fan that draws the air sample past a UV lamp where the compounds in the air sample are ionized. The sample produces an ion current proportional to mass and thus concentration. An electronic signal is then displayed on an analog or digital output.[3] The sensor consists of a sealed UV light source that emits photons with an energy level high enough to ionize many organic and some inorganic compounds.

There are two basic designs of PIDs. In one, the lamp is contained within the instrument housing and one set of electronics applies to all lamps. In the other design, the lamp is incorporated into a probe and there is a separate set of electronics for each probe/lamp. The result is a difference in the cost as well as the frequency of replacement. The all-in-one design can be held in one hand. The limit of detection for most contaminants is approximately 1 ppm.

The HNU 101 is one of the best known PID survey instruments. Its design is the type where the lamp and its electronics are contained in the probe. A chamber in the probe exposed to a UV light source contains a pair of electrodes: one a bias electrode and the second a collector electrode. When a positive potential is applied to the bias electrode, a field is created in the chamber. Ions formed by the absorption of photons are driven to the collector electrode. The current produced is then measured and the corresponding concentration is displayed on a meter directly in ppm.[4]

$$RH + h\nu \longrightarrow RH^+ + e^-$$

The HNU has three scales: 0–20 ppm, 0–200 ppm, and 0–2000 ppm. There are three lamps available for the HNU: 9.5 eV, 10.2 eV, and 11.7 eV. According to the manufacturer, the linear range for most compounds is from 0.1 ppm to 600 ppm. Measurements from 600 pm to 2000 ppm generally are not linear and may read lower than actual concentrations.[5]

The instrument can be used continuously for 10 hours without a strip chart recorder and 6 hours with one. Because of concerns regarding the effects of humidity on PID measurements, a second model was introduced by HNU that has a slightly different design. While the lamp is still installed inside the probe, rather than using a fan to pull in samples, it has a positive displacement pump and a different orientation of the sample inlet to the ion chamber, which tends to prevent condensation.

An example of a microprocessor-based general-survey PID is the Photovac Microtip (Fig. 9-4). Since the growth in the use of PID-based general survey analyzers is one of the largest areas in sampling applications, it is likely that in the future most instruments will incorporate microprocessors. This instrument is the type where the lamp is contained inside the instrument housing. Most units of this type offer five different lamps: 8.5, 9.5, 10.2, 10.6, and 11.7 eV. Microprocessor-based units have several functions allowing for adjustment of parameters such as calibrations, concentration readout, data logging, lamp selection, and communication with printers and computers. The initial setup requires selection of such values as the concentration readout, type of logging interval, and type of display interval. For example, readings can be averaged from one time per second to once every 4 minutes. The type of lamp installed in the instrument must be reprogrammed if a different lamp is installed or the calibration will not be accurate. In a microprocessor-based PID such as a Microtip the microprocessor can subtract background from the signal and multiply this signal by a response factor previously obtained by calibrating with a gas of known concentration.

The PID has a large linear range and is a good general detector for organic and some inorganic gases. These instruments can detect aliphatic and aromatic hydrocarbons, halohydrocarbons, alcohols, ketones, aldehydes, ethylene oxide, arsine, phosphine, total reduced sulfur components including hydrogen sulfide, and glycol ether solvents. In general, the PID has a lower response to compounds, such as short-chain hydrocarbons, methyl cyanide, ethylene, carbon tetrachloride, and methyl alcohol. The lamps used in these devices do not ionize major components of air, such as oxygen, nitrogen, carbon monoxide, carbon dioxide, and water, thus eliminating interferences from these compounds and allowing measurements to be made in ambient air.

As with other general survey instruments, the

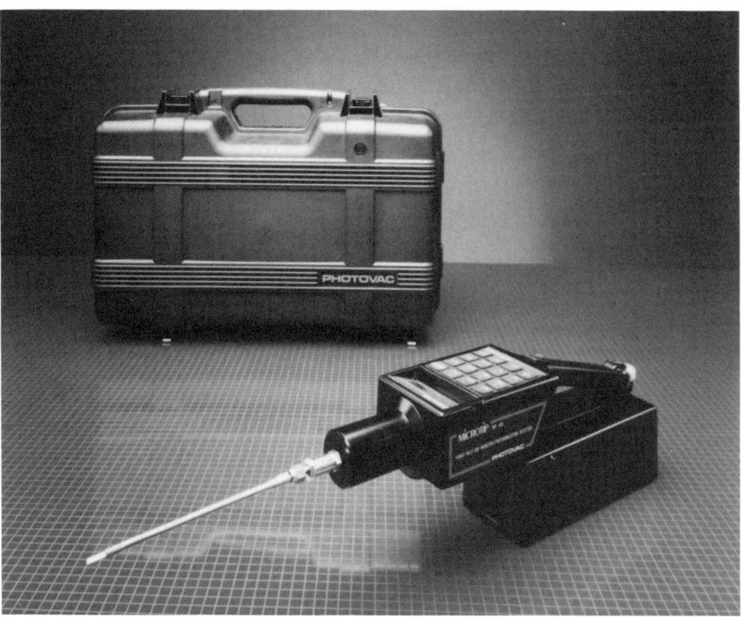

Figure 9-4. The Microtip, a microprocessor-based photoionization detector. (Courtesy Photovac, Inc.)

TABLE 9-4. Relative Photoionization Sensitivities for Gases (if instrument is calibrated to benzene)

*Relative Sensitivity**	*Examples*
10	Aromatics, such as benzene, xylene, toluene
10	Aliphatic amines, such as diethylamine
5–9	Unsaturated chlorinated hydrocarbons, such as trichloroethylene, dichloroethylene
7–9	Ketones, such as MEK, MIBK, acetone, cyclohexanone
3–5	Unsaturated hydrocarbons, such as acrolein, propylene, cyclohexanone, allyl alcohol
3–5	Sulfide compounds, including hydrogen sulfide, methyl mercaptan
1–3	Paraffinic hydrocarbons with 5 to 7 carbons, hexane, heptane, and so on
<1–0	Ammonia and paraffinic hydrocarbons with 1 to 4 carbons, ethane, methane

*Relative sensitivity is the meter reading when measuring 10 ppm of the listed gas with the HNU, with the 10.2-eV probe calibrated for 10 ppm of benzene, span plot setting at 9.8 for direct reading of benzene (*HNU Manual*).

response varies depending on what compounds are being sampled. As mentioned previously, if only one chemical species is present, the PID can be set to quantitatively respond to that chemical. However, the PID will not quantitatively respond to a mixture unless the ionization potentials of all the chemicals in the mixture are the same, because the instruments have a different sensitivity to compounds with different ionization potentials (Table 9-4). As a rule, the PIDs are more sensitive to complex compounds and less sensitive to simpler ones.

Some key instrument parameters affecting PID sensitivity include lamp intensity, lamp seal, and flow rate through the detector. In terms of chemical structure, PID sensitivity depends on the following parameters: carbon number, functional group, and type of bond. Sensitivity increases as carbon number increases. The following sensitivity relationships have been observed[6]:

alkanes < alkenes < aromatics
alkanes < alcohols < esters <aldehydes < ketones
cyclic compounds > noncyclic compounds
branched compounds > nonbranched compounds
fluorine-substituted < chlorine-substituted < bromine-substituted < iodine-substituted

As noted, the ionization source in the PID is a gaseous discharge lamp. These lamps contain a low-pressure gas in which a high-potential electrical current is passed. By varying the composition of the gas in the detector lamp, manufacturers are able to produce lamps with different energy (ionization potential) levels. Lamps are manufactured in several energy levels: 8.5, 9.5, 10.0, 10.2, 10.6, 11.4, and 11.7 electron volts (eV) (Table 9-5).[7] However, since lamps must be purchased from the manufacturer of the instrument, not all lamps are available for all instruments.

Currently 11.7 eV is the highest-energy lamp commercially available that has the widest sampling range, and therefore it is the most logical one to use; however, its lamp window is manufactured from lithium fluoride due to the need to transmit such high-photon energy, and this material degrades rapidly in the presence of humidity. This high-photon energy also produces radiation

TABLE 9-5. Types of Gases Used in PID Lamps

Gas	Energy (eV)	Wavelength (ηm)	Window
Oxygen	9.5	130	MgF_2
Krypton	10.0	123.58	MgF_2
Hydrogen	10.2	121.6	MgF_2
Nitrogen	10.6	113	LiF
Helium	11.4	108	LiF
Argon	11.7	105	LiF

damage to the lithium fluoride. As a result, the lamp can be unreliable and often needs repair. The 10.6-eV lamp has a magnesium fluoride window that is much more stable than the lithium fluoride; therefore, lamps in this range tend to be used more often. The 11.7-eV lamp is useful for compounds with ionization potentials (IPs) greater than 10.7 eV, which includes many chlorinated compounds.[3]

For stability and useful lives of the lamps, the noble gases (krypton, helium, and argon) are much more stable and would be more applicable for continuous use. The nonnoble gases (oxygen, nitrogen, and hydrogen) are more appropriate for intermittent use, and their lives can be greatly extended if they are turned off between use. Another concern relating to stability is the choice of the window. For example, the lithium fluoride window will discolor though use, which would further reduce the transmittance and, thus, the sensitivity of the detector.[7]

Lamps can detect some compounds with higher IPs than the lamp's ionization energy (eV), but it has been noted that the sensitivity decreases rapidly as the IP of the analyte exceeds the eV of the lamp.[8] However, there is enough of a response to certain compounds with higher IPs for any given lamp to indicate that the cutoff point for any lamp is not well defined.[3] For example, in theory, a sampler would expect a lamp with a 10.2-eV cutoff to detect only compounds with ionization potentials (IP) of 10.2 eV and less; however, in reality it is not the case as lamps are categorized by their major emission line and as a result instruments can often detect compounds with IPs above the rating of the lamp.[7] A review of the krypton lamp spectrum shows that krypton has two distinct emissions. The major one is at 123.58 ηm, and the minor one of about 10% is at 116.49 ηm. These wavelengths correspond to 10.0 eV and 10.6 eV. Thus, the 10.0-eV krypton lamp can detect ethene, which has an IP of 10.52, although the response for ethene is much less than that for materials that have an IP less than 10.0 eV, such as benzene. Although the hydrogen lamp is rated at 10.2, materials that have ionization potentials of up to 10.9 eV can be seen because of the relatively broad spectrum of this lamp.[7] The sensitivity is also a function of the lamp window, since magnesium fluoride transmits a greater amount of radiation, as compared to the lithium fluoride window.[7]

Lamps of different energies also have different sensitivities to a given compound. For example, an 11.7-eV lamp has only one tenth the sensitivity to benzene that a 10.2- or 10.6-eV lamp has, even though it has a much broader detection capability than the other lamps. Each lamp energy has different responses to compounds other than what has been used for calibration. Thus, if the response of a 10.2-eV lamp is set to 100 ppm of benzene to be 100 ppm, then its response to 100 ppm of acetone would be 63 ppm.

Selectivity can be introduced to some degree by the use of different energy lamps. For example, the 9.5-eV lamp allows for the selective detection of aromatics in the presence of alkanes or oxygenated hydrocarbons, and the determination of mercaptans in the presence of hydrogen sulfide. Since aromatics, amines, and organic sulfur compounds have low IPs, these compounds can be determined with greater selectivity using a 9.5-eV lamp instead of a 10.2-eV lamp. A 9.5-eV lamp has a high sensitivity for the amines but will not detect ammonia, which has an IP of 10.15 eV.[9] The 10.9-eV lamp has a wider range of response than the 10.2-eV lamp and can detect formaldehyde, formic acid, and other compounds that are difficult to measure with most instruments.

PIDs are sensitive to humidity extremes, possibly due to the lamp fogging and to the fact that the IP for water vapor is 12.61. It has also been theorized that water molecules may collide with a photoionized contaminant is molecules inside the detector and deactivate them. In one test, the presence of 90% relative humidity (RH) caused a PID instrument to take twice as long to stabilize than at 0% RH. When compared to measurements at 0% RH, the response of this instrument to the same concentration was much lower at 90% RH. In another test, it was found that while stable readings were attained after 10 seconds in dry atmospheres, 5 minutes were required to stabilize when organics at 90% RH were sampled. It was also concluded that 90% RH decreases the response of the 10.2-eV lamp by a factor of 2 for most compounds as compared with the response under dry conditions.[10] For the 11.7-eV probe, humidity is especially a problem due to the sensitivity of the lamp material to water vapor.

Calibration

All instruments are calibrated in the factory by adjusting a span setting (potentiometer) that controls instrument sensitivity by varying the gain on the amplifier, so that the instrument will read directly for a defined concentration of a specific vapor. However, as the lamp fatigues or becomes contaminated, the factory calibration becomes inaccurate, so each instrument must be recalibrated regularly by adjusting the span and zero settings.[5]

If the compounds most likely to be present are known, calibrating the instrument to one of these will make the readings in the field more relevant. Another strategy which is sometimes used is to adjust the response of an instrument calibrated with one gas to the sensitivity needed for another gas. This can be done using response factors to multiply the reading or utilizing the span adjustment instead. Although most PIDs are calibrated with isobutylene, the instrument is usually adjusted to respond as if this gas was benzene (55 ppm of benzene is equivalent to approximately 100 ppm of isobutylene if using a 10.2-eV lamp). In this case, the sampler can consult charts that give the response factor for a different compound and multiply the result accordingly. The response factor also depends on the lamp intensity (Table 9-6).[5] In general, the higher the response factor multiplier, the lower the actual response of a given lamp to a compound. The use of a direct calibration provides more accuracy than a calculation involving response factors. The choice depends on the options available.

One approach is to use a background zero

TABLE 9-6. Comparison of the Relative Responses of the 9.5- and 10.2-eV Lamps in an HNU 101

Compound	*10.2 (eV)*	*9.5 (eV)*	*9.5/10.2 Response × 100 (%)*	*IP (eV)*
Xylene	112	112	100	8.4
Benzene	100	100	100	9.2
Styrene	100	100	100	8.5
Toluene	100	100	100	8.8
Phenol	75	77	102	8.5
Aniline	35	39	111	7.7
Pyridine	30	22	73	9.3
Methyl ethyl ketone	57	29	51	9.5
Acetone	63	6.5	10	9.7
Methyl methacrylate	30	<6	—	—
Heptane	17	<2	<10	10.1
Hexane	22	0	0	10.2
Ammonia	2	0	0	10.2

with a fixed-point scan, a method good for determining whether a response to a sample is actually different from that expected from ambient air. In this type of calibration the instrument is zeroed in a noncontaminated area so as to obtain an ambient clean air or background zero reading while the span is kept at the same fixed point, usually in the middle range.[3]

Another option is background zero, followed by a relative response to a calibrant span. Zeroing is done as described previously; then the instrument is connected to a Tedlar bag containing a standard calibration gas, such as 100 ppm of isobutylene in air. There are two variations for this method. The span may be adjusted to give a reading of 100 so that all measurements are recorded as equivalent to the isobutylene signal, or it may be adjusted to read directly in ppm for another compound, such as benzene.

In situations where only one compound is present (in this example, benzene), the instrument's reading will be actual ppm values for that compound as long as the response factor for the desired gas or vapor (to that lamp) is known. This method may be chosen rather than calibrating directly with a concentration of the contaminant of interest because the calibration gas may not be readily available or may be very toxic. The sampler may also want to do several scans using different calibrations of the instrument. However, the most accurate method would be to calibrate the instrument with the compound of interest.[3]

Under some working conditions it may be impossible to obtain clean background air, or alternatively, very accurate low-level results are necessary. In such cases, one approach is to use a "zero air zero" with a specific gas calibration. One application is when high-toxicity compounds are being sampled. In these cases, a Tedlar bag of zero-grade air is used to obtain an accurate zero followed up with exposure to a standard of the compound of interest to adjust the span setting. This procedure assures that the instrument is calibrated at both ends of the scale. The zero air is essential, because if the instrument is zeroed with contaminated air and subsequently encounters a cleaner atmosphere, negative readings could result.[3]

Clear plastic overlays that incorporate the response of a compound to a specific calibration gas are available for at least one PID model that allow the user to avoid modifying the span setting for compounds other than the calibration gas. The concentration scale on the overlay corrects for regions of nonlinearity for each compound, and also indicates if there is significant nonlinearity at very low concentration ranges.

Procedure

1. Identify the probe by the lamp label and connect it or input the correct codes on the unit to identify which lamp is inside. If there is any doubt that the readout may not have been updated since the lamp was changed, the unit should be opened to physically inspect the lamp.

2. Set the span pot to the proper value for the probe being calibrated.

3. Check the IP of the calibration gas to be used. The IP of the calibration gas must be near or below the IP of the lamp.

4. Set the unit on standby. In this position the lamp is off and no signal is generated. Set the zero point with the zero set control.

5. For calibrating for general survey work on lower ranges, such as 0 ppm to 20 ppm, only one calibration point is required. Turn the switch to the proper concentration range and note the readout. Adjust the span control settings as required to read the ppm concentration of the standard. For each lamp, compound, and concentration combination a different span setting will be required. Recheck the zero on the readout and adjust if necessary. A two-point calibration is given: zero and the gas standard concentration.

6. For calibrating on the higher ranges, such as 0–200 ppm and 0–2,000 ppm, the use of two or more concentrations is recommended. At a minimum, use one that will give a response of 70% to 85% of full scale and other for 25% to 35% of full scale. First calibrate with the highest standard, using the span control setting. Then

calibrate with the lower standard, using the zero adjust. Repeat these several times to ensure that a good calibration is obtained.

Maintenance

During periods of operation, dust or other foreign matter can be drawn into the instrument, forming deposits on the surface on the UV lamp or ion chamber. This condition is indicated by meter readings that are low, erratic, unstable, nonrepeatable, or drifting, or show apparent moisture sensitivity. When exposed to levels of gases and vapors higher than an instrument's detection capability, it is possible to saturate an instrument. Some contaminants are "sticky" and will remain in the instrument for long periods of time. The instrument should be inspected monthly for this condition, if in regular use. Cleaning is described below.

Procedure

1. Very carefully disassemble the probe or instrument and remove the lamp and ion chamber.

2. During the course of normal operation a film will build up on the window of the UV lamp. First check the lamp window for fouling by looking at the surface at an incident angle because deposits, films, or discoloration may interfere with the ionization process. The rate at which the film develops depends on the type and concentration of the gas and vapor being sampled. As a guide clean the window following every 24 hours of operation. For magnesium-fluoride lamps, the windows are cleaned by rubbing gently with lens tissue that has been dipped in a detergent solution. If further cleaning is needed, use the cleaning compound supplied by the manufacturer and spread it evenly over the surface with a lens tissue. Wipe off the compound and rinse the surface with warm water (27°C), or use a damp tissue to remove all traces of grit or oils and any static charge that may have built up on the lens. Dry carefully with clean tissue. For lithium-fluoride lamps, cleaning is done using chlorinated solvents. Never clean this lamp with water or any water-miscible solvents, such as methanol or acetone, as they will damage the lamp. Cleaning compounds, unless specifically recommended for this lamp, should not be used either.

3. Inspect the ion chamber for particulate deposits. If present, the chamber should be cleaned. A tissue or cotton swab, dry or wetted with methanol, can be used to clean off any stubborn deposits but nothing else. Make sure the chamber is absolutely dry prior to reassembly.

4. Reassemble the probe or instrument and check the analyzer's operation by zeroing and calibrating the instrument.

5. If these steps do not work, the lamp may require replacement.

Field Measurements

While there are differences in the operation of all instruments, the following section contains some general information on survey PIDs. Be alert for sources of electromagnetic fields, such as power lines, transformers, and radio wave transmissions, since these may affect readings. Never look directly at the light source from closer than 6 inches or less without wearing proper eye protection; continued exposure to UV light will damage the eyes. Prior to field use, the instrument should be calibrated for the range of interest. If the range is unknown, calibrate the instrument at three points over the instrument's range.

The length of the probe that can be used with a PID depends on the instrument's flow rate. The flow rate should not be decreased by the diameter or length of the probe or the pump's lifetime may be affected. To minimize sample absorption and reactions that cause or probe memory, most probes are constructed of Teflon or stainless steel.

As a field check a magic marker can be used to see if the instrument is responding. Never use automobile exhaust to check the status of an instrument as it contains condensable organic compounds and particulates that can deposit in the probe and foul the lamp.

If an initial hazardous waste site investigation, walk the perimeter of the site first and record measurements. The PID has a relatively long response time, and therefore the sampling professional must walk slowly when using this type of analyzer, or areas of high gas and vapor concentrations may be entered inadvertently before the instrument has a chance to respond. The PID is also highly directional and must be held close to a source in order to get an accurate reading. High humidity and wind will affect the accuracy of the meter.

If a reading is unstable a lower span setting may be necessary. Sampling in a windy location can also cause the reading to jump, so the inlet should be sheltered in these situations. If the chemical concentration in the air is fluctuating, so will the readout.

Procedure

1. Make sure prior to field use that the instrument is fully charged. During charging of an instrument where the lamp is contained in the probe, it is important to have a probe attached or the unit will not fully charge and will not show the battery check. The charger should be disengaged from the wall before disconnecting from the instrument. With a full charge, most instruments should last 8 hours, although colder temperatures can decrease this time.

2. Depending on the energy (eV) of the lamp, the span may have to be changed. Set the span to the value specified for the lamp on the calibration gas cylinder if available.

3. Adjust the zero to zero of the instrument. If the span adjustment setting is changed after the zero is set, the zero should be rechecked and adjusted if necessary. Wait 15 to 20 seconds to assure that the zero reading is stable.

4. Set the instrument to the appropriate operating range. Start with the highest position and then switch to the more sensitive (lower) ranges. In this position, the UV light source should be on. If the lamp is installed in a probe, check the end of the probe for the purple glow of the lamp. This works on some units but not on all.

5. On instruments where a single scale is used for all concentration ranges, be aware of the range on which the instrument is set at all times, or confusion in what concentrations are being measured may result, potentially exposing individuals to high concentrations. Position the end of the intake probe in close proximity to the area or item to be sampled because of the low flow rate associated with most instruments. Use slow sweeping measurements to avoid by-passing problem areas.

6. If using the PID to screen drum contents or other containers in which there may be high concentrations of chemicals, set the instrument on the highest-range setting first and move gradually closer to the top of the open bung hole. If wanting to approximate exposure concentrations, take a measurement at shoulder level. Do not stick the instrument's probe directly into bore holes, drum bungs, or other situations unrealistic of exposures as doing so could saturate the instrument and most likely will require the use of a scale for which the instrument is not calibrated. Saturating an instrument can result in having to wait long periods of time to flush it out, or may cloud up the lamp during a time when sampling is critical.

7. If high levels of dust are present attach a filter to the front of the inlet to prevent dust from entering the detection chamber and causing incorrect readings. Always have at least one spare lamp along on surveys in the event that the lamp is scratched during cleaning, damaged by deposits of nonvolatile compounds, or by physical shock such as dropping the instrument or it simply wears out.

Collection of Follow-Up Samples

Some PIDs have an exhaust port that allows collection of an integrated sample on a sorbent tube. An integrated sample is collected for laboratory analysis to confirm identities of the suspected airborne contaminants because the PID is a nondestructive detector. To collect such a sample, one or two charcoal tubes are hooked to the exit port of the sampler. If the flow rate is high, two tubes may be necessary to split the airflow of the unit so that half goes through each tube, although only one has to be analyzed.[11] The flow rate should be calibrated with the tubes "in line" prior to use. For more information on flow rate calibrations, see Appendix A.

Flame Ionization Detectors

Hydrocarbon analyzers that utilize an FID are carbon counters; therefore, their response to a given quantity of a typical C_6 hydrocarbon is six times that of methane, as long as the flow rate is the same. The FID is essentially a stainless steel burner in which hydrogen is mixed with the incoming sample in the base of the unit; combustion air or oxygen is fed in and diffused around the jet through which the hydrogen gas mixture flows to the cathode tip where ignition occurs. When organic compounds are introduced into the flame, positively charged, ionized carbon fragments form that are collected by a negative platinum loop (the collector electrode). An electric field exists between the conductors surrounding the flame and the collecting electrode that drives the ions to the collecting electrode. As the positive ions are collected, a current corresponding to the collection rate is generated on the input electrode.

The current carried across the electrode gap is proportional to the number of ions generated during the burning of the sample and thus the concentration. Different organic compounds will ionize to a different but repeatable extent in the flame, allowing the FID to analyze by the mechanism of breaking bonds as the following reaction indicates:

$$RH + O_2 \longrightarrow RH^+ + e^-$$
$$\longrightarrow CO_2 + H_2O$$

The organic vapor analyzer (OVA) is the most common general survey FID analyzer in use. The method used in the OVA to introduce the sample to the hydrogen flame modifies the typical ion formation process, so that it has a response to nearly all commonly encountered hydrocarbons of between 50% and 150% of the response of methane. This instrument is designed to operate within a wide range of relative humidity, 5% to 95% (Table 9-7).

TABLE 9-7. Relative Response of the OVA to Different Chemicals* (if calibrated to methane)

Compound	*Relative Response (%)*
Acetaldehyde	25
Acetic acid	80
Acetone	60
Acetonitrile	70
Acetylene	225
Acrolein	27
Acrylonitrile	70
Allyl alcohol	30
Allyl chloride	50
Aniline	38
Benzene	150
Benzyl chloride	60
Bromoethane	75
1-Bromopropane	75
n-Butane	63
1,3-Butadiene	28
n-Butanol	55
iso-Butanol	70
sec-Butanol	65
tert-Butanol	105
Butene	55
n-Butyl acetate	80
n-Butyl acrylate	60
2-Butyl acrylate	70
n-Butyl formate	50
2-Butyl formate	60
n-Butyl methacrylate	60
2-Butyl methacrylate	80
Carbon tetrachloride	10
Chlorobenzene	200
Chloroform	65
1-Chloropropane	75
2-Chloropropane	90
Cumene	70
Cyclohexane	85
Cyclohexanone	110
n-Decane	75
o-Dichlorobenzene	50
p-Dichlorobenzene	113
1,1-Dichloroethane	80
1,2-Dichloroethane	80
trans-1,2-Dichloroethylene	50
Dichloromethane	100
1,2-Dichloropropane	90
1,3-Dichloropropane	80
Diethylamine	75
o-Diethylbenzene	66
Diethyl ether	18
Diethyl ketone	80
Dimethyl ether	21

TABLE 9-7. *(continued)*

Compound	*Relative Response (%)*
Dimethyl formamide	34
Dimethyl hydrazine	20
p-Dioxane	30
Dipropylamine	110
Ethane	80
Ethanethiol	30
Ethanol	25
Ethrane	150
Ethyl acetate	65
Ethyl acrylate	40
Ethyl benzene	100
Ethyl bromide	75
Ethyl butyrate	70
Ethyl cellosolve	22
Ethylene	85
Ethylene dibromide (1,2-dibromoethene)	50
Ethylene dichloride (1,2-dichloroethane)	60
Ethylene oxide	70
Ethyl formate	40
Ethyl methacrylate	70
Ethyl propionate	65
Freon 11	10
Freon 12	15
Freon 21	72
Freon 22	42
Freon 113	90
Freon 114	110
Halothane	30
Heptane	75
Hexachloro-1,3-butadiene	95
Hexane	75
Hexanol	60
Hexene	70
Isobutene	82
Isoprene	50
Isopropyl alcohol	65
Methane	100
Methanol	12
Methyl acetate	41
Methyl acrylate	40
Methyl chloride	78
Methyl cyclohexane	100
Methyl cyclopentane	80
Methyl ethyl ketone	80
Methyl isobutyl ketone	100
Methyl methacrylate	50
4-Methyl-2-pentanone	100
Methylene chloride	90
Methyl ethyl ketone	80
Methyl methacrylate	50

TABLE 9-7. *(continued)*

Compound	*Relative Response (%)*
Methyl propyl ketone	65
Methyl isopropyl ketone	90
4-Methyl-2-butanone	90
Monomethyl hydrazine	16
Nitrobenzene	98
Nitromethane	34
1-Nitropropane	60
2-Nitropropane	70
Nonane	90
Octane	80
n-Pentane	65
Pentanol	40
Phenol	54
Piperidine	70
Propane	70
Propanoic acid	32
n-Propanol	41
2-Propanol	65
n-Propyl acetate	75
Propyl bromide	75
n-Propyl ether	65
n-Propyl formate	50
Pyridine	128
Styrene	80
1,1,1,2-Tetrachloroethane	100
1,1,2,2,-Tetrachloroethane	100
Tetrachloroethylene	70
Tetrahydrofuran	40
Toluene	110
1,1,1-Trichloroethane	105
1,1,2-Trichloroethane	85
Trichloroethylene	70
1,2,3-Trichloropropane	73
Triethylamine	70
Vinyl acetate	50
Vinyl chloride	35
Vinylidene chloride	40
o-Xylene	116
m-Xylene	111
p-Xylene	116

*Relative response (%)

$$= \frac{\text{Concentration of organic vapors on readout}(100)}{\text{Actual concentration}}$$

Therefore,

Actual concentration

$$= \frac{\text{Concentration of organic vapors on readout}(100)}{\text{Relative response in }\%}$$

There are three models of OVAs available with two different readouts: 1 – 1000 ppm and 1 – 10,000 ppm on a logarithmic scale. The unit with the lower range is used for boundary line monitoring and hazardous waste work while the log scale model is used to measure fugitive emissions, which are discussed in the chapter on Survey Preparations and Performance.

General survey FIDs respond to almost all organic compounds, but the response is greatest to aliphatic hydrocarbons. While FIDs do not respond to atoms other than carbon, other atoms can alter an instrument's sensitivity to carbon by altering the chemical environment of the carbon atom. For example, compounds containing oxygen, such as alcohols, ethers, aldehydes, and esters, and nitrogen-containing compounds, such as amines, amides, and nitriles, have a lower response than that observed for hydrocarbons. It can also detect halogenated hydrocarbons, such as trichloroethylene, chloroform, and 1,1,1-trichloroethane, although its sensitivity to these is low compared to most combustibles. The FID is insensitive to water, inert gases, and inorganic compounds and has a negligible response to carbon monoxide and carbon dioxide, which due to their structure do not produce appreciable ions in the detector flame.

Instrument characteristics, such as sensitivity, are usually given as methane equivalents because of the high sensitivity to short-chain hydrocarbons. Any FID-based instrument will have slightly different responses for other organic vapors relative to the gas it is calibrated for, usually methane.

Identification of gases to aid in determining suspected cases of arson is another application wherein the instrument is used to screen materials at the site to target those most likely to be useful for analysis at a crime laboratory. In arson surveys first the entire area is scanned, and then the uncovered debris is surveyed, layer by layer. The most common compounds involved in arson are commercially available volatile liquids, such as acetone, alcohol, gasoline, paint thinner, mineral spirits, or other petroleum-based chemicals.

When using an FID in the initial site investigation of landfills, the sampler should be aware that these sites, whether used for sanitary or hazardous waste, have appreciable levels of methane present due to decomposition of organic material. By using a charcoal filter to scrub incoming samples for organics, in these situations the sampler can determine to some degree whether the response is primarily due to methane, because it does not collect on charcoal.

Calibration and Maintenance

The OVA (Fig. 9-5), is calibrated by the manufacturer with methane. The preference for methane is based on its availability, linearity on the calibration curve, and cost. Methane calibration standards are available in a wide variety of concentrations and are less expensive than most other hydrocarbon gases. The OVA has a linear response to methane over its entire dynamic range (10,000 ppm), whereas for other gases, such as hexane, it does not. The instrument is calibrated by adjusting a gas select setting and a trim pot that controls instrument sensitivity, so that it will read directly for a defined concentration of a specific vapor. As the FID filament becomes contaminated, the factory calibration becomes inaccurate, so the instrument must be recalibrated regularly.

Calibration may be accomplished using a single standard of methane in air in the range of 90 ppm to 100 ppm. This is often adequate for field survey work. If the primary intent is to use the OVA in situations where very low levels will be encountered, such as screening samples for headspace vapors, it would be better to calibrate the instrument for two to three levels over the range of interest. It may be best to have more than one instrument if it is necessary to use it for widely disparate measurements, such as perimeter monitoring and drum sampling.

Never connect a gas cylinder directly to the probe. If the regulator flow is greater than the inlet flow rate of the OVA, a direct connection could result in pressurization of the analyzer, resulting in an inaccurate calibration and possibly causing damage to the system. Instead fill a sample bag with the cylinder, then draw from the bag into the instrument. It is important not to draw a vacuum on the inlet as well.

For instruments with several measurement ranges, calibrations should be done with known concentrations for each range. For example, to

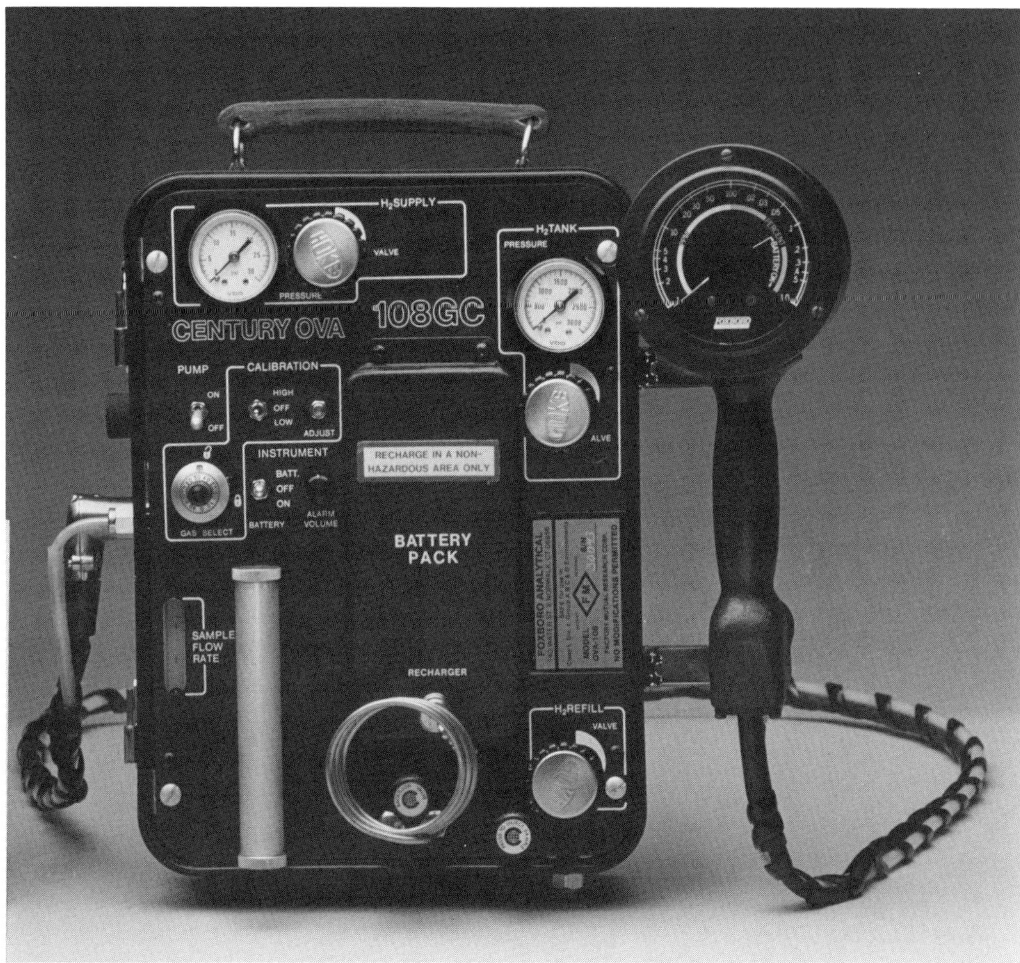

Figure 9-5. The organic vapor monitor, a flame ionization detector. (Courtesy The Foxboro Company.)

calibrate the OVA 128, which has ranges of 1 ppm to 10 ppm, 1 ppm to 100 ppm, and 1 ppm to 1,000 ppm, samples of methane in air in the following concentration ranges are used: 7 ppm to 10 ppm, 90 ppm to 100 ppm, and 900 ppm to 1,000 ppm.

There are two hydrogen cylinders on an OVA: a high-pressure main cylinder and a low-pressure supply tank. A supply of analytical grade hydrogen is needed for recharging. When charged to 100 psi, the hydrogen supply will provide 8 hours of service. Compressed hydrogen cannot be shipped by air, therefore, a supply of compressed hydrogen must be obtained locally. This should be prepurified or zero-grade hydrogen that has been certified to have <0.5 ppm total hydrocarbon (as methane) contamination.

There are two filters designed to prevent dust from getting into the instrument along with sampled air. One is located in the sampling tube, and the second filter, actually a porous stainless steel filter cup, is behind the sample inlet connector inside the instrument's case. If the unit is saturated by sampling too high a concentration, contamination can be trapped in the porous bronze sample filter. The air sample lines can also become contaminated and cause high readings.

A potential problem associated with OVAs is

leaks in the air sample pumping system. These leaks can result in dilution or loss of sample, causing low readings and slow response times. The flow rate rotameter can be used to check for air leaks. Cover the end of the sampling probe and observe the gauge to see if the ball goes to the bottom, indicating no airflow. Next, cover the screen over the exhaust port and again observe the gauge for lack of airflow. Failure of the ball to go to the bottom when the inlet is blocked indicates a leak in the system between the probe and the pump inlet or the inlet check valve.[12]

Special shipment is required if the FID analyzer is to be filled with hydrogen gas during shipment under Department of Transportation (DOT) regulations. Exemptions have been provided to various firms by the DOT that allow the instrument to be shipped on passenger aircraft, however a formal application must be submitted to the DOT.

Field Operation

For most users, the most difficult step is lighting the flame. Although it is suggested that the unit can be used after a warm-up period of only 10 minutes, up to 30 minutes is often better. The hydrogen cylinder should remain turned off during this time, but the instrument's power switch should be turned on so the filament is allowed to heat up. A full battery charge is critical to light the flame. Firmly depress the red igniter button and count off for the full 6 seconds recommended by the manufacturer. A little "pop" will be heard and the meter needle will move upscale, usually pegging briefly and then returning to zero. Do not depress the igniter for more than 6 seconds. If the flame does not ignite, wait 1 minute and try again.

If using the audible alarm it must be set for the range on which measurements are going to be made. As an example, if making survey measurements on a site where 5 ppm has been selected as the level to upgrade from an air purifying mask (level c) to a supplied air respirator (level b), then the alarm should be set for this level. Bear in mind that near freeways or on landfills where methane and other hydrocarbons may be present, and due to other hydrocarbon sources, background alone may equal or exceed this amount, so the 5 ppm should be set at a level that is in excess of background.

The background reading on linear OVAs is determined by zeroing the meter with the flame out and noting the rise in the meter readout after the flame is lit. On logarithmic-scaled units the background reading is observed on the meter at all times. This reading will vary depending on the geographical location where the instrument is being used; however, a typical background reading is 3–5 ppm (as methane). This consists of 1–1.5 ppm from ambient air methane and 2–4 ppm from contamination in the hydrogen fuel. Over a period of time fuel contamination can create deposits on the burner face that will cause a constant background when the flame is lit. If the OVA is to be used in the 0–10 ppm range, it should be "zeroed" in an area with very low background. For zeroing the instrument in contaminated air a charcoal filter should be used.

As a field check a butane lighter (button pushed but not lit) can be used to see if the instrument is responding. Never use automobile exhaust to check the status of the instrument as it contains condensable organic compounds and particulates that can deposit in the probe, partially plug the filters, and possibly damage the detector.

Survey the areas of interest while listening for the audible alarm. For ease of operation, use the strap and carry the unit on the shoulder opposite the hand carrying the probe. Walk slowly or a high concentration could be encountered before the instrument responds, since it does not respond fast enough to detect vapors at a rapid walking pace.

In general, the probe should be positioned at chest level while taking readings. If trying to approximate an individual's exposure, hold the probe near the person's breathing zone. If taking a headspace vapor measurement above a bulk sample, start on the highest range and be careful not to suck any liquid up the probe.

Two sampling probes are available: a close area sampler and a telescoping probe. The telescoping probe is most useful for leak detection and for sampling near situations where the sampler is unsure of the concentration, such as drums. Make sure the probe extension is firmly seated against the flat seals in the probe handle and in the tip of the telescoping probe or the sample inlet might be cut off.

COMPARISON OF FID AND PID FOR GENERAL SURVEY USE

As has been discussed, the differences between the detection principles of the PID and the FID will impact measurements. Since the FID uses a hydrogen flame as the means to ionize organic vapors, it responds to virtually all compounds that contain carbon-hydrogen or carbon-carbon bonds. The PID can respond to some inorganic compounds and most but not all organic compounds.

The FID response (sensitivity) to most organic vapors is based on breaking chemical bonds, which requires a set amount of energy. For the PID, the response is based on the IP of a compound and the ease in which an electron can be ionized (displaced) from the molecule. This mechanism is variable and highly dependent on the individual characteristics of a particular substance, resulting in a highly variable response factor for most organics that are ionizable. Therefore, in general, large sensitivity shifts are not seen between different substances when using an FID as compared to those seen with a PID.[9]

The relative responses of the FID and PID to a given compound also vary due to the fact that different compounds are commonly used to calibrate these instruments. PIDs are commonly calibrated with isobutylene and adjusted to respond as if this compound was benzene, and the FIDs are calibrated to methane. This difference sometimes causes problems if trying to compare results from simultaneous measurements of an FID with those of a PID.

An advantage of PIDs is that they are generally easy to use compared with many others, including FID units. The controls are usually limited to an on/off switch, a zero adjustment, a span adjustment, and a range adjustment for meters to read different ranges of concentrations. The zero and span controls are used to adjust zero for ambient background conditions and to adjust the sensitivity of the instrument depending on the response factor to the compounds being sampled. There is no gas reservoir to keep filled and no flame to start. In addition, in the general survey mode these instruments can be made to respond to a narrower range of compounds than FID monitors in that the lamps can be changed to allow for some selectivity.[3]

In a comparison test of an FID and a PID at relative humidities of 0% and 90%, a number of notable differences were observed.[5]

1. The relationships between the span setting and the meter reading for the FID was linear, whereas that for the PID with a 10.2-eV lamp was exponential, the result being less precision for the PID instrument than the FID.
2. The FID stabilized faster than the PID and was affected much less by 90% RH. The presence of 90% RH significantly decreased the sensitivity of the PID, but not the FID.
3. The FID had a wider linear range than the PID.
4. As long as the reference vapor was humidified appropriately, both detectors responded in an additive manner to the simultaneous presence of several organic compounds; the FID was more precise than the PID.

INTERPRETATION OF GENERAL SURVEY MEASUREMENTS FOR HEALTH HAZARDS

The limitations of each instrument are important. General survey instruments should not be used where a compound with an extremely low IDLH might be present, for example, acrolein, which has a TLV of 0.1 ppm and an IDLH of 5 ppm. Similarly, monitoring for compounds with TLVs of 1 ppm and less is better done with instruments capable of greater specificity and sensitivity. FID-based instruments that measure up to 1,000 ppm are linear throughout the entire scale. Although a PID can measure concentrations from 1 ppm to 2,000 ppm, its response is not linear over this entire range. Generally measurements greater than 600 ppm are considered higher than the readout value due to the nonlinearity of the instrument.

General survey instruments do not differentiate between specific chemicals, as has been discussed previously. Since their response varies, depending on what compounds are being sampled, readings are not specific for any one compound, unless the situation is such that the compound present is the only one and the instrument has been calibrated with that chemical. Since the types of compounds that the instrument can potentially detect are only a fraction of the chemicals potentially present in a spill incident or on a hazardous waste site, a zero reading does not necessarily signify the absence of air contaminants.

All readings in a contaminated area should be compared to upwind, background readings. If used for OSHA compliance measurements, the PID and FID are only allowed for use as screening instruments. A negative reading on a PID may indicate that a sample has fewer total ionizables than the zero reference air. It is an advantage to use both instruments when methane may be present, since the response of the PID to toluene and gasoline has been shown to be decreased by as much as 30% in the presence of 0.5% methane (10% of the LEL).[13]

If an FID is used along with a PID in situations involving clay soil, cement, or asphalt, which are diffusion barriers allowing biogenic levels of methane to build up, this will allow a better estimation of what is present. In this situation, if a measurement of a particular contaminant is desired, it may be necessary to use a gas chromatograph instead. One example is soil gas measurements. Another alternative would be the use of integrated measurement techniques, such as charcoal tubes, also sometimes used for soil gas measurements.

On uncontrolled hazardous waste sites (sites where the hazards and potential levels are unknown), the FID and PID are often used to determine what level of personal protection is required by providing data for setting arbitrary concentrations at which workers would either upgrade or downgrade their respiratory protection. This concept is called the setting of action levels.

On a site where the hazards are unknown, level C equipment (half mask respirator minimum) should be worn whenever levels exceed background and up to 5 ppm. Level B self-contained breathing apparatus (SCBA) is required for levels exceeding 5 ppm and up to 500 ppm above background. Level A protection (impermeable bodysuit with SCBA) is required whenever levels exceed 500 ppm over background. While some may disagree with this practice, it is better than doing nothing and allows some leeway for workers to do their jobs without having to wear respirators, which are often uncomfortable.

COMBINATION MEASUREMENTS OF EXPLOSIVE AND HEALTH HAZARD LEVELS: SOLID-STATE SENSORS

A growing area of detection is based on semiconductors, also sometimes called solid-state sensors. Currently much research is being done in this area and these instruments are likely to present many options in the future because these sensors are inexpensive and can be applied to a wider variety of detection needs than many other sensors.

They utilize a principle similar to that of catalytic combustion, but rather than using a catalyst to speed up the reaction at a lower temperature, they have a semiconducting material. As semiconductor characteristics can be altered by "doping" with impurities, and they are small and cheap to make, many more varieties may exist in the future. In a sense, these oxide coatings might also be described as catalysts; however, they can be designed to selectively accelerate the oxidation of only certain compounds, thus providing more specificity than a conventional catalytic combustion sensor.[14]

Many combustible gases, including hydrocarbons, alcohols, ethers, ketones, and esters, have reducing properties that allow for detection via semiconductors. These sensors also can be used to detect nitro and amine compounds as well as halogenated hydrocarbons, and a number of inorganic gases such as ammonia, carbon monoxide, hydrogen, hydrogen cyanide, and hydrogen sulfide.

Many of these instruments can detect both combustible gas concentrations and health hazard concentrations of certain chemicals often with the same sensor. Where more specificity is desired, different sensors are installed in the same instrument for each application. The combustible gas sensor generally is deliberately left unmodified so that it will detect the maximum number of hydrocarbons, while the other sensor, often termed the toxics sensor, is engineered to be as selective as possible for compounds such as carbon monoxide and hydrogen sulfide. Semiconductor sensors require a more complex circuitry than catalytic combustion and thermal conductivity sensors.

Selectivity is made possible through different mixtures of oxides and selected temperatures of operations. Thus, the semiconductor coatings both enhance and accelerate oxidation processes as high as 20 times that of its conductivity in air. Because semiconductor sensors depend on intergranular resistance, each sensor can be expected to differ slightly in its initial characteristics.

Metallic oxide semiconductor (MOS) sensors are one of the most common devices. MOS sensors are very small and consist of a metal oxide film made up of mixed metal oxides (iron, zinc, tin) coated on a heated ceramic substrate fused or wrapped around a platinum wire coil. A heater and collector are imbedded along with two electrodes into a metal oxide material in the sensor cell and a voltage is applied between the electrodes. Inside the cell two matched elements, one active and one inactive, are installed in a Wheatstone bridge circuit similar to the configuration previously discussed for hot wire combustion sensors. Like other semiconductor sensors, the signal or sensing mechanism of MOS sensors is based on changes in the concentration of electrons that in turn alters the conductivity of the semiconductor.

Absorbed oxygen from the ambient air establishes an equilibrium reaction with conduction layers of the metallic oxide, giving a base level of electron conduction through the sensor. The partial pressure of oxygen in air is nearly constant; heating the bead to a fixed temperature results in an amount of oxygen being absorbed in a stable manner. When a combustible or otherwise reducing gas such as carbon monoxide or hydrogen sulfide adsorbs on the sensor surface of the heated semiconductor bead, it reacts with the adsorbed oxygen layer and an electron charge transfer occurs between the surface of the semiconductor and the adsorbed molecule because of the difference in electron energy between them.

As gas molecules displace oxygen molecules, the surface of these sensors is altered, causing changes in its resistance. When the gas disappears, the sensor returns to its original resistance level. In order to promote the reactions of contaminants with the sensor surface, the temperature on the sensor surface (usually between 150°C and 300°C) must be carefully controlled by applying a specific voltage across the heating coil. The most widely used semiconductors in commercial gas sensors include SnO_2, ZnO_2, WO_3, and NiO.[15]

There are two types of MOS monitors, based on how the signal from the sensor is handled by the controller electronics and on how the monitors are calibrated. Some monitors electronically linearize the output from the sensors, and others simply use nonlinear scales to compensate for not linearizing the sensor signal electronically. The monitors with a linearized output require calibration at two different concentrations. Those with nonlinear scales need only be calibrated at one concentration. The disadvantage of a linearized output is that at lower concentrations (<5 ppm) these monitors show little or no response. It has been suggested that electronic linearization results in better span stability and reproducibility of measurements.[16]

By being very selective with the metal oxide coatings and the Wheatstone bridge temperature on MOS sensors, the number of interferents can be minimized by reducing the response of the sensor to other gases and vapors. For example, the same sensor can be used for detecting either carbon monoxide or methylene chloride. For methylene chloride detection, the temperature is increased to the point where carbon monoxide molecules burn off faster than the sensor can detect them.

The Enmet CGS-100 Tritechtor (Fig. 9-6) is an example of an instrument that takes advantage of this concept by allowing the user to adjust the temperature for different uses. One calibration is used for hydrogen sulfide and/or carbon monoxide (assuming that both compounds are unlikely to be present at the same time during measurements) and another for methylene chloride. In both cases, the same sensor, with a heater voltage change, is used; when the heater voltage is increased on an MOS sensor, the temperature of the sensor is correspondingly increased. The sensor also must be calibrated with the gas to be monitored following the adjustment in sensor temperature.

The AIM monitor has a diffusion tin oxide semiconductor sensor for hydrocarbons and a microprocessor. It looks like a wand or a "space sabor" (Fig. 9-7). The microprocessor in the unit allows storage of calibrations of five different levels for as many as 33 different industrial gases at several concentrations at the factory. Some compounds have calibrations for only toxicity (IDLH in ppm levels), others for flammability (LEL), and some for both. Compounds for which a specific calibration has been done include acetone, acetylene, ammonia, benzene, butane, butylene, diethylamine, dimethylamine, ethane, ethanol, ethyl acetate, ethyl chloride, ethyl ether, ethylene, ethylene chloride, hexane, hydrogen, hydrogen sulfide, methane, methanol, methyl ethyl ketone, methylene chloride, octane, pentane, propane, propanol, propylene, sulfur dioxide, toluene, and xylene.

Compounds are selected by punching in a number on the instrument's keyboard that triggers a display of what calibrations are available. A TLV button is incorporated into some models that will integrate the "exposure" over a period of time as long as the instrument is kept on a specific setting for a given chemical. The readout is presented as a range of the IDLH or LEL: 0–9, 10–19, and 20–29. An alternative readout is the analog to digital value indicating the absolute response of the instrument. The digital value is a three-digit number ranging between 200 and 250 in clean air depending on the specific characteristics of the sensor. Exposure to gas will increase this value proportionately.

Colored light emitting diodes (LEDs) flash to signal different alarm levels as well as to indicate whether an instrument is in the general sensing mode or not. The AIM instruments store the maximum value detected for each measurement up to the number of events a particular model is capable of storing. This "peak measurement" storage capability allows the unit to be lowered into a tank or other confined space. Sampling data can be recalled on the unit's display, downloaded to a PC-compatible serial printer, or loaded into AIM report software. Response time for toxic/combustible measurements is 0.5 second and the limit of resolution is 10%.

Calibration checking and sensor replacement/recalibration kits are available from the manufacturer. Instruments are precalibrated in the factory and a list of calibration points for all chemicals is available from the manufacturer. Since the instrument is nonlinear, it is important that the user check the calibration at several points. The microprocessor of the AIM is programmed to compensate for the nonlinearity of the sensor.

Catalytic combustion, as discussed in the section on combustible gas indicators, has also been modified to provide detection of both LEL and ppm ranges in a single instrument. The Bachrach TLV Sniffer is an example of this type of instrument. Calibrations on both scales are required and the scale readings must be divided by a factor, say 20, to get low ppm results.

Even when conditioned electronically to respond to a particular gas, the MOS sensor will still respond to a variety of other gases and vapors, mainly hydrocarbons. For example, a sensor designed to measure carbon monoxide may also be sensitive to nitrogen oxide, sulfur dioxide, hydrogen sulfide, and unburned hydrocarbons. However, MOS sensor response to gases other

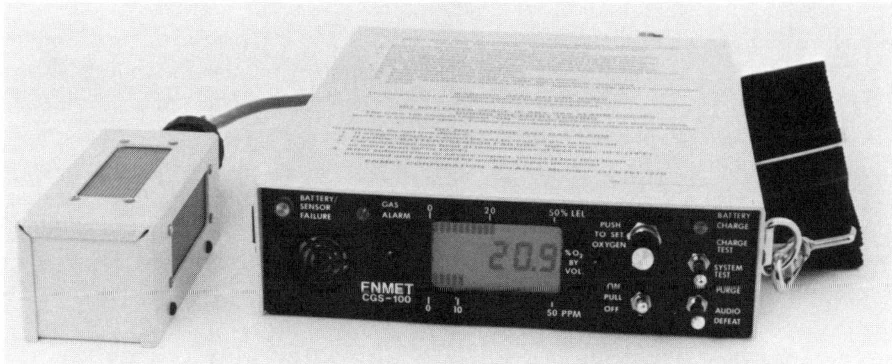

Figure 9-6. An MOS-based monitor capable of measuring combustible gas and carbon monoxide/hydrogen sulfide. (Courtesy Enmet.)

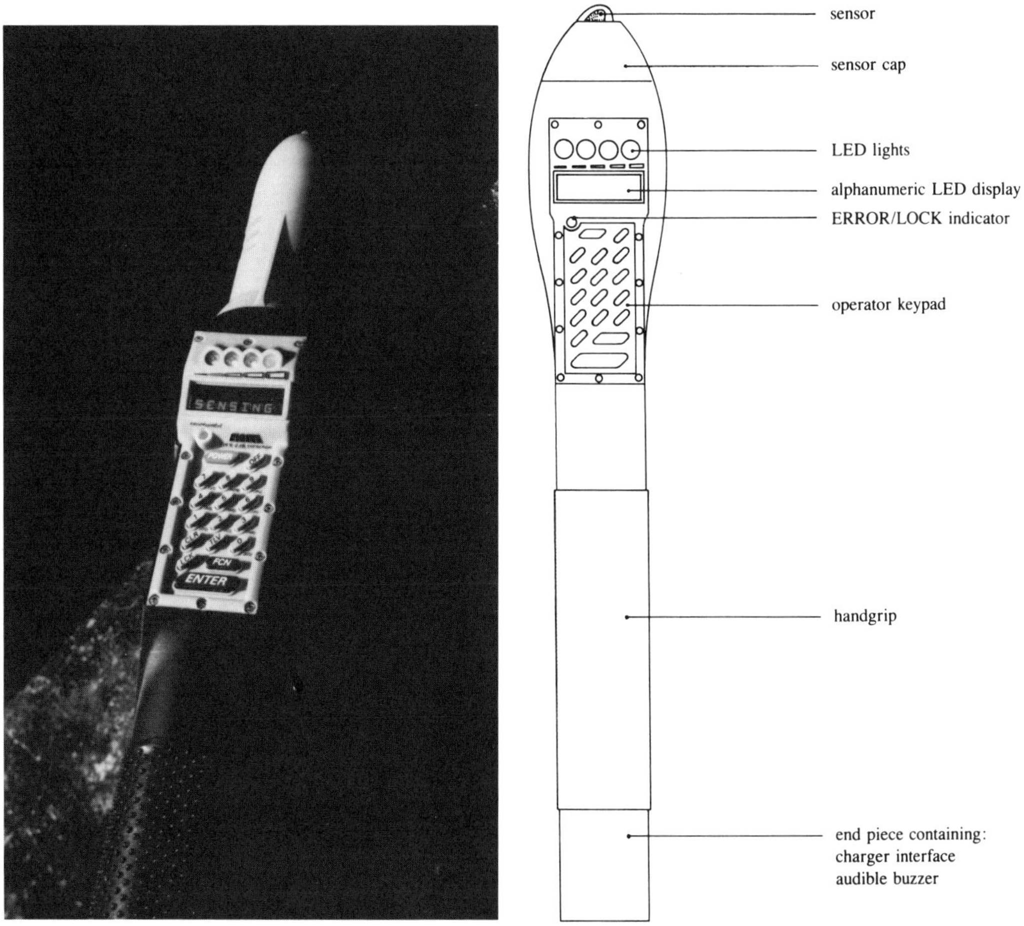

sensor
sensor cap
LED lights
alphanumeric LED display
ERROR/LOCK indicator
operator keypad
handgrip
end piece containing:
charger interface
audible buzzer

Figure 9-7. The AIM monitor, calibrated to measure 17 compounds. (Courtesy AIM Safety.)

than those for which it is calibrated will not result in accurate readouts of concentrations of such gases. For example, a sensor calibrated to activate an alarm in a concentration of 20% methane may also alarm in the presence of a different concentration of gasoline.

After repeated exposures to gases, MOS sensors can saturate. These sensors can also be attacked by corrosive gases, such as chlorine, bromine, and fluorine, and other compounds similar to those found in cigarette smoke.[17] Most sensors will not recognize the presence of inert gases, unless the concentrations are sufficient to displace the level of oxygen to below 10% in air. This potential for saturation and susceptibility of certain deleterious compounds is the reason most MOS detectors have a purge switch to "burn" off contamination on the sensor. Eventually the sensor will saturate and the purge will not be effective and the sensor will need to be replaced. In general, if an instrument begins to read high in clean air, or if the unit is unused for several weeks or months, it must be purged by powering it up in a clean, well ventilated environment for 1 to 3 hours. During this time it may alarm as contaminants burn off the sensor. Regular replacement is needed, regardless of the concern about saturation, because the reactive material on the sensor's surface gets used up over time.

MOS sensors cannot be operated in an absence of oxygen, such as can occur when monitoring confined spaces. A minimum of 10% relative humidity is generally required for accurate calibrations and field measurements; high humidity may adversely affect the sensors. Solid-state sensors also are affected by temperature in that their resistance increases in elevated temperatures, so these instruments are not desirable for making measurements on hot processes.

The major advantages of solid-state sensors are their simplicity in function, small size, and low cost. The major disadvantages are lack of stability, lack of reproducibility, and lack of selectivity as well as insufficient sensitivity for certain purposes. Due to the elevated sensor temperature, irreversible reactions with gaseous impurities in the atmosphere can affect stability and cause baseline drift.[14]

Solid-state devices with external (passive) sensors should not be stored in a contaminated environment, because the sensor may be "fouled" due to its susceptible location outside the instrument's case. Examples are closets with oily rags, operations where airborne contaminants are generated, and chemical storage areas near sources of combustion such as gas stoves and motors. Oil mist and dust can coat the interior of these sensors.

Calibration and Maintenance

There are two different types of calibrations required for solid-state units: the basic laboratory calibration and the field check. During the field check, the unit is exposed to a single concentration of gas for each sensor. If the unit fails the field check, a full calibration is required. For a full calibration of a solid-state sensor, generally a special apparatus is required. It is usually available from the manufacturer along with the calibration gases.

Sometimes multiple calibrations can be done on a single solid-state sensor. For example, the AIM gas detector can be calibrated for several different chemicals and concentrations at a time, and then these calibrations can be stored in the microprocessor's memory for later recall (Fig. 9-8). When a gas calibration is selected in the field, the unit will respond accurately to that gas as long as it is the only one present.

Other types of sensors can be recalibrated for use in measuring different atmospheres. For example, the ENMET CGS-100 can be calibrated to respond selectively in either confined spaces or petrochemical plants. The first calibration uses hydrogen sulfide and carbon monoxide because they represent the most likely gases to be encountered in a confined space. In the petrochemical calibration, 100 ppm of methylene chloride is used because it adjusts the circuitry and MOS sensor to respond to low levels of general hydrocarbons. In other special applications the same sensor has been calibrated for jet fuel, gasoline, or ammonia. Calibration is also required following sensor replacement, and whenever the sensor heater voltage is changed.

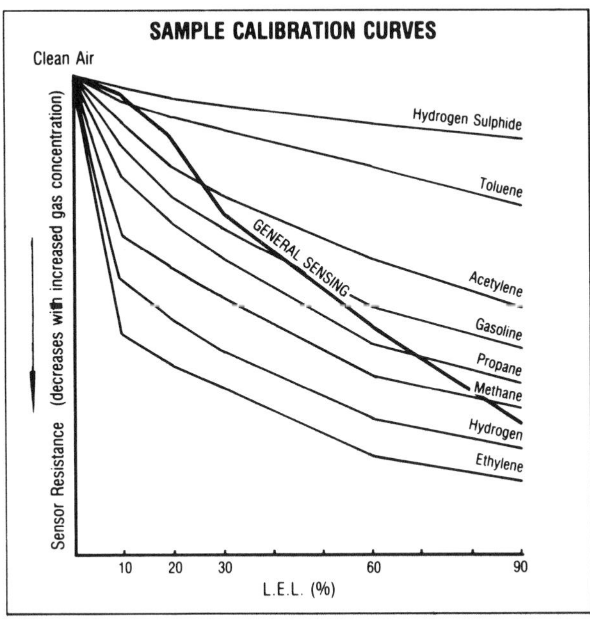

Figure 9-8. Calibration curves for the AIM. (Courtesy AIM Safety.)

Oxygen-dependent solid-state sensors should not be calibrated with compounds that are mixed with inert gases, such as nitrogen rather than air, because the sensor will give an inaccurate alarm response when used in the field. If a solid-state sensor should accidently fall into a tank of liquid or become submerged in some other way, the instrument will require repair before it can be used.

Calibration of MOS sensors can also be affected by humidity. A lack of water vapor can desensitize the MOS sensor, and since most cylinder gases are extremely dry, they may require some humidification. On the other hand, excessively high humidification will also affect the performance of the MOS sensor. Thus, care must be taken to calibrate instruments with MOS sensors within the humidity range specified by the manufacturer, as well as at the humidity level expected to be encountered in the field.[16] The span gas flow rate is critical when calibrating these instruments as an excessive flow rate might cool the sensor temperature, the result being that the instrument would read high.[17]

Procedures for full calibration are more extensive than those for the field check and include exposing each solid-state sensor to at least two different concentrations of gas, as well as checking the sensor heater voltage with a voltmeter. The use of two to four different gas concentrations over the measurement range of the instrument is important, because MOS-type sensors do not have a linear response. If the unit is used as a warning device only, then a single calibration at the alarm concentration is considered sufficient.

The actual calibration procedure requires adjusting first one and then a different gas concentration in sequence, until each time the instrument is exposed to one of these concentrations, no further adjustments are required. Several cycles are usually required; thus, calibration adjustments on solid-state instruments can be somewhat complicated and time-consuming over those required for other types of general survey instruments. Since these analyzers are nondestructive, the operator should be aware that the calibration gases are released from the instrument during calibration. The following procedure recommended by the Enmet Corporation is typical of that used for most instruments.

Procedure

1. Turn on the instrument and purge (burn sensor) for approximately 10 minutes. The purge switch may have to be taped in position for this step.

2. After purging, allow sensors to stabilize for 6 minutes.

3. Check sensor heater voltages. Generally this check requires that the sensor or instrument cover be removed. The manufacturer's manual should indicate the location of the test points. Usually the probe attached to the black or ground lead goes to the ground test point and the positive lead's probe (red) goes to the sensor test point. The manual will indicate what voltage range to expect; if the unit is out of range, a screwdriver is used to adjust the proper screw. The screw is actually a potentiometer. If the voltage cannot be adjusted to within the manufacturer's specified operating range, the sensor will have to be replaced.

4. The next step often involves turning on a calibration switch and checking the same voltages at specified points on the main circuit board.

5. The humidifier bottle should now be filled approximately halfway and recapped. Handling these units can be difficult. The assembly should not be allowed to tip, because if water runs into the gas supply line, it may eventually reach the sensors and damage them during calibration.

6. The unit is now ready to expose to calibration gas mixtures with the humidifier in line. The regulator valve on the cylinder is opened just enough to allow for the gas to flow at the rate specified by the manufacturer, often 0.9 cfh to 1 cfh, which is a very low flow rate. If the gas flows too fast, it may force water from the humidifier into the gas supply line to the sensors.

7. Once the gas has flowed to the sensors for at least 3 minutes, adjustments can be made to the correct potentiometer (screw), if needed. Sometimes as many as three different potentiometers will have to be adjusted in this step (reference, gain, alarm).

8. When changing concentrations, especially if going from a higher concentration to a lower one, the sensor may have to recover for 10 minutes.

Use

Solid-state sensors designed to detect low concentrations (ppm) of gases usually have the slower response times. For example, when testing a confined space for toxics, such as carbon monoxide and hydrogen sulfide, always allow at least 3 minutes of sampling time to fully test any atmosphere prior to entry. For more information on confined space testing, see the section on confined space measurements in the chapter on Survey Preparations and Performance. If using a solid-state instrument to measure combustible gases, always measure for oxygen first, since some MOS units are not accurate in an oxygen-depleted atmosphere.

Some manufacturers recommend using exhaled cigarette smoke as a field check of a carbon monoxide sensor. However, in addition to carbon monoxide, cigarette smoke contains many other hydrocarbons and particulates that can damage or contaminate the sensor surface. Therefore, do not operate these instruments with these sensors in the presence of cigarette or other smoke.

Do not use solid-state instruments in fast-moving air currents, such as when tanks are being ventilated with blowers or around conduits or ventilation diffusers. High air flow rates can cool electronically heated MOS sensors and interfere with a true gas signal. Do not attempt gas measurements in fast moving air currents such as conduits or ventilation diffusers as the high flow rate may cool the warm metallic oxide sensor element and interfere with a true gas signal.

It is important to identify and operate within the humidity range specified for the instrument, and during sampling, humidity should be monitored since wide fluctuations in humidity during measurements may affect accuracy.[16] For continuous monitoring applications in excess of 3 hours, MOS devices often require an external source of humidity. It may consist of a bottle of distilled water installed in-line with the intake port. If a filter is incorporated into the inlet of the instrument, it should be removed and reconnected in front of the bottle, because it will trap the water vapor intended for the sensor. When moving between areas where large-scale temperature changes can occur, allow a few minutes for the instrument to stabilize between measurements.

REFERENCES

1. McKinnon, G. P., ed. *Fire Protection Handbook*, 15th ed. Batterymarch Park, Quincy, MA: National Fire Protection Association, 1986.
2. American Industrial Hygiene Association. *Manual of Recommended Practice for Combustible Gas Indicators and Portable, Direct Reading Hydrocarbon Detectors.* Akron, OH, 1980.
3. Smyth, R. T., and D. Bingham. *The Use of Nonspecific PID Monitors for Remedial Investigations and Field Studies.* HazTech International Conference Proceedings, Denver, Colorado, August 11–15, 1986.
4. *HNU Manual*, HNU Corporation.
5. Lee, I. N., et al. Additivity of detector responses of a portable direct-reading 10.2 eV photoionization detector and a flame ionization gas chromatograph for atmospheres of multicomponent organics: Use of PID/FID ratios. *AIHA J.* **48**(5):437–441, 1987.
6. Langhorst, M. L. Photoionization detector sensitivity of organic compounds. *J. Chromatogr. Sci.* **19.** February 1981.
7. Technotes: PID—Different Ionization Sources. Thermo Environmental Instruments, Franklin, MA.
8. Burroughs, G. E., and J. L. Woebkenburg. *Effectiveness of Real-Time Monitoring.* ACGIH: Advances in Air Sampling, pp. 243–250, Cincinnati, OH, 1988.
9. Driscoll, J. N., and J. H. Becker. *Industrial Hygiene Monitoring with a Variable Selectivity Photoionization Analyzer.* American Laboratory, November 1979, pp. 69–76.
10. Barsky, J. B., et al. An evaluation of the response of some portable direct-reading 10.2 eV and 11.8 eV photoionization detectors, and a flame ionization gas chromatograph for organic vapors in high humidity atmospheres. *AIHA J.* **46**(1):9–14, 1985.
11. *OVA 580 Manual.* Thermo Environmental Instruments, Franklin, MA.
12. Organic Vapor Analyzer, The Foxboro Co., Foxboro, MA.
13. Nyquist, J. E. et al. Decreased sensitivity of photoionization detector total organic vapor detectors in the presence of methane. *AIHA J.* **51**(6):326–330, 1990.
14. Gentry, S. J. *Catalytic Devices, Chemical Sensors*, pp. 259–274. Routledge Chapman and Hall, 1987.
15. Stetter, J. *Instrumentation in Synfuel Industry.* Annals of the American Conference of Governmental Industrial Hygienists, Cincinnati, 1984.
16. Smith, J. P., and S. A. Shulman. An evaluation of H_2S continuous monitors using metal oxide semiconductor sensors. *Appl. Ind. Hyg.* **3**(7):214–221, 1988.
17. American Industrial Hygiene Association. *Manual of Recommended Practice for Portable Direct-Reading Carbon Monoxide Indicators.* AIHA, Akron, OH, 1985.

CHAPTER 10

Monitoring Instruments for Many Specific Gases and Vapors: GC and IR

A single field instrument capable of instantaneously measuring and identifying any and all compounds encountered during a survey would be ideal, but is not possible utilizing current technology. Instruments are available that can offer both specificity and the ability to monitor for many different compounds with only a few adjustments. A gas chromatograph (GC) can separate and identify components of a mixture. An infrared spectrophotometer (IR) can be tuned so that it is specific for many different compounds, although it can detect only one at a time. There are, however, new developments that promise more capabilities in the near future.

The portable GC is used for environmental sampling for hazardous waste work, and community and industrial monitoring. Portable GCs equipped with multiple detectors have also been used by the EPA as screening air devices to determine hot spots and potential interferences, and to quantify volatile organic compounds (VOCs) prior to setting up fixed site samplers. They are used for occupational monitoring primarily to collect compounds such as benzene, which must be sampled at very low levels. The IR is used almost exclusively for occupational monitoring because it is better suited for indoor use. Leak detection is a common application of the IR because of its ability to monitor for gases such as halothane, nitrous oxide, and ethylene oxide.

PORTABLE GAS CHROMATOGRAPHS

Portable GC as used here refers to any gas chromatograph capable of being used in the field under battery power. A portable GC consists of, at a minimum, an injection system, a separator in the form of a column, and a detector (Fig. 10-1). There are two basic injection systems: An air or liquid sample is injected using a syringe through a septum, or a pump pulls an air sample directly into a sampling loop. From the injection port or sampling loop the sample is carried through a column where the mixture is separated into individual components. Packings and coatings in columns physically interact with the components in a sample to slow their passage through the column. The interaction between the injected sample containing one or more compounds in the vapor phase and a liquid phase, which is coated either on the walls in the case of a capillary column or on the packing in the case of a packed column, separates the components in a given mixture. Since each compound will interact differently and is slowed down to a different extent, the net

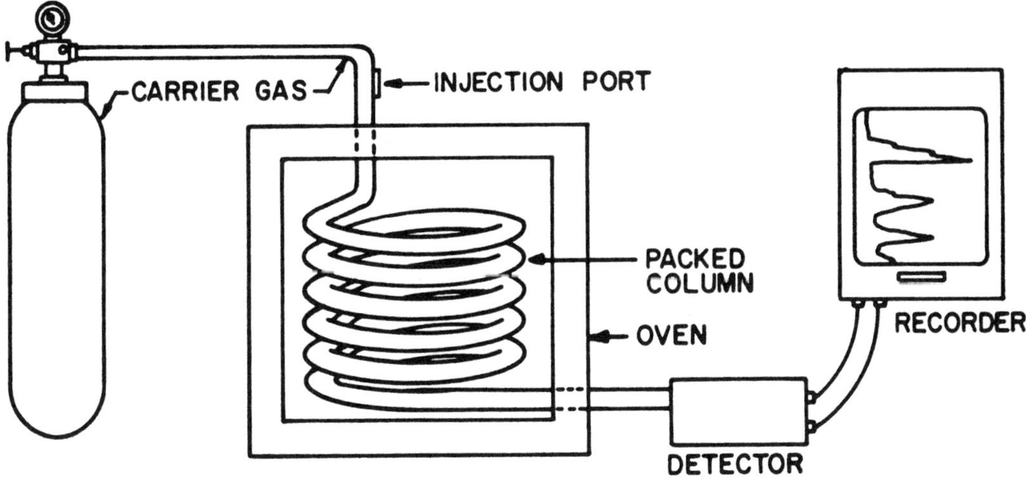

Figure 10-1. Block diagram of a typical gas chromatograph. (From *The Industrial Environment—Its Evaluation & Control*, NIOSH, Washington, D.C., 1973, p. 259.)

effect is a separation of the gases in such a way that each component elutes from the column and is detected at a different time, known as its retention time (Fig. 10-2). The stronger the interaction between the compound in the vapor phase and the liquid phase, the more strongly the movement of the compound will be slowed by the column, and therefore the longer its retention time. The carrier gas can be high-purity air, hydrogen, helium, or nitrogen.

After the sample leaves the column, it goes directly to the detector.

If a strip chart recorder is hooked up to an instrument, the actual peaks can be seen. Peaks can be of different sizes even if each compound is present in a calibration mixture at the same concentration, since detectors have different sensitivities (responses) to different chemicals.

The retention time is primarily dependent on the type of packing material, the length of the

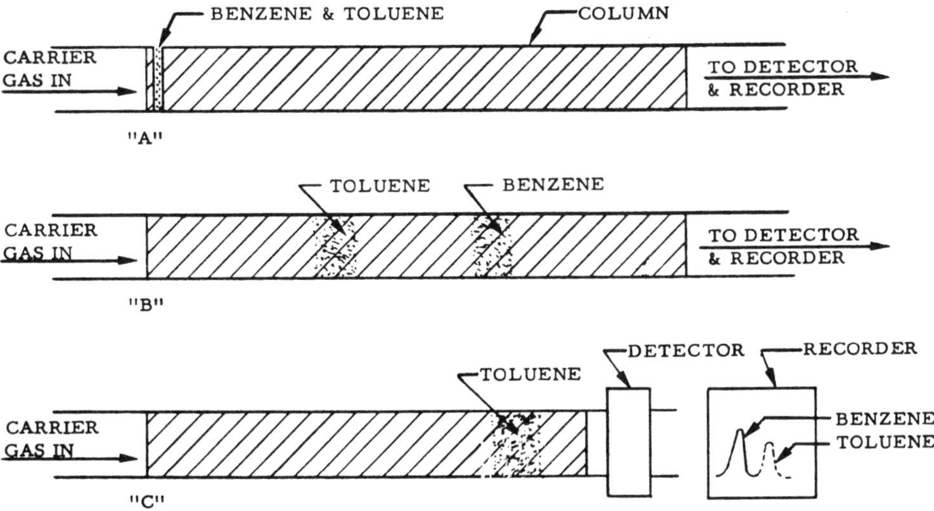

Figure 10-2. Chromatogram/pictorial separation of benzene and toluene. (Courtesy The Foxboro Company.)

column, the flow rate of the gas carrying the mixture through the column, and the temperature range of the system. These variables can be changed to allow better resolution of the components in a mixture. Lighter compounds are retained roughly in proportion to their molecular mass. Increases in temperature will shorten retention times as well as increase carrier gas flow rate.

The peak area of the sample when compared to the peak area in the calibration is used to calculate the concentration of each compound, and the shape of the peak often provides information about the individual compounds (Fig. 10-3). Most GCs have the capability to automatically integrate the electronic signal that the peak represents; however, if this technology is not available manual methods can be used, such as calculating the area using the ½ (base × height) formula for triangle areas. A basic book on gas chromatography will describe this technique and others.

Portable GCs are the best instruments to use for the identification of mixtures and unknowns. One of the most popular environmental applications is soil gas monitoring. Because of their sturdiness they can be used as fixed-point continuous monitoring systems, particularly in hazardous waste work. If sampling is done outdoors, it is important that wind speed and direction be incorporated into the sampling plan. Generally, field GCs are limited to volatile compounds, due to the time necessary for lower boiling point compounds to get through the chromatographic column and because of the differences in field conditions compared to the stability of the laboratory.

Generally, portable GCs have not been used for Occupational Safety and Health Administration (OSHA) compliance measurements, nor as a substitute for The National Institute of Occupational Safety and Health (NIOSH) integrated sampling methods that have been validated for personal exposure measurements primarily because so few instruments have been available until recently. It is likely that occupational sampling applications for portable GCs will increase because NIOSH has developed two methods that rely on the use of the portable GC: one for ethylene oxide and one for trichloroethylene, and more are likely to follow.

GC Selection

Once the decision to purchase a portable GC has been made, selection of the type of column and detector will be based on the specific application for which the instrument is to be used. While there are some GC setups that can analyze more compounds than others, there is no single configuration that will sample any and all compounds. For example, a setup to monitor for unknown compounds in a situation where high concentrations (ppm range) were expected would enable an instrument to identify aromatics, straight chain hydrocarbons, ketones, and some chlorinated compounds if a flame ionization detector (FID) were installed. If the primary concern was low levels of benzene, then a different column and a photoionization detector (PID) would be installed in a unit. However, this setup would not be used for compounds other than aromatics and hydrocarbons.

Other sampling needs would necessitate other choices. It is not uncommon for a portable GC to be set up and dedicated to a single application; however, should sampling needs change, the instrument can often be reconfigured to accommodate such changes. The number of options available for a given model of portable GC will determine its flexibility for changes in its configuration. Increased flexibility is often proportional to increased cost and decreased portability. Modifications such as installing new compounds and detectors, and programming a microprocessor-based instrument must generally be done in a laboratory by experienced personnel.

Some portable GCs have only a manual port, which means that in order to inject samples a gas syringe must be used. Others have a pump and sampling loop that can be used to automatically inject samples from ambient air as well as to calibrate the instrument at preprogrammed intervals from a gas cylinder, which can be installed on many units. The loop system is considered inherently more accurate than the syringe, but the volume injected is fixed for loops. Changing sample volumes requires a change of the loop. Syringes are available in many sizes, and injection volumes can be varied between 10 μl and 1 ml easily. Both loop and syringe injection sys-

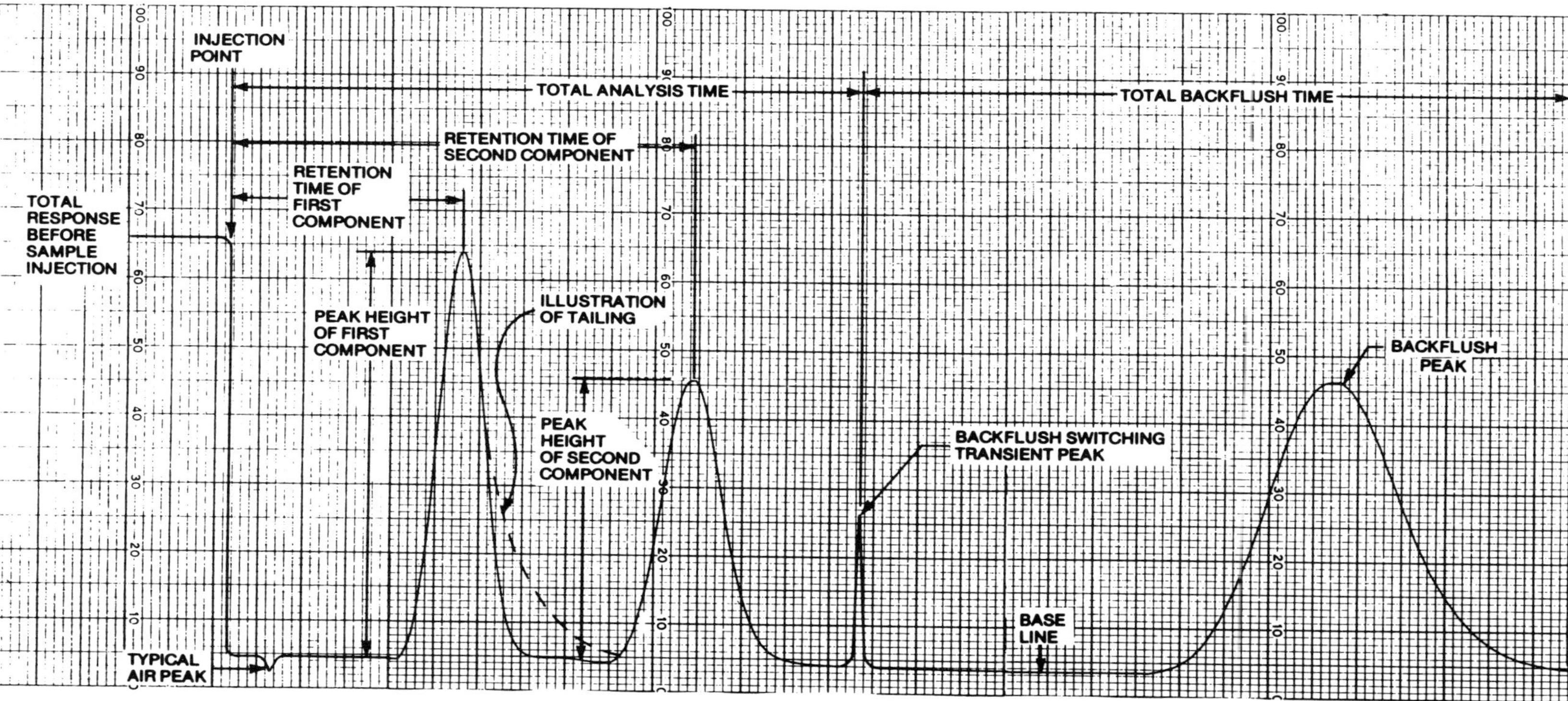

Figure 10-3. A typical chromatogram. (Courtesy The Foxboro Company.)

tems may require purging to reduce memory effects from residues of low-boiling compounds retained in these systems.[2]

The selection of appropriate column length and packing materials are important considerations in maximizing column efficiency. Columns come in two different types: packed or capillary (Fig. 10-4). The packed column consists of a solid support coated with a liquid stationary phase (gas/liquid chromatograph) or simply a solid adsorbent (gas/solid chromatography). The capillary column is also called a wall-coated open tubular column and consists of a liquid stationary phase coated or bonded to a specially treated glass or fused silica tubing. Fused silica tubing is most commonly used because of its physical durability. Bonded (or cross-linked) columns are used in preference to coated columns because of the greater operating temperatures that can be obtained. The typical linear flow velocity through both capillary and packed columns is about 30 cm/sec.[3] Table 10-1 lists commonly used stationary phases for GC.

The packed column can accept a larger injec-

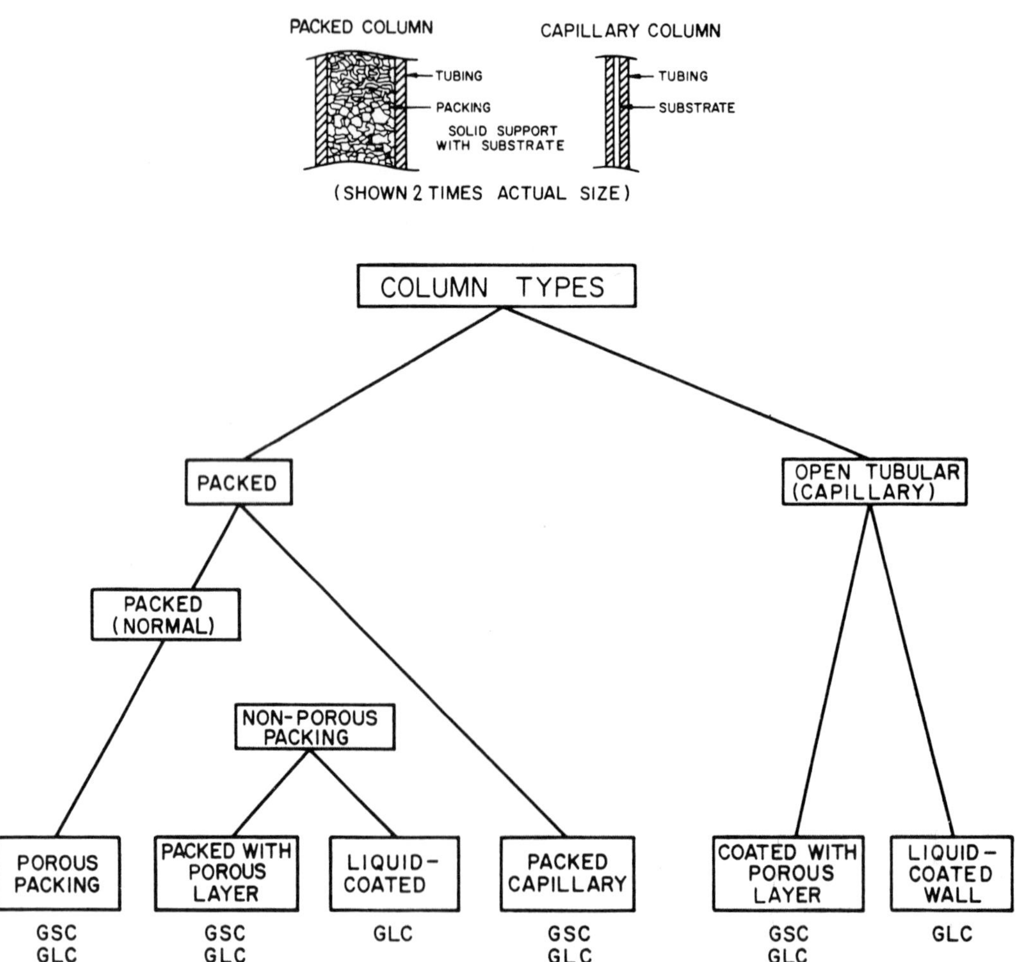

Figure 10-4. Packed and capillary GC columns. (From *The Industrial Environment—Its Evaluation & Control*, NIOSH, Cincinnati, OH, 1973, p. 262.)

TABLE 10-1. Commonly Used Stationary Phases for GC

	Applications
Liquid Phases	
SE-30, OV-1 (methyl silicones)	Hydrocarbons, chlorinated hydrocarbons
OV-1, SE-54 (methyl/phenyl silicones)	PAHs, chlorinated pesticides, hydrocarbons
Carbonax 20M (polyethylene glycol)	Polar compounds such as esters, alcohols
FFAP, SP-1000 (polyethylene glycol terephthalate)	Phenols, volatile acids
Solid Phases (Packed Columns Only)	
Chromosorb, Porapak series (styrene/divinylbenzene polymers)	Volatile alcohols, ketones, hydrocarbons, halocarbons (boiling points 30 – 100°C)
Carbon molecular sieves (Carbosphere, Spherocarb, Carbosieve)	$C_1 - C_5$ hydrocarbons
Porous silica (Unibeads, Porasil A)	$C_1 - C_5$ hydrocarbons

tion volume, but the capillary column has better separating power. As an example, Figure 10-5 shows the difference between an analysis of gasoline vapor using a packed column and a capillary column. Capillary columns are used with isothermally heated systems, and packed columns are used with unheated systems. It should also be noted that columns for portable GCs may require special shapes and sizes, material for construction, and hardware specific to a given instrument. The choice of columns that can be used will depend on the type of instrument. In most cases, columns are referred to by the type of packing inside, the coating, if appropriate, and their length.

Most field GCs come with a single detector, although at least one unit is on the market that allows detectors to be interchanged. Detectors vary according to their selectivity, sensitivity, and linearity. There are a number that can be used in GCs, but the most commonly used detectors for portable GCs are the FID, PID, and electron capture detector (ECD). Table 10-2 lists characteristics of commonly used detectors. The choice of detector will depend on the chemicals for which monitoring is needed, as well as on the nature of any other contaminants present and the sensitivity required. While humidity can bother most direct reading instruments, the portable GC is

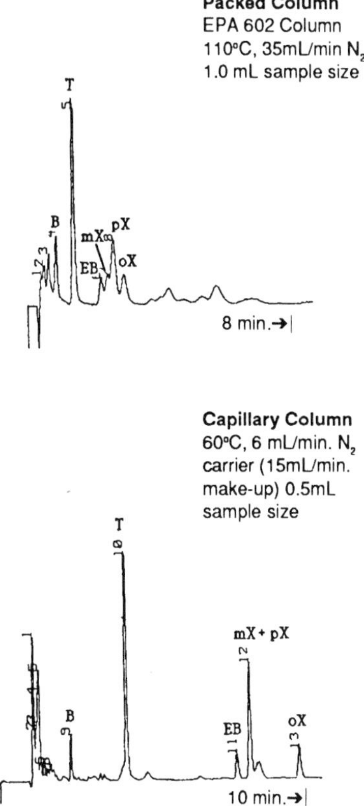

Figure 10-5. Gasoline vapor analysis on packed and capillary columns. (Courtesy HNU Systems, Inc.)

TABLE 10-2. Types of GC Detectors

Type	*Principle*	*Compounds Detected*
Flame ionization detector (FID)	A stainless steel burner in which hydrogen is mixed with the sample stream; combustion air feeds in and diffuses around the jet (burner) where ignition occurs. The current carried across the gap of a platinum-loop collector electrode is proportional to the number of ions generated by burning the sample. Not as sensitive as many other detectors. Large linear range. Response is relatively uniform from compound to compound. Detection limit is 0.01 ppm.	Almost all organics; response is greatest to hydrocarbons; decreases with increasing substitutions of other elements: O, S, Cl. Not sensitive to water, the permanent gases, and most inorganic compounds.
Photoionization detector (PID)	A sealed UV light source emits photons with an energy level high enough to ionize the compounds in the sample. In a chamber exposed to the light source containing a pair of electrodes, ions formed by the absorption of photons are driven to the collector electrode where a current is produced that is proportional to the concentration. Large linear range; high sensitivity; selectivity can be introduced by using different energy lamps. Response varies from compound to compound. Detection limit: 1–100 ppb.	Organics, some inorganics, lower response to low-molecular-weight hydrocarbons. Can detect aliphatic, aromatic, halogenated hydrocarbons. Arsine, phosphine, hydrogen sulfide. Sensitive to water. No response to methane. Especially good for aromatics.
Electron capture detector (ECD)	Utilizes a radioactive source such as ^{63}Ni to supply energy to the detector in the form of β radiation. Intensity of the electron beam arriving at a collection electrode is monitored. When an electron-capturing species passes through the cell the intensity of the electron beam decreases, sending out an electronic signal. Response varies widely from compound to compound. Highly sensitive and selective. Potential problems include excessive heat, which could vaporize some of the source; "aging" of the coated foil, which must be replaced; gradual loss of the source's activity. Detection limit: 0.1–100 ppb.	Compounds containing halogens, cyano or nitro groups; ECD has a minimal response to hydrocarbons, alcohols, and ketones.

less susceptible to it than other instruments such as the IR. In general, the FID is water resistant, while the PID is more susceptible.

The use of a heated column system greatly increases the flexibility of the analysis. Unheated columns suffer from a number of problems; the most critical is thermal drift, meaning changes in the retention time of a given compound over time because of changes in the ambient temperature. It can be compensated for by increasing the calibration intervals between samples, but must be done very carefully in order to obtain accurate data. Generally, portable GCs with heated systems are only capable of a single, uniform column temperature, and therefore are referred to as isothermal. In general, unheated columns are best for compounds with short retention times (high volatility) and for routine monitoring situations where the same compounds are sampled regularly. The retention time for any compound to elute from the column is a function of the column's temperature. As a rule of thumb, the retention time doubles for every decrease in column temperature of 30°C. An increase in temperature will decrease column retention time and vice versa.[2]

As a comparison, laboratory GC systems use temperature programing, meaning a range of increasing temperatures exists over the length of a single column. This capability allows for analysis of very volatile compounds, as well as semivolatile, heavier-molecular-weight compounds in the same sample. Many laboratory-grade GCs have two separate and different types of columns that are used in cases where interferences due to coeluting compounds are suspected. If the peaks for a given compound on the separate columns are found at retention times that can be correlated, the compound is considered to have been identified.

In summary, items to consider when selecting a GC include the following:

Column variety available: Increasing column length is proportional to increasing retention times. The use of capillary versus packed columns and type of packings and coatings (phases) available for these columns depend on the manufacturer. Customized columns are available in many cases.

Types of compounds to be sampled: Identify these according to molecular weights, solubility, and boiling points (volatility). Nonpolar compounds (alkanes, halogenated alkanes) are separated best by nonpolar stationary phases, and polar columns (alcohols, ketones, and esters) work best with polar phases.

Backflushing capability: This capability is important if semivolatile compounds may be encountered. This topic is discussed further in the calibration section.

Heated versus unheated columns: This factor determines the consistency of retention times, environments in which the GC can be used, and time for analysis. Temperature and location inside (or outside) the GC unit are important.

Dedicated versus interchangeable detectors: Different compounds can be detected best by different detectors, so a need to use an instrument in a wide variety of situations will be best served by an instrument with interchangeable detectors. Fulfilling this need, however, can require considerable expertise and increased costs.

Length of battery power: The need for extra battery units is impacted by the field sampling time. Heated columns usually require additional power, such as generators or line voltage.

Air sampling pump versus manual injection only: The pump allows for continuous operation of the GC.

Heated or unheated injection ports: A heated injection port along with a column that can be heated allows semivolatile compounds to be sampled and allows for the injection of liquid samples.

Microprocessor capability: This capability determines whether the unit is programable and if data can be stored and transferred.

Carrier gas: The type as well as the flow rate is important. In remote areas replacement air may be difficult to get. The higher the flow rate, the faster the gas is used up.

Field calibration setup: Automatic calibration capability eliminates the need for manual calibrations.

Data recording capability: Internal data recording and storage capability decreases the need for a separate unit and makes transport and setup of the GC in the field easier.

Technical advice: The availability of training and troubleshooting expertise from the manufacturer will minimize problems in the field.

Types

An example of a field GC is the OVA with GC (Fig. 10-6). This unit has an ambient air column and an FID detector, and it uses "scrubbed" air as a carrier gas.[3] Samples can be injected through a septum or through a gas sample valve. The amount of air injected is small, 0.25 ml. A significant advantage is its cost. A thermal desorber unit with stainless steel sorbent tubes can be used to inject integrated samples into the GC column. It allows collection of long-term air samples on Tenax or charcoal sorbent tubes; desorption; and quantification of the compounds on the GC. The OVA with the GC option is intended for applications where there are a limited number of compounds of interest and the compounds are normally known. Under these conditions, the operator must know the retention time and peak height characteristics of these compounds under specific operating conditions. The OVA must be calibrated using the compounds of concern with the exact setup planned for use in the field, includ-

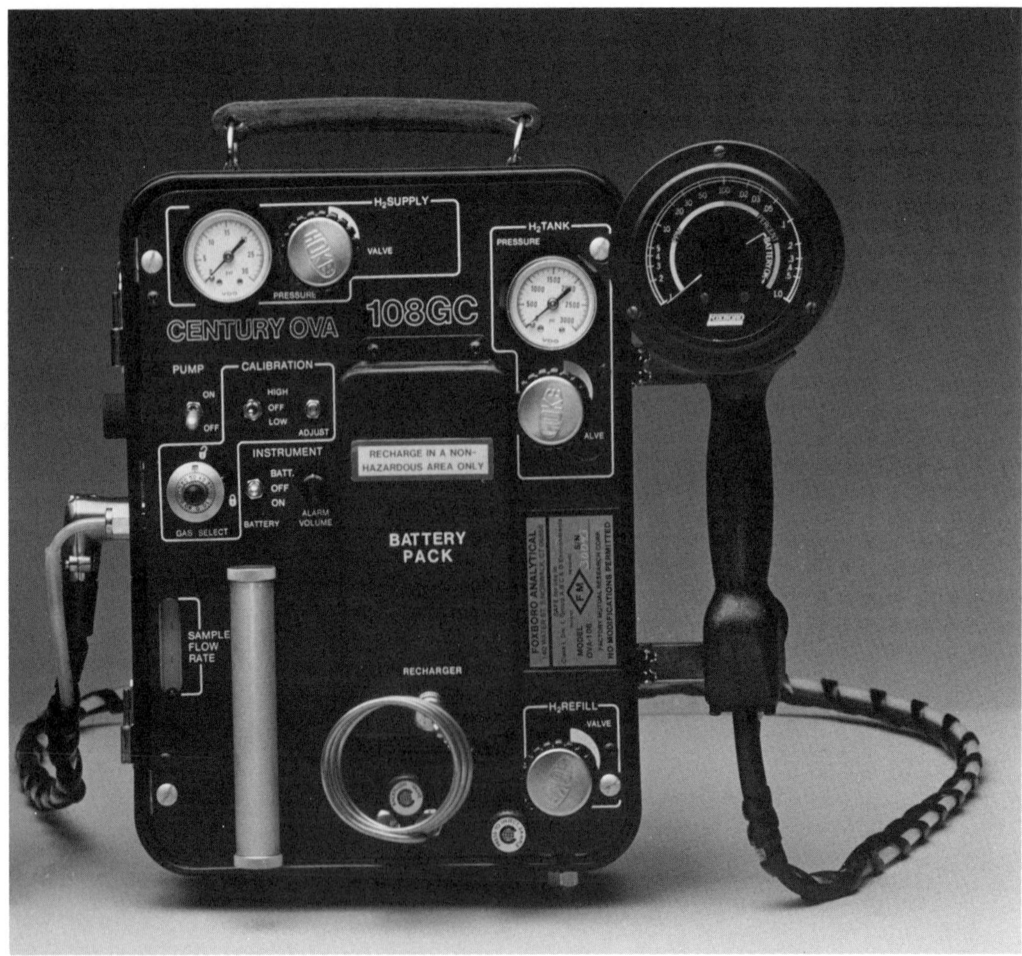

Figure 10-6. Organic vapor analyzer with GC. (Courtesy The Foxboro Company.)

ing the column type, expected field temperature range, and expected concentration range. In the field, periodic calibrations must be done. A limitation of this model is that it can only be used for relatively light, volatile compounds and the susceptibility of the exposed columns to changes in ambient temperature.

An example of a state-of-the-art GC that really is portable is the Photovac 10S series (Fig. 10-7). This unit utilizes a PID detector only. It is capable of detecting concentrations of benzene as low as 1 ppb. Several models are available, but only the upper-end models have microprocessors to convert the area of the peak to a concentration. The

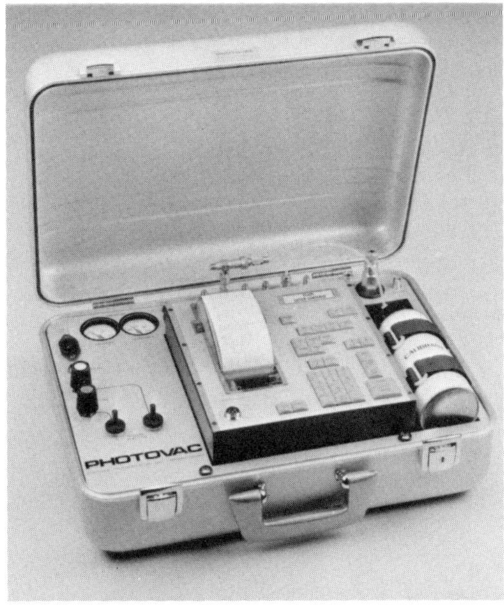

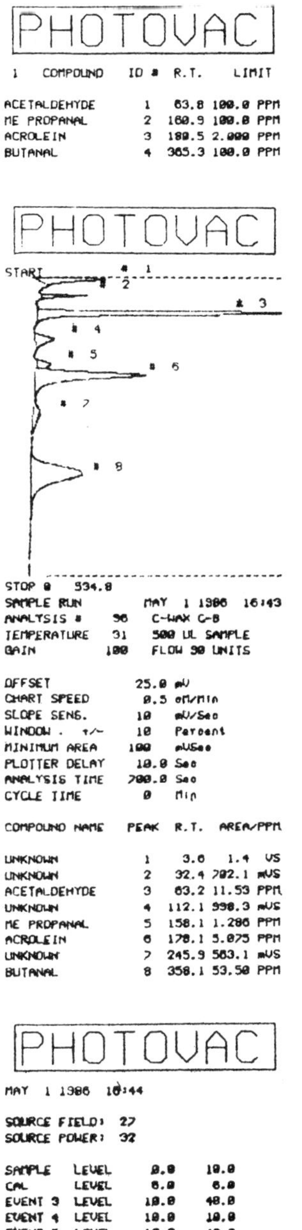

PHOTOVAC

1 COMPOUND	ID #	R.T.	LIMIT
ACETALDEHYDE	1	63.8	100.0 PPM
ME PROPANAL	2	160.9	100.0 PPM
ACROLEIN	3	189.5	2.000 PPM
BUTANAL	4	365.3	100.0 PPM

PHOTOVAC

STOP @ 534.8

SAMPLE RUN MAY 1 1986 16:43
ANALYSIS # 98 C-WAX C-8
TEMPERATURE 31 500 UL SAMPLE
GAIN 100 FLOW 90 UNITS

OFFSET	25.0	mV
CHART SPEED	0.5	cM/Min
SLOPE SENS.	10	mV/Sec
WINDOW . +/-	10	Percent
MINIMUM AREA	100	mVSec
PLOTTER DELAY	10.0	Sec
ANALYSIS TIME	700.0	Sec
CYCLE TIME	0	Min

COMPOUND NAME	PEAK	R.T.	AREA/PPM
UNKNOWN	1	3.6	1.4 VS
UNKNOWN	2	32.4	702.1 mVS
ACETALDEHYDE	3	63.2	11.53 PPM
UNKNOWN	4	112.1	998.3 mVS
ME PROPANAL	5	158.1	1.286 PPM
ACROLEIN	6	178.1	5.075 PPM
UNKNOWN	7	245.9	583.1 mVS
BUTANAL	8	358.1	53.50 PPM

PHOTOVAC

MAY 1 1986 16:44

SOURCE FIELD: 27
SOURCE POWER: 32

SAMPLE	LEVEL	0.0	10.0
CAL	LEVEL	5.0	5.0
EVENT 3	LEVEL	10.0	40.0
EVENT 4	LEVEL	10.0	10.0
EVENT 5	LEVEL	10.0	40.0

Figure 10-7. Field portable gas chromatograph (photo) and GC printout. (Courtesy Photovac.)

lower-end models must hooked up to a strip chart and a manual integration of the area of the column is required in order to determine the actual concentrations. From a practical standpoint, only the automatic units will be truly portable, because the addition of a strip chart recorder will limit the mobility of the unit as well as increase the time for identification, since between runs the user will have to mark the chart with the times for each peak. The Photovac 10S70 has a computer and modem incorporated, allowing the instrument to be programed for continuous monitoring and automatic transmission of data over telephone lines. This type of operation is extremely useful in situations where continuous monitoring is needed, but requires the availability of electricity and a dedicated phone line on site as well as a terminal and modem at a second site to receive the data. The manufacturer does provide the ability to upgrade every model, which would allow a company to gradually invest the capital required to have the state-of-the-art capabilities offered by the high-end units.

The HNU 311 is a portable GC (Fig. 10-8) that incorporates some features of laboratory GCs. It has a heated injection port that permits injection of both gaseous and liquid samples into the unit. Columns of three different diameters (0.25, 0.32, and 0.53 mm) are available. Columns are installed inside an isothermal oven that can be heated up to 200°C if the instrument is hooked to line power or a portable generator. Four different lamps are available for the PID detector: 8.3, 9.5, 10.2, and 11.7 eV. Microprocessor capabilities and a printout of the analysis are also incorporated into this unit.

GCs with microprocessor and storage capabilities permit the user to create a library allowing multiple calibration data for many different compounds to be stored in memory. Then in the field a lesser number of compounds can be used for intermittent calibration, the assumption being

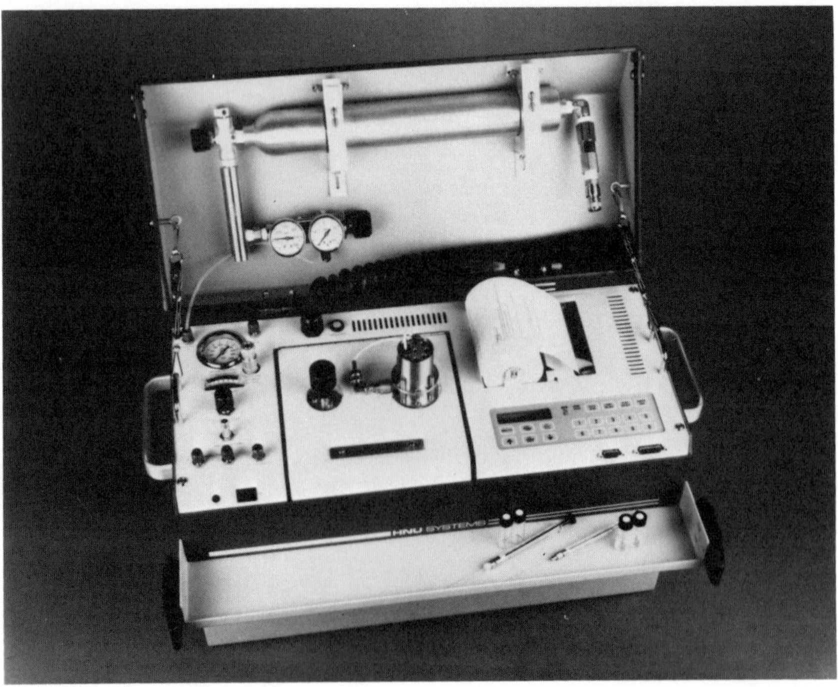

Figure 10-8. Field portable gas chromatograph. (Courtesy HNU Environmental Monitoring Instrumentation.)

that if their retention times stay the same so will those of the others. GCs are available with preconcentrators for sampling lower airborne concentrations as well as dilution setups for higher concentrations to prevent saturation of the detector or contamination of the column.

Advantages and Limitations

The technology of GCs is increasing rapidly, making them more precise than instruments of the past. An advantage of a field GC is the ability to measure samples without having to collect them on separate media, ship, store, and desorb them.

The advantage of being able to use the instrument directly in the field to try and identify unknown components of mixtures is very valuable, because it eliminates to some degree the need to store and ship samples, as well as allowing sampling during intermittent releases. Integrating methods as have been discussed in prior chapters tend to be quite specific for a narrow range of compounds, and therefore trying to find an unknown using these can require many repetitive and expensive samples and time, whereas screening samples with the GC first can narrow down the number of suspected compounds. While humidity affects many direct reading instruments, the portable GC is less susceptible than other instruments.

The foremost limitation for the use of portable GCs is the skill required for their setup and use. A novice with no instrument experience or chemistry background will have a difficult time using this type of instrument. However, training is usually available with the purchase of most units, and once the units are set up, technicians will be able to use them in the field with relative ease as long as they are provided with procedural information.

Calibration and Maintenance

Calibration will decrease the likelihood of errors due to potential differences in ambient temperatures between the laboratory and field if an unheated column is used. Calibration should be done both in the laboratory prior to the survey when the GC is set up and in the field between measurements. Calibration should always be done as close to the time of actual use as possible using a mixture of gases rather than a single compound.

Calibration standards may be of two types: commercially available certified gas mixtures or standards in bags that have been prepared from a known gas, or liquid aliquot of the desired compound plus a carefully metered volume of diluent, usually air or nitrogen. It is strongly advised that the preparation of standards be done in the laboratory as a practice exercise before attempting it in the field, and that the ability to produce accurate bag standards is verified in the laboratory by comparison with other standards. For more information on gas and vapor calibration standards, see Appendix B.

In the laboratory a calibration graph should be plotted for each contaminant(s) to be sampled, using at least three repeat analyses of at least five different concentrations.

The range of standards should include the expected ambient concentration at the field site. If it is not known, preliminary samples should be taken to determine the range of expected concentrations. Three duplicate injections of each standard should be made, and the average value should be used for the calibration graph. The expected field concentration should be within the range in which the contaminant has a linear response on the instrument. In general, three to five points over the expected sampling range are sufficient.

Detector drift averaged over the expected time period(s) should be determined as well as the ability of the GC column to separate the contaminant of interest from other substances known or predicted to be present in the field samples.

If experience shows that detector drift is significant over the period of time that samples will be taken, the number of injections of calibration standards over the course of sampling should be increased. Even in the event that drift does not seem to be a problem, occasional injections of standards throughout the sampling period are needed to assure that the instrument is operating as expected.

If the injection system operates by means of a loop, the calibration curve should be in the form of concentration versus peak height (obtained from the strip chart recorder tracing). If a gas-tight syringe is used, the calibration curve is more conveniently of the form of amount (concentration, in units of weight per volume, times volume injected) or volume (ppm times injection volume) versus peak height. A list follows of information to record during GC calibration.[2]

1. Column type and serial number
2. Temperatures of column and ambient air
3. Chart speed in units of distance per time
4. Carrier gas flow rate through the column in ml/min
5. Concentration of each compound in ppm
6. Sample volume
7. Recorder scale in terms of ppm per unit of deflection
8. Range scale of the instrument
9. Instrument serial number

Backflushing is a technique used to clear a column of high-boiling components much more rapidly than if they were allowed to flow completely through the column. The primary purpose of backflushing is to clear the column of heavy compounds so they do not contaminate the column and interfere with subsequent analyses by creating unwanted peaks or high background levels. Since field GCs are limited to shorter column lengths, lower temperatures, and shorter analytical time than laboratory GCs, most field GCs provide this option.

Backflushing requires two columns, a precolumn and an analytical column, usually hooked together in series. At the junction of the two there is a port where carrier gas can be programmed to automatically inject, thus forcing heavier compounds back out the precolumn while the lighter (more volatile) compounds of interest make the transition to the analytical column. Once the volatile sample compounds have cleared the precolumn, those remaining are simply flushed out. Some GCs backflush to ambient air and others to the detector. Backflushing can significantly shorten analytical time. It is important to select the correct time for the back or reverse flush of carrier gas to begin, or compounds of analytical interest may be lost or undesirable compounds may reach the analytical column. The time interval for backflushing is usually slightly longer than the time that has passed from the point of injection to the point at which the backflush valve was used.

Any column can become contaminated with compounds having long retention times. If contamination of a column is suspected, installing a new column is one way to check. Contaminated columns can be purged by backflushing the column after every analysis, purging the column with clean air for a period of time, or baking columns in a drying oven at a temperature recommended by the manufacturer in a drying oven while passing nitrogen through the column. The baking time and temperature of the oven depends on the type of column being cleaned. Purging between samples increases the time between analyses.

When changing columns, a conditioning period is required. If the ends of the column have been kept capped during storage, the amount of conditioning time is reduced. When installing the column, avoid touching the ends, as contamination may result. Also, ensure that the fittings are tight to avoid carrier gas leaks.

General maintenance during operation for GCs involves the replacement of the carrier and calibration gas and paper and pens if using a printer/plotter. Many systems have refillable gas cylinders that must be charged before going to the field. Others may need external carrier gas cylinders and regulators. Tubing used to connect gas cylinders and GCs should be made of inert materials such as Teflon or stainless steel. Other materials such as rubber or tygon can offgas, causing contamination of the sample. After changing carrier gas most instruments require a period of adjustment, since the new gas is likely to have a different background impurity level than the old one. It is important to check the carrier gas requirements in the manual of instructions for that particular portable GC. It is also important to consider the appropriate shipping regulations concerning compressed gases if a survey is planned at a distant location. Generally it is best to have carrier gas shipped to the destination ahead of time or purchased locally, since there are many restrictions on shipping compressed gas cylinders.

Field Operation

A rule of thumb for selecting chemicals to sample when using a GC with an unheated column is that compounds whose boiling points are greater than 180°C are poor candidates in that their retention times are too long. The type of column will also determine which chemicals can be monitored in any given survey. The most important aspect of making GC measurements is repeatability in retention times. This can be achieved through maintaining a constant column temperature and flow rate, or by the use of standards injected before and after each analysis. In contrast to laboratory GCs, most portable GCs do not generally have heated injection ports. Therefore, field samples must be a gas or vapor at room temperature.

Winter temperatures may preclude using a GC with an unheated column outdoors. Instruments with exposed columns will be most affected by ambient temperature; however, in some portable GCs the unheated column is recessed inside the instrument, thus providing it with some insulation from temperature changes. Like other direct reading instruments, portable GCs should never be left in a car overnight, or delays in sampling can result due to the need to warm up the instrument and the column. High-purity air is commonly used as the carrier gas.

Depending on what is present in a sampled mixture, adjustments to a method (type of column, carrier gas flow rate) will usually allow for different retention times among these compounds. Conditions are usually selected to promote retention times of less than or equal to 10 minutes, because the peaks of late-eluting compounds will be poorly formed (broad lumps), which makes calculating their true area and thus the concentration difficult. An increase in carrier gas flow rate will decrease retention times. Faster flow rates allow for detection of slow-moving compounds, but at some point resolution of the compounds with early peaks (small molecules) will be lost. Also this increases consumption of the carrier gas.

Due to the changes that can occur in field use, such as going from indoor to outdoor locations, in order to use a portable GC for quantitative measurements it must be calibrated with the compound of interest under the exact same conditions in the field, so the peaks can be compared correctly. The GC should be set up and turned on for a period long enough to allow it to stabilize. While this time will vary depending on the type, on whether the column is heated, and on other factors, it should be known ahead of time. One suggested period is 40 minutes.[4] Monitor the ambient temperature. If the variation is too great, a shelter may be needed. This is especially important when using an unheated column. One range of temperatures recommended for ambient air monitoring is 10–35°C.[12]

Once an instrument is set up, sampling is done by injecting ambient air periodically into the portable GC and obtaining concentrations by comparison of peaks with the calibration graph, or by viewing the printout if the GC has its own computer. The size of the sample collection syringe should be the same as that used to perform the calibrations, so that the concentrations will be comparable. The frequency of sample collection is based on the time necessary for the peak(s) of interest to appear (retention time) plus the time necessary for any other contaminants to elute from the column. As it is seldom possible to know the exact composition of the atmosphere to be sampled beforehand, the time between samples will usually have to be determined in the field.

During field use a strip chart recorder should be used as well as a peak integrator (unless the integrator also produces a chromatographic tracing). A record of the GC setup and each measurement should be kept on a sampling data sheet (Fig. 10-9).

A qualitative field check for a given compound's response and retention time on a GC can be useful. One technique for doing this involves placing a 10-ml syringe inside a vial of the compound of interest, touching the needle into the vapor phase above the liquid (headspace). Briefly, for approximately 1 second, withdraw the syringe and then draw back the plunger to 1 ml and inject. This is usually sufficient to check the response of the GC without saturating the detector.[2]

For instruments that are somewhat heavy and

Plant ______________________________ Date ______________
Location ______________________________
General information

Source temperature (°C)	________	Columnar temperature:	
Probe temperature (°C)	________	Initial (°C)/time (min)	________
Ambient temperature (°C)	________	Program rate (°C/min)	________
Atmospheric pressure (mm)	________	Final (°C)/time (min)	________
Source pressure (¹¹Hg)	________	Carrier gas flow rate (ml/min)	________
Absolute source pressure (mm)	________	Detector temperature (°C)	________
Sampling rate (liter/min)	________	Injection time (24-hour basis)	________
Sample loop volume (ml)	________	Chart speed (mm/min)	________
Sample loop temperature (°C)	________	Dilution gas flow rate (ml/min)	________
		Dilution gas used (symbol)	________
		Dilution ratio	________

Components to be analyzed	Expected concentration

Suggested chromatographic column ______________________________
____

Column flow rate ______________ ml/min Head pressure ______________ mm Hg
Column temperature:
Isothermal ________ °C
Programmed from ____ °C to ____ °C at ____ °C/min
Injection port/sample loop temperature ____ °C
Detector temperature ____ °C
Detector flow rates: Hydrogen ____ ml/min.
head pressure ____ mm Hg
Air/Oxygen ____ ml/min.
head pressure ____ mm Hg
Chart speed ________ inches/minute
Compound data:

Compound	Retention time	Attenuation

Sampling considerations
Location to set up GC ______________________________
____
Special hazards to be considered ______________________________
____
Power available at duct ______________________________
Power available for GC ______________________________
Plant safety requirements ______________________________
____
Vehicle traffic rules ______________________________
____
Plant entry requirements ______________________________
____
Security agreements ______________________________
____
Potential problems ______________________________
____
Site diagrams. (Attach additional sheets if required).

Figure 10-9. GC field sampling data sheet.

not very portable, especially if hooked to a strip chart recorder, one technique is to use a syringe. Syringe sampling is a simple and convenient way of sampling areas that are remote. Since it is a one-time analysis, the concentration of the contaminant(s) must be estimated in advance so that the proper size sample can be collected. An injection volume as small as 10 μl can be used if the concentration is expected to be several hundred ppm, while a 1.0-ml syringe is more appropriate if a concentration on the order of 1.0 ppm or less is expected.[5]

A sample is drawn into the syringe from the area or source to be sampled. Several flushings of the syringe should be used before the final sample is taken to ensure a representative sample. At this point, the sample is carried to and injected through the septum of the GC. The septum provides an airtight seal for the injection of the needle installed. If a GC with an automatic air sampler (loop) is used alternately with syringe samples, the syringe injection must be the same size as the air sample volume for that particular GC.

Portable GCs are most commonly operated by collecting grab samples such as syringe collection; however, some are capable of continuous monitoring (automated). Important parameters in continuous monitoring are the cycle time (how often each analysis is performed during the monitoring) and the frequency of calibration during sampling. The determining factor in selecting analysis time is the retention time of the last compound and a margin. Instruments that do not have heated columns require increased numbers of calibrations due to the increased drift.

Samples can also be collected by using a personal air sampling pump to fill a gas bag and then injecting a sample from the bag into the GC. In this case, the sample concentration reflects the average of the concentrations present over the filling time of the bag as opposed to an instantaneous concentration that direct injections of ambient air would represent. Bag samples can be collected over an 8-hour day or a few minutes. At 1 L/min, 60 seconds of sampling will yield enough sample for several replicate analyses. By selecting a low flow rate and a larger bag, such as a 10-ml/min flow rate and a 5-liter bag, an 8-hour time-weighted average (TWA) sample can be obtained. The maximum volume is limited only by the size of bags available and the space in which to store them.

Procedure[5]

1. Start the GC instrument and recorder and allow them to warm up according to the manufacturer's instructions.

2. Set the sensitivity (gain) at the most likely value to be used, and get a straight baseline on the strip chart recorder. Try to select a realistic setting.

3. Calibrate the instrument either with a standard mixture of the gases to be analyzed from a purchased mixture in a gas cylinder or a standard made up in a gas bag. If a cylinder is used, it can be connected directly to the GC in some cases or used to fill a bag. A gas syringe is used to pull the sample from inside of the bag. Set conditions so that peaks are at least 50% of full scale. Make sure carrier gas flow rates are maintained; if not, results will vary.

4. Select the sampling mode either for pumped or syringe samples.

5. If the instrument has a built-in pump, it can be taken directly to the area where sampling is planned and turned on to collect the sample. In some cases tubing can be attached to the GC inlet to draw samples from a distance. Alternatively, the instrument can be set up in a clean area and a gas syringe can be used to pull the sample. The sample should be taken immediately back to the instrument for injection. Use syringes that have not been used previously for sample collection.

6. Integrated samples involve the use of Tedlar bags. Evacuate a clean sample bag using the inlet port of a personal sampling pump. To reduce memory effects and contamination, use only previously unused sample bags. Attach the sample bag to exhaust outlet of the personal sampling pump using a minimum amount of Tygon tubing. Pump the air using a rate calculated to fill $\leq$80% of the sample bag's capacity during the sampling period.

7. Manual injections require practice for consistency. Depressing the plunger on the syringe

should be a smooth, rapid motion. GCs that perform calculations may limit the time in which an injection can be made once the instrument has received a signal that an analysis is about to begin. Other injection problems can result from a leaky plunger or septum or block in a needle. Never use the same syringe to inject both calibration mixtures and samples. At least three replicate samples should be collected and analyzed for each sample of interest. For more information on the use of syringes, see Appendix B.

8. Test syringes for contamination periodically by filling them with clean air and analyzing the contents. Bags should be analyzed within 4 hours after completion of sampling. Most bags have a septum on one side that will allow a sample to be removed with the gas syringe.

Ethylene Oxide Measurements on a Portable GC[6]

Procedure[6]

1. Set up a portable GC with a PID using the following parameters: column = 1.2 m × 3 mm OD PTFE, packed with Carbopak BHT 40/100 mesh. Carrier gas flow rate = 15 ml/min. Also assemble portable strip chart or integrator, battery charger, regulator, and any other peripherals necessary for individual instruments. Allow the instrument to warm up and equilibrate for 30 minutes.

2. Check the sampling equipment to prevent contamination. Use different syringes for sampling and for standard preparation. Identify each syringe with a unique number. Segregate bags used for sample collection from those used for calibration standards.

3. Calibrate the GC daily in the field. Prepare bag standards by adding a known volume of ethylene oxide to a known volume of clean air in a bag. By adding a known volume in μl to a specified number of liters of air, a known concentration in ppm will be created. *Note:* Because ethylene oxide is a flammable gas, the shipment of the compressed gas must comply with 49CFR 171-177 regulations regarding shipment of hazardous materials.

4. Evacuate a 5-L to 10-L bag completely by drawing the air out with a large 1-L to 2-L syringe.

5. Draw clean air (or oxygen or nitrogen) from a supply cylinder into the syringe for measured transfer into the bag. Alternately, if a clean air supply is not available, draw room air through charcoal sorbent into the syringe. Repeat until the bag contains 5 L of air.

6. Add a known amount of pure ethylene oxide or standard ethylene oxide mixture to the bag by means of a gas-tight syringe. *Example:* Using a gas-tight syringe, take 50 μl from a cylinder of pure ethylene oxide and inject it into 5 L of air to create a 10-ppm standard. Alternately, 200 μl of a 27% v/v ethylene oxide mixture (e.g., 88/12 w/w Freon 12 and ethylene oxide) can be added to the 5 L of air to obtain a 10.8 standard.

7. Allow the bag to equilibrate, with occasional kneading, for at least 5 minutes.

8. Analyze aliquots of various sizes to establish a calibration graph. For each point three replicate samples should be done. For example, using a high instrument attenuation (low sensitivity), injections of 0.2 ml, 0.4 ml, 0.6 ml, 0.8 ml, and 10 ml might be possible. These amounts would correspond to injections of 2 ηl, 4 ηl, 6 ηl, 8 ηl, and 10 ηl. On a more sensitive attenuation, injections of 0.02 ml, 0.04 ml, 0.06 ml, 0.08 ml, and 0.10 ml would be typical. Results will vary from instrument to instrument, and from time to time on the same instrument.

9. Plot ηl of ethylene oxide versus peak height or area if the GC cannot do this automatically. This plot should be a straight line.

10. Periodically throughout the day, check the calibration by repeating some of these injections. Ideally, each sample would be bracketed, before and after, with injections of standards, although this situation is seldom practical. Some GCs are capable of doing automatic periodic calibrations with programming.

11. Collect samples by drawing air from the contaminated area directly into a syringe or filling bags for an integrated TWA sample.

12. For syringe sampling, draw air directly into a gas syringe. Collect syringe samples by first purging a gas-tight syringe several times with clean air to remove any residual ethylene oxide from previous samples; then draw air into the syringe at the time and location of interest. If larger syringe samples are desired, such as 20 ml of air, it will be necessary to transfer some of the sample to the smaller syringe being used in the chromatograph. It is essential that the sizes of all grab sample injections be the same if concentrations are to be compared. A rubber cap placed over the end port of the larger syringe can serve as a septum for the smaller (500 μl) syringe.

13. For integrated air samples for TWA determinations, a clean bag of plastic or other material must first be evacuated. A personal air sampling pump is used at the highest airflow available.

14. Attach the plastic bag via tubing to the outlet of a personal sampling pump and pump air from the contaminated area into the bag at a rate calculated to fill the bag over the sampling period. This rate will be between 20 ml/min and 500 ml/min. Terminate sampling before the bag is 80% full. The pump's flow rate must be within ±5% of the initial setting throughout the sampling period.

15. Analyze the bag sample within 2 hours after completion of the sampling to minimize loss of analyte by adsorption and permeation as follows: Fill a gas-tight syringe, purged several times with sample, from the sample bag. Then empty it to the desired volume, and inject that volume into the chromatograph with a quick firm motion. Record the number of the syringe. Use replicate analyses to determine the repeatability of the analysis. If no estimate of concentration is available, use an injection volume 10 μl to 25 μl at a high attenuation to reduce the possibility of column and detector overload. Depending on the results of this injection, larger volumes and/or ore-sensitive attenuations may be selected.

16. Calculate the concentration of ethylene oxide from the calibration graph (ηl) and the injection volume (ml):

$$\text{ppm} = \frac{\mu\text{l (gas)}}{\text{L (gas)}} = \frac{\eta\text{l (gas)}}{\text{ml (gas)}}$$

Interpreting Measurements with the Portable GC

As noted previously, it is possible for more than one compound to have the same retention time, and in certain circumstances where unknowns may be present, it will mean that a peak appearing at or near a given retention time is not 100% confirmation of a given compound's presence. However, if the instrument is operating properly, a lack of peaks will usually mean there are no compounds present at the detection level of the instrument. The best way to confirm the presence of a given compound is to collect additional samples for analysis in the laboratory on a GC with two different columns, each of which is capable of resolving the mixture of interest and doing sequential analyses. As this is difficult to do in most portable GCs where the results are questionable, bag samples should be collected and taken to a laboratory for analysis. The interpretation of a chromatogram requires the use of calibration reference data that have been generated through testing.

In general, the more triangular and symmetrical a peak, the easier it is to do an analysis. However, many GC peaks have "tailing." Quantitation based on peak height is acceptable as long as the area under the tail is small compared with the total peak area. The OVA GC and many other field GCs rely on peak heights for quantitation.

If peak area measurements are desired, the areas may be measured using an integrator on the GC's output signal. Other manual methods may also be used, such as counting squares, weighing curves, or calculating the area using the formula for triangles: ½ (base × height). Under a given set of operating conditions the retention time is characteristic of that particular substance and can be used to identify specific compounds.

Copies of chromatograms should be saved as a permanent part of the field measurement file. When complex mixtures such as gasoline are analyzed, it may be desirable to record the backflush peak for future reference and peak area comparison. If the instrument is capable of doing a total gas or vapor concentration reading, this reading should also be recorded. It is useful for determining relative response factors.

INFRARED SPECTROPHOTOMETERS

IR analyzers are based on the principle that many compounds absorb light in the wavelength of the IR spectrum. Detection of specific compounds is due to their selective absorption of infrared radiation at specific wavelengths to give energy transitions between vibrational and rotational energy levels of molecules in their electronic ground state. This selective absorption provides specificity by allowing for selection of an analytical wavelength at which potential interferences do not absorb. The larger the number of molecules through which the IR beam passes, the greater the absorption. The amount absorbed by any compound is a function of both the concentration of material in the sample (molecules per length of beam travel) and the distance through which the IR beam passes. The IR source is often a Nichrome wire resister. Many of the vibrational energy transitions observed in IR spectra are characteristic of certain functional groups. For example, the ketone group (C=O) has an absorption band from 5.60 to 5.90 μm, depending on the characteristics of the compound with which it is affiliated.[2]

Instruments fall into two categories: dispersive and nondispersive. Dispersive instruments use gratings or prisms to disperse the transmitted beam of radiation. The dispersed light is focused to a spectrum in the plane of the vertical exit slit. Rotation of the dispersing device causes the spectrum to move across the face of the exit slit, and ultimately, the detector. Dispersive instruments are used in the laboratory and frequently use dual beams of IR radiation for analysis.[7] Most field instruments are nondispersive and do not incorporate prisms or gratings, so the total radiation from the IR source passes through the sample.

Most instruments consist of an IR source, a circular variable filter to select IR radiation of different frequency intervals, and a detector. They may have fixed or variable path length and wavelength, and most are operable only on AC power, although 4-hour batteries are available for some units. Field instruments are single beam devices and are used almost exclusively for quantitative analysis, although in some cases it is possible to do some qualitative analysis by running a scan. The degree of expertise required to operate IRs varies. In some cases, the user must set the analytical parameters, such as wavelength, slit width, and path length. In other instruments, a preprogrammed microprocessor automatically makes these adjustments. Table 10-3 lists specific IR absorption bands.

The importance of selecting the proper peak wavelength for IR analysis cannot be ignored. For example, there are three absorption maxima that can be selected from trichloroethylene's spectrum: 10.58, 11.77, and 12.78 μm. However, the

TABLE 10-3. Specific Infrared Absorption Bands

Grouping	*Absorption Band (μm)*
Alkanes	
CH_3-C-, -CH_2=	3.35–3.65
Alkenes	
-CH=CH_2	3.25–3.45
Alkyne	
-C=C-	3.05–3.25
Aromatic hydrocarbons	3.25–3.35
Substituted aromatics	6.15–6.35
Alcohols	
-OH	2.80–3.10
Acids	
-COOH	5.75–6.00
Aldehydes	
-COH	5.60–5.90
Ketones	
-C=O	5.60–5.90
Esters	
-COOR	5.75–6.00
Chlorinated compounds	
-C-Cl	12.80–15.50

Source: NIOSH: *The Industrial Environment, Its Evaluation & Control.*

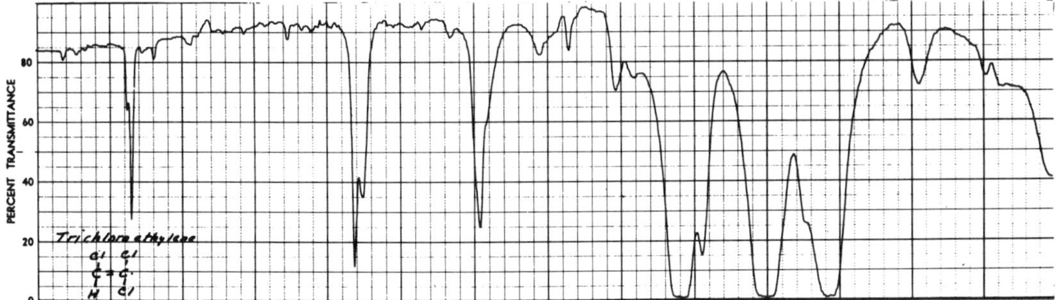

Figure 10-10. Infrared spectra of trichloroethylene. (From *The Industrial Environment—Its Evaluation & Control,* NIOSH, Cincinnati, OH, 1973, p. 234.)

10.58-μm peak overlaps with freon 113 and cannot be used if this compound is present and the 12.78-μm peak is of low intensity and overlaps with water peaks. Therefore, in many cases there is only one peak that will provide good data.[8] Figure 10-10 shows the infrared spectra of trichloroethylene.

Not all IRs are the same. IR analyzers come in a variety of levels of sophistication, the simplest being an instrument dedicated to a single compound. Multiple-compound instruments are available for the field as well as for fixed station monitoring. A number of different field instruments are available. All are capable of continuously sampling ambient air. The measurement units will be in absorbance and percent transmittance or ppm. The trend is to read out in ppm, although at least one model using the absorbance mode, the MIRAN 1A, is still in wide use.

The majority of field instruments operate in the near-IR region of the electromagnetic spectrum, generally between 2.5-μm and 16-μm wavelengths. This wavelength range encompasses both the group frequency region (2.5–8 μm) and the fingerprint region (8–19 μm). Neither oxygen nor nitrogen interfere with IR measurements. Every organic molecule has a characteristic IR spectrum, and this information can be used for identification and for quantifying the amount of material present.

IR systems are useful for leak detection. Many industrial manufacturing facilities use toxic gases in closed systems; however, any system can malfunction and leak. Figure 10-11 demonstrates the use of an IR to monitor for ethylene oxide leaks in a hospital sterilizer. Other portable dedicated IRs are designed to be set up and used as stationary monitors for certain gases, such as carbon monoxide. These are not useful for leak detection, but they can be moved and set up in different areas with ease. Employee exposure monitoring can also be done with these units to a degree; however, since the measurements are made at fixed positions, these systems do not replace personal sampling but supplement it.

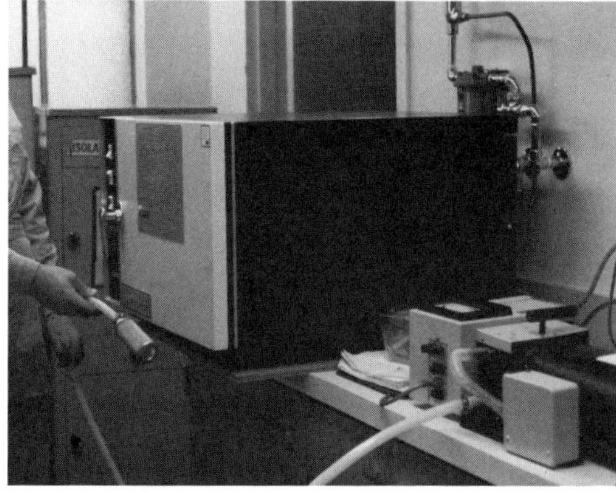

Figure 10-11. Using a Foxboro MIRAN 1A infrared analyzer to monitor for leaks.

The IR is often used to measure carbon dioxide for indoor air problems. By detecting increases in carbon dioxide levels over time in various areas of a building, over a full day the sampler can determine the efficiency of the heating, ventilating, and air conditioning (HVAC) system to provide enough fresh air and air changes. Anesthetic gases, such as nitrous oxide, halothane,

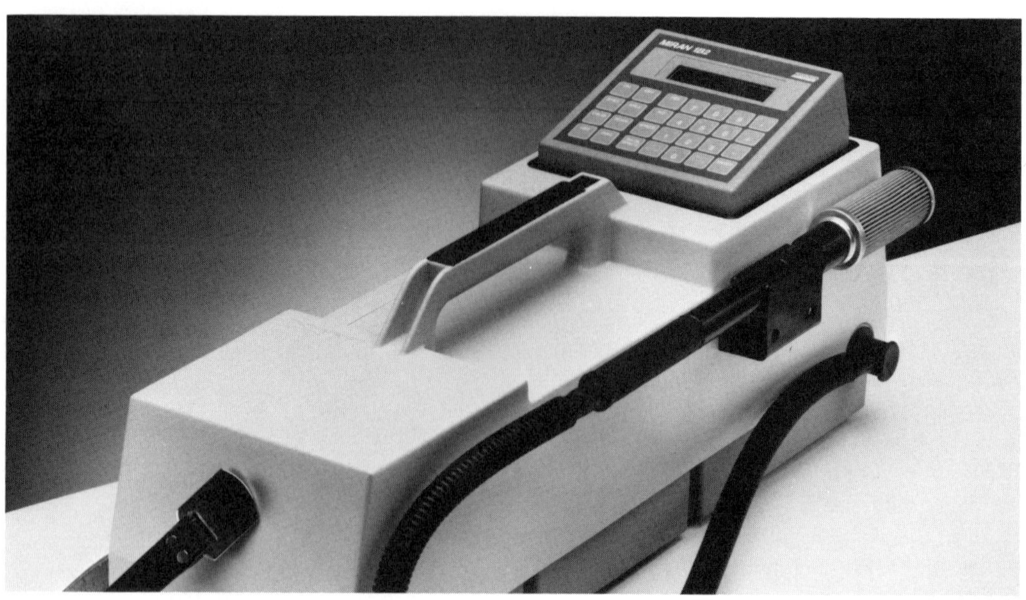

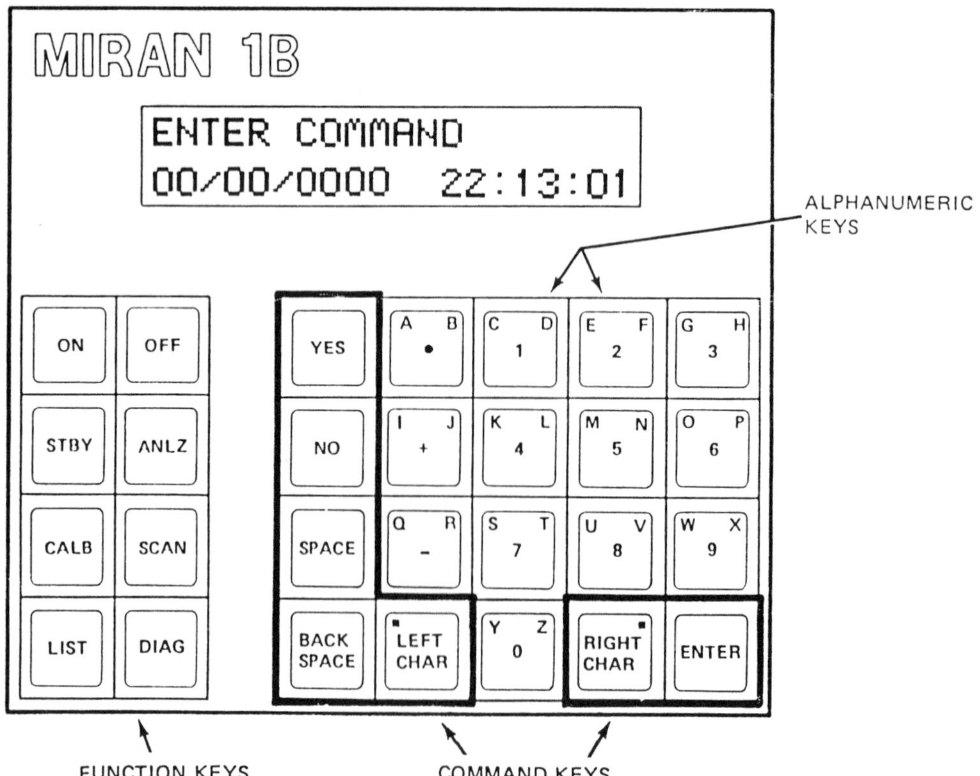

Figure 10-12. The MIRAN 1B infrared analyzer *(top)* and its alphanumeric keyboard *(bottom)*. (Courtesy The Foxboro Company.)

enflurane, penthrane, and isoflurane, are also a common application. In this case, the IR is used to detect any leaks in the gas supply and scavenging system. The determination of residual fumigants, such as ethylene dibromide, chloropicrin, methyl bromide, and phosphine, used to disinfect fruits and grains, is another application for IR.

The best-known infrared analyzers are the MIRANs (miniature infrared analyzers). These range from dedicated instruments for a specific gas to those capable of being "tuned" to monitor for many individual compounds, one at a time. In some cases, tuning is done by interchanging filter and meter packages that allow up to 300 different compounds to be monitored separately. In other cases, it is done by changing the wavelength and path length of an instrument.

The MIRAN 1B (Fig. 10-12) is designed for field measurement of gases and vapors. It is a microprocessor-controlled single-beam IR that utilizes interactive programming to prompt the operator through available choices and functions. The instrument has a touch-sensitive alpha numeric keyboard.

Included in the analyzer's software is a diagnostics test routine that is run each time the analyzer is turned on. Every time the instrument is unplugged and plugged in, it must go through its entire warm-up cycle, approximately 15 minutes. For quick measurements it can be carried around, but in cramped spaces and for making lots of measurements it generally needs to be on a cart. It is necessary to wait 15 seconds and to the hold the probe in place in order to get a measurement. The clock display is helpful in this respect, allowing the sampler to watch the seconds to be sure enough time has passed. The instrument reads out to 0.1 ppm.

In the MIRAN 1B, infrared energy is emitted from a nichrome wire source through a light pipe assembly. The light is then directed to the filter wheel that allows energy at the selected wavelength to pass into the gas cell. The sample is drawn into the cell by the integral air pump at a rate of 25 Lpm to 30 Lpm. The sample absorbs IR energy from the beam, and the amount of absorption is measured by the detector, amplified and converted to concentration units by the electronics, and transmitted to the liquid crystal display. See Table 10-4 for compound data for the MIRAN 1B2.

The analyzer also has a variable path-length gas cell that is adjustable from 0.75 M to 20.25 M at discrete path-length settings. The greatest sensitivity is obtained at the longest path length of 20.25 M, where measurements below 1 ppm can be made for some gases according to the manufacturer. There are five path lengths available on the 1B, 0.75, 2.25, 6.75, 12.75, and 20.25 M. These are automatically selected by the instrument depending on which concentration range is used for sampling. The slit setting is fixed at 1.25 mm. The sample cell window is made of silver bromide. The instrument converts absorbance to concentration, and the display reads out in ppm or absorbance, contrary to previous models that required extrapolation of concentrations from absorbances using a graph. Since infrared absorbance is nonlinear for most compounds, all data must be corrected in order to get a straight line. The 1B corrects automatically.

According to the manufacturer, a recent innovation in the MIRAN 1B series is the sealing of the nichrome wire source. As a result of this enhancement, the unit is more stable than previous units and can detect lower concentrations of many gases and vapors that absorb energy between 2.5 μm and 14.5 μm in the IR spectrum. For example, according to the manufacturer, the sensitivity for ethylene oxide has been increased from 1 ppm to 0.2 ppm. To date, no studies have been published on these improvements. Prior instruments suffered from errors caused by tilting during operation, and sealing minimizes this problem. The 1B2 model has this improvement.

On the most sophisticated level, instruments are available that can measure five or more compounds simultaneously and print out concentrations at regular programmed intervals at multiple sites. For example, the MIRAN multipoint monitoring system can provide coverage for a complete facility. An IR system can monitor as many as 24 points, with each point being as far as 1,000 feet from the monitor. These systems are capable of automatically calculating the concentration of each measured component and com-

TABLE 10-4. Compound Library for the MIRAN 1B2

Compound	*Wavelength (μm)*	*Path Length (m)*	*Min. Detectable Concentration (ppm)*
Acetaldehyde	9.26	20.25	1.1
Acetic acid	8.72	20.25	0.3
Acetone	8.48	2.25	0.6
Acetonitrile	9.68	20.25	7.6
Acetophenone	10.70	20.25	0.3
Acetylene	3.05	20.25	1.6
Acetylene tetrabromide	8.99	20.25	1.2
Acrylonitrile	10.67	20.25	0.6
Ammonia	10.95	20.25	0.5
Aniline	9.53	20.25	0.2
Benzaldehyde	8.58	20.25	0.3
Benzene*	9.93	20.25	2.2
Benzyl chloride	9.54	20.25	2.5
Bromoform	8.96	20.25	0.4
1,3-Butadiene	11.10	2.25	3.9
Butane	10.40	20.25	5.1
Butyl acetate	8.33	2.25	0.6
n-Butyl alcohol	9.70	12.25	0.3
Carbon dioxide	4.72	0.75	10.2
Carbon disulfide	4.70	20.25	4.8
Carbon monoxide	4.76	20.25	2.1
Carbon tetrachloride	12.76	20.25	0.1
Cellosolve acetate	8.89	6.75	0.3
Chlorobenzene	9.40	20.25	0.4
Chlorobromomethane	8.39	6.75	1.5
Chlorodifluoromethane	9.20	0.75	1.1
Chloroform	13.12	2.25	1.1
Cresol	8.88	20.25	0.2
Cumene	9.90	20.25	2.3
Cyclohexane	3.41	2.25	0.7
Cyclopentane	11.40	20.25	9.0
Diborane	3.83	20.25	0.6
m-Dichlorobenzene	9.47	20.25	0.3
o-Dichlorobenzene	13.55	20.25	0.6
p-Dichlorobenzene	9.30	20.25	2.5
1,1-Dichloroethane	9.50	20.25	0.4
1,2-Dichloroethylene	12.30	0.75	5.3
Dichloroethyl ether	9.05	20.25	0.09
Diethylamine	8.99	20.25	1.1
Dimethylacetamide	10.10	20.25	0.4
Dimethylamine	8.79	20.25	1.2
Dimethylformamide	9.36	20.25	0.2
Dioxane	9.06	6.75	0.3
Enflurane	8.96	20.25	0.03
Ethane	12.20	20.25	10.6
Ethanolamine	12.93	20.25	2.9
Ethyl acetate	8.32	0.75	2.1
Ethyl alcohol	9.67	2.25	2.3
Ethyl benzene	9.90	20.25	3.0
Ethyl chloride	10.50	6.75	3.4
Ethylene	10.70	20.25	0.5
Ethylene dibromide	8.68	20.25	0.6

TABLE 10-4. *(continued)*

Compound	*Wavelength (μm)*	*Path Length (m)*	*Min. Detectable Concentration (ppm)*
Ethylene dichloride	8.37	20.25	1.3
Ethylene oxide	3.30	20.25	0.7
Ethyl ether	9.03	0.75	2.8
Formaldehyde	3.56	20.25	0.5
Formic acid	9.36	20.25	0.2
Freon 11	10.96	2.25	11.2
Freon 12	9.30	20.25	0.1
Freon 13B1	8.54	0.75	0.4
Freon 21	9.50	0.75	2.3
Freon 112	9.90	0.75	12.8
Freon 113	8.70	0.75	3.4
Freon 114	8.67	0.75	1.7
Halothane	12.46	20.25	0.3
Heptane	3.40	0.75	3.5
Hexane	3.39	0.75	3.9
Hydrazine	10.67	20.25	0.6
Hydrogen cyanide	3.03	20.25	1.7
Isoflurane	8.84	20.25	0.04
Isopropyl alcohol	8.94	6.75	1.5
Isopropyl ether	9.12	0.75	4.1
Methane	7.70	20.25	1.0
Methoxyflurane	12.10	20.25	0.2
Methyl acetate	9.7	2.25	2.3
Methyl acetylene	3.0	12.75	2.0
Methyl acrylate	8.57	20.25	0.1
Methyl alcohol	9.70	6.75	0.7
Methylamine	3.36	20.25	1.9
Methyl bromide	7.60	20.25	2.3
Methyl cellosolve	9.62	20.25	0.3
Methyl chloride	13.59	20.25	3.0
Methylene chloride	13.47	0.75	10.0
Methyl iodide	3.36	20.25	1.8
Methyl mercaptan	3.38	20.25	1.5
Methyl methacrylate	8.80	2.25	1.0
Morpholine	9.20	20.25	0.7
Nitrobenzene	11.94	20.25	0.9
Nitromethane	3.37	20.25	4.5
Nitrous oxide	4.68	12.75	0.4
Octane	3.40	6.75	0.3
Pentane	3.39	0.75	4.6
Perchloroethylene	11.10	6.75	0.2
Phosgene	11.98	20.25	0.1
Propane	3.37	0.75	6.5
Propyl alcohol	9.60	6.75	0.8
Propylene oxide	12.16	20.25	1.1
Pyridine	9.90	20.25	8.6
Styrene	11.10	20.25	0.5
Sulfur dioxide	9.00	20.25	1.9
Sulfur hexafluoride	10.80	6.75	0.02
1,1,2,2-Tetrachloroethane	8.60	20.25	1.4
Tetrahydrofuran	9.40	6.75	0.6
Toluene	13.89	2.25	9.2

(continued)

TABLE 10-4. *(continued)*

Compound	*Wavelength (μm)*	*Path Length (m)*	*Min. Detectable Concentration (ppm)*
Total hydrocarbons	3.39	0.75	3.9
1,1,1-Trichloroethane	9.39	2.25	1.2
1,1,2-Trichloroethane	10.90	20.25	0.9
Trichloroethylene	10.84	6.75	0.4
Vinyl acetate	8.42	20.25	0.1
Vinyl chloride	11.30	20.25	0.8
Vinylidine chloride	9.40	20.25	0.3
Xylene	13.20	20.25	2.0

Source: The Foxboro Co., Foxboro, MA.

*Minimum detection level exceeds current permissible exposure standard; therefore, recommended for gross leak detection only.

Analytical wavelength: The analytical wavelength has usually been chosen as that of the strongest band in the spectrum that is free from interference due to atmospheric water and carbon dioxide. The listed wavelengths are approximate and should be peak-picked if extra accuracy is required. If more than one IR-absorbing material is present in the air in significant concentration, the use of another analytical wavelength may be necessary.

Path length: Those indicated have been chosen for optimized readings at the exposure limits.

Minimum detectable limit: The concentration that would produce an absorbance equal to twice the peak-to-peak noise of the MIRAN 1B. All minimum detectable limits are based on a path length of 20.25 meters.

pensating as required for the interference resulting from the presence of other compounds in the area being monitored.

Advantages and Limitations

The advantage of IR is that units can be purchased that are specific for given chemicals, or tunable units can be purchased and used for many different compounds with a minimum of setup time. In the field they are easy to use and provide stable operation.

Detection limits depend on the absorption coefficient of the compound at a given wavelength or frequency, since IR absorption coefficients are generally several orders of magnitude lower than those for visible or ultraviolet absorption. In order to compensate for this difference, portable instruments incorporate long, often variable, path-length cells, typically 1–20 m.[9]

The limit of detection for many compounds is in the range of 1 ppm to 20 ppm using an instrument with a 20-m path length, although a more realistic minimum value may be 3–5 ppm due to background problems in most cases. Maximum concentrations are commonly in the low percent range.

The usefulness of IR analyzers for monitoring complex mixtures is limited because overlapping peaks produce an additive response, making concentrations appear higher than they actually are. A wavelength that is relatively unique to a chemical is selected for monitoring and ideally there should be no others that interfere. However, this is often not the case, and a review of the chemicals likely to be present during monitoring should be done to assure that an interferent will not be a problem. Interferences depend on the contaminant being measured. For example, chlorinated hydrocarbons will all absorb at approximately the same wavelength. This problem can be partially minimized by taking measurements at a secondary absorption wavelength for substance confirmation. A primary interferent is water vapor. The effect of water vapor can be minimized by passing the air sample through silica gel or a similar drying agent, maintaining constant humidity in the sample and calibration gases by refrigeration, saturating the air sample and calibration gases to maintain constant humidity, or using narrow-band optical filters in combination with some of the other measures.[10] Carbon dioxide can also be an interference, although its effect at concentrations normally present in ambient air is minimal. According to the American Society for Testing and Materials (ASTM), 750 ppm of carbon dioxide may give a response equivalent to 0.5 ppm carbon monoxide.[10] Some instruments have microprocessors that allow them to correct to some degree for interferences.

Calibration

A multipoint calibration is required when an analyzer is first purchased, when the analyzer has had maintenance that could affect its response characteristics, or when the analyzer shows drift in excess of specifications as determined when the zero and span calibration is performed. A zero and span calibration is required before and after each sampling period, or if the analyzer is used daily. The flowmeter should be calibrated as well when the analyzer is first purchased, when it is cleaned, and when it shows signs of erratic behavior.[10]

Once calibration plotting curves are prepared for a given compound, and concentrations can be obtained by measuring the absorbance at the analytical wavelength and reading the concentration from the point where the absorbance intersects the curve. Most calibration curves of this type will have some curvature, so it is best to use a plot prepared with three or four data points, rather than using a single-point and a straight-line approximation. If only a single calibration concentration is available, it may be used to set the span of the instrument. The measurements will be most accurate near this concentration point and less accurate at other values due to curvature.

Manual calibration of most long path-length IR instruments is done using a closed-loop system. Small amounts of contaminant (on the order of microliters) are added to a fixed volume sample chamber (generally 2–6 L) without measurably affecting the pressure of the system, creating a known concentration on the order of parts per million (ppm). User calibrations can be somewhat difficult to do and require considerable practice in using the closed-loop system and becoming familiar with the instrument in order to consistently generate reliable accurate calibrations. Calibration of IRs can also be done using known concentrations in gas cylinders.

Care must be taken during calibration of IRs, because these are nondestructive instruments; thus, any contaminant that enters the instrument will come out essentially unchanged, allowing the user to be exposed to the calibrant gas. The exhaust from the analyzer should be vented to a laboratory hood or other exhaust to remove contaminant, both during calibration and analysis, to prevent buildup in the surrounding environment.

The first step in setting up an IR analyzer to measure a gas is to select the wavelength and path length. Wavelengths are chosen to maximize analytical sensitivity while minimizing interferences from other vapors commonly present in the workplace or atmosphere where this contaminant is typically found. For example, if measuring styrene vapor in a chemical plant where acrylonitrile butadiene styrene (ABS) polymer is manufactured, then 1,3-butadiene is likely to be present. Since this compound absorbs at the most frequently chosen analytical wavelength for styrene, an alternative wavelength must be chosen. On the other hand, if doing a survey in a fiberglass boat hull construction facility where acetone is commonly used as a cleaning solvent for styrene, other vapors would not be a concern as they are not likely to have any absorption peaks at wavelengths that would interfere with styrene.

Since ambient air always contains water vapor and carbon dioxide, the analytical wavelengths selected must reflect this makeup. Some gases have a limited number of possible analytical wavelengths, and occasionally a wavelength must be chosen where water vapor can interfere. This situation is associated with the 5-μm to 8-μm region where strong water vapor absorbance occurs. To reduce the interference of any water vapor, try to zero the instrument with air that has about the same humidity as the sample.

If an IR is overloaded from being exposed to

too high a concentration, inaccurate calibrations and negative or suppressed readings at certain concentrations can result. If the absorbances for any of the calibration standards go above one absorbance unit, switch to a shorter path length or use a weaker absorption band.

There are many compounds for which IRs can be used. In the case where the instrument has not already been set up to monitor for a given compound, the manufacturer can often provide information to select appropriate wavelengths and path lengths for setting up IRs to monitor many different gases and vapors. When the path length in a variable path length instrument, such as a MIRAN 1A or 1B, is changed, the instrument should be recalibrated.

Closed-Loop Calibration for the IR

When using the closed-loop technique, the user introduces pure samples through a septum with a liquid or gas syringe. The sample is circulated through the system by means of the closed-loop pump. Using the closed-loop system for the calibration of organic compounds with low vapor pressures may result in errors due to adsorption on the walls of the closed-loop system. The result of using a calibration curve based on these data would be to overestimate concentration during surveys. Therefore, even though closed-loop calibration systems are not recommended for compounds with vapor pressures less than 15 mm Hg (25°C), from the standpoint of using the instrument the errors may not be significant in certain types of surveys.[11]

Procedure[12]

1. Determine the sample volume required for the desired concentration limit. For gases the volume to be injected is calculated using

$$\text{ppm} = \frac{V_1}{V_2}$$

where

V_1 = volume of gas to be injected
V_2 = volume of closed-loop calibration system.

For liquids (at atmospheric pressure and 23°C) the volume to be injected is calculated using

$$\text{ppm} = \frac{V_1(d)(24.45 \times 10^3)}{MV_2}$$

where

V_1 = liquid sample volume in μl
d = liquid density in g/ml
M = molecular weight
V_2 = volume of closed-loop calibration system in L
24.45×10^3 = number of μl of vapor per millimole of analyte at standard temperature and pressure (STP)

For other temperatures recalculate 24.2 using

$$\frac{0.08205\ T}{P}$$

where T = temperature and P = pressure.

2. Introduce clean air into the sampling chamber using clean tank air or room air drawn through the zeroing cartridge.

3. Connect tubing from the "out" connector of the closed-loop pump to the input port of the analyzer. Connect tubing from the "in" connector of the pump to the "cal port" connector of the analyzer. Turn the sample valve to the "calibrate" position.

4. Turn on the closed-loop pump.

5. Inject the quantity of gas or liquid needed to fill the sample cell with the desired concentration. Inject equal increments to cover a full range of concentrations. For example, if the maximum concentration is 200 ppm, and the number of calibration points is to be five, then each injection should increase in concentration by 40 ppm. Frequently replace the silicone rubber disks, called *septa*, used in the injection ports of closed-loop calibration systems. After a number of injections they may not reseal properly.

6. Following each injection, flush the closed-loop system and cell with clean air or nitrogen.

For direct-reading scales, the obtained reading should agree closely (±5%) with the concentration injected into the closed-loop system. When calibrating for other concentration ranges using the general-purpose scale, add the steps below to the basic closed-loop calibration procedure.

7. Adjust the expansion controls so that the full-scale concentration value reads exactly 100 divisions on the meter scale.

8. Disconnect the pump and flush the cell with "zero" air.

9. Inject three or four increments of sample corresponding to approximately 20% of the full scale value. The zero does not have to be reset after each injection. Record the meter reading for each injection.

10. Prepare a plot of meter reading versus concentration, with meter reading on the *y* axis and concentration on the *x* axis. Unknown concentrations can then be found using this curve and a meter reading.

Given the microprocessor capability of some instruments, calibration has been simplified in that certain parameters are programed into the instrument at the factory to allow for an electronic calibration check. The MIRAN 1B's memory contains a library of calibration data for 116 gases and vapors (see Table 10-4). It is a permanent library and cannot be altered by the user, although a second user-programmable library is also available. If the information becomes outdated, the instrument must be sent back to the factory. Most 1B's are not calibrated against all 116 compounds at the factory; instead, the filter wheel is calibrated. The internal electronic calibration system on the MIRAN 1B allows for field calibration without the need for calibration gas which is sufficiently accurate for performing screening samples. Although this automatic (factory-provided) calibration of the 1B saves time, it is not as accurate as calibration with a standard. The accuracy of the factory calibration is rated at ±15% under the manufacturer's specifications.

A multipoint calibration can be plotted by repeated injection of small aliquots of analyte (thereby building up the concentration in the sample cell), and taking a reading of instrument response after each addition. It also has provisions for the storage of up to 10 user-calibrated gases in a separate library that can be reprogramed and changed as the user desires. In order to add gases, the closed-loop calibration system must be used. The MIRAN 1B also must be periodically checked after every 1,100 hours of operation, or more frequently, using the "automatic" path-length calibration mode. A "calibration error" will be indicated if the instrument is out of synch.

Although the 1B analyzer automatically fits a calibration curve to gas standards that the user supplies, and a special curve-fitting routine takes care of any nonlinearity in the data, problems can develop. The sampler should always check to see what linear (*P*) and quadratic (*Q*) constants the MIRAN 1B calculates from the data. If the absorbances measured by the 1B are too high, which would occur if the cell path length was too long, a negative *P*-coefficient can result. Then during measurements negative concentrations will be displayed until a certain concentration threshold is reached. In this case, the correction involves reducing the path length and recalibrating. A similar problem can occur if a negative *Q*-coefficient is calculated by the instrument.[13]

Due to its electronic components and its precisely aligned system of mirrors, the IR is particularly susceptible to damage during shipment unless properly packaged. Knobs can also be moved during shipment. Therefore, most manufacturers recommend using the original packing container whenever shipping these instruments.

Field Operation

There are three primary types of sampling modes for IR analyzers: bag sampling, leak testing, and continuous monitoring. In bag sampling, discrete breathing zone, area, or process samples can be collected in bags as described in the chapter on Integrated Sampling for Gases and Vapors using personal sampling pumps. The flow rate of the sample pump is not critical, as long as it is constant over the duration of the sample. Following sample collection, the bag is transported to the location of the IR analyzer where it is connected to the inlet of the sample cell. An internal or auxiliary pump pulls the sample from the bag, through the sampling chamber, where it displaces the ambient air. When the instrument's response to the sample in the bag reaches a maximum, a reading is taken. The concentration of the contaminant can be determined from comparing the plot of instrument response versus the calibration curve concentrations, or directly from the readout if it is in ppm. Another application is to collect exhaled breath in bags and analyze it on the IR. For more information on this sampling method, see the chapter on Biological Monitoring.

Bags can also be analyzed to identify the presence of unknowns on IRs with scanning capabilities. The scan of the unknown atmosphere is compared with a scan of contaminant(s) expected to be present to see if any unidentified peaks are present. The instrument should be hooked up to a strip chart recorder during this analysis.

When sampling in situations where the atmosphere is unknown, one technique is to operate the IR at 3.4 μm (the C—H bond stretching frequency) using the maximum possible path length. In the field, the response under these conditions should be reported as equivalents of hexane whatever compounds have been used to calibrate the instrument over the desired concentration range using the closed-loop method.[4]

In survey instruments designed to detect leaks, an extended sample probe of up to 6 feet in length made of inert tubing with a dust filter attached is used to make spot measurements at many different locations, such as door seals, sumps, lines, and fittings. If the filter becomes clogged, the sampling flow rate will decrease, resulting in a slower instrument response. Replace the filter cartridge when the response time has increased noticeably, or after 400 hours of use.

In continuous monitoring, the instrument is stationed to collect samples at a location of interest. A pump continuously draws sample into the sample cell, and the absorbance is continuously monitored on the instrument readout and recorded on a strip chart or stored on a datalogger for printout later. If measurements exceed the range for which the instrument was calibrated, it must be remembered that the response of IR analyzers is not linear over the total range of the instrument, particularly at the ends of the scale; therefore, extrapolation past calibration points is not advised. If much higher concentrations than anticipated are identified, monitoring may need to be repeated following calibration at higher concentrations.

If the process being monitored is a constant operation with little variation, readings can be taken with the monitor every 30 minutes. If the process is variable, it may be necessary to take the readings as often as every 5 minutes or continuously as described above.

In compliance sampling with the MIRAN, calibration should be done using either a premixed gas in a cylinder or the closed-loop calibration system. For personal sampling, Tedlar gas bags capable of collecting a minimum of 6 L and personal sampling pumps equipped with exhaust ports are needed. Care must be taken to assure sufficient air to purge the sample cell during analysis: for example, 6 L are required for a MIRAN 103 and 8 L for a MIRAN 1B2. In certain environments, a drying tube, Teflon sampling line, particulate filter, or zero gas filter may be necessary.

In general, where ambient air is drawn directly into the instrument all sampling should be done with the sample hose and particulate filter attached to prevent dust and particulate matter from entering the cell and accumulating on the mirrors and windows, thus interfering with the performance of the instrument by increasing background levels.

IRs should be used outdoors with great care because cooler temperatures and humidity can cause condensation on the mirrors and windows. If the instrument has been in a cold environment, condensation may take place in the cell if the sample is drawn into it before the instrument has warmed up properly. If the mirrors do become fogged, permanently high background readings can result. Some mirror materials are more susceptible to humidity. For example, sodium chloride windows cloud more easily than those made of silver bromide. The mirrors on some instruments can be periodically cleaned, but not all of them. For example, the mirrors on the MIRAN 1B are gold-plated and very delicate, and therefore repair is very expensive. In addition, if the seal on the housing is broken, the warranty may be voided for many instruments. The cell ports in IRs should be covered with protective plastic caps when the IR is not in use to prevent dust and moisture from contaminating the optics. An alternative is to use a charcoal cartridge as described below.

Zeroing IR analyzers is a critical step. If the sampler zeros in an atmosphere containing the compound of interest or an interference, and then analyzes air that contains less of the same compound, a negative reading will occur in the MIRAN 1B. This reading can occur when the zero gas filter has become saturated, or if the sampler zeros on humid air and subsequently analyzes air that is drier than the zero. A charcoal canister, sometimes termed a zeroing cartridge or zero gas filter, must be placed over the probe for proper zeroing. This cartridge is essentially the same as the organic vapor cartridges used in respirators. Note that charcoal has a limited ability to absorb water vapor. These cartridges should be stored in a sealed plastic bag to prevent exposure to contaminated atmospheres when not in use. Saturation of the cartridge depends on the nature and concentration of the contaminants that are pulled through it. If a new cartridge produces an appreciably lower analyzer reading, the old cartridge should be discarded. Replace the cartridge at least once a year. If an instrument is zeroed with a contaminated cartridge, a negative value may occur when making measurements, the result being low values for all measurements because the ambient background readings may be lower than the concentration in the cartridge used for zeroing. Whereas some sources recommend using nitrogen for zeroing, others recommend using room air to account for normal humidity conditions.[8] Carbon monoxide readings tend to read low on the MIRAN 1B if the instrument was *not* zeroed on room air because the instrument assumes the presence of constant levels of carbon dioxide in carbon monoxide samples.[10] If the moisture is not subtracted as background, results will appear higher than they are. In dark areas liquid crystal displays are hard to read from certain angles as well as from directly overhead.

Bag Sample Collection for Nitrous Oxide Using an IR[14]

1. Set up a long path-length portable IR analyzer to the following parameters: $\lambda = 4.48 - 4.68$ μm, path length $= 0.5 - 40$ M. The path length varies with pressure and instrument, but 12.75 M is used for the MIRAN 1B. Allow the instrument to warm up and equilibrate for 15 minutes.

2. Perform an on-site multipoint calibration at five or more concentrations over the range of 10 ppm to 1,000 ppm. *Note:* Because nitrous oxide supports combustion, the shipment of the compressed gas must comply with 49 CFR 171-177 regulations regarding shipment of hazardous materials. Calibration for nitrous oxide can frequently be done using material available on-site when surveys are done in hospitals, dental offices, and veterinary operations, because generally the nitrous oxide in use is of sufficient purity.

3. Zero the instrument while recirculating uncontaminated air through the sample cell. If the area where the instrument is being calibrated is serviced by the same ventilation system as the area to be monitored, it will be necessary to obtain a source of uncontaminated air (or nitrogen or oxygen) for zeroing the instrument.

4. Inject a known volume of nitrous oxide into the sample cell with a gas-tight syringe through tubing, or by using a septum attached to the sample cell. Calculate the concentration of nitrous oxide in the sample cell:

$$C_s = \frac{\text{volume of } N_2O \text{ injected } (\mu l)}{\text{volume of cell (L)}} = \text{ppm}$$

When the instrument reading stabilizes, record meter or recorder deflection.

5. Prepare a calibration graph of ppm (C_s) versus meter deflection.

6. Select one of the following sampling modes according to the desired form of the data: ambient air or integrated air samples for TWA determinations.

7. Move the instrument to the first area to be sampled. A probe with a sample line can be used to allow more flexibility. Thus, the instrument can be placed nearby and the probe can be used to move, for example, around the doors of a sterilizer, or held over the shoulder of an employee while opening the door of the sterilizer to get a peak measurement.

8. For ambient air, turn on the instrument pump and record the readings on the meter. If

data are to be expressed as a TWA concentration, a data recorder will have to be used. A strip chart recorder can be used instead of the data recorder, but the levels must be mathematically averaged.

9. Pump air to be analyzed through the sample cell to purge the cell. Typically, two to three cell volumes are necessary. When output stabilizes, record the meter reading as a measurement.

10. Use the calibration graph to determine the concentration by moving along the horizontal axis to the meter deflection, and moving to the point on the line that corresponds to this concentration. The ppms for this point on the line can then be read on the vertical axis.

11. During each day's operation, periodically recheck the calibration by repeating measurements with the calibration gas at three or more points on the graph.

12. For integrated air samples for TWA determinations a clean bag of plastic or other material must first be evacuated. It can be done using a personal air sampling pump at the highest airflow available.

13. Attach the plastic bag via tubing to a personal sampling pump and pump air from the contaminated area into the bag at a rate calculated to fill the bag over the sampling period. This rate will be between 20 ml/min and 500 ml/min. Terminate sampling before the bag is 80% full. The pump's flow rate must be within ±5% of the initial setting throughout the sampling period.

14. Analyze the bag sample within 2 hours after completion of the sampling to minimize loss of analyte by adsorption and permeation.

Interpreting Measurements with Infrared Analyzers

Readouts on IRs can be in three different forms: absorbance, percent transmission, and parts per million (ppm). Ppms require no further adjustment as long as the ambient temperature and humidity do not exceed the operating range of the instrument. The percent transmission (%T) readout is used with a calibration curve to determine the concentration of the sample. The absorbance (A) scale is related to the %T scale by $A = -\log T$ and is not a linear scale as is the %T scale. It too must be compared to a calibration curve in order to determine the actual concentration of the compound being sampled. The calibration curves must be prepared in advance using the same control settings (pathlength, wavelength, slit width) which will be used during measurements and the instrument calibrated immediately prior to sampling to assure the curve is still valid.

Given the difficulties in interpreting IR data qualitatively in that using a spectrum to identify a compound requires considerable IR experience as well as access to a library of spectra of various compounds, and the fact that certain compounds—such as carbon monoxide, carbon dioxide, and water vapor, among other common air contaminants—need to be subtracted, the most common use of the IR in the field is quantifying known specific gases and vapors rather than identification of compounds.

THE NEXT STEP

As sophisticated and multipurpose as the instruments described in this chapter may seem, there are developments that will allow even better sampling in situations where mixtures exist. These involve the use of portable mass spectrometers and Fourier transform infrared spectroscopy (FTIR) for much more flexibility in making measurements to detect contaminants.

The mass spectrometer produces charged ions consisting of the parent ion and ionic fragments of the original molecule. These ions are sorted by their mass/charge ratio. The mass spectrum is a record of the numbers of different kinds of ions and their molecular weights. This allows for qualitative identification of the composition of mixtures of organic compounds.

In FTIR, the sensitivity (lower limits of detection) and speed of analysis are much increased over that of conventional IR measurements. The Fourier transform is a mathematical process applied to the infrared data. The integration of microprocessors into the units allows the complicated Fourier transform calculations to be done in seconds and for interferences from compounds such as water and carbon dioxide to be automatically subtracted. In addition, concentrations can be computed by reference to lab spectrum of a known amount of a pollutant stored in a library in the memory of the instrument.[15] A computer program can be used to resolve overlapping peaks to identify contaminants within mixtures. It has been used to quantitate up to 16 compounds simultaneously.[16]

The sample system of the FTIR and the conventional IR are essentially the same, with both depending on the use of a gas cell with internal mirrors so that the cell volume can be kept low and the path length of the light can be long. However, the FTIR has only one moving part, which is the mirror in the interferometer. Most conventional IR systems have several moving parts, including the filter, slits, and beam chopper. The FTIR rejects interferences from stray light whereas the conventional IR is strongly affected by stray light. The FTIR has an increased beam area over conventional IR, and IR energy is allowed to reach the detector, thus allowing for the greater sensitivity of this method.

With either instrument the sample can be drawn into the sample cell either continuously or one sample at a time. If the sample is drawn continuously, the result is an average of the composition and concentration of the air contaminants during the sampling period. Finally, the FTIR monitors all of the wavelengths of light simultaneously whereas conventional IRs monitor only one wavelength at a time. However, the FTIR will not replace the need for the conventional IR, because there are many situations where the sensitivity of this instrument is all that is needed for a particular job. Currently the cost for FTIR is prohibitive for it to be used in situations other than research, but in the future it is likely that instruments will be available for common sampling use.[14]

REFERENCES

1. National Institute of Occupational Safety and Health. *The Industrial Environment, Its Evaluation & Control.* NIOSH, Cincinnati, OH, 1973.
2. *Photovac 10S50 Operating Manual.* Photovac International, Inc., Huntington, NY.
3. *OVA 108 Operating Manual.* The Foxboro Co., Foxboro, MA.
4. *Guide to Portable Instruments for Assessing Airborne Contaminants at Hazardous Waste Sites.* World Health Organization, Geneva, Switzerland, 1988.
5. Environmental Protection Agency. *Monitoring Methods for VOCs.* EPA, Washington, D.C., 1988.
6. National Institute of Occupational Safety and Health. *NIOSH Manual of Analytical Methods, Method 3702,* 3d ed. P. M. Eller, ed., NIOSH: Cincinnati, OH, 1984.
7. Willard, H. H., L. L. Merritt, and J. A. Dean. *Instrumental Methods of Analysis,* 4th ed. D. Van Nostrand, New York, 1965.
8. Levine, S. P., et al. Advantages and disadvantages in the use of Fourier transform infrared (FTIR) spectrometers for monitoring airborne gases and vapors of industrial hygiene concern. *AIHA J.* **4**(7):180–187, 1989.
9. Daisy, J. Real time portable organic analyzers. *Advances in Air Sampling.* American Conference of Governmental Industrial Hygienists, Cincinnati, OH, 1988.
10. American Society for Testing and Materials: Method D3162. ASTM, Philadelphia, PA.
11. Sammi, B. S. Calibration of MIRAN gas analyzers: Extent of vapor loss within a closed loop calibration system. *AIHA J.* **44**(1):40–45, 1983.
12. *MIRAN 1 B Operating Manual,* The Foxboro Co., Foxboro, MA.
13. Burroughs, G. E., and W. J. Woodfin. Report on a microprocessor-based infrared analyzer. *J. Appl. Ind. Hyg.* **3**(12):R-2–R-4, 1988.
14. National Institute of Occupational Safety and Health. *NIOSH Manual of Analytical Methods, Method 6600,* 3rd ed., P. M. Eller, ed., NIOSH, Cincinnati, OH, 1984.
15. Xiao, H., S. P. Levine, J. B. D'Arcy, G. Kinnes, and D. Almagner. Comparison of the Fourier transform infrared (FTIR) spectrometer and the miniature infrared analyzer (MIRAN) for the determination of trichloroethylene (TCE) in the presence of Freon-113 in workplace air. *AIHA J.* **51**(7):395–401, 1990.
16. Li-Shi, Y., and S. P. Levine. Evaluation of the applicability of Fourier transform infrared (FTIR) spectroscopy for quantitation of the components of airborne solvent vapors in air. *AIHA J.* **50**(7):360–365, 1989.

CHAPTER 11

Real Time Sampling Methods for Aerosols

Real time field instruments for measuring aerosols can determine total mass, respirable mass, total particle count, or particle size distribution, although there is no one instrument that can perform all of these measurements. These instruments can be used to measure short-term exposures such as can occur when a worker dumps bags, or to compare one operation with another and break down an operation into specific exposures associated with various sequential tasks.

In the direct reading mode, aerosol detection instruments are useful as screening devices to estimate total or respirable dust levels. Typical situations where use of these instruments is applicable are development of a sampling strategy during the walk-through of a suspected dusty workplace to determine areas of highest potential employee exposure; determination of abatement progress in dust control; determination of effectiveness of dust control measures; identification of dust-generating sources; and evaluation of work practices.

Many have outputs that can be connected to a strip chart recorder or data logger. Thus, the instrument may operate unattended while it monitors the work process. In some cases, a personal dust monitor with a data logger is worn by a worker. These systems can be used to provide a time-dependent profile of exposure over a work period, as well as an integrated TWA exposure. In addition, workers can be made aware of aerosol concentrations generated during their tasks, so that they can modify their work practices to reduce exposure. While instruments generally measure only total aerosol samples, a cyclone containing a filter can often be placed at the exhaust of an instrument to collect a respirable sample similar to the method used for integrated measurements.

Real time aerosol monitors are not a replacement for classic integrated filter sampling methods, but rather a complement. A common use is screening measurements to identify the areas or positions where levels are likely to be the highest to select the best sites for traditional sampling measurements for 8-hour TWAs. In certain situations, such as hazardous waste sites, direct reading instruments along with suitable data recorders are preferred to traditional filter samples. However, prior to this filter samples are used to field calibrate these aerosol monitors.

While the measurement of aerosols may seem straightforward, it is actually more complex and difficult than the measurement of gas and vapor molecules, because aerosols are subject to many factors, such as size, settling velocity, and shape, and the same factors relating to the dynamics of particle collection, already discussed in the chapter on Integrated Sampling for Aerosols also apply to making measurements of aerosols with real time devices.[1] As a consequence, there is a variety of instruments available for aerosol measurements, based on two different types of detection methods for aerosols: The Piezobalance analyzes particle deposits on a crystal substrate, and light-scattering instruments analyze the aerosol as it moves past the sensor.

Some instruments measure particle mass directly, whereas others count particles and convert this amount to mass based on reference values programmed into the instrument. Another difference is whether an instrument measures total or respirable dust. When set up to measure respirable dust, U.S.-made instruments follow the criteria of the American Conference of Governmental Industrial Hygienists (ACGIH) previously described in the chapter on Integrated Sampling for Aerosols. Some monitors (passive) do not use pumps to pull the aerosol into the sensing chamber, whereas others (active) depend on them. Studies have shown that when the same aerosol is measured using both active and passive methods, the passive instrument will show more deviations in the results.[2] One explanation may be homogenization of the aerosol during active sampling, and when an aerosol is collected as a result of natural air convections, concentration variations may show up more readily.

Limitations of these instruments include their nonspecificity for any particular aerosol meaning they cannot distinguish between types of dusts. This factor can be a problem in situations where highly toxic dusts, such as lead, cadmium, and cobalt, may be present. Nonspecificity also impacts concentration measurements when an instrument has been calibrated with one type of aerosol and is used to measure another. At least one instrument has been designed to specifically detect particles in the shape of fibers.

The aerosol monitoring instruments described within this chapter were selected because they are easily obtainable for use in the field. However, there are other more sophisticated and expensive units available, and the instrument of choice will depend on the specifics of the sampling situation.[3]

LIGHT-SCATTERING MONITORS

Light-scattering aerosol monitors are also called nephelometers or aerosol photometers. In light-scattering aerosol monitors of the nephelometric type, an instrument continuously senses the intensity of light from the combined scattering of the population of particles present within the sensing volume at a given angle relative to the incident beam as they pass through a sensor cell of defined volume. As the number of particles increases, the light reaching the sensor increases. The scattered light detected by the photodetector is transformed into a voltage proportional to the light intensity.

The light source can be monochromatic, such as light-emitting diode or laser, or a broad-wavelength light source, such as a tungsten filament lamp. The choice of light source has more to do with the ability to control the light output level than with the wavelength of the output. The detector is generally a solid-state photodiode but can be a photomultiplier tube. The scattered light received from a single particle depends on the size and shape of the particle, the refractive index of the particle, the wavelength of the light, and the angle of scatter.[4] Thus, the amount of scattered light received from the multiple particles in a cloud when the wavelength and angle of scatter are fixed will vary with the concentration, the particle size distribution, and changes in refractive index associated with the aerosol's composition. The relationship between the measured light scattered by a dust and its mass also depends on dust density.[4] The response to light-scattering measurements for larger particles is influenced more by their surface area than their mass.[5]

The components of photometers that have the greatest influence on their readings are the light source, scattering angle, and preselector. Of these, the scattering angle may be the most important. The angle of scattering is defined with respect to the beam of light passing through the aerosol in the detection volume. The smaller the value of this angle, the more the detection is weighted toward larger particles. For many instruments, the scattering angle is 90°, which provides the greatest sensitivity for small particles. Some instruments use near forward light scattering with scattering angles of 12° to 20°. This results in smaller variations based on the optical properties of the aerosol (refractive index and absorption coefficient) than with the other photometers. In general, systems using a 45° scattering angle may give the greatest accuracy. Forward scattering instruments give the best results for absorbing particles, but they suffer from a lack of sensitivity because of the difficulty of reducing the stray background radiation from the source and are considered less sensitive to changes in refractive index than other photometers, which use other sensing angles such as 30°, 45°, and 90°.[6,7,8]

Factors influencing the sensitivity, or response. of the light-scattering aerosol monitors are the geometry of the optical system, wavelength (distribution) of the light source, spectral sensitivity of the light detector, particle-size distribution, and physical properties of the particles to be measured. Figure 11-1 shows different types of light-scattering monitors.

The RAM-1 is an example of a portable survey dust monitor that utilizes right-angle light scattering. A pulsed light-emitting diode works in combination with a silicon detector to sense the light scattered over a forward angle of 45° to 95° from particles traversing the sensing volume. The RAM-1 can measure over three ranges of concentrations, 0–2 mg/M^3, 0–20 mg/M^3, 0–200 mg/M^3, allowing it to be used to detect both health hazard levels and explosive dust levels. It can be used with or without a cyclone at the inlet. With the cyclone, the RAM-1 measures respirable dust. The unit has four selectable measurement periods: 0.5 sec, 2 sec, 8 sec, and 32 sec. Each increase in the time for measurement halves the internal noise fluctuations. However, longer time measurements provide more stable readings at the expense of the ability to follow rapid fluctuations of the aerosol concentration.

The sampling flow rate is normally 2 Lpm; however, it can be adjusted from 1 Lpm to 3 Lpm. The particle-size range of the instrument is 0.1 μm to 20 μm in diameter. The RAM-1 has been used to do boundary line monitoring on hazardous waste sites where dusty activities are in progress, such as moving contaminated soil to aerate or screen it, as well as for other typical applications discussed at the beginning of this chapter. For continuous, unattended monitoring, a data logger is needed.

By closing the inlet valve so that clean, filtered air is passed through the sensing system of an instrument, a zero check can be accomplished. Zeroing should be done in clean air, which includes the absence of cigarette smoke.

A secondary clean airstream of 0.2 Lpm provides continuous flushing of filtered air over all of the critical optical surfaces of the RAM-1. Both of the cartridge filters are replaceable. A drying column is provided to protect the RAM-1 optics from water vapor condensation in case the instrument is moved from a cold to a warm environment. The desiccant material in this drying column must be replaced periodically.

Water droplets do not appear to significantly affect RAM-1 results, as the RAM-1 inlet has been tested and found to be effective in removing them, due to their size, from the airstream before they enter the instrument. Electromagnetic fields have also been determined not to affect the instrument.[9]

The mini-RAM is a miniaturized version of the RAM-1 with no active sampling pump. It is designed to be a portable personal monitoring device for respirable dusts that can be mounted on an employee. During passive sampling it depends on air currents, convection, and work movement to push the aerosol through the detection region. With a flow-through adapter, the unit can be converted to an "active" sampler by hooking it up to a personal sampling pump that will pull 2 Lpm through the sensing chamber and placing a cyclone at the inlet. The use of the personal sampler adapter permits concurrent mini-RAM readings and filter collection of the same aerosol to calibrate the unit for a specific aerosol, or to determine both concentration and chemical composition of the aerosol. The mini-RAM is worn on the belt of the worker so it samples from the waist area. During normal operation, the display indicates the aerosol concentration in mg/M^3 and the displayed reading is updated every 10 seconds.

A microprocessor is incorporated into the unit to process the signal from the light-scattering detection circuit, control the measurement sequence program, compute concentration averages, keep records of elapsed time, perform automatic zero correction, control auto-ranging, drive the liquid-crystal display, store average concentration values, obtain timing and identification information, sense battery and overload conditions, sequence playback of stored information, and provide alarm signals. The instrument is capable of giving both the 8-hour TWA and the projected TWA. It can save data in its internal memory for up to seven shifts or seven different measurements.

Stored information playback can be accomplished either by using the display on the instrument or through the digital output jack. A data logger can be hooked up and then fastened to the mini-RAM for use during personal monitoring. This significantly increases the weight of the sampling unit and requires some adjustments, such as the use of duct tape in order to hook the units together.

Background drift can occur in the mini-RAM if decay of the internal coated surfaces in the optical sensor occurs or if there is particle deposition on the optical surfaces, since this will cause a change in absorbance.[3] At high flow rates, turbulence in the chamber may reentrain deposited particles, thus adding particles to those that are being transmitted through the chamber. It has been noted that use of a cyclone in the inlet of the mini-RAM might prevent many larger particles from entering the chamber, thus reducing this effect. Momentary overloads can be caused by the insertion of an object into the sensing chamber or sudden exposure to sunlight.

The HAM is a light-scattering monitor in the shape of a hairdryer that collects near-forward scattering (10° to 15°) light from light-emitting diodes (830 ηm incident beam). The HAM samples passively and detects total particles from 0.5 μm to 12 μm in size over three measurement ranges: 0–2 mg/M^3, 0–20 mg/M^3, and 0–200 mg/M^3. Measurements can reflect either 1-sec or 5-sec averages. A digital readout is on the back of the unit where the user can easily see it while making measurements. For most processes the HAM's signal is proportional to the mass concentration if the mass mean diameter (MMD) of the particles lies within the range of 0.3 μm to 2 μm. While for those particles larger than 2 μm, the instrument responds to the projected surface area of the particle. The unit is accurate to within ±25%. If the particle composition and size distribution are constant, and the instrument is calibrated for the specific process, greater accuracy is possible.

Calibration adjustments are made based on the density of the dust. If substantial differences in particle MMD exist, such as can occur with process variations that can change the MMD drastically to lower values, then the mass concentration may be overpredicted by the HAM.[3]

Light-scattering methods have distinct advantages: Measurements can be made without interfering with the particles and can easily be converted into electrical signals to be transferred to data loggers. Since most light-scattering monitors provide a linear response over a broad concentration range, they are especially useful in monitoring applications where response time is important and exact agreement with gravimetric mass is not important.[10] Another advantage of direct light-scattering aerosol sensing is that the rate at which air passes through the sensor does not influence the concentration readout, but does influence the response time. Because they actually measure particle count and not mass, aerosol photometers should be used to determine mass concentration only under conditions in which the optical properties, density, and size distribution of the particles are constant.[11]

As noted, light-scattering aerosol monitors are sensitive to size-distribution changes, refractive index changes, humidity, and composition of the aerosol being measured. Most instruments will have some detection losses, as some dust particles may fail to reach the detector. However, these losses are generally small, unless there are strong air currents blowing across the sampling path or the particles are highly charged. Some aerosols can be difficult to measure with light-scattering devices, such as welding fume particles that have

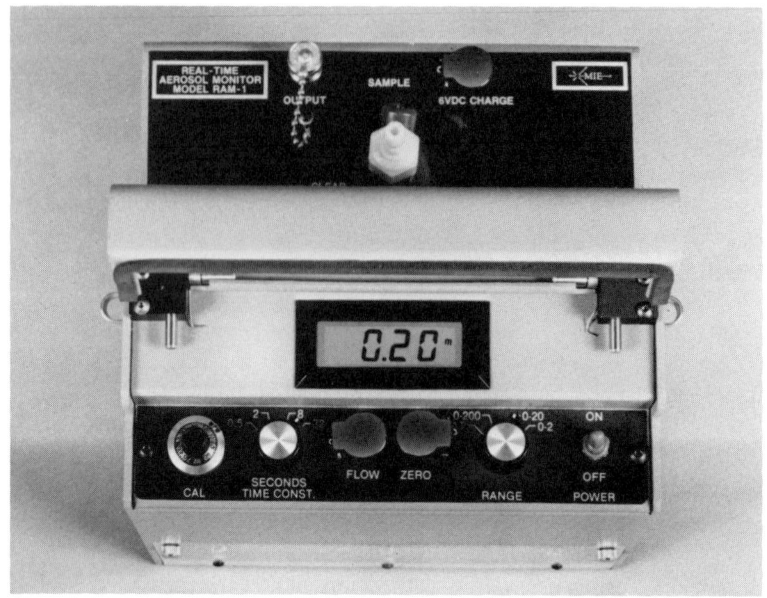

A

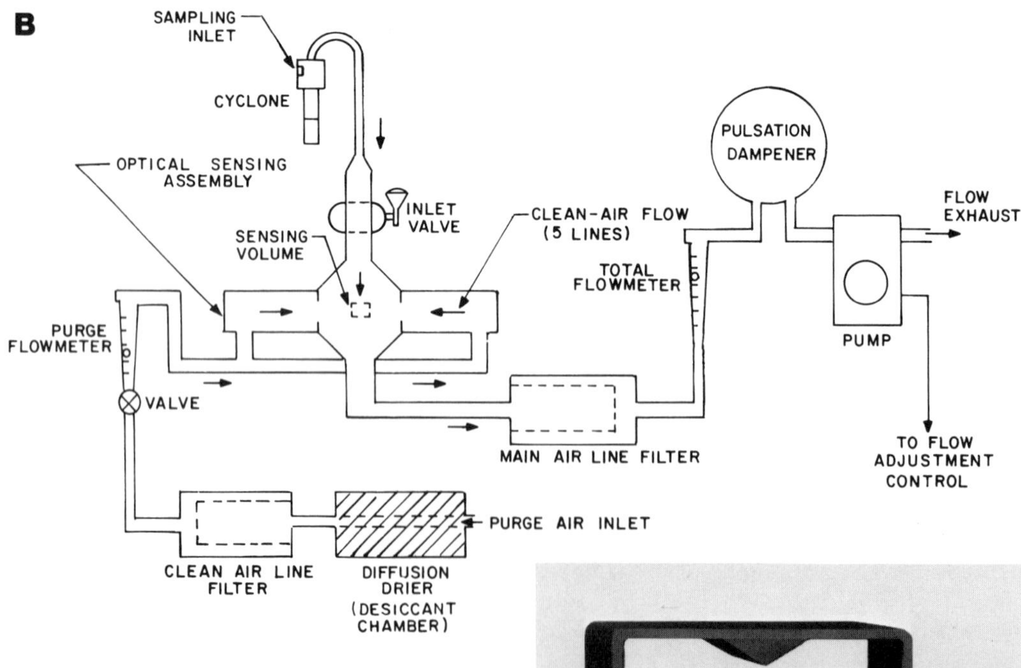

C

Figure 11-1. Portable survey real-time aerosol monitors. *A:* RAM-1 dust monitor; *B:* schematic of the RAM-1 flow system; *C and D:* personal monitoring with the Mini-RAM; *E:* typical Mini-RAM printout format. (Courtesy MIE, Inc.)

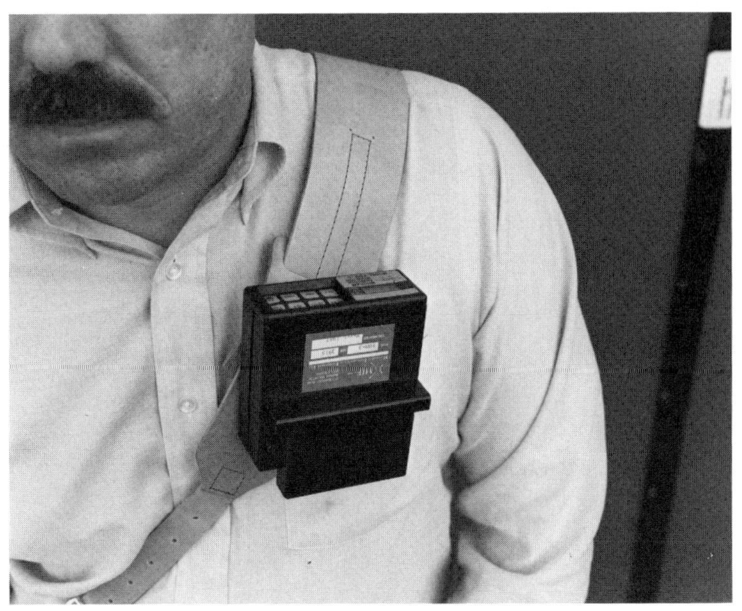

D

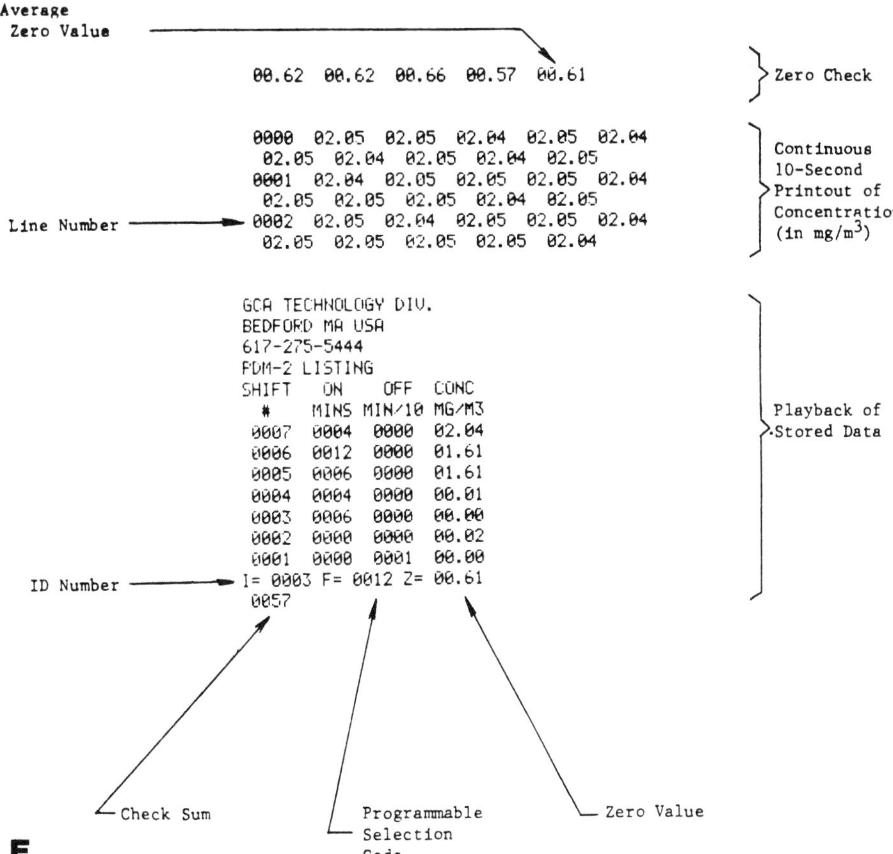

Average
Zero Value
00.62 00.62 00.66 00.57 00.61
Zero Check
0000 02.05 02.05 02.04 02.05 02.04
02.05 02.04 02.05 02.04 02.05
0001 02.04 02.05 02.05 02.05 02.04
02.05 02.05 02.05 02.04 02.05
0002 02.05 02.04 02.05 02.05 02.04
02.05 02.05 02.05 02.05 02.04
Line Number
Continuous
10-Second
Printout of
Concentration
(in mg/m3)
GCA TECHNOLOGY DIV.
BEDFORD MA USA
617-275-5444
PDM-2 LISTING
SHIFT ON OFF CONC
MINS MIN/10 MG/M3
0007 0004 0000 02.04
0006 0012 0000 01.61
0005 0006 0000 01.61
0004 0004 0000 00.01
0003 0006 0000 00.00
0002 0000 0000 00.02
0001 0000 0001 00.00
I= 0003 F= 0012 Z= 00.61
0057
Playback of
Stored Data
ID Number
Check Sum
Programmable
Selection
Code
Zero Value
E

a tendency to stick together, forming agglomerates over time.[5]

Calibration of Light-Scattering Monitors

Light-scattering monitors are calibrated both in the factory and in the field. The purpose of the factory calibration is to ensure that an instrument is operating properly and responding the same as other similar instruments while the field calibration is to improve the specificity of the instrument for dust generated by a given process. Most light-scattering monitors are factory calibrated by comparing the instrument response in a well-defined aerosol to measurements by the gravimetric method. During factory calibration, the rate at which particles are detected, expressed in counts per minute (cpm), is related to one or more concentrations of a particular calibration aerosol. For a given aerosol, a certain value of cpm's will be equal to a given mg/M^3. In most cases, the instrument's response is modified to read directly in mg/M^3. The calibration aerosol is the manufacturer's choice. As examples, the factory calibration of the RAM-1 is done with Arizona road dust with the calibration control set at 5.0. The HAM is factory calibrated with either oil mist or a special cement dust.

Field calibration for a given process is carried out by side-by-side sampling to compare the TWA instrument readings directly with field measurements using a cyclone for respirable measurements or filter cassette for total dust measurements (NIOSH Methods 0600 and 0500) to establish an instrument—gravimetric ratio. The object of side-by-side sampling with direct reading dust monitors and gravimetric methods is to generate calibration factors that adjust for differences in the type of dust being measured compared to that being used to calibrate the instrument. To do this, the average mass concentration determined on filters is compared to the average mass concentration determined by the aerosol photometer. When the methods are in agreement, the ratio of these measurements is 1:1. However, when measurements differ in the field, this ratio will change.

In some cases, rather than using a ratio, the gravimetric measurement is compared to an instrument's voltage output. An example of a value obtained in a HAM study on a hand-held sander was 5.4 mg/M^3/V.[12] It is important to note that since any calibration factor determined this way will be an average calibration factor, because no industrial source produces dusts with constant particle-size distributions or optical properties, actual results may differ somewhat.

Adjusted levels are then obtained by multiplying the photometer readings by the appropriate calibration factors for each type of dust. The most likely causes of differences between gravimetric measurements and light-scattering instrument results are cyclone orientation and differences in dust type. Of these, cyclone orientation has been found to be the most significant. Orientation also impacts the velocity of particles as they enter the instrument's inlet.

Some instruments are equipped with an internal light-scattering board with which the instrument can be checked and readjusted to the original calibration. A field check to see if the instrument is still within the manufacturer's calibration can be done using a reference scatterer that can be inserted into the sensing volume in order to check the calibration and overall operation of the instrument. For the HAM this is done by inserting the reference element and adjusting a potentiometer to achieve a known and traceable reading on the digital display. The mg/M^3 value of this reference element is specific to each instrument. This check does not compensate for differences in the calibration aerosol and that being measured, so the field calibration against NIOSH methods is still needed.

Therefore, scattered-light instruments provide a reliable indication of aerosol mass concentration if the instruments are calibrated with the same aerosol to be measured and the size distribution of this aerosol stays constant. Since this is rarely the case in the field, an additional field calibration is needed to compensate for the fact that once an instrument is calibrated for a given dust, its response to other dusts may be different. For the highest accuracy quantitative measurements, it is necessary to calibrate instruments with an aerosol similar in refractive index and particle size to the one being measured, because aerosols from a different source, of different composition, or different size distribution can produce photometer responses differing by more than a factor of ten.[3] The size distribution of dust particles is also important in evaluating the response of an instrument for a specific dust. As an example, the response of a RAM-1 to a volcanic ash aerosol is shown in Figure 11-2. When compared to the ACGIH respirable dust curve the curve of this aerosol has a similar shape but is composed of larger size particles.[3]

Curve A in Figure 11-3 demonstrates the difference in response of a RAM-1 to equal concentrations of different sizes of particles.[13] Each point was calibrated with a monodisperse aerosol. From this it is obvious that changes in particle size can produce large changes in response. Curve B shows the response of the instrument using various aerosols that have a broad size distribution and plotted according to MMD. The result is a broader curve. According to the manufacturer, calibrating for MMD rather than a specific diameter decreases the huge variations in response as shown in curve A.

One use for calibration graphs such as these, if enough is known about the dust being monitored, is a better estimation of actual concentration. For example, using the RAM with its basic calibra-

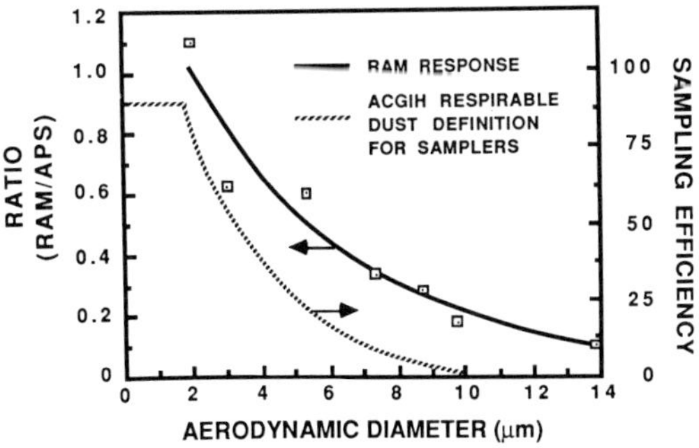

Figure 11-2. Size-dependent response of the RAM-1 for a volcanic ash aerosol compared with the ACGIH respirable dust curve. (From Baron, P. A., Modern real-time aerosol samplers, *Appl. Ind. Hyg.* 3(3):97–103, 1988.)

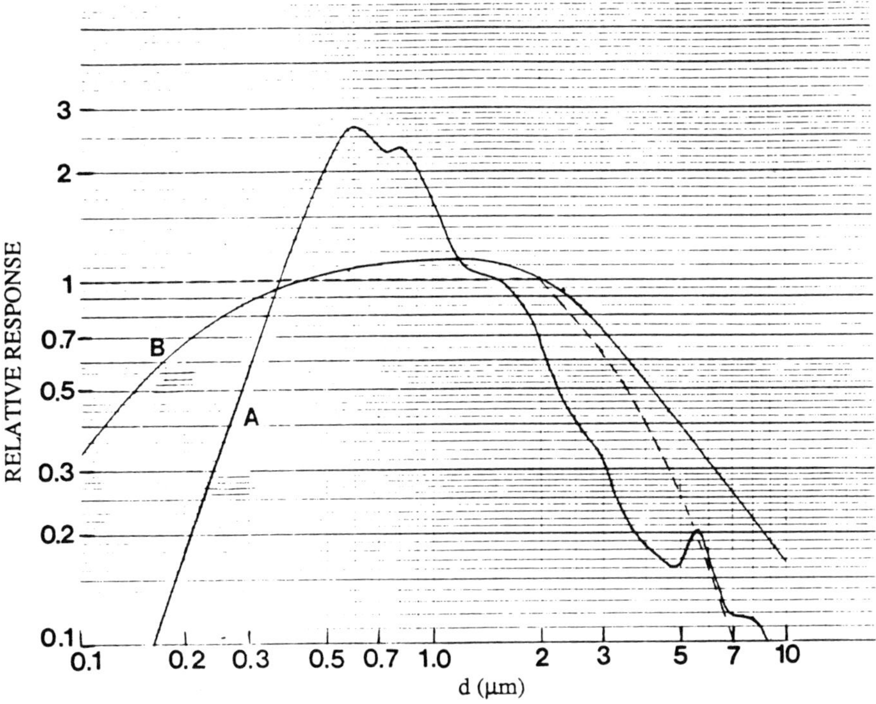

Figure 11-3. Calibration curves for a light scattering instrument using dusts of specific diameters (curve A) and using the MMD of dusts with a broad size distribution (curve B). (From MIE Technical Note No. 2, Bedford, MA, March 1990.)

tion to an aerosol of Arizona road dust with an MMD of 2 μm to monitor a dust with an MMD of 1 μm would result in readings 13% higher as shown in curve B.

Should a light-scattering analyzer become contaminated by overloading or sampling too high a concentration of oil or other particulates, the critical optical surfaces should be cleaned. Also, if sensor calibration and zeroing procedures result in settings far from the factory-set values, contamination should be suspected. Erratic behavior in mass concentration readings can also be caused if fibrous material (or spider web) is deposited in the particle sensing chamber and interacts with the light beam. Use a small flashlight and check for contamination of the chamber by shining it through the sampling ports. For some instruments cleaning can be done by the user. When cleaning optical surfaces, do not rub or attempt to polish. Dip a cotton tip applicator in freon or alcohol and clean the surfaces with a single very lightly applied motion.

Use of Light-Scattering Instruments

Light-scattering devices cannot be used to discriminate between different types of aerosols, since these instruments will respond to any and all aerosols present in the sensor cell. For example, if lead pigment is to be measured but there are dusts present, the photometer would respond to all of the dust as if it were lead. However, if the sampler is relatively sure that the contaminant of interest is the major component of a dust, the instrument will give results that are representative of this compound.

Care should be taken when measuring aerosols in high-humidity situations. An example is in cotton mills and carpet manufacturing, where water mists are deliberately introduced to decrease airborne dusts, mostly for the purpose of minimizing the potential for dust explosions. Water droplets can exist in the air for extended periods of time and will be detected by a photometer.

FIBROUS AEROSOL MONITORS

The MIE fibrous aerosol monitor (FAM) is a direct reading instrument designed to measure airborne concentrations of fibrous materials, such as asbestos and fiberglass, with a length-to-diameter (aspect) ratio greater than three. Ambient air is drawn into the FAM by a diaphragm pump into an air intake that is a short section of vertical tubing that is bent in a shape designed to exclude large particles and to prevent access to an internal laser beam (Fig. 11-4).

Concurrent with its transport through the sensing chamber, each fiber is illuminated by a continuous-wave He-Ne laser beam aligned parallel to the direction of airflow. Electro-optical sensors detect fiber oscillations as the aerosol passes through a rapidly oscillating high-intensity electrical field that induces the fibers to oscillate rapidly as well. An electric field quadrupole provides alternate excitation, causing the fiber to scatter laser light.

The scattering pulses resulting from the induced fiber oscillation are detected by a photomultiplier tube located alongside the sensing chamber. The fiber count is digitally displayed during the sampling period, and the concentration reads directly in fibers per cubic centimeters (f/cc) at the end of the cycle. Sample times of 1, 10, 100, or 1,000 minutes can be selected.

The width and length parameters for fiber detection are determined when the instrument is initially set up. There are two comparator outlets: one for the length:width ratio signal and the other for the output signal from the peak of the pulse. The ratio signal is entered into a comparator whose reference voltage can be adjusted on the instrument panel to select different fiber-length thresholds. The output signal from the peak of the pulse is fed into another comparator in order that only signals whose amplitude exceeds a selected minimum are accepted, thus protecting against low-signal (noise) ambiguities. The two comparator outputs are connected to a discrimination mode selector switch that has two positions: amplitude and ratio plus amplitude. When the switch is placed at the ratio plus amplitude position, only fibers producing pulses meeting the predetermined criteria of ratio and amplitude are counted.[14] The fiber length can be set at the 5-μm criterion conventional for characterizing asbestos fibers. The criteria for fiber detection are particles with a length-to-width (aspect)

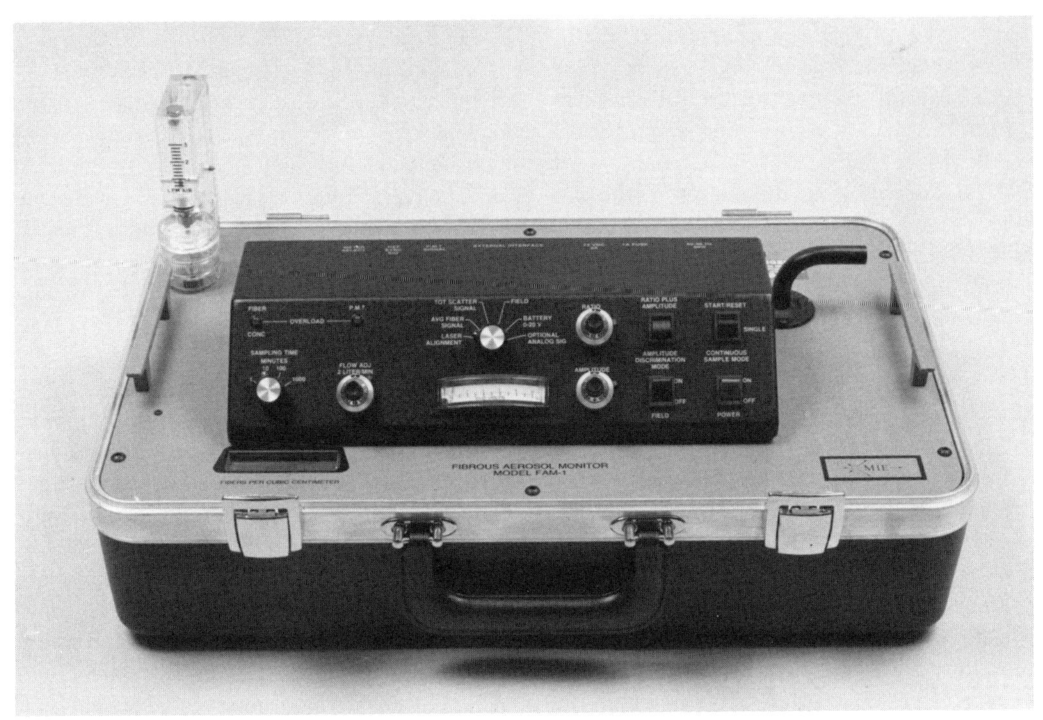

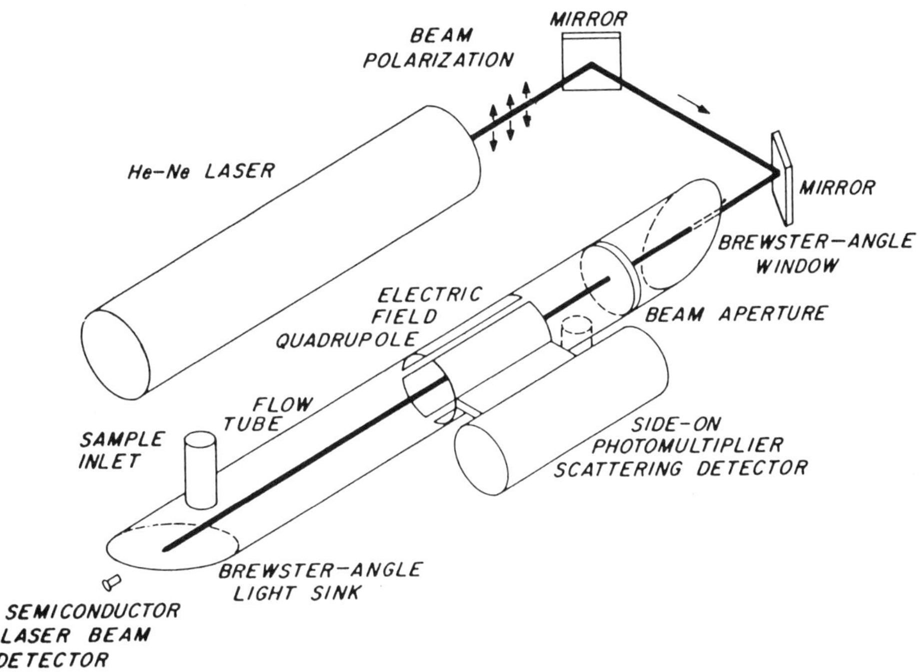

Figure 11-4. Fibrous aerosol monitor (photo) and a diagram showing optical configuration. (Photo and diagram courtesy MIE, Inc.)

ratio of at least 3:1. Fiber curliness and curvature effects have been studied, and are not considered significant due to the powerful straightening force exerted by the aligning electric field.[15]

Like other light-scattering instruments, the FAM is susceptible to differences in the field aerosol compared to the calibration aerosol.

The FAM has been evaluated in several field studies. In a test against chrysotile asbestos in an asbestos mine, it was set to count fibers equal to, or longer than, both 4 μm and 5 μm for sampling durations of 90 minutes.[16] Standard 37-mm MCE filter samples were collected alongside the unit. In this study, the FAM undersampled when the minimum detectable fiber length was set to 5 μm. In another study, with anthophyllite asbestos, side-by-side sampling was done in a test chamber. In this test, the FAM performed within ±25% of the reference membrane filter fiber count method. In a more recent study, the FAM was tested against a mixture of asbestos and nonasbestos materials, more representative of a typical fireproofing found in a building. These FAM measurements were comparable to the NIOSH counting method when the unit was set to detect fibers greater than 4 μm. However, when only fibers greater than 5 μm were counted, the FAM overstated the airborne concentration. Overall, the FAM gave higher measurements as compared to the standard phase count microscopy (PCM) method. One hypothesis for this situation was the presence of curly chrysotile fibers and the physical differences between the FAM, and the "B" counting rules specified by NIOSH Method 7400.[14]

The limit of the FAM's detection is fibers ≥0.2 μm in diameter according to the manufacturer. The FAM cannot replace microscopy as an analytical method for detection of fibers, but it does provide instant on-site information as to whether fiber levels in the air are rising, falling, or remaining constant. The limit of detection by PCM for asbestos is fibers 0.3 μm in diameter. Figure 11-5 shows a field data comparison of the FAM with filter asbestos fiber measurements.

FAMs cannot differentiate between fibers of different materials, such as cotton, paper, or asbestos. It is also possible that particles hooked in a chainlike formation might be counted as a fiber. High levels (5 – 10 mg/M^3) of nonfibrous irregularly shaped dust particles that are >3 μm may cause excess counts and a corresponding increase in the indicated fiber concentration. A total dust concentration (including fibers) greater than 10 mg/M^3, a fiber concentration greater than 25 fib/cc, or extremely large particle sizes can overload the instrument. Fiber clustering and clumping and fiber ends splitting into fibrils may not be detected.

A virtual impactor is available from the manufacturer to use as a precollector to minimize interference from high levels of nonfibrous dust particles. The impactor will also collect large-diameter fibers, such as fiberglass, carbon, and organic fibers that are usually nonasbestos. Because most asbestos fibers have diameters of less than 1 μm, the impactor will retain typically only 20% to 40% of these fibers, but at the same time it will retain most particles larger than 3 μm.

A correction factor of approximately 1.3 should be applied to FAM asbestos fiber counts when using this virtual impactor. Fiberglass filters rather than membrane filters must be used with the virtual impactor due to the high pressure drop of membrane filters. Under typical use, this filter should be replaced every 6 months or when the secondary flow on the FAM, normally 0.2 Lpm, changes (decreases) significantly.

Calibration and Maintenance of FAMs

The FAM is factory calibrated with amosite asbestos. The manufacturer recommends that this calibration be repeated by the factory once a year, or after every 500 hours of operation in the field. For a field check, clothes can be brushed and a count taken.

An adjustment can be made to FAM measurements to compensate for the instrument's inability to measure fibers less than 0.5 μm in width. A number of air samples that are representative of the material to be sampled are collected on a filter cassette and analyzed by transmission electron microscopy (TEM) and a

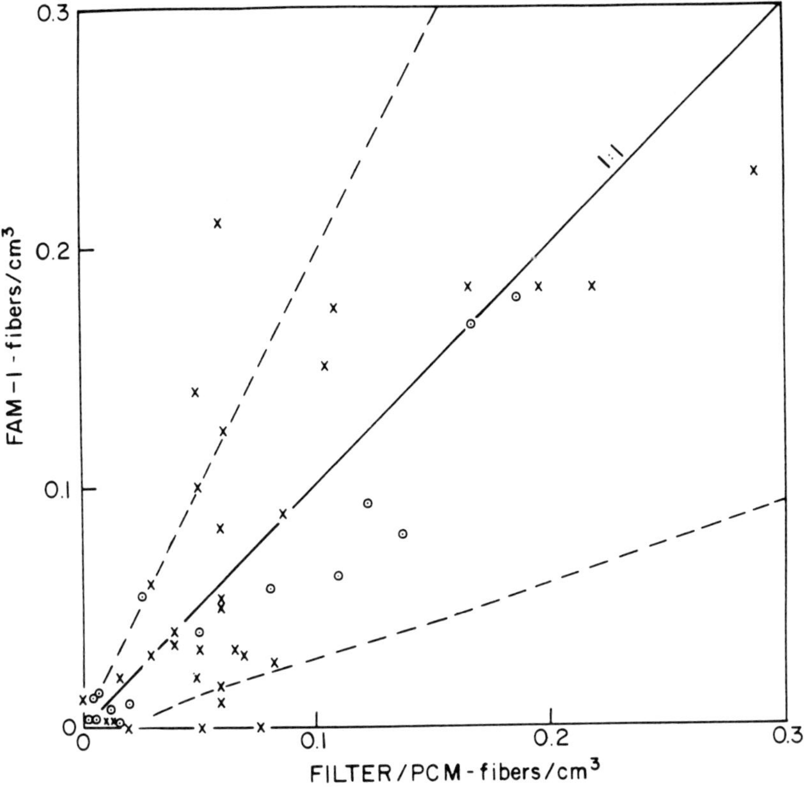

Figure 11-5. Field data comparing the FAM with filter (PCM) asbestos fiber measurements: ⊙, amsite; x, chrysotile. The error band (dashed lines) is based on NIOSH Method 7400. (Courtesy MIE, Inc.)

ratio of thin-to-thick fibers is developed that depends on the material being sampled and the degree of processing it has undergone. Once this ratio is established, the FAM readings are multiplied by this factor.

Although field calibrations are not possible, it is possible to do voltage, laser alignment, signal, and flow rate checks. These should be done prior to and after each use. The actual alignment of the laser should be performed only by trained individuals. The 2-milliwatt He-Ne laser can damage the eye if the sampler looks directly at it. The ratio, amplitude, and gain are factory set and should not be readjusted in the field.

The optical system is sensitive to shock and vibration and should be protected from jarring during transportation. Tight packing is necessary because excessive movement will cause misalignment of the laser.

Use of the FAM-1

The FAM is best utilized as a screening instrument for the presence of airborne fibers. It can also be used to evaluate the effectiveness of engineering controls during monitoring and follow-up inspection. During asbestos abatement projects the FAM will provide constant monitoring outside the containment area.[15] It can be used for checking for releases in the event of tears in the containment plastic walls and to supplement quarterly surveys for general building air for operations and maintenance (O&M) programs. It is not designed to replace other monitoring that may be required for legal or compliance purposes. The unit is best suited for area screening samples at a fixed location where a 110-VAC power supply is available.

A standard 25-mm cassette typical of that used for asbestos monitoring must be placed between the exhaust of the unit and the flowmeter. This cassette can be sent to a laboratory for qualitative analysis to confirm the presence, but not the amount, of asbestos, because the results are not comparable with the measurements of the instrument. Extension tubing should not be attached to the inlet of the FAM because fibers will stick to the walls of the tubing, resulting in gross sampling error. Low-humidity conditions may cause fiberglass fibers not to oscillate, and may result in erroneous readings.

In the field, when using the FAM-1 for asbestos monitoring, at least three samples for PCM or TEM analyses should be taken concurrently: one at the start of work, another midway during the work period, and another just prior to clearance. The sampling period should be the same for both the FAM and the filter samples, and sample inlets should be within 4 inches of each other to assure comparability. Areas in close proximity to air movers (fans), ventilation duct exhausts, and other sources of strong localized air motion should be avoided during these tests.

In general, to increase the reliability of FAM results, the number of samples and the sampling time should be increased. For example, for a statistical precision of ±25% at the 95% confidence level that a reading exceeds 0.1 fib/cc, a minimum of six 10-minute runs can be performed.[18]

Another application for abatement monitoring is identification of "hot spots" for inside containment. For example, boiler rooms are particularly prone to areas of high concentration, because they are so congested with tanks, pipes, and other equipment, making a uniform negative air interior difficult. When a "hot spot" of high-fiber counts is found within the work zone, it normally indicates that the negative air system's flow is allowing eddies of air in certain spaces in a way that allows these fiber levels to accumulate.[15]

Another use of the FAM is to monitor an abatement contractor's performance when a third-party representative, such as an asbestos project manager, is unavailable, for example, on graveyard shifts. A similar use would be to monitor employee work practices. The FAM has also been used in monitoring fiber levels during cleanups of accidental releases of asbestos material. For example, in industrial operations boiler upsets and emergency shutdowns that release loose asbestos insulation from steamlines and boiler walls can occur. Often these releases are very short-term, and conventional filter sampling would "flatten" out the peak levels that occur due to the time needed for collection. Another application is to test new technologies to see if fiber levels are increased or decreased.

In asbestos abatement monitoring, usually two FAMs are used: one outside the containment area and one inside.[15] Often the outside unit is hooked to a strip chart recorder or data logger to allow continuous monitoring of the concentrations. If the inside unit is hooked to a strip chart recorder and run off of AC current, the wires must be hooked up overhead in the work zone to prevent them from being coated with debris or disconnected during work activities.

One recommendation that has been established for FAM monitoring is to have the inside FAM make a minimum of at least two 10-minute measurements per hour, and the outside FAM perform at least four 10-minute tests per hour.[15] In situations where a building is to remain partially occupied while removal is done, the outside FAM should be run continuously and the information recorded. The inside unit may have to be shut off periodically and moved to avoid damage from water spray, falling clumps of material, and swinging tools. The computer circuitry is durable under routine conditions, but is very susceptible to damage by water and impact.

If possible, when used outside the containment, the FAM should be set up in close proximity to the personnel whose environment is to be monitored. The instrument should also be kept at least 3 feet above the floor if possible. In some cases, a third unit is also used as a "rover." For example, in a very large building random air checks can be taken on various floors during a removal operation. It is not recommended that the FAM be used for clearance monitoring on abatement projects, because of the legal implications of these samples. Asbestos monitoring is described in more detail in the chapter on Specific Sampling Situations.

PARTICLE MASS MEASUREMENTS WITH THE PIEZOBALANCE

The Piezobalance utilizes a different method for particle measurements than those previously discussed. The Piezobalance measures respirable mass by depositing the particles that penetrate the precollection impactor on a piezoelectric sensor by electrostatic precipitation. The impactor separates out nonrespirable ($>10 \mu m$) particles. As the aerosol stream is drawn into the impactor chamber, its velocity is accelerated by passage through a small-diameter nozzle directed at the impactor surface. Large particles, due to their greater inertia, maintain their trajectories and strike the impaction surface, where they are captured by stop-cock grease. Smaller particles, with less inertia, follow the flow and reach the sensor. Ideally, the impactor collects all particles larger than its aerodynamic 50% cutoff diameter. For example, the impactor in the TSI 3500 respirable mass monitor has an operable cutoff diameter of 3.5 μm. The combination of flowing air and the electric field generated by the precipitator needle carries the particles to the electrode surface of the piezoelectric crystal. The high-adhesive force between the particles and the crystal's surface causes the particles to stick firmly to the surface.

Initially, with the sensing crystal clean, the output of the instrument is related to the difference between the reference crystal frequency and the sensing crystal frequency. The reference crystal frequency is a fixed value for each instrument, equal to f_1. As particles precipitated onto the sensing crystal, the crystal frequency changes in direct proportion to the mass of the collected particles. The rate of change of the resonant frequency of the quartz crystal sensor is directly proportional to the mass of material deposited on it. The frequency difference, the final frequency f_2 less initial frequency f_1, is directly proportional to the total particle mass. Therefore, the quartz transducer on which the electrostatic precipitator collects the respirable dust electronically weighs the deposit and calculates particle concentration, C, from the basic formula

$$C = 0.333 \frac{f_1 - f_2}{t_2 - t_1} \text{ in mg/M}^3$$

where t_1 = initial time and t_2 = final time. *Note:* With few exceptions, t_2 is either 24 sec or 120 sec. Figure 11-6 shows the Piezobalance and how its sensing system operates.

The Piezobalance in one series of tests was more sensitive to quartz than to concrete dust. Accuracy is considered to be $\pm 10\%$ of a reading or ± 0.01 mg/M^3. Precision varies with the type of aerosol. Side-by-side Piezobalance and integrated filter measurements have been compared for a variety of aerosols, including tobacco smoke, and results agreed within 10%, although the Piezobalance underestimated tobacco smoke by as much as 15% relative to the filter measurements.[19] Changes are linear up to a maximum change in frequency of 1,000 Hz. The sensitivity of the crystal was found to decline with particle size.[5]

Compared to light-scattering devices, the Piezobalance contains a larger number of sensitive components necessary for operation. For example, the quartz crystal is very delicate. The corona needle is also a sensitive feature of the instrument and has been known to experience a current drop following only 5 to 10 measurements.[5] The impactor in the inlet must be greased.

Limitations to be aware of when using the Piezobalance include possible nonuniform sensitivity of the crystal's surface and temperature and

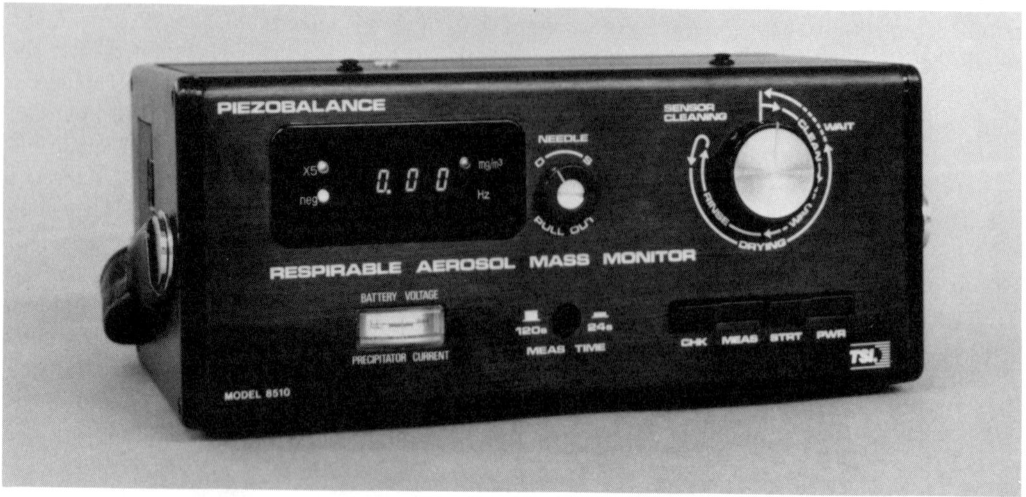

Figure 11-6. The Piezobalance monitor and its sensing system: the respirable aerosol monitor (photo) and diagrams *(at right)* of the system operation. (Courtesy TSI, Inc.)

humidity effects.[20] For example, both humidity and high levels of wood dust have been known to lead to many instrument stoppages.[5] These instruments may have a decreased sensitivity to particles greater than 10 μm in size, due to the ease with which the crystal's surface may be overloaded in this situation.[6]

Calibration and Maintenance

Calibration of Piezobalances to increase accuracy in the field is similar to that of light-scattering instruments using side-by-side instrument:integrated filter measurements. During the factory calibration, using arc welding fume, every Piezobalance is adjusted to a mass sensitivity of 180 Hz/μg, which is equivalent to 0.00278 mg/M^3/(Hz/2 min). Once a new ratio is established from a field calibration, it can be used to multiply this mass sensitivity factor to establish a new one for the dust of interest.

Calibration measurements are generally performed over 30-minute periods running the instrument alongside a Teflon membrane filter with a pore size <1 μm in a cyclone, at the standard flow rate of 1.7 Lpm for cyclones, and six periodic measurements with intervening sensor cleanings must be made with the Piezobalance at its flow rate of 1 Lpm. After several trials with the same aerosol, the Piezobalance mass sensitivity is adjusted to compare with the cyclone results. This requires practice and agility on the part of the sampler if all six measurements (including sensor cleaning) are to occur within the 30-minute measurement period. The adjustment is made by removing the aerosol inlet panel and changing a number of white dip switches, each of which represents a different value when in the on position.

The sensor should be cleaned as soon as possible following use of the instrument. Cleaning is particularly important if the unit will be stored for a period of time (several weeks). As mentioned previously, in dusty environments the sensor may require cleaning between runs. The sensor cleaning sponges should be cleaned at least once a day or whenever 20 sensor cleanings have been made. In extremely dry areas, the sponges may have to be re-wetted between cleanings. If the sponges are dried out, wet them prior to removal. Table 11-1 shows approximate number of measurements that can be performed between sensor cleanings.

The Piezoelectric Microbalance Technique

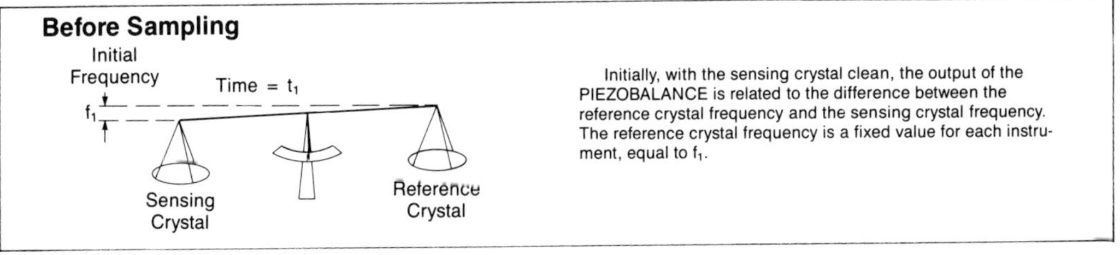

Before Sampling
Initial Frequency
Time = t_1
f_1
Sensing Crystal
Reference Crystal
Initially, with the sensing crystal clean, the output of the PIEZOBALANCE is related to the difference between the reference crystal frequency and the sensing crystal frequency. The reference crystal frequency is a fixed value for each instrument, equal to f_1.

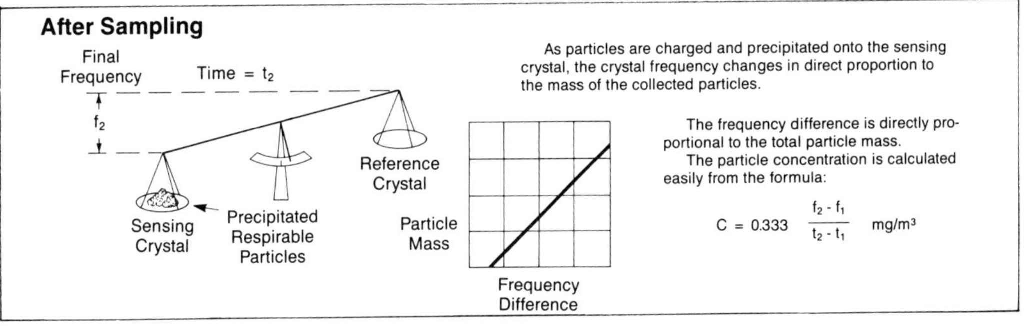

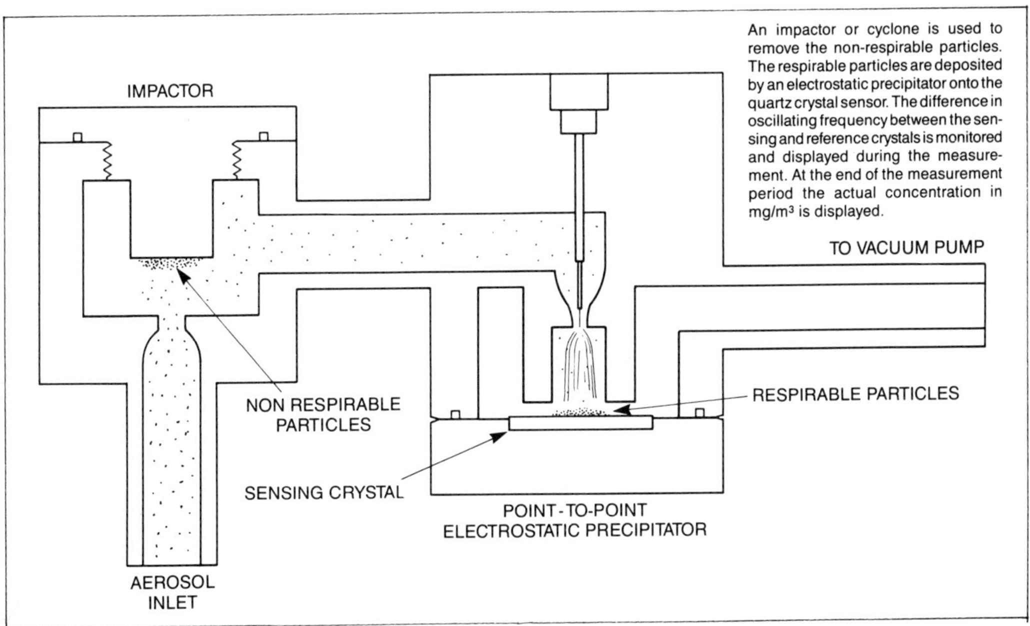

Figure 11-6. *(Continued)*

TABLE 11-1. Approximate Number of Readings Between Sensor Cleanings

Concentration mg/M³	*Measurements Before Cleaning*	*Total Continuous Measuring Time (min)*
0.1	5	10.0
0.2	5	10.0
0.5	5	10.0
1.0	2, 5*	5.0
2.0*	5*	2.5
5.0*	2*	1.0
10.0*	1*	0.5

*Using 24-second measurement mode.

The crystal must be cleaned regularly during runs. To determine how often cleaning must be done, the sensor reading should be compared with the basic crystal frequency, which typically varies between 1,200 Hz and 1,400 Hz. The sensor must be cleaned if the frequency readout on the instrument exceeds the basic frequency by 1,000 or more. It should also be cleaned whenever extra accuracy is important, for example, when measuring aerosol concentrations expected to be less than 0.1 mg/M^3. The manufacturer recommends that the precipitator needle assembly be cleaned and the vacuum pump be checked for air leaks every 2 weeks.

The impactor should be checked daily for particle deposits. In very dusty or smoky sampling environments, cleaning may have to be done hourly. This involves taking the impactor out of the instrument housing, inspecting the impaction surface, and applying a light coat of stopcock grease to the instrument's surface before reassembly.

In summary, common concerns in obtaining valid calibration data for Piezobalances are listed below.[21]

1. Cleaning the sensor takes practice. It takes nearly 5 minutes for an experienced operator to take a 120-second measurement and clean the sensor.
2. The filter cassette should be stationed as closely to the inlet of the Piezobalance as possible in order to avoid the large differences that can occur in aerosol concentrations over very small distances.
3. If the respirable mass concentration is fluctuating strongly during the calibration, longer sample times or more calibration tests will be needed to obtain statistically valid determinations.
4. At least 100 μg must be collected on a filter, because electrobalances generally have an error of about $\pm 5\ \mu$g. The flow rate should be verified during the measurement period and following it.
5. The Piezobalance impactor must be removed during calibration.
6. Flow rates should be measured and compared to the manufacturer's stated value, which is usually 1 Lpm, in order to assure accuracy of measurements. The flow rate should be measured using a bubble buret or any other rotameter. A water pressure drop of $\leq$1 cm at a flow rate of 1 Lpm should be adjusted until it is 1.00 $\pm$ 0.05 Lpm. For more information on flow rate calibration, see the appendix on Pumps and Flow Rate Calibrations.

Use of the Piezobalance

A significant amount of experience is required to operate the Piezobalance accurately, due to maintenance requirements already described as well as the limited sampling periods, which require the sampling professional to be in constant attendance in order to collect a sufficient number of samples. Measurement times can be either 120 seconds for environments where levels of 0.01–2.0 mg/M^3 are expected, or 24 seconds for levels of 2.0–10.0 mg/M^3. When starting up the instrument, always make certain the sensor reading is near the basic frequency. If the precipitator current displayed on the front of the meter does not rise and stabilize, the sensor cleaning knob may not be in the correct position.

When measurements begin, the meter will display the frequency change of the sensor, which represents the accumulation of particles. At the end of the measurement period, the concentration in mg/M^3 is displayed. For most Piezobalances, if the MEAS button is depressed while the power is on, the sensor is collecting particles regardless of what is on the display. Since this situation could lead to overloading, the MEAS button should only be pressed when actual measurements are desired.

There is a method to override the time limitation of the measurement periods so that concentrations in the 0.005 mg/M^3 to 0.05 mg/M^3 (very low) range can be measured with greater accuracy. This method involves extending the measurement time by using a stopwatch to manually time the measurement and recording the frequencies that are displayed at the beginning and end of a 5- to 20-minute period. The concentration can then be calculated using the basic equation described earlier.

COMPARISON OF AEROSOL MONITORING INSTRUMENTS

The instruments are not capable of differentiating between different types of dusts regardless of their chemical composition; therefore, the readout represents the total mass of all dusts present with particles within the size range measured by the instrument. It is the Occupational Safety and Health Administration's (OSHA) policy that aerosol monitoring instruments are to be used for screening purposes only and they cannot be used for generating citations.

In a series of tests designed to measure cigarette smoke, mini-RAM, HAM, and TSI 3500 Piezobalance monitors were used. A comparison of the results indicated that the mini-RAM overestimated the results and the HAM underestimated them, relative to the Piezobalance and reference method, which was integrated collection on a filter cassette and gravimetric analysis. Figures 11-7 and 11-8 compare aerosol monitors. However, it should be noted that light-scattering devices allow continuous measurements and are simpler to calibrate and use than a Piezobalance. As with all monitoring situations, there are tradeoffs when selecting among instruments.

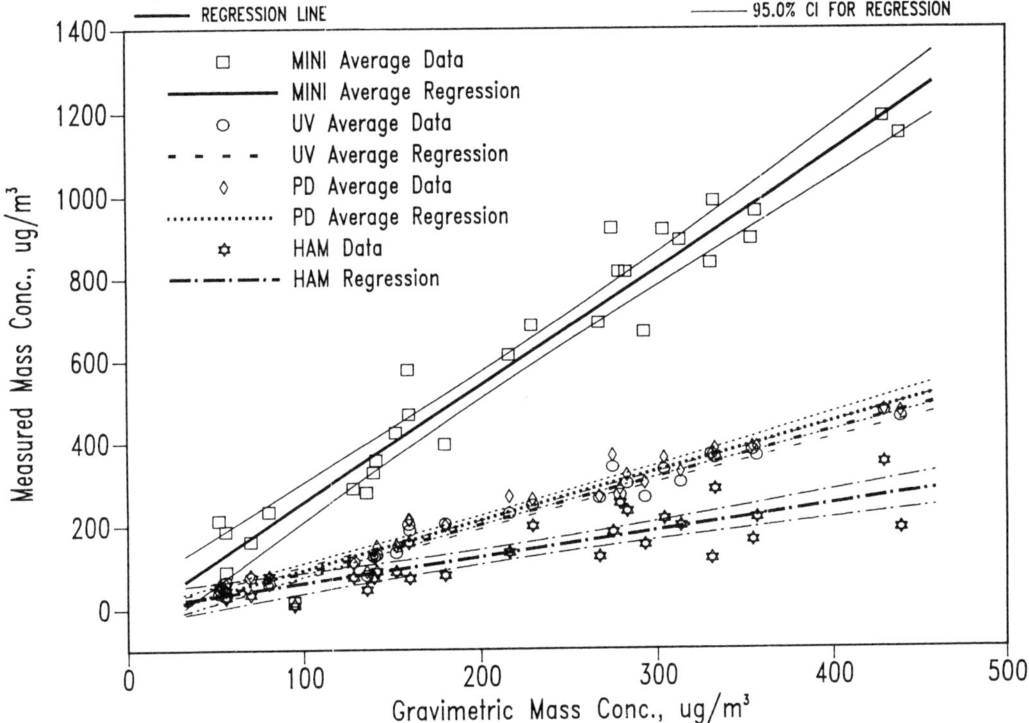

Figure 11-7. Comparison of aerosol measurements made with the RAM-1, Mini-RAM, HAM, and Piezobalance. (From Ingebrethsen, B. J., et al., A comparative study of environmental tobacco smoke particulate mass measurements in an environmental chamber, *JAPCA* **38**:413–417, 1988.)

Figure 11-8. Comparison of selected real-time aerosol monitors. (From Baron, P. A., Modern real-time aerosol samplers, *Appl. Ind. Hyg.* 3(3):97–103, 1988.)

Instrument	*Response Time for Readout*	*Measured Parameter*[1]	*Important Aerosol Properties*[2]	*Particle Size Range (μm)*
RAM-1	Continuous	Light scattering (45°–95°)	Density Size distribution Refractive index	0.1–20
Advantages: Continuous readout; respirable and total dust measurements Disadvantages: Calibration changes with dust type Accuracy: ±0.1 of full scale or 0.005 mg/M^3 Factory Calibration: Arizona road dust Concentration Range: 0.01–200 mg/M^3				
Mini-RAM		Light scattering (45°–95°)	Density Size distribution Refractive index	0.1–10
Advantages: Personal monitoring Disadvantages: Passive collection Accuracy: ±0.02 mg/M^3 Factory Calibration: Arizona road dust Concentration Range: 0.01–100 mg/M^3				
HAM	1–5 sec	Light scat. (10–15°)	Density Size distribution Refractive index	0.5–12
Advantages: High sensitivity, digital readout Disadvantages: Best for particles 0.3–2 μm, passive collection Accuracy: ±25% Factory Calibration: Oil mist or cement dust Concentration Range: 0.001–200 mg/M^3				
Piezobalance (TSI 3500)	0.5–2 min	Mass	Particle size	0.02–10
Advantages: Direct measure of mass, measures fumes Disadvantages: Frequent sensor cleaning required, not simple to use Accuracy: ±10% or ±0.01 mg/M^3 Factory Calibration: Welding fume Concentration Range: 0.01–10 mg/M^3				
FAM-1	1–1000 min	Light scat. lasers (parallel to airflow)	Fiber length Size of fibers Fiber diameter and count	
Advantages: Specific for fibers Disadvantages: Cannot distinguish between types of fibers Accuracy: ±25% Factory Calibration: Amosite fibers Concentration Range: 0.01–10 f/cc				

[1]Physical property that produces instrument signal.
[2]Parameters that significantly affect the accuracy of the instrument.

REFERENCES

1. Knutson, E. O., and P. J. Lioy. Measurement and presentation of aerosol size distributions. In *Air Sampling Instruments,* 6th ed., ACGIH, Cincinnati, OH, 1983.
2. Willeke, K., and S. J. Degarmo. Passive versus active aerosol monitoring. *Appl. Ind. Hyg.* **3**(9):263–266, 1988.
3. Baron, P.A. Modern real time aerosol samplers. *Appl. Ind. Hyg.* **3**(3):97–103, 1988.
4. Schnakenberg, G., and B. Chilton. Direct reading systems. *ACGIH Ann.* **1**:272–290, 1981.
5. Kusisto, P. Evaluation of the direct reading instruments for the measurement of aerosols. *AIHA J.* **44**(11):863–874, 1983.
6. Swift, D. L. Direct reading instruments for analyzing airborne particles. In *Air Sampling Instruments,* 6th ed., ACGIH, Cincinnati, OH, 1983.
7. Hodkinson, J. R., and J. R. Greenfield. Response calculations for light-scattering aerosol counters and photometers. *Appl. Opt.* **4**:1463–1474, 1965.
8. Gucker, F. T., and D. G. Rose. The response curves of aerosol particle counters. In *Proceedings of the 3rd National Air Pollution Symposium.* National Air Pollution Control Association, Pasadena, CA, 1955, pp. 120–130.
9. Page, S. J., and R. A. Jankowski. Correlations between measurements with RAM-1 and gravimetric samplers on longwall shearer faces. *AIHA J.* **45**(9): 610–616, 1984.
10. Smith, J. P., et al. Response characteristics of scattered light aerosol sensors used for control monitoring. *AIHA J.* **48**(3):219–229, 1987.
11. Lehtimaki, M., and J. Keskinen. A method of modifying the sensitivity function of an aerosol photometer. *AIHA J.* **49**(8):396–400, 1988.
12. O'Brien, D. M., et al. Acquisition and spreadsheet analysis of real time dust exposure data: A case study. *Appl. Ind. Hyg.* **4**(9):238–243, 1989.
13. MIE Technical Note No. 2. *Particle Size Dependence of MIE Dust/Smoke Monitors.* MIE, Bedford, MA, March 1990.
14. Phanprasit, W., et al. Comparison of the fibrous aerosol monitor and the optical fiber count technique for asbestos measurement. *Appl. Ind. Hyg.* **3**(1):28–33, 1988.
15. Natale, A., and H. Levins. *Asbestos Removal and Control: An Insider's Guide to the Business.* Source Finders, Cherry Hill, NJ, 1984.
16. Lilienfeld, P., and M. A. Trudeau. A comparison: The GCA model FAM fibrous aerosol monitor to the NIOSH recommended procedure for asbestos sampling and microscope counting tests performed by five Quebec asbestos companies in August and September 1978. *Asbestos* **6**(2):4, 1979.
17. Iles, P. J., and T. Sheldon-Taylor. Comparison of a fibrous aerosol monitor (FAM) with the membrane filter method for measuring airborne asbestos concentrations. *Ann. Occup. Hyg.* **30**(1):77, 1986.
18. *FAM-1 Operations Manual.* MIE, Bedford, MA.
19. Ingebrethsen, B. J., et al. A comparative study of environmental tobacco smoke particulate mass measurements in an environmental chamber. *JAPCA* **38**:413–417, 1988.
20. Lundgren, D. A., et al. Aerosol mass measurement using piezoelectric crystal sensors. In *Fine Particles in Gaseous Media,* Howard E. Hesketh (ed.). Ann Arbor Science, Ann Arbor, MI, 1978.
21. *Instruction Manual, Piezobalance Model 3500.* TSI, Inc., St. Paul, MN, 1979.

PART IV

Sampling with Detector Tubes and Monitoring for Agents Other Than Chemicals

CHAPTER 12

Sampling of Gases and Vapors Based on Color Change

COLORIMETRIC TUBES

Sampling media that change color when contaminated air is pulled through them have been available for 50 years, and can provide the ability to do direct reading measurements while in the field for a wide variety of gases and vapors.[1] Detector tubes are also termed colorimetric indicating tubes, or Draeger tubes, after a manufacturer who has been offering them for many years. A typical sampler is a glass tube filled with a solid granular material, such as silica gel, that has been coated with a chemical that reacts to cause a color change when air containing a specific chemical or specific group of contaminants moves through the tube. The primary use for these devices is for occupational sampling, since they are not sensitive enough to detect the low levels of contaminants needed for environmental detection.

Color change media occupy their own slots as sampling categories in that they share certain aspects of both integrated and real time techniques. Like integrated sampling methods there are many tubes that are specific for a given chemical or chemical family; however, no laboratory analysis is required. They share the ability to provide fast on-site results with real time instruments and for that reason are often lumped into categories with real time instruments called direct reading devices.

Detector tubes are available for short-term measurements, often termed grab sample measurements, or long-term measurements; however, different tubes are used for each type of sampling. In the case of a short-term measurement, the ends are broken off of a fresh tube and a specific amount of air is drawn through the tube using a hand pump, during which the tube will change color if there is enough of a particular chemical present to be detected by the tube. Chemical reaction times are 1 to 2 minutes. Long-term samples for 4 to 8 hours are collected on color change media using both active and passive methods. As with integrated samples, active methods involve the use of a portable battery-powered pump or hand-powered pump. Passive detectors that change color with exposure to contaminants can be in the shape of a tube or a "badge" with a large spot or square of reacting material used as the sampling surface. Tubes designed for active sampling cannot be interchanged with those designed for passive sampling. In all cases, the detection principle is essentially the same but modifications have been made for a given application.

Tubes detect via one of two mechanisms: concentration being related to the amount of material stained, called length of stain, or concentration related to the degree of color change as compared to a standard. Most detector tubes are indicating, that is, they have the calibration scales marked on them that correspond to various concentrations when a specified volume is pulled through the tube. A variety of types of tube construction exist: Tubes can be designed with or without a prelayer, require priming with a liquid reagent contained within an ampule inside the tube, or can be used in series for measurements. The type of construction and reagents used depends on the degree of specificity required in sampling; type of compound being detected and its chemical characteristics; and the concentration range of interest. Regardless of tube construction, the tube in which the measurement is made is called the indicator tube.

Detection by indicator tubes is based on chemical reactions that take place in the tubes with the contaminant of interest (Table 12-1). The reagents used inside the tubes are specific for each tube, and the reactions used for detection of a specific gas can vary among manufacturers. For example, the Draeger 10/b carbon monoxide tube reacts carbon monoxide with iodine pentoxide, selenium dioxide, and fuming sulfuric acid to produce a brown-green color. The Gastec 1La carbon monoxide tube that measures a similar range of concentrations reacts carbon monoxide with potassium palladosulfite to generate metallic platinum and a brown color. Sometimes in the course of a measurement tubes emit smoke or heat up. Some tubes require oxygen for proper reactions to take place.

The simplest configuration is a tube containing one layer designed to detect the chemical of interest. Long-term tubes and passive collectors are all of this type as are a few grab sampling tubes such as ozone tubes.

Some grab sampling tubes contain a prelayer that absorbs interferents, such as moisture or other similar compounds to that being measured, and reacts with the gas or vapor of interest to convert it to a more suitable reacting compound, or reacts with the gas or vapor to produce a different gas or vapor that can be measured by the second section. The prelayer in tubes designed to detect benzene removes toluene and xylene because these compounds would also react with the indicating layer due to their similarity in structure to benzene. When gases and vapors, because of their low reactivity, are not easily detected, a reaction with very powerful chemical reagents to break down these nonreactive compounds into other more readily detectable substances is used. Reactors may be in the form of a second glass tube or ampules contained in plastic tubing. Preconditioning tubes, sometimes called equilibrium tubes, are sometimes needed to prime a reaction tube. The contaminated atmosphere is pulled through the assembled preconditioner-reactor tubes to activate the chemical in the reactor tube; then the preconditioner tube is removed and the indicating tube is attached to the reactor tube to make the measurement.

Most tubes have scales printed on their exteriors. Concentration scales are the most common and are used in grab sampling tubes. Usually these are in ppm, although certain tubes designed to measure high concentrations (such as carbon dioxide) are in percent. If there are no deviations from the manufacturer's basic instructions, the concentration can be read directly off the tube with no adjustment. In the case where a grab

TABLE 12-1. Color-Change Reactions in Selected Dosimeter Tubes

Gas Measured	*Reaction*	*Color Change*
Carbon monoxide	$CO + K_2Pd(SO_3)_2 \rightarrow Pd + CO_2 + K_2SO_3$	Yellow – black-brown
Hydrogen sulfide	$H_2S + Pb(CH_3COO)_2 \rightarrow PbS + 2CH_3COOH$	White – dark brown
Sulfur dioxide	$SO_2 + BaCl_2 + H_2O \rightarrow BaSo_3 + 2HCl$	Green – yellow
Hydrogen cyanide	$HCN + HgCl_2 \rightarrow Hg(CN)_2 + 2HCl$	Yellow – red

Source: Roberson, R. W., et al. Performance testing of Sensidyne/Gastec Dosi Tubes for CO, H_2S, SO_2 and HCN. *AIHA Conf. Presentation*, May 21, 1985.

sampling tube has more than one range, there may be two scales on a tube, each corresponding to a different number of strokes of a grab sampling pump. Long-term tubes usually require a calculation or comparison to a chart of time-concentration curves to determine concentrations.

New calibration scales for detector tubes are printed for each individual production lot by most manufacturers to eliminate variations in tube diameters, precision of packing, and reagent quality or reactivity. Some tubes have the name of the detected chemical printed on the exterior as well, but more often it is the part number.

Selection of detector tubes (Fig. 12-1) depends on the chemical for which monitoring will be done and the concentration range of interest. The largest variety of detector tubes is in the grab sampling category. Only a few chemical-specific tubes are available for long-term sampling and most are for inorganic compounds. Often tubes for the same chemical will come in two or more measurement ranges. The sampler will need to know in advance if multiple measurement ranges are available on the same tube or if different tubes are required for each range. Also, if low levels are of the greatest concern, a tube should be selected with a narrow measurement range or it will be too difficult to determine the results with accuracy. For example, a tube whose range is 1 ppm to 50 ppm would not be the best choice for a compound like sulfur dioxide whose PEL is 2 ppm, since most tubes with this range would be marked at 5-ppm intervals.

The manufacturer's instructions for the specific detector tubes to be used are usually included as an insert in the tube box. Prior to starting, read these for information on interferences and relative standard deviations for the specific tube, as well as the number of strokes, time between strokes, and time necessary for color development, temperature, humidity, and atmospheric pressure effects.

Another reason it is important to review the literature first is to understand the function of each tube layer. The indicating layer may be in the middle between the prelayer and a layer to be used for color comparison or at the end of the

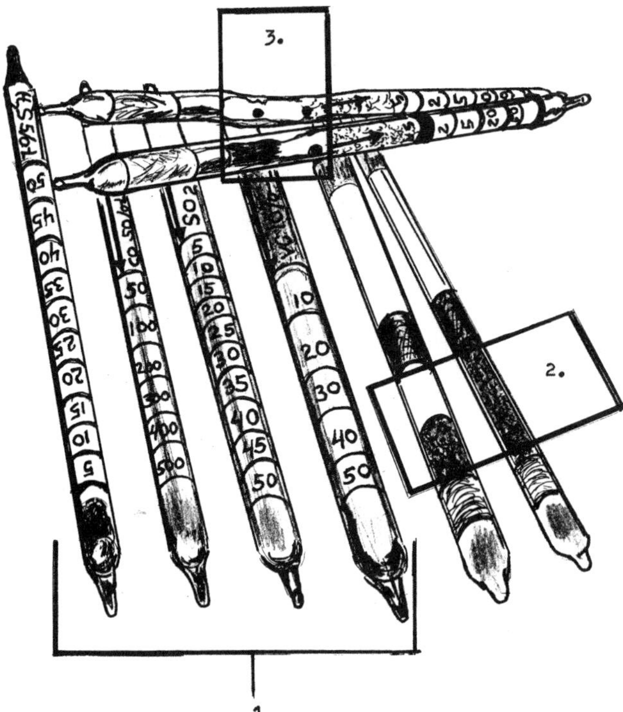

Figure 12-1. Types of detector tubes: (1) length-of-stain tubes; (2) color comparison tubes; (3) ampule tubes before and after breaking ampule.

tube. Prelayers often change color and should not be confused with the indicating layer.

Some tubes, such as those containing ampules, require special procedures not typically done with other tubes. These are broken just before the tube is used. An example is the Draeger tube for toluene diisocyanate, which is a long tube with two separate ampules and an indicating layer. In these situations it is important to follow the specific sequence indicated in the manufacturer's instructions and practice with tubes before attempting measurements in the field. The portion of the tube containing the ampule is covered in plastic. The tube should be bent gently until the ampule breaks, making sure it is pointing in the right direction so the reagent reaches the correct layer. Sometimes the indicating layer changes color when a reagent reaches it, and in other cases one reagent causes a change in color and the next one bleaches out the color.

High humidity can cause discolorations in these tubes. Some tubes must be held in a certain position, for example, the Draeger Methyl Bromide 5/b has to be held vertically upward. The time for a stroke varies depending on the tube's resistance.

Detector tubes have been used in the past for confined-space testing for both toxic and explosive gases as well as oxygen deficiency, although they have mostly been replaced with real time instruments for this use. They are currently used on hazardous waste sites, for measurements of leaks and spills, for personal monitoring (long-term tubes and dosimeters), and for "grab" samples in a variety of situations. A new application that has been proposed is to use detector tubes to measure permeation resistance of personal protective clothing.[2] In some cases, electronic real time instruments have replaced the use of detector tubes; however, the tubes still offer a wider variety of specificity than the real time instruments available.

A device to draw air through 10 detector tubes at a time in order to allow for rapid detection of compounds on hazardous waste sites has been developed (Fig. 12-2). It can be assembled using

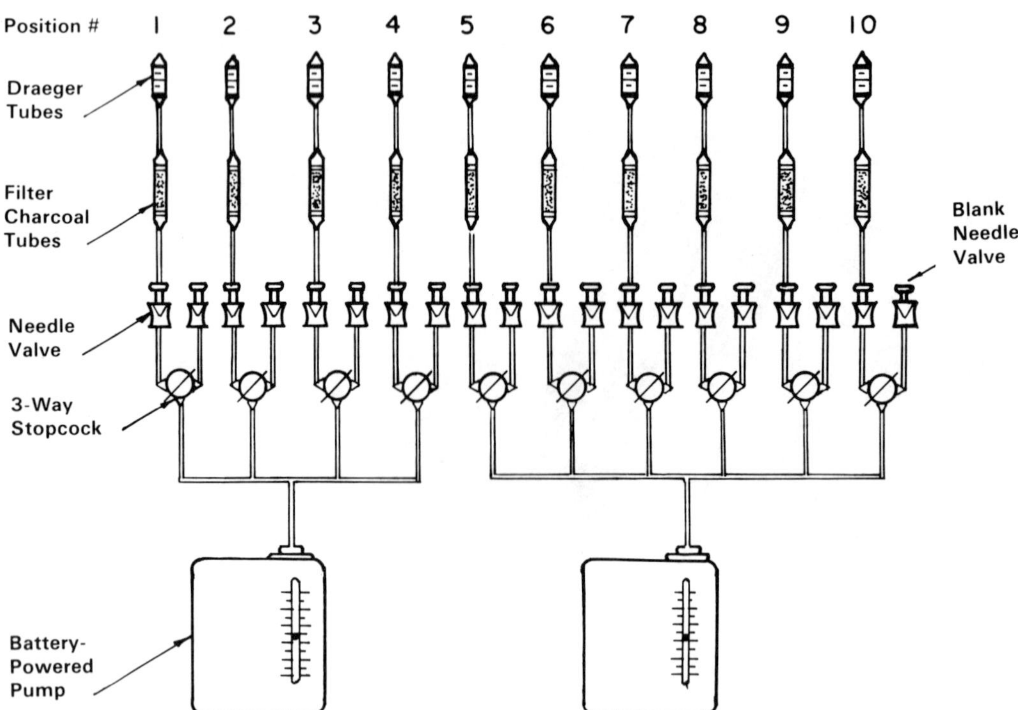

Figure 12-2. Schematic of simultaneous direct reading indicator tube system. (Reprinted with permission from *Am. Ind. Hyg. Assoc. J.* **44:**617, 1983.)

two battery-operated, intrinsically safe personal air sampling pumps and a series of 20 paired-needle valves for sampling and bypass are attached to three-way stop-cocks. This method overcomes the lengthy time period required to make 10 successive measurements for different tubes. Considerable setup time is required in order to assure that each valve provides the proper volume required for each individual tube. A stopwatch is used to determine when the proper sample volume has been pulled into each tube. At this point, the user switches the flow of contaminated air to the blank or bypass valve by turning the stop-cock.[3]

Detector tubes are best used to screen areas to determine what locations should receive more in-depth sampling. They should only be used for determining compliance when no other sampling method is readily available. Only NIOSH (National Institute of Occupational Safety and Health)-certified detector tubes should be used for compliance testing according to the Occupational Safety and Health Administration (OSHA) policy; however, this program is no longer in existence and the number of certified tubes is few compared to the number currently available.[4] They may also be used concurrently with integrated samples to trace sources of exposure and track variations in exposure levels throughout the work shift.

Quality Control

The most important elements of quality control in detector tubes are the purity of the reagents used, grain size of the gel or adsorbant, method of packing the tubes, moisture content of the gel, uniformity of the tube diameter, and proper storage precautions to preserve the shelf life.[5] Many factors, such as interferences, packing of the granules/reagent, airflow patterns, actual air volume, and sampling rate, can affect the final reading. Precision, accuracy, and reproducibility vary with the age, conditions of storage, and lot-to-lot manufacturing of these tubes.

As noted, NIOSH had a detector tube certification program until 1982 when it was suspended. The Safety Equipment Institute (SEI), a private, nonprofit organization funded by safety equipment manufacturers, recently started its own detector tube certification program. Tube testing is performed by laboratories that have been accredited by the American Industrial Hygiene Association (AIHA) using similar protocols to the NIOSH program. Currently the types of tubes that have been certified are for ammonia, benzene, carbon dioxide, carbon monoxide, chlorine, hydrogen cyanide, hydrogen sulfide, sulfur dioxide, trichloroethylene, and nitrogen dioxide.[6] The program includes an evaluation of the sampling pump used by each system, and testing is done for volume accuracy and leakage. The tubes are exposed to a specified concentration of test gas, and then a six-person team analyzes each tube's corresponding length of stain. Following certification, a manufacturer can use the SEI certification mark on the product.[7] Each lot of detector tubes is supposed to be separately calibrated by the manufacturer so that the scale markings on the tube or chart are accurate; therefore, to use a calibration scale or chart from another lot may lead to errors. In one study, significant calibration errors were associated with tubes for carbon monoxide, carbon dioxide, nitrogen oxides, and sulfur dioxide.[8]

It is considered a good practice to verify the manufacturer's calibration for each new lot of tube received. Detector tubes whose expiration date has passed should not be used without testing in a test atmosphere. For more information on generating concentrations of gas and vapors, see Appendix B.

Length of stain tubes are generally considered preferable to color change tubes, since color charts can fade with time or not provide realistic color comparisons.[9] Grain size of the coated granules in the tubes may vary among tubes and within tubes. Among tubes, the difference will be determined by the size of the tube. A fine grain size is said to provide a uniform distribution of airflow through the tube along with sharp demarcation lines.[2] Within a given tube, variations in particle size can lead to striations in the stain and cause an indefinite (fuzzy) stain line.

It has been recommended by NIOSH, on the basis of detector tube certified accuracy standards and other considerations, that the methods of measurement should have an accuracy to a confidence level of 95%. Table 12-2 lists accuracies of detector tube measurements.

TABLE 12-2. Accuracy of Detector Tube Methods of Measurement

Concentration	*Required Accuracy*
Above PEL	±25%
At or below PEL and above 50% of the PEL	±35%
At or below the 50% of the PEI	±50%

Temperature and Humidity

Chemical reactions occurring in detector tubes are temperature- and humidity-dependent. Most manufacturers specify a range of temperatures over which their tubes will work, and if a tube is a different temperature than that for which it was calibrated, the reading must be adjusted.

While some manufacturers indicate operating ranges for many tubes of 10% to 90% RH, this may be optimistic. In one study of tubes for carbon monoxide, nitrogen oxides, sulfur dioxide, formaldehyde, and hydrogen sulfide, it was found that errors for some tubes exceeded NIOSH criteria at 50% relative humidity (RH) and higher, whereas others were only affected at 90%. Elevations in temperature to 4.4°C and 49°C also affected most tubes examined in the study.[10]

At least one large manufacturer of detector tubes (Draeger) routinely reports humidity operating limits in units of absolute water vapor concentration (mg/L) instead of percent RH. From a survey of operating limits for Draeger detector tubes, upper water vapor concentrations of 12 mg/L and 15 mg/L and lower water vapor concentrations of 3 mg/L and 5 mg/L were the most common limits specified.[11] Tubes with absolute humidity ranges of 5 mg/L to 12 mg/L may be used only between 22% and 52% RH at a temperature of 25°C, or between 16% and 40% RH at a temperature of 30°C. Some areas of the United States in the summer may exceed these limits and in winter humidities may be lower than these limits.

In most industrial situations the temperature and humidity are at tolerable levels, and the majority of detector tube units are not adversely affected. However, sampling professionals should be aware that scale calibrations when printed on tubes may not be accurate outside of the temperature and pressure range specified by the tube manufacturer and they should also observe the correction factors specified for temperature and humidity.

Interferences

The primary disadvantage of detector tubes is their propensity to interferences, both positive and negative. For example, many tubes used for chlorine will also react to hydrogen sulfide, ammonia, nitrogen dioxide, ethylene, or halides as well, with the result being an increase in the concentration due to these positive interferents. If any of these compounds are present in the air during sampling, the result will not be specific for chlorine. Negative interferences manifest themselves as no color change when an interferent is present. Sulfur dioxide, for example, when present in an atmosphere being measured for hydrogen sulfide, will produce a lower reading than the concentration actually present.

Sometimes manufacturers specify that an interferent will cause a different discoloration of a tube than that specified for the contaminant of interest. For example, chlorine causes a bluish stain in one tube, and nitrogen dioxide will turn the same tube pale yellow. Therefore, a discoloration significantly different from what is predicted should be considered suspicious. Table 12-3 provides examples of interferences in dosimeter tubes.

The best prevention for spurious results due to interferences is for the sampling professional to evaluate this potential prior to sampling. A review of the reagents used in the detector tubes along with the reactions can be useful. Most manufacturers provide this information. Tolerable concentrations of interferents for each tube should be reviewed along with allowable ranges and corrections for temperature, pressure, and relative humidity.[9]

Shelf Life

Shelf life, as an expiration date, is generally stated on each box of tubes. It is the period of time within which the calibration accuracy of the tubes can be expected to remain within ±25%. Most indicator tubes deteriorate within a year or two. It is a common practice to refrigerate these tubes in order to extend their shelf life, but because the

TABLE 12-3. Interferents for VaporGuard Dosimeters

Dosimeter	Temperature Range (°F)	Concentration Range (ppm)	Interferents
Carbon monoxide	0–125	0–250	Hydrogen sulfide, acetylene
Ammonia	40–90	0–125	Acid gases (negative), amines
Sulfur dioxide	40–90	0–25	Acid gases
Nitrogen dioxide	32–120	0 20	Halogens
Hydrogen sulfide	32–120	0–50	None listed
Carbon dioxide	32–120	0–4%	None listed

Source: Courtesy of Mine Safety Appliances, Inc.

speed of most chemical reactions is sensitive to temperature, the tubes must be warmed to ambient conditions prior to use if the tubes are going to perform according to the manufacturer's specifications. Refrigeration can assure that the tubes are not exposed to high concentrations of heat, which might decrease the shelf life.

Small variations in impurities, such as moisture content, may have a large effect on the shelf lives of different batches; therefore, every batch of tubes should be labeled by the manufacturer with a lot number and an expiration date.

For all these reasons, prior to use detector tubes should be examined visually for obvious problems, such as partial discoloration of a layer. If tubes change color during storage, some decomposition has probably occurred, and the tubes should be discarded regardless of whether time remains until the expiration date. Multilayer tubes often have a shorter shelf life than other tubes, due to the potential for diffusion between the layers. Tubes incorporating a preliminary reaction zone to convert the contaminant being sampled to a derivative also frequently have a shorter shelf life than others, because of diffusion of chemicals between layers and consequent deterioration.[9]

GRAB SAMPLE MEASUREMENTS USING DETECTOR TUBES

The primary application for grab sampling measurements is screening to see if contaminants are present as a preliminary to more complex and accurate methods, or these measurements are used in situations where other methods do not exist or cost considerations are important; detector tubes generally cost much less than real time instruments.

Due to the concerns of sampling on hazardous waste sites in recent years, the interferences from which the tubes suffer have been advantageous. A sampling scheme (Fig. 12-3) has been developed for use in identifying unknown chemicals to some degree. It is most useful in situations where individuals are already wearing protective gear, or when a spill or the contents of a drum must be identified quickly. This sampling method would not be useful to determine the necessity to don or upgrade personal protective gear.

Inaccessible locations, such as manhole pits, can be sampled by using a flexible tube attached to the pump and putting the detector tube at the other end. The sample should be drawn directly into the detector tube, meaning the detector tube should be lowered into the hole attached to the end of the tubing, and not with the tubing in front of the tube, since contaminants might adsorb onto the tubing wall and result in a reduced reading. When using the extension tubing, attaching it to a metal probe will allow for measurements to be taken in areas other than just the middle of the space. For more information on confined space monitoring, see the chapter on Specific Sampling Situations.

Detector Tube Pumps for Grab Sampling

There are two different types of pumps available: the bellows variety and the piston type. Although in principle all grab sample tubes appear to do

Detection of unknown substances by means of DRAEGER detector tubes*

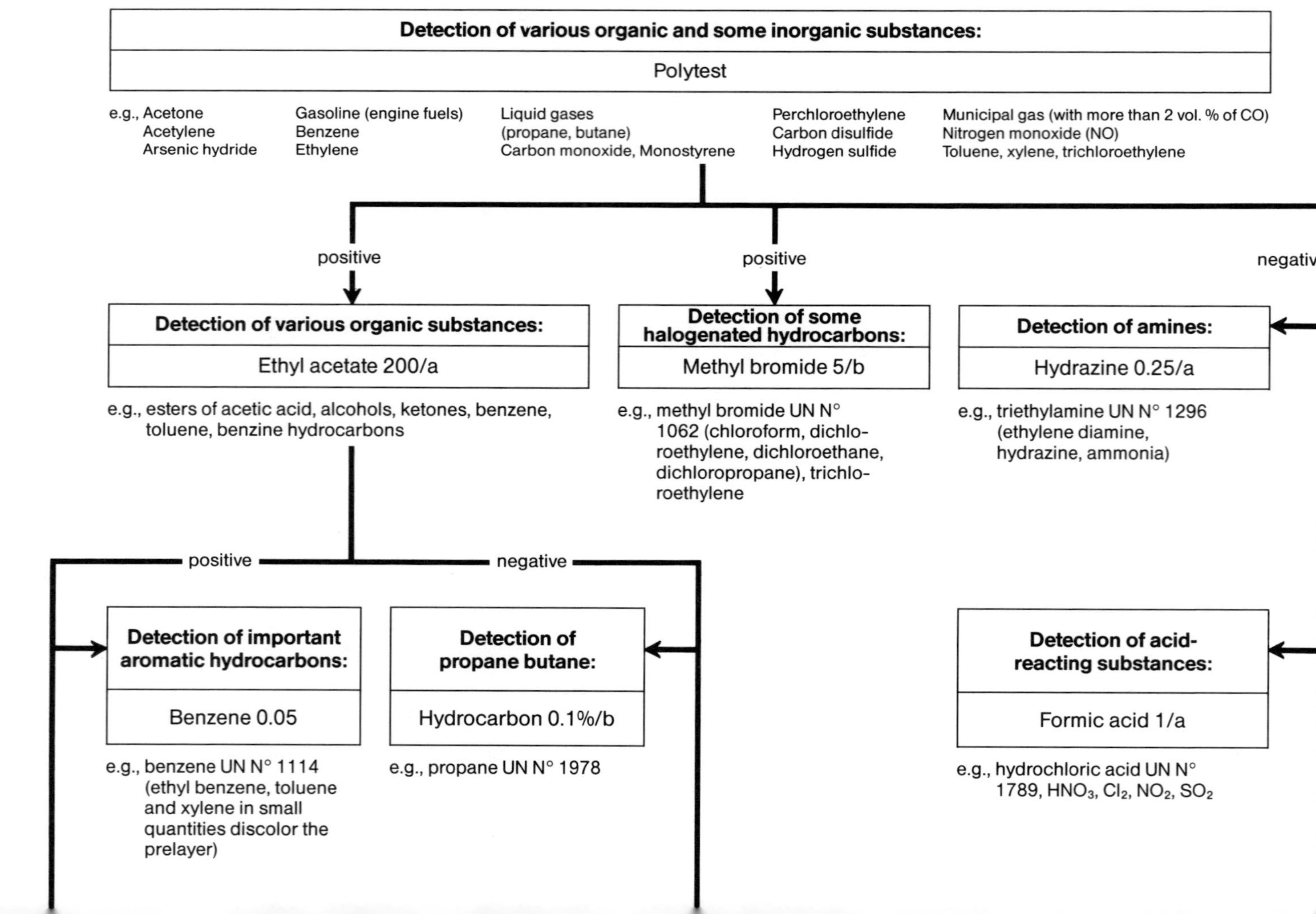
Detection of various organic and some inorganic substances:
Polytest
e.g., Acetone
Acetylene
Arsenic hydride
Gasoline (engine fuels)
Benzene
Ethylene
Liquid gases
(propane, butane)
Carbon monoxide, Monostyrene
Perchloroethylene
Carbon disulfide
Hydrogen sulfide
Municipal gas (with more than 2 vol. % of CO)
Nitrogen monoxide (NO)
Toluene, xylene, trichloroethylene
positive
positive
negative
Detection of various organic substances:
Ethyl acetate 200/a
e.g., esters of acetic acid, alcohols, ketones, benzene, toluene, benzine hydrocarbons
Detection of some halogenated hydrocarbons:
Methyl bromide 5/b
e.g., methyl bromide UN N° 1062 (chloroform, dichloroethylene, dichloroethane, dichloropropane), trichloroethylene
Detection of amines:
Hydrazine 0.25/a
e.g., triethylamine UN N° 1296 (ethylene diamine, hydrazine, ammonia)
positive
negative
Detection of important aromatic hydrocarbons:
Benzene 0.05
e.g., benzene UN N° 1114 (ethyl benzene, toluene and xylene in small quantities discolor the prelayer)
Detection of propane butane:
Hydrocarbon 0.1%/b
e.g., propane UN N° 1978
Detection of acid-reacting substances:
Formic acid 1/a
e.g., hydrochloric acid UN N° 1789, HNO3, Cl2, NO2, SO2

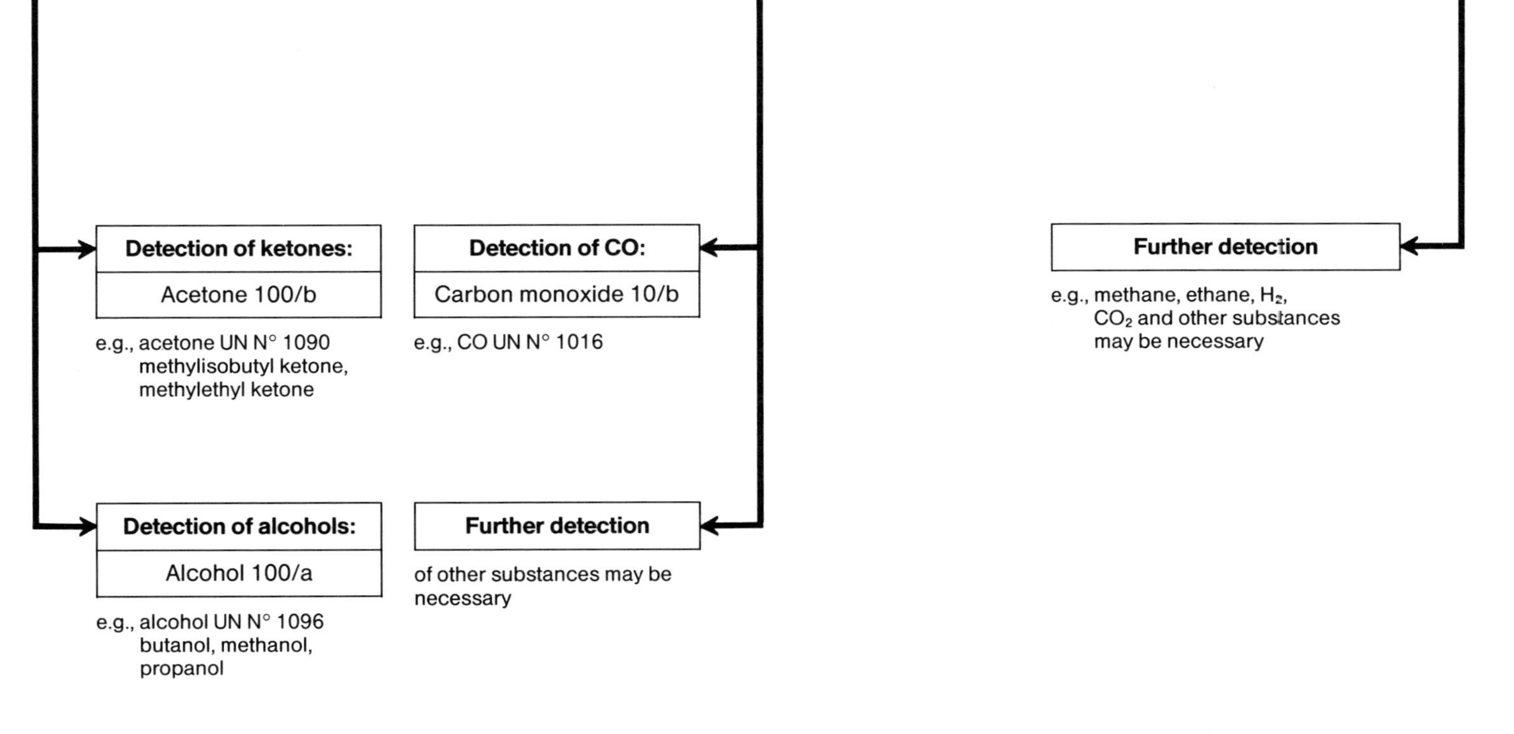

*Important: This sample taking schedule refers to a selection of substances which occur frequently in practice. Other situations may necessitate another sequence of measurements and, the case being, the use of additional detector tubes, or measurements according to other procedures must be carried out.

Figure 12-3. Sampling scheme for chemical identification of unknown gases and vapors. (Courtesy National Draeger, Inc.)

the same thing, that is, change color when a certain minimum concentration of a chemical passes through them, using tubes from one grab sample pump from a different manufacturer may lead to errors in measurements.

Each manufacturer has its own design, manufacturing techniques, and quality control. For example, some piston-type pumps use orifices to control flow rate and other piston-type pumps use only the resistance of the granules packed in the detector tube to limit flow rate. Bellows-type pumps also utilize the tube packing for resistance. Since the method of controlling flow rate, and thus the total volume pulled with each stroke, varies among manufacturers, differences in suction pressure per unit time and flow rate through the tube over time can be expected. Flow rate determines the absorption rate for the chemical reactions occurring in the detector tubes to produce the color change and length of stain.

With a bellows pump the stroke is finished when the bellows has completely opened, often indicated by a chain straightening. Usually these pumps are calibrated to pull 100 ml of air per stroke and 2 to 10 strokes are required; therefore bellows pumps use larger tubes than piston pumps and are sometimes more difficult to interpret. Stroke counters are very useful for bellows-type pumps, because as noted some measurements require 10 or more strokes and it is easy to get distracted during this period. However, the counter position must be checked with each stroke, because if not pushed hard enough, it will not register the stroke. There is some evidence to indicate that the squeeze bellows type does not always deliver reproducible volumes, and therefore the accuracy is less than that provided with the piston type of pump.

A variation on the bellows pump is the Kwik-Draw sampling pump from MSA that can be operated with one hand. It has an automatic stroke counter and can also be outfitted with an end of stroke indicator. An advantage is its ability to use both types of tubes: the wider ones designed for bellows pumps and the thin ones used with piston pumps. Since the minimum volume of air pulled by the Kwik-Draw pump is 100 ml, it cannot be used for the high range on some tubes.

On some piston pumps an orifice is selected to vary the volume delivered. On others the pump handle is marked with the volume the pump will pull when the handle is locked in a certain position. Most piston pumps pull 100 ml volume of air, although at least one has four calibrated volumes: 25, 50, 75, and 100 ml. A choice of volume is made by rotating an indexed head with four orifices. Some piston pumps also have stroke counters that may be automatic and change with every stroke, or they may be changed manually by the user. Another option is the flow finish indicator, which is often a button that pops up when the full volume is pulled, thus eliminating the need for a stopwatch to time the stroke. Handles are designed to lock open to allow for the full air volume to be pulled into the tube. Some pumps have openings to break off tube ends, which is a convenient feature.

A multiorifice sampling pump will accurately control both the volume of air sampled and the rate of airflow during a test. The specific manufacturer's instructions will list the required orifice setting and the volume of air required for each tube.

Some manufacturers offer pyrolyzers as accessories for piston pumps. A pyrolyzer is a hot-wire instrument operated by batteries that is attached to pumps and is used for certain gases and vapors, such as freons and other chlorinated compounds, that require thermally induced decomposition to release substances that can be measured. The purpose of the pyrolyzer is to break down difficult-to-detect compounds into other compounds that are more easily detected. A sample is drawn across the heated filament of the pyrolyzer, which decomposes the contaminant, thereby releasing a measurable gas. The decomposition products then pass through the detector tube, causing a change in color relative to the concentration. Instructions for its use and maintenance are given in the manufacturer's instruction manuals. The pyrolyzer is also useful for organic nitrogen compounds, one of the products of breakdown being nitrogen dioxide, which is easily monitored.

Maintenance is important on detector tube pumps. For example, piston-type pumps may bind, or blow by check valves can become inoperable. Elastic components may lose elasticity. Pumps should be accurate to ±5% of their stated volume according to a currently published method.[12] NIOSH's certification requirement was that leakage should not exceed 3% per minute of the pump's total volume. There are two types of maintenance that should be performed on grab sampling pumps to assure accuracy in measurements: leakage tests and flow rate calibrations.

Leakage Test

A leakage test should be performed on each pump when first purchased, after an extended period of idleness, and periodically during use in accordance with the manufacturer's instructions in order to minimize erroneous readings due to air leaks around the seals. At least one manufacturer recommends a leak test after every 10 to 12 samples. If a pump has multiple settings, the procedure should be repeated for each setting.

Procedure

1. If the pump has multiple orifices, turn the rotating head through several revolutions, finally stopping at the highest position.

2. Insert an unopened detector tube into the tube holder. Line up the guide marks, usually dots on the pump shaft and stopper, and pull back the handle all the way. Most pumps lock automatically. Unlock the handle by rotating one-quarter turn and pull the handle back, locking the piston in the farthermost position.

3. Wait 2 minutes and release the pump handle. In order to protect the pump from breakage, do not release the handle and allow it to spring back; instead hold it and release it slowly. When the handle is released the piston should spring back all the way to 0 cc. If it does not, the pump is not leaktight and the amount of leakage is indicated by the handle as it comes to rest. If there is greater than 5 cc of leakage in 2 minutes, or whatever time is required for the stroke, the accuracy of any concentration determination will be subject to error.

For a bellows pump, squeeze it as far as it will go.

4. Rotate the handle to release the locking mechanism. When released, the piston should return to the 0-cc position, indicating the pump is satisfactory for use. If the handle remains out more than 5⁄16 inch (approximately 5 cc), it indicates excessive leakage that could occur either between the piston and cylinder or at the inlet in the rotating head.

5. If the handle fails to return to the starting position, leakage is indicated. If the handle stops at more than 6 mm from the 0-ml stop, the pump should not be used until the leakage has been corrected.

If excessive leakage is found, it usually takes place either at the pump inlet or between the piston and cylinder walls. Leakage in the cylinder can usually be corrected by cleaning and relubrication. Leakage at the inlet may result from a poor seal between the detector tube and the rubber inlet tip, or the flange of the rubber inlet tip and the pump body.

Volume and Flow Rate Calibration

Calibrate detector tube pumps for proper volume and flow rate measurement quarterly. If a pump has more than one orifice, a calibration should be performed for each orifice. For most types of pumps, the flow rate is not constant. Instead it is high initially and slower toward the end of the cycle, which is not a problem as long as the total volume is pulled through the pump in the time period specified for measurement. However, the flow rate determines the adsorption rate for chemical reactions to occur in some tubes. Generally, flow rate is only a concern with piston pumps. For bellows pumps, volume is the primary concern. One of the reasons the pumps lose their volume is due to the buildup of smoke particles released during the tubes' reactions. These particles are pulled into the interior of the pumps and gradually build up, causing low flow rates and possibly inaccurate measurements.

Procedure

1. Test the detector tube pump for leakage following the procedures previously described.

2. Connect the inlet of a detector tube pump to the top of a 200-ml bubble buret using Tygon tubing. For procedures on the use of bubble burets see Appendix A. With limiting orifices, use a suitable adapter, such as a short piece of glass tubing. Do not put a detector tube in line for these pumps. Pumps that rely on the resistance of a detector tube to govern airflow must have a tube in line during measurements.

For the piston pump continue as follows.

3. Select an orifice and set the pump on it, if the pump has orifices.

4. Dip a bubble buret into bubble solution and pull the pump handle slightly. Pull the pump handle all the way out for a full pump stroke and time the bubble front from the start until it comes to rest. Note where the soap bubble stops. The difference between the initial and final points is the sample volume. The volume divided by the time is the flow rate.

5. Review the literature for the pump being used to determine the number of minutes to wait after pulling the pump handle for the full volume of the pump to be pulled.

For the bellows pump, continue as follows.

6. Squeeze the pump, making sure the bellows is completely and evenly pressed. Allow the pump to open on its own, measuring the volume as described in step 4. Repeat two more times and average the three measurements.

7. Compare your findings with the manufacturer's literature for allowed variation in total volume and the time required to draw the total volume. Lubricate the pump if the volume error is greater than 5% and repeat the volume calibration.

8. Also, using the same procedure for tubes with multiple orifices, check the volume for one-half and one-quarter pump stroke.

9. If the error is greater than ±5% the pump should be repaired.

Try to determine what other compounds are likely to be present during sampling and compare these with the list of interferents. The tube should also be selected for the concentration range of interest; usually this is the OSHA permissible exposure limit (PEL). The time for a stroke on a piston pump to be completed varies depending on the tube's resistance. Some tubes require as long as 4 minutes for a stroke and others require ½ minute. This can be important. For example, in order to reach a sensitivity of 1 ppm, at least one tube for benzene requires five strokes, each 2 minutes in length; thus a single measurement requires holding the sampling pump in place for 10 minutes.

In many cases, the sensitivity of a reading can be improved by taking two to four additional strokes and dividing the final result by the total number of strokes. For example, consider a reading well below the first concentration mark of 10 ppm that is obtained with one stroke. It could be estimated to be between 5 ppm and 10 ppm of that tube. What if two additional strokes are collected and a reading of 20 ppm is now obtained, which can be read more clearly? This value can be divided by 3 to obtain an average result of 7 ppm, which is more specific than reporting a 5-ppm to 10-ppm range.

In the case of a negative result following a measurement, most manufacturers recommend reusing tubes, although in general the accuracy may suffer with subsequent measurement. The best time to do this is when several measurements are being made for screening purposes and the tube is immediately reused.

One of the most common mistakes made during grab sampling is allowing the tube to move away from the point being sampled, as the sampling professional's arm tires. The tube must remain positioned during the entire sampling period in order to make sure all the air sucked across comes from the same point or source.

Figure 12-4. Bellows pump and tubes. (Courtesy National Draeger, Inc.)

Using a Bellows Pump

Procedure

1. Break off the ends of a fresh detector tube (Fig. 12-4) by inserting each end in the tube breaker hole usually located in the pump head. Dispose of the broken glass in a safe place. In the case where the tube consists of a two-tube system, connect the ends of the two tubes using the rubber tubing supplied after breaking the tube tips. Insert the tube(s) securely into the pump inlet, taking care that the arrow on the tube points toward the pump.

2. Squeeze the bellows pump and wait for the chain straddling the bellows to become straight or for the end of the flow indicator to change color. Bellows-type pumps must be compressed completely flat to get an accurate sample. Both hands may be needed, since equal pressure must be applied to both ends of the pump. If only one hand is used, the entire bellows may not collapse, causing an inaccurate volume to be pulled into the tube. Repeat the suction process as often as specified in the tube's operating instructions. Make sure if using the stroke counter that it moves with each stroke; otherwise the count will be off.

3. When drawing air into a tube, keep your eyes on the tube to note any color change or stain development. For example, if using a hydrogen sulfide tube with a sensitivity range of 0.5 ppm to 1.5 ppm, requiring 10 strokes, and the concentration is higher than 1.5 ppm, the tube will fully color with a fewer number of strokes. Do not stop until the full number of strokes has been pulled unless the tube saturates as described above. If this happens, put in a new tube and pull one-half as many strokes, in this example five, and multiply the reading by two.

4. Record readings immediately after using tubes, since some changes in the coloration will occur with time, possibly including a complete loss of coloration. Where there is a gradation of color change, the end point should be taken as that point where the slight discoloration can just barely be made out from the unused portion. If the end point occurs at an angle, which often happens when using these wider tubes, estimate the concentration for each side of the tube and average them.

Using Piston Pumps

Procedure

1. Initially, make sure the pump handle is all the way in. Align the index marks on the handle and back plate of the pump (Fig. 12-5).

2. Pull the handle straight back (without turning) to the sample volume required (full or partial pump stroke). The piston will automatically lock in this position. On some models the handle can be locked on either one-half (50 cc) or a full (100 cc) stroke. The time required for the pump to draw its full volume is important and is specified in the tube's literature. If the handle is retracted before enough time has passed, the volume will be smaller than is required, the result being a lower reading than the concentration that is actually present. For the next stroke, if needed, unlock

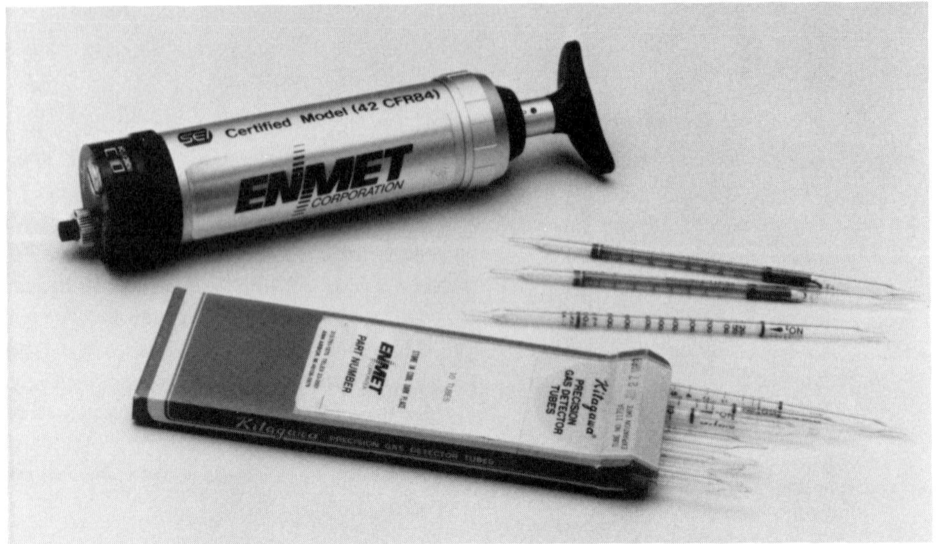

Figure 12-5. Piston pump and detector tubes. (Courtesy Enmet Corp.)

the pump handle by making a one-quarter turn, return it to the starting position, and pull again. The detector tube should remain in place until all the strokes have been made. Some piston-type pumps have multiple orifices that allow different air volumes to be selected.

3. With a multiple orifice an air sample is taken as follows: The proper orifice is selected by sliding the locking button on the rotating head forward and turning the head to the designated index number on the flow control plate. When the center line of the locking button is adjacent to the index mark, release the button. The spring-loaded button will then lock the rotating head in this position.

4. To exhaust the pump rotate the handle 90 degrees, thereby releasing the locking mechanism, and push in the handle all the way. Rotate the handle and realign the index marks for the next stroke or test.

LONG-TERM DETECTOR MEASUREMENTS

Long-term detector tubes are hooked up to the same personal battery-operated sampling pumps as are used for integrated gas and vapor sampling. The change in color is translated to an integrated exposure value using a short calculation to convert the microliters measured on the tube to ppm based on the total volume (in liters) collected. They can be used for 8-hour time-weighted average personal samples on employees. As noted, the calibration scales on long-term tubes are usually in microliters (μl).

When selecting these tubes, pay attention to the maximum number of hours a given tube can be used. Also, they require extremely low flow rates, in the range of 10 ml/min to 20 ml/min, and not all pumps can achieve these. For more information on personal sampling pumps see Appendix A. Sample times vary from 1 hour to 4 hours. In one study, it was determined that the manufacturer's recommendations for sampling time and volume did not necessarily fit the capacity of the indicator layer. The sampling rates, total volume, and sampling time suggested by the manufacturer differ for most tubes. The response time for the indicating layer to change color varies for different tubes. The volume sampled should be corrected for pressure and temperature. Long-term tubes are subject to the same cross sensitivities as short-term tubes, and may be more greatly influenced by humidity because the exposure time is longer.[13]

Normally tubes must be used for the length of time specified and the minimum volumes required by the manufacturer if the results are to be used quantitatively. For a screening technique, when it is suspected that very low levels of a contaminant might be present—such as sulfur dioxide, which is capable of rusting metal at levels far below OSHA PELs or the odor threshold—the tubes can be run for up to 24 hours. In this case, the pumps should be run off the battery charger and the results are only qualitative, that is, suggestive that a compound might or might not be present. The tubes should be checked periodically to see if any color change is occurring. During sampling the tubes should be kept in a vertical position. The setup is essentially the same as that used for gas and vapor sampling with sorbent tubes discussed in the chapter on Integrated Sampling for Gases and Vapors, although the flow rates used are much lower than those for sorbent tubes.

Although their accuracy is not as high as the accuracy of tubes analyzed in the laboratory, there are situations where the use of long-term detector tubes can be an advantage. In situations where there is a high degree of employee concern over a potential exposure, these tubes give immediate results that are more representative of employee exposure (integrated) than a series of grab samples using the short-term tubes. Long-term detector tubes can be very useful in situations where a gas or vapor is suspected, in that they can be used as a screening device. They can detect much lower levels than other detector tubes, because they can collect samples for a longer period of time. In a situation where the concern is what is present, rather than how much of a given compound is airborne, these tubes can be used for 8- to 12-hour screening samples.

It can be more difficult to interpret results in these cases due to differences in color changes when contaminants are present at very low levels. It is especially important to remember that these tubes still suffer from the same problem with interferences as the ones used for grab samples. Long-term or long-duration tubes are available for a limited number of compounds, but for some of these chemicals the only other available measurement methods are impingers or real time instruments, for example, ammonia, and the use of long-term tubes is necessary.

Procedure

1. Prior to use, test each batch of tubes with a known concentration of the air contaminant that is to be measured. Observe the manufacturer's expiration date closely, and discard outdated tubes. Refer to the manufacturer's data for a list of interfering materials.

2. Break off the ends of a tube. Calibrate a pump at the flow rate indicated, usually 10 ml/min to 20 ml/min, with a proper length of tubing and a long-term tube in place using a bubble buret.

3. Break off the ends of a fresh tube if there is any possibility of contaminants being present during calibration. Put the pump on a worker or in the area to be sampled. Turn on the pump and record the time or the stroke number if using a stroke volume pump.

4. At the end of the sampling period, turn off the pump and record the time. If the instructions for the tube require that fresh air be pulled through the tube (many long-term tubes require this), take the tube to an area where clean air is present and pull the required amount of air through it. If only a small amount is required, such as 500 ml, then a suction pump such as a bellows pump can be used; otherwise, the battery-operated pump must be used.

5. Read the tube immediately after the last amount of air is pulled through it. Read the length of stain in a well-lighted area. Read the longest length of stain if stain development is not sharp or even (conservative estimate). Many tubes will retain their stain for a period of time if capped off at the ends.

PASSIVE COLOR-CHANGE MONITORS

Dosimeter Tubes

Passive detector, length-of-stain tubes, or stain-length passive dosimeters, as they are also known, have been developed. Their primary advantage over the color-change method is their ease of use in that they are simply opened and hung in place for a period of time. The basic concept is that increasing exposure times results in longer stain lengths. The stain length is thus related to the product of exposure time and ambient concentration.

A length-of-stain dosimeter tube consists of a glass or plastic tube containing an inert granular material or a strip of paper that has been impregnated with a chemical system that reacts with the gas or vapor of interest. As a result of this reaction, the impregnated chemical changes color. The granular material or paper strip is held in place within the glass tube by a porous plug of a suitable inert material. In some cases, a scale is printed on the glass tube to make it easy to read the length of stain of the reacted chemical; in other instances, a scale is provided for comparison of the length of stain. During use, the dosimeter tube is held in a lightweight, plastic holder. The tube holder protects the dosimeter during use and also helps to minimize effects of air currents on performance. The holder has a clip that allows it to be fastened to a collar or pocket during personal sampling, or to some appropriate object during area sampling. These tubes have been developed for only a few chemicals. They are designed to be exposed and read after a certain number of maximum hours.

The dose is derived from length-of-stain dosimeters by reading the length of stain from the calibration curve provided with the tubes. The dose from colorimetric units is obtained by comparing the color intensities provided by the manufacturer with the color of the unit. Other tubes are marked in ppm-hr. At the end of the sampling period, this value is divided by the number of hours in the sampling period to obtain a time-weighted average (TWA).

Some dosimeters are designed to be used over several days, for example, units placed in homes to measure formaldehyde. In this case, they often have a calibration for 1 day's exposure and for 7 days' exposure. These devices are usually designed for screening, that is, determining what type of follow-up measurements should be taken. Results are not considered to be precise.

The sampling rate of colorimetric dosimeter tubes is very slow, on the order of 0.1 ml/min; thus, the "starving" effect of static air, which is a concern for many passive samplers, is not significant for these devices, so that air velocity is not critical. However, a stream of high-velocity air should not be permitted to flow directly into the open end of the tube.

The primary disadvantages are a dependence on face velocity for contaminant collection, need for temperature correction, and sampling or diffusion rates that must be determined independently for each compound. The specificity of these dosimeters is limited by the specificity of the color-change reaction, just as it is for other color-change tubes. Other problems include the possibility of adsorption into the granules and incomplete reaction as the diffusion front passes. The latter could be caused by slow diffusion of the gas or vapor into the pores of the sorbent to react with the indicator sorbed into the pores. For more information on these concepts, see the section on passive collectors in the chapter on Integrated Sampling for Gases and Vapors.

These tubes are susceptible to interferents just as other color-change units are (see Table 12-3). Since these units have maximum end-point exposure constants, high concentrations will result in a rapid saturation of color. The accuracy of dosimeter tubes is generally within ±25%, although some tube types may vary from this percentage, and specific tube accuracy may vary from lot to lot and among manufacturers.

In one laboratory study on passive colorimetric dosimeter tubes for carbon monoxide comparing three commercially available units, it was found that under certain circumstances the tubes tended to overestimate the carbon monoxide levels at higher concentrations and underestimate them at lower concentrations with the best accuracy at 100 ppm. It was important to monitor for humidity.[14]

For sampling, the dosimeter is either "broken" open or removed from its protective wrapper, and the unit is placed vertically on the person's lapel or in a given area and the time is recorded. At the end of the sample period, the time that the unit is removed is also recorded. The length of stain or the color intensity is compared with the standard or measure provided by the manufacturer, often a graph of concentration curves corresponding to stain length at a given time. By finding the stain length on the *y* axis and drawing a line to intersect with the point on a curve that corresponds to the number of hours for which monitoring was done, the concentration for the sampling period can be determined.

VaporGuard tubes from MSA are one example of a dosimeter. The user breaks off one end of the tube, inserts the tube into a holder, and attaches it to the sampling point for up to 8 hours. To determine the gas concentration the user measures the length of the color stain in millimeters from the scale printed on the tube. A calibration graph provides the average gas concentration as a function of stain length and sampling time.

"Spot Plates" and Tapes

Other variations on passive colorimetric sampling are spot plates or badges (Fig. 12-6). They have indicator strips or buttons designed to change color when a critical accumulation of a given gas is reached. The color of the badge is compared to a chart to determine the integrated concentration. Some have the colors corresponding to various concentrations printed directly on them, whereas others must be compared to a color card.

An example of a color-change indicator for hydrogen sulfide, the H_2S DETECTOR, uses a tape that is chemically impregnated with lead acetate and can detect concentrations up to 20 ppm. This chemical reacts with hydrogen sulfide to change the white surface of the tape to lead sulfide, which darkens to a tan-brown. This color is proportional to the concentration of hydrogen sulfide. After 3 minutes, this tan-brown color is then compared to one of five color references located beside the sampling window. These types of detectors can be sensitive to sunlight and high temperatures and should be protected. The use of chemically impregnated tapes as a method of detection in real time instruments is discussed in the chapter on Monitoring Instruments Dedicated to a Single Chemical.

The most important criterion for using tape-based monitors is their need for humidity. Often just breathing on the tape several times will provide enough humidity to activate the tape. Usually constant observation is required, as after a period of time, usually a few minutes, the tape dries out and a new section must be exposed to restart the sampling period.

Some models have white plastic cards with a sensitized or reagent-impregnated pad. The pads are usually covered with plastic to protect them from premature exposure. The "dose" is evaluated by comparing the exposed pad or plate to color standards provided by the manufacturer. For quality control a part of the pad or "spot" should also be kept covered as a control during sampling.

The Carbon Monoxide "Dead STOP" is an example of this type of indicator (Table 12-4). A tan indicator button is located in between two brown squares. The button turns darker upon exposure to carbon monoxide. According to the manufacturer, the discoloration by carbon monoxide depends on the amount of exposure time as well as the concentration. After the badge is removed from its enclosure, it will last up to 30 days unless it is exposed to high concentrations of carbon monoxide during use.

Caution in using these detectors is important. Uneven color changes can occur when the impregnated surface contains large granules. Sometimes only small areas change color on the detection surface, making it difficult to determine what concentration was present. Problems in printing can change the color of the standards that the manufacturers provide. Badges must be shielded from both direct sunlight and rain. The color change for "spot" plates should be recorded immediately following exposure, because the stains can change with time. Prolonged exposure of some pads to light, in the absence of contaminant, will cause the stain to fade.

Badges should not be used to warn employees that levels in excess of the PEL are present. Real time instruments with audible alarms are designed for these situations. In any event, badges should be checked regularly by the wearer to determine if any changes in the color have occurred.

INTERPRETATION OF COLOR-CHANGE MEASUREMENTS

As described in Table 12-2, general certification requirement for the accuracy of tubes is ±25% of the true value when results are at one to five times the PEL, and ±35% at one-half the PEL. Therefore,

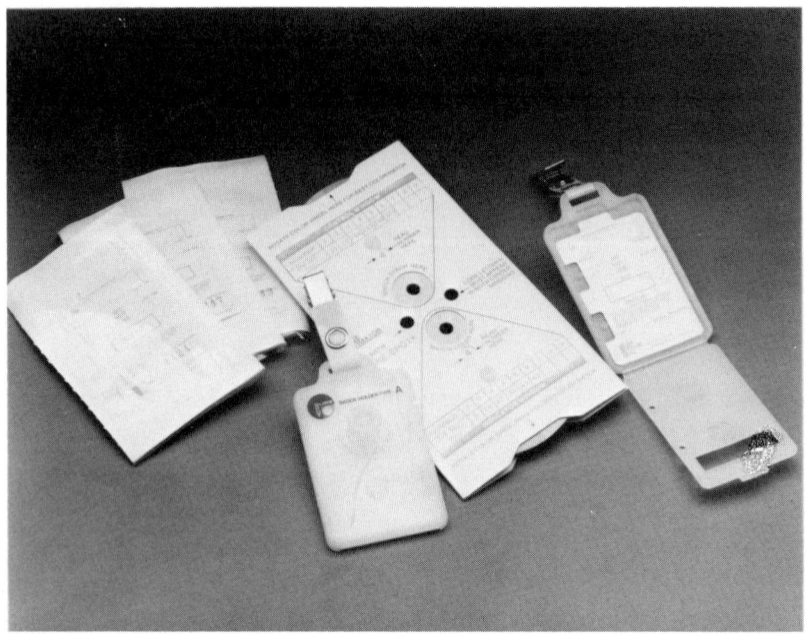

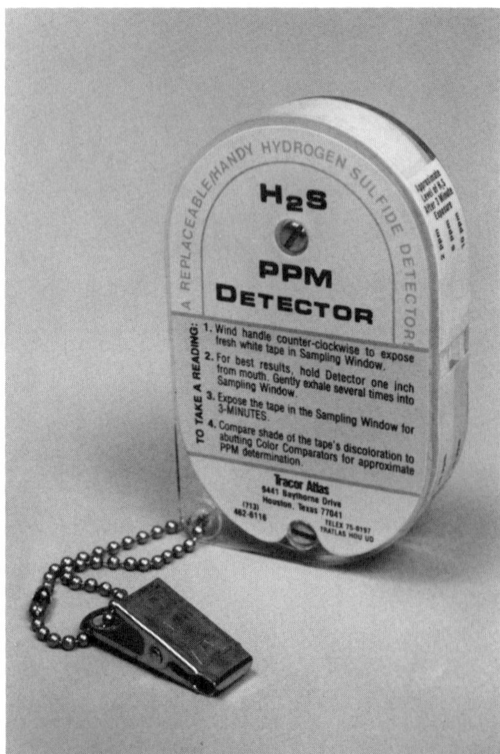

Figure 12-6. Types of colorimetric badges. *Top:* Sure-Spot dosimeter badge (courtesy GMD Systems, Inc.). *Bottom left:* H_2S detector, and *bottom right:* Dead-Stop badge (both courtesy Tractor Atlas, Inc.).

TABLE 12-4. Color Changes of the Carbon Monoxide "Dead STOP" Badge

Concentration (ppm)	*Coloration*	*Minutes*
10–20	No change	60
40–50	Slight darkening	15
95–105	Light to medium copper	5–10
150–160	Medium to dark copper with brown specks	8–20
195–215	Dark copper to brown shade with dark specks	1 8
400–500	Brown to dark brown with black specks	2–10

it is best to present data from color-change tubes as a range of values to reflect this. For example, short-term samples indicating 100 ppm of toluene should be presented as 75 ppm to 125 ppm to take into account the likelihood of 25% error at the TLV level.

The largest relative standard deviation reported by the manufacturer (for the exposure range) should be used in calculating the exposure level. For example, if a reading of 100 ppm is obtained for a substance with a detector tube whose relative standard deviation is reported as being 25% to 35%, the reported reading should be 100 ppm − (100 × 35%) = 65 ppm.

It is a good practice to apply a 50% safety factor to results obtained when detector tubes are used for confined space measurements.

Most manufacturers claim that corrections for temperature and/or atmospheric pressure changes are not required in order to directly compare sampling results to PELs for ambient temperatures (also tube) of 0°C to 40°C and for an RH range of 20% to 90%. However, where the detecting reagent is abnormally sensitive to temperature or humidity, the tube reading must be corrected according to the table or chart supplied with each box of tubes. In addition, tube readings are proportional to absolute pressure, and for altitudes above 5,000 feet, gas law corrections must be applied. For more information on this calculation, see the Overview of Integrated Sampling chapter.

Where there is a gradation of color change, the end point is generally taken as the end of any discoloration, even if it can just barely be made out. If the end of the stain is at an angle, take the reading of the longest and shortest discoloration and get the average.

Some color stains fade or change with time, so tubes must be read immediately following exposures. Sometimes the color change is subtle, such as that of the Draeger nitrogen oxide 0.5, which is a tube that changes from pale gray to blue-gray. Depending on the concentration the tube is exposed to, the actual color change may vary from what is predicted. The visual judgment required to determine the length of stain or color change depends strongly on the color perception of the observer and the lighting conditions. Individuals with color vision problems may find it difficult to use detector tubes. Exposed tubes must be read in an area with daylight or incandescent illumination. Fluorescent lighting may not be good for matching certain colors. A tube being reused because of negative results with prior measurements should be followed up with a fresh tube immediately to confirm a positive result.

Detector tubes are portable, are less expensive on a per sample basis compared to conventional samples that must be analyzed in an analytical laboratory, provide immediate on-site results, and are available for a wide variety of compounds. They are relatively simple to use, and the instrumentation is simple enough that the maintenance compared to other types of direct reading instrumentation is minimal. However, as has been discussed, training is necessary so that measurements can be performed properly or the results may be misleading. Therefore, ease of use should not be confused with the need to read instructions and be aware of the limitations of color-change measurements.

REFERENCES

1. Littlefield, J. B., W. P. Yant, and L. B. Berger. A detector for quantitative estimation of low concentrations of hydrogen sulfide. *US BOM Report*, vol. 3276. Washington, D.C.: Department of the Interior, 1935.
2. Sarner, S. F., and N. W. Henry. The use of detector tubes following ASTM Method F-739-85 for meas-

uring permeation resistance of clothing. *AIHA J.* **50**(6):298–302, 1989.

3. King, M. V., P. M. Eller, and R. J. Costello. A qualitative sampling device for use at hazardous waste sites. *AIHA J.* **44**(8):615–618, 1983.
4. OSHA. *Industrial Hygiene Technical Manual.* Washington, D.C.: U.S. Department of Labor, March 30, 1984.
5. Keenan, R. G. Direct reading instruments for determining concentrations of aerosols, gases, and vapors. In *The Industrial Environment, Its Evaluation and Control.* Cincinnati: NIOSH, 1973.
6. Wilcher, F. E. SEI gas detector tube certification. *Appl. Ind. Hyg.* **3**:R-7–R-8, 1988.
7. Wilcher, F. E. SEI—Committed to high standards of safety. *Appl. Ind. Hyg.* **3**(7):F-7–F-11, 1988.
8. Carlson, D. H., M. D. Osborne, and J. H. Johnson. The development and application to detector tubes of a laboratory method to assess accuracy of occupational diesel pollutant concentration measurements. *AIHA J.* **43**(4):275–285, 1982.
9. Saltzman, B. E. Direct reading colorimetric tubes. In *Air Sampling Equipment,* 6th ed. Cincinnati: ACGIH, 1983, pp. T-1–T-29.
10. McCammon, C. S., W. E. Crouse, and H. B. Carroll. The effect of extreme humidity and temperature on gas detector tube performance. *AIHA J.* **43**(1):18–25, 1982.
11. Stock, T. The use of detector tube humidity limits. *AIHA J.* **47**(4):241–244, 1986.
12. Standard Practice for Measuring the Concentration of Toxic Gases or Vapors Using Detector Tubes. ASTM D-4490-85.
13. Jentzson, D., and D. A. Fraser. A laboratory evaluation of long term detector tubes: Benzene, toluene, trichloroethylene. *AIHA J.* **42**(11):810–823, 1981.
14. Hossain, M. A., and B. E. Saltzman. Laboratory evaluation of passive colorimetric dosimeter tubes for carbon monoxide. *Appl. Ind. Hyg.* **4**(5):119–125, 1989.

CHAPTER 13

Radon Measurements

Radon comes from the radioactive decay of uranium. There are several isotopes of radon, but ^{222}Rn is the most common. Radon-222 is widely distributed through the environment due to its chemical inertness, gaseous state, and relatively long half life. Radon can be found in high concentrations in soils and rocks containing uranium, granite, shale, phosphate, and pitchblende. Radon may also be found in soils contaminated with certain types of industrial wastes, such as the by-products from uranium or phosphate mining.

Radon gas has a half life of 3.8 days and decays in several steps to radioactive particles, called radon daughters. These are, in sequence, isotopes of ^{218}Po, ^{214}Pb, ^{214}Bi, ^{214}Po and, finally, stable ^{210}Pb. Two daughter products, ^{217}Bi and ^{218}Po as well as ^{222}Ra decay with the emission of alpha particles.

Radon can be immobilized in rock. Radon atoms originating from this source may migrate within the rock and subsequently dissolve into ground or surface waters.[1] Consequently, radon can occur in the soil and in water. Direct sources of radon include tap water and seepage of radon gas and natural gas.

Because radon is a gas, it can move through small spaces in the soil and rock on which dwellings are built (Fig. 13-1). Radon in the soil can enter buildings in two ways: (1) by passive diffusion through the air-pore system in the soil, through dirt floors, cracks in concrete floors and walls, floor drains, sumps, joints, and tiny cracks or pores in hollow-block walls; and (2) by pressure-

Figure 13-1. Major routes of radon entry into buildings.

driven flow, which is created by thermal effects. The thermal effects are due to the pressure differential that builds up whenever there are air masses of different temperatures. During the heating season, this buildup gives rise to a pressure gradient, with the net effect that the radon gas is actually pulled into the building as the hot air rises.

Loose, sandy soil promotes radon transfer, whereas clayey, compacted, wet, or frozen soil inhibits gas flow. The soil properties that most influence radon diffusion are soil moisture regime, depth to bedrock, and porosity. Although soil water decreases radon permeability, it can increase chemical weathering; facilitate the transport of the mobile, oxidized form of uranium (hexavalent); and allow more radon isotopes to be emanated in the interstitial pore space of rock instead of remaining embedded into the solid matrix during decay. A characteristic of the upward flow of radon, as well as other gases, is that it is highly variable because geological material tends to be in horizontal formations and due to anisotropic development of soil horizons.[2] Table 13-1 lists risk classification guidelines for radon in soil gas.[3]

In some situations, radon may be released from construction materials, for example, a large stone fireplace or a solar heating system in which heat is stored in large stone beds. Recently the potential for radon exposure in schools has become an issue.[4]

Radon can also enter water in private wells and be released into the air when the water tap is on. Activities and appliances that spray or agitate heated water, such as showers, dishwashers, and clothes washers, create the largest releases of waterborne radon. The airborne contribution of radon from water is usually small compared to that from soil gas. Usually radon is not a problem with large community water supplies.

The primary populations exposed to radon are individuals living in residential housing and uranium and phosphate miners. Another potential source of occupational exposure to radon is natural gas. During fractionation at processing plants, radon may be concentrated in the liquified petroleum gas (LPG) product stream. Radon can contaminate the interior surfaces of plant machinery and when equipment is repaired, exposures can occur.[5] Safe levels for occupational exposures are higher than exposures of concern in the community where exposure can be 24 hours.[6]

The primary adverse health effect associated with radon exposure is lung cancer from inhalation of the radon decay products. Radioactive substances emit three principal types of radiation: alpha, beta, and gamma. The primary hazards associated with radon decay are alpha particles that do not penetrate through the skin. Although they act at very short range, if they are released in the lungs they can do significant cellular damage, including causing lung cancer. Radon gas concentrations are expressed in picocuries per liter (pCi/L) of air or water and radon daughters in working levels (WL). One WL is any combination of radon progeny in 1L of air whose ultimate decay will produce 1.3×10^5 MeV of alpha energy. Figure 13-2 is a radon risk evaluation chart.[3]

TABLE 13-1. Risk Classification Guidelines for Radon in Soil Gas

Classification	*Soil Type and Radon Concentration*
High-risk	Uranium-rich granites, pegmatites, phosphates, and alum shale Highly permeable soils, such as gravel and coarse sand Radon cencentration in soil gas >1,350 pCi/L
Medium-risk	Rocks and soils with low or normal uranium content Soils with average permeability Radon concentration in soil gas from 270 pCi/L to 1,350 pCi/L
Low-risk	Rocks with very low uranium content, such as limestone, sandstone, and basic igneous and volcanic rocks Soils with very low permeability, such as clay and silt Radon concentration in soil gas <270 pCi/L

ENVIRONMENTAL COLLECTION METHODS FOR RADON IN AIR

Several different measurement methods can be used to determine radon concentrations. In prac-

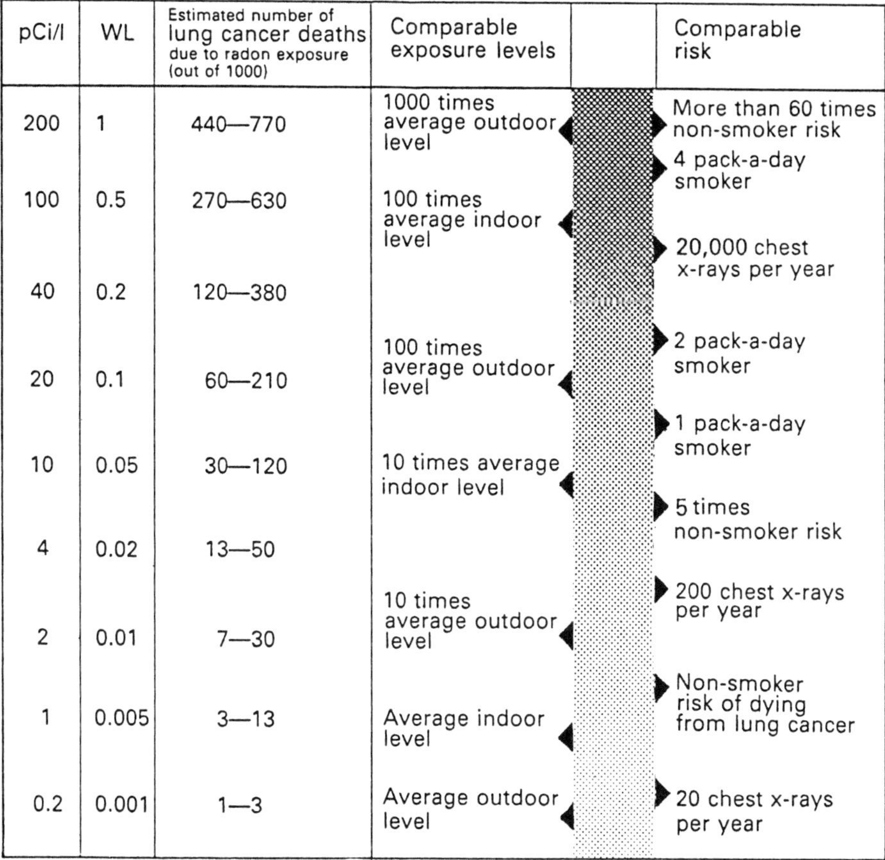

pCi/l
WL
Estimated number of lung cancer deaths due to radon exposure (out of 1000)
Comparable exposure levels
Comparable risk
200
1
440—770
1000 times average outdoor level
More than 60 times non-smoker risk
4 pack-a-day smoker
100
0.5
270—630
100 times average indoor level
20,000 chest x-rays per year
40
0.2
120—380
100 times average outdoor level
2 pack-a-day smoker
20
0.1
60—210
1 pack-a-day smoker
10
0.05
30—120
10 times average indoor level
5 times non-smoker risk
4
0.02
13—50
10 times average outdoor level
200 chest x-rays per year
2
0.01
7—30
Non-smoker risk of dying from lung cancer
1
0.005
3—13
Average indoor level
0.2
0.001
1—3
Average outdoor level
20 chest x-rays per year

Figure 13-2. Radon risk evaluation chart.

tice, the choice of a method is often dictated by availability. If alternate methods are available, then the cost or the duration of the measurement may become the deciding factor.

Sampling methods for radon are designed to detect either radon or its decay products, radon daughters, and the basis for the measurements is the detection of alpha particles from radon decay. When ionizing radiation measurements are discussed often the Geiger counter is the first instrument that comes to mind. However, this detector does not usually provide accurate measurements of low-level alpha particles, so other methods are used instead.

The techniques for measuring radon decay and its decay products are different. There are three main decay products whose activities are measured: ^{218}Po, ^{214}Pb, and ^{214}Bi. Generally the measurements combine the counts from all three. There are three basic types of samples: short-term, continuous, and grab. Short-term sampling is the most common, and involves the use of charcoal canisters and alpha-track detectors. These samples are generally used for screening measurements. Continuous sampling is done with real time instruments and is often used to determine the effectiveness of controls or for very long-term monitoring. Grab sampling, the most involved method, is generally reserved for calibrating other methods and collecting occupational samples. It is sometimes called scintillation flask, Kusnetz, or modified Tsivoglou. Table 13-2 provides an overview of sampling methods.

Short-term screening is useful for comparative

TABLE 13-2. Overview of Radon Measurement Methods

Instrument	*Sampling Times*
Charcoal canister	2 to 7 days
Alpha-track detector	3 months (or less if laboratory uses adequate lower limit of detection)
Radon progeny integrating sampling unit (RPISU)	100 hours minimum, 7 days preferred
Continuous Working Level monitor	6 hours minimum, 24 hours or longer preferred
Continuous radon monitor	6 hours minimum, 24 hours or longer preferred
Grab Working Level	5 minutes
Grab radon	5 minutes

measurements in various locations in the same building or in different buildings. Results of remediation activities can be verified by doing short-term measurements. Long-term measurements will give a more valid measurement of actual radon concentrations in that seasonal and other variations will be averaged out.

The equilibrium factor (EF) is an index of the extent to which radon gas has reached equilibrium with its daughters. This index is useful to describe the efficiency of air exchange, especially in confined spaces such as caissons or basements. The EF is defined as:

$$EF = \frac{100\ WL}{Rn}$$

where WL = working level and Rn = radon gas concentration. A high EF reflects a lack of air exchange, leading to rapid equilibrium between radon gas and its daughters. The maximum EF value of 1 could be obtained in an unventilated area.

Problems encountered when measuring indoor radon and radon decay product concentrations include variability due to nonstandardized procedures, different house conditions prior to and during measurement, seasonal and other weather conditions, and different interpretations of the results. For example, in the case of a house next to a lake, rising water level in early summer helps drive radon into the basement through the sump hole. In the winter the lake is frozen and the water table is considerably lower. Because of seasonal trends several measurements are required over a year to determine the actual radon levels in a building. It has been suggested that a 24-hour measurement once a month would be sufficient to determine the seasonal trend and annual average for most buildings.[7]

Measurements should be made under "closed-house" conditions. To a reasonable extent, windows and external doors should be closed for 12 hours prior to and during the test. Normal entrances and exits are permitted, but should be limited to brief opening and closing of doors. In addition, external–internal air-exchange systems, such as high-volume attic and window fans, should not be operating.[8]

In northern climates where the average daily temperature is less than 4.4°C, measurements should be made during the coldest months of the winter season. In southern areas that do not experience extended periods of cold weather, an attempt should be made to identify if there are time periods when closed-house conditions normally exist. Air-conditioning systems that recycle interior air can be operated during the closed-house conditions.

Measurements of 3 days or less should not be conducted if severe storms with high winds are predicted. Severe weather will affect the measurement results in a variety of ways. A high wind will increase the variability of radon concentration due to wind-induced differences in air pressure between the house interior and exterior. Rapid changes in barometric pressure increase the chance of a large difference in the interior and exterior air pressures, therefore changing the rate of radon influx. Measurements should also not be made if remodeling, changes in the heating, ventilating, and air-conditioning (HVAC) system, or other modifications that may influence the radon concentration during the measurement period are planned.

Schools and other public buildings should conduct measurements on the weekends, so that closed conditions can be more easily satisfied. Ventilation systems should not be shut down or operated at a reduced rate during the time when the measurement is made.[7]

A sampling site must be selected for stationary monitors where they will not be disturbed during the measurement period, which is at least 20 inches above the floor with the detector's top face at least 4 inches from other objects and well away from exterior house walls. The unit should sit in open air that people might breathe rather than in a closet or a drawer. All measurements should be made away from drafts caused by HVAC vents, doors, windows, and fireplaces. The influence of changes in ventilation and on radon and radon decay product concentrations will be reduced.

Samples should be collected in areas permanently occupied (office space, break areas) and areas routinely visited (mechanical areas, storage areas, work shops, garage areas) by employees of the building on floors directly in contact or directly over (crawl spaces) the ground. Rooms should be selected that are expected to have the lowest ventilation rates, such as interior rooms with no windows and tight doors. Avoid locations near excessive heat or in direct, strong sunlight and areas of high humidity. Detectors can be placed close to, but not directly in, inaccessible areas, such as closets, sumps, crawl spaces, or nooks within the foundation. Detectors should be placed inside the mixing chambers in HVAC systems where the return air from the ground level mixes with outside air before recirculation to other areas of the building. In gymnasiums or buildings with large open rooms, detectors can be placed every 2,000 square feet. A knowledge of the building's profile (structure, mechanical areas, utilities) will assist in identifying the sampling areas. Review below-grade (basement) floor plans to help determine sampling positions.

Documentation is critical so that data interpretations and comparisons can be made. It includes:

- Start and stop times, and the date(s) of the measurement
- Information about how the standardized conditions have been satisfied
- Exact location of the monitor, including a floor plan of the building
- Any other useful information, including type of building, heating system, existence of a crawl space, smoking habits of the occupants, operation of humidifiers, air filters, electrostatic precipitators, or clothes dryers

The primary types of short term passive monitors for radon are charcoal canisters and alpha track detectors. There are a number of variations on these. Some are designed for shorter sampling periods.

Charcoal Canisters

Charcoal canisters are passive integrating devices that measure radon gas directly. They consist of containers most often shaped like a respirator cartridge or round canister with a screw top filled with a measured amount of activated charcoal. In one version, one side of the container is fitted with a screen that keeps the charcoal in but allows air to diffuse into the charcoal when the cover is removed. When the canister is prepared by the supplier, it is sealed until it is ready to be used. Another type of charcoal detector is shaped like a large pouch into which room air can diffuse. The top of the container is usually perforated to allow collection without loss of charcoal. Some have a filter to keep out radon-decay products. All charcoal adsorbers should be stored in airtight containers when not being used for sampling.

Charcoal canisters are generally set out for 4 to 7 days, although some recent EPA guidelines recommend only 2 days.[7] Sealed environmental conditions in the testing area (or building) for 12 hours prior to and following the test are required for accurate results. Laboratory analysis is performed using a sodium iodide gamma-scintillation detector to count the gamma rays emitted by the radon-decay products on the charcoal. Gamma rays of energies between 0.25 MeV and 0.61 MeV are counted. The detector may be used in conjunction with a multichannel gamma spectrometer or with a single-channel analyzer with the window set to cover the appropriate gamma-energy window.

A correction must be made for the reduced sensitivity of the charcoal due to adsorbed water. It may be done by weighing each canister when it is prepared, and then reweighing it when it is returned to the laboratory for analysis. Any weight increase is attributed to water adsorbed on the charcoal. The weight of water gained is correlated to a correction factor that should be empirically derived. The correction is unnecessary if the charcoal canister configuration is modified to significantly reduce the adsorption of water, and if the user has experimentally demonstrated that, over a wide range of humidities, there is negligible change in the collection efficiency of the charcoal.

Advantages of charcoal canisters include

- Low cost per canister
- No special skills required for use
- Convenient to handle and install
- Unobtrusive when installed
- No external source of power needed
- Precise results with proper analytical techniques

Disadvantages include

- Some charcoal adsorbers are more sensitive than others to temperature and humidity.
- Canisters can only be used for short-term testing, that is, 7 days or less.

The primary use of canisters is as screening devices; due to the short sampling period their accuracy is lower than other passive methods. The downside of short-term monitoring is that it may detect an unrepresentative peak or a valley of radon concentration, causing a false sense of alarm or security depending on the situation.

Calibration

Duplicate canisters should be placed in enough rooms or buildings to monitor the precision of the canisters, that is, approximately 10% of the number of sites monitored in a month, or 50 duplicates, whichever is less.[7] The duplicate canisters should be shipped, stored, exposed, and analyzed under the same conditions.

Procedure

1. Prior to sampling, make sure for 12 hours before and during the 4- to 7-day measurement period that windows and external doors are kept closed, except for normal entry and exit. Fans or ventilation systems that use outside air, such as attic fans, must not be operated.

2. Put a unique number on the canister.

3. Record on the sampling data sheet the type of building in which the measurement is being made: single-family home, multifamily dwelling, business, school, other.

4. Generally, if the building being sampled has a basement, this place is the best to put the canister. Garages root cellars, and crawl spaces have too much ventilation and are undesirable. If there is no basement, place the canister in any room on the lowest floor of the house, except in a bathroom, kitchen, or porch. Record the floor of the building where the measurement is being made.

5. Within the selected room, the canister should not be in a location frequently exposed to noticeable drafts of an open door, window, fireplace, and so forth. The canister should be exposed to the same air that the building's occupants are breathing. It should be placed on a table or shelf at least 2 feet above the floor, and should be in the open air, not in a closet, drawer, or cupboard. Record the room and the exact location in which the measurement is to be made: bedroom, family room, office, living room, unfinished basement, classroom, or other.

6. Record the date and time, and unscrew the cap or lift off the tape to start the measurement. Some canisters are designed to be exposed for 48

hours (2 days), and others for 72 hours (3 days), 96 hours (4 days), or 168 hours (7 days). A deviation from the schedule of up to 6 hours is acceptable as long as the actual time of termination is documented.

7. At the end of the sampling time, replace the cap or the tape on the canister. Record the date and time and immediately send the canister to the laboratory.[7]

Alpha-Track Detectors

The alpha-track (AT) method measures radon. It is considered a short-term passive monitor. An AT detector is a small sheet of special plastic material (usually polycarbonate) enclosed in a container with filter-covered opening. Alpha particles that strike this plastic cause microscopic markings while other types of radiation pass through without causing any changes. At the end of the measurement period, the detectors are returned to a laboratory, where the plastic is placed in a caustic solution that accentuates the damage tracks so they can be counted using a microscope or an automated counting system. The number of tracks per unit area is correlated to the radon concentration in air, using a conversion factor derived from data generated at a calibration facility.

These are passive devices and require no external source of power. Some AT devices look like large pill boxes; others look like clear plastic drinking glasses. Alpha particles emitted by the radon-decay products strike the plastic and produce submicroscopic damage tracks.

When these detectors are used according to guidelines in the EPA protocol, they are left for periods of up to 3 months for screening and 12 months for follow-up measurements.[7] At the end of the desired testing period, they must be sent to an analytical laboratory for processing and evaluation. The plastic from the detectors is placed in a caustic solution that enlarges the damage tracks so they can be counted using a microscope or an automated counting system. The number of tracks per unit area is converted to an average radon gas concentration in air for the measurement period.

These passive detectors require no special skills to use, and can measure the long-term average concentrations over a 12-month period, which is the optimal measure of long-term concentrations. During this period, closed-house conditions do not have to be satisfied. The long measurement period, a 3-month minimum, limits the situations for which these devices can be used.

Many factors contribute to the variability of AT detector results, including differences in the detector response within and between batches of plastic, nonuniform platcout of decay products inside the detector holder, differences in the number of tracks used as background, variations in etching conditions, and differences in readout. The variability in AT detector results decreases with the number of net tracks counted, so counting more tracks over a larger area of the detector will reduce the uncertainty of the result. In addition, use of duplicate AT detectors will reduce error.

The sensitivity of an AT detector system is dependent on the area of the detector that is counted for alpha tracks (Table 13-3).[7] At a minimum, the detector should provide an adequate sensitivity at 4 picocuries per liter (pCi/L).

Advantages of AT detectors include

- Low cost per detector
- Easy to handle and install
- Unobtrusive when installed
- No training required for use
- No external power source needed
- Ability to measure the integrated average concentration over a 12-month period, which is the optimal measure of long-term concentration

TABLE 13-3. Precision of Alpha-Track Detectors

Number of Net Tracks Counted	*2 Sigma Error (%)**
4	100
6	82
10	63
15	52
20	45
50	28
75	23
100	20

*This is the minimum error for the number of net tracks indicated; the absolute error is dependent on the actual number of background tracks counted.

Disadvantages include

- Relatively long measurement period needed, minimum of 3 months
- Large inherent variability (precision errors), especially at low concentrations, if the area of the detector that is counted is small

Maintenance and Calibration

Determination of a calibration factor requires exposure of AT detectors to a known radon concentration in a radon exposure chamber. These calibration exposures are to be used to obtain or verify the conversion factor between net tracks per unit area and radon concentration. AT detectors should be exposed in a radon chamber at several different radon concentrations, or exposure levels similar to those found in at least three tested buildings. A minimum of 10 detectors should be exposed at each level. The period of exposure should be sufficient to allow the detectors to achieve equilibrium with the chamber atmosphere. A calibration factor should be determined for each batch of detector material received from the material supplier.

AT detectors should be used as soon as possible after delivery from the supplier. If the storage time exceeds more than a few months, the background exposures from a sample of the stored detectors should be assessed.

Always review the manufacturer's instructions furnished with the sampler. As with all passive detectors, measurements should be collected in an area away from drafts that can be caused by heating and air conditioning systems, openings such as doors and windows, or ventilation from cracks and openings in exterior walls. A rule of thumb is the placement of a detector for each 2,000 square feet of floor space. In basements note walls that penetrate the solid and place one detector here. Except in this case, avoid locating the detector near concrete or masonry walls.

Procedure

1. Remove the protective covering by cutting the edge of the bag, or remove it so that it can be reused to reseal the detector at the end of the exposure period. The monitoring period begins when the detector is exposed.

2. Inspect the detector to make sure the front is intact and has not been physically damaged in shipment or handling.

3. It is usually convenient to suspend the detector from the ceiling or on a joist. Some have adhesive strips for this purpose. A thumb tack can also be used. The detector should be positioned at least 8 inches below the ceiling. However, it can also be placed upright with holes showing at least 2–3 feet above the floor on an exposed surface.

4. Record the serial number of the detector in a log book along with a diagram showing the location in the room where the detector was placed. If during the exposure period it is necessary to relocate the detector, make certain it is noted in the log book, along with the date it was relocated and recorded on the diagram.

5. At the end of the measurement period, the detector should be inspected for damage or deviation from the conditions entered in the log book at the time of installation. Any changes should be noted. The date of removal is entered on the data form for the detector and in the log book. The detector is then resealed using the protective cover provided with the detector, sometimes an adhesive seal, or a bag with the correct serial number for that detector. If a bag is used, the open edge of the bag is folded several times and resealed with tape. If the bag or cover has been destroyed or misplaced, the detector should be wrapped in several layers of aluminum foil and taped shut. After retrieval, the detectors should be returned as soon as possible to the analytical laboratory for processing.[7]

Comparison of Charcoal and Alpha-Track Detectors

Measurements made with AT detectors give a better estimate of average radon concentration than charcoal canisters. AT detectors are better integrating devices, and are not as affected by fluctuating radon levels. AT detectors do not require immediate analysis. Unlike charcoal canisters, no radon decay occurs in the AT detector once the measurement is taken. Therefore, the time between when the radon measurement is completed and when the devices are shipped is not as critical, allowing large numbers of detectors to be handled.

AT detectors cost about one-third more than charcoal canisters. They can be tampered with to yield inaccurate results. The likelihood of tampering is increased due to the long period required for measurement.

Grab Sampling Methods

The major differences in various grab sampling methods are the total time period required for sampling and analysis, the capability of determining exposure concentration at the work site, and the amount of routine maintenance and calibration requirement of an instrument. Grab sampling methods measure concentrations of radon gas or radon decay product concentrations.

Some detector systems can sample both radon and radon decay products simultaneously. Several samples can be collected and analyzed each day. Due to the short measurement periods, the samples may not be representative of the typical concentrations to which people may be exposed. Careful control of the environment to assure closed conditions is required for 12 hours prior to and following the test. Because of the highly reactive nature of radon decay products, WL measurements are more susceptible to sampling error than radon gas measurements.

While an individual grab sample may be quite accurate in representing the concentration of radon or radon decay products at the moment of sampling, it is usually a poor indicator of the long-term average concentration because of the inherent variability of the radon concentration in homes. For this reason, grab sample results should be interpreted with caution..It is especially important to adhere to closed hose conditions for 12 hours prior to grab sampling measurements.

Grab sampling is useful for sites where high concentrations are suspected because of the need to implement control measurements as soon as possible. They can also be used as diagnostic tools to trace the probable cause of elevated levels in a building. Multiple grab samples are preferred to single grab samples, especially when making decisions as to the need for remedial action. Grab samples are not recommended for follow-up measurements because of their poor correlation with long-term averages. Advantages of grab sampling include

- Results are quickly obtained.
- Equipment is portable.
- Both radon and its decay products can be measured at the same time.
- Several samples can be run per day.
- Conditions during measurement are known to the sampling professional.

Disadvantages include

- Since the measurement period is very short, the results may not be representative of the long-term average.
- Sampling requires considerable skill.
- House conditions must be under careful control for 12 hours prior to measurements.
- Cost of a system is relatively expensive.

Critical factors in assuring accuracy when collecting grab samples include the proper calibration of radiation detectors and pumps, filters that precisely fit the equipment, and accurate maintenance of the flow rate during the sampling period. It is also important to prevent contamination of the pump, counting equipment, and filters.

Grab Sampling Methods for Radon

Grab samples for radon are very quick determinations of the concentration collected in scintillation cells holding 100 to 2,000 cubic centimeters (cc) of air. In this method a sample of air is drawn into these cells that have a zinc sulfide phosphor coating on the inner surfaces with a clear window at one end. To take a measurement, the cell is evacuated, taken to the sampling location, opened to allow room air to rush in, and then sealed to trap the radon inside.

The cell is counted within 4 hours after filling to allow the short-lived radon daughter products to reach equilibrium with the radon. In the laboratory, the cell is put into contact with a photomultiplier tube and tiny light flashes (scintillations) produced by radon decay products striking the cell's zinc-sulfide-coated interior are electronically detected. The number of light flashes is proportional to the radon in air concentration. Correction factors are applied to the counting results to compensate for decay during the time between collection and counting to account for decay during counting.

For accurate measurements it is necessary to standardize cell pressure prior to counting, because the path lengths of alpha particles are a function of air density. The counting system, consisting of the scaler, detector, and high-voltage supply, must be calibrated. The correct high voltage is determined via a plateau. Each counting system should be calibrated before being put into service, after any repair, or at least once per year. Also, a check source or calibration cell should be counted in each system each day to demonstrate proper operation prior to counting any samples.

An accurate calibration factor must be obtained for each counting cell. This is done by filling each cell with radon of a known concentration and counting the cell to determine the conversion factor of counts per minute per picocurie. The known concentration of radon may be obtained from a radon calibration chamber or estimated from a bubbler tube containing a known concentration.[8]

Procedure

1. Prior to collection of the sample, counter efficiency must be verified and a background measurement must be taken.

2. Evacuate the cell (Lucas type) to at least 25 inches of mercury, attach the filter to the cell, and open the valve, allowing the cell to fill completely with air. Allow at least 10 seconds for the cell to completely fill. To assure a good vacuum at the time of sampling, the cell may be evacuated using a small hand-operated pump in the room being sampled. It is a good practice to evacuate the cell at least five times, allowing it to fill completely with room air each time. Make sure the air to be sampled flows through the filter each time. If it can be demonstrated that the cells and valves do not leak, the cells can be evacuated in a laboratory.

3. With a double-valve, flow-through-type cell, attach the filter to the inlet valve and a suitable vacuum pump to the other valve. The pump may be motor-driven or hand-operated. Open both valves and operate the pump to flow at least 10 complete air exchanges through the cell. Stop the pump and close both valves.

4. Record all pertinent sampling information after taking the sample, including the date and time, cell number, name of person collecting the sample, and any other significant conditions within the building or notes regarding weather conditions.

5. After the cells have been counted and the data recorded, the cells must be flushed with nitrogen to remove the sample. Flow-through cells should be flushed with at least 10 volume exchanges at a flow of about 2 Lpm. Cells with single valves are evacuated and refilled with nitrogen at least five times. The cells are left filled with nitrogen and allowed to sit overnight before being counted for background. If an acceptable background is obtained, the cell is ready for reuse.[8]

Grab Sampling for Radon Decay Product Measurements

Grab sampling methods for radon daughters involve drawing a known volume of air through a filter and counting the alpha activity during or following sampling. They are collected by drawing a known volume of air through a filter using an air sampling pump for very short sampling periods, usually 5 minutes. The radon decay products, if present in the air, are collected on the filter. Filters are counted for alpha particle emissions during mathematically determined periods after the sample is collected. There are one-count methods where the alpha activity is determined over a single sampling period and two-count methods involving two sampling periods with the ratio of results used for the calculation. Another method known as the three-count derives radon daughter concentrations from the relative changes in the measurements taken at three 30-minute intervals. A single count method is known as the Kusnetz method and a three-count method is known as the Tsivoglou method. The Kusnetz method is the most common.

There are two methods commonly used to do grab sampling measurements of radon decay products: Kusnetz and modified Tsivoglou. The primary difference between the two is in the analytical phase. For a 5-minute sampling period (10–20 liters of air) on a 25-mm filter, the sensitivity using either method is approximately 0.005 working level.

TABLE 13-4. Kusnetz Factors

Time (min)	*K(t)*	*Time (min)*	*K(t)*
40	150	66	98
42	146	68	94
44	142	70	90
46	138	72	87
48	134	74	84
50	130	76	86
52	126	78	78
54	122	80	75
56	118	82	73
58	114	84	69
60	110	86	66
62	106	88	63
64	102	90	60

Kusnetz Procedure

1. Place a new filter in the holder prior to entry to the building where sampling will be done. Care should be taken to avoid puncturing the filter and to avoid leaks.

2. Start the pump and the clock simultaneously. Note the flow rate and record it on the sampling data sheet. Collect up to 100 liters of air on a filter over a 5-minute sampling period. The sampling time should be carefully monitored and recorded.

3. Remove the filter from the holder using forceps and carefully place it facing the scintillation phosphor. The side of the filter on which the decay products were collected must face the phosphor disc. The chamber containing the filter and disc should be closed and allowed to adapt to the dark prior to starting the count.

4. The total alpha activity on the filter must be counted at some time between 40 and 90 minutes after the end of sampling. Counting can be done using a scintillation counter to obtain gross alpha counts for the selected period. Counts from the filter are then converted to disintegrations using the appropriate counter efficiency.

5. The disintegrations from the decay products collected from the known volume of air can then be converted into working levels using the appropriate Kusnetz factor (Table 13-4) for the counting time utilized and the following calculation:

$$WL = \frac{C}{K(t)VE}$$

where

WL = working level
C = sample cpm − background cpm
K(t) = factor determined from Table 13-4 for the time from the end of collection to the midpoint of counting
V = total sample air volume in liters (flow rate × time)
E = counter efficiency in counts per minute (cpm) or disintegrations per minute (dpm)

Tsivoglou Procedure

1. Collect a sample on a filter using the same method as described in the Kusnetz procedure.

2. Remove the filter using forceps and carefully place it facing the scintillation phosphor. The side of the filter on which the decay products were collected must face the phosphor disc. The chamber containing the filter and disc should be closed and allowed to adapt to the dark prior to starting counting. If the counter used has been

slow to dark adapt, the counting procedure should be done in a darkened environment.

3. The filter must be placed into the counting position very quickly, since the first of the three counts must begin 2 minutes following sampling. Count the filter at the following intervals following collection: 2–5, 6–20, and 21–30 minutes.

4. Using the following equations that are based on a 3.11-minute half life for ^{218}Po, calculate concentrations of radon daughter products for each interval[8]:

$$C2 = 1.(0.16746\, G1 - 0.0813\, G2 + 0.0769\, G3 - 0.0566\, R)/(F)(E)$$

$$C3 = 1.(0.00184\, G1 - 0.0209\, G2 + 0.0494\, G3 - 0.1575\, R)/(F)(E)$$

$$C4 = 1.(-0.0235\, G1 + 0.0337\, G2 - 0.0382\, G3 - 0.0576\, R)/(F)(E)$$

$$WL = (1.028 \times 10^{-3})(C2 + 5.07 \times 10^{-3}) \times (C3 + 3.728 \times 10^{-3})\, C4$$

where

C2 = concentration of ^{218}Po in pCi/L
C3 = concentration of ^{214}Pb in pCi/L
C4 = concentration of ^{214}Po in pCi/L
F = flow rate in Lpm
E = counter efficiency in cpm/dpm
G1 = gross alpha counts for 2–5-minute interval
G2 = gross alpha counts for 6–20-minute interval
G3 = gross alpha counts for 21–30-minute interval
R = background counting rate in cpm
WL = working level

Continuous Monitors

Continuous radon and continuous WL monitoring measurement methods are similar in that they both use an electronic detector to accumulate and store information related to the periodic (usually hourly) average concentration of radon gas or radon-decay products. The primary difference between methods is that a continuous WL monitor samples the ambient air collected on a filter cartridge at a flow rate of about 0.1–1.0 Lpm, while a continuous radon instrument samples the air after it has passed through a filter. Radon WL instruments are generally faster than radon instruments (0.5 hours versus 3-hour response time).

A continuous monitor consists of three parts: a pump, a filter to remove dust and radon-decay products, and a scintillation cell. These monitors sample the radon in the ambient air that is collected in a scintillation cell after passing through a filter that removes dust and radon-decay products. As the radon in the cell decays, ionized radon-decay products plate out on the interior surface of the scintillation cell. The alpha particles that are produced strike the coating on the inside of the scintillation cell, causing scintillations to occur that are detected by a photomultiplier tube. The resultant electronic signals are processed and the data are either stored in the memory of the continuous monitor or printed on paper tape by the printer. These units can be the flow-through cell type or the periodic-fill type. In the flow-through cell type, air continuously flows into and through the scintillation cell. The periodic-fill type fills the cells once each preselected time interval, counts the scintillations, then begins the cycle again.

The At Ease (Fig. 13-3) radon monitor is an example of a portable continuous monitor using a silicon detector that provides short-term and long-term measurements of radon and radon daughter concentrations. Ambient air diffuses through a filter into the detection chamber. As the radon gas decays alpha particles are counted using a solid state detector. A microcomputer records the pulses and computes the radon concentration from an internal calibration factor. The measurement range is from 0.1 to >1000 pCi/L. The unit is capable of measuring ±25% accuracy or 1 pCi/L, whichever is greater. The calculated lower limit of detection as stated by the manufacturer is 1 count per hour per pCi/L.

The device has three indicator lights and two push-button controls. The red light indicates measurement at or above threshold. The green light indicates a measurement below threshold, and the yellow light blinks to indicate detection of an alpha particle. There is also a cumulative readout button. By pushing the button marked "average," the digital display will light and give the average long-term radon gas concentration for the entire period since the last time the memory was cleared. If the "current" button is pushed, the display indicates the average radon gas concentration for the last 12-hour period.

A data logger/detector that can store up to 90 data points in memory is also available. This amount represents the average radon gas concentration measured during each interval. The unit requires AC current, but if power is interrupted internal batteries supply power for up to 7 hours. Each concentration value is graphically represented in the same order, since the device is equipped with tamper-resistant features that inform the operator of any movement or power disconnect that may invalidate the test.

A printer is available for the data logger unit that will provide hard-copy reports on demand. It prints the average radon gas concentration for each measurement interval in sequential order. A continuous radon monitor requires electrical power. It can be used to look at changes in radon concentrations over periods of minutes to hours to days.

Errors that may occur with continuous monitors include inherent errors caused by measuring events instead of alpha energy, absorption errors that affect the system's sensitivity and inherent error, statistical errors from counting random events, calibration errors, airflow rate errors, radon daughter plate out error, instrument instability and nonlinearity errors, and external background activity errors.[9]

Most often monitors are turned on or programmed for the desired operating time—a minimum of 6 hours for screening and 24 hours for follow-up measurements. At the end of the test, the monitor must be retrieved and the results must be analyzed by a skilled instrument operator.

In an EPA evaluation the At Ease radon monitor was subjected to varying conditions of relative humidity and temperature, and in each case no significant change in sensitivity or accuracy was observed.[10] The average coefficient of variation in this evaluation was 8.3%. The unit passed the evaluation. There may be problems with consistency in measurements if the reset button is not pushed long enough for the unit to completely clear and reset. One way to do this is to take a second reading after clearing to verify the unit's memory has been cleared. Each instrument should have its respective calibration factor verified at least once a year.

Figure 13-3. The At Ease continuous radon monitor. (Courtesy Nuclear Associates.)

Maintenance and Calibration

These monitors must be calibrated in a known radon environment to obtain the conversion factor used by the electronics to convert count rate to radon concentration. After every 1,000 hours of operation, units should be examined to check the background rate by purging with clean, aged air or nitrogen. A second check for background count rate is done by operating the instrument in an outdoor or other low-radon environment. Twice a year the unit's response to a known radon concentration should be measured, and the flow rate of the pump should be measured.

Use

The continuous monitor should be programmed to run continuously, recording the hourly integrated radon concentration measured and, if applicable, the total integrated average radon concentration. The sampling period should not be less than 24 hours.

Prior to and after each measurement, test the monitor to verify that the correct input parameters and the unit's clock are set properly and to verify that the pump is operating properly.[8]

Continuous Working-Level Monitors

As noted, the continuous working-level (WL) monitor samples the ambient air by filtering airborne particles as the air is drawn through a filter cartridge at a low flow rate of about 0.1 Lpm to 1 Lpm. An alpha detector, such as a diffused-junction or surface-barrier detector, counts the alpha particles produced by ^{218}Po and ^{214}Po as they decay on the filter. The detector is normally set to detect alpha particles with energies between 2 MeV and 8 MeV. The event count is directly proportional to the number of alpha particles emitted by the radon-decay products on the filter. Total counts over a specific time period can also be measured.[8]

An example of this type of instrument is the Thomson and Nielsen Radon WL Meter (Fig. 13-4). The Radon WL Meter operates on the same basic principles as most other active WL meters. Air containing radionuclides is sampled through a slot in the top of the instrument via a small continuously operating pump and is deposited on a standard 0.8-μm MCE filter. Alpha particles are detected by a semicustom digital-integrated circuit (silicon detector) and counted with their total displayed as alpha counts. The air sampling and alpha counting are controlled by a timer that can be preset to operate for 0.5–8 hours, or optionally, 24 hours. Background gamma radiation does not interfere as the instrument counts alpha particles only. At the end of the air sampling period, the total alpha count is displayed on the readout and the WL is manually calculated using the alpha counts and the instrument's calibration factor. The instrument can be operated with AC for long periods or with a battery for shorter periods.

These instruments require a relatively short measurement period. Hourly results can track the variation of concentrations present. Most models are very precise, and results are available on-site following the measurements.

Prior to a survey, careful control of conditions are necessary to assure the environment has been sealed for 12 hours prior to following the test. Because of the highly reactive nature of radon-decay products, WL measurements are more susceptible to sampling error than radon gas measurements.

Advantages of continuous radon monitors and continuous WL monitors include

- Relatively short measurement durations—a minimum of 6 hours for screening, 24 hours for follow-up measurements
- Hourly results that allow identification of variations of building concentrations
- Small precision error
- Result available on site

Disadvantages of these monitors include

Figure 13-4. The Radon WL Meter continuous radon monitor. (Courtesy Thomson and Nielsen Electronics Ltd.)

- Higher cost than most other methods
- Heavy and awkward to move some models
- Extensive calibration requirements utilizing a radon calibration chamber
- Trained operators necessary

Procedure for Maintenance and Calibration

Calibration is done by the manufacturer in a calibration facility and a calibration factor is printed on a label on the back of the instrument. The EPA recommends recalibration by participation in a semiannual laboratory intercomparison program in which the meter's response is compared to a known radon-decay product concentration. Recalibration is also necessary after repair or modification of the instrument. A checkout should be performed prior to use to check the detector calibration and the pump performance.

1. Turn over the instrument and open the filter holder compartment. Remove the cap from the filter holder and replace the filter with a capped alpha-emitting check source using tweezers so as to prevent damage from occurring to the active surface of the source and to avoid contact of the source with the skin. The check source used must be the one supplied by the manufacturer of the instrument.

2. Set the timer to 30 minutes and turn on the instrument but not the instrument's pump.

3. Note the alpha counts displayed on the readout at the end of the 30-minute period and divide this count by 30 to determine the calibration check factor in cpm. Compare this factor with that provided by the manufacturer of the instrument. Generally they should agree within 10%.

4. The flow rate is checked using the methods discussed in Appendix A. The calibration factor is used to convert the number of counts into average milli-working-level (mWL) using the following calculation:

$$mWL = \frac{N}{(T - 0.5)(CF)}$$

where

N = total number of counts at the end of the sampling period
T = sampling time in hours
CF = calibration factor in counts per hour per mWL
1 WL = 10^3 mWL

After every 100 hours of operation, the unit should be checked to measure the background count rate using the procedures that may be identified in the operating manual for the instrument. Twice annually the unit's response should be compared to a known radon-decay product concentration using a calibration chamber and the flow rate of the pump.

Procedure for Use

The continuous WL monitor should be programmed to run continuously, recording the hourly integrated WL measured and, when possible, the total integrated average WL. The sampling period should not be less than 24 hours for most purposes. The longer the operating time, the smaller the uncertainty associated with the measurement result. The integrated average WL over the measurement period should be used as the measurement result.[8]

Prior to a survey, careful control of conditions is necessary to assure the environment has been sealed for 12 hours prior to and following the test. Because of highly reactive nature of radon-decay products, WL measurements are more susceptible to sampling error than radon gas measurements.

1. Prior to use the continuous WL monitor should be tested to verify that a new filter has been installed and the input parameters and clock are set properly. The detector performance should be checked daily using a check source, often ^{214}Am.

2. The following information should be recorded during measurements: date and time of the start and finish of the measurement period; whether the conditions during the measurement period were standard or if exceptions occurred; the exact location of the measurement on a floor plan of the building; type of building and number of stories; type of heating system; existence of a crawl space or basement; occupants' smoking habits; operation of humidifiers, dehumidifiers, air filters or electrostatic precipitators, and clothes dryers.

3. As with all radon measurements, closed-house conditions must prevail. Measurements should not be taken near drafts caused by heating, ventilation and air conditioning vents, doors, windows, and fireplaces. The measurement location should also not be close to the outside walls of the building. The instrument should be placed at least 20 inches from the floor.

4. For long-time averaging calibration the instrument should be set to run for a minimum of 8 hours. The calibration factor is used to convert the number of counts into average mWL by performing the calculation shown in the preceding section on maintenance and calibration of continuous WL monitors.

5. For short-time screening, the timer can be set to between 0.5 and 4 hours, although 1 hour is most commonly used. This mode of operation can be used to do comparative measurements in various locations in the same building or in different buildings. The equation used for long-time averaging is not accurate when used for

short-term screening; instead the best use of the results is to compare various areas in order to target the best areas to perform long-term measurements.

6. Turn over the instrument and remove the filter holder. Remove the cap holding the filter and put a clean 25-mm filter inside. Do not touch the alpha particle detector located below the filter holder. It has a thin Mylar-film window that is easily penetrated. Any contact with the detector may destroy it.

7. Set the timer to the desired sampling time.

8. Turn on the power and check the display for the "000000" readout. Turn on the pump.

9. At the end of the sampling period, note the alpha counts and calculate the WL using the equation shown in the preceding section on maintenance and calibration of continuous WL monitors.

OCCUPATIONAL AIR SAMPLING FOR RADON

The primary situations where occupational exposure to radon is measured are in uranium and phosphate mines. Other hard rock mines where radon daughter exposure may be a problem are iron, zinc, fluorspar, and bauxite. In uranium mines the exposure usually depends on the quality and amount of uranium ore present and the effectiveness of the ventilation system.[11] The primary methods used to measure occupational exposure to radon daughters in mines are grab samples collected in various area rather than a continuous integrated personal sample.

Radon samples in mines are initially on the exhaust air using procedure specified by the MSHA standards.[12] If greater than 0.1 WL is detected in the exhaust air of a uranium mine, sampling representative of the breathing zone of miners must be conducted every 2 weeks at random times in all active working areas. This includes stopes, drift headings, travelways, haulage ways, shops, stations, lunchrooms, magazines, and any other place or location where persons work, travel, or congregate. Higher levels increase the sampling period to weekly. The monitoring schedule is required until levels decrease to less than 0.1 WL. In mines other than uranium mines, radon also has to be sampled in the exhaust air, and if levels exceed 0.1 WL, samples representative of the breathing zone of miners must be taken every 3 months.

When relying on grab samples, a sampling strategy must take into account the typical variability in radon daughter concentrations that can occur in any given area. Furthermore, once the samples are collected a correlation must be made for individual occupations to account for the times spent in various areas often within a single work shift. A statistical approach is used to attempt to randomize the results as much as possible. Work stations in close proximity are clustered together so that they can all be sampled on the sample day. Grab samples are collected at each station in a day, but the sequence in which the samples are collected is rotated.

Sampling is done on 2 different days within a given 2-week period with a provision for additional repeat next-day sampling whenever results exceed 0.14 WL. Calculations of results include average work shift concentrations for a given day or 2 sequential days. By apportioning the length of time an individual spends in a given location with the measured concentration in that area, an estimated value of the working-level month (WLM) is determined. A WLM is the exposure that takes place to radon daughters for 170 hours in a given month. The time spent by each worker at a given workplace can be determined from the time sheets and can be used to calculate individual monthly exposure to alpha. Gamma radiation measurements using Geiger counters are often done simultaneously to see if this is a significant contribution to the exposure.[11]

Occupational grab sampling techniques use the Kusnetz count method or the instant WL monitor. Grab sampling methods for radon daughters are described in the section on grab sampling methods in this chapter.

Continuous monitoring systems are rarely used for occupational sampling, since they are usually stationary systems and their primary use is for indicating problems with the ventilation system and identifying exposure sources. While personal sampling dosimeters for radon daughters exist, they are not commonly used in occupational measurements due to problems with calibration and precision of results. Where there are high thorium concentrations, special sampling techniques may be required in order to accurately monitor the contribution from radon daughters.[13]

COLLECTION METHOD FOR RADON IN WATER

Since radon is a dissolved gas when it is in water, it can escape if sampling is not done carefully. Other types of contamination that can release radiation in water are iodine, strontium, radium, and uranium.

Procedure

1. Use evacuated glass vials with rubber septa and syringes to collect the samples. Bladder pumps and tubing, a bailer with dual-check valves, a bottom emptying device, or a positive displacement pump can also be used. For sampling, 40-ml glass vials with Teflon liners must be used. Do not use suction, airlift, or peristaltic pumps to collect samples of water suspected of containing radon.

2. Prior to sampling each point, label two vials with a unique sampling number and record the number on the sampling data sheet. For duplicate samples, and "A" and "B" designation added to the number can be used.

3. Make sure the liner is in the correct position in the cap with the Teflon-coated septum, which is usually white on the outside; when the lid is screwed on, it should point toward the sample.

4. For evacuated vials, load the canister, lower it down to the depth of the screen, and let the messenger fall. The vial will fill itself.

5. If pumps or bailers are used, minimize air contact with sampled water while filling two vials to a high meniscus, or almost to overflowing. While filling the vial, try to minimize turbulence and air—water contact. Cap the vial immediately and screw the cap on snug (Teflon side down). There should be no air space at the top of the vial.

6. Invert the vial and tap it on a hard surface to determine if any air bubbles remain inside. If there are any, refill the vial(s). If samples without bubbles are not possible to collect, open the vial and add more sample. Reclose the vial and again inspect it for air bubbles. Continue to work with the sample until no bubbles are present.

7. Fill two vials from each sampling point. Do not filter samples for radon even if the water is silty.

8. Store samples in a cooler at 4°C upside down, but do not allow them to freeze. Keep a thermometer in the cooler with the samples. Keep the trip blank and any field blanks with the samples.

9. Send samples to a laboratory and request an analysis as a gross alpha count.

INTERPRETATION OF RADON MEASUREMENTS

Radon measurements may be reported in two ways. One is to report the concentration of the radon gas itself. The unit used is pCi/L. The other method is to report units that represent the energy released by the chain of products produced when radon gas radioactively decays. This unit is the WL. One WL is any combination of radon progeny in one liter of air whose ultimate decay through ^{214}Po will produce 1.3×10^5 MeV of alpha energy. It is the amount of alpha energy delivered by short-lived radon daughters in equilibrium with 100 pCi of ^{222}Ra. The most common devices readout in pCi/L while the unit most directly related to dose and health effects is the WL. Table 13-5 lists radon standards and guidelines.

Residential and Other Buildings

The current EPA guideline for radon is 4 pCi/L, or a 0.02 WL. If the test results are less than these, follow-up measurements are generally not necessary. If the result is >0.02WL to 0.1 WL, or >4

TABLE 13-5. Radon Standards and Guidelines

Organization	*Maximum Radon Level (pCi/L)*	*Applicability*
Mine Safety and Health Administration (MSHA)	16	Regulatory requirement that applies only to miners
National Council on Radiation, Protection, and Measurement	8	Guideline for protecting the general population
ASHRAE	5.4	Guideline for commercial buildings and residences
EPA	4	Guideline for protecting the general population

pCi/L to 20 pCi/L, follow-up measurements should be performed utilizing at least a 1 week measurement period during each of the four seasons, or exposing one detector for a full year.

If the result is >0.1 WL, or >20 pCi/L, follow-up measurements should commence immediately. Detectors should be exposed for no longer than 1 week. Doors and windows should be closed during this testing period. If the follow-up tests confirm initial tests, control measures are necessary.

Occupational Measurements

The interpretation of occupational radon samples depends on the current MSHA standard. NIOSH recommends that miners' exposures be limited to 1.0 WLM per year and that the average concentration of radon daughters in any work area should not exceed 1/12 WL during any single work shift.

Water Samples

As a rule of thumb, 10,000 pCi/L of radon in incoming (tap) water is equivalent to 1 pCi/L of radon in indoor air. If the gross alpha count is high, radium and uranium should be sampled for. For more information on collection of samples for these elements, see the chapter on Bulk Sampling Methods. Table 13-6 gives recommended actions based on screening results.

PERFORMING FOLLOW-UP MEASUREMENTS

The purpose of follow-up measurements is to estimate the long-term average radon or radon-decay product concentrations in general-use/living areas with sufficient confidence so that the need for remedial action can be estimated.

Follow-up results of 4 pCi/L or 0.02 WL are

TABLE 13-6. Recommended Actions Based on Results of Screening Measurements

Measurement Result	*Recommended Action*
>1 WL or 200 pCi/L	Perform short-term follow-up measurements and consider short-term actions to reduce the radon levels as soon as possible
0.1–1 WL 20–200 pCi/L	Perform short-term follow-up measurements within several months
0.02–0.1 WL 4–20 pCi/L	Perform follow-up measurements over the next 12 months
<0.02 WL or 4 pCi/L	Relatively low probability of significant health risk from concentrations in general living areas; follow-up measurements generally unnecessary, but can be made for confirmation

TABLE 13-7. Follow-up Measurement Periods to Estimate Annual Averages

Instrument	*Result >20 pCi/L**	*Result <20 pCi/L†*
Alpha-track detector	3-month intervals	12-month intervals
Charcoal canister	2–7 days	Four measurements every 3 months
Continuous working level	24 hours	Four 24-hour measurements every 3 months
Continuous radon	24 hours	Four 24-hour measurements every 3 months

*Made under closed-house conditions.
†Made under normal living conditions.

considered average or slightly above average for structures by the EPA, and no remedial action is necessary. In general, it is difficult to lower exposures much below these levels with existing remedial actions. If the follow-up measurement results are between 4 pCi/L and 20 pCi/L, or 0.02 WL and 0.1 WL, remedial actions should be undertaken within a few years.

If the results of the follow-up measurements are between 20 pCi/L and 200 pCi/L, or 0.1 WL and 1.0 WL, remedial actions should be taken within several months. If short-term measurements were used under closed-house conditions to achieve these results, the average annual concentration may be overestimated. However, this alternative is preferable to allowing exposures at high levels to continue for an additional 12 months. If follow-up measurements indicate levels to be at 1.0 WL, or 200 pCi/L, immediate remedial actions should be implemented.

These measurement guidelines must be followed:

1. Measurements, should be made in each level of the building that is frequently used as a living, classroom, or work area.
2. Measurements should be made in the most frequently occupied room of each level. For example, in residences a bedroom is a good choice, because most people spend more time in their bedrooms than in any other room of a house.
3. If children use or live in the building, also perform measurements in the areas where they spend most of their time.
4. Do not make measurements in kitchens, because they often contain stove exhaust systems and cooking contributes to particle levels. Bathroom air measurements are also not useful because of the high humidity often present and the small amount of time usually spent in these rooms. Bathrooms can be used for sampling radon in water.

Measurements should also be made after control measures have been installed to determine the effectiveness of these mitigation techniques. Once satisfactory levels have been obtained, follow-up measurements should only be done if physical changes occur, such as new construction, repair and alterations to the building, or settling of the foundation. Table 13-7 lists follow-up measurement periods to estimate annual averages.

REFERENCES

1. McManus, T. N., and J. W. Smith. A simple compact system for the extraction of radon from water samples. *AIHA J.* **48**(3):276–286, 1987.
2. Boyle, M. Radon testing of soils. *Environ. Sci. Technol.* **22**(12):1397–1399, 1988.
3. USEPA. Interim Guide to Radon Reduction in New Construction. Draft, January 1987.
4. USEPA. Radon Measurements in Schools: An Interim Report. 520/1-89-010, 1989.

5. Summerlin, J., and H. M. Prichard. Radiological health implications of lead-210 and polonium-210 accumulations in LPG refineries. *AIHA J.* **46:**202–205, 1985.
6. Gesell, T. F. Occupational radiation exposure due to ^{222}Rn in natural gas and natural gas products. *Health Phys.* **29:**681–687, 1985.
7. Thomson, I., and T. K. Nielson. A New Portable WL Meter for Indoor Radon. Presented at the 32nd Annual Meeting of the Health Physics Society, July 5–9, Salt Lake City, Utah.
8. Ronca-Battista, M., et al. Interim Indoor Radon and Radon Decay Product Measurement Protocols. EPA 520/1-86-04, February 1986.
9. Droullard, R. F. Instrumentation for Measuring Uranium Miner Exposure to Radon Daughters.
10. USEPA, Office of Radiation Programs, Las Vegas Facility. *Operational Evaluation of the AT EASE Model 1020 Radon Monitor.*
11. National Institute of Occupational Safety and Health. *Criteria for a Recommended Standard to Radon Progeny in Underground Mines.* NIOSH, Cincinnati, 1987.
12. 30 *CFR*. Chapter 1, MSHA, Radiation—Underground Only.
13. Phillips, C. R., and H. Leung. Working Level Measurement of Radon Daughters and Thoron Daughters by Personal Dosimetry and Continuous Monitoring.

CHAPTER 14

Sampling for Bioaerosols

Bioaerosols, meaning airborne particles derived from microbial, viral, and related agents, come in a wide variety of sizes, shapes, and classifications. Bioaerosol sampling is a specialized field utilizing its own terminology. Textbooks on microbiology can be useful for learning more about microorganisms, while texts on immunology discuss antigens and hypersensitivity. Following is a list of some terms associated with bioaerosols:

Antigens: Foreign materials that once in the body generate a response resulting in the production of antibodies.

Autotrophic organisms: Organisms that have the ability to synthesize carbohydrates.

Biogenic volatiles: By-products released by microorganisms that are volatile and are often responsible for odors.

Endotoxins: A lipopolysaccharide found in the outer membrane of gram-negative bacteria.

Facultative anaerobes: Organisms capable of growing in the presence or absence of air.

Gram-positive bacteria: Bacteria that stain purple when exposed to crystal violet dye.

Gram-negative bacteria: Bacteria that do not react with crystal violet stain but do turn red when exposed to safranine dye.

Mycotoxin: Fungal metabolite, many of which are carcinogenic, such as aflatoxin.

Parasitic organisms: Organisms that utilize living material for food.

Saprophytic organisms: Organisms that utilize dead organic material for food.

Thermophilic bacteria: Bacteria that grow at elevated temperatures (above 55°C).

TNTC: Too numerous to count.

CFU: Colony forming unit.

Bioaerosols can cause two basic conditions: infections and allergies. Infections are generally the result of multiplication and growth of microbes inside humans while allergies are the result of exposures to antigens. Not all infectious organisms cause pathogenic diseases in humans, but those that can are of concern. Well-known diseases associated with occupational exposures include anthrax, Q fever, and brucellosis. Diseases for which concerns are increasing are those associated with health workers and include AIDS and hepatitis.

Antigens are capable of stimulating the production of antibodies that produce allergic diseases. Allergic reactions are the result of an antigen producing a response from the immune system.

Hypersensitivity disease is another term for the allergic reactions produced by these agents. These include hypersensitivity pneumonitis, allergic asthma, and allergic rhinitis. Sources of airborne antigens include bacteria, fungi, pollen, insect body parts, and skin scales (dander) and saliva of mammals.[1] In these situations antibody assays on blood from affected individuals may be performed in conjunction with monitoring for bioaerosols

Sometimes it is not the microbe itself that produces the harmful effect but the fact that it produces a toxin. Botulism is an example wherein the botulinum toxin is the responsible agent. When release of a toxin is involved, the organism can produce a disease without extensive multiplication or dissemination throughout the body.

Just as there are factors that can predispose individuals to the health effects caused by chemicals, certain persons are also at increased risk when exposed to bioaerosols if they are over 50 years old, drink alcohol excessively, smoke, or have preexisting respiratory disease or other illnesses such as diabetes or kidney disease.

Bioaerosols can exist in both viable (living) and nonviable states. Viable microorganisms such as bacteria, fungi, yeasts, and molds originate from sprays or splashes of media, from the agitations of dusts, and from sneezes and coughs of which only the small particles ($<10\mu m$) remain in the air long enough to travel any distance. Examples of nonviable agents that are occasionally sampled include pollens and insect parts. Grains, clusters of cells, and skin scales are much larger-sized particles ($10-50\mu m$) than bacteria and viruses[2] (Fig. 14-1). Spores, which can be formed by fungi and certain bacteria, can be both viable and nonviable and are capable of causing disease in both forms. Most techniques attempt to sample for only viable particles as these can be cultured so that they multiply, making identification easier.

The specialized characteristics of viable agents require specialized sampling instruments in order to preserve the organisms for laboratory culture, which is the primary means of identification. Their fragility and temperature, moisture, and nutrient needs are the primary considerations when selecting a sampling device. While passive air sampling is simple and can be done by setting out plates containing culture media, it is not as effective as the use of active techniques involving the use of pumps. There are two basic methods for collecting these air samples: (1) specialized instruments and (2) air sampling trains incorporating a personal air sampling pump, rotameter and media, such as is used for integrated chemical sampling. The specialized instruments can be used to house culture media and therefore in most cases are preferred to integrated sampling techniques. Area air samples are more commonly collected for bioaerosols than personal samples, regardless of the type of situation being monitored, due to the need to house culture media inside of instruments specialized for sampling viable bioaerosols.

Most air sampling for bioaerosols is related to occupational exposures in hospitals, laboratories, and research facilities; certain industrial operations, such as brewery fermentation, cotton preparation and ginning, wool sorting, hemp handling, and sawmills; and agricultural operations, including hay preparation and the use of biological insecticides and wastewater and sewer treatment facilities. A current interest is exemplified by a recent survey for a biological insecticide, *Bacillus thuringiensis,* to characterize exposures to personnel during a large-scale spraying application.[3] Workers in some sectors of the food industry where fruits and vegetables are processed may also be affected. As an example, slicing sugar beets was found to generate exposures to bacteria originating in soil during beet growth.[4] Other examples of operations where bioaerosol sampling might be done include pharmaceutical manufacturing plants, clean rooms in semiconductor manufacturing plants, animal laboratories, and food processing plants.

Historically most environmental monitoring has consisted of collecting bulk samples of water, especially drinking water and wastewater. Indoor air surveys in buildings incorporate both indoor and outdoor air samples for various agents.

Sampling usually attempts to determine whether the agents are being generated from a source within a building rather than from an outside source where they naturally occur. It has been noted that the majority of fungal spores found indoors are derived from outdoor sources, such as dead or dying plant and animal materials,

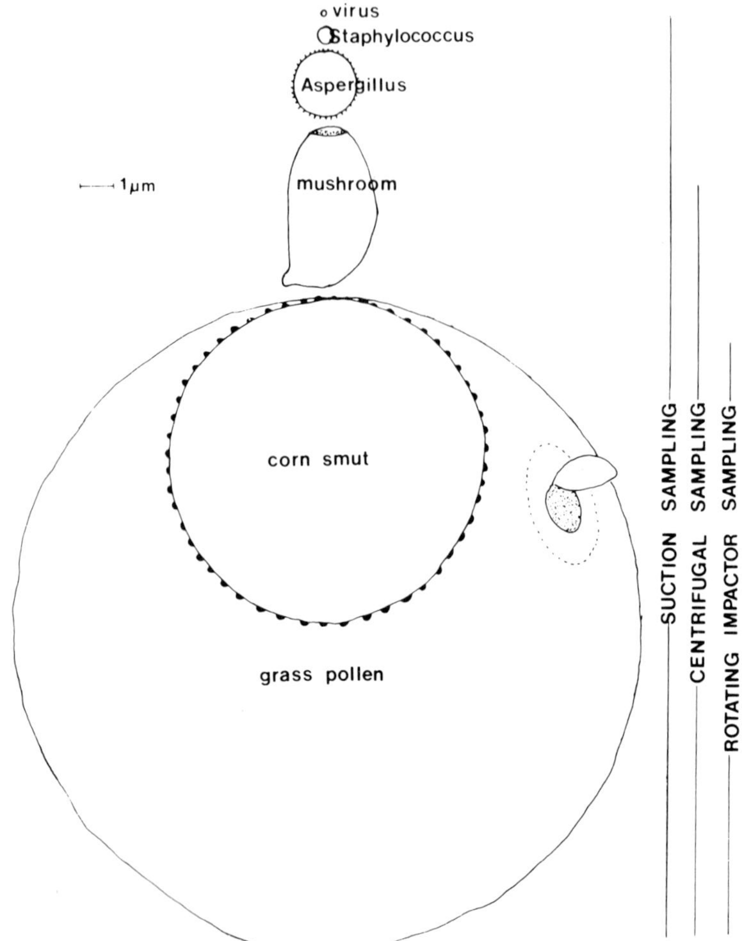

Figure 14-1. Relative sizes of microbial agents. (From Burge, H. A., and W. R. Solomon. Sampling and analysis of biologic aerosols. *Atmospheric Environment* **21**(2):451–456, 1987.)

while the primary source indoors is human bacteria shedding. Sources of microbes include organic materials; humidifiers; vaporizers; heating, ventilating, and air conditioning systems (HVAC); as well as their associated equipment, such as cooling towers.

Another situation of growing interest that incorporates both occupational and environmental exposures is bioremediation of contaminated soils and waters using specially engineered "super bugs." These techniques have proved very successful for treating petroleum compounds. In these situations, the release of volatile organic compounds (VOCs) is also a concern, so a boundary line monitoring strategy would need to incorporate both sets of agents. Industrial wastewater treatment has incorporated the use of various microbes for many years.

Air sampling for bioaerosols may be combined with sampling for chemical agents in some situations such as indoor air surveys and exposures to wood dust or bark. In some situations, monitoring is performed for chemicals where there are metabolic by-products of the organism, such as endotoxin, released by certain bacteria. Biological monitoring may also need to be performed. For example, blood and urine samples have been collected in cases of suspected Legionnaires' disease, since *Legionella pneumophila* infections cause the release of antigens to the urine.[5] Surface contamination may also be a concern.

BACTERIA

Bacterial infections are the most commonly seen in humans, such as those that occur in minor wounds and scratches. Diseases related to bacteria that are found in occupational exposures include those caused by anthrax, also called wool sorter's disease, transmitted by handling imported goat hair, wood, and hides; *Brucella canis* infections (brucellosis) from the contaminated blood of slaughtered animals; and Leptospira-induced disease (leptospirosis), associated with farm animals, dogs, and rodents, which is spread through contact with infected urine, animal tissue, or water.[6]

Other bacteria of concern in sampling include staphylococcus and streptococcus, which are carried by humans and cause infections under the right conditions; Pseudomonas, which cause pneumonia; and bacillus, which is associated with hypersensitivity pneumonitis. Thermophilic bacteria of concern include Thermoactinomyces, known to cause hypersensitivity; and Micropolyspora, Thermomonospora, and Saccharomonospora.[7] Another exposure of concern is the Salmonella bacteria responsible for food poisoning. In this case, ingestion rather than inhalation is the route of exposure. While not strictly occupational in nature, this may be a concern in indoor air investigations.

Rickettsiae are intracellular parasites in fleas, ticks, and lice that are considered to be bacteria. The tick is the most common reservoir and tick bites are the primary route of transmission. Rickettsiae do not appear to produce symptoms of disease in their hosts, but if they are transmitted to humans, a severe and often fatal infection may result. The major rickettsial disease of humans is epidemic typhus. Other human diseases are Rocky Mountain spotted fever and Q fever, both transmitted by ticks, and scrub typhus, normally transmitted by mites to field mice, but also transmissible to humans. The chlamydias, other specialized bacteria, are carried in birds and the primary disease they are associated with, ornithosis, is transmitted by inhalation of dried discharges and droppings of birds.

Most bacteria are 1 mm to 5 mm in size and are classified through their nutritional and growth requirements, shape (round or rod), ability to utilize certain compounds, oxygen needs, metabolites produced, and the type of stain they take in the Gram stain test (positive or negative). There are four main factors that influence bacterial growth: temperature, nutrient supply, availability of stagnant water, and presence of particulate matter to bind to. Temperature is the most important factor for many bacteria. While some organisms, called thermophilic, can exist at elevated temperatures, most bacteria exist in a range of 20°C to 45°C. Nutrients can include nitrites; by-products of other microbials since the growth products of one organism may support another; certain chemical additives often used for water treatment; nonmetallic materials such as insulation, rubber parts, gaskets, and sealants; and atmospheric contamination by dusts and insects. In order to select the proper sampling medium, it will be necessary to know what the specific characteristics are of the bacteria suspected of being present. It has been suggested that Gram-positive bacteria are more likely to survive during air sampling than gram-negative.[1] Gram-negative bacteria produce endotoxins that are pyrogenic and induce local inflammatory responses.[8] It is currently thought that endotoxins may be responsible for a number of diseases in workers.[9] Some bacteria, like fungi, can produce spores whose characteristics are discussed further in the section on fungi.

Bacteria are widely present in the environment and the presence of certain odors, slime, and foam on a surface are often an indication of bacterial growth. Certain types live in the human body while others live outdoors on vegetation; therefore, when sampling indoors, those bacteria associated with humans will predominate while outdoor samples will contain mostly the other type. Bacteria are spread primarily through inhalation, although bacteria that are very small are often dispersed on skin scales.[10] It has been estimated that 7 million skin scales are shed per minute by humans, each containing an average of 4 viable bacteria.[11] Bioaerosols are usually associated with water, and an undisturbed source of water or a humid environment is highly conducive to growth. When water is aerosolized, droplets range in size, but larger droplets can evaporate and become smaller, thus increasing the potential for inhalation. Taps, showers, whirlpool spas, and cooling towers are all sources. When bacteria attach to particles, they are often protected against the environment.

The primary tools for collecting bacteria for culture are impingers and cascade impactors. Screening samples can be collected with a slit or centrifugal impactor.

Coliforms

Coliform bacteria are a concern in drinking water; therefore, potable water is collected and analyzed for these microbes. Fecal coliforms and *Escherichia coli* are not harmful in themselves, but their presence in water intended for potable uses is serious as it is indicative of contamination by sewage or animal waste. This is generally a result of a problem with water treatment of the pipes that distribute the water and indicate that the water may be contaminated with organisms that can cause disease. Environmental Protection Agency (EPA) standards are based on the presence or absence of total coliforms in a sample rather than on the coliform density.[12] The likelihood of finding coliforms in a water supply is based on monitoring frequency. Monitoring frequency for total coliforms in community water systems is based on the population served and to some degree on whether the system uses surface water or groundwater.

A total coliform sample is considered invalid by the EPA if the sample produces a turbid culture in the absence of gas production using an analytical method where gas formation is examined, produces a turbid culture in the absence of an acid reaction, or exhibits confluent growth or produces colonies TNTC with an analytical method using a membrane filter.[13] If a routine or repeat sample is total coliform-positive, a follow-up analysis on the culture must be done to determine if fecal coliforms are present. A test for *E. coli* in lieu of fecal coliforms is acceptable.

Legionella Pneumophila

Legionnaires' disease became a concern when it became apparent that aerosol from contaminated cooling towers could enter fresh air vents and spread this agent to the HVAC system in buildings. *Legionella pneumophila,* the agent in this case, is a rod-shaped, slow-growing, gram-negative bacterium. Ideal growth conditions require a temperature of 35–45°C and a pH of 6.9–7.0. The bacterium is ubiquitous in the environment. It can coexist with amoebae and can survive and grow on blue-green algae. The vast majority of outbreaks have been associated with Legionella sero group 1, but other sero groups, if detected, are also of concern, and control measures should be initiated if they are detected. Legionella sero group 1 has been isolated from a variety of surface and potable aquatic habitats. It is viable in tap water for more than a year. Hot-water tanks, and in particular their bottom sediments, are excellent media for its survival and proliferation.

Cooling towers are especially susceptible to contamination, since their primary function involves inducing large amounts of air into large amounts of flowing water; thus, they act as air scrubbers, washing out dust, debris, pollen, insects, and plant materials. These bacteria have also been known to build up in water softeners.

Most sampling for Legionella is done by collecting bulk samples of suspected sources. One source considers heavy contamination by Legionella to be counts greater than 10 colony-forming units per liter in a bulk sample.[14] Sampling should be done in both suspected and background areas. As a precautionary measure, bulk samples of suspect sources of Legionella should be collected every 6 months. Chemical analyses of makeup water and system water should be performed monthly. Samples should not be collected following cleaning or immediately after startup. It is best to sample in the middle or end of each 6-month period during normal operation. The best places are system water dead spots or slow-flowing areas; however, these should be selected so they are not near incoming fresh water or biocide treatment points. Testing should precede but not follow slug feed of biocide.

In the course of evaluating the potential for this microorganism to be the source of an outbreak, the operational procedures and facilities are considered as important as the sources of microbial contamination and dissemination. An investigation for Legionella in cooling towers would focus on the following areas[14]:

Temperature: If the normal operating temperature of the water is greater than 30°C, there is a high risk of bacterial growth.

Contamination: Nonmetallic materials such as washers, coating, gaskets, linings, and sealants in the system can harbor bacterial growth.

Stagnation: If the water is left standing for more than 5 days at a time, there is a high risk of bacterial growth.

Particulate matter: The sump can accumulate sludge, debris, scale and bacterial growth.

Aerosol generation: If there is a significant amount of spray the likelihood of spread is increased, especially if there is any possibility of aerosol escaping from the cooling tower.

Susceptible populations: If the cooling tower is associated with any buildings or sites occupied by susceptible people, such as hospitals or schools, the risk is increased.

Other factors of importance include whether a responsible person is in charge of the cooling system, the type of training provided to the staff responsible for its upkeep, and the availability of adequate record keeping.

If a routine test for Legionella is positive, the cooling tower should be cleaned immediately. Following startup, the system should be resampled within seven days. As Legionella require soluble iron for growth, control of corrosion is an important factor. Often it involves adjusting the pH of the water. Biocides are often used to kill Legionella and other waterborne microbes, but their effectiveness depends on controlling water chemistry including pH, alkalinity, and cycles of concentration (i.e., addition periods).

Endotoxin

Endotoxin is a distinct lipopolysaccharide (LPS) found in the outer membrane of gram-negative bacteria, and it varies among bacterial types. Endotoxin can be present in several forms: the bacteria, fragments of bacteria membranes incorporated in dust, as well as the endotoxin molecule itself. Endotoxin is considered highly toxic and is suspected of causing pulmonary impairment in humans. Endotoxin has been implicated as having a significant role in the development of byssinosis from cotton dust exposures.[15] It has been found in agricultural, industrial, and office environments.

Endotoxin in air has been sampled using filters attached to personal sampling pumps. Bulk samples are useful when endotoxin is suspected. These must be collected in oven-baked glassware and must be analyzed promptly.[1]

The most common analytical method for endotoxin is known as the Limulus assay. Testing results are often reported as picograms per meter cubed (p/M^3), which is an extremely small amount of material. LPS is inactivated by filter media, including 5.0-mm PVC, 1.0-mm Teflon, 0.45-mm MCE and Polyflon, the result being reduced concentrations when testing is conducted.[16] It is of note that currently there are a number of variations both in the Limulus assay and in the extraction technique used to remove endotoxin from the filter media; therefore, comparison of results from different laboratories will depend on how similar their analytical techniques are.[17]

Given the fact that an agreed upon method that eliminates the problem of loss of sample on filters has not yet evolved, the best approach to sampling is to collect background and source samples and compare the results rather than attempt to associate the hazard with some type of standard. The same method should be used for all samples and similar volumes should be collected.[16]

FUNGUS AND MOLDS

The fungi class includes yeasts, mold, mildew, and mushrooms. Soil is the most common habitat of the fungi, although many of the primitive fungal groups are aquatic. Fungi occur on the surface of decaying plant or animal materials in ponds and streams or grow on top of aqueous industrial fluids such as metal-working coolants. They are common in grain-handling facilities, paper mills, fruit warehouses, and agricultural environments as well as indoor air environments. Fungal species commonly encountered include Aspergillus, which is ubiquitous in the soil and air, especially in agricultural products and in standing water, whose spores are known to cause a variety of pulmonary effects; and Histoplasma and Cryptococcus, found in bird droppings. Penicillium is a mold that grows on damp organic materials and standing water and is associated with hypersensitivity pneumonitis. *Candida albicans* is a yeast that is ubiquitous and known to cause Candidiasis, a disease of the skin and mouth that occurs in dishwashers, cooks, cannery workers, and others who frequently have their hands in food-contaminated water. In immunosuppressed individuals it can have systemic effects. Other fungi of concern are Alternaria, Aureobasidium, Chaetomium, Cladosporium, and Mucur.[7]

Fungal-related diseases can be divided into two types: mycosis and mycotoxicosis. Mycosis represents a variety of toxic effects, including dermatitis, hypersensitivity pneumonitis, and some systemic diseases that result from an infection by the organisms themselves. Mycotoxicosis is produced by metabolites of various fungi and causes diseases such as toxic aleukia and yellow rice disease.[18]

Occupations associated with exposure to fungi include sawmill, sugarcane, and cork workers as well as jobs where seeds and textile fibers are handled. Other work environments conducive for the growth and sporulation of fungi are farming, grain handling, mushroom cultivation, insect rearing, and pharmaceutical manufacturing.[18]

Like other microbes, fungi have specific nutritional requirements that vary among the species and produce metabolic products, a classic example being penicillin produced from the mold penicillium. Fungi are also dependent on having water present. The presence of a moldy odor is suggestive that fungi are growing.[1]

A unique stage of some fungus' life cycle is the spore stage consisting of a wide variety of shapes in a very broad size range (<2 mm to >100 mm).[19] In this stage they form a durable coating over the exterior and become dormant. Spores can be classified by size, morphology, and color, allowing them to be categorized into different taxonomic groups. Since spores are relatively hardy structures, they can survive in dry environments and become airborne when disturbed. Fungal spores are released into the air either by mechanical means, such as wind or other agitation, or biologically by specialized (active) spore discharge mechanisms usually occurring during periods of high relative humidity.[19] When airborne, fungal spores tend to travel as single units.[10]

Certain foods such as peanuts and animal feed contain fungal spores that begin growing and producing aflatoxins when environmental conditions (time, temperature, moisture, nutrients, and pH) are favorable. Aflatoxins are a group of chemically similar compounds known to be acutely toxic and carcinogenic at low doses, and are metabolites of two common fungi: *Aspergillus flavus* and *Aspergillus parasiticus*. If fungal spores are suspected, water reservoirs should be identified and bulk samples should be collected at all suspected sources.[1] Suitable niches for growth and sporulation include stored food, house plants, air conditioners, humidifiers, cold air vaporizers, books and papers, carpets, and damp areas.

The primary air sampling tools used for fungi spores are slit impactors and filters. Screening samples can be collected with a centrifugal impactor.

VIRUSES

Viruses represent a unique class of agent and are different from cellular organisms. A virus alternates in its life cycle between two phases: one extracellular and the other intracellular. In its extracellular phase, a virus exists as an inert, infectious particle, or virion. A virion consists of one or more molecules of nucleic acid, either

DNA or RNA, contained within a protein coat, or capsid. In its intracellular phase, a virus exists in the form of replicating nucleic acid, either DNA or RNA. Viruses utilize the host cell for replication (reproduction) and thus are intracellular parasites. In the extracellular phase, some viruses are quite stable and resistant to heat and light.

Viral infections may be acquired from vectors such as needles or from handling of animals or animal products and from humans. Laboratory-acquired infections may result from needle sticks; animals; clinical or autopsy specimens; or contaminated glassware. Diseases include rabies, cat-scratch disease, and viral hepatitis (both serum and infectious).[6]

Viruses survive best in situations where high humidity and moderate temperatures are present. Situations where water containing a virus is being aerosolized are especially conducive to viral multiplication. Table 14-1 describes factors affecting survival and dispersion of bacteria and viruses in wastewater aerosols.

Collection of viruses often requires very specific techniques, although some have been collected on filters.[20] Viruses have also been collected on the slit sampler and the multistage cascade impactor.

Viruses are usually measured as either infectious units or total particle numbers.[1] While it is important to be aware that viruses may be a cause of an outbreak of illness, it is unlikely that air sampling would be useful in most situations, due to complicated analytical techniques usually requiring that a live species be injected and the degree of specialization necessary to perform these analyses. Instead, a more common technique is to identify the symptoms associated with a suspect virus and determine if the disease is present through a physician's clinical evaluation.[1]

Collection of viruses often requires very specific media, although some have been collected on filters.

OTHER MICROORGANISMS

Spirochetes have a unique cell structure relative to other bacteria in that they have very long, wormlike bodies. Thus, they can swim in liquid media and are found in mud and water.

Mycoplasmas, the smallest known cellular organisms, are a large and widespread group. The first member of this group to be identified was the agent of bovine pleuropneumonia. Many

TABLE 14-1. Factors That Affect the Survival and Dispersion of Bacteria and Viruses in Wastewater Aerosols

Factor	*Comment*
Relative humidity	Bacteria and most enteric viruses survive longer at high relative humidities, such as those occurring during the night. High RH delays droplet evaporation and retards organism die-off.
Wind speed	Low wind speeds reduce biological aerosol transmission.
Sunlight	Sunlight, through UV radiation, is deleterious to microorganisms. The greatest concentration of organisms in aerosols from wastewater occurs at night.
Temperature	Increased temperature can also reduce the viability of organisms in aerosols, mainly by accentuating the effects of RH. Pronounced temperature effects do not appear until a temperature of 80°F (26°C) is reached.
Open air	It has been observed that bacteria and viruses are inactivated more rapidly when aerosolized and when the captive aerosols are exposed to the open air than when held in the laboratory.

other species have been isolated from humans and other vertebrates, where they occur as parasites on moist mucosal surfaces. These organisms contaminate tissue cultures, and mycoplasmas have been found in hot springs and other thermal environments, in which case a bulk sample is useful.

Protozoa are large in size compared to other microorganisms and include amoeba. Most protozoa are encountered as a result of contaminated water reservoirs. Naegleria is a protozoan that can infect humans and Acanthamoeba releases antigens that cause hypersensitivity pneumonitis.[6] Protozoa may pose an exposure in wastewater treatment (sewage) workers. Sampling is labor-intensive and problematic, so generally a bulk sample of a suspected source is more useful than an air sample for these microorganisms.[1]

SAMPLING METHODS AND STRATEGIES

Although microbials have certain characteristics that make them difficult to collect and analyze, the collection methods are similar to those used to collect other types of particulates. The basic strategy is to collect the material and demonstrate the presence of microbials through growth in various culture media. While there are some standard devices used for collection, there are no widely accepted standards for sampling. Therefore, the best way is to review the literature and select a sampling method based on the specific situation under evaluation. Selection should consider sampling location(s), type of agent, expected concentration, analytical methods, and organism characteristics. The type of agent will dictate whether results will be expressed as the total number of viable particles (colony-forming units) or the total number of individual viable cells.

Selection of an air sampling method will depend on the type of organism being sampled as well as on the expected concentration. Viable bacteria and fungi including yeasts and molds are most commonly sampled using agar culture media contained in a variety of different sampling instruments or impingers containing nutrient solutions. Nonviable agents such as spores are usually collected on filters. A rule of thumb is to limit the use of culture plate samplers to environments where concentrations are expected to be less than 10,000 organisms/M^3.[1] Filters have been recommended for situations where high concentrations of microorganisms are suspected.[21] The selection of culture media and conditions such as temperature and humidity can determine which organisms grow and thus are collected and which do not, because organisms vary significantly in culture requirements. Media are usually selected for their ability to culture virtually every organism of a certain type that might be present for screening samples or to selectively grow only certain species for identification samples.

Malt extract agar is recommended for the general detection of fungi while agar containing casein peptone, soy peptone, and sodium chloride is used similarly for bacteria.[7] Trypticase soy agar has been used to collect and support the growth of both fungi and bacteria.[22] In order to make a culture media more specific, antibiotics and other compounds are often added, to inhibit the growth of certain microbes. For example, inhibitory mold agar will promote fungi growth. Rose-Bengal-Streptomycin has also been recommended as a useful medium for fungal sampling as it yields high colony counts and impedes colony spreading; however, it is sensitive to direct sunlight and must be shielded when sampling outdoors.[18] All media should be prepared within days of use. Poured plates should be stored upside down and any condensate that accumulates in the lid should be shaken out before use.

Ideally samplers to be used for viable microorganisms should be sterilized before each use. Most samplers commonly used with culture media can be swabbed with isopropyl alcohol or bleach before each use. All glass impinger fluid should be sterile and the impinger should be rinsed with sterile fluid before use. Table 14-2 describes samplers recommended for viable bioaerosols.[7]

Analytical methods are direct, meaning the organism or its toxin is identified, or they are indirect in that a skin bioassay or immunological assay on blood for an allergic reaction to that organism is performed and considered evidence that an individual has been exposed. Direct methods include microscopy that may be optical, scanning electron microscopy (SEM), or epifluorescence

TABLE 14-2. **Samplers Recommended for Viable Bioaerosols**

Sampler	*Principle of Operation*	*Sampling Rate (Lpm)*	*Rec. Sample Time (min)*	*Min. CFU Detected*
Slit impactor	Impaction onto adhesive-coated slide	10	15	N/A
N-6 Single-stage impactor	Impaction onto agar in 100-mm culture plate	28	1	35
2-Stage impactor	Impaction onto agar in 2 – 100-mm culture plates	28	1	35
Filter cassettes	Filtration	1–2	15–60	8–33
Glass impingers	Impingement into liquid	1.5–2.5	30–60	5–25
Centrifugal impactor	Impaction onto agar strips	40	½	50

Source: ACGIH BioAerosols Committee. Guidelines for assessment and sampling of saprophytic bioaerosols in the indoor environment. *Appl. Ind. Hyg.* **2**(5):R-10–R-16, 1987.

(requires that the sample be stained) and culturing followed by a variety of chemical tests for specific properties of microbes. Biochemical tests can be performed to identify metabolites, especially those considered toxic.

Prior to sampling, an appropriate laboratory should be selected for analyses. Many clinical laboratories specialize in medical samples such as throat cultures and other human cultures and do not have experience in analyzing environmental air samples. Some laboratories specialize in mycology, the study of fungi, while antigen, endotoxin, and virus sampling often are still research techniques, and arrangements for analysis must be made with laboratories conducting research in a given area. Often a review of the literature will provide information on laboratories specializing in various techniques. Some universities allow private samples to be analyzed by microbiology departments.

Colony counts should be done according to the manufacturer's specifications for the sampling instrument used. If using a new laboratory, these specifications should be provided along with the samples. Samples for bioaerosols associated with allergic illnesses, such as hypersensitivity pneumonitis and asthma, must be sent to a laboratory proficient in growing and identifying environmental molds.[23] Clinical hospital laboratories have little experience in identifying environmental fungi and therefore may not be the best analytical choice.

Sampling professionals should be aware that not all microbes can be cultured, and some are extremely small and difficult to see in an optical microscope. In the case where sampling is being done for a specific strain, a nonspecific method may lead to confusion in the situation where other interfering strains may also be present. Identification of specific strains can be time consuming and difficult for laboratories and is not always necessary to identify what controls are needed. In some cases, rather than conducting extensive sampling for bioaerosols, it will be better to identify possible sources of stagnant water and continuously damp organic materials and correct the source of humidity. Any organic material may support mold growth when wet, including leather, cotton, paper, carpets, furniture, and furniture stuffing. The determination of which way to proceed is up to the sampling professional.

Sampling strategies for bioaerosols depend on the situations being sampled. When identifying the source of disease outbreak, sampling must be done as soon as possible following the outbreak. In this case, the best strategy may be to have affected individuals examined by a physician and specific sampling should be performed to identify the source of exposure once likely pathogens have been identified.

Viable sampling indoors will be affected by the sedimentation rate of airborne particles in the case of passive samplers, HVAC filtration units, and seasonal variations in loads and types of microorganisms entrained from the outdoor air.[24] More variations will occur when samples are collected outdoors rather than indoors, due to differing conditions, such as climate changes and unpredictability of the wind. Other variations can also affect sample results. For example, in one study on bacteria in grain dust, it was found that *Enterobacter agglomerans* was the predominant species in warm months, but in winter it was Pseudomonas and Klebsiella species.[8]

One strategy suggested for situations where a bioaerosol is suspected but sampling must be performed for a broad range of organisms is as follows[10]: Collect screening samples with a slit impactor, followed by sampling with at least three cascade impactors, each containing a different type of culture medium for bacteria, and filters for spore collection and examination. Identify specific bioassays or biochemical tests for any suspect toxins and allergic reactions, and have exposed individuals tested.

For indoor air studies, sampling should be done during work periods when occupants are present and also when the building is empty. A concern is that during off-periods, such as evenings, ventilation systems are often set to 100% recirculation air whereas during the workday a certain percent of outdoor air is utilized. In the indoor environment, samples should be collected in the supply and return air of rooms housing the affected occupants, as well as in other rooms where occupants have no complaints. Outdoor samples should be taken close to the intakes of HVAC systems when they are open and, if applicable, in the vicinity of potential bioaerosol sources such as cooling towers, stored organic material, and dense vegetation. Air samples should also be collected at an outdoor site remote from obvious sources. It is essential to sample at different times during the day and, if at all possible, on different days. Seasonal sampling may be required if the sampler is dealing with a past event. The following sequence is one option for collecting samples for indoor air situations[23]:

1. Sample before the HVAC system is turned on and before occupants arrive for work.
2. Sample before occupants arrive for work at a time when the HVAC system is operating.
3. Sample at the time of maximal occupancy with the HVAC system on.
4. Sample after occupants leave with the HVAC system off.
5. Sample in relation to normal changes in energy conservation that affect the HVAC system, or before and during artificial manipulation of the HVAC system.

Humidity or the lack of it is important to measure when sampling microbials for indoor air exposures, since virtually all microbes require a consistent source of moisture for growth. While growth can be supported at between 25% and 75% relative humidity (RH), greater than 70% is considered optimal.[11] Since buildings are often equipped with humidifying facilities along with heating and cooling systems, these can become infested with a variety of molds and other microbials that can release spores to the ventilation air when the systems cycle from wet to dry operation, so bulk samples should also be collected.[25] Indoor air surveys are discussed in greater detail in the chapter on Specific Sampling Situations.

PASSIVE AIR SAMPLING

Passive methods were the first attempt to collect airborne microorganisms and are still used. In setting out culture plates, otherwise known as settling plates or gravity sampling, dishes of culture media or adhesive coated glass slides are used that are placed in various areas and left open for air exposure for a number of hours. Then they are incubated and the number of colonies is counted. Extreme care must be taken to locate them away from supply and exhaust (return) vents, windows, sitting areas, and heaters. The best application is as a screening technique when instruments for active air sampling are unavailable. Sensitivity to air currents and human infections is a disadvantage of this method. Problems with this approach include the fact that

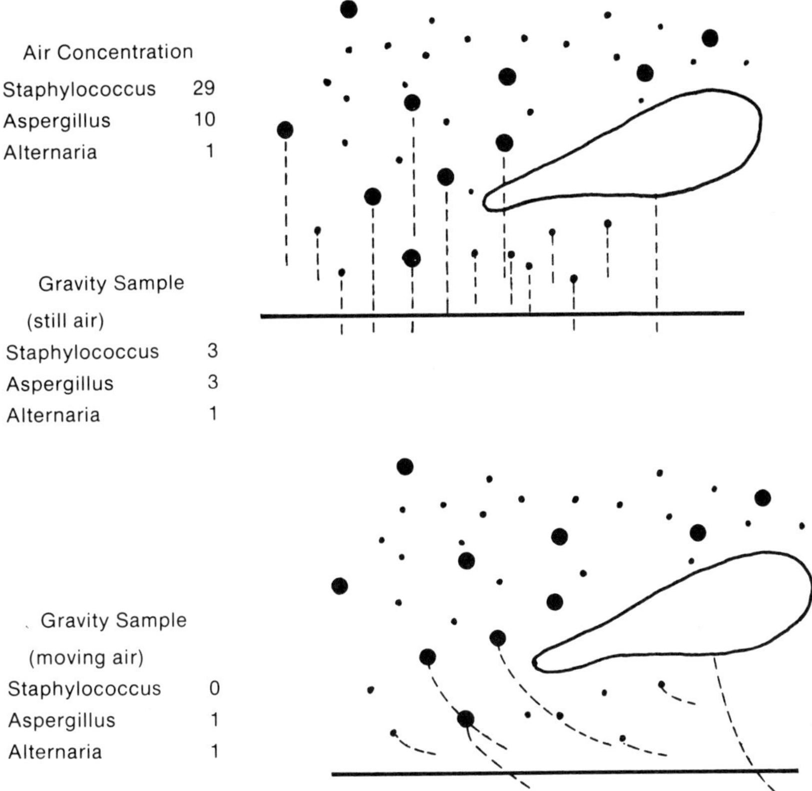

Figure 14-2. Problems with settling methods. (From Burge, H. A., and W. R. Solomon. Sampling and analysis of biologic aerosols. *Atmospheric Environment* **21**(2):451–456, 1987.)

heavier particles settle out first; thus, they predominate in the results and air currents create turbulence that can reduce collection efficiencies for all particles. Figure 14-2 demonstrates problems with settling methods.

If settling plates are left near HVAC air vents, open windows, or other sources of air currents, the large:small particle ratio will change even more. Also they are susceptible to outdoor contamination if left near open windows and to human contamination if too close to areas where persons are present.

ACTIVE AIR SAMPLING

Active air samplers fall into two categories: (1) inertial impactors that include specialized instruments and (2) impingers, rotating impactors, and filters. Different methods offer different advantages. Most active air sampling methods rely on impaction. The Burkard slit sampler and the Andersen cascade impactor are examples of inertial impactors and the Biotest is a rotating impactor. Impingers collect material in a liquid.[10] Filtration is another method used for collecting bioaerosols. Collection techniques using impingers and filters have been discussed at length in the Integrated Sampling for Gases and Vapors and the Integrated Sampling for Aerosols chapters, respectively.

The slit sampler and rotating impactor are generally used for screening areas in that they have short sampling times and allow for a wide variety of organisms to be sampled for over a short period of time either because the sampled surface (in the case of the slit sampler) can be examined directly under a microscope or a variety of sampling strips containing different culture media can be used sequentially.

Collection by impaction onto an agar surface makes it possible to count viable particles and particle clumps as colony forming units (CFU), whereas the vigorous motion occurring in an impinger causes the clusters of cells to break apart and makes it possible to count total (individual) viable cells.[2]

Although generally not a problem for active sampling indoors, wind speed and direction must be considered when sampling outdoors. Incident airflow should be parallel and going in the same direction as the suction flow of the sampler.[10] All sample volumes should be adjusted to provide a sensitivity near 50 CFU. Sample times at each site should be varied from this maximum time/volume to prevent overloading, especially when culture plate samplers are used, and to allow for the logarithmic variability that these organisms display.[7] Compared to the collection of chemical samples, sampling periods for bioaerosols are much shorter, especially when viable organisms are being collected. Table 14-2 details sampling characteristics for various types of bioaerosol samplers.

Inertial Impactors

Andersen Impactors

Inertial impactors, such as the Andersen multistage cascade impactor, can be used to separate viable particles into respirable and nonrespirable particles. Impactors for viable microbe sampling are available from several manufacturers. One of the most well known is the Andersen sampler. The Andersen viable sampler was first recommended as a standard microbial sampler in around 1970. Inertial impactors, like all samplers relying on collection on culture plates, are used for situations where airborne counts are expected to be low, such as for indoor sampling.

There are three sizes of Andersen impactors designed for collecting microbials: 6-stage, 2-stage, and single-stage impactors (Fig. 14-3). An 8-stage sampler is also available; however, generally it is used only in extensive studies where a wide range of size distributions are desired. Over the years, modifications to the original 6-stage sampler have resulted in various methods. S-6 refers to the use of the 6-stage sampler with the top 5 stages left on during sampling, but with a single culture dish in the last stage. N-6 is used to refer to a method using only one stage of the Andersen 6-stage sampler. In this case, only stage number 6 is in place with an agar collection surface and base plate. It has been replaced by the introduction of a single-stage unit for N-6 sampling. When impactors other than the Andersen are used, their collection efficiency relative to this unit should be known, because the Andersen instrument has been in use for so long. This can sometimes be done by consulting the literature. If the efficiency is significantly different, the data should be adjusted. A personal-size impactor has been used for personal microbial sampling. The collection efficiency of this unit was tested against that of the Andersen in one study and it compared well.[27] The principle of collection is the same as it is for other impactors. For more information on the use of this device, see the chapter on Integrated Sampling for Aerosols.

The basic mechanism for collection by single and multistage inertial impactors is as follows (Fig. 14-4): As air is drawn through the sampler, it passes through the orifices and strikes the surface of the agar medium. As the air passes through the orifices, the velocity of the air increases in a manner that is inversely proportional to the area of the orifice: The smaller the orifice, the higher the velocity of the air. When the velocity imparted to a particle entrained in the airstream is sufficiently great, its inertia will overcome its aerodynamic drag, and the particle will leave the turning airstream to be impacted on the agar surface. If the particle does not achieve a velocity sufficient to cause impaction, it remains entrained in the airstream and proceeds to the next stage.

Each stage in the 6-stage Andersen impactor contains a plate with 400 orifices. The orifice size is constant for a given stage, but decreases for each following stage. The 6-stage sampler separates particles into six aerodynamic ranges (Table 14-3). Directly below each stage is a glass petri dish containing a nutrient agar medium. The stages are held together by three spring clamps. Air is pulled through at a rate of 1 cfm.

Each stage of the two-stage unit contains 200 tapered orifices (Fig. 14-5). The diameter of the orifices on the first stage is 1.5 mm, and on the second stage it is 0.4 mm. Directly below each

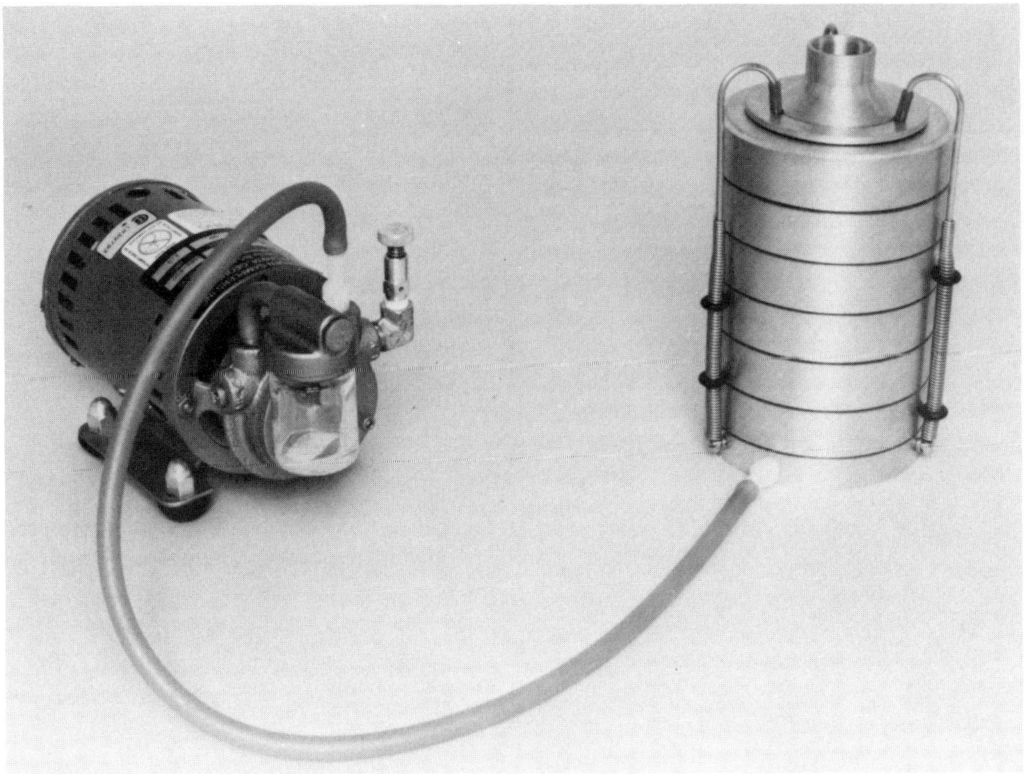

Figure 14-3. Andersen samplers for microbials. *Top:* a six-stage sampler; *left:* a single-stage sampler. (Courtesy Andersen Instruments, Inc.)

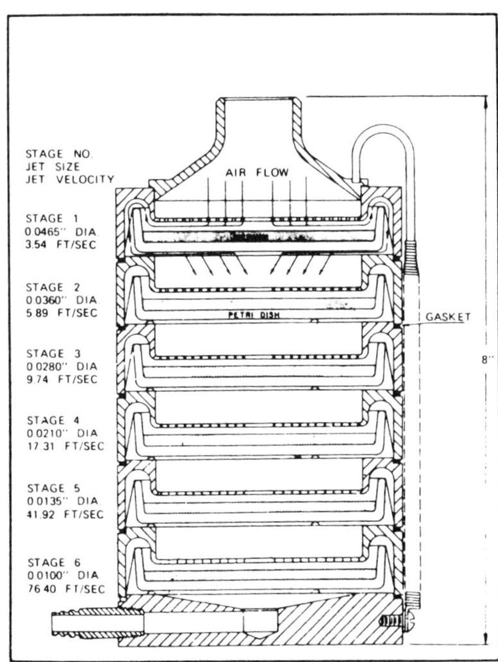

Figure 14-4. Cross section of the six-stage viable impactor stages showing the direction of air flow. (Courtesy Andersen Instruments, Inc.)

TABLE 14-3. Particle-Size Ranges for the Six-Stage Andersen Sampler

Stage	*Orifice Diameter (mm)*	*Range of Particle Sizes (μm)*
1 (top)	1.18	7.0 and above
2	0.91	4.7–7.0
3	0.71	3.3–4.7
4	0.53	2.1–3.3
5	0.34	1.1–2.1
6 (bottom)	0.25	0.65–1.1

Figure 14-5. Interior of a stage in the two-stage viable impactor. (Courtesy Andersen Instruments, Inc.)

stage is a standard 100-mm by 15-mm disposable plastic petri dish that will hold various culture mediums. The stages are held together by three dowel pins with Teflon caps. The particles on the first stage are considered nonrespirable, and those on the second stage are respirable. This sampler also requires a flow rate of 1 cfm.

When compared with the 6-stage unit, the 2-stage model has been determined to have decreased counts.[28] It has been suggested that smaller particles of 0.65 μm to 1.1 μm (0.00065 mm to 0.0011 mm) may not be collected by the 2-stage model due to its larger orifice sizes, whereas they are collected by the fifth and sixth stages of the 6-stage model. It may only be a problem for certain types of microbes of very small size. Another observation is that the 2-stage sampler has one half the number of orifices on each stage as the 6-stage sampler. Therefore, the result is multiparticle impaction through the same orifice. Since each orifice point is only counted as one CFU, the result in the 2-stage readings would be consistently lower than the 6-stage readings.

The single-stage sampler is useful in situations such as large office buildings where it is important to sample simultaneously in many areas of a building in order to determine the relationship of the ventilation system and its variables with levels of viable microbes in various areas. Two advantages of the S-6 and the N-6 methods are that shorter sampling times and less collection media are needed. There is a tendency for the very small holes of the lower stages to become occluded, however, and for the single collection plate to be overloaded. With the S-6 method, there also is more potential for wall losses on the upper stages.[28]

In one study involving 6-stage Andersen samplers to collect airborne bacteria and fungi, 4 minutes was the ideal sampling time for a satisfactory count for total microorganisms to assure that a sufficient number of microorganisms had reached all six stages.[29] For respirable samples, defined as the amount of microorganisms that deposit on stages 3 through 6, 20 minutes was required. The average flow rate was 28.3 Lpm. One method for decontamination of Andersen samplers in between uses is to wash with 70% ethanol and follow up with acetone to dry.[29] Excess cleaning solution should be collected for proper disposal.

Most inertial impactors for bioaerosol sampling are constructed of aluminum. Stainless steel substrates (plates) are often used for viable particles because they can be sterilized more easily (wrapped in paper and autoclaved) or placed in boiling water for 10 minutes and stored in sterile petri dishes.

All impactors that use solid media produce data in CFUs. Corrections must be made for multiple impactions at single holes. No more than one fourth of the holes should have multiple impactions, especially for fungus counts, to avoid inhibition effects resulting from overcrowding.

Procedure for Use of the Andersen Sampler

1. Six petri dishes, each containing 27 ml of the agar medium appropriate for the microorganisms that may be encountered, are placed in the instrument and the high-volume pump calibrated at 1 cfm is turned on. The plates are then removed, inverted in their covers, incubated, and counted.

2. Air should not be drawn through the sampler unless the petri dishes are inplace; otherwise, dirt may lodge in the small holes of the lower stages.

3. If it is desired to know the number of viable cells in each particle, as well as the number and size of particles collected by the sampler, duplicate samples should be collected. One set of plates is used for particle counts and the other set for cell counts. The cell count may be obtained by immediately washing the collected material from each plate into a flask, shaking the flask vigorously, and pouring the contents through a membrane filter.

4. The plates are counted by selecting a number of fields, and then counting the total number of colonies in these fields. In stages three through six, the colonies conform to the pattern of jets, and are counted by either the positive-hole method or by the microscope method. The positive-hole method is essentially a count of the jets that delivered viable particles to the petri plates, and the conversion of this count to a particle count by the use of a positive-hole conversion table.

5. CFUs are calculated by the following:

$$\frac{\text{CFU}}{\text{M}^3} = \frac{\text{adjusted number of colonies on plate}}{\text{total volume of air sampled in M}^3}$$

Slit Samplers

The Burkard sampler is an example of a slit sampler, also considered a type of inertial impactor for collecting screening samples of bioaerosols (Fig. 14-6). The unit is a battery-powered air sampler that can collect area samples directly on microscopic slides. An impaction orifice is located on top of the unit. A slide is slipped in through the side of the sampler with its collection surface facing up and air is pulled into the unit at a rate of 10 Lpm. The batteries must be fully charged to assure this flow rate.

Figure 14-6. Burkard personal volumetric air sampler (a slit impactor). (Courtesy Burkard Mfg. Co., Ltd.)

The Burkard sampler is useful for identification of specific sources of microbes. As an example, in an indoor air situation where occupants on only one side of a building are complaining, samples can be collected in several representative rooms in a number of different areas for comparison over the same day. Each sample must be sealed in a container immediately following collection to avoid contamination.

Clean glass slides can be used without adhesive if the final sample will not be preserved. However, an adhesive such as Vaseline is often used. Heating slides coated with Vaseline will make it easier to cover a slide with a smooth coating of material. Vaseline as an adhesive works well with oil immersion. A mixture of Vaseline and hexane (1:5 w/v) has also been found useful.[26] Gelvatol, a water-soluble plastic, is recommended by the manufacturer. A number of factors are involved in the selection of an adhesive for the slides: (1) stickiness, since the surface should be as wet as possible, especially for sampling spores; (2) weather resistance, since fog or high humidity will cause adhesives to slough off or to be emulsified; (3) the type of microscopy to be used on the samples including stains and mountants.

Maintenance involves cleaning the sampling orifice to remove contaminants and obstructions to airflow.

Procedure

1. Prepare the glass slide with a suitable adhesive.

2. Rotate the upper ring assembly until the red dots are in line; this exposes the aperture for the glass slide. Insert the glass slide through the aperture, adhesive side up, until the end of the slide is firmly against the stop inside the unit. Close the sampling chamber by rotating the upper knurled ring by at least 1 inch. Make sure the slots are fully covered.

3. Turn on the unit. When warmed up, it draws 10 Lpm. Normal sampling time is 15 minutes.

4. At the end of the sampling period, open the ring and push the slide out.

Rotating Impactors

The Biotest sampler (Fig. 14-7) is an example of a rotating impactor. A multibladed impeller draws air into the sampler from up to 40 cm away, depositing microbes on a plastic strip containing agar culture media lining the inside of the impeller housing. After the samples are collected, the strips are incubated and the colonies are counted. The sampler has an average rotational speed of 4,096 rpm, with an accuracy of ±2% according to the manufacturer. This device is a compact field sampler that is battery-operated, and samples a large volume of air in a short time. The present design is inefficient for collection of small particles ($\leq 5\ \mu$m).[2]

An advantage of this sampler is its ability to collect samples to screen for a number of different organisms by collecting sequential samples on strips that contain different types of agar. The following types of agar are recommended by the manufacturer:

TSA-Agar for total counts
Rose-Bengal-Agar for yeasts and molds
Mannitol-Salt-Agar for staphylococci
MacConkey-Agar for coliform bacteria
TSA-Penase-Agar for total count in air-containing penicillins and semisynthetic penicillins

Due to its principle of operation and the geometric properties of the impeller drum, the Biotest sampler has special volume characteristics. It is therefore necessary to differentiate between the total volume sampled (sampling volume) and the volume relevant for separating the particles (separation volume). The separation volume is 40 Lpm and is the basis for calculating the number of organisms per air volume. The sampling volume is 280 Lpm (based on 4,096 rpm). By virtue of the high centrifugal force, the particles in the rotating ring of air are forced outward and impacted onto the surface of the nutrient medium. However, this separation takes place only from one part of the sampling volume. Calculations based on the angle velocity, viscosity of the air, diameter and density of a relevant particle, and the sampling volume are used to determine the separation volume. At least one study has been conducted on the use of the Biotest sampler.[30]

The unit cannot be calibrated; instead the angle of the impeller blades is to be checked either weekly or, if less often, prior to each sampling. With use, the angles of the impeller blades may become altered, and this correction is required. The impeller is placed into the calibration set so that the teeth interlock. By pressing the units firmly together, the position of the individual blades is checked and corrected.

The sampling time should be measured monthly. The 8-minute sampling time is the one most often checked. The manufacturer recommends that it not exceed or be less than 480 ± 9.6 seconds, which is an accuracy of ±2%. If the result is not within this limit, the batteries should be changed. If the result is still not within this

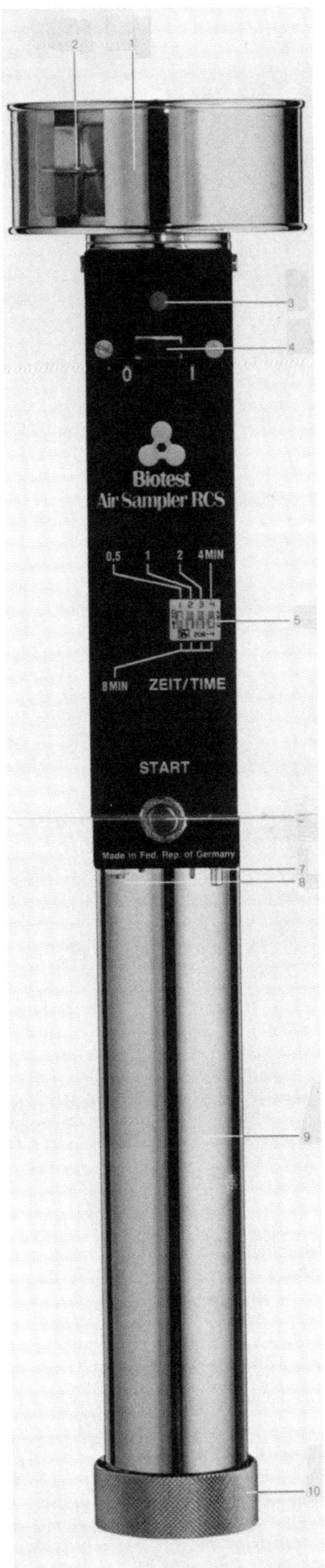

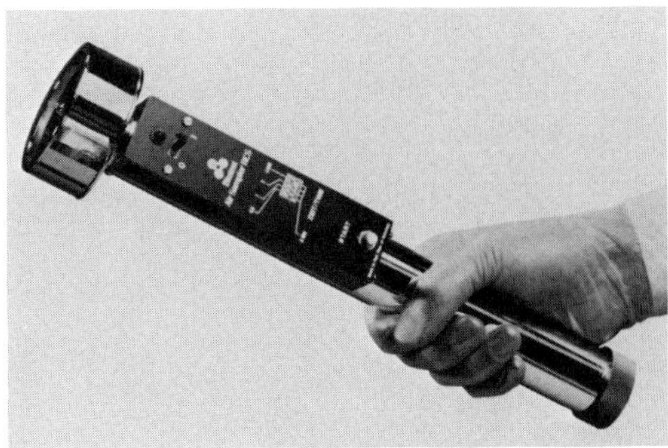

Figure 14-7. The Biotest RCS centrifugal air sampler. *Left:* schematic showing sampler and control panel. 1. Open end drum housing with impeller blades; 2. slot for insertion of the agar strips; 3. control light for "on-off" and battery check; 4. main switch; 5. timer; 6. start button; 7. remote control input; 8. AC adapter input; 9. battery housing; 10. screw off cap for battery housing. *Above:* sampler in use. (Courtesy Biotest Diagnostics Corp.)

limit following this change, the unit must be sent in for electronic calibration.

The primary application of the Biotest is as a probe for hot spots, in order to screen areas to determine those with the highest concentrations.[31] Like all samplers relying on collection on a culture plate, the Biotest is limited to situations where airborne counts are likely to be low, which is generally for indoor sampling.

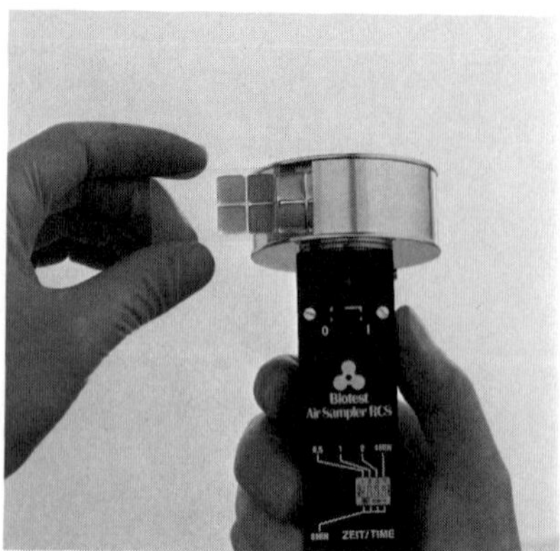

Figure 14-8. Preparing the Biotest sampler for use. (Courtesy Andersen Instruments, Inc.)

Procedure for Use of the Biotest Sampler

1. Prior to use, the impeller drum should be removed from the unit and sterilized. It can be done by boiling the unit in water or soaking it in isopropyl alcohol.

2. Select the sampling time. The instrument provides five selections for sampling times. Table 14-4 indicates sample time settings for the Biotest. The sampler should not be used more than 8 minutes for any one agar strip as excessive desiccation and microbial death will result. Sampling time is selected by setting the switches in a variety of configurations.

3. Remove the agar strip from the wrapper and inspect it for contamination. Do not use a strip with growths on it. Handle it by the edges only.

4. Insert the strip into the slot in the open-end drum with the agar surface facing toward the impeller blades (inside). Continue inserting until the slot is closed by the strip and the tab protrudes about 2 cm (Fig. 14-8).

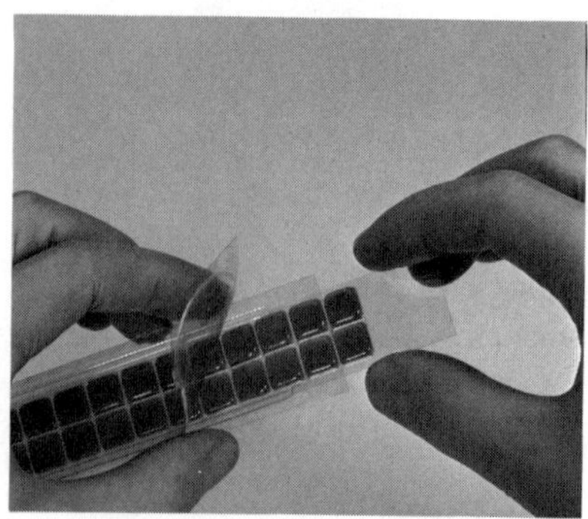

Figure 14-9. Removing the agar strip from the Biotest sampler. (Courtesy Andersen Instruments, Inc.)

TABLE 14-4. Sampling Time Settings for the Biotest

Selector	*Position*	*Time*	*Volume (liters)*
1	ON	30 seconds	20
2	ON	1 minute	40
3	ON	2 minutes	80
4	ON	4 minutes	160
All	DOWN	8 minutes	320

5. Switch on the instrument by moving the main switch to 1. Check that the control light is on.

6. Press the start button to begin sampling. Depending on what is being sampled, such as the vent of an HVAC system, it is important to keep the unit placed correctly. Even though it is light, after 8 minutes it can be difficult to hold in place. After the drum stops turning, move the main switch to off.

7. Remove the agar strip by holding its tab and gently pulling it out of the impeller drum (Fig. 14-9). Replace it in its original wrapper with the agar surface facing away from the lid (face down). Seal the wrapper by sliding the plastic

seal over the top to prevent the media from drying out during incubation. Label the holder immediately using an indelible felt tip pen or grease pencil with a unique number.

8. Calculations are done following the count.

$$\text{CFU/L} = \frac{\text{colonies on agar strip}}{40 \times \text{sampling time in minutes}}$$

$$\text{CFU/M}^3 = \frac{\text{colonies on agar strip} \times 25}{\text{sampling time in minutes}}$$

$$\text{CFU/ft}^3 = \frac{\text{colonies on agar strip} \times 0.708}{\text{sampling time in minutes}}$$

Impingers

Liquid impingment collects microbes directly into a liquid, providing some protection for the microorganisms, and allows immediate initiation of repair of any damage that may have been caused by the collection process itself. Impingers are described in the chapter on Integrated Sampling for Gases and Vapors. The sample liquid can be processed to detect a wide range of aerosol concentrations. It can be sampled directly or filtered with all particles removed from the total volume. Impingers can be used in situations where high concentrations of microorganisms are suspected as the solutions can be diluted prior to culturing.[1] An advantage of having the sample in a liquid is that several types of media can be inoculated from the suspension or several plates of the same medium can be inoculated and incubated under different conditions.[27] Aggregates of cells that would grow as single colonies on an impactor sampler are broken up in the impinger liquid, allowing each cell to develop into a separate colony. Calculation of inhaled doses from this measurement of air concentration may be preferred for some disease agents, especially those for which a small number of organisms constitutes an infective dose. A spillproof impinger has been used successfully to collect microbials with the same efficiency as the standard all-glass impinger and showed more efficient recovery of hardy spores.[2]

Impingers are recommended for bacteria collection as long as appropriate media are used. In general, impingers are useful for the recovery of soluble materials, such as mycotoxins, antigens, and endotoxins, and for sampling aerosols of bacteria and viruses that require gentle handling.[10] All-glass impinger fluid should be sterile and the impinger should be rinsed with the sterile fluid prior to use.

Procedure[24]

1. Prepare a sampling train with a portable pump and a sterile glass midget impinger containing 10 ml of sterile phosphate buffer solution.

2. Collect a sample at a flow rate of 2.0 Lpm ± 0.5. Sampling time will be dependent on an estimated bacterial concentration.

3. After sampling, cap the impinger openings and return to the laboratory within 30 minutes.

4. Aspirate the sample through an opened Millipore sterile clinical field monitor (catalog no. MHWG037HO) with a 0.45-μm pore-size gridded filter inside.

5. Rinse the empty impinger with 10 ml of sterile phosphate buffer solution, and aspirate this washing through the filter.

6. Add 1 ampoule of bacterial media to the filter by aspiration.

7. Cap the filter cassette, and incubate the cassette sample in a 30°C incubator for 2 to 3 days.

8. Count each colony using a stereoscopic microscope.

9. Calculate the volume of air sampled, and report as CFU/M^3.

Filtration

Filters in plastic cassettes attached to a personal air sampling pump can be used to collect nonviable dusts such as spores, but viable cells will most likely be dehydrated and killed by the large volumes of air that pass over the filter in the course of sampling. Filters are also useful for sampling highly contaminated environments. Examples include sawmills during wood chip handling, grain elevators, and straw handling in pig houses.[21] Spores that have a tougher cell wall and are not actively metabolizing can be collected on gelatin and cellulose ester and polycarbonate filters. The analytical method often dictates what type of filter can be used. One study evaluated three different analytical methods for fungi collection on filters.[21] Filters were analyzed by SEM and light microscopy for direct counts and epifluorescence microscopy, a staining technique. The material collected on filters can also be resuspended and cultured on different agar media, and analyzed for mycotoxins, endotoxins, or specific allergens. Polycarbonate, 0.4-μm filters in 37-mm cassettes are used to collect spores to be examined by SEM or epifluorescence microscopy. Cellulose ester 0.8-μm membrane filters were used for samples for light microscopy and for culturing. The flow rate for all samples was 1.5 Lpm and samples were collected closed-face.

An advantage of gelatin filters is that they can be used to provide an estimate of both total number of bacteria-laden particles and total number of bacteria present. A gelatin filter placed on warm, moist agar readily dissolves and is absorbed. Intact particles are deposited on the agar, and if viable, they grow into visible colonies, one colony per particle, even if a particle carries more than one bacterium. A gelatin filter dissolved in water releases collected particles, and clusters of bacteria adhering to particles are dispersed by vigorous shaking. Dilution of the liquid avoids the problem of an overloaded filter, resulting in colonies too numerous to count (TNTC).[2] When polycarbonate filters are used, spores can be washed from the filter and centrifuged to form a pellet that is then examined in a haemocytometer under a microscope.[7]

Culturing filters allows for identification of both bacteria and fungi, although different types of agar must be used. Since counts usually involve inspection of only a portion of the filter, the loading characteristics are important. Accuracy is affected by uneven distributions of particles on filters and the difficulty of counting small spores close to the limits of microscopy resolution.

Filter counting procedures must deal with the fact that a typical sample may contain single spores, aggregates of spores, and other particles carrying microorganisms.[21] If a filter is washed so the sample can be resuspended, spores may be lost.

Plastic cassettes used to contain filters should be sterilized prior to use. Filtration is recommended for fungi collection but not for bacteria.

Procedure[24]

1. Prepare a sampling train with a personal air sampling pump and a sterile Millipore clinical field monitor (catalog no. MHWG037HO) containing a 0.45-μm gridded filter.

2. Collect a sample at a flow rate of 2.0 Lpm $\pm$ 0.5. Sampling time will depend on estimated fungal concentration.

3. Return the sample to the laboratory within 30 minutes.

4. Open the cassette and aspirate 10 ml of sterile phosphate buffer solution through the filter. This wetting step helps to distribute the media added next.

5. Add fungal media, by aspiration, from one Millipore ampoule, to the opened filter cassette.

6. Cap the cassette and place the sample in a drawer for 3 to 5 days to incubate at room temperature.

7. Count each individual colony that has grown during the incubation period. The use of a magnifier or stereoscopic microscope is helpful. Ideally, if the sampling time has been estimated correctly, the colonies should be easy to read with no overcrowding.

8. Calculate the volume of air sampled and report as CFU/M^3.

SURFACE SAMPLING

Slides can be used to "wipe" an area to see if microbials are present in dust. Slides can also be pressed onto surfaces. Properly prepared, these will allow the organisms to be preserved. Scotch tape imprints have been used to collect spores for qualitative identification of the presence of molds and other compounds. This technique has also been used for rapid identification of mold colonies growing on organic materials.[32] Cotton swabs can be used to collect air-conditioning unit condensate as well as other suspect surface materials.[22] Sterile swabs should be used to wipe the surface of interest. The swabs are then stroked against the surface of a culture plate containing agar; plates are incubated and then examined.

BULK SAMPLING

Bulk samples are useful for qualitative identification of microbials, since the sample can be cultured and examined directly under a microscope. When bacteria are fragile and difficult to culture, such as Legionella, bulk samples of suspect sources are preferred to air samples.[1] Bulk samples are often collected of cooling tower water because of the association with Legionnaires' disease. Deteriorating insulation and other organic materials such as fiberglass can also develop microbial growths. Filters in HVAC systems as well as ducts often become heavily laden with dust, which can be collected and cultured for fungus and molds that may have been present as spores. An example of an industrial situation where bulk samples can be useful is microbial contamination of water-based metal-working fluids, which can occur if the concentration of biocide decreases upon standing or through dilution with other solutions.

Microbial growth on surfaces can often be readily identified, such as the powdery black or blue-green growths of molds. In these cases, if a specific identification of species is needed, a bulk sample will be useful. If slime is found in any reservoirs, such as those associated with humidifiers and refrigerators, it should be sampled by scraping the material and placing it in a glass culture plate. Slime growth has been found on the surfaces of water spray pumps, on masonry wall surfaces located between water spray systems and fans, and on the floor in the vicinity of the fans. In addition to microbes, slimes can contain flagellates, nematodes, and mites. Wetted surfaces that occur in the air-handling units of the HVAC systems designed for cooling may also encourage slime growths. Fan coil units can also develop slime growths.

Many construction materials can supply sufficient nutrients for microbial growth.[33] HVAC filters and carpet samples are often collected and submitted for analyses. Damp organic materials that can support microbial growths include leather, plastics, and nylon. The presence of a film of soap or grease on any surface increases the likelihood of microbial contamination. Usually debris is washed out of these materials and the wash is cultured. These materials can be collected in ziplock plastic bags.

Microorganisms, particularly fungi, have been found in settled dust from floors, equipment, or ledges.[34] Settled dust exposures could result in significant worker exposures during sweeping and cleaning. Bulk samples of dust are useful for spore analysis or for the presence of antigenic materials such as bacteria, fungi, or insect body parts. If histoplasma or crytococcus, are suspected, then a bulk sample of bird droppings should be collected. If the birds are nesting in air intakes, this is especially important.[1]

If water is suspected as the source of contamination, bulk samples should be collected for culturing in addition to air samples. Typical samples might be potable and nonpotable water. Other sources for bulk water samples include water from piping dead ends, faucet aerators, sink taps, gasket materials, and joint sealants. Sterile pipets can be used to collect water samples. Bulk water samples are usually filtered according to standard methods.[13] The filters are then placed on plates containing agar and are incubated.

Tap water is frequently collected for *E. coli* counts as this organism is used as an indicator for contamination by sewage microbes by the EPA. OSHA regulations also require the availability of potable water for drinking, and washing hands and eating utensils at work. Testing for this organism is regularly done on community water supplies.

Procedure for Collecting Water Samples for Microbials[13]

1. Samples should be collected in capped bottles that have been cleansed, rinsed, and sterilized.

2. Keep the sampling container closed until collection time. Remove the cap with care to prevent soiling and do not handle the cap or

neck of the bottle. Instead hold the bottle near the base.

3. Fill the bottle without rinsing and replace the cap immediately. During collection, leave ample air space in the bottle to allow mixing of the sample in the laboratory. At least 100 ml of water should be collected for microbial analysis.

4. Record the date and time of collection along with the ambient temperature on the label of the sample.

STORAGE AND SHIPMENT OF SAMPLES

Prompt shipment and proper storage of samples following collection is critical for viable organisms. A rule of thumb is to get viable samples to a laboratory within 24 hours.[1] During storage the sample temperature should be maintained as close as possible to that of the source.[13] Samples taken on solid or liquid media will begin to grow almost immediately after collection. To prevent growth of bacteria that grow best at temperatures between 20°C and 55°C on media designed for high-temperature bacteria, refrigerate samples or begin incubation at the proper temperature immediately.[7] Blanks should be included of sampler and/or culture medium. Containers should be sterilized prior to use and shipping packages should insulate samples from environmental stresses, especially heat and freezing cold.

Impactor plates and filters should be protected from bright light and maintained at refrigerated temperatures for 12 hours to 36 hours until transported to the laboratory where they can be incubated.

INTERPRETATION OF RESULTS

No threshold limit values (TLVs) or other official standards exist for levels of microbials in an air sample. Results of bioaerosol sampling will be presented differently depending on how samples were collected. Air samples are reported as CFU/M^3. Specialized sampling is often reported in terms of the entity collected, for example, if only spores were sampled, results would be reported as spores/M^3, or if pollen was sampled, results would be reported as grains/M^3. Impinger samples are reported as CFU/ml while bulk samples are reported as CFU/g in the case of dusts and CFU/cm^2 for carpet. Viruses are reported as infectious units describing the ability of the organisms to infect, multiply, and produce new virus. CFU data must be interpreted with care because many fungus spores travel in chains and many bacteria are carried on other larger particles that produce only 1 colony for 5 to 100 cells or spores. The designation TNTC on a report indicates that sample results were too high to be effectively counted. The EPA defines TNTC as a situation where the total number of bacterial colonies exceeds 200 on a 47-mm diameter membrane filter for coliform detection.[12] If blanks are positive, sampling must be repeated.

When using culture techniques for collecting air samples, most results will be less than the actual number present due to losses in viable organisms during sampling as well as the fact that other organisms will not grow.[1] Some organisms have specialized growth needs: Growth inhibitors produced by one microbe may reduce levels of other organisms; overcrowding can reduce recoveries due to both soluble inhibition factors and "contact" suppression by adjacent growth points; filters can allow microbes to dry out and die; and there are limited periods when organisms are viable when airborne.[10] Results of cultures can be affected by the viability of the organisms; choice of nutrient medium; conditions for growth, especially temperature; and interactions between different organisms.[21]

In indoor air investigations, the best standard is the comparison of the outside air at the intakes of a building to the indoor air to determine whether there is indoor amplification of particular organisms. The percent of outdoor air allowed into the building and the percent of air that is recirculated must also be considered. For fungus, a rule of thumb is that indoor counts should be less than one half of outdoor levels present when HVAC systems are on.[1] The presence of any one fungus in levels exceeding 500/M^3 indoors when comparison samples collected outdoors have no detectable levels of this strain or low levels of fungus can lead to a presumption of an indoor source. Levels of fungal spores indoors should be less than one third of those collected on simultaneous outdoor samples or a significant amount of contamination may exist.[7]

A rule of thumb is that overall levels of a bioaerosol should be at least ten times greater in the area suspected of contamination than in the area considered background.[1] For example, a study in southern California determined that a common outdoor count for molds is 1000 to 1500.[32]

According to one source, if the total count of microorganisms at an affected person's area equals or exceeds 10,000/M^3, remedial action is necessary.[23] Another source suggests it is a good practice to attempt to identify any viable microorganism recovered from air in levels greater than 75 CFU/M^3.[7] It has been suggested that a level of viable microorganisms in excess of 1×10^3 viable particles per M^3 indicates that the indoor environment may be in need of investigation and improvement.[7] High levels of human source bacteria (gram-positive) in an indoor environment indicate overcrowding and/or inadequate ventilation. High levels (>500 CFU/M^3) of gram-negative bacteria or Bacillus species in an indoor air situation suggest contamination of a source, since these organisms are not usually present indoors.[1] Actinomycetes bacteria are usually associated with agriculture and their presence in other situations is suspect.[1] The presence of thermophilic actinomycetes in indoor environments at levels above 500/M^3 has been associated with outbreaks of hypersensitivity lung illness.[23]

One method for evaluating the contribution of outdoor air to indoor bioaerosols and wipe samples is rank-order assessment.[23] Identify taxa at experimental sites and at control (outdoor) sites. Calculate relative abundance of each taxon. If locations have the same flora, the taxa should fall into a similar order of abundance. Individual taxa are listed in descending order of abundance for indoor sites and outdoor controls. For example, if one fungus is much higher indoors than outdoors, even though others are found, it is likely that it may be the source of the problem. Usually results (in CFU/M^3) are set up in a tabular form for comparison:

Indoors:	microbe A >	microbe B >	microbe C
	(2000)	(400)	(100)
Outdoors:	microbe C >	microbe D >	microbe B
	(400)	(100)	(50)

In this example, microbe A is the suspect source in the indoor air. As a precaution, it should be noted that in some ventilation systems filtration may remove only large spores from the outside air, and thus change the rank order of smaller spores in the interior. A combination of quantitation and rank order assessment is usually necessary.

The type of organisms detected often depends on the type of operation being sampled. For example, bacteria are more prevalent than fungi in straw handling operations, grain elevators, and pig houses whereas fungi are more prevalent than bacteria in wood-related operations such as wood chip handling and sawmills.[21] Fluctuations in results from sampling situations where animals are suspected as the source can be due to variations in feeding activities, maintenance, and cleaning of animal housing facilities.[29]

The presence of a microbe in a bulk sample of water or other material does not mean it is airborne unless a means of aerosolizing the microbe can be identified. If a growth is positive, it may or may not be the source of disease. There are often several different strains of a given organism and some are pathogenic to humans whereas others are not. Other guidelines are 1×10^6 fungi/g in dusts from HVAC filters, and 1×10^5 bacteria or fungi/ml in stagnant water or slime are indicative of excessive microbial contamination.[24]

REFERENCES

1. ACGIH Committee on Bioaerosols: *Guidelines for the Assessment of Bioaerosols in the Indoor Environment.* ACGIH, Cincinnati, OH, 1989.
2. Macher, J. M. and M. W. First. Personal air samplers for measuring occupational exposures to biological hazards. *AIHA J.* **45**(2):76–83, 1984.
3. Elliott, L. J., R. Sokolow, M. Heumann, and S. L. Elefant. An exposure characterization of a largescale application of a biological insecticide, *Bacillus thuringiensis. Appl. Ind. Hyg.* **3**(4):119–122, 1988.
4. Forster, H. W., et al. Investigation of organic aerosols generated during sugar beet slicing. *AIHA J.* **50**(1):44–50, 1989.
5. Muraca, P. W., et al. Legionnaires' disease in the work environment: Implications for environmental health. *AIHA J.* **49**(11):584–590, 1988.
6. National Institute of Occupational Safety and Health. *Occupational Diseases: A Guide to Their Recognition.* Cincinnati, OH: NIOSH, 1977.
7. ACGIH Bioaerosols Committee. Guidelines for assessment and sampling of saprophytic bioaerosols

in the indoor environment. *Appl. Ind. Hyg.* **2**(5):R-10–R-16, 1987.

8. DeLucca, A. J., and M. S. Palmgren. Seasonal variation in aerobic bacterial populations and endotoxin concentrations in grain dusts. *AIHA J.* **48**(2):106–110, 1987.
9. Clark, S., et al. Airborne bacteria endotoxin and fungi in dust in poultry and swine confinement buildings. *AIHA J.* **44**(7):537–541, 1983.
10. Burge, H. A., and W. R. Solomon. Sampling and analysis of biologic aerosols. *Atmos. Environ.* **21**(2):451–456, 1987.
11. Burge, H. A. Indoor sources for airborne microbes. In *Indoor Air and Human Health,* H. B. Gammage and S. V. Kaye (eds.) Chelsea, MI: Lewis Pubs. Inc., 1985.
12. Environmental Protection Agency. *National Drinking Water Regulations.* CFR 40. Chap. 1, Subchapter D, Part 141.
13. American Public Health Association. *Standard Methods for the Examination of Water and Waste Water,* 16th ed. APHA, et. al., New York, 1985.
14. DuBois Chemicals, Ltd. *Reducing the Risk of Legionnaires' Disease.* 1990 management report, England, 1990.
15. Fischer, J., et al. Environment influences levels of gram-negative bacteria and endotoxin on cotton bracts. *AIHA J.* **43**:290–292, 1982.
16. Milton, D. K., et al. Endotoxin measurement: Aerosol sampling and application of a new Limulus method. *AIHA J.* **51**(6):331–337, 1990.
17. Jacobs, R. R. Airborne endotoxins: An association with occupational lung disease. *Appl. Ind. Hyg.* **4**:50–55, 1989.
18. Morring, K. L., W. G. Sorenson, and M. D. Aitfield. Sampling for airborne fungi: A statistical comparison of media. *AIHA J.* **44**(9):662–664, 1983.
19. Burge, H. A. Fungus allergens. *Clin. Rev. Allergy* **3**:319–329, 1985.
20. Chatigny, M. A. Sampling airborne microorganisms. In *Air Sampling Instruments,* 3rd ed. Cincinnati, OH: ACGIH, pp. E-1–E-9.
21. Eduard, W., et al. Evaluation of methods for enumerating microorganisms in filter samples from highly contaminated occupational environments. *AIHA J.* **51**(1):427–428, 1990.
22. McJilton, C. E., et al. Bacteria and indoor odor problems—three case studies. *AIHA J.* **51**(10): 545–549, 1990.
23. ACGIH BioAerosols Committee. Airborne viable microorganisms in office environments: Sampling protocol and analytical procedures. *Appl. Ind. Hyg.* **1**(4):R-19–R-23, 1986.
24. Morey, P. R., et al. Environmental studies in moldy office buildings: Biological agents, sources and preventive measures. *Am. Conf. Gov. Ind. Hyg. Ann.* **10**:21–35, 1984.
25. Furst, M. W. Air sampling. *Appl. Ind. Hyg.* **3**(12):F-20, 1988.
26. Burkard Manufacturing Company. *Manual for Use of the Burkard Sampler.* England.
27. Macher, J. M., and H. C. Hansson. Personal size-separating impactor for sampling microbiological aerosols. *AIHA J.* **48**(7):652–655, 1987.
28. Macher, J. M. Positive-hole correction of multiple-jet impactors for collecting viable microorganisms. *AIHA J.* **50**:561–568, 1989.
29. Cormier, Y., et. al.: Airborne microbial contents in two types of swine confinement buildings in Quebec. *AIHA J.* **51**(6):304–309, 1990.
30. Macher, J. M., and M. W. First. Reuter centrifugal air sampler: Measurement of effective air flow rate and collection efficiency. *Appl. Environ. Microbial.* **45**:1960–1962, 1983.
31. Smid, T., et al. Enumeration of viable fungi in occupational environments: A comparison of samplers and media. *AIHA J.* **50**(5):235–239, 1989.
32. Kozak, P. P., and J. Gallup. Endogenous mold exposure: Environmental risk to atopic and nonatopic patients. In *Indoor Air and Human Health,* H. B. Gammage and S. V. Kaye (eds.). Chelsea, MI: Lewis Pubs. Inc., 1985.
33. Brief, R. S., and T. Bernath. Indoor pollution: Guidelines for prevention and control of microbiological respiratory hazards associated with air conditioning and ventilation systems. *Appl. Ind. Hyg.* **3**(1):5–10, 1988.
34. Burge, W. R., et al. Measurements of airborne aflatoxins during the handling of 1979 contaminated corn. *AIHA J.* **43**(8):580–586, 1982.

PART V

Air Sampling Decisions: Gathering Background Information

CHAPTER 15

Biological Monitoring

Biological monitoring measurements of bodily fluids or breath are considered better measurements of exposure than the values measured in the air since they represent an individual's actual absorbed dose. Biological monitoring is the trend for the 1990s, and the number of compounds for which methods are available can be expected to increase.[1] Currently it is almost exclusively applied to occupational exposures.

There are many specific purposes for conducting biological monitoring. Many of the Occupational Safety and Health Administration's (OSHA) substance-specific standards require monitoring of urine or blood for compliance if certain airborne levels are exceeded. Biological monitoring can assist in evaluating the contribution from dermal and oral exposures and in the absence of hazardous airborne levels, whether skin or oral exposures are occurring. In the case of an emergency release, an assessment of the degree of exposure can be done (but not detection of whether an acute exposure has occurred as discussed in the section on limitations later in this chapter). It can also be used to monitor the effectiveness of personal protective equipment (PPE). For example, in one study it was determined that a respirator cartridge was overloaded due to high mandelic acid levels in the urine of a worker exposed to styrene in a fiberglassing operation.[2]

Another use of biological monitoring is to assess the effectiveness of engineering controls or work practices to limit uptake of environmental chemicals. Many workers' compensation claims for overexposure to chemical agents could be simplified if biological samples were collected either during the time of the incident or when the employee complains. Since this procedure is generally not the case, the result is that claims investigations for occupational exposures are generally drawn out and rely on indirect methods to determine whether the exposure actually took place.

Blood and urine are sampled by collecting the bulk fluid. Breath sampling is done using basic air sampling techniques such as Tedlar bags, sorbent tubes, and a specialized glass pipet. Analysis of blood and urine is always done in the laboratory and real time instruments are sometimes used for breath samples. Respirators with specially adapted cartridges have also been used for breath collection, although as of now there are no biological standards based on this technique.

Biological monitoring has often been called the ultimate personal sampler, because when properly used it can assess worker exposure to industrial chemicals by all routes, including skin absorption and oral indigestion. Specifically, biological monitoring includes the measurement of the absorption of an environmental chemical in an individual. There are two basic types of biological monitoring used to develop methods: direct measurement of a chemical or a metabolite in biological media, or indirect measurement by quantifying a nonadverse biological effect related to the exposure.

Nonadverse or subcritical effects have been defined as effects that do not impair cellular function, but are still evident by means of biochemical or other tests, and as a precursor of a critical effect, a subcritical effect may be a more useful measurement than others in preventing harmful exposures.[3] Direct measurements are done more often than indirect ones and include such tests as blood lead as an assessment of chronic lead exposure, measurement of urinary phenol for assessment of exposure to benzene, and measurement of carboxyhemoglobin in blood as an index of carbon monoxide or methylene chloride exposure. Examples of indirect measurements include cholinesterase inhibition by organophosphate compounds and inhibition of a delta-amino levulinic acid dehydratase (ALAD) (a red blood cell enzyme) by inorganic lead.

Another type of test is the measurement of a specific antibody to determine if exposure to an allergen has taken place. In one study individuals who had become sensitized to pigeons were identified by testing for a reaction of a specific IgG antibody in their blood with pigeon serum proteins.[4] The limitation of direct and indirect measurement methods is that although they measure exposure, they do not quantify the amount that has reached a target organ, such as the liver in the case of exposures to chlorinated solvents. Since under most circumstances this method is not possible, other substances such as blood, urine, and breath are monitored instead.

The basis for biological monitoring is pharmakinetics (also sometimes called toxicokinetics when used to describe the effects of nondrugs or toxic compounds), which is the study of the process of uptake, distribution, and elimination of substances from the body. Once in the body, chemicals are metabolized and excreted or stored in an organ, fat, or bone. The liver is the primary site for metabolism. Some of the chemical may be excreted unchanged, such as occurs when solvents are inhaled and partially eliminated in the breath.

Accumulation of a chemical such as lead in the blood or in the hepatobiliary intestinal loop can also occur. Once in the blood, chemicals can bind to plasma or other proteins or circulate unbound. The unbound chemical is usually the entity that is responsible for toxic effects unless the target organ is the blood.

Each individual represents a unique biological system (Table 15-1). Workers differ in age,

TABLE 15-1. Sources of Pharmacokinetic Variability

Absorption	*Distribution*	*Metabolism*
Route	Body size	Genetic factors
Physical form	Body composition	Age and sex
Solubility	Protein binding	Environment (pollution, diet, habits)
Physical work load	Physical work load	Chemical intake (alcohol, medications)
Exposure concentration	Exposure concentration	Physical activity (pulmonary ventilation, blood flow)
Exposure duration	Exposure duration	Protein binding
Skin characteristics		Life-style (smoking)
		Exposure level

Source: Droz, P. O. Biological monitoring I: Sources of variability in human response to chemical exposure. *Appl. Ind. Hyg.* **4**(1):F-23, 1989.

body build, weight, fitness, physiological and nutritional status, and habits such as smoking and alcohol consumption, all of which can affect their intake. Work rate has also been shown to affect intake in a given exposure situation. As an example, concentrations of mandelic acid in urine due to styrene exposure decrease when light work rather than heavy work is performed.[2] Metabolizing enzymes can be induced by smoking, medication, and absorption of other chemicals present in the environment or diet. The simultaneous absorption of chemicals in relatively high dosages, as in alcohol consumption, drug intake, or simultaneous occupational exposure to other chemicals, can slow down the metabolic rate by inhibition of the metabolizing enzyme system.[5] For example, alcohol (ethanol) consumption during or shortly following an exposure period to toluene will increase toluene levels in the blood.[6]

As a result, when the same group of workers is exposed to the same airborne concentration of a given chemical, each individual will have differences in the uptake, absorption, biotransformation, and elimination of that compound. This will produce differences in the amount of active metabolite(s) reaching their target organs.[7]

Some types of organic compounds can undergo extensive biochemical transformations during metabolism. In these situations, it is a metabolite that is usually monitored rather than the individual compound, since there is often more than one metabolite. Selection of the proper one to monitor is important. The metabolite must be specific to the exposure and must accumulate in biological fluids in quantities proportional to the dose of the original compound in order for monitoring to be effective. An example is the determination of 2,5-hexanedione in urine following exposure to *n*-hexane. Often compromises are necessary and the metabolite does not always meet this criterion perfectly. In the case of *n*-hexane, while there are no natural (endogenous or diet) sources of 2,5-hexanedione, it is also a metabolite product of methyl *n*-butyl ketone (MBK).[8]

The biologic half life (BHL) determines whether a chemical is rapidly excreted or stored in the body (Table 15-2). This value is specific for a given chemical and can be determined for a chemical's residence time in the entire body or in a specific tissue, in the case where a chemical is stored. The formula for computing a BHL is given by[9]:

TABLE 15-2. Biologic Half Lives

Chemical	*Half Life (hours)*
Acetone	4
Aniline	2.9
Benzene	3
Carbon disulfide	0.9
Dieldrin	365
Ethylbenzene	5
Hexane	3
Lead	840
Pentachlorophenol	33
Phenol	3.5
Toluene	1.5
Xylene	3.6

$$\text{BHL} = 0.693\frac{(V_d)}{(F)}$$

where V_d = distribution volume in liters and F = rate of blood flow to the tissues.

The distribution volume depends on the volume of the tissue (or body) and the tissue-blood partition coefficient. As with other measurements of this type, the greater the BHL value, the longer a chemical will reside in the body. There are very few BHL's that have been determined when compared to the number of chemicals of concern.

Another calculation that can be used to estimate the uptake of a chemical by a worker through inhalation uptake is based on

$$(\text{RV})(\text{EC})(\text{LE})$$

where RV = respiratory volume, EL = environmental concentration, and LE = length of exposure.

Typical respiratory volumes and an example of this type of calculation have been previously described in the chapter on Hazards.

Biological exposure indices (BEIs) are developed by the American Conference of Governmental Industrial Hygienists (ACGIH) to use as guidelines for biological monitoring. BEIs are based on a variety of information: mechanisms of absorption and elimination, including biotransformation of the substance; all possible toxic effects, but especially the most significant effect, which is generally observed at the lowest exposure level; field studies in which biological measurements are compared with exposure levels; controlled laboratory studies with volunteers for providing pharmacokinetic data; and computer modeling to do simulation studies for extrapolation and matching of BEIs and threshold limit values (TLVs).[10] BEIs, as with all occupational standards, are not developed to protect the fetus or children, or the elderly, but rather the healthy working population. Some BEIs are to be applied as group rather than individual tests. Urinary lead and urinary phenol (for benzene exposure) are measurements that due to wide variations to be expected in the test results are examples of this requirement.[11,12] Some BEIs are recommended only for confirmation of another test: Ethyl benzene and benzene in expired air are both used to confirm the results of urine tests.[12,13] Confirmatory tests are generally recommended when a metabolite can be generated by more than one source.

As BEIs are tied to a given airborne exposure limit, the TLV, the criterion for a metabolite, may depend on a minimum airborne exposure. The BEI for monitoring toluene exposure via the concentration of hippuric acid in urine is not considered useful below an airborne toluene concentration of 50 ppm due to interferences from nonoccupational sources of hippuric acid including sodium benzoate, a food preservative, and many fruits.[6]

Sometimes biological monitoring is termed medical monitoring; however, this should not be confused with health or medical surveillance that consists of periodic (often annual) exams and tests since its purpose is to identify whether an adverse effect has occurred such as increased protein in urine from exposure to hepatotoxic chemicals.[7] Biological monitoring is meant to be preventive in that the tests are selected to identify whether an overexposure has occurred but not an adverse effect. These changes and concentrations that are measured are temporary in nature and should be representative of events that occur long before those changes measured in conventional health surveillance tests such as liver function, blood counts, and x-rays. Clinical pathology tests can be difficult to interpret as normals fall into a wide range.[14] Therefore, if a result is near the end of a range it is difficult to tell if damage has occurred or not. Changes are easiest to detect when monitoring has been conducted on the same individual over a long period of time, often years.

Another way to view the difference is to consider that the aim of biological monitoring is to assess whether a significant dose has been absorbed, whereas many medical monitoring tests are aimed at determining whether a disturbance in physiology as a result of the absorbed dose has occurred. For example, liver function tests are used to indicate abnormalities in function rather than the mere presence of a compound. While an abnormality in such a test does not mean the organ is compromised, it does indicate some change may be taking place. Also, there are often two levels of concern when biological monitoring is performed: one being of statistical significance, and the other of clinical concern, because at this point the individual may develop a disabling illness.

Finally, the types of tests appropriate for biological monitoring can vary including those that provide useful information for medical surveillance. For example, an increase in cadmium levels in urine may be indicative of an adverse effect on the kidneys (saturation of the detoxification mechanism), whereas cadmium in blood correlates much better with recent exposure; therefore, it is the preferred test for biological monitoring.[15] In some cases, biological monitoring is preferred to air sampling. For example, studies of occupational exposures to agricultural (primarily organophosphate pesticide-related) workers have found that inhalation exposures contribute much less to the absorbed body burden than dermal exposures.[16] Other situations include the assessment of individual work practices, determination of the effectiveness of PPE, and assessment of the contribution of oral and dermal exposures.

ADVANTAGES AND DISADVANTAGES

Biomonitoring also has its limitations. While it is useful for preventing adverse effects that might result from chronic exposures, it will not help prevent acute (peak) exposures since it is not done routinely, for example, daily. The same tests can be used, however, to measure the extent of exposure as an aid to identifying proper medical treatment following acute or emergency exposures. Biomonitoring is not useful for assessing exposures to substances that exhibit toxic effects at the site of first contact and these substances are poorly absorbed. Such is the case for primary lung irritants.[7] Biomonitoring is only useful when the relationship between airborne (and in some cases, dermal) levels, internal dose, and adverse effect is known for a specific chemical.

A major limitation to widespread use of biological monitoring is the lack of detailed information on the fate of industrial chemicals in humans. Most of the toxicokinetic data available are from experimental animal studies and must be extrapolated to humans. Attempts are being made to use modeling to develop more data.[17]

Biological monitoring, unlike environmental monitoring, should be considered a medical procedure since by definition the specimen comes directly from a human. Permission, as well as trained and in some cases (blood) licensed personnel, may be required for collection of biological samples. This rule can be a problem in some situations, for example, when workers are spread out over a wide geographical area.

As air monitoring standards decrease, biological monitoring methods may need to change, since metabolites must be detected at increased levels of instrument sensitivity. Sometimes this requires a change in metabolite. As an example, a standard of 50 mg/L of phenol in urine can be reliably used to detect benzene overexposures associated with an airborne standard of 10 ppm, but for airborne standards of 1 ppm and less the background levels of urinary phenol that range up to at least 20 mg/L in nonoccupationally exposed individuals will mask any changes due to occupational exposures.[10,12]

There are situations in which air monitoring is a better measure of exposure. In atmospheres where the airborne concentration is rapidly fluctuating, the amount of material actually absorbed by an individual will be less than that measured by a continuous monitoring sampler next to the breathing zone. The number of available biological monitoring methods is limited, and for many substances only air monitoring methods exist. However, given the limitation involved in collecting, analyzing, and interpreting air and biological samples, the best program is to use air sampling, wipe sampling, and bulk sampling in conjunction with biological monitoring to assess not only the true dose a worker is receiving but also the sources of exposure.

METHOD SELECTION

Considerable advance planning is necessary when biological monitoring is contemplated to aid in making sure that the appropriate test, time for monitoring, and properly trained, and in some states, licensed, personnel are available for sample collection. The type of exposure (chronic, acute, intermittent), the major route of exposure, and a knowledge of the process and all chemicals in use are also important considerations when contemplating biological monitoring. The proper medium to sample, the frequency of sampling, the parameters to measure, additional data to collect, and the need for air sampling will be determined by the goals and objectives.

In general, blood serum, urine, and breath sampling require different sample preparation and involve completely different collection methodologies. The limit of detection for the test should be sensitive (low) enough to differentiate exposed from nonexposed workers, and if exposure limits have recently been lowered, existing methods should be reviewed for adequacy. Also, when selecting a test it must be remembered that the fate of any chemical in the body will depend on its volatility, polarity, and chemical and biological stability.

Selection of the test method is important since different tests can represent different types of exposure, for example, the ALAD test is useful for detecting early lead exposure, and the free erythrocyte protoporphyrin (FEP) or zinc protoporphyrin (ZPP) tests indicate chronic (long-term) exposure to lead.

While standards such as BEIs often specify when samples are to be collected, in some cases it is up to the sampling professional. Timing is important because levels of some rapidly metabolized compounds decrease significantly in hours. Therefore, a measurement immediately following the shift is an indicator of the most recent exposure, and samples collected prior to a shift, 16 hours away from work, reflect the average exposure of the prior workday.[10]

Sample collection times often depend on the BHL of a compound. For example, toluene if sampled in breath must be collected immediately following the end of exposure due to its short half life and rapid clearance from the lung.[6] The nature of the operation and availability of workers for sampling will also affect the timing.

An understanding of the chemicals used in the process can help interpret results. For example, the same metabolite can be a product of more than one substance. Mandelic acid is the metabolite of styrene and ethyl benzene, while 2,5-hexanedione can be produced by *n*-hexane and methyl *n*-butyl ketone. The same biochemical change can be induced by different compounds, for example, methylene chloride is metabolized to carbon monoxide in the body, thus creating carboxyhemoglobin just as a carbon monoxide exposure does. As described, interferences can occur as a result of diet, drugs, alcohol, disease, or other workplace chemicals.

Labeled containers with the proper preservative or anticoagulant must be used, and immediate shipping procedures must be identified. All laboratories that analyze biological specimens must be licensed as clinical laboratories under either a federal or state program. Several professional organizations have established accreditation programs for clinical laboratories, including the College of American Pathologists and the American Association of Blood Banks.

Blood Sampling

Most chemicals capable of causing systemic effects are generally transported by the blood. The blood is often considered the most useful biological medium to monitor, since it generally provides an accurate, although indirect, measurement of the level of most toxic agents in target organs or tissues. Blood samples are usually collected if a urinary or other noninvasive test (breath) is not applicable. This can occur when airborne exposure concentrations will not produce a sufficient amount of a key metabolite in urine to be detected, when a compound does not undergo significant biotransformation, or when urinary metabolites are not specific to that compound.

Chemicals that enter the blood can be found at different concentrations depending on whether the site of entry is measured or what blood vessel is sampled. Concentrations in venous and arterial blood differ with the same exposure, and capillary blood concentrations will resemble those found in arterial blood. Therefore, blood collected from an area where dermal exposure has occurred will have higher levels of a contaminant than blood from another location.

Measurements on blood can be done on whole blood, plasma, or serum, depending on the contaminant. Chemicals can concentrate in red blood cells or plasma, or can be found in equal concentrations in both. For example, when measuring a cholinesterase in organophosphate-exposed workers, both red blood cell and plasma cholinesterase—two different enzymes—are determined, because with some pesticides, such as Demeton, plasma cholinesterase is affected by exposure long before red cell cholinesterase.[18]

Blood samples also require an anticoagulant, usually heparin. Vacuum tubes, the most common method of collection, are coated on the inside with the anticoagulant. The color of the cap indicates what has been added to the tube and the color coding is standardized. Blood samples cannot be frozen but can be stored at 4°C for 5 to 7 days.[6] Preservatives such as sodium fluoride can also be added and storage in the dark at 4°C is recommended for some samples, such as those collected for carboxyhemoglobin analysis.[19]

Blood is usually collected from the cubital vein or finger or earlobe capillaries. Cubital vein blood contains the same concentration of contaminant as the muscles. Capillary blood resembles the concentrations found in arterial blood.[7]

The primary advantage of sampling blood is the relatively small amount of variation in its composition relative to its effect on concentrations of chemicals as well as the fact that the sampling technique is simple and straightforward. The measurement of compounds in blood is much more specific than that of metabolites in urine and is subject to much less interference. However, sampling blood is also an invasive technique that causes discomfort among the subjects and thus most individuals are reticent to regularly provide samples. In particular, routine blood sampling in the field is impractical. Venipuncture requires trained personnel who are not usually available under these conditions. Also, the importance of cleanliness to prevent sample contamination makes field collection difficult. There is a concern among both analytical personnel and those qualified to collect blood samples about the potential for AIDS (acquired immune deficiency syndrome) and hepatitis exposure. Also, the samples must be carefully stored and handled or deterioration will occur.

General Method for Sampling Blood

1. If more than 0.5 ml blood is needed for the test, a venous blood sample is required.

2. For less than 10 ml of unclotted blood, one of the following anticoagulants should be used: 20 mg of potassium or sodium oxalate, 50 mg of sodium citrate, 15 mg of disodium ethylene diamine tetra-acetic acid (EDTA), or 2 mg of heparin. The anticoagulant should be dispersed in a concentrated solution along the bottom wall of the tube, and then desiccated. Care must be taken with vacutainers, because the rubber stoppers may contaminate the blood with low levels of organics.[18] If the sample is to be analyzed for metals, acid-washed, metal-free glass containers should be used for collection.

3. The skin should be washed prior to sampling, first with soap and water, and then with isopropanol.

4. Gently rotate the specimen container to mix.

5. Do not freeze whole blood or hemolysis will occur.

Blood Sampling for Polychlorinated Biphenyls in Serum[20]

1. Collect 20 ml to 25 ml of whole blood by venipuncture using a 30-ml glass syringe.

2. After the blood has clotted, centrifuge for 10 minutes at 2,000 rpm. Transfer the serum to a 16-mm by 150-mm culture tube with a Teflon-lined screw cap using a sterile, disposable pipet.

3. Ship the serum in an insulated container with ice to keep the samples at 4°C.

4. Freeze samples upon arrival at the laboratory until analysis.

Blood Sampling for 2-Butanone, Toluene, or Ethanol[21]

1. Collect 5 ml of venous whole blood in a vacuum tube containing heparin. Invert the tube several times to mix.

2. Ship samples in polyfoam packs containing bagged ice or refrigerant.

Urine Sampling

Urine is suitable for monitoring hydrophilic chemicals, metals, and metabolites. It is noninvasive and relatively easy to analyze. Urine consists of 90% to 98% water, and in healthy individuals it is sterile. The balance is solids consisting of many inorganic and organic compounds. Total daily urine output of an average adult varies between 600 ml and 2,500 ml. The volume at any given time depends on the time of day, diet, temperature, and humidity.

There are three types of urine samples: collection of all urine voided during a 24-hour period; collection of a single sample, often called "spot" collection; and pooling of several spot samples over the course of a day. The 24-hour sample is the most desirable, since the levels best represent actual exposures; however, this sample is the most difficult to obtain. Pooled specimens, representing several collections during the day on the same individual, are also representative of exposure but are more likely to be contaminated. A single spot sample is the easiest to obtain, but the least representative of exposures.[22]

In the case of 24-hour urine samples, the volume of the urine is measured using a graduated cylinder, followed by saving some of the sample in sealed vials for analysis. Other measurements such as specific gravity and pH are often done.

Most exposure measurements are based on a single sample. Due to the variation in urine volume over the day, urine samples must be corrected for dilution. Since the excretion rate of the solids is relatively constant, whereas that of the water is not, measurements are generally adjusted to the specific gravity of the solids or creatinine, a metabolic product of skeletal muscles, whose excretion rate is relatively steady. Specific gravity can vary up to a factor of 10 over a day while creatinine excretion varies much less.[23] Results are expressed as mg/L when measurements are adjusted for specific gravity or in grams per gram of creatinine. NIOSH recommends correcting urine to a specific gravity of 1.024.[24] Since other factors have also been used for specific gravity corrections, when comparing results of one sample to another or to a standard, or samples from different laboratories, it is important to make sure that the same correction factor was used. Sampling professionals should also be aware that creatinine excretion is affected by kidney function. The concentration of creatinine is also dependent on age, sex, and muscle mass.[25]

Urine samples are often preserved using thymol or acid to prevent bacterial degradation and then stored at 4°C until analysis. Frozen specimens are stable for greater than a month while specimens at 4°C are stable for 3 to 4 weeks.[6]

One strategy in situations where the metabolite selected for analysis also occurs in nonexposed individuals is to collect samples of urine after a period of nonexposure such as a weekend to use as a background for those collected following an exposure period.[25] Fallout from work clothing or contaminated hands can contaminate a urine sample.

General Method for Sampling Urine[26]

1. The subject is asked to empty the bladder and record the time. For end-of-shift sampling, this procedure should be done 3 hours before the end of the shift.

2. The subject is provided with a 500-ml (or 200–300-ml) glass container with a wide mouth and a Teflon-lined screw-top lid inside a securely sealed bag, and asked to collect the next void. Do not discard any sample; leave this decision to the laboratory. If volatiles must be collected, a 50-ml container with a screw-top lid should be used. Ask the subject to fill the container completely. It must be immediately sealed following collection to prevent losses. These containers should be able to endure changes of temperature during transportation and storage that might result in overpressure. The time of this next collection should be recorded. The container should be labeled with a unique number.

3. For 24-hour collections, subjects void into a 4-L plastic jug with a large screw cap. In between collections, the jug is stored in an ice-pack, cooled Styrofoam chest.

4. For storage periods of 5 days or less, refrigerate; for longer periods, freezing is necessary.

5. The total volume of urine collected divided by the total elapsed time between voids represents the urine output. Compensate for dilution by correcting for either specific gravity or creatinine. Samples with a specific gravity <1.01 are too dilute and sampling should be repeated. If creatinine is to be measured, it should be done on the same sample as the chemical measurement.

Collection of Urine Samples for Benzene Exposure[27]

1. Collect 50 ml to 100 ml of urine in a 125-ml polyethylene bottle containing a few crystals of thymol. Close the bottle immediately after sample collection and swirl gently to mix.

2. Collect two urine samples for each worker, one prior to exposure and one after. A representative number of workers should be sampled. Submit individual samples as a group.

3. Collect and pool urine samples from nonexposed workers to use as background phenol levels.

4. Freeze the urine and ship in dry ice in an insulated container. Submit for analysis for phenol.

Exhaled Air (Breath) Sampling

Generally, when doing measurements on exhaled breath the compound representing the exposure rather than its metabolite(s) is monitored, because metabolites are usually not volatile and therefore are not excreted through the lung.[28] Breath analysis is suitable for monitoring volatile solvents, carbon monoxide, and other gases excreted through the lungs. Once inhaled, volatile compounds can pass through the alveoli in the lungs to the bloodstream very rapidly. Gas (air) in the alveolar region of the lungs is almost in equilibrium with the arterial blood gases. Compounds that are poorly soluble in water and fat, compounds that are poorly metabolized, and compounds that have a high vapor pressure are poorly retained in the lungs.[29] For other compounds factors that influence the concentration passing from the alveoli to the blood are solubility in blood, breathing rate, solubility in fat, duration of exposure, and biotransformation. The less soluble a compound is in the blood, the higher the concentration in the alveolar air. As a general rule, the ratio of alveolar air to outside air provides some guidance in predicting the degree of solubility. Compounds such as hexane with a ratio greater than 0.5 (0.8 – 1.0 is best) are poorly soluble in blood while the opposite is true when this ratio is less than 0.5, as is the case for methyl ethyl ketone and toluene. Breath analysis works best on compounds with low blood solubility; these compounds are eliminated into expired air unmetabolized.[7]

The concentration of a volatile compound in the exhaled breath is directly related to the blood concentration, and is dependent on the total amount absorbed, the time passed since absorption, and the rate of elimination from the body. An example of the typical decrease in breath concentration as time passes following exposure can be determined from a study done using trichloroethylene (TCE) during which subjects were exposed to 100 ppm for a 7.5-hour period (Fig. 15-1).[30]

Solubility can also affect BHL. In one study concentrations of methylene chloride, a poorly soluble compound, leveled off in alveolar air during the work week, staying relatively constant, while concentrations of toluene, which is highly soluble (in blood), increased throughout the week.[17]

Breath sampling is not as simple as it might appear. When a breath of an exposed person is exhaled it contains some plain air and some contaminated air. This is because the total exhalation, or mixed expired air, as it is called, is not the same as air from the alveolar region. After inhalation the contaminant is concentrated in this alveolar region while the air from the upper respiratory tract, sometimes called the "dead space," does not contain contaminant. In quiet breathing, a healthy human inhales and exhales about 500 ml of air 15 times per minute. When the total air exhaled during normal breathing is collected, the concentration in the sample represents the gas mixture displaced from the dead space (150 ml) and from the alveoli (350 ml).[22] When the air is collected only at the end of an expiration phase, the sample is called end-exhaled air.

Ideally, the concentration of vapors and gases in end-exhaled air equals the concentrations in alveolar air as this space is in equilibrium with concentrations in arterial blood.[31] Benzene provides an example of the difference in concentrations that can be expected in a sample of mixed-exhaled breath versus end-exhaled breath for the same exposure. If the airborne concentration is 10 ppm, benzene in mixed-exhaled air is expected to be 0.08 ppm while for an end-exhaled breath the concentration would be 0.12 ppm.[12]

During exposure, the end-exhaled concentration is smaller than the mixed-exhaled concentration, but the difference between them diminishes with the length of exposure. After exposure, the relationship reverses and the mixed-exhaled concentration is approximately two-thirds of the end-exhaled concentration. The difference tends to be smaller during increased physical activity, and generally increases with age.[28]

There are a number of factors that can affect the concentrations of chemicals in exhaled breath. As work activity increases, the amount of highly blood-soluble chemicals (such as toluene) eliminated in the breath will increase, but there will be minimal effects on poorly blood-soluble compounds.[17] Postexposure activity and obesity also affect concentrations in the breath. In the case of obesity, these individuals may develop higher exposures than others. Increased activity after work has decreased levels when compared to individuals who rested.[17] The breathing rate of subjects is also a source of fluctuations: Some individuals breathe normally, others hyperventilate, and still others breathe very slowly.

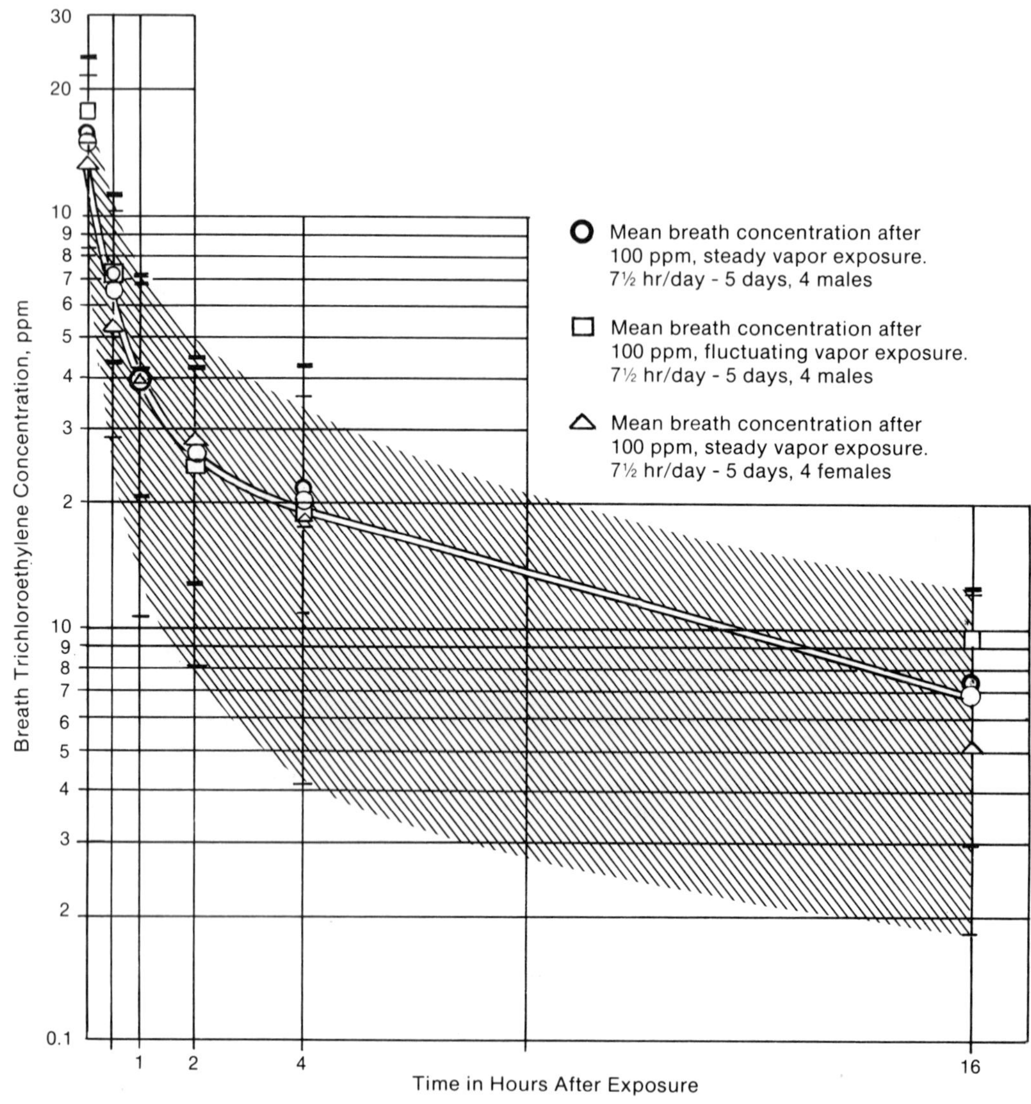

Figure 15-1. Trichloroethylene breath-decay curve: mean and range breath concentrations after 100 ppm vapor exposure for 7.5 hours/day for 5 days with 12 subjects. The shaded area encloses the mean ±25 standard deviation. (From *Biologic Standards for the Industrial Worker by Breath Analysis: Trichloroethylene,* NIOSH, Cincinnati, OH, 1974.)

In addition to the physiologic variations, there are other considerations. Storage of breath samples after collection can be a problem, especially if there is any possibility of adsorption onto the walls of the bag. Samples must be collected in a "clean" area, meaning that no workplace contaminants are present. With these factors to consider, it may come as no surprise that it is more difficult to obtain good reproducible samples with breath than with blood or urine. However, since the technique is noninvasive and can be repeated without causing significant discomfort to the subject, it may be useful for screening exposures, accident investigations, and providing supporting data for other measurements. Table 15-3 lists gases and vapors for which breath screening is currently recommended.

Breath samples represent different aspects of

TABLE 15-3. Gases and Vapors for Which Breath Screening Is Currently Recommended

Compound	*Infrared Wavelength (μm)*
Ammonia	10.77
Benzene	9.62
Carbon dioxide	4.27
Carbon monoxide	4.58
Carbon tetrachloride	12.60
Chlorobenzene	9.16
Chloroform	12.95
Dichlorodifluoromethane (freon)	10.85
Ethanol	9.37
Ethyl benzene	9.70
Ethylene oxide	11.48
Ethyl ether	8.75
Fluorotrichloromethane	11.82
Furfural	13.27
n-Hexane	3.40
Methanol	9.45
Methyl chloride	3.35
Methylene chloride	13.10
Perchloroethylene	10.92
n-Propyl alcohol	10.47
Styrene	12.90
Toluene	13.75
1,1,1-Trichloroethane	9.20
Trichloroethylene	11.78

exposure depending on how the sample is collected. Therefore, there are a number of approaches to take when it comes to breath sampling. Depending on the sampling technique, breath measurements reflect either the instantaneous blood levels of the contaminant in the body (a single breath) or the average blood level (multiple breaths) during the sampling period. Multiple breath samples are generally considered more representative of actual body concentrations than single breaths.

One approach is to use the breath-holding technique that consists of having the subjects hold their breath for 5 seconds to 30 seconds before exhaling into the sampling device. The content of the air can differ here as well, depending on whether the lungs were full of air or partially empty before starting.[22] Other aspects that have been considered include the use of end tidal volume where the breath is sampled only at the end of an expiration. Depending on whether the expiration is normal or forced, the content of the breath may differ.

A variation involves using a mouthpiece with a Y tube in which one tube of the Y functions as an exit bypass for that portion of the breath that would dilute the sample, and the other tube is attached to the inlet valve on the bag. During collection the valve on the bag is open, but the exit bypass prevents sample from entering the bag until the subject has exhaled 60% of lung capacity. At this point, a thumb is placed over the fixed end of the Y to shunt the balance of the exhalation into the bag. This is repeated until 100l to 1L has been collected. The bag is stored in a shaded area for analysis.[32]

End-Exhaled Air Sampling Using a Bag

1. Select a direct reading instrument that is specific for the compound to be measured. Often these are electrochemical, infrared, or GC-based units. For more information on their operation, see the chapters on Monitoring Instruments Dedicated to a Single Chemical, and Monitoring Instruments for Many Specific Gases and Vapors: GC and IR. The instrument should be hooked up to a data recording device, such as a strip chart recorder or data logger. Prior to sampling, take a baseline measurement in the area where the instrument is stationed. If several samples are being processed, take a background after every 10 runs.

2. Explain the technique to the employees and have them practice during a briefing session prior to sample collection. Attach a fresh piece of Tygon tubing to the valve of a Tedlar bag.

3. Timing of sampling is important. Preshift samples should be collected 16 hours after the previous shift in clean air and prior to the next shift. Postshift samples should be collected immediately following the shift. During-shift samples should be collected as soon as the employee leaves the work area.

4. Have the employee take several deep breaths, then hold a deep breath for 25 to 30 seconds.

5. The employee should then exhale half of this breath and blow the remaining half of the

breath into a Tedlar bag. Close the valve on the bag immediately.

6. If using an analyzer with a pump attach the bag to the inlet and record the reading; otherwise use a syringe to withdraw a sample from the septum on the bag and inject it directly into a properly set-up GC.

7. Subtract the baseline reading from the final level. The result represents the change in gas concentration due to the exposure.

Carbon Monoxide Breath Sampling in Bags[33]

1. Have subjects hold their breath for 20 seconds and then discard the first portion of the expired breath and collect the last portion in a 5-liter Saran bag.

2. The smoking habits of the person being sampled should be obtained (especially for that day). The normal carboxyhemoglobin level of the blood ranges from 0.5% to 2%. A person who smokes one pack of cigarettes per day can be expected to have an average level of 5% carboxyhemoglobin, and a two-pack-per-day smoker can be expected to have an 8% to 9% level. In terms of impacting the results, one pack a day yields 30 to 35 ppm while two to three packs per day will produce as much as 45 to 50 ppm background CO in exhaled breath.[19]

3. Have subjects hold their breath for 20 seconds, and then discard the first portion of the expired breath and collect the last portion in a 5-L Saran bag.

4. Bags are immediately hooked up to a direct reading carbon monoxide monitor, such as an Ecolyzer.

5. Correlate exposure duration, carboxyhemoglobin, and carbon monoxide concentration.

Breath Sampling with Sorbent Tubes[22]

1. Breath sampling should be carried out in an uncontaminated area. Set up a sampling train that is connected in the following order: expired air bag, silica gel or charcoal tube, ascarite tube, personal air sampling pump calibrated to 500 ml/min. The choice of tube depends on the contaminant being collected. For guidance, consult the NIOSH air sampling method.

2. Instruct the worker to take a normal inspiration, expel a small amount, and then direct the rest of the expiration into a heated bag through a piece of tygon tubing and close the valve on the bag.

3. Attach the tubing on the bag of expired air to the inlet end of the sampling tube. While keeping the bag warm, 1,000 ml of air (2-minute sample) from the bag is collected using the pump.

4. The time of sampling, the time of the end of work, the ambient pressure, and the temperature of the sampling bag should be recorded.

5. After sampling, the tubes are again separated and capped. Care should be taken to use the same plugs for the ascarite tube if it was preweighed with the plugs on.

6. One silica gel tube and one ascarite tube should be handled the same way, but without drawing air through, for laboratory blanks.

7. Label each tube with a unique number and send to the laboratory in a refrigerated container.

Pipet Breath Sampling[30]

The glass pipet method is best for those applications in which highly sensitive detection methods will be used. A 50-ml glass tube with screw caps on each end is used. This technique is useful for situations where a large number of repetitive samples must be collected.[34]

1. Alveolar breath samples are obtained from each subject prior to exposure and immediately following exposure. Duplicate background samples are also collected by unscrewing the caps and allowing the tube to lie for several minutes.

2. The samples are collected in duplicate for each subject. One end of the pipet (Fig. 15-2) is

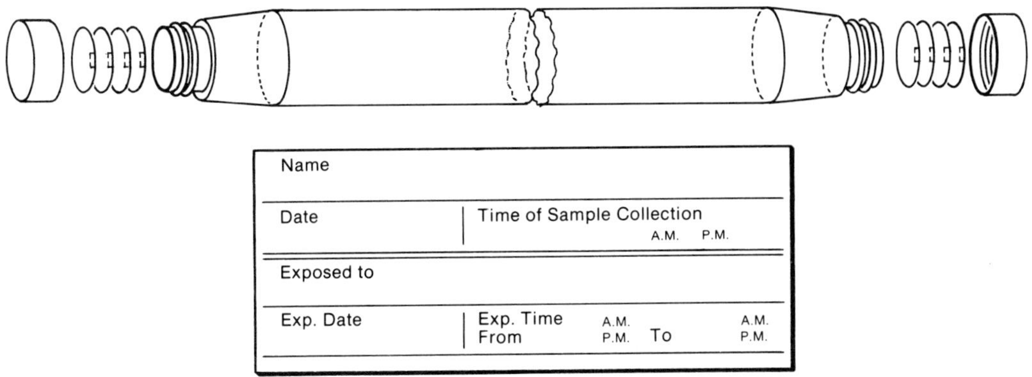

Figure 15-2. Typical breath sampling pipet. (From *Biologic Standards for the Industrial Worker by Breath Analysis: Trichloroethylene*, NIOSH, Cincinnati, OH, 1974.)

sealed with the lips, flushing the chamber with three exhaled breaths, and then, after holding a fourth breath for 30 seconds, exhaling through the pipet chamber so that the end-tidal portion can be collected.

3. The distal cap is secured while the subject is still expiring through the pipet chamber. After the distal cap is secure, the proximal end of the pipet is removed from the mouth and quickly sealed with a fingertip.

4. The proximal cap is then screwed shut. Screw the caps tightly; leaks in caps are the most common source of sample loss with this technique.

5. While it is desirable to construct individualized breath decay curves for each subject, more often these results are compared to standard curves.

Other Methods for Breath Sampling

Analysis of a respirator filter worn by a worker would give a good integrated sample of the air that would have reached the lungs, although the exact air volume sampled can only be estimated. The mask must be properly fitted, thus ensuring that all inhaled air will pass through the filter system. An example of this type of sampler has been evaluated for breath sampling.[35] It is based on a half-face, dual-cartridge respirator whose inhalation ports are fitted with standard air-purifying elements. On the exhalation port is mounted a special cartridge for sampling exhaled air. Exhaled air is directed through the respirator's exhaust port and then through the two-part cartridge. The first section contains four layers of activated charcoal cloth, and the second section contains a molecular sieve adsorbant whose weight gain of water is proportional to the volume of air exhaled.

The sampling procedure consists of having the subjects first wear the respirator without the special sampling cartridge for 7 to 10 minutes to allow air temperature and relative humidity (RH) inside the mask to stabilize. Then the sampling cartridge is attached, and the subject continues to breathe at rest for 15 to 20 minutes. Following exposure, the charcoal cloth is desorbed and analyzed by GC, and the molecular sieve is weighed before and after sampling and corrected using the following calculation:

$$MG = \frac{0.0268EV}{0.73}$$

where MG = moisture given in grams and EV = expired volume in liters.

Breath concentrations are corrected for the dilution effect of the respirator cavity dead space. Correction is done by measuring carbon dioxide at the mouth and at the respirator outlet port.

Another modification that has been used on a mask is to put a special funnel over the front.[36] The stem of a funnel is shortened and plugged, and two holes 12 mm in diameter and 6 mm apart are drilled midway between the base and the apex of the funnel. By firmly taping this modified funnel with holes directed downward to the retaining ring of the respirator, the design of the mask emulates air entering the nostrils. The purpose of the funnel is to prevent direct impaction of large particles on the filter and to regulate particle size. A multilayer filter consisting of 32 layers of surgical gauze stapled to an alpha-cellulose respirator has been found to be efficient for trapping both sprays and dust.[37] The filter pads are preextracted with the same solvent to remove any contaminants that might later interfere with analysis. A filter is bagged and sent to the laboratory with the exposed filters to serve as a blank. Following exposure, the entire pad is extracted in a Soxhlet apparatus. The amount of contaminant removed from the pad and the amount rinsed from the inside of the funnel are considered the exposure. The exposure can then be calculatead in mg/hr.

TABLE 15-4. Normal Ranges for Trace and Minor Elements in Hair

Element	*Concentration Range (μg)*
Aluminum	20–40
Arsenic	2.0–3.0
Cadmium	1.0–2.0
Calcium	200–600
Chromium	0.50–1.50
Cobalt	0.2–1.0
Copper	12–35
Iron	20–50
Lead	20–30
Lithium	0.10–0.80
Magnesium	25–75
Manganese	1.0–10
Mercury	2.5–5.0
Molybdenum	0.10–1
Nickel	1.0–2.0
Phosphorus	100–170
Potassium	75–180
Selenium	3.0–6.0
Sodium	150–350
Vanadium	0.5–1.0
Zinc	160–240

Hair and Nail Sampling

Hair and nails are usually not suitable for use as biologic indicators, because they are frequently contaminated.[10] However, hair has been used for monitoring exposures to such heavy metals as arsenic, lead, cadmium, and mercury. One problem is the wide spread of data when attempting to come up with average hair values for the normal individual from exposures to typical environmental contaminants in food and water (Table 15-4).[38] Another concern is the likelihood of concentration gradients in a single hair.

Human hair consists of approximately 80% protein and 15% water, with smaller amounts of lipid and inorganic materials. The water content of hair varies directly with the ambient RH. One consideration with selecting hair as a sampling medium is that hair does not grow continuously, but over periods of alternating activity. These periods include active growth, a resting stage, and a transitional phase. During the active growth phase, the follicles of the scalp produce hair at a rate ranging from 0.2 mm/day to 0.5 mm/day. Growth rates and the duration of activity are highly variable, and are dependent on such factors as the individual's age, race and gender, and the anatomical location of the hair and the season. However, it is generally assumed that human hair grows about 1 cm/month. Levels of various trace elements in hair are affected by various disease states, such as anemia, cirrhosis, and epilepsy.[39] Nonoccupational sources that can contribute to the types and concentrations of metals found in hair are the diet, especially fruits, vegetables, and grains, and hair dyes.[23]

It has been suggested that hair analysis would be most useful when a regular testing program is conducted on an individual over several years.[23] Test data used for comparison should be based on the same cleaning procedures. Detergent wash followed by a distilled water rinse is the most common method for cleaning hair, although other means, such as solvents, have also been used. The biggest concern in sampling hair is an inability to identify when the exposure occurred and the potential for exterior and nonoccupational contamination.

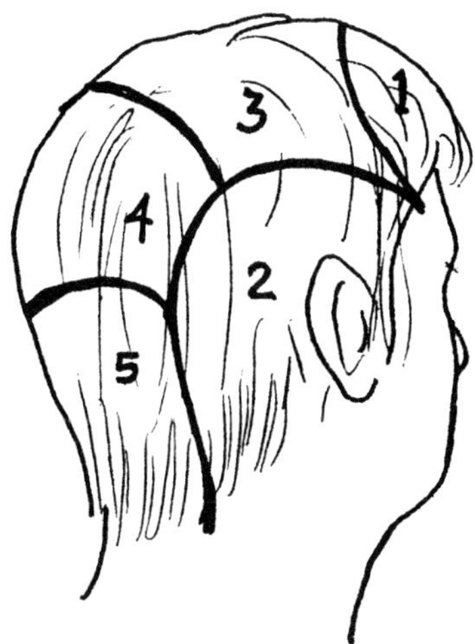

Figure 15-3. Regions of the scalp: 1. frontal, 2. temporal, 3. vertex anterior, 4. vertex posterior, 5. nape. (Reprinted by permission of VCH Publishers, Inc., 220 East 23rd St., New York, NY 10010, from *Hair Analysis, Applications in the Biomedical and Environmental Sciences*, by S. A. Katz and A. Chatt, 1988, p. 75.)

The International Atomic Energy Agency protocol recommends that at least 100 individual hairs be taken for an analysis: 5 to 10 strands from 10 to 20 different sites on the scalp. If analyses are to be compared between individuals, the lengths of hair sampled must be the same. Scalp hair can be divided into five regions: frontal, temporal, vertex anterior, vertex posterior, and nape (Fig. 15-3).[39]

Collection Method for Hair[22,38]

1. Collect from the back of the head at the nape of the neck. Use clean stainless steel or plastic scissors and have clean, washed and dried, hands.

2. Measure lengths of 1–2 inches (5–10 cm) from the area closest to the scalp outward, and cut as close to the scalp as possible. This hair represents the most recent growth.

3. Collect 0.5–1 gram of hair (1–2 tablespoons), and place in a plastic bag for shipment.

4. Interview the subject regarding the type of shampoo, hair colorants, and other hair cosmetics in use, such as mousse, sprays, and conditioners, including brand names.[21]

5. When submitting the sample to the laboratory, indicate from where the hair was collected (portion of the scalp) and include an empty plastic bag as a blank in the event contamination is suspected.

INTERPRETATION OF RESULTS

Interpretation of data is difficult, because there are very few standards relative to the types of biomonitoring that can be done for chemical exposures. When ACGIH BEIs are available, they should be used. The acceptable daily intake (ADI) may be a useful standard where no BEIs exist.

Results obtained are often dependent on time of collection, specimen integrity, drug or chemical interference, and methodologies. Concentrations in blood and breath reflect the most recent exposure if the sample is collected during the work shift or represent an integrated exposure (for an entire period) if performed 16 hours after exposure stops. For cumulative compounds such as hexachlorobenzene concentrations represent the body burden.[7]

When attempting to compare the results of blood and breath samples, it is important to note that alveolar air represents arterial blood concentrations while blood samples are collected from veins. During exposure, tissue uptake of a contaminant decreases its concentration in the blood so the concentration in veins is less than that in the arteries.[7]

An analysis of very dilute urine samples (specific gravity less than 1.010 or creatinine concentration less than 0.3 g/L) is not likely to be accurate and should be repeated.[7] It is important to review test data carefully and make sure that the proper analysis has been requested and performed. In the case of styrene, results may be given as either mandelic acid or total mandelic acid. In the latter case, they represent the total concentration of both mandelic and phenylglyoxylic acids, both the metabolites of styrene.[25]

Results of biological monitoring of chemicals with a long biological half life are indicators of long-term exposure, and do not correlate well with the current time-weighted average (TWA) inhalation exposure. Examples are metals, such as lead, and cadmium or chlorinated compounds, such as DDT and PCBs. On the other hand, biological "determinants" with a short half life are indicators of the most recent exposure, and do not correlate well with the TWA exposure if the air concentration fluctuates widely. Examples are blood and expired air measurements in samples taken at the end of a shift. Factors such as the chemical's half life and the volume of body fat, and variability of metabolism among individuals as well as differences in workloads are also important to consider when making correlations between airborne exposure and biological results.

Nonoccupational exposures must also be taken into account. For example, smokers have significantly higher carboxyhemoglobin than nonsmokers, which must be considered when interpreting results of measurements for employees exposed to carbon monoxide. Correlations must also be made with the observed levels of "natural" pollutants in the environment, as bioaccumulation of these pollutants can occur, resulting from long-term exposure to low levels in the environment. One approach to this problem is to collect specimens from nonexposed control subjects at the workplace, and to use this group mean as a baseline to interpret the values from an exposed population. If a nonexposed control group cannot be obtained from the workplace, the use of an outside control group should be considered.

REFERENCES

1. Corn, M. Strategies for prospective surveillance. In *Advances in Air Sampling*. ACGIH, Cincinnati, OH, 1988.
2. Bowman, J. A., J. L. Held, and D. R. Factor. A field evaluation of mandelic acid in urine as a compliance monitor for styrene exposures. *Appl. Occup. Environ. Hyg.* **5**(8):526–535, 1990.
3. Friberg, L. T. The rationale of biological monitoring of chemicals—with special reference to metals. *AIHA J* **46**(11):633–642, 1985.
4. McSharry, C., et al. Seasonal variation of antibody levels among pigeon fanciers. *Clin. Allergy* **13**:293–299, 1983.
5. Droz, P. O. Biological monitoring I: Sources of variability in human response to chemical exposure. *Appl. Ind. Hyg.* **4**(1):F-20–F-24, 1989.
6. Lowry, L. K. Review of biological monitoring tests for toluene. In M. K. Ho and H. K. Dillon (eds.), *Biological Monitoring of Exposure to Chemicals. Organic Compounds.* Wiley, New York, 1987, pp. 99–109.
7. Bernard, A., and R. Lauwerys. General principles for biological monitoring of exposures to chemicals. In M. K. Ho and H. K. Dillon (eds.), *Biological Monitoring of Exposure to Chemicals. Organic Compounds.* Wiley, New York, 1987, pp. 1–16.
8. American Conference of Governmental Industrial Hygienists. *n*-Hexane—recommended BEI. In *Documentation of the Biological Exposure Limits.* ACGIH, Cincinnati, OH, 1986.
9. Brugnone, F. Monitoring of individual exposures to organic volatile compounds by analysis of alveolar air and blood. In M. K. Ho and H. K. Dillon (eds.), *Biological Monitoring of Exposure to Chemicals. Organic Compounds.* Wiley, New York, 1987, pp. 59–71.
10. Fiserova Bergerova, Thomas V. Development of biological exposure indices (BEIs) and their implementation. *Appl. Ind. Hyg.* **2**(2):87–92, 1987.
11. American Conference of Governmental Industrial Hygienists. Lead—recommended BEI. In *Documentation of the Biological Exposure Limits.* ACGIH, Cincinnati, OH, 1986.
12. American Conference of Governmental Industrial Hygienists. Benzene—recommended BEI. In *Documentation of the Biological Exposure Limits.* ACGIH, Cincinnati, OH, 1986.
13. American Conference of Governmental Industrial Hygienists. Ethyl benzene—recommended BEI. In *Documentation of the Biological Exposure Limits.* ACGIH, Cincinnati, OH, 1986.
14. Meeks, R. G. Clinical pathology: A role in biological monitoring. In M. K. Ho and H. K. Dillon (eds.), *Biological Monitoring of Exposure to Chemicals. Organic Compounds.* Wiley, New York, 1987, pp. 85–90.
15. Thomas, V. Five years of the biological exposure indices committee. *Appl. Ind. Hyg.* **3**(10):F-26–F-28, 1988.
16. Yeary, R. A.: Urinary excretion of 2,4-D in commercial lawn specialists. *Appl. Ind. Hyg.* **1**(3):119–121, 1986.
17. Fiserova-Bergerova, V. Simulation model as a tool for adjustment of BEIs to exposure conditions. In M. K. Ho and H. K. Dillon (eds.), *Biological Monitoring of Exposure to Chemicals. Organic Compounds.* Wiley, New York, 1987, pp. 27–57.
18. Lowry, L. K., et al. Biological monitoring III: Measurements in blood. *Appl. Ind. Hyg.* **4**(3):F-11–F-13, 1989.
19. American Conference of Governmental Industrial

Hygienists. Carbon monoxide—recommended BEI. In *Documentation of the Biological Exposure Limits.* ACGIH, Cincinnati, OH, 1986.

20. NIOSH. *Manual of Analytical Methods, Method 8004,* 3rd ed. NIOSH, Cincinnati, OH, 1984.
21. NIOSH. *Manual of Analytical Methods, Method 8002,* 3rd ed. NIOSH, Cincinnati, OH, 1984.
22. Hill, R. H., et al. Sample collection. In T. J. Kneip and J. V. Crable (eds.), *Methods for Biological Monitoring: A Manual for Assessing Human Exposure to Hazardous Substances.* American Public Health Association, Washington, DC, 1988.
23. Waritz, R. S. Biological indicators of chemical dosage and burden. In L. G. Cralley and L. V. Cralley (eds.), *Patty's Industrial Hygiene and Toxicology. Vol. III: Theory and Rationale of Industrial Hygiene Practice. 3B: Biological Responses.* Wiley, New York, 1985, pp. 175–312.
24. NIOSH. *Criteria for a Recommended Standard . . . Occupational Exposure to Benzene.* NIOSH, Cincinnati, OH, 1974.
25. van Hemmen, J., and G. de Mik. Biological monitoring of solvents, no panacea. In M. K. Ho and H. K. Dillon (eds.), *Biological Monitoring of Exposure to Chemicals. Organic Compounds.* Wiley, New York, 1987, pp. 73–84.
26. Rosenberg, J., et al. Biological monitoring IV: Measurements in urine. *Appl. Ind. Hyg.* **4**(4):F-16–F-21, 1989.
27. NIOSH. *Manual of Analytical Methods, Method 8305,* 3rd ed. NIOSH, Cincinnati, OH, 1984.
28. Fiserova-Bergerova, Thomas V., et al. Biological monitoring II: Measurements in exhaled air. *Appl. Ind. Hyg.* **4**(2):F-10–F-13, 1989.
29. Bardodej, Z., J. Urban, and H. Malonova. Important considerations in the development of biological monitoring methods to determine occupational exposures to chemicals. In M. K. Ho and H. K. Dillon (eds.), *Biological Monitoring of Exposure to Chemicals. Organic Compounds.* Wiley, New York, 1987, pp. 17–28.
30. NIOSH. *Biologic Standards for the Industrial Worker by Breath Analysis: Trichloroethylene.* NIOSH, Cincinnati, OH, 1974.
31. West, J. B. Ventilation. In *Respiratory Physiology—the Essentials.* Williams and Wilkins, Baltimore, MD, 1974.
32. Prevost, R. J., et al. Biological monitoring of exposures to chemical vapors released in marine operations. In M. K. Ho and H. K. Dillon (eds.), *Biological Monitoring of Exposure to Chemicals. Organic Compounds.* Wiley, New York, 1987, pp. 179–195.
33. Stewart, Richard, et al. Rapid estimation of carboxyhemoglobin level in fire fighters. *JAMA* **235**(4):390–392, 1976.
34. Linch, A. L. *Biological Monitoring for Industrial Chemical Exposure Control.* CRC Press, Cleveland, OH, 1974.
35. Morgan, M., et al. Design and laboratory evaluation of a breath sampling respirator for organic solvent biological monitoring. *Appl. Ind. Hyg.* 3(2):41–46, 1988.
36. Batchelor, G. S., and K. C. Walker. Health hazards involved in the use of parathion in fruit orchards of north central Washington. *AMA Arch. Ind. Hyg. Occup. Med.* **10**:502–529, 1954.
37. Durham, W. F., and H. R. Wolfe. Measurement of the exposure of workers to pesticides. *WHO Bull.* **26**:75–91, 1962.
38. Sheldon, L., et al. (Research Triangle Institute, EPA, NC). Chemicals identified in human biological medium. In *Biological Monitoring Techniques.* Noyes Pubs., Park Ridge, NY, 1986 (adapted from MineraLab, Inc., 22455 Maple Court, Hayward, CA 94541).
39. Katz, S. A. and A. Chatt. *Hair Analysis: Applications in the Biomedical and Environmental Sciences.* VCH, New York, 1988.

CHAPTER 16

Surface Sampling Methods for Dermal Exposure

There are a variety of reasons why surface contamination, especially removable surface contamination, may have to be sampled. Many toxic materials can gain entry into the body via ingestion and, in some instances, via penetration (absorption) through the skin. Air sampling will not measure exposure to chemicals capable of penetrating the skin. This exposure must be considered in addition to inhalation. However, for substances that can rapidly penetrate the skin, wipes of the skin are not recommended; instead, biological monitoring is the prescribed method. Toxic materials may reach the skin through fallout from sprays, via direct immersion into the chemical, as a result of accidental spills onto the body, and from contact with contaminated surfaces.

Field studies on crop workers have shown that in some cases dermal exposure predominates over inhalation exposure. In these situations, vapors represent the easier of the two types of hazards to manage, because respirable concentrations tend to be relatively homogeneous throughout the field. In general, this case will be true for compounds with relatively low vapor pressures, such that at maximum vapor concentrations the airborne levels are close to the permissible exposure limits (PELs), but rarely exceed them. For other compounds with high vapor pressures, such as ethylene dibromide, airborne concentrations can be significant for several days following application. Table 16-1 compares respirable and dermal exposures to parathion during citrus harvesting.

Surfaces that may contact food or other materials such as chewing tobacco, gum, and cigarettes that are ingested or placed in the mouth may be a source of contamination to hands and fingers. Heated contaminated surfaces may produce toxic products of combustion as well as contaminate hands and fingers (Fig. 16-1).

Worker activities that can result in dermal exposure include splashes when working with or near liquids, wiping with contaminated rags, contact with contaminated tools or surfaces, and immersing parts in cleaning solvents. Ingestion should

TABLE 16-1. Comparison of Respirable and Dermal Exposures to Parathion During Citrus Harvesting

Exposure	*Week 1 (mg)*	*Week 2 (mg)*	*Week 3 (mg)*
Total dermal	0.382	0.246	1.030
Total respiratory	0.008	0.009	0.011

Source: Popendorf, W. J., and J. T. Leffingwell. Regulating OP pesticide residues for farmworker protection. *Res. Rev.* **82:**125–201, 1982.

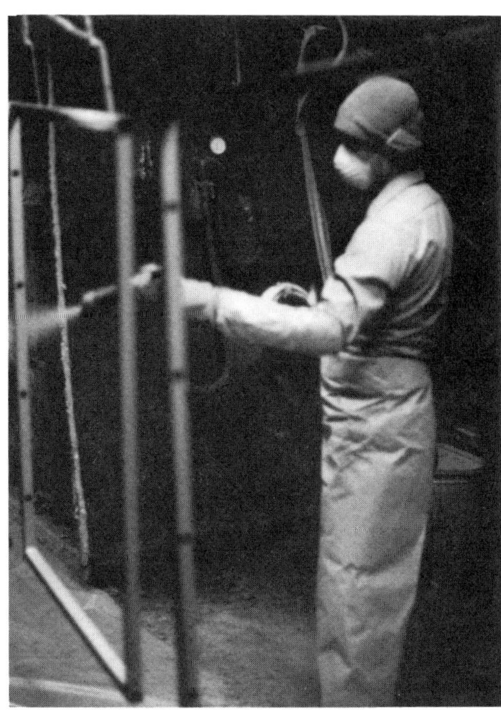

Figure 16-1. Dermal exposure situations. Cleaning a mixing tank *(left)* with solvents can lead to skin exposure and working near paint-spray robots *(right)* can yield high vapor concentrations.

be included in an assessment for skin exposure. Hand contact with chemicals and transfer to the mouth during such activities as smoking, eating, and cosmetics are a concern. Contaminated clothing that is worn during the workday or during several workdays can be a source of exposure. The carrier solvent in a mixture, such as a pesticide formulation, can also affect the skin permeation rate either by retaining a chemical or by facilitating its penetration.

Surface sampling is primarily done to assess occupational exposures. There are two types of methods: those that assess worker exposure indirectly by monitoring for surface contamination and those that assess the degree of contamination on the worker's skin. Indirect methods involve using filters, gauze pads, or swabs to wipe surfaces; instruments to "sniff" surfaces where compounds are volatile; and adhesive tape to lift dust from surfaces. Direct methods are often experimental and have used gauze patches, charcoal pads, direct washing of the skin with a solvent, and fluorescence detection. Sampling professionals intending to try monitoring a worker's skin are cautioned that incorrect application or use of direct techniques may result in injury to the worker. However, since dermal exposures are not currently assessed sufficiently, these methods are provided in the hope of increasing the awareness of sampling professionals.

Surfaces can also be sampled for microbial contamination. This type of sampling is discussed in the chapter on Sampling for Bioaerosols.

Given that many chemicals can penetrate or injure the skin, one way of estimating the potential for skin exposure is to collect wipe samples when evaluating situations where these types of contaminants may pose a hazard. Selection of compounds for dermal exposure studies depends on their ability to cause systemic toxicity by penetration of the skin, dermal irritation, or sensitization. Other terms for wipe sampling are "swipe" and smear sampling. All are terms used to describe the techniques used for assessing surface contamination.

The effectiveness of personal protective equipment (PPE) such as gloves against penetration by contaminants can be determined in part by wiping the interior of the gloves as well as the skin that was protected. The effectiveness of decontamination procedures can be determined. Accumulated toxic dusts, such as lead and beryllium, may become resuspended in air, and thus may contribute to airborne exposures. Some compounds, such as arsenic and lead, pose a greater threat via ingestion than others.

The assessment of dermal exposure must also consider that volatilization of a portion of the initial deposit before complete absorption can occur, as well as the impact of workplace factors on affecting dermal absorption rate, such as humidity (skin hydration), dermatitis, abrasion, and protective clothing. Occlusion, wherein materials are trapped next to the skin by coverings such as barrier creams or PPE, will not only enhance the likelihood of the material penetrating or irritating the skin but will cause the exposure to last longer by restricting evaporation and removal by other mechanisms.

Important factors that influence permeation include the particular compound's solubility, its oil/water partition coefficient, its molecular size, concentration gradient, presence of skin injury or change in the condition of the stratum corneum, species differences (when extrapolating animal toxicity data to humans), skin area of application, and the degree of contact between the absorbable material and the skin.[1]

WIPE SAMPLING

Chemical Selection

Chemicals that can penetrate or injure the skin are the primary types of compounds for which surface contamination is a concern. Compounds such as the heavy metals that do not penetrate the skin but can be transferred to the mouth if an eating area is contaminated are also good candidates for surface sampling. Examples of heavy metals are lead, arsenic, and cadmium.

Chemicals for which the American Conference of Governmental Industrial Hygienists (ACGIH) or OSHA has a "skin" notation (Table

TABLE 16-2. Chemicals That Have a TLV Skin Notation

Acetonitrile
Acrylamide
Acrylic acid
Acrylonitrile
Aldrin
Allyl alcohol
Allyl glycidyl ether (AGE)
4-Aminodiphenyl
Aniline
Anisidine
Azinphos-methyl
Benzidine
Bromoform
2-Butoxyethanol
n-Butyl alcohol
Butylamine
tert-Butyl chromate
o-*sec*-Butylphenol
Captafol
Carbon disulfide
Carbon tetrachloride
Catechol
Chlordane
Chlorinated camphene
o-Chlorobenzylidene malononitrile
Chlorodiphenyl (PCBs)-42% and 54% chlorine
β-Chloroprene
Chlorpyrifos
Cresol
Cumene
Cyanides
Cyclohexanol
Cyclohexanone
Cyclonite
Decaborane
Demeton
Diazinon
2-N-Dibutylaminoethanol
3,3′-Dichlorobenzidine
Dichloroethyl ether
Dichloropropene
Dichlorvos
Dicrotophos
Dieldrin
2-Diethylaminoethanol
Diethylene triamine
Diisopropylamine
Dimethyl acetamide
Dimethylaniline
Dimethyl 1,2-dibromo-2,2-dichloroethyl phosphate
Dimethylformamide
1,1-Dimethylhydrazine

TABLE 16-2. *(continued)*

Dimethyl sulfate
Dinitrobenzene
Dinitro-O-cresol
Dinitrotoluene
Dioxane
Dioxathion
Disulfoton
Endosulfan
Endrin
Epichlorohydrin
EPN
Ethion
2-Ethoxyethanol
2-Ethoxyethyl acetate
Ethyl acrylate
Ethylene chlorohydrin
Ethylene dibromide
Ethylene glycol dinitrate
Ethylenimine
n-Ethylmorpholine
Fenamiphos
Fenthion
Fonofos
Furfural
Furfuryl alcohol
Heptachlor
Hexachlorobutadiene
Hexachloronaphthalene
Hexafluoroacetone
Hexamethyl phosphoramide
Hydrazine
2-Hydroxypropyl acrylate
Isooctyl alcohol
Isophorone diisocyanate
n-Isopropylaniline
Malathion
Manganese cyclopentadienyl tricarbonyl
Mercury vapor, aryl and inorganic compounds
Mercury, alkyl compounds
Methacrylic acid
2-Methoxyethanol
2-Methoxyethyl acetate
Methyl acrylate
Methyl acrylonitrile
Methyl alcohol
n-Methyl aniline
Methyl bromide
O-Methylcyclohexanone
2-Methylcyclopentadienyl manganese tricarbonyl
Methyl demeton
4,4′-Methylene *bis*(2-chloroaniline)
4,4′-Methylene dianiline

TABLE 16-2. *(continued)*

Methyl hydrazine
Methyl iodide
Methyl isobutyl carbinol
Methyl isocyanate
Methyl parathion
Mevinphos
Morpholine
Naled
Nicotine
p-Nitroaniline
Nitrobenzene
p-Nitrochlorobenzene
Nitroglycerin (NG)
n-Nitrosodimethylamine
Nitrotoluene
Octachloronaphthalene
Parathion
Pentachlorophenol
Phenol
Phenothiazine
p-Phenylene diamine
Phenylhydrazine
Phorate
Picric acid
Propargyl alcohol
Propyl alcohol
Propylene glycol dinitrate
Propylene imine
Sodium azide
Sodium fluoroacetate
Sulfotep
TEPP
1,1,2,2-Tetrachloroethane
Tetraethyl lead
Tetramethyl lead
Tetramethyl succinonitrile
Tetryl
Thallium, soluble compounds
Thioglycolic acid
Tin, organic compounds
o-Toluidine
m-Toluidine
o-Toluidine
p-Toluidine
1,1,2-Trichloroethane
Trichloronaphthalene
Trichloropropane
2,4,6-Trinitrotoluene (TNT)
Triorthocresyl phosphate
Vinyl cyclohexene dioxide
m-Xylene, α, α'-diamine
Xylidine

16-2), or substance that has a skin LD_{50} (to rabbits) of 200 mg/kg or less are considered to have significant skin penetration properties.

A class of irritants for which wipe sampling is commonly done is amines, since they can stick to surfaces. Some amines can also be sensitizers. The degree of irritation can vary. For example, hydrogen fluoride molecules actually "burrow" through the skin to the bone. In some cases, the impact of a corrosive chemical such as concentrated sodium hydroxide can be very severe and can cause significant burns to the skin's surface. The following is a list of chemicals that are skin irritants:

Acrolein
Allyl alcohol
Allyl glycidyl ether
Ammonia
Ammonium chloride fume
n-Butyl acetate
Caprolactam, dust and vapor
Chlorine
Chloroacetyl chloride
o-Chlorobenzylidene malononitrile
Diethylamine
Ethyl benzene
Glutaraldehyde
2-Hydroxypropyl acrylate
Methyl 2-cyanoacrylate
Phosphoric acid
Potassium hydroxide
Propylene glycol monomethyl ether
Sodium bisulfite
Sodium hydroxide
Thioglycolic acid
1,2,4-trichlorobenzene
Triethylamine
Tetrasodium borate salts

A special case is allergic sensitization of the skin, such as that caused by isocyanates and cobalt, where exposure can result in an individual becoming unable to work around the material and workers needing to change jobs or leave a facility entirely. Another type of effect is photosensitization that results from exposure to certain compounds such as petroleum asphalt fumes and ultraviolet (UV) light, usually from the sun. The following chemicals are examples of skin sensitizers:

Captafol
Cobalt metal, fume and dust
Isophorone diisocyanate
Phenothiazine
Phenyl glycidyl ether
Picric acid
Subtilisins
Toluene-2,4-diisocyanate

In selecting chemicals to sample for ingestion potential, their oral toxicity is important. In general, compounds with oral LD_{50}'s (to rats) of 50 mg/kg and less are high candidates, but those with higher oral LD_{50}'s (meaning less toxic) must be evaluated on a case-by-case basis.

Chemicals for which wipe sampling is ineffective even though they can penetrate the skin include volatile solvents such as benzene and *n*-butyl alcohol, because their evaporation from surfaces can be rapid. Therefore, sampling will not provide adequate quantitation. Most gases (e.g., bromine and boron trifluoride) do not redeposit on surfaces, so sampling for them is ineffective; however, some gases (e.g., arsine and stibine) may revert to their metallic form after contact with surfaces and remain. In some cases, biological monitoring such as collecting urine samples from benzene-exposed workers for phenol analysis can be an effective means of determining whether skin absorption is a significant concern for contaminants that are difficult to sample.

Sampling Materials

Generally there are two types of filters recommended for taking wipe samples[2]:

1. Glass fiber filters (37 mm) are usually used for materials analyzed by high-performance liquid chromatography (HPLC) and often for substances analyzed by gas chromatography (GC).
2. Cellulose (paper) filters are generally used for metals, and may be used for anything not

analyzed by HPLC. For convenience, the Whatman smear tab may be used.

Gauze is sometimes used rather than filter papers. An example is for PCB sampling. Gauze is generally extracted using the Soxhlet technique to assure its purity prior to use.

Acids and bases (or alkalies) can be detected by their reaction with pH paper. For example, ammonia and amines are very basic and can be detected by a high pH. This method can also be used as a first identification if an unknown spill is detected. Acids turn litmus (pH) paper blue, and bases turn it pink.

Purity of solvents is important. Although they need not be spectroscopic grade, they should be of sufficient purity not to contaminate the sample. For example, distilled water should be used rather than tap water. Solvents are not required, but they can enhance collection if appropriate for the contaminant under investigation. Solvents other than distilled water should not be used to sample direct skin (Table 16-3). Filters are fragile and when using solvents extreme care must be used or they will be damaged.

TABLE 16-3. Solvents Used in Wipe Sampling

Compound	*Solvent*
PCBs	Hexane
Aromatic amines	Methanol
4-Aminodiphenyl Azinphos methyl Toxaphene DDVP Diazinon Dieldrin Dinitrotoluenes Lindane Malathion Parathion	Ethylene glycol
4-Aminopyridine Aniline Anisidine Benzidine Heptachlor Nitroglycerin Pentachloronaphthalene TEDP Tetrachloronaphthalene O-Toluidines *o*-Toluidene Trinitrotoluene (TNT)	Isopropanol
General	
Metals and salts	Distilled water
Low-chain hydrocarbons	Distilled water
Bases	Dilute acids
Amines	Dilute acids
Hydrazines	Dilute acids
Acids	Dilute bases (detergents)
Phenols	Dilute bases (detergents)
Thiols	Dilute bases (detergents)
Nonpolar hydrocarbons	Organic solvents

Sampling Methods

Direct evaluation of dermal exposure usually is far more complicated than that of airborne exposures due to variability in deposition rates onto the body, the effect of clothing, the duration of actual skin contact with the chemical, and the importance of time in the retention and permeation of the chemical through the skin. A number of methods for assessing surface and personal contamination have been developed, including wet and dry smears, adhesive tape sampling, and skin-washing techniques. Wipe sampling is also sometimes referred to as smear sampling. Wipe sampling variables include the degree of pressure applied, accuracy of selecting the area to sample, types of wipe media used, and the physical nature of the contaminated surface (porous versus nonpermeable or smooth) since particles will be deposited in surface cracks and crevices where they are not removed by the wiping.[3]

For some types of surface contamination that do not respond well to manual collection, such as mercury, direct reading instruments like a mercury sniffer may be used. Other examples are general survey instruments, such as an organic vapor analyzer for organic vapor contamination. In this case, it is important to remember that unless the instrument is calibrated with the solvent in question, the results are all relative to the material with which it was calibrated.

It can be difficult to determine whether equipment or instruments are clean after they have been used in a contaminated area, such as on a hazardous waste site or for sampling of concentrated materials. Wipe sampling can be useful in determining whether these procedures are effective. It is often useful to sample the sump where the decontamination water is being collected, such as happens on hazardous waste sites. See the chapter on Bulk Sampling Methods for information on this method.

Wipe sampling is inappropriate for porous surfaces, and would absorb the compound of interest, such as PCBs. These include wood and asphalt. Instead, a bulk sample should be collected in these situations, emphasizing collection of the surface (1-cm deep) material.

The patch technique has been the most widely used method for estimating dermal exposure, but there are concerns that it has never been properly validated. Its primary limitation is the assumption that exposures are uniform over various body parts. Since patches generally cover 6% or less of the body, exposures could be grossly overestimated or underestimated if droplets hit or miss the patch in situations where spray applications are being monitored.

General Procedure [2]

1. Preload a group of vials with appropriate filters. Make sure the vials have labels.

2. Always wear clean impervious gloves when doing wipes. Disposable gloves are preferable, and can prevent contamination of future samples as well as to oneself. A clean set of gloves should be used with each individual sample. The selection of gloves will depend on the contaminants being sampled and the types of solvents in use.

3. Prepare a diagram of the area or room(s) to be wipe sampled along with locations of key surfaces.

4. Label the sample vial with the place where the sample is being collected and a unique number. Withdraw the filter or other media from the vial. If a damp wipe sample is needed, moisten the filter with distilled water or other solvent as recommended for the contaminant being sampled.

5. Wipe approximately 100 cm^2 of the surface to be sampled. The purpose is to have consistent areas to compare. Even if standards are not available, samples should be comparable for determining which areas are the most contaminated.

6. Without allowing the filter to contact any other surface, fold the filter with exposed sides against each other and then fold it again. Put the filter in its sample vial, cap the vial, and place a corresponding number at the sample location on the diagram. Some substances, such as benzidine, must have solvent added to the vial as soon as the wipe is placed inside.

7. Take notes as well as including any further descriptions that may later prove useful when evaluating sample results, for example, employees' names if personal protective equipment is being wiped.

8. At least one blank filter should be folded and put into a vial. Clean gloves and remember that no contact with surfaces is important. Also provide a sample of the solvent used as a blank in case it is needed.

9. Samples that can evaporate must be contained within airtight bottles or samples will be lost. Shipping wipe samples containing solvents may be a problem and may delay receipt at the laboratory from normal air times. For example, hexane, which is used to sample PCBs, is highly flammable and cannot be shipped by air.

Wet Wipe Test for Arsenic

This method is appropriate for arsenic and arsenic-containing inorganic compounds.[2]

Procedure

1. Using a clean, disposable glove, remove a 7-cm (2¾-in.) diameter Whatman 41 filter from the box.

2. Moisten the filter with distilled water. Use a dry filter if sampling for a liquid residue of arsenic trichloride or arsenic trifluoride.

3. Select a sampling area that is at least 100 cm^2.

4. With the leading edge of the filter slightly raised, wipe the surface in a back and forth and up and down motion. A 10-cm by 10-cm wire frame can be used as a guide.

5. Pick up the filter paper, place it on a clean sheet of paper, fold the contaminated side inward, and then make one more fold to form a 90° angle in the center of the filter.

6. Place the filter, angle first, into a glass vial and close the lid tightly.

Modified Wet Wipe for Aromatic Amines

MOCA (4,4′-methylene bis(2-chloroaniline)) is an example of an aromatic amine that is also a potent animal carcinogen used as a curing agent for isocyanate-containing polymers. An example is urethane molds that are often used in high-precision aluminum foundries. Surface contamination can occur while mixing the urethane with this catalyst. Residual material could be contacted by workers brushing against bench tops, opening doors, or handling containers. Wipe samples would include all surfaces in the area where MOCA is stored, used, or handled in its pure form, such as mixing benches, door knobs, personal protective equipment like gloves and aprons.[2] Also, if it is suspected that the employees are not being required to wash after using the material, or that other modes of contamination may be present, then wipes may also have to be collected on lunchroom surfaces or other surfaces where other employees may be exposed. It is not uncommon for the mixing area to be separate from the molding area. In the OSHA standard it is required that areas where employees may be exposed to MOCA be "regulated areas," that is, areas where entry is restricted and where certain controls, such as personal protective equipment and training, are required. Once the material is mixed into the urethane, curing should begin, and the exposure to residual isocyanate is a greater concern than the small amounts of catalyst that are generally used in these mixtures.

Procedure

1. Follow steps 1 through 3 of the General Procedure.

2. Wipe approximately 100 cm^3 with a Whatman 42 filter that is 7 cm (2.8 in.) in diameter after moistening its center with 5 drops of methanol.

3. After wiping the sample area, apply 3 drops of fluoroescamine (a visualization reagent) to the contaminated area of the filter. Also place a drop of this reagent on an area of the filter that has not contacted the wiped surface. This area becomes a blank, and using the same filter allows for better comparison.

4. Allow 6 minutes for the reaction, then irradiate the filter with a 366-nm ultraviolet (UV) light. Compare the color development of the contaminated area with the area designated as the blank. A color change to yellow is a positive for contamination with the following amines: MOCA, benzidine, 3,3′-dichlorobenzidine, alpha-naphthylamine, beta-naphthylamine, and 4-aminodiphenyl.

5. Additional samples should be collected and sent to a laboratory for confirmation. Send a vial containing a blank filter along with the wipe samples.

Wipe Test Using Gauze for Polychlorinated Biphenyls

Other situations where PCB wipe sampling may be needed include testing decontamination procedures on waste sites where PCB cleanup is in progress and in offices where leaks of PCB-containing ballasts in fluorescent lamps have occurred. Another situation is to determine whether permeation has occurred through personal protective equipment. If the surface to be sampled is smooth and impervious, a wipe sample will be effective for PCBs. It should be noted that cleaning solutions should be analyzed prior to disposal for PCB levels in excess of 50 ppm in order to determine whether they must be disposed of as a hazardous waste. For more information on collecting bulk samples of water and sending them in for analyses, see the chapter on Bulk Sampling Methods.

Procedure

1. Use 3-inch by 3-inch gauze pads that have been Soxhlet extracted with hexane. Prior to sampling, wet each pad with 8 ml of pesticide grade hexane.

2. Put on a phthalate-free glove prior to each sample.

3. Mark off a surface area of 100 cm^2 and wipe the surface with the hexane-saturated pad initially in a horizontal direction using a forward and backward motion. Do a second wiping of the surface using a clean portion of the same gauze pad in the vertical direction with the same forward and backward motion.

4. Place the gauze pad in a brown glass sample container equipped with a Teflon-lined lid.

Dry Wipe Test

Procedure

1. Using the tip of the thumb, wipe a 2.4-cm-diameter filter paper disk in a "Z" or "S" pattern over a representative portion of the surface to be sampled. The length of the wipe should be 50 cm. The pressure-bearing portion of the filter paper disk will be about 2 cm wide; therefore, the area of the surface that is sampled will be approximately 100 cm.^{2}

2. Avoid contacting excess dirt when wiping an area.

Swab Test

Analysis of these samples is qualitative, but will reflect the general degree of surface contamination.

Procedure[4]

1. Assemble the following materials.

Cotton swab, wooden stem
Acetone, "distilled-in-glass" Nanograde, or other proper solvent
Hexane, pesticide grade
Isooctane, pesticide grade
Metal clamp

Containers:

Glass-stoppered glass jar
10-ml cone-shaped bottom vial with Teflon-lined screw cap
2-dram glass vial with Teflon-lined cap
Amber glass bottle, 1 pint
Plastic Nalgene bottle, 1 quart

Protective equipment:

Butyl rubber gloves
Plastic disposal bag

2. Wipe off a square area of approximately 0.24 M^2 on the surface to be sampled. Alternatively, mark off five 2-inch-diameter circles distributed at the four corners and center of a 1-M^2 area for building surfaces, or one 2-inch-diameter circle for vents and other surfaces. If sampling an area of known contamination, select an area of 4 in.^{2} in the center of the contamination.

3. While holding a swab in a clean metal clamp, saturate it with 20 ml to 30 ml of a 1:4 acetone/hexane mixture. Continue holding the swab in the clamp while wiping the sampling area back and forth in a vertical direction, applying moderate pressure. Wipe several times. Turn over the swab and wipe back and forth in the horizontal direction.

4. Alternatively, dip a swab in a 2-dram vial containing 1.5 ml of acetone or other solvent, and swab one circle at a time, dipping the swab in the solvent before and after each circle is swabbed.

5. Wrap each swab in aluminum foil and place the wrapped swab in a clean, labeled glass container with a Teflon-lined lid, and close the cap until extraction and analysis can be performed. When all circles have been swabbed, dip a swab into the solvent, wrap it in foil, and put it into a jar labeled "blank." Tightly seal the acetone-containing vial with a Teflon-lined cap.

6. Preserve the collected samples and blank at 4°C in a refrigerated box.

7. When resampling a surface after decontamination, position the sampling grid 6 inches to the right of the initial sampling points, or if movement to the right is restricted, position 6 inches downward.

OTHER SURFACE SAMPLING METHODS

"Sniff" Test

For volatile contaminants, a general survey monitor with a photoionization detector (PID) or a flame ionization detector (FID) can be used. The probe is moved around the article to see if there are any increases in levels over background. For a discussion on how to use these instruments, see the chapter on General Survey Instruments for Gases and Vapors. The following is an example of how this test can be used to identify mercury contamination.[5]

Procedure

1. Enclose the suspect material in a polyethylene bag or close-fitting airtight container for 8 hours at room temperature (76–80°F [24–30°C]), or place the bag in an oven set at 125°F ± 5°F (52°C ± 2°C), for 1 hour.

2. Make a small slit in the bag and sample the air with a mercury vapor analyzer.

3. If the mercury vapor concentration is greater than 0.01 mg/M^3, the material is contaminated.

Surface Dust Contamination

This procedure is particularly useful for collecting samples from rafters and beams. One application is in asbestos abatement work where residual materials on areas such as window sills, pipes, and floor cracks are suspected.[6] This method could also be used to determine if residual dust, such as lead, is present on a worker's clothing, and therefore could be contributing to the worker's exposure.

Tape samples are used primarily for surface asbestos contamination, but can be used for other dusts as well. It has been suggested that adhesive tape may be more efficient in taking samples from rough surfaces than filter paper.[3]

Procedure

1. Calibrate a personal air sampling pump at 2 Lpm. Attach a tube and a filter in a cassette. Depending on the analysis, the type of filter will vary. Another short length of tubing is attached to the inlet of the cassette.

2. Start the pump and use the exposed end of the second piece of tubing like a vacuum cleaner.

3. Move the sample collector through the work area, stopping at 10 sites or more that represent the most likely places where dust might collect. Each sampled spot should be marked with an "X" on a diagram, and the specific area should be described on a list, such as window sill, left sleeve, cuff, top of pipe.

4. The amount of time spent sampling each site will be determined by filter loading. If the dust is readily visible, the filter will rapidly become loaded.

METHODS THAT DIRECTLY ASSESS WORKER EXPOSURE

A number of techniques to directly monitor for dermal exposure on the worker have also been tried. These include a patch type of dermal dosimeter made of gauze or charcoal cloth, a skin wash technique (most appropriate for chemicals with low rates of dermal absorption), urinary excretion of chemicals readily absorbed, and fluorescence of selective chemicals.[7] Common areas to sample are forearms, hands, wrists, feet, ankles, and neck (Table 16-4 and Fig. 16-2). The neck is generally exposed and is not washed during the course of the day as the hands are.

By placing the pad under the clothing, gauze and charcoal pads can be used to monitor exposure of exposed skin as well as skin exposed through clothing permeation. The pad is attached to an elastic band, and slid onto an arm or leg or pinned to underwear. Hand or face exposure can be monitored by attaching these pads, often called dosimeters, to thin cotton gloves or a hat. Outside clothing exposure can be monitored by attaching the pads using strips of Velcro.[8]

TABLE 16-4. Surface Areas of the Body

Body Part	*Surface Area (%)*
Whole body	100.0
Face	3.5
Hands	4.4
Forearms	6.5
Back of neck	0.6
Front of neck and "V" of chest	0.8

Source: From Durham, W. F., and H. R. Wolfe. Measurement of the exposure of workers to pesticides. *WHO Bull.* **26:**75–91, 1962.

Gauze Patches or Pads

Patch or pad samplers have been primarily used to estimate exposures to pesticides, because of the significant potential for skin contact during reentry after spraying crops. These absorbant gauze pads can be attached to the hands, face,

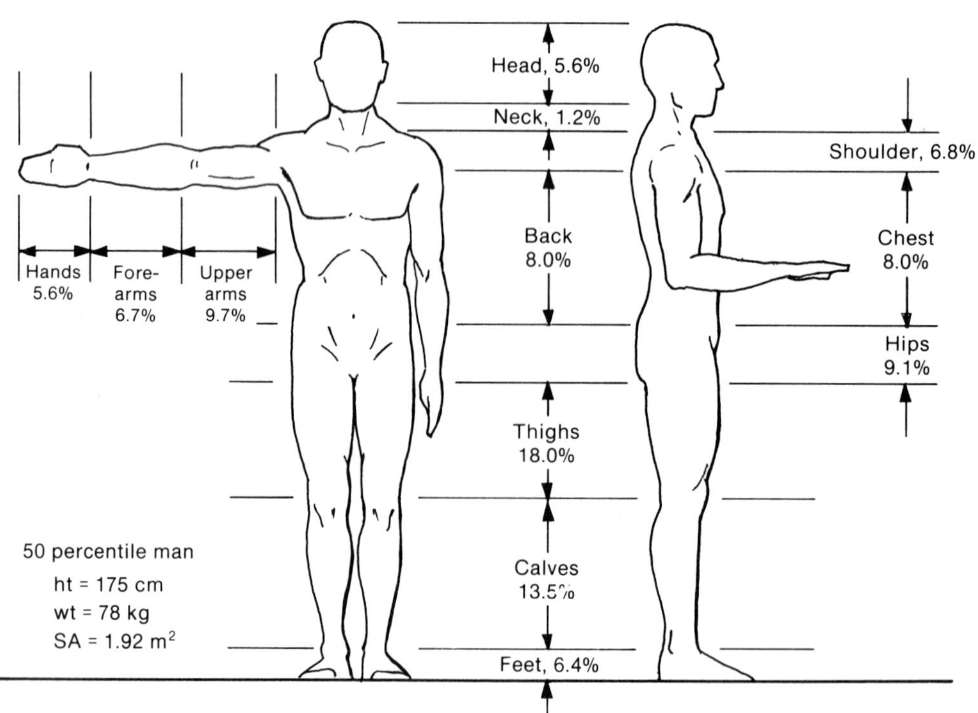

Figure 16-2. Total surface areas for humans. (From Popendorf, W. J., and J. T. Leffingwell, Regulating OP pesticide residues for farmworker protection, *Res. Rev.* **82:**125–201, 1982.)

and other parts of the body that are normally unclothed. When placed underneath personal protective equipment, they can be used to estimate the degree of protection afforded by this clothing. Harvesters have been sampled extensively using this method, because they are almost literally bathed in contaminated dust.

A limitation of this technique is the necessity of assuming that the area covered by the pad is representative of the entire body being sampled. The pads have several advantages in that they are small, portable, inexpensive, and passive. They tend to absorb and retain materials like spray mists or particulates that have low vapor pressures. One limitation to accurate dose estimates using these pads is that exposures must be relatively uniform between monitored and unmonitored points of the body.

Another disadvantage to gauze pads is that because they do not physically resemble the skin, adsorption, absorption, and retention of volatile or even nonvolatile materials by the cotton gauze may not be similar to the skin. There are a wide range of skin variations both between and within individuals due to such factors as hairiness, ease of sweating, wrinkling, callouses, and smoothness.[1]

The composition of a gauze pad for dermal sampling is several (often 16) layers of gauze or alpha-cellulose material, an impervious backing to prevent loss of material, and tape or other means to hold the pad together and attach it to the wearer. Both polyethylene and aluminum foil have been used as backings. Surgical tape, safety pins, or the use of an overgarment, such as a vest containing special pockets, have been used for attachment. However, ideally to measure dermal dose the pads should be worn under the clothing against the skin and on exposed skin.[9]

Procedure

1. Pads of alpha-cellulose have been used for measuring sprays, and pads made from 32 layers of surgical gauze and backed by filter paper have been used for measuring dusts. The gauze pads should be taped around the entire circumference. Whatever material is selected must be preextracted (Soxhlet) with the same solvent to be used during analysis. Pads and gauze should be free of ink, oils, and other markings. The size of the pads is optional, although 5-cm by 5-cm gauze (exposed area) patches appear to be a useful size.

2. For transport, pads should be placed between folds of Whatman filter paper.

3. Secure the pad with surgical tape to the area where monitoring is to be done.

4. After use, each pad should be placed in an airtight, labeled glass container.

5. For analysis, the tape is cut away from the pad, leaving a 5-cm by 5-cm square. This 25-cm^2 piece is put through a Soxhlet extraction, and the extractant is analyzed. Results are reported in $\mu g/cm^2$.

Charcoal Pads

Charcoal pads have been used to monitor dermal exposure to volatile compounds, such as organic solvents and fumigants. The pads consist of a commercially available charcoal cloth[10] cut into 8-cm by 5-cm squares. Four layers are then sewn together and a protective covering of a thin polyester fabric ("Horizon East" Japan) is applied, and then a baking of a strong polyethylene-coated belting material is provided.

These dosimeters have been recommended for monitoring highly toxic materials that are either harmful to the skin or can permeate the skin readily.[11] Materials with vapor pressures of 10 mm Hg and greater should be monitored only where ventilation effectively reduces vapor concentration. Chemical volatility also affects the usefulness of these dosimeters in that a liquid deposit on the skin is subject to air evaporation as well as absorption by the skin. Therefore, the more volatile a compound, the shorter will be the contact period on the skin, thus decreasing dermal absorption.

Collection techniques will be similar to those used for gauze pads; however, sample handling following exposure is different. The units must be stored at −20°C or less until analysis to minimize evaporation.

TABLE 16-5. Retention Efficiencies for Charcoal Cloth

VP (mm Hg):	0.3				1.0				4.0				10				15			
RL (μL):	10	25	60	100	10	25	60	100	10	25	60	100	10	25	60	100	10	25	60	100
RH(%)																				
30	94	93	92	92	92	90	89	88	87	85	83	82	83	81	77	75	81	78	75	72
50	91	90	89	88	88	86	84	83	81	78	75	73	75	71	66	63	72	67	61	58
70	89	87	85	84	84	82	79	77	75	72	67	64	67	62	55	51	62	56	49	43
90	87	84	82	81	81	78	74	72	70	75	60	56	60	52	44	38	53	45	34	26

Following collection, a correction must be made, because charcoal has a much higher adsorptive surface area than human skin, and the amount of vapors adsorbed onto the charcoal pad will be several times higher than the amount that could be adsorbed onto a similar area of human skin. Also, the total amount of chemical collected on the dosimeter would include depositions from both the liquid phase and the vapor phase in the case of materials that are being sprayed. The following calculation can be used to determine the initial liquid deposit to the charcoal pad.

$$\text{Initial liquid deposit} = \frac{\text{total amount} - \text{adsorbed vapors}}{RE/100}$$

where the total amount is obtained from the desorption and analysis of the cloth in the dosimeter; *RE* is the retention efficiency (see Table 16-5); and the amount of adsorbed vapors is unknown.

Table 16-5 lists estimated retention efficiencies for charcoal cloth patches as a function of vapor pressure (VP), retained liquid (RL), and relative humidity (RH).[11]

Skin Wash

Another technique involves wiping the skin directly. Skin washing has been used to estimate dermal exposures to organophosphates by using alcohol to wash unabsorbed residues from workers' skin and clothing. This technique may be more representative of actual exposure because the skin is subjected to more variables (e.g., sweating and changes in body temperature) than are external collectors such as pads. However, the effectiveness of this method is severely limited by the rapid rate with which the skin absorbs and retains certain compounds, such as organophosphates. Therefore, skin rinsing may underestimate the potential dose, because only the unabsorbed or slowly absorbed fractions of a pesticide formulation may be collected.[1] Solvents should be selected so that they are not irritating or readily absorbed by the skin. In addition, they should not facilitate skin penetration of the material being sampled.

Procedure

Prior to exposure, the area of the skin to be monitored must be cleansed.[11]

1. Swabs consist of two 8-ply 10-cm by 10-cm surgical gauze squares that have been extracted with the same solvent to be used for analyses. The squares are folded twice to form 5-cm by 5-cm squares and then stapled shut.

2. Place the swabs in a glass jar saturated with 95% ethanol (do not use 99.9% ethanol because it may contain benzene). The lid of the jar should be lined with aluminum foil or Teflon to prevent contamination of the gauze with residue on the lid. The jar should be kept shut until use.

3. In the field, open the jar and grasp a swab at the stapled edge using ring forceps or another similar device. Allow the excess ethanol to be squeezed off by rubbing the pad against the inside of the jar.

4. Continuing to hold the pad using forceps, the skin area selected is "washed" by rubbing the cloth back and forth over the surface with light pressure. Each area selected should be swabbed up to four times.

5. Pads for the same area can be combined into one labeled jar as a single sample.

Alternate Procedure

1. A prelabeled 12-ml vial is affixed to a sampling cup and the cup is held against the surface of the skin to be sampled (Fig. 16-3).

2. Cyclohexane is sprayed through the cup onto the skin from an aerosol can. The cup will drain the skin washings into the sample bottle.

3. Sampling is terminated when the level of cyclohexane in the 12-ml vial reaches the sampling cup drip tube. This process happens very fast, in approximately 10 seconds.

4. The vial is then removed and sealed with a Teflon-lined screw cap and labeled. The cup is cleaned between samples by rinsing with a small amount of solvent from the can, and the rinse is discarded into a waste container.

5. A blank is also created by spraying 10 ml of solvent directly into a vial.

Fluorescence Monitoring

Fluorescence has been used to identify dermal contamination in two situations: when compounds such as polynuclear aromatics (PNAs) possess a natural fluorescence and when fluorescent compounds have been deliberately added to contaminants. A UV light is used to visualize the contamination in both cases. Eye damage from excessive exposure to UV radiation both to the sampling professional and to the worker is a concern when using this technique and therefore protective glasses must be worn. Interferences include deodorants, coal-tar shampoos, and other cosmetic preparations that can also fluoresce. Therefore, when samples are obtained, the sampling professional should ask what cosmetics or shampoos have been used in order to determine potential interferences.

Scanning a surface with a UV light for fluorescence has been done for detecting PNAs exposure derived from raw synthetic crude oils and their distillate fractions and coal tars (coal gasification facilities and cleanup sites); however, this technique is likely to be replaced by fluorescent tracers due to the potential problems related to the deliberate exposure of humans to carcinogenic UV light, as well as the synergistic effects of UV and PNAs.

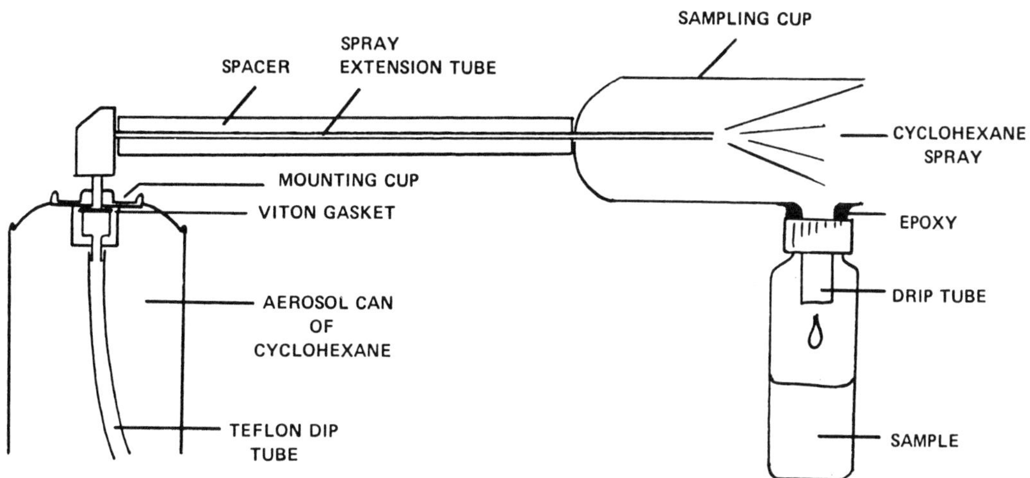

Figure 16-3. Skin was sampling cup. (Reprinted with permission from *American Industrial Hygiene Association Journal* 43:474, 1982.)

The technique consisted of scanning the body of a worker with a hand-held UV lamp ("black-light" lamp) and examining the fluorescence in a dark room. The contaminated areas generally exhibit a fluorescence signal brighter than that of the skin. Background luminescence of human skin has to be corrected for. Natural background skin fluorescence can vary from 10% to 15% among individuals. Calluses, dried skin, and fingertips usually exhibit a fluorescence intensity greater than other parts of the hands. Deposits of coal dust and coal tars can decrease luminescence of the skin; however, they are visible to the naked eye in natural light. It has been recommended that if this type of monitoring is done, a data file should be developed for each individual based on a "background" fluorescence scan.[13]

A variation of this technique is to wash the skin with a solvent such as cyclohexane and then analyze the solvent with a fluorimeter as discussed in the section on skin washing.[14] An instrument of this type fitted with a background-nulling feature has been found the most useful for skin measurements, since it can compensate for normal skin variations, providing a direct reading of the net fluorescence signal. The records developed for each individual worker should be used for individual calibration.[15] The best use of this instrument is in qualitative detection of contamination or to verify that decontamination procedures are adequate.

The addition of fluorescent compounds to mixtures has been used during pesticide application to "trace" leaf coverage, equipment efficiency, droplet size, and "drift" of spray clouds. The EPA has used fluorescent compounds to investigate visual warning methods for farmworker exposure to pesticide residues.[16]

Methods have been developed for the use of fluorescent tracers as quantitative indicators of pesticide exposure during mixing and application. One approach utilizes gauze pads, which are discussed in another section in this chapter, to collect the material. The basis is the mixing of the fluorescent material with the pesticide to produce a "tracer" effect, assuming quantitatively similar behavior of the two compounds from mixing to deposition on the skin surface. A concern is the ability of the tracer and the formulation being applied to be compatible due to solubility and mixing characteristics. This will determine the likelihood of partitioning of the tracer from the rest of the formulation during the application process. Problems will most likely be the result of a combination of solubility differences and incomplete mixing by the spray apparatus. This procedure has the potential to quantify worker exposure as well as to evaluate the effectiveness of protective clothing under actual field conditions. It also provides a measure of the effect of personal hygiene and work practices upon exposure. It can also be used for worker training in order to enhance awareness of the ramifications of not using the correct handling procedures.[17]

Fluorescent tracers were used in a study designed to see if workers could be motivated to use good work practices that minimize the potential for skin contamination if they could actually see the contamination on their skin. Sawmill employees involved in treating lumber with chlorophenols or handling the treated wood were monitored. Diisodium fluorescein was added daily to the mixing tanks supplying Permatox, a chlorophenol-containing wood preservative. Daily visualization of skin and clothing was done using long wavelength UV lamps. As a result, it was concluded that employees were motivated to practice better hygiene and work practices while on the job: Their levels of exposure, as measured during urine monitoring, decreased over the period of the study.[18]

Fluorescent tracers are not yet widely used. One problem is a lack of nontoxic fluorescent tracer compounds that would be compatible with a wide variety of contaminants.[19] The major disadvantage of this technique when compared to the patch method is its relative complexity in the field along with high cost compared to that of the patch tests. The stability of the UV illumination and the instrument's performance must be monitored. The ratio of pesticide to tracer deposition on the skin's surface must be determined through field sampling in each survey. In some cases, clothing penetration may be different for the two compounds, requiring a correction factor. The problem of quenching is inherent to any method measuring dermal fluorescence. When excess fluorescent material is deposited on the skin, the fluorescence produced is no longer proportional to deposition. Therefore, areas of high concentration may be underestimated.[20]

EVALUATING SAMPLE RESULTS

As discussed before, there are no standards related to skin exposure; each situation must be handled on a case by case basis. The type of health effects caused, the contribution of skin absorption or ingestion to the total dose, and air sample results must be considered when evaluating the results of wipe samples. For example, detection of lead on the surfaces of an office associated with a radiator shop or lead battery shop means that individuals other than production workers are potentially exposed. Exposure could include younger people, for example, teenagers working after school, or office workers. For wipe samples, it must be remembered that quantitation is related to the specific area that was sampled. Since it is generally impossible to wipe all areas of a surface, other portions could have higher levels of contamination.

In general, for OSHA compliance wipe sampling is necessary to establish the presence of a toxic material posing a potential absorption or ingestion hazard. A citation for an ingestion hazard can be issued when there is reasonable probability that in areas where employees consume food or beverages (including drinking fountains) a toxic material may be ingested and subsequently absorbed. A citation for exposure to materials that can be absorbed through the skin or can cause a skin effect such as dermatitis shall be issued where appropriate personal protective equipment is necessary but not worn. Neither of these citations require any air sampling in addition to wipe sampling. There are two primary considerations when issuing a citation of an ingestion or absorption hazard, such as a citation for lack of protective clothing: The first consideration is whether a health risk exists as demonstrated by a potential for an illness, such as dermatitis, and/or the presence of a toxic material that can be ingested or absorbed through the skin or in some other manner. The second is if there is a potential that the toxic material can be ingested or absorbed, meaning that it can be present on the skin of the employee, and can be established by evaluating the conditions of use and determining the possibility that a health hazard exists.

There is always the possibility that false negative results, that is, nondetection of existing contamination, will occur because surface contamination is not removed by a wipe sample. It can be due to the selection of the wrong solvent or permeation of the contamination into a porous material. Dirt and grease on surfaces will adhere to wipes, obscuring sample results and making analyses impossible or difficult.

The conditions of use must be documented by taking both qualitative and quantitative results of wipe sampling into consideration when evaluating the hazard. In order to support the issuance of a citation the following considerations must be met:

The potential for ingestion or absorption of the toxic material must exist.

The ingestion or absorption of the material must represent a health hazard.

The toxic substance must be of such a nature and exist in such quantities as to pose a serious hazard. The substance must be present on surfaces that have hand contact, such as lunch tables and cigarettes, or on other surfaces that present the potential for ingestion or absorption of the toxic material if contaminated, for example, a water fountain. The PPE or other abatement means would be effective in eliminating or significantly reducing exposure.

Exposure is dependent on the total amount absorbed, the amount available, and the rate of absorption. There are a number of mechanisms that affect absorption of chemicals through the skin. In some cases, a portion of the initial deposition volatilizes before complete absorption can take place. Other factors include skin hydration from sweating, preexisting dermatitis, abrasions, and wearing contaminated work clothing for extended periods of time. In addition, the absence or presence of adequate warning properties, such as visual changes to the skin, irritation, or corrosive properties that would make a worker more prone to wash the skin rapidly, thus limiting the exposure period, are important. Another situation that can occur is occlusion, for example, when wearing contact lenses in concentrated atmospheres. It can enhance exposures by restricting the natural processes of evaporation and removal of the contaminant. The same phenomenon can occur under clothing.

Commonly used in pesticide applications, the carrier solvent in a mixture can affect the skin permeation rate either by retaining a chemical or by facilitating its penetration. For example, a compound like dimethyl sulfate would allow a poorly absorbed chemical to be absorbed much faster through the skin.

If the deposition rate is less than the absorption rate, then the deposition rate will be the limiting factor. Under these conditions, it might be expected that the majority or a high fraction of the deposition would be absorbed, depending on the absorption rate coefficient. If the deposition rate is greater than the absorption rate, then the absorption rate will be the limiting factor, for example, immersion of the skin into chemicals with low absorption potential, or deposition of larger quantities of these chemicals by splashing.[7]

Whereas oral doses have historically been based on the weight of the exposed individual (mg/kg), skin exposure should be related to total body surface area. It has actually been suggested that all doses should be based on surface area rather than weight, because interspecies variation would be reduced (for purposes of comparing animal tests to humans) as well as human variations.[21]

Skin contact can be minimized by the use of PPE such as gloves, aprons, and boots; replacing PPE when worn or damaged and decontaminating it each day; practicing good hygiene including washing hands at lunch and breaks; wearing fresh work clothing each day; not smoking, eating, or drinking on the job; and washing contaminated skin immediately.[18]

REFERENCES

1. Webster, R. C., and H. I. Maibach. Cutaneous pharmacokinetics: Ten steps to percutaneous absorption. *Drug Metab. Rev.* **14:**169–205, 1983.
2. OSHA. *OSHA Industrial Hygiene Technical Manual.* 1984.
3. Chavalitnitikul, C., and L. Levin. A laboratory evaluation of wipe testing based on lead oxide surface contamination. In *Fundamentals of Analytical Procedures in Industrial Hygiene.* AIHA, Akron, OH, 1987.
4. Rosbury, K. D. Handbook: *Dust Control at Hazardous Waste Sites.* EPA-540/2-85/003.
5. Arizona Instrument Co., Tempe, AZ.
6. Natale, A., and H. Levins. *Asbestos Removal and Control: An Insider's Guide to the Business.* Source Finders, Cherry Hill, NJ, 1984.
7. Popendorf, W. J. Workshop: Predicting workplace exposure to new chemicals. *Appl. Ind. Hyg.* **1**(3): R-11–R-13, 1986.
8. Cohen, B. M., and W. Popendorf. A method for monitoring dermal exposure to volatile chemicals. *AIHA J.* **50**(4):216–223, 1989.
9. Durham, W. F., and H. R. Wolfe. Measurement of workers to pesticides. *World Health Organ. Bull.* **26:**75, 1962.
10. MDA Scientific, Glenview IL.
11. Cohen, B. M., and W. Popendorf. A method for monitoring dermal exposure to volatile chemicals. *AIHA J.* **50**(4):216–223, 1989.
12. Keenan, R. R., and S. B. Cole. A sampling and analytical procedure for skin-contamination evaluation. *AIHA J.* **43**(7):473–476, 1982.
13. Vo-Dinh, T., and R. B. Gammage. The lightpipe luminoscope for monitoring occupational skin contamination. *AIHA J.* **42**(2):112–120, 1981.
14. Keenan, R. R., and S. B. Cole. A sampling and analytical procedure for skin contamination evaluation. *AIHA J.* **43:**473–476, 1982.
15. Vo-Dinh, T. Evaluation of an improved fiberoptics luminescence skin monitor with background correction. *AIHA J.* **48**(6):594–598, 1987.
16. Johnson, D. E., L. M. Adams, and J. D. Millar. *Sensory Chemical Pesticide Warning System, Part I. Experimental, Summary and Recommendations.* San Antonio, TX: Southwest Research Institute, 1975. EPA-540/9-75-029.
17. Fenske, R. A., et al. A video imaging technique for assessing dermal exposure. II. Fluorescent tracer testing. *AIHA J.* **47**(12):771–775, 1986.
18. Bentley, R. K., S. W. Horstman and M. S. Morgan. Reduction of sawmill worker exposure to chlorophenols. *Appl. Ind. Hyg.* **4**(3): 69–74, 1989.
19. Dubelman, S., and J. E. Cowell. Biological monitoring technology for measurement of applicator exposure. In *Biological Monitoring for Pesticide Exposure.* American Chemical Society, Washington, D.C., 1989.
20. Fenske, R. A. Validation of environmental monitoring by biological monitoring. In *Biological Monitoring for Pesticide Exposure.* American Chemical Society, Washington, D.C., 1989.
21. Klaassen, C. D. and J. D. Doull: Evaluation of safety: Toxicologic evaluation. In *Casarett and Doull's Toxicology: The Basic Science of Poisons.* Doull, J., Klaassen, C. D., and Amdur, M. O., (eds.). Macmillan, New York, 1980.

CHAPTER 17

Bulk Sampling Methods

Although the collection of bulk samples is very simple, few understand the usefulness of these types of samples. The incorporation of bulk samples into a sampling strategy can often make the difference between a successful or unsuccessful sampling effort. Bulk sample results can assist in making air sampling decisions, because laboratory methods can identify their constituents. Bulk samples can be collected from air, soil, water, chemicals, and many other media, such as carpet filters from heating, ventilating, and air-conditioning (HVAC) systems. Chemicals, including chemical wastes, are the most common materials from which a bulk sample is collected.

PURPOSE

Bulk samples are collected for both occupational and environmental purposes. When used to supplement air sampling for occupational exposures, samples are most often of process materials. Reasons for collecting occupational bulk samples include providing support for certain types of air sampling, such as silica sampling. An employer may have recently changed suppliers for a solvent and some of the old material may still be present in unlabeled containers in the shop. Occupational samples are also collected if interferences are suspected or for use as an analytical reference. Environmental samples are generally of chemical wastes, soil, sludge, or water. The Environmental Protection Agency (EPA) has specific definitions for materials in which environmental samples are collected. The term *solids* is applied to soils, sludges, sediments, liquids, and other bulk materials. Sludges are further defined as semidry materials ranging from dewatered solids to high-viscosity liquids and can be found on the bottom of creeks, ponds, and tanks or in any other place where solids can settle out of a body of liquid.

Water and soil are sampled on hazardous waste sites or during environmental audits for real estate transfers. Bulk soil and water samples are useful for identifying contaminants for which air sampling must be conducted during remediation work. Air samples are often collected along the site boundaries to determine if there is a potential for community exposure and on site for occupational exposures. These are performed using integrated and real time techniques as discussed in previous chapters. It is important to note that any situation where extensive samples of soil and/or surface or groundwater are collected requires the involvement of a geologist, hydrogeologist, or engineer, since an understanding of soil and groundwater characteristics and chemistry is required. Discharge water sampling to meet EPA requirements requires special techniques and knowledge as well.

Sampling of tap water is often performed in conjunction with environmental audits for real estate transfer and in evaluations for contamination of potable water sources as required by the Occupational Safety and Health Administration (OSHA). The most common compounds for which analysis is performed on water are lead, volatile organic compounds, and radon.

Bulk samples of air can be collected in bags, evacuated containers, or sorbent tubes and then analyzed to identify hazardous constituents, thus allowing an air sampling strategy to be developed for quantitation of specific compounds. Bags are used to collect gases and vapors for high-resolution analysis of trace constituents when sampling in open fields or vapor wells.[1] Drums and tanks are the most common types of chemical containers that are sampled. Both concentrated chemicals and wastes are stored in drums and tanks. Samples may be collected to identify the composition of the material or to determine if a waste meets any of the EPA criteria set for characterization of hazardous wastes under RCRA regulations. A tank may be scheduled for cleaning and the composition of the sludge inside may be uncertain. In this situation a bulk sample will assist in selecting adequate personal protective equipment and other controls to make sure workers who enter the tanks to remove the sludge are protected.

Analytical methods differ depending on the composition of the sample. For soil and water samples, EPA analytical methods are generally used. For bulk chemical samples the techniques may be from OSHA, NIOSH, EPA, or other source. If the material is classified as a waste, EPA methods are generally used. Samples of bulk air are most commonly analyzed using gas chromatography for separation and a mass spectrograph for detection. If collected on a charcoal tube, a NIOSH analytical method will be used whereas for other situations EPA methods are more likely. When interpreting the results of analysis of a bulk sample containing volatile components, the sampling professional should be aware that the percent composition of the mixture released to the air may be different from that remaining in the container. The airborne mixture will generally reflect a larger percent of the volatiles than were present in the bulk sample.

SAMPLE COLLECTION STRATEGIES

Most commercially available solid sampling devices are steel, brass, or plastic. Stainless steel is considered one of the most practical materials. Some devices are plated with chrome or nickel. They are not advisable to use, since scratches and flaking of the plating can drastically alter the results of analysis. Sample containers used to collect chemicals should be compatible with the material to be sampled. Polyvinyl chloride sample bottles can be used for acids and bases and other water-soluble materials. Glass, preferably with a safety plastic coat, should be used for hydrocarbons and solvents. Bakelite tops with Teflon seals should be used with glass bottles. As a general rule, equipment used to sample hazardous wastes should be disposable. If not, it must be carefully decontaminated between each sample to prevent contaminating subsequent samples. In general, metal sample containers should not be used to collect samples of liquid chemicals or wastes.[2]

Strategies for bulk sampling rarely involve collection of a single bulk sample. At a minimum, several samples are more likely to be collected and mixed together, thus creating a composited sample. If only surface samples are to be collected, then a design must be developed to maximize collection of a representative sample. Generally this involves the use of a grid. The number of samples and distance between samples depends on the surface area to be sampled and other factors such as the cost of analyses. When depth of sampling is involved for samples that are homogenized, as are many liquids, three samples are often sufficient, each one at a different depth and then combined. In other cases a grid is developed for each of several levels. Table 17-1 describes sampling points recommended for waste containers.

Two general types of design are possible for most types of bulk sampling situations where

TABLE 17-1. Sampling Points Recommended for Most Waste Containers

Container Type	*Sampling Point*
Drum, bung on one end	Withdraw sample through bung opening.
Drum, bung on side	Lay drum on side with bung up. Withdraw sample through the drum opening.
Barrel, fiberdrum, buckets, sacks, bags	Withdraw samples through the top of barrels, fiberdrums, buckets, and similar containers. Withdraw samples through fill openings of bags and sacks. Withdraw samples through the center of the containers and to different points diagonally opposite the point of entry.
Vacuum truck	Withdraw sample through open hatch. Sample all other hatches.
Waste pile	Withdraw samples through at least three different points near the top of the pile to points diagonally opposite the point of entry.
Storage tank	Sample from the top of the sampling hole.
Soil	Divide the surface area into an imaginary grid. Sample each grid.

multiple samples are required: grid designs and random designs. A grid system uses a regular pattern, either rectangular or triangular, to determine regular or random sampling points. A circular pattern of sampling around a central point may also be used. Random designs have some disadvantages compared to grid designs in that random designs are more difficult to implement in the field, since the sampling professional must be specifically trained to generate random patterns on-site, and since the resulting pattern is irregular. Grid designs (Fig. 17-1) are more efficient in that they are certain to detect a sufficiently large contaminated area, whereas many random designs are not. For grid sampling, there are equations that may be used to determine grid intervals and the number of samples in a given area.[3] For sites larger than 3 acres,

$$GI = (A\pi/GL)^{0.5}$$

For sites smaller than 3 acres,

$$GI = \frac{(A/\pi)^{0.5}}{2}$$

where GI = grid interval, A = area to be sampled, and GL = grid length.

Compositing is often done to save on analytical costs. Several samples are collected and mixed together to form one sample (Fig. 17-2). Situations where compositing is often done are tank, drum, waste pile, soil, and water sample collection. The disadvantage of sample compositing is the loss of specific data for specific levels or areas, but this loss is usually offset by having a more representative estimate of concentration. Samples can be composited in the field or the laboratory, although compositing is most often done in the laboratory so the samples can be recomposited in a different combination or individual analysis of specific samples can be performed. The actual compositing of samples requires that they be homogenized. This procedure varies according to the type of waste being composited and the parameters to be measured.

The applicability of compositing is dependent on whether individual specimens contain sufficient material to form a composite and leave enough material for individual analyses if needed.

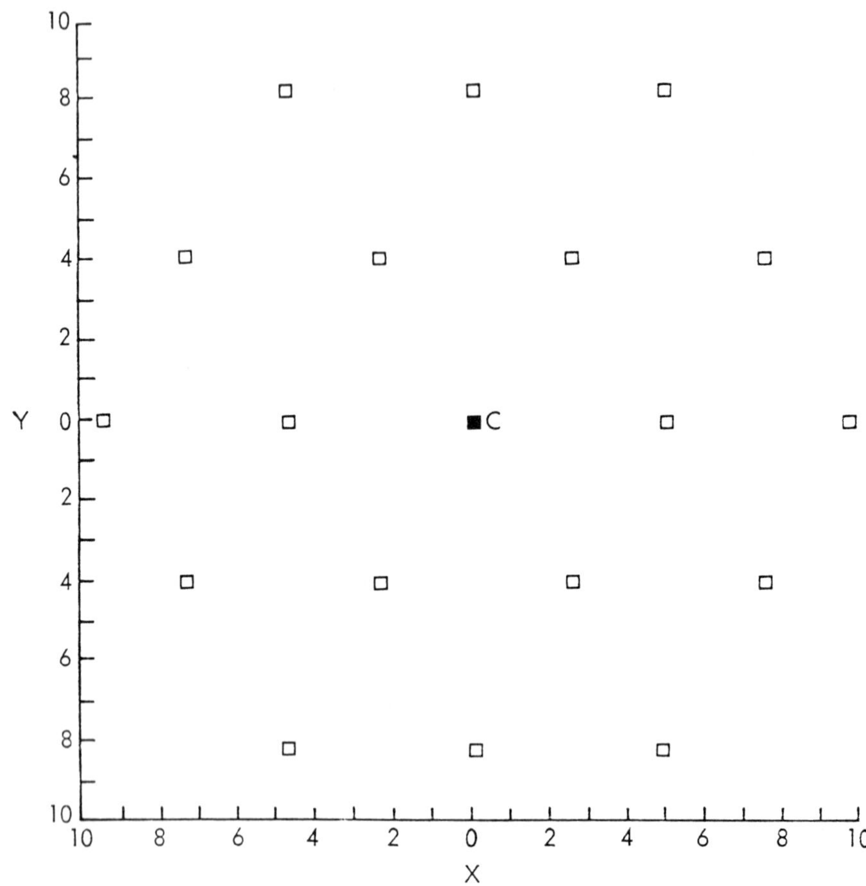

Figure 17-1. Location of sample points in a 19-point grid. The outer boundary of the contaminated area is assumed to be 10 feet from the center (C) of the spill site. (From EPA. *Verification of PCB Spill Cleanup by Sampling and Analysis.* EPA-560/5-85-026, August 1985.)

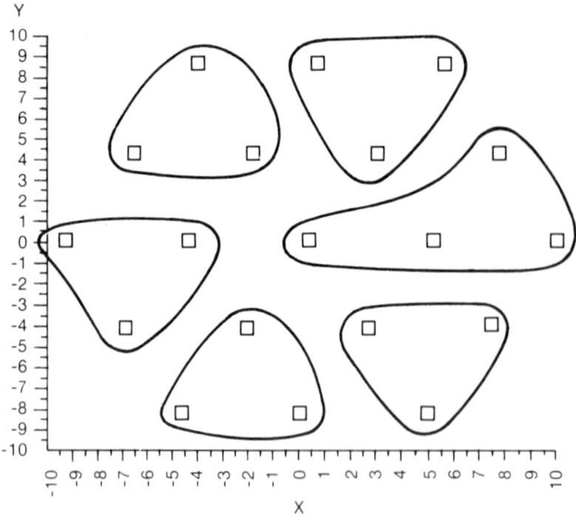

Figure 17-2. Plan showing a six-group compositing design. (From EPA. *Verification of PCB Spill Cleanup by Sampling and Analysis.* EPA-560/5-85-026, August 1985.)

Other limiting factors are the analytical method and cleanup criteria. As a general rule, the EPA cleanup level divided by the number of composited samples should not exceed the minimum detection level of the method. Cleanup levels are generally set by individual states or the EPA and information on specific chemicals can be obtained from the state environmental agencies or the EPA.

CONTAINERS AND SHIPPING

The most important factors to consider when choosing containers for bulk samples are compatibility with the material sampled, cost, resistance to breakage, and volume. Containers must not distort, rupture, or leak as a result of chemical reactions with constituents of waste samples. Therefore, it is important to have some idea of the properties and composition of the material sampled.

All samples should be labeled with a unique sampling number and identified as such on the laboratory request form. It is best to label the bottle or bag first and then collect the sample. If the samples are related to air samples, then an exact reference including their number, should be made to them on the analytical request sheet.

In general, leakproof glass containers are best since they will not react with most chemicals. Brown glass is needed for photosensitive waste. However, polyethylene containers can be used for most dusts (but not soils). Bulk samples are generally collected in vials of 20 ml to 50 ml in volume with Teflon liners in screw lids. A recommended container is a 20-ml scintillation vial with a PTFE-lined cap. Specific chemicals for which glass vials should be used include aromatic compounds, chlorinated hydrocarbons, and strong acids. Generally, large quantities of material are not necessary for analyses. The shorter the time between the collection of a sample and its analysis, the more reliable will be the analytical results. Before the samples are removed from the sampling site, a check should be made to ensure that the tops are correctly and securely fastened. Decontaminate the outside of each sample container thoroughly before packing for transit.

Care should be taken when storing or shipping samples in plastic bottles or bags, since gases and vapors can diffuse in and out of these materials. Exhaust components from transportation vehicles, including jet engine exhaust and gasoline vapors, may contaminate the samples during shipment. Containers should be airtight.

The most important rule in shipping bulk samples is to ship them by themselves, although similar types of bulk samples may be sent together. Bulk samples should never be sent in the same package as air samples. If a material safety data sheet is available, it is a good practice to send it along with the bulk sample to the laboratory.

If a sample is flammable, combustible, corrosive, or poisonous, the shipping options are limited to a land carrier. Department of Transportation (DOT) labeling requirements may have to be observed.[4] For example, certain chemicals are not allowed to be mailed as bulk samples: nitric acid, gasoline, perchloric acid, benzoyl peroxide, class A poisons, aniline, chloropicrin, organophosphate pesticides. Shipping containers for liquids should be cushioned within another sealed waterproof container with absorbent material sufficient to take up all leakage or breakage of the sample container. It is generally a good practice to ship containers of liquids by surface mail.

Sample preservation is sometimes required, especially for water samples. Methods of preservation are relatively limited, and are generally intended to retard biological action, to retard hydrolysis of chemical compounds and complexes, and to reduce volatility of constituents. Preservation methods are limited to pH control, chemical addition, refrigeration, and freezing. Storage at a low temperature, usually 4°C, is perhaps the best way to preserve most samples until the next day. Use chemical preservatives only when they are shown not to interfere with the analysis being made. When they are used, add them to the sample bottle initially so that all sample portions are preserved as soon as collected.

In the case of volatile bulk samples, consideration should be given to shipping the samples on dry ice or with a bagged refrigerant inside an ice chest. The carrier should be consulted about possible restrictions on the amount of dry ice accepted; carriers usually accept 5 pounds or less. Specific package labels are usually required when dry ice is used. Drawbacks include obtaining dry ice and the need for adequate refrigeration in order to cool Blue Ice enough to get it thoroughly frozen. Occasionally a sample may require a preservative other than cooling. Usually this requirement is limited to environmental samples undergoing a certain type of analysis.

PERSONAL PROTECTION

Many materials from which bulk samples are collected are in a concentrated form; therefore, when collecting samples it is important for the sampling professional to wear personal protective equipment. The type of equipment needed will vary depending on whether liquids that might splash the eyes or skin are being handled, or whether toxic dusts such as asbestos are being collected. The selection of proper gloves will vary depending on the characteristics of the material, for example, whether the material is corrosive or is a solvent. Respirators should have the proper type of chemical cartridge or filter for the contaminant(s) being sampled. When sampling unknown materials, it is especially important to have proper personal protective equipment available and to use it.

Sampling of drums and tanks may require the use of self-contained breathing apparatus (SCBA) or other supplied-air respiratory protective equipment, such as pressure-demand type airline equipment. This equipment requires individuals to be trained and regularly drilled in its use. These drums and tanks may hold hazardous materials and special procedures are often required.

Sampling professionals should be aware of the hazards of sniffing samples to determine if any chemicals are present, since some chemicals can be hazardous at very low levels. For example, benzene, a carcinogen, has an odor threshold of 12 ppm,[5] which is much higher than the exposure levels considered safe. Also some compounds at levels of environmental concern do not have a significant odor, and thus the lack of odor is an invalid criterion for not analyzing the sample.

BULK AIR SAMPLES

Bulk air samples can be collected in bags, evacuated containers, or on sorbent tubes. Another variation of sampling bulk air involves using a direct reading instrument to identify levels of volatiles above a bulk sample of a solid or liquid.

Generally, bulk air samples are collected for the purpose of qualitative analysis, that is, to see what contaminants might be present. An example is a situation where occupants are concerned about odors in a building where a carpet was recently installed and the exact identity of suspected airborne hydrocarbons is unknown. Generally reactive gases, such as nitrogen oxides, hydrogen sulfide, and isocyanates, cannot be collected in this manner unless analysis is to be done in the field immediately after sampling, because these gases can react with dust particles, moisture, sealing compounds, glass, or metal, thus altering the sample's composition.

During hot processes, decomposition or offgassing of products that are part of the formulation can occur. In addition, process intermediates, which are temporary products that are either extracted during production or react into other compounds during the manufacturing process, are often of interest. Sometimes these are given off by an interim step, for example, in a reaction vessel that has local exhaust.

The use of bags to collect air samples for personal exposures was discussed in the chapter on Integrated Sampling for Gases and Vapors. Bulk air samples can also be used to detect concentrations of gases and vapors in exhaled breath. For more information on this sampling technique, see the chapter on Biological Monitoring.

Headspace Vapor Samples

Headspace vapor sampling is often done on samples of soil, water, and mystery liquids as a screening technique. It is done by putting the material into a container, capping it, and using a general survey monitor such as a photoionization detector (PID) or flame ionization detector (FID) to "sniff" the sample. Figure 17-3 illustrates the use of this technique in the field. For soil and groundwater samples, it is sometimes used as a way of

Figure 17-3. Headspace vapor sampling with an HNU photoionization monitor.

screening samples to select the most contaminated to send to the laboratory, thus minimizing laboratory fees.

Headspace gases are the accumulated gaseous components found above solid or liquid layers in closed vessels. Bulging, stainless steel, lined, or other special designated drums are likely to contain headspace gases. These gases may be the result of volatilization, degradation, or chemical reaction and concentrations are generally high. Therefore, precautions must be taken. Since higher concentration ranges are to be expected, real time monitoring instruments with higher-range scales or a sample dilution system should be used in order to prevent saturation or deterioration of the detector. The term *headspace vapor* is used to refer to a field screening technique that is a different method than when the term is used to describe a technique that often involves heating a sample in a closed container, collecting any accumulated gas from above the sample, and injecting it directly into a gas chromatograph.

A method for taking a headspace vapor sample over a bulk soil or water sample is to put the material in a cup and cover it with cellophane. Puncture the cellophane with a probe and take a reading with a PID or FID analyzer. When monitoring soils for organic vapor content, conditions such as volatility of the sample, porosity of the soil, general weather conditions, and the ambient background hydrocarbon concentration must be taken into consideration.

Most vessels can be sampled though small hatches or openings. Fully sealed vessels require special techniques, such as the use of remote control units to drill openings, use of safety screens for explosion protection, and vessel pressure monitoring, and these vessels should not be disturbed unless the sampling professional has received special training in this area. In some cases, such as headspace vapor sampling on large tanks, an extension tube will be useful for collecting samples at various depths. However, if the depth of the liquid is unknown, this is not advisable since liquid may be sucked up the tubing into the instrument. A preliminary scan of the external seams, edges, or any corroded areas of the vessel with a PID or FID may indicate if any vapors are present.

Sorbent Tubes

Sometimes integrating methods using sorbent tubes are used to collect high-volume air samples for qualitative analysis. For most volatile and semivolatile organic compounds, a sample can be collected using a charcoal or Tenax tube connected to a personal sampling pump and collecting air at 1 L/min for several hours. Although the sample is likely to exhibit breakthrough, it does not matter since sampling professionals are primarily interested in what substances are present rather than their exact concentrations. In some situations where the concentrations remain constant and the air is stagnant, such as might occur in a confined space, a grab sample might be sufficiently representative of the actual concentrations present. However, the sampling professional should always exercise caution when interpreting the results of this type of sampling. For more information on collection techniques using sorbent tubes, see the chapter on Integrated Sampling for Gases and Vapors.

Bag Sampling for High-Resolution Analysis

Bags are often used to collect gases and vapors for high-resolution analysis of trace levels. This method is commonly used for environmental sampling in open fields or vapor wells. The sampling apparatus consists of a 5-gallon open-top container with a sealable lid, a Teflon stop-cock, Teflon bulkheads, Teflon tubing, and a sampling pump.[1] Flexible bags can also be used for grab samples (Fig. 17-4).

Sample collection bags can also be constructed of a number of synthetic materials, including polyethylene, Saran, Mylar, Teflon, and Tedlar. Tedlar is produced without plasticizers, thereby eliminating the outgassing problem common to other materials. For a discussion of bag materials and sampling issues, see the chapter on Integrated Sampling for Gases and Vapors.

Procedure[1]

1. A 5-L or 10-L Tedlar sampling bag is placed inside the container and connected to the small section of Teflon tubing attached to the bulkhead.

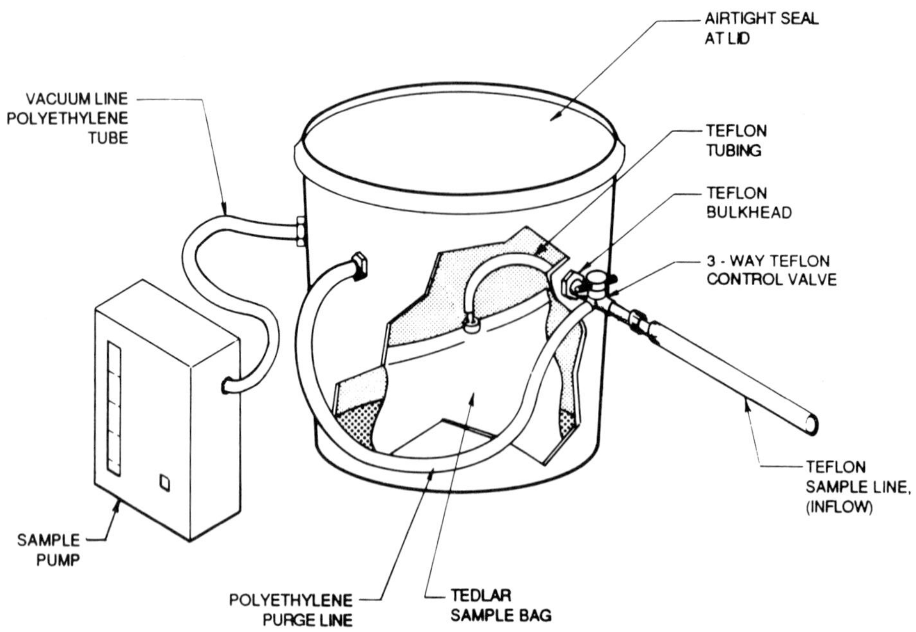

Figure 17-4. Air sampling system for high-resolution analysis. (Reprinted with permission from *American Industrial Hygiene Association Journal* **50**:A-591, 1989.)

The lid, which requires a gasket to ensure an airtight seal, is placed tightly on the container.

2. Prior to collecting an air sample, the stop-cock should be positioned to close off the line to the sampling bag. The pump is then turned on and the air in the sample line is purged. If necessary, the air between the stop-cock and the sample bag can also be purged by putting a small amount of ultra-high pure nitrogen in the bag and pulling it out through the sample line.

3. The air sample is collected by opening the stop-cock to allow air to be pulled through both lines into the plastic container. In this arrangement, air is drawn through the sample line into the bag, due to the vacuum that is created by the sampling pump onto the sampling apparatus.

4. After the sampling is completed, the stop-cock is returned to the original position. In this position, the air in the sample bag is closed off and simply redirected through the sampling apparatus. The pump is then shut off, the lid is opened, and the sample bag is removed. If the contaminants of interest are photochemically reactive, the sample bag immediately must be placed into a light-occlusive container.

Evacuated Containers

In many cases, the sampler simply wants to collect a quick sample of air from a given area. It might be during the heating of a tire or during some other short-term operation. Evacuated flasks and cans can be used for this collection. Evacuation-type devices are sealed vessels with only one entry in which a vacuum has been created. These types of devices can be used to collect gross gas contamination, such as methane or sewer gas. In general, a limitation is that the concentrations of contaminants must be high enough to be detected in a small sample volume.

Evacuated cans similar to those that hold shaving creme are available commercially. By pushing in the top of the can, the seal is broken, and an inlet allows air to enter. Vacu-Sampler (Fig. 17-5) from MDA Scientific, Inc. is an example. These ridged, aerosol-type cans are completely evacuated and then backfilled with nitrogen to a partial vacuum of 20 inches of mercury in the factory. When the sample is collected at standard temperature and pressure, each can will collect 123.3 cubic centimeters of vapor or gas. Analysis is generally done by gas chromatography (GC). During analysis the nitrogen background is elimi-

Figure 17-5. The Vacu Sampler, a bulk air sampler. (Courtesy MDA Scientific, Inc.)

nated using a dilution correction factor that takes into account temperature during sampling as well as analysis. In order to collect general area samples the following procedures are used.

Procedure

1. Break the tape over the top cover and remove the safety cover, thus exposing the actuator button. Do not use Vacu-Samplers if the tape has been broken.

2. Hold the sampler firmly, with the index finger over the actuator button. Position the sampler close to where the sample is to be collected. Since the sample will be qualitative rather than strictly quantitative, the closer to the source the better. Caution: Heated sources may cause evacuated containers to explode.

3. Depress the actuator button firmly and hold it down for 10 seconds, allowing the atmosphere to enter the can.

4. Immediately record the sample data, including the ambient dry bulb temperature on the label of the Vacu-Sampler.

5. Replace the safety cover.

BULK SAMPLES OF SOLID OR LIQUID CHEMICALS

When choosing to sample bulk chemicals there can be a number of different goals, but most commonly the goals are to provide supplementary information for air samples, to identify the constituents of a mixture, to identify a suspected contaminant, or to characterize a hazardous waste. This is generally done when all other means of obtaining information regarding the chemical composition of a process material have been exhausted, for example, the material safety data sheet (MSDS) isn't specific enough, the manufacturer has claimed a trade secret exemption, or a source of contamination is suspected. Mixtures in which the constituents are unknown are often wastes and can be found in any type of container as well as in soil and water. The composition of the material as well as its container will determine the method of collection. The most common chemical containers sampled are drums and tanks for liquids and sacks, bags, and piles for dusts. In each case, a different type of sampler is required. The collection techniques are discussed further in each of the specific sections for sampling these containers.

In situations where air sampling is being conducted for certain compounds used in specialized processes, laboratories may request a sample to use as a standard or to be certain it is of the same composition as their standard. Bulk samples are used to confirm the presence of free silica in respirable air samples and to assess the presence of other substances that may interfere in their analysis. The following are examples of chemicals for which bulk samples should always be collected along with air samples:

Chlorinated camphene
Chlorinated diphenyl oxide
Chlorodiphenyl
Gasoline
Hydrogenated terphenyls
Kerosene
Mineral oils
Naphtha
PCBs
Petroleum distillates
Stoddard solvent
Turpentine

Never include bulk samples in the same shipping container as air samples or they may contaminate one another.

In cases where the mixture is identified, such as kerosene, a contaminant of concern may be benzene, because the kerosene comes from a bulk supplier who may use the same bulk tank for storage of several grades of petroleum products without cleaning in between. Perhaps waste oil is being burned in a heater. Many contaminants could be present in this case, including polychlorinated biphenyls (PCBs), arsenic, lead, antimony, and other heavy metals that would result in PCB decomposition products or in the metals being released to the workroom air upon burning of this oil. Other situations where PCB contamination may exist include capacitors used in the ballasts of older fluorescent lamps, hydraulic systems on machines that handled hot metals, and hydraulic elevators. Although PCBs are present in lamps in small amounts, they can leak or release hazardous decomposition products in a fire.

Fugitive and Other Dusts

Fugitive dusts are not containerized. They may be on the rafters or other surfaces in a manufacturing operation or contained in a building material whose composition is suspected of having a hazardous component. Settled dust samples collected from rafters or near a worker's job site are considered representative of the airborne dust to which workers are being exposed. Usually rafter samples are used to determine the average size distribution and composition of settled airborne dusts.[6] Common contaminants found in dusts include lead in processes where it is heated, such as battery manufacturing and radiator repair, because of its tendency to settle over a widespread area, and tremolite, a type of asbestos often present in talc, because it tends to form in conjunction with talc deposits. A process material might also be sampled to determine the composition of the material before it is airborne. Sampling dusts and other dry materials in containers are discussed in the next section. Samples of suspect asbestos-containing materials utilize specialized techniques that are discussed in the chapter on Specific Sampling Situations.

When used in conjunction with air samples, the method of collection as well as the source of dust samples must be selected carefully in order to be representative of the situation in which the material is collected. If a bulk sample of the raw material used in the manufacturing process is sampled, it may not represent the particle size ranges likely to be present following processing, thus allowing a better estimation of the respirable hazard. The next choice would be to fill a bottle with settled dust from a rafter near the operation of interest. This would be representative of the changes in particle size due to the processing of the raw material. An even better choice might be to attach a cyclone containing a filter cassette to a personal air sampling pump and "vacuum" up the dust from the rafter or near the process. This would collect the particles that were 10 μm and less on the filter. If the material collected in the bottom of the cyclone is weighed, as well as that on the filter, an estimation of the percent of respirable dust present can be made. For more information on the use of cyclone collectors, see the chapter on Integrated Sampling for Aerosols.

Another concern is exterior paint on tanks, piping, and other metal surfaces that may be sandblasted or cut using oxyacetylene torches. Most paint used for these applications contains varying amounts of lead that will become airborne during the aformentioned procedures. A section of the surface should be scraped down to bare metal. If any detectable levels of lead are present in the paint, precautions should be set up for workers. At a minimum, air sampling should be done during the work to determine what levels of lead are present.

Bags, Sacks, Fiberdrums, and Waste Piles Containing Dry Materials

Laboratory scoops or triers, also called sampling probes or tubes, can be used to sample dry bulk material in containers. A typical sampling trier (Fig. 17-6) is a long tube usually made of stainless steel cut in half lengthwise with a sharpened tip. In some models an ejector is incorporated into the sampler for easier removal of the sample. The ejector snaps into the probe's slot, sliding up

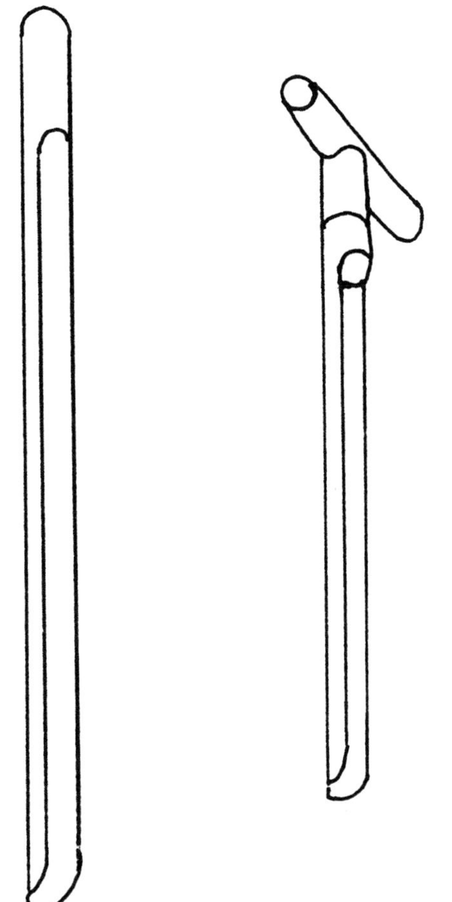

Figure 17-6. Sampling triers. (From EPA. *Characterization of Hazardous Waste Sites—A Methods Manual: Vol. II. Available Sampling Methods.* EPA-600/4-83-040, September, 1983.)

the probe as it is inserted into the material. When the trier is pulled up, the ejector slides down the probe and the sample is pushed out. Triers are preferred when the granular material to be sampled is moist or sticky and are also used to sample surface soil. Scoops are preferred when the material to be sampled is dry, granular, or powdered. The scoop or trier can be used to collect samples of solids at depths greater than 3 inches. The sampling depth is determined by the hardness and types of solids being collected.[2] Dry powdered or granular waste in containers such as barrels, fiberdrums, sacks, or bags. Since these wastes tend to generate airborne particles when the containers are disturbed, containers must be opened slowly.

Trier samplers are generally used to sample waste piles. A waste pile can range from a small heap to a large aggregate of wastes. The wastes are predominantly solid and can be mixtures of powders, granules, and large chunks. Hazardous materials stored in waste piles are usually of a small granular size, such as sand and dust. Waste piles should be approached from upwind due to the potential for dust emissions. Sampling should be designed to disturb the pile as little as possible, minimizing the amount of dust released.

In waste piles, the accessibility of the waste for sampling is usually a function of pile size. Ideally, piles containing unknown wastes should be sampled using a three-dimensional simple random sampling strategy. This strategy can be employed only if all points within the pile can be accessed. In such cases, the pile should be divided into a three-dimensional grid system, the grid sections are assigned numbers, and the sampling points are then chosen using random-number tables. If sampling is limited to certain portions of the pile, the collected sample will be representative only of those portions, unless the waste is known to be homogeneous.[2]

When using a trier to sample waste piles, the sampler must be decontaminated in between piles or disposed of after each set of samples. A number of core samples, depending on the size of the waste pile, must be taken at different angles and composited in order to obtain a sample that upon analysis[9] will give average values for the hazardous components in the waste pile.

Procedure

1. Barrels, fiberdrums, and cans must be positioned upright for sampling. If possible, sample sacks or bags in the position in which they are found, since standing them upright might rupture them. For waste piles, determine the number and locations of sampling points.

2. Collect a composite sample from the container with a trier or stainless steel scoop. Insert the trier into the material at a 0° to 45° angle from the horizontal. This orientation minimizes the spillage of sample from the trier. Rotate the trier two or three times to cut a core of material.

3. Slowly withdraw the trier, making sure the slot is facing upward. It might require tilting of the container. Withdraw samples through the top of barrels, fiberdrums, buckets, and similar receptacles. Withdraw samples through fill openings of bags and sacks. Withdraw samples through the center of the receptacles and to different points diagonally opposite the point of entry. Transfer the sample into a suitable container with the aid of a spatula and/or brush.

4. If composite sampling is desired, repeat the sampling at different points two or more times, and combine the samples in the same container with the sample from step 3 and cap it. For waste piles, sample through at least three different points near the top of the pile to points diagonally opposite the point of entry, and composite these in the same sample container.

5. Wipe the sampler clean and store it in a plastic bag for subsequent cleaning.

6. Where there is more than one container of wastes on a site, segregate the containers by waste type and sample according to a table of random numbers. For more information on random numbers, see the chapter on Survey Preparation and Performance.

Drum Sampling

The 55-gallon drum has become synonymous with hazardous waste disposal. Since it is the most frequently used container for storage and disposal, it is also one of the most common fixtures sampled and analyzed.[7] Because of the drum's frequent appearance at hazardous waste dump sites, and the number of problems associated with sampling it, special precautions and techniques must be followed. Structurally sound drums present the least amount of risk of rupture during mechanical handling. A responsible, experienced member of the sampling party should determine whether the drums can be opened or punctured and sampled in place. This decision will be based on the extent of cleanup in case of rupture, the danger involved in a rupture, and other factors such as the size of the site and drum spacing. Drum sampling can be one of the most hazardous activities to worker safety and health because it often involves direct contact with unidentified wastes and concentrated materials.

In some situations the relative composition of a waste may be known yet it still requires a sample. For example, drums of mineral oil waste generated by a process such as metalworking very likely contain dissolved amounts of the metals being tooled. In the case of carbide tools, this might include cobalt and tungsten. Therefore, drums used to store waste material will most likely require sampling before a determination of the proper category for this waste can be achieved. Printing processes often end up with waste mixtures of solvents and inks. A representative sample of the waste has to be collected to determine the ranges of the components in order to submit it to a disposal facility.

The following situations require extraordinary procedures and are beyond the scope of this book. Contact specialists whenever these are present: drums that may contain shock-sensitive wastes, drums containing radioactive wastes, drums containing laboratory packs (laboratory wastes), and buried drums. Identifiers of high-risk drums include drums with bulged heads most likely due to internal gas formation, drums with bulges on the side or bottom most likely due to freezing and expansion of the contents, and drums that have been deformed due to mishandling.

All tools used for drum opening should be of the nonsparking variety. Often these tools are made of brass. For drums that may contain unknown hazardous wastes, remote-controlled devices such as pneumatically operated impact wrenches, hydraulically or pneumatically operated drum piercers, or a backhoe equipped with a bronze spike to penetrate the drum top are preferable. Do not use picks, chisels, and firearms to open drums. Reseal openings as soon as possible to minimize vapor release. Decontaminate equipment after each use to avoid mixing incompatible wastes and contaminating future samples.

Drum sampling is generally done using a containerized liquid waste sampler (COLIWASA) (Fig. 17-7) readily available from scientific and hazardous waste supply houses. It is especially useful for sampling wastes that consist of several

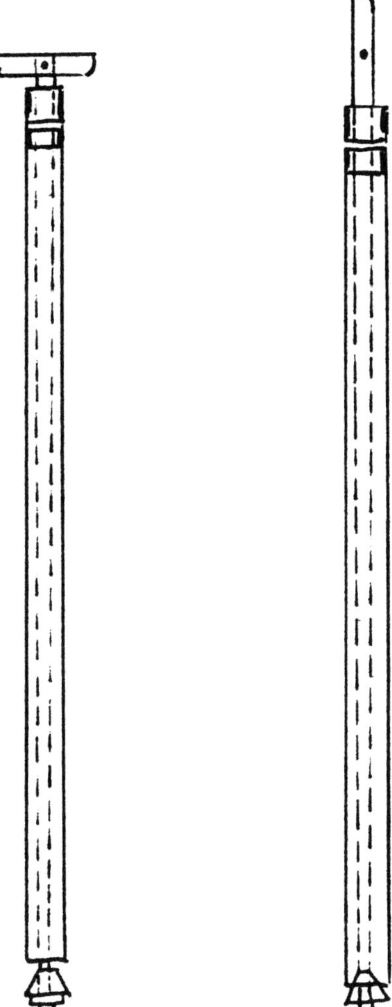

Figure 17-7. COLIWASA. (From EPA. *Characterization of Hazardous Waste Sites—A Methods Manual: Vol. II. Available Sampling Methods*. EPA-600/4-83-040, September 1983.)

immiscible liquid phases. It is made of glass or plastic and can be either disposable or reusable. One reusable model is made of borosilicate glass and has a sample capacity of 200 ml. A general-purpose disposable model has a slightly tapered bottom that accepts a glass inner tube with a blown ball at the end. The ball can be transferred to a sample bottle with minimal loss. Tubes come prescored to permit components to be snapped in half for easy disposal. The glass COLIWASA is used to sample hydrocarbons such as solvents and all other containerized liquid wastes that cannot be sampled with the plastic COLIWASA except for strong alkali and hydrofluoric acid solutions. PVC COLIWASAS are best for acids, bases, or other water-based substances. The COLIWASA can also be used to sample free-flowing liquids and slurries contained in shallow tanks, pits, and similar containers.

Make sure COLIWASAS are clean between successive samples or use a different tube for each sample. To clean, use a long-handled brush, rags, and a solvent and clean the tube inside and outside. Proper PPE must be worn during decontamination of the sampling device. Using a separate brush and rags, wash next with soap and water and then rinse with clean water. Before using the tube to sample again, inspect it for signs of deterioration. All wash materials must be decontaminated or disposed of properly. Any solids should be removed and the sampler drained before using it for additional sampling. Check to be sure that the sampler is functioning properly.

Procedure

1. Develop a sampling plan. Research background information about the wastes. Determine which drums should be samples. Select the appropriate sampling devices and containers. Develop a sampling plan that includes the number, volume, and locations where samples are to be taken. When there is more than one drum of wastes to be sampled at a site, segregate the drums according to waste types and use a table of random numbers to determine which drums to sample. For sampling of known wastes a composite sample should be collected by pulling samples from at least 10 percent of the containers. Drums should be selected at random, taking care not to sample only outside drums.[8] For more information on random numbers, see the chapter on Survey Preparations and Performance.

2. Develop standard operating procedures for opening drums, sampling, and sample packaging and transportation Based on available information about the wastes and site conditions, have a trained health and safety professional determine the appropriate personal protection to be used during sampling, decontamination, and packaging of the sample.

3. Choose a drum whose bung is up. Drums with bungs on the ends would be positioned upright. Drums with bungs on the side should be lying on the side with the bungs upright. Use a manual nonsparking hand wrench and slowly loosen the large bung to allow any gas pressure to escape. If the bung cannot be removed, the drum should be punctured.

4. Select a clean plastic or glass COLIWASA for the liquid waste to be sampled, and assemble the sampler. Check the sampler to make sure it is functioning properly. Adjust the locking mechanism if necessary to make sure the neoprene rubber stopper provides a tight closure. Put the sampler in the open position by placing the stopper rod handle in the "T" position and pushing the rod down until the handle sits against the sampler's locking block. While wearing appropriate PPE, slowly lower the sampler into the liquid waste at a rate that permits the levels of the liquid inside and outside the sampler tube to be about the same. If the level of the liquid in the sampler tube is lower than outside the sampler, the sampling rate is too fast and will result in a nonrepresentative sample.

5. When the sampler stopper hits the bottom of the waste container, push the sampler tube downward against the stopper to close the sampler. It should be noted that this sampler will not sample the bottom 1 to 2 inches of the drummed material, nor will it sample solids. Lock the sampler in the close position by turning the "T" handle until it is upright and one end rests tightly on the locking block. Slowly withdraw the sampler from the waste container with one hand while wiping the sampler tube with a disposable cloth or rag with the other hand. Dispose of the cloth into an appropriate container.

6. Secure the container to keep it from tipping over before transferring the sample. The container to receive the sample should have a mouth wide enough for the COLIWASA to fit into it and large enough to hold the volume contained in the COLIWASA. Carefully discharge the sample into a labeled sample container by slowly pulling the lower end of the "T" handle away from the locking block while the lower end of the sampler is positioned in a sample container. Continue to add other samples in the same way if this is to be a composite sample. Mix the contents of the random samples together after each addition. Cap the container in between samples. Most laboratories require only a pint of liquid to conduct analytical tests. Return any excess material to the drums.

7. After the last sample has been added to the bottle, making sure it is securely closed, invert the sample bottle a few times to check for leaks. Regardless of whether visible leaks are detectable or not, wipe the bottles with rags to remove any wastes on their outside.

8. Unscrew the "T" handle of the COLIWASA and disengage the locking block. Clean the sampler on site or store the contaminated parts of the sampler in a plastic storage tube for subsequent cleaning.

9. Mark the drum with paint or other indelible marking compound so that it can be identified later. Put a plastic drum cover over the drum, or reinstall the bung, to prevent any liquid such as rain from entering. Do not use the cover as the drum marker as it may be blown off by the wind.

Procedure Using Glass Tubing

The simplest method for sampling drums is to use a length of glass tubing to collect samples.[9] The tubing is normally 122 cm long and ranges in size from 6 mm to 16 mm inside diameter with larger tubes used for more viscous fluids. The tubing is broken and discarded in the container after the sample has been collected. This is a quick and relatively inexpensive means of collecting concentrated samples of containerized materials. Nonviscous materials are more difficult to retain in the tubing. The sampling professional should be protected against splashes, wearing at a minimum neoprene gloves, butyl rubber apron, and a faceshield.

1. Remove cover from container to be sampled.

2. Insert glass tubing almost to the bottom of the container, keeping at least 30 cm of tubing above the top of the container.

3. Allow the liquid in the drum to reach its natural level in the tube. Hold a thumb (must

have gloves on) over the exposed end of the tube or put a rubber stopper in it.

4. Carefully remove the tube from the drum and insert the uncapped end into the sample container. Remove thumb or stopper to allow the sample to flow into the bottle. Cap the sample container and make sure it is labeled with a unique sample number.

5. Repeat sample collection until sufficient volume is collected. Discard the glass tube in the drum, breaking it in such a way that all of it is inside the drum.

6. Replace the cover on the container.

Tank Sampling

For tanks, procedures similar to drum sampling are followed. Sampling above-ground storage tanks can require great dexterity. Usually it requires climbing to the top of the tank through a narrow vertical or spiral stairway while wearing protective equipment and carrying sampling paraphernalia. At least two persons must always perform the sampling: one to collect the actual samples; the other to stand back, usually at the head of the stairway, and observe, ready to assist or call for help in case of problems.

When opening a tank hatch, make sure that excess pressure from stored volatiles has been vented. Often underground storage tanks have vent pipes. If there is any doubt, call in experts prior to sampling. Guard manholes to prevent personnel from falling into the tank.

A bomb sampler is a composite sampling device for storage tanks. Bomb samplers are available in stainless steel and consist of a cylindrical reservoir chamber, a weighted plunger that seals the chamber at the bottom, and a cable attachment for suspending the apparatus and activating the sampling device. These samplers typically hold 500 ml of sample.

Procedure

1. Collect one sample each from the upper, middle, and lower sections of the tank contents with a weighted bottle sampler.

2. Combine the samples in one container and submit the container as a composite sample for analysis.[7]

SOIL SAMPLING

Soil samples can provide useful information about the exposures to personnel who might have skin contact with the soil, or about the potential for dust clouds to occur (fugitive emissions) due to activities such as grading or excavating or the wind. For volatile materials trapped in soil, airborne vapors will be a concern as well. Generally these tend to increase during excavations.

In general, soil samples are taken in a grid pattern over the entire site to ensure a uniform coverage of the site. As noted in the section on grid sampling, there are elaborate statistically designed patterns for sampling soils. Soil sampling increments are often 6 inches for the upper 2 feet of soil and every 12 inches below a depth of 2 feet.[3] The depth will depend on the degree of contamination and it is usually the depth at which the cleanup criterion (concentration) is met.

Background samples also must be collected to allow for statistical comparison of the natural condition to the potentially contaminated area. They can be used to differentiate between contaminants and materials naturally occurring in the soil, such as heavy metals. A few contaminants such as formaldehyde and phenol may be naturally produced, but their concentrations in soils are typically very low and near or below detection limits. Background samples should be taken in areas not affected by the hazardous waste units or by the site itself. They may be taken away from the site, but the sampling location should be as close as possible to the site. Background samples should be taken from soil depths and soil horizon materials similar to those of the potentially contaminated area. The location and depth of background samples must be indicated. As many as 10 background samples from each soil strata may be needed.[3]

Soil sampling intervals and total depth may be dependent on several factors, including soil type and hydraulic conductivity; suspected magnitude of surface contamination; physical state of the waste and its mobility; height of liquid head at the ground surface; length of time that the waste was present at the site; relative toxicity of the waste; and location of the waste management unit (indoors or outdoors).[1] Many of these decisions require the expertise of geologists and engineers. Soil types vary considerably; therefore, it is important to maintain a detailed record during sampling, especially listing locations and depth sampled, and characteristics such as soil grain size, color, and the presence of an odor.

Soil samples for underground storage tank (UST) closures should be taken beneath the tank invert and as close to the tank as possible. Excavations should only be entered after the sides have been properly sloped, shored, or braced. Near-surface soil samples should also be taken in loading/unloading areas adjacent to all storage tanks to determine whether soil contamination from spills has occurred.[3]

Figure 17-8. Scoop. (From EPA. *Characterization of Hazardous Waste Sites—A Methods Manual: Vol. II. Available Sampling Methods.* EPA-600/4-83-040, September 1983.)

Surface Soil Samples

When samples of surface soil, sand, or sediment are to be collected, scrape samples from the surface can be collected. Using a 10-cm by 10-cm template to mark the area to be sampled, the surface should be scraped to a depth of 1 cm with a stainless steel trowel or similar implement. The yield should be at least 100 g soil. If more sample is required, expand the area but do not sample deeper. Use a disposable template or thoroughly clean the template between samples to prevent contamination of subsequent samples. The sample should be scraped directly into a precleaned glass bottle. If it is free-flowing, the sample should be thoroughly homogenized by tumbling. If not, successive subdivision in a stainless steel bowl should be used to create a representative subsample.

In some cases, such as sod, scrape samples may not be appropriate. For these cases, core samples not more than 5 cm deep should be taken using a soil coring device such as the corers discussed in the section on Sludge Sampling. These core samples should be well homogenized in a stainless steel bowl by successive subdivision. A portion of each sample should then be removed, weighed, and analyzed.

The simplest method of collecting surface soil samples for analysis is with the use of a trowel or scoop. The laboratory scoop has a curved blade and a closed upper end to permit the containment of material. Scoops (Fig. 17-8) come in different sizes and makes and the size will depend on the amount of material that needs to be collected. The stainless steel laboratory scoop is the preferable choice for sampling. Identical sample amounts for a composite sample are difficult to collect with this sampler. If undisturbed sections of soil are needed, a flat, pointed trowel can be used to cut a block of the desired soil rather than the scoop.[9]

Procedure

1. Divide the area to be sampled into an imaginary grid on the site map with number codes in two perpendicular directions. Then mark the grid on the ground with numbered stakes and flags.

2. If it is not known whether contamination is present, use a calculator or random-number table to choose random numbers; two numbers are required to choose a sampling site, and these can be used to number the sample.

3. If discolored areas of soil exist, which strongly suggest chemical contamination, plan to sample each area as well as the grid.

4. Do not kneel, squat, or otherwise touch the ground, because contaminants could be transferred from boots to clothing. Use a ground cover such as plastic for kneeling or placement of equipment. Do not lean or sit on trees or equipment.

5. Carefully remove the top layer of soil to the desired sample depth with a spade.

6. To sample up to 8 cm deep, collect samples with a scoop or spoon. Dig down to the desired depth; then use a spatula to take a "channel" sample composited over the vertical interval of interest.

7. Sample each section of the grid, and combine appropriate samples into composite samples. Do not combine if sampling in highly contaminated areas.

8. Generally a stainless steel sampler is used to sample volatile organic compounds (VOCs) in soil. If sampling for VOCs, do not composite, mix, or aerate the soil because volatilization may result, causing loss of sample. Instead quickly fill a 40-ml vial with soil, wipe its rim, screw the septum cap tightly, and keep the vial on ice. Fill the sample container completely to eliminate any headspace. Cover both ends with aluminum foil, and put plastic caps over the foil. Put the sample into the cooler immediately.

9. Decontaminate the auger and bucket before moving on to the next sample site. Following collection, transport the samples to the laboratory as soon as possible, generally within 24 hours of sampling.

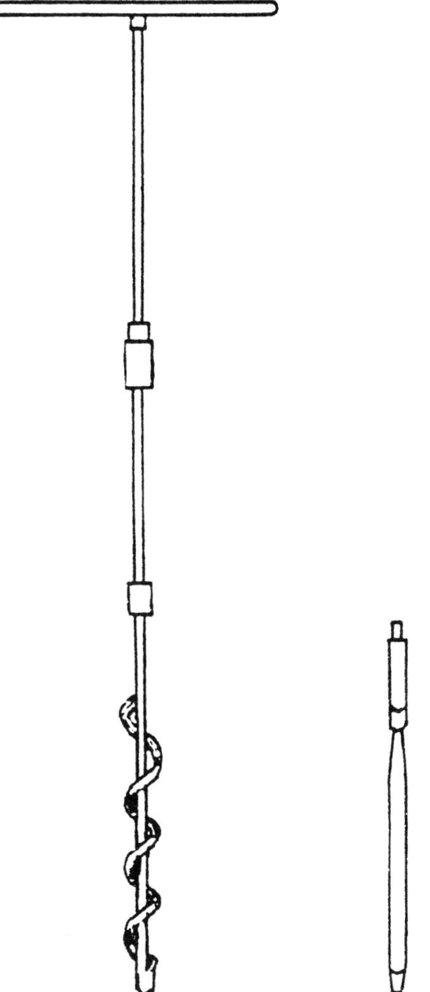

Figure 17-9. A screw auger. (From EPA. *Characterization of Hazardous Waste Sites—A Methods Manual: Vol. II. Available Sampling Methods.* EPA-600/4-83-040, September 1983.)

Subsurface Soil Sampling with Augers and Thin-Tube Samplers

The auger boring is the most common method of soil investigation and sampling. The soil auger can be used for both boring the hole and for bringing up samples of the soil.

The auger (Fig. 17-9) consists of an auger bit at the end of a cylinder, a series of drill rods, and a "T" handle. A threaded coupling at the top of the auger allows different sizes of extension tubing to be attached. Bits are generally made of high-carbon alloy steel and are available in a variety of configurations for different soil types such as sandy and clayey. Screw augers are used for collecting smaller soil samples. Augers are available in different size diameters. In order to push the auger through the soil, techniques that are easier than sheer physical strength are available. The simplest modification is using a mallet with a rubber or urethane head to pound the auger into the ground. Other modifications include foot-operated "jacks" on the extension tubing that take the stress off the arms, back, and shoulders.

This system can be used in a variety of soil conditions. It can be used to sample both from the surface or to a depth of 6 meters. The presence of rock layers and the collapse of the borehole usually prohibit sampling at depths in excess of 2 meters. Since this sampler destroys the structure of cohesive soil and does not distinguish between samples collected near the surface or toward the bottom, it is not recommended when an undisturbed soil sample is desired.[9]

Procedure

1. Wearing appropriate PPE, select the necessary sampling points and remove unnecessary rocks, twigs, and other nonsoil materials.

2. Attach the auger bit to a drill rod extension and further attach the "T" handle to the drill rod.

3. Bore a hole through the middle of an aluminum pie pan large enough to allow the blades of the auger to pass through. The pan will be used to catch the sample brought to the surface by the auger.

4. Spot the pan against the sampling point. When using the auger, twist it down to the requisite depth, collecting all auger scrapings in a bucket for a composite. Note that augers tend to cross contaminate between depths.

5. The auger bit is used to bore through the hole in the middle of the aluminum pie pan to the desired sampling depth and then it is withdrawn. Transfer the sample collected in the pie pan and the sample adhering to the auger to a labeled container. Spoon out the rest of the loosened sample to the same container.

6. The auger tip is then replaced with thin-wall tube sampler and the proper cutting tip is installed.

7. Carefully lower the tube into the borehole and gradually force the tube into the soil. Care should be taken to avoid scraping the borehole sides. Hammering of the drill rods to facilitate coring should be avoided as the vibrations may cause the boring walls to collapse.

8. The tube is then withdrawn and the drill rods are unscrewed. The cutting tip and tube are then removed from the device.

9. Discard the top of the tube sample, which represents any material collected by the tube before penetration of the layer in question. Place the remaining sample into a labeled container and cap the container.

10. Brush off and wipe the sampler clean or store it in a plastic bag for subsequent cleaning. Using a stainless steel scoop, collect the desired quantity of soil, transfer the sample to a labeled container, and cap the container.

Sampling of Sludges and Sediments

Sludges are defined as semidry materials ranging from dewatered solids to high-viscosity liquids. Frequently sludges form as a result of settling of the higher-density components of a liquid. In this instance, the sludge may still have a liquid layer above it. The primary characteristic of a sludge is that the material is completely saturated with liquid. Sediments are the deposited material underlying a body of water.[2]

Corers (Fig. 17-10) are used for collecting samples of most sludges and sediments. They collect essentially undisturbed samples that represent the profile of strata that may develop in sediments and sludges during variations in the deposition process. Depending on the density of the substrate and the weight of the corer, penetration to depths of 30 inches can be attained. Improvements on the basic core sampler are available. One model features liners so that many samples can be collected with the same corer. The ends of the liner are capped following collection and a new liner is inserted. Liners are available made of plastic, stainless steel, aluminum, and Teflon. A butterfly valve has been incorporated into one model. When the core sampler is pushed into the sludge, the valve opens, allowing the material to enter the liner, and upon removal the valve closes to prevent loss of sample. Another model fea-

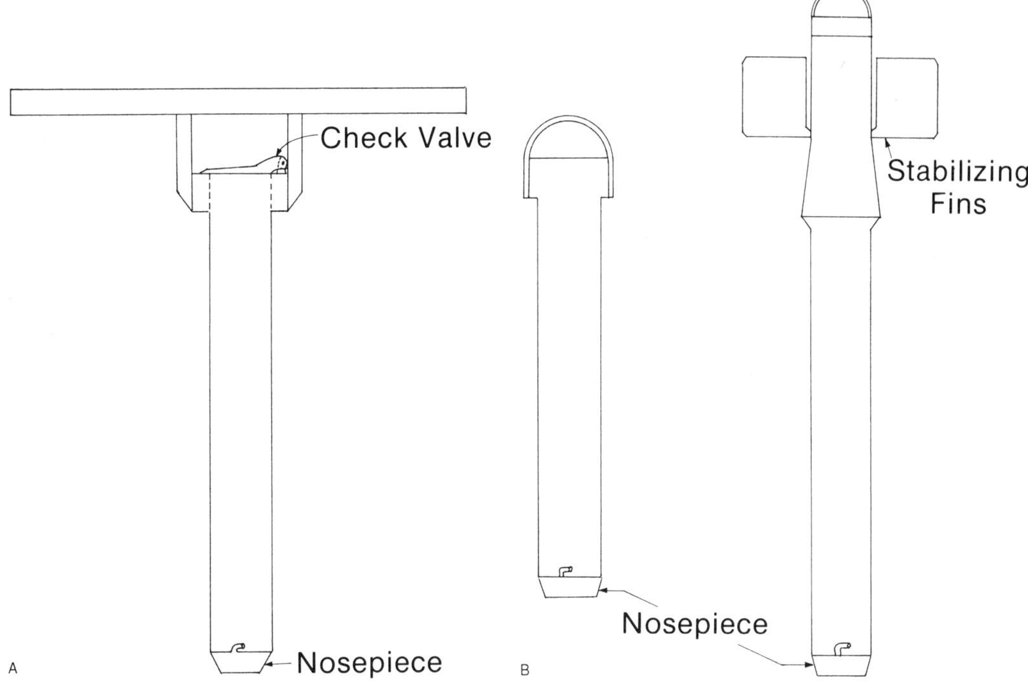

Figure 17-10. Corers. A: Hand corer. B: Gravity corers. (After EPA. *Characterization of Hazardous Waste Sites—A Methods Manual: Vol. II. Available Sampling Methods.* EPA-600/4-83-040, September 1983.)

tures an auger bit for easier penetration of denser sludges. Some hand corers can be fitted with extension handles that will allow the collection of samples underlying a shallow layer of liquid.

A gravity corer is a metal tube with a replaceable tapered nosepiece on the bottom and a ball or other type of check valve on the top. The check valve allows water to pass through the corer upon descent but prevents washout during recovery. The tapered nosepiece facilitates cutting and reduces core disturbance during penetration.

Procedure[7]

1. Inspect the corer for proper cleanliness.

2. Attach a precleaned corer to a length of sample line. Solid-braided 3⁄16-inch nylon line is sufficient, although ¾-inch nylon line is easier to grasp during hand hoisting. Secure the free end of the line to a fixed support to prevent accidental loss of the corer. Allow the corer to free fall through the liquid to the bottom.

3. Force in the corer with a smooth continuous motion. Twist the corer and then withdraw it in a single smooth motion. An alternative is to use the gravity corer.

4. Retrieve the corer with a smooth, continuous lifting motion. Do not bump the corer since sample loss may result. Remove the liner from the corer, cap the ends, label, and wrap in foil. If a liner is not being used, transfer the sample from the corer into a labeled sample bottle and cap tightly. Refrigerate the sample in an ice chest until it reaches the laboratory.

WATER SAMPLING

Water is commonly sampled as a result of environmental audits for real estate transfer, to identify whether contaminants are being transferred to decontamination water, to identify potential contaminants in indoor air investigations, and for the OSHA requirement that potable water be used by workers for washing and cleaning utensils. This may involve sampling drinking water from a tap, chiller, or drinking fountain or private well. Sumps in basements and for decontamination systems are also frequently sampled.[10] Some situations such as stagnant water in a basement or HVAC ducts and requirements for tap water may involve sampling for micoorganisms. This type of sampling is discussed further in the chapter on Sampling for Bioaerosols. Likewise, radon in water is discussed in the chapter on Radon Measurements.

Lead is the most common metal of interest in drinking water samples, and procedures for collection of these samples are well established.[11] Other metals can be more difficult to sample and often require preservatives. Therefore, prior to sampling for metals other than lead, the specific method should be reviewed for that element.[12]

VOCs are also frequently sampled for in water. The most common VOCs found in water are chlorinated and aromatic compounds. These might be encountered in situations associated with leaking underground storage tanks. High levels of trihalomethanes are often the result of organic compounds reacting with chlorine in drinking water systems. Vinyl chloride and methane have been found in water contaminated by nearby sanitary landfills. The biggest problem associated with sampling for VOCs is their volatility, resulting in loss of sample due to evaporation during collection. Sampling procedures for these compounds are designed to minimize agitation and contact of the sample with air. Samples must be kept cool (4°C) to avoid degradation of organics.

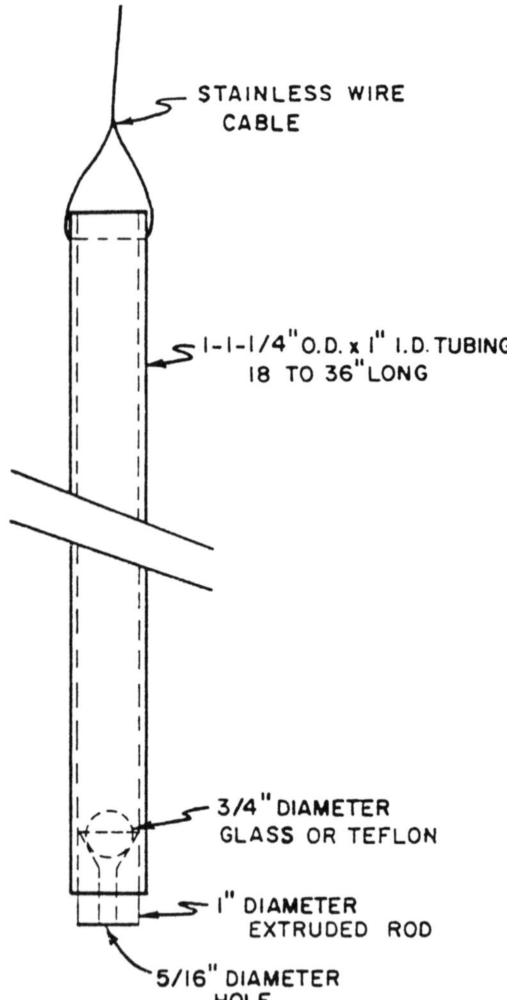

Figure 17-11. Check valve bailer. (From EPA. *Characterization of Hazardous Waste Sites—A Methods Manual: Vol. II. Available Sampling Methods*. EPA-600/4-83-040, September 1983.)

Bailers are one of the most common samplers for water and can be constructed from a wide variety of materials in various designs. The conventional bailer consists of a weighted bottle or basally capped length of pipe attached to a cable or cord that fills from the top as it is lowered into the well or sump to retrieve a sample. Check-valve bailers (Fig. 17-11) have a valve located at the base that allows them to fill from the bottom. When the bailer is pulled, the valve closes, retaining the sample as the bailer is raised to the surface. Bailers can be constructed of plastics, Teflon, or borosilicate glass; the need dictates the material, although plastic is the most common. Glass bailers resist chemicals and scratching better than most plastics and therefore are useful for collecting samples for analysis for organics. Teflon bailers are often used for groundwater monitoring. Since plastic bailers are relatively inexpensive, they can be disposed of after sampling a source. Bailers allow samples to be recovered with a minimum of aeration if care is taken when lowering the bailer. The primary disadvantages of bailers are limited sample volume and the inability to collect discrete samples for various depths.[9]

The following is a general procedure for collecting samples of water from standing bodies of water, sumps, and private wells using a bailer.[9] It is not designed for groundwater sampling situations, since these require more sophisticated techniques[13] as well as the involvement of geologists, hydrogeologists, and engineers. There are also specific EPA methods for collection and analyses of water from standing sources for chemicals that may prove useful to review.[12,14]

General Procedure

1. Gather all necessary equipment, making sure it has been decontaminated from prior uses.

2. Record observations including type of water being sampled, location, incongruities about the surface such as scum, and presence or lack of an odor.

3. Evaluate the area around the sampling point for possible contamination from external sources such as a loosely sealed gasoline can or solvent drum, automobile, or exhaust from nearby processes. Make sure the rinse water for the samplers is not near any of these sources.

4. Place a plastic tarp on the ground next to the sampling site to prevent equipment from touching the ground. Gloves and other PPE should be donned prior to collecting samples.

5. Prior to sampling each point, label sample bottle with unique sampling number and record it on the sampling data sheet.

6. Attach a bailer to a cable or line. Gradually lower the bailer until it contacts the source. Allow the bailer to sink and fill with a minimum of surface disturbance.

7. Slowly raise the bailer. Do not allow the bailer line to contact the ground.

8. Tip the bailer to allow slow discharge from the top to flow gently down the inside of the sample bottle with a minimum of entry turbulence. Repeat collection until a sufficient volume is obtained.

9. Filter samples if the procedure requires it. Wipe the exterior of the container, label it with a unique sample number, and record this on the sampling data sheet.

10. Put labeled sample into a plastic bag and seal. Put bagged sample into clean metal can with top. Fill can with vermiculite and attach lid. Put container into cooler. Change outer gloves to avoid contaminating the next sample or other equipment during the next phase of the work.

11. Generally field blanks are collected after the samples have been collected. Other quality control samples may be collected as well, such as duplicate samples or splitting a single sample for separate analyses by different laboratories.

12. Store samples and any blanks in a cooler with a thermometer to make sure the temperature is maintained at 4°C.

13. If possible, deliver to the laboratory on the same day that the samples were collected. If this is not possible, store samples in a refrigerator overnight and deliver the next day. Shipping water samples is undesirable due to the potential for delays, which can result in warming of the samples and rough treatment that can cause containers to leak.

Tap Water Samples

The most common purpose for sampling tap water is for lead detection. The primary sources of lead contamination are lead-containing pipes and plumbing system components and lead-containing solder and flux used in drinking water distribution lines and home plumbing. These sources can release lead into the water supply, especially in areas with particularly corrosive water and where lead solder is less than 5 years old. Some municipalities have water supply systems made of copper pipe with lead solder. Older cities, such as Chicago, specified the use of lead pipe in their building codes at certain periods. Generally, it was used in smaller services, such as the 2-inch pipe used in residences. Lead levels are usually highest in water that has the longest contact time with the plumbing. The highest levels are usually found in faucets that are seldom used, or from first-draw samples in the morning.

Through the early 1900s, it was a common practice in some areas of the country to use lead pipes for interior plumbing. Also, lead piping was often used for the service connections that join building piping to the public water system mains. Plumbing installed before 1930 is most likely to contain lead.

Since the 1930s, interior plumbing has usually been constructed from galvanized iron or copper pipe. Copper pipe was usually soldered with 50% tin and 50% lead solder until recently when many city building codes changed to require lead-free solder. Although relatively little solder is exposed to the interior of the pipe on a well-made joint, significant amounts of lead can leach into the water from lead solder joints. The degree of lead released from soldered joints depends on the pH of the water; the higher the pH, the more likely it is that lead will be released. Newly installed plumbing is affected most.

The configuration of the interior plumbing can vary depending on the layout of the building. Therefore, it is useful prior to sampling to study the blueprint(s) of the building's plumbing system. Locate service intakes, headers, laterals, fixture supply pipes, drinking water fountains, central chiller units, storage tanks, riser pipes, and different drinking water loops. In multistory buildings water is elevated to the floors by one or more riser pipes. Water from the riser pipes is usually distributed through several different drinking water loops and in some buildings may be stored in a tank prior to distribution. In single-story buildings water comes from the service connection via main plumbing branches, often called headers. These in turn supply water to laterals. Smaller plumbing connections from the laterals and loops supply water to the faucets, drinking fountains, and other outlets. Water within a plumbing system moves downstream from the source, for example, from the distribution main in the street. Figure 17-12 shows suggested sample sites in single-level and multistory buildings.[11]

There are different techniques for sampling tap water. One of the most common is a first-draw sample collected after overnight stagnation; others involve collecting random daytime samples and flushed samples. First-draw samples generally have the highest lead concentrations, but are the least convenient to collect. Samples collected after flushing a system result in the most consistent values, but the data reflect the *minimum* exposure of the water to lead. Random daytime samples are most representative of lead levels in the water being consumed, but are the most variable, so a greater number of samples are required. Sampling variables include the flow rate from the tap and the volume collected. If the flow rates vary from tap to tap, it may be difficult to compare samples from different taps. The volume must be consistent from sample to sample or changes in water quality may not be detected. Analysis of drinking water samples for lead should be done by a state-certified laboratory using EPA methods.[14]

Sample Collection for VOCs in Drinking Water

Procedure

1. Prior to sampling each point, label two vials with a unique sampling number and record it on the sampling data sheet. For duplicate samples an "A" and "B" designation added to the unique number can be used.

2. Turn on the tap and minimize air contact with sampled water while filling 2 vials to overflowing. Make sure the cap liner is in the correct position with the Teflon liner, which is usually white, on the outside; when the lid is screwed on, it should point toward the sample. Cap the vials immediately, screwing on the caps tightly.

3. Fill two vials from each sampling point.

4. There should be no air space at the top of the vials. To test this, invert each vial and tap it on a hard surface to determine if any air bubbles remain inside. If there are any, refill it and again inspect for air bubbles. Continue to work with the sample until no bubbles are present.

5. If samples are being collected for compliance, split samples are usually taken, meaning each sample is usually divided between the facility owner and the regulatory agency. In the case of VOCs, it is best to fill all vials from the same tap.

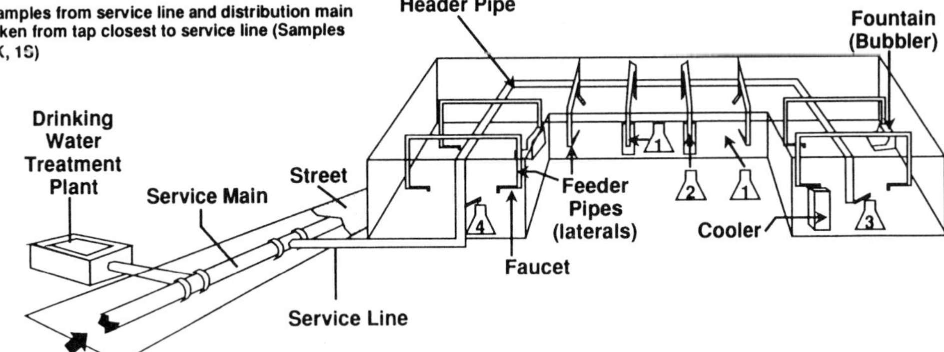

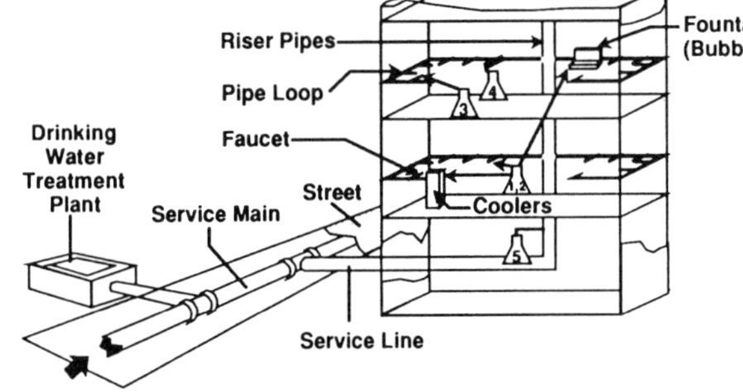

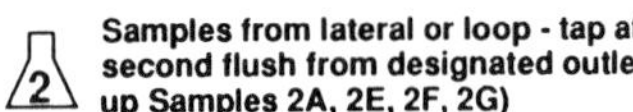

Figure 17-12. Suggested sampling sites for lead in water in a single-level building *(top)* and in a multistroy building *(bottom)*. (From EPA. *Lead in School's Drinking Water.* EPA-570/9-88-001, January 1989.)

Sample Collection of Lead in Drinking Water

Sampling sites most likely to show lead contamination include areas where the plumbing is used to ground electrical circuits; areas where corrosive water having low pH and alkalinity is distributed; areas of low flow and/or infrequent use, such as where water is in contact for a long time with sediments or plumbing containing lead; areas containing lead pipes or areas of recent construction and repair in which lead solder or materials containing lead were used; and water coolers identified by the EPA as having lead-lined storage tanks or other parts containing lead.

The procedure for collecting samples involves collecting screening samples of outlets that provide drinking water, and doing follow-up sampling of outlets that showed significant lead levels in the initial screening in order to identify the sources of lead in the drinking water.[11]

Procedure

1. Determine the source of the water and the type of piping used to carry the water both for the municipality and the building being sampled. Individual properties as well as townships can rely on wells for water. Pipelines may be polyvinyl chloride (PVC), copper with lead–tin solder, cast iron, asbestos concrete (A/C), and galvanized pipe.

2. The faucet chosen for sampling must be a supply without filters, softeners, or heaters. This sample consists of water that has been in contact with the fixture and the plumbing connecting the faucet to the lateral. It is representative of the water that may be consumed at the beginning of the day or after infrequent use. It should be collected before any water is used for the day.

3. Prior to collecting samples, lines must be flushed sufficiently to ensure that the sample is representative of water supply, taking into account the diameter and length of the pipe to be flushed and the velocity of flow. For volatile organics, let the water just dribble into the vial to avoid loss of volatiles. Do not use plastic containers for volatiles. For bacteria, be aware that contamination may be in any part of the system: the well or other source, pipes, filter medium, or faucet. These possibilities can impact the sampling strategy and frequently call for multiple samples. For more information on sampling for bacteria, see the chapter on Sampling for Bioaerosols.

4. Remove the aerator if there is one. Aerators are usually found on the kitchen and bathroom faucets. Collect the water immediately after opening the faucet without allowing any water to run into the sink. Fill the container with 250 ml of water.

5. Follow-up samples should be taken from those water faucets where test results indicate lead levels over 20 ppb. The follow-up sample is representative of the water that is in the plumbing upstream from the faucet. Take this sample before any water is used. Let the water from the faucet run for 30 seconds before collecting the sample. Collect 250 ml of water.

Interpreting the Results

1. If the lead level in the first sample is higher than in the second, the source of lead is the water faucet and/or the plumbing upstream from the faucet.

2. If the lead level in the second sample is very low–close to 5 ppb–very little lead is coming from the plumbing upstream from the faucet. The majority or all of the lead in the water is from the faucet and/or the plumbing connecting the faucet to the lateral.

3. If the lead level in the second sample significantly exceeds 5 ppb (e.g., 10 ppb), lead may be contributed from the plumbing upstream from the faucet.

Service Connections and Distribution Mains

The most common sample collected at these points is for lead analysis.[11]

Procedure

1. A first-morning sample should be collected.

2. Open the tap closest to the service connection.

3. Let the water run and feel the temperature of the water. As soon as the water changes from warm to cold, collect the sample. Because water usually warms slightly after standing in the interior plumbing, this colder water sample represents the water that had been standing just outside of the building and in contact with the service connector. Fill the sample container with 250 ml of water.

4. Allow the water to run for an additional 3 minutes and then collect a second sample. Fill this container with another 250 ml of water. This sample is representative of the water that has been standing in the distribution main.

Interpreting the Results

1. If the lead level of the service connection sample significantly exceeds 5 ppb (e.g., 10 ppb), and is higher than in the second sample of the distribution main, lead is being contributed from the service connector. Check for the presence of a lead service line. In the absence of a lead service connector, lead goosenecks or other appurtenances containing lead in line with the service connection may be the sources of contamination. The distribution main is rarely the source of contamination.

2. If the lead level of the service main sample significantly exceeds 5 ppb, lead in the water may be attributed to the source water, sediments in the main, or possibly lead joints used in the installation or repair of cast iron pipes. If the water supplied is from a well, a lead packer in the well may also contribute lead to the water.

3. If the lead level of both samples is low, close to 5 ppb, very little lead is picked up from the service line or the distribution main.

Interior Plumbing

Sampling interior plumbing should proceed systematically upstream from the initial follow-up sample sites. The goal is to isolate sections of the interior plumbing that contribute lead to the water by comparing the results of these samples with results of previous samples.

Procedure for Laterals[11]

Laterals are the plumbing branches between a fixture or group of fixtures, such as taps and water fountains.

1. Open the tap that has been designated as the sample site for the lateral pipe.

2. Let the water run for 30 seconds before collecting the sample. The purpose of flushing the water is to clear the plumbing between the sample site and the lateral pipe, which will assure collection of a representative sample. Fill the container with 250 ml of water.

Interpreting the Results

1. If the lead level exceeds 20 ppb, collect additional samples from the plumbing upstream (the service line, the riser pipe, the loop or header supplying water to the lateral). High lead levels may also be caused by recent repairs and additions using lead solders, or by sediments and debris in the pipe. Debris in the plumbing is most often found in areas of infrequent use, and a sample should be sent to the laboratory for analysis.

2. If the lead level of this sample is the same as the lead level in a sample taken downstream from the sample site, lead is contributed from the lateral or from interior plumbing upstream from the lateral. Possible sources of lead may be the loop, header, riser pipe, or service connection.

3. If the lead level is very low—close to 5 ppb—the portion of the lateral upstream from the sample site and the interior plumbing supplying water to the lateral are not contributing lead to the water.

4. If the lead level significantly exceeds 5 ppb, and is less than the lead level in a sample taken downstream from the sample site, a portion of the lead is contributed downstream from the sample site.

Procedure for Loops and/or Headers[11]

A loop is a closed circuit of a plumbing branch that supplies water from the riser to a fixture or a group of fixtures. A header is the main pipe in the internal plumbing system of a building. The header supplies water to lateral pipes.

1. Locate the sampling point farthest from the service connection or riser pipe on a floor.

2. Open the faucet and let it run for 30 seconds before collecting the sample. The purpose of flushing the water is to clear the faucet and plumbing between the sample site and the loop and/or header pipe, thus assuring collection of a representative sample. Fill the container with 250 ml of water.

Interpreting the Results

1. If the lead level is over 20 ppb, collect additional samples from the plumbing upstream supplying water to the loop or header. Compare the sample results with those taken from the service line or the riser pipe that supplies water to the loop and/or header. High lead levels may also be caused by recent repairs and additions using lead solders, or by sediment and debris in the pipe. Debris in the plumbing is most often found in areas of infrequent use, and a sample should be sent to the laboratory for analysis.

2. If the lead level in either sample is equal to the lead level in a sample taken downstream from the sample site, the lead is contributed from the header or the loop and from the interioı plumbing upstream from the header or loop. Possible sources of lead may be the loop, header, riser pipe, or service connection.

3. If the lead level in either sample is close to or equal to 5 ppb, the portion of the header or loop upstream from the sample site and the interior plumbing supplying water to the loop or header are not contributing lead to the drinking water. The source of lead is downstream from the site.

4. If the lead level in either sample significantly exceeds 5 ppb, and is less than the lead level in a sample taken downstream from the sample site, a portion of the lead is contributed downstream from the sample site.

Procedure for Riser Pipes[11]

A riser is the vertical pipe that carries the water from one floor to another.

1. Open the tap closest to the riser pipe.

2. Let the water run for 30 seconds before collecting the sample. The purpose of flushing the water is to clear the faucet and plumbing between the sample site and the riser pipe to assure collection of a representative sample. Collect 250 ml in the sample container.

Interpreting the Results

1. If lead levels exceed 20 ppb, collect additional samples from the plumbing upstream from the riser. High lead levels in the riser pipes may also be caused by recent repairs and additions using lead solder.

2. If the lead level in the sample is equal to the lead level in a sample taken downstream from this site, the source of lead is the riser pipe or the plumbing and service connection upstream from the riser pipe.

3. If the lead level in the sample is close to or equal to 5 ppb, the portion of the riser pipe and plumbing upstream from the sample site and the service connection are not contributing lead to the water. The source of lead is downstream from the sample site.

4. If the lead level in the sample significantly exceeds 5 ppb, and is less than the lead level in a sample taken downstream from the sample site, a portion of the lead is contributed downstream from the sample site.

Drinking Water Fountains

Drinking water fountains have three basic configurations: (1) a self-contained water cooler equipped with its own cooling system that receives water from the building's supply; (2) a system of drinking fountains that receive water directly from the building's plumbing system via a central chiller (Fig. 17-13); and (3) a drinking fountain that receives water directly from the building's plumbing system.

For sampling, valves to the water fountains should not be closed to prevent their use, since minute amounts of scrapings from the valves will be collected, resulting in higher than actual lead levels. All samples should be collected with the

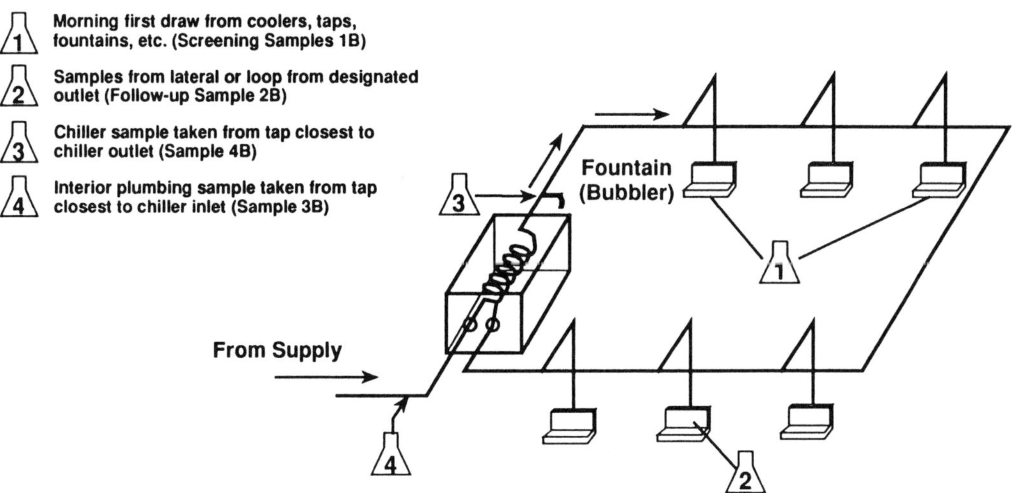

Figure 17-13. Suggested sampling sites for water supplied from a central chiller to water fountains. (From EPA. *Lead in School's Drinking Water.* EPA-570/9-88-001, January 1989.)

taps fully open.[11] A drinking water cooler is considered lead-free when no component that comes into contact with drinking water contains more than 8% lead, and no solder, flux, or storage tank interior that may come into contact with drinking water contains more than 0.2% lead.[11]

Sampling Bubblers with or without a Central Chiller

This sample is representative of the water that may be consumed at the beginning of the day or after infrequent use. It consists of water that has been in contact with the bubbler valve and fittings and the section of plumbing closest to the outlet of the unit.

Procedure

1. Take this sample before any water is used. Collect the water immediately after opening the faucet without allowing any water to run into the sink. Fill the container with 250 ml of water.

2. Follow-up samples should be taken from water fountains where test results indicate lead levels over 20 ppb. These consist of samples that are representative of the water in the plumbing upstream from the bubbler. Take them before any water is used. Let the water from the fountain run for 30 seconds before collecting the sample. Fill the container with 250 ml.

Interpreting the Results

1. If the lead level in the first sample is higher than in the follow-up, a portion of lead in the drinking water is contributed from the bubbler.

2. If the lead level in the follow-up sample is very low—close to 5 ppb—very little lead is picked up from the plumbing upstream from the outlet. The majority or all of the lead in the water is contributed from the bubbler.

3. For a bubbler without a central chiller, if the lead level in the follow-up sample exceeds 5 ppb, lead in the drinking water is also contributed from the plumbing upstream from the bubbler. If the lead level in the follow-up sample exceeds 20 ppb, sampling from the header or loop supplying water to the lateral should be done to locate the source of contamination. For a bubbler with a central chiller, if the lead level in the follow-up sample significantly exceeds 5 ppb,

the lead in the drinking water may be contributed from the plumbing supplying the water from the chiller to the bubbler, from the chiller, or from the plumbing supplying water to the chiller.

4. If the lead level in the follow-up sample exceeds 20 ppb, sampling from the chiller unit supplying water to the lateral to locate the source of contamination should be done.

Sampling Water Fountains with Coolers

There are two types of water coolers: wall-mounted coolers and free-standing coolers. Water in the cooler is stored in a pipe coil or in a reservoir. Refrigerant coils in contact with either of these storage units cool the water. Sources of lead in the water may be the internal components of the cooler, including a lead-lined storage unit; the section of the pipe connecting the cooler to the lateral; and/or the interior plumbing. Some water coolers have storage tanks lined with materials containing lead. Sediments and debris containing lead on screens or in the plumbing frequently produce significant lead levels. Lead solder in the plumbing can also contribute to the problem.[11]

This sample is representative of the water that may be consumed at the beginning of the day or after infrequent use. It consists of water that has been in contact with the valve and fittings, the storage unit, and the section of plumbing closest to the outlet of the unit.

Procedure

1. Take the sample before any water is used. Collect the water immediately after opening the faucet, without allowing any water to be wasted. Fill the sample containers with 250 ml of water.

2. Follow-up samples should be taken from water coolers where test results indicate lead levels over 20 ppb. These water samples are representative of the water that is in contact with the plumbing upstream from the cooler. Take these samples at the end of the day, letting the water run from the fountain for 15 minutes before collecting them. Flushing for 15 minutes is necessary to ensure that no stagnant water is left in the storage unit. Fill the containers with 250 ml of water.

3. This sample must be taken the morning following the sample collected in step 2. Because the water in the cooler should be flushed the afternoon prior to collecting this sample, it is representative of the water that was in contact with the cooler overnight, not in extended contact with the plumbing upstream. Take this sample before any water is used. Collect the water immediately after opening the faucet, without allowing any water to be wasted.

Interpreting the Results

1. If the lead level in the sample collected in step 3 is higher than the level in the sample collected in step 2, the water cooler is contributing lead to the water.

2. If the lead level in the sample collected in step 3 is higher than the level in the sample collected in step 2, and the lead level in the sample collected in step 1 is higher than the level in the sample collected in step 2, the plumbing upstream from the water cooler may also be contributing lead to the water.

3. If the lead level in the sample collected in step 3 is identical or close to the level of the sample collected in step 2, the water cooler probably is not contributing lead to the water.

4. If the lead level in the sample collected in step 1 is higher than the level in the sample collected in step 3, and if the lead levels in the samples collected in steps 2 and 3 are close or identical, the plumbing upstream from the cooler and/or the plumbing connection leading to the cooler are contributing lead to the water.

5. If the lead level in the sample collected in step 2 is in excess of 10 ppb, and is equal to or greater than the lead levels in the samples col-

lected in steps 1 and 3, the source of the lead may be sediments contained in the cooler storage tank, screens, or the plumbing upstream from the cooler.

Verifying the Source of Lead

1. Take a 30-second flushed sample from a tap upstream from the cooler. If a low lead level is found in this sample, the source of lead may be sediments in the cooler or the plumbing connecting the cooler to the lateral, or lead solder in the plumbing between the taps.

2. If the flushed samples from the upstream outlets have lead levels in excess of 5 ppb, then the cooler and the upstream plumbing may both contribute lead to the water.

Confirming Whether the Cooler Is a Source of Lead

1. Turn off the valve leading to the cooler. Disconnect the cooler from the plumbing and look for a screen at the inlet.

2. Remove the screen. If there is debris present, check for the presence of lead solder by sending a sample of the debris to the laboratory for analysis.

3. Some coolers also have a screen installed at their bubbler outlet. Carefully remove the bubbler outlet by unscrewing it. Check for a screen and debris, and have a sample of any debris analyzed.

4. Some coolers are equipped with a drain valve at the bottom of the water reservoir. Water from the bottom of the water reservoir should be sampled, and any debris should be analyzed. Collect 250 ml in the container.

5. Collect a sample from the disconnected plumbing outlet. Take the sample before any water is used. Collect this fraction as soon as possible after disconnecting the plumbing from the cooler.

Interpreting the Results

1. If the lead level in the last sample collected from the disconnected plumbing outlet is less than 5 ppb, lead is coming from the debris in the cooler or the screen.

2. If the lead level in the last sample is significantly higher than 5 ppb, the source of lead is the plumbing upstream from the cooler.

REFERENCES

1. Hamaan, M. E. An air sampling collection apparatus for high resolution air analysis. *AIHA J* **50**(8):A-591–A592, 1989.
2. U.S. Environmental Protection Agency. *RCRA Inspection Manual*. EPA, Washington, DC, September 1982.
3. State of Illinois Environmental Protection Agency. *Instructions for the Preparation of Closure Plans for Interim Status RCRA Hazardous Waste Facilities*. Springfield, IL, March 2, 1989.
4. U.S. Code of Federal Regulations, Title 49, Parts 100–199.
5. Notice of intended changes for benzene TLV. *Appl. Ind. Hyg.* **5**(7):453–463, 1990.
6. First, M. W. Air sampling. *Appl. Ind. Hyg.* **3**(12):F-20, 1988.
7. USEPA. *Test Methods for Evaluating Solid Waste*, SW 846, 3rd ed., 1986.
8. Chichowicz, J. Hazmat samples: Proper packaging, labeling and shipping are critical. *Hazmat World*, June 1989, pp. 60–61.
9. Environmental Protection Agency. *Characterization of Hazardous Waste Sites—A Methods Manual: Vol. II. Available Sampling Methods*. EPA 600/4-83-040, EPA, Washington, DC, Sept. 1983.
10. Rosbury, K. D. *Handbook: Dust Control at Hazardous Waste Sites*. EPA-540/2-85/003.
11. Environmental Protection Agency. *Lead in School's Drinking Water*. EPA 570/9-88-001, Washington, DC, January 1989.
12. American Public Health Association, American Water Works Association, and the Waste Pollution Control Federation. *Standard Methods for the Examination of Water and Wastewater*, 15th ed. APHA, AWWA, and WPCF, 1980.
13. Environmental Protection Agency. *Practical Guide for Ground-Water Sampling*. EPA/600/2-85/104, EPA, Washington, DC, 1985.
14. Environmental Protection Agency. *Methods for Chemical Analysis of Water and Wastes*. EPA-600/4-79-020, US EPA, Cincinnati, OH, March 1979.

PART VI

General Information for the Sampler

CHAPTER 18

Survey Preparations and Performance

Surveys vary in complexity and purpose. They range from the relatively straightforward identification of significant chemicals to sample for in a well-characterized industry, such as a foundry, to the often difficult and complex identification associated with a hazardous waste site. Different types of surveys often dictate different approaches. For example, an insurance survey usually involves limited sampling, and consultants generally must operate within the contractual scope of the proposal accepted by their client. A health and safety or environmental professional assigned to a single industrial facility usually performs repetitious sampling for the same contaminants. In consulting and insurance work, rarely is there opportunity to collect enough samples for high statistical accuracy; mistakes requiring a survey to be redone can be costly, especially if extensive travel or time expenditures are involved.

SURVEYS INITIATED FROM WITHIN THE WORKPLACE

Insurance Surveys

Loss control and industrial hygiene surveys are generally designed to assist an employer to minimize the potential for occupational disease claims under the employer's workers compensation policy. An insurance carrier may request that an industrial hygiene survey be done because the employer is in a high-risk classification; otherwise the policy holder may request one. The size and nature of an operation can also determine whether or not an industrial hygiene survey will be done. Other survey issues vary with the particular employer's situation, for example, employees may be complaining of discomfort in the course of their work. Usually air sampling surveys are provided annually as part of the insurance service. These annual surveys are usually not comprehensive since only a few individuals are sampled for one or two contaminants. The reports are often brief and may be in memo form. Table 18-1 lists industries with potential hazardous exposures.[1]

Workers compensation claim investigations result in another type of insurance survey. Often the employer requests sampling of a specific job function be done to determine whether there is actually an exposure or possible overexposure. The main problem with these surveys is that they often take place some time after the individual's employment. Changes in the process, materials, and chemicals may have occurred since that time. Consequently, it may be difficult to assume that the exposure evaluated by the survey is significantly related to the claim.

TABLE 18-1. Industries with Potential Hazardous Exposures, 1985*

SIC Code	*Description*	*Total Number of Production Workers*
20	Food and kindred products	1,118,000
21	Tobacco manufacturers	48,000
22	Textile mill products	607,000
23	Apparel products	945,000
24	Lumber and wood products (except furniture)	584,000
243	Millwork, veneer, and plywood	190,000
245	Building and mobile homes	56,000
249	Miscellaneous wood products	64,000
25	Furniture and fixtures	394,000
26	Paper and allied products	512,000
27	Printing, publishing, and allied industries	789,000
28	Chemical and allied products	578,000
281	Industrial inorganic chemicals	72,000
282	Plastics and Synthetics	114,000
283	Drugs	206,000
284	Soap, cleaners, and cosmetics	148,000
285	Paints, varnishes, and lacquers	31,000
286	Industrial organic chemicals	160,000
287	Agricultural chemicals	37,000
289	Miscellaneous chemical products	54,000
29	Petroleum refining and related industries	109,000
291	Petroleum refining	82,000
295	Paving and roofing materials	20,000
30	Rubber and plastic products	607,000
307	Miscellaneous plastic products	435,000

Response to a Complaint or Injury

Surveys are also done in response to employee complaint or injury. An added difficulty in this context arises from poor relations between management and labor. Subjective employee complaints and injuries resulting from a nonroutine activity or unknown exposure source can be most difficult to investigate. Maintenance and repairs are often the cause of unexpected situations. An example of the latter factor can occur in some production plants where the process sewers are open and may be linked throughout the entire facility. If two reactive components get into the sewer at the same time and meet downstream, it can be difficult to determine the source of the resulting problem.

Evaluation of Engineering and Process Controls

Health and safety are important on an industrial site. The most common and the most expensive employee injuries are usually safety-related, such as hand and back injuries. Consequently, there is concern at such locations about the lost workday injury rate and the latest safety innovation programs. One of the best ways to prevent problems is to review a potential change in a process for problems before the change is implemented. Sampling surveys are often done to establish the need for engineering controls, such as local exhaust ventilation, enclosures, or other modifications to the processor equipment. Follow-up surveys can determine the effectiveness of these controls and

TABLE 18-1.
(continued)

SIC Code	Description	Total Number of Production Workers
31	Leather and leather products	137,000
311	Leather tanning and finishing	12,000
32	Stone, clay, glass, and concrete products	451,000
33	Primary metal industries	612,000
34	Fabricated metal products	1,084,000
35	Machinery, except electrical	1,307,000
36	Electrical and electronic machinery, equipment and supplies	1,300,000
37	Transportation equipment	1,257,000
38	Instruments	391,000
39	Miscellaneous manufacturing	264,000
40	Railroad transportation	264,000
45	Transportation by air	
47	Transportation services	
49	Electrical, gas, and sanitary services	729,000
5093	Scrap and waste materials	
5161	Chemicals and allied products	
5191	Farm supplies	
5198	Paints, varnishes, and supplies	
55	Auto dealers and service stations	1,886,000
72	Personal services	
73	Business services	3,863,000
75	Auto repair, services and garages	614,000
7641	Reupholstery and furniture repair	
7692	Welding repair	
80	Health services	5,607,000

*Identified by OSHA.[1]

determine if further modifications are necessary. The Occupational Safety and Health Administration (OSHA) requires that sampling be conducted whenever there has been a change in production, process, engineering, or other types of controls, or personnel changes that might impact exposures.

In sampling to evaluate controls, the ability of the hood to contain air contaminants can often be resolved by comparing concentrations measured immediately outside of the hood near the source with those in the general work area. A significant difference between a general work area sample and a sample collected near a poorly controlled emission source indicates that the source is a significant contributor to contaminant concentrations.

SURVEYS INITIATED FROM OUTSIDE THE WORKPLACE

OSHA Surveys

There are two types of OSHA inspections: compliance and consultation. The compliance survey is better known. Upon request, the less familiar consultation branch of OSHA will generally assist employers free of charge. But many employers are reluctant to invite OSHA into their operation for fear of dispute over whether an attempt to correct a problem has been genuine. Working for OSHA compliance may involve adversarial situations in which inspectors are viewed as police who issue citations.

Many companies have a policy requiring side-by-side sampling to be done whenever an OSHA compliance inspection takes place. Generally, the company's sampler is also instructed to try to find discrepancies in OSHA's techniques.

NIOSH Health Hazard Evaluations

In unusual health hazard situations, the National Institute of Occupational Safety and Health (NIOSH) may be asked to conduct health hazard evaluations. These in-depth surveys usually involve both an industrial hygienist and a physician. Sometimes epidemiologists are also involved. When an exhaustive study is warranted because of a significant change in a manufacturing process that changes exposures, engineers and other professionals may be involved as well.

BASELINE AND ROUTINE MONITORING SURVEYS

An approach commonly used in medical monitoring is the estimation of baseline exposures. Baseline exposure assessments may be done for job classes, tasks, and individuals. Such sampling is often directed at decreasing worker compensation exposures. Once enough samples to estimate a baseline have been collected, periodic monitoring is done to detect any change in exposures. This approach minimizes overlooking any high-hazard job or activity where the need is to get the most objective measurement possible from a group of employees, all of whom very likely have similar exposures. Such sampling is often done to establish large data bases on exposure histories associated with certain types of occupations. These data are useful in the event of workers compensation claims. In such cases, the sampling is usually randomized using a random-numbers table.

For certain high-toxicity chemicals, or in potential overexposure situations, employees are regularly monitored. Lead and benzene are examples of chemicals for which ongoing monitoring programs are often established. Additionally, certain OSHA standards, such as those for benzene and lead, require annual monitoring whenever employees' exposures exceed the action level.

A routine monitoring program differs from sampling to determine if exposures are within OSHA levels. OSHA regulations usually require an employer to sample a representative number of employees and jobs to determine if an exposure over the permissible exposure limit (PEL) or action level exists. After this sampling, no monitoring is required unless there has been a change in production, process, control equipment, personnel, or work practices that may result in additional exposures, or when the employer has any reason to suspect that a change may result in new or additional exposures.

Routine monitoring programs are designed to eventually characterize virtually any chronic exposure that may occur. One method is termed the exposure zone method. Once all the potential agents have been categorized, they are broken down into various chronic categories, such as central nervous system (CNS), asphyxiant, and carcinogen. In the case of chemicals without known toxic effects, monitoring may consist of reviewing the literature rather than actual sampling. Often there is no established sampling method for these chemicals, so accurate sampling is difficult. Sometimes sampling and analytical methods must be developed for a specific chemical. A recommended exposure limit may also be developed. This is how most of the current OSHA PELs came into existence. Other chemicals are scheduled for annual, semiannual, quarterly, or other periodic sampling. A certain percent of employees in each zone is sampled following this schedule.[2]

If the purpose of a survey is part of an ongoing monitoring program, or to determine if a change has been effective, it is important to either use the same method as previously used or to make sure that the new method is comparable with the old one. One way is to do side-by-side sampling with each method. An example of problems that can happen occurred when the asbestos sampling method changed from PC&M 239 to NIOSH 7400. Some studies indicated that the results were lowered by a factor of three,[3] making prior sampling results difficult to compare with the new method. In these situations, it is necessary to determine any differences between methods.

HAZARDOUS WASTE

Hazardous waste sites are potentially more dangerous than any other environment in which sampling may be needed. There are usually large amounts of concentrated and often unknown materials at such sites; such materials are often in open pools or deteriorating containers. Special personal protective equipment is often necessary. Consequently investigations, sampling, and other activities at a hazardous waste site are usually much more difficult and time-consuming than routine environmental sampling.

Under OSHA recommendations, several types of monitoring are performed at hazardous waste sites. Upon initial entry, general survey air monitoring is done to identify any immediately dangerous to life and health (IDLH) conditions, exposures exceeding health standards, or other dangerous conditions, such as the presence of flammable atmospheres or oxygen-deficient environments. Subsequently, periodic or continuous monitoring is to be done when any of these situations may exist or when there is indication that exposure levels may have increased since prior monitoring. Examples of situations where additional monitoring may also be required include work on a different portion of the site, handling of contaminants other than those previously identified, initiation of a new type of operation, and safeguarding employees at the greatest risk of exposure during cleanup operations.

EMERGENCY RESPONSE

Emergency monitoring situations differ from most surveys in several ways. Since acute rather than chronic effects are the primary concern, the required detection sensitivity is less stringent. However, in many cases, the exact compounds of interest may not be immediately known and therefore a broad range of monitoring techniques must be used to make sure the toxic compounds are detected. In addition, the development of a monitoring strategy is difficult due to the need for immediate response.

INDOOR AIR AND COMMUNITY COMPLAINTS

Nuisance complaints arising from the general public are primarily related to noxious odors, eye irritation, or in some cases, more serious illnesses. Monitoring activities are often geared to identifying mystery contaminants. Often the levels of contaminant measurement are extremely low due to the highly sensitive nature of many individuals, especially when compared to standards for workplace exposures.

Tight building syndrome assessments involve sampling exposures of individuals who may or may not be the client's employees. The survey may be at the request of the building management or in a government building. The cause of irritation or illnesses may be due to outside pollution being entrained into the building's ventilation system.

OTHER SURVEY SITUATIONS

Surveys in mines are covered by the Mine Safety and Health Administration (MSHA) regulations. Consequently, sampling methods and monitoring requirements may vary somewhat from those of OSHA. Mines can present greater hazards than work in typical industries. The environment is often more dusty, wet, and noisy as well as often being underground. These surveys can require special preparations, equipment, and techniques.

Sometimes the problem does not involve exposures to people, but deterioration of equipment, such as rusting metal in a warehouse. Some chemicals such as sulfur dioxide are capable of oxidizing metals at levels far below either their odor threshold or the capability to elicit health effects.

The purpose of sampling to detect unknowns is to document previously unknown high exposures and then determine what activities present exposures. A typical situation is where contaminants may periodically be high, but employees rarely work in these areas. An example would be sulfur dioxide sampling in a back room associated with a water treatment process. In some situations, rather than attempting to identify the actual compounds present, a nonspecific instrument is used to screen for the presence of a wide variety of gases or vapors. This screening is commonly done on hazardous waste sites where there is often not enough time to characterize all unknown components before they get into the air. A variation is the measurement of combustible gases where the screening is done to determine potentially explosive levels rather than health hazard concentrations.

Stack monitoring studies generally require specialized techniques and experience. Exhaust sources can range from pipes less than 1 foot high to 100-foot stacks. The sampling professional might be asked to do some sampling related to emissions suspected to arise from a stack, vent pipe on a reactor, storage vessel, hospital incinerator, or on personnel suspected of being exposed to these types of emissions. In these situations, the sampling professional may have to determine qualitatively or quantitatively the types of contaminants and the amounts prior to making decisions as to the likely sources of exposure to personnel. Another ancillary type of situation already suggested is airborne contamination of one property from emissions occurring on another. The concern may be levels of a contaminant so low that the primary effect is corrosion of machinery rather than existence of a health hazard.

SAMPLING STRATEGIES

In general, an acceptable strategy is one that prioritizes needs, optimizes resources, is readily implemented, and is cost effective. Traditionally, sampling done for limited periods (such as one or two days) has concentrated on worst-case exposures in that sampling decisions are based on the periods when production is highest, the areas where highest concentrations are used, or the jobs involved with the most toxic compounds. If the primary aim of air sampling is to prevent worker overexposure, a sampling strategy should be developed that takes into consideration the nature of the contaminant, such as its biological half-life, as well as setting up a statistical estimation of the actual exposure. The biologic half-life is one way of determining the likelihood of a compound remaining in the body. The longer the half-life, the more likely it is that the compound will accumulate in the body.

It has been suggested that in order to have air sampling results more closely reflect the potential for the body burden to be exceeded, the minimum sampling time should be equal to 0.3 times the biological half-life.[3] This subject is discussed at greater length in the chapter on Biological Monitoring. There are at least three types of substances: (1) chemicals with a long half-life, such as coal dust, which will result in cumulative biological effects over a lifetime of exposure; (2) chemicals that produce long-term effects because they have exposure thresholds above which there is a dose rate dependency; and (3) compounds that have short half-lives, such as chlorine, but cause irritation and acute effects in high levels.[4]

The type of toxic effect can also impact the sampling strategy. For example, grab sampling using a series of short samples can provide useful information as to whether high levels of compounds that are irritants are present.[5]

Sampling strategies will also differ depending on whether sampling for chronic or acute exposures. Situations that can pose a potential for acute exposure to fast-acting chemicals at relatively low levels are sampled over shorter periods; the most dangerous situations such as chlorine and hydrogen sulfide in a confined space require continuous real time monitoring. There are two basic strategies for sampling acute occupational exposures: Collect samples during periods of suspected highest exposure or collect short samples at regular intervals in order to minimize bias and identify unpredictable peaks.[6] The sampling period should be short in order to identify peaks and should also be a realistic reflection of the actual exposure period. Integrated sampling techniques are best for chronic exposures to cumulative poisons such as lead and mercury.[7]

The placement of monitoring stations containing instruments for environmental area monitoring of gases and vapors in ambient air to measure exposures to communities will also vary depending on the study. For chronic effects studies, locations of monitoring stations should be chosen to include high, intermediate, and low pollution exposure areas. For acute effects only one station may be needed in the area of the source of the highest exposures, but often multiple monitors operating simultaneously upwind and downwind from the source are very valuable and efficient.

The purpose for sampling will dictate whether area or personal samples are collected. Area samples are useful for evaluating "what if" situations where a contaminant is suspected but not known, and are often used to document conditions, but should not be considered representative of employee exposures. Indoor air surveys for tight building syndrome are an example where area samples are more commonly collected than personnel samples, since there is a need to do clearance sampling to document the effectiveness of asbestos removal inside a containment area and to do sampling outside the same enclosure during removal to assure that no leaks occur. During hazardous waste operations area real time monitoring is done more often than personnel integrated sampling because of the concern for the presence of unknown chemicals at unknown concentrations. Area samples are also useful for determining the actual source of exposure when many possible ones exist or for identifying leaks. A problem when doing source monitoring is the temptation to put the pump or instrument as close to the source as possible: If the source is too hot, it can damage the instrument.

Most often it will be impractical to monitor every employee or area, so the exposures of a smaller number are assumed to be representative of the whole. Collection of personal samples involves special techniques over and above those necessary to simply run equipment. When putting pumps on workers, always provide a brief description of the purpose for the sampling. Tell workers how the equipment works, what compounds are being sampled for, and that it is important to conduct their work normally. Also tell the employees what to do if a problem arises with the sampling equipment (pump shuts off, sampling media fall off) or what to do if they have to leave early (family emergency).

Sometimes the nature of the sampling, such as the use of impingers where the employee should be told not to bend over to prevent liquid from being sucked up the pump, requires a slight modification in work habits. Record the person's name, position(s), social security number (sometimes), and lengths of breaks and lunches on a field sampling data sheet. Ask the location of lunch and breaks. It is not advisable to sample during lunch unless lunches are eaten in areas where exposure exists. In most cases, the pump should simply be turned off prior to lunch, and turned on again after lunch. It is not usually necessary to remove equipment unless the employee leaves the company premises.

Care should be taken to ensure that contamination of the collection medium does not occur. Therefore, the sample should be removed or capped off. Generally it is preferable to remove the sample medium and replace it after lunch. If the employee plans to leave the facility for lunch, the pump may have to be removed. If the employee remains in the work area, the pump can be left on. Care must be taken to ensure that the sampling inlet is not covered up during sampling. It can occur when a coat or jacket is donned after the monitor is placed on the worker. Tubing attached to pumps should be secured so that it does not catch on equipment or get in the way of the employee's movement. Duct tape is often used because it is strong, attaches to clothing, and does not leave marks. Spring-loaded metal clips attached to the tubing can also be used.

Once the sampling purpose and corresponding strategy have been decided, the sampling period must be selected. This time depends on three main factors: the length of the workday or the task being performed; the expected concentration; and the standard for comparing the data. Wherever possible, it is good to characterize exposures on all shifts or situations since variations often occur. The day shift may be the busiest, whereas frequently the grave and swing shifts have fewer employees. However, lack of supervision on evening shifts can lead to lapses of good work practices and exposures can increase.

The work shift is generally defined as 8 hours. When collecting samples to compare to an 8-hour TWA standard such as OSHA's, sample periods must reflect this. A full shift sample can be a combination of longer integrated (2–4 hours) samples and short-term samples for STELs. An 8-hour TWA can still be calculated as long as most of the day was sampled. The best way to identify an unknown concentration is to collect a series of STEL samples over a workday while attempting to identify sources. Short-term air samples are also collected when concentrations are being evaluated to compare to STELs or excursion standards, or when a task is short in length. For example, many mixing operations involve opening bags and dumping them. The complete time required to pick up the bags, open them, dump them, and dispose of the empty bags may take less than 15 minutes. Generally it is difficult to sample for shorter or longer periods of time because of limitations in the sampling and analytical methods.

The ideal method of measuring compounds with ceiling standards would be continuous monitoring equipment with cumulative display capability as well as having an alarm to warn when the ceiling has been exceeded. However, since few instruments exist that are chemical specific, compared to the number of chemicals that have ceilings, an alternative is to use integrated methods and collect a series of shorter samples, although these are not necessarily 15 minutes in length. These are then added together to develop an 8-hour TWA. Also, any given sample can be compared to the TWA-C, because none of the averages should exceed it. Occasionally, in some peak exposure situations the exposure may last 20 to 30 minutes, or the minimum sampling volume may require more than 15 minutes to collect a sample to meet laboratory detection limits; then a sample must be collected over a longer period.

Not only do concentrations vary over a typical day but they fluctuate when one day is compared with another. If a single sample is collected over a full day, then only the variation between different days will be apparent, because the daily fluctuations are averaged out. On the other hand, if a number of very short-term samples, sometimes termed grab samples, are collected, they will reflect both types of fluctuations. From a practical point of view, it is usually impossible to continuously sample all areas at all times, so instead a number of small samples are collected and generalizations are made from them. The more samples collected, the more confidence the sampling professional will be able to put in the reliability of the sampling data. However, while statistical accuracy increases with increasing numbers of samples for each task collected over a day, these are often outweighed by the significant increases in analytical costs. A middle ground must often be reached as to the number of samples taken.

Outdoor monitoring is often done using area monitoring techniques. There are two primary situations where outdoor air monitoring may be required over the course of an occupational health investigation: indoor air problems and hazardous waste site monitoring, including boundary line sampling. The number of points actually sampled outdoors will depend on the average wind speed and direction, the prevailing wind direction, the locations of greatest concern nearest the site boundaries (residences, schools), and the locations of the areas with the highest concentrations of hazardous materials on the site.

Several pieces of information must be known in advance to plan a statistical sampling strategy, including the size of the work force to be sampled, the accuracy of the sampling and measurement method to be used, and the confidence the sampler wishes to have in predicting the exposure of the work force. For example, to determine with 90% confidence that at least one worker from a workplace subgroup will be in the top 10% of the exposures occurring in the group, the number of employees to sample would be chosen from a table set up for these parameters. Other tables are available for confidence limits of 95% and for the top 20% of exposures (see Table 18-2).

Random sampling is often favored by NIOSH and OSHA. Random sampling can be useful in cases involving nonroutine exposures or when an attempt is being made to identify an unknown

TABLE 18-2. Minimum Sample Size for Including at Least One High-Risk* Employee (90% Confidence Level)

Size of employee group:	1	2	3	4	5	6	7	8	9	10	11–12	13–14	15–17	18–20	21–24	25–29	30–37	39–49	50
Minimum number of measured employees:	1	2	3	4	5	6	7	7	8	9	10	11	12	13	14	15	16	17	18

*High risk is defined as exposure in the highest 10%.

exposure. The best situations are those requiring several samples over a lengthy period. A generally accepted method for obtaining a random sample is through the use of a table of random numbers. To obtain 100 random air samples, over a period of several weeks, for a given area, over different shifts and time periods, to investigate the potential for a particular exposure, the sampler would assign a number to each of the samples intended for collection and then draw 100 random numbers from a table to determine the period in which that sample would be collected. The periods to be included are those that correspond to the selected number in the random table. Follow the instructions below.

1. Determine the total number of samples from Table 18-3.
2. Starting with the number 1, assign a number in sequence to each sample.
3. Select the smallest random number table with more numbers than sample numbers.
4. Starting at the top of column one, select each random number that fits into the total number of samples. For example, if the total number of samples is 35, then all random numbers greater than 35 would be passed over. Stop when a total of 35 random numbers have been selected from the table.
5. Draw a line after the last random number used. It will be the starting point for the next set of samples.
6. A sequence for 12 samples selected from a random number table might end up with the following numbers: 6, 7, 10, 13, 5, 1, 3, 9, 2, 12, and 8. Thus, sample number 6 will be collected first, sample number 7 second, and so forth.

It should be kept in mind that reading down a list of random numbers is analogous to drawing numbered chips from a bowl. The numbers are random; therefore, it is not necessary to skip around the pages. However, the sampler should mark lightly in pencil the last number used and continue from that point when the next set of random numbers is needed. If the random number desired is a three-digit number, then the first column should be combined with the first digit of the second column to create all three-digit

TABLE 18-3. Random Number Table

Column 1	*Column 2*	*Column 3*	*Column 4*	*Column 5*
6	46	89	65	33
7	43	83	21	96
59	14	71	70	31
68	51	9	41	74
73	22	54	2	35
10	3	28	38	25
81	17	90	82	49
50	57	45	29	78
23	15	84	37	62
55	18	85	98	32
69	58	80	12	77
66	91	30	27	16
48	19	99	56	26
13	63	52	86	92
94	72	39	53	97
5	60	61	8	95
1	24	76	75	44
64	93	40	11	36
20	42	87	4	100
34	47	88	79	67

Prepared by the Office of Data Analysis, 09/02/82/.

random numbers. If it is a four-digit number, then the first two columns are combined to create all four-digit random numbers.

Systematic sampling is a practical approximation of random sampling in some cases. For example, to draw a systematic sample of 100 units from a file containing 1,000, the sampler would select only one random number between 1 and 10, and proceed by selecting the unit in the position corresponding to this number and every tenth unit thereafter. In situations where this type of sampling is applicable, it eliminates the need for looking up all but one of the random numbers in the table.

As another tool in deciding how many samples to collect, the sampler should note that statistics are a way of increasing the consistency of sample results as well as optimizing the number of samples to collect, although properly applying statistics generally requires a substantial amount of data. A problem is that few sampling professionals have a good knowledge of the use of statistics or the confidence to apply them. Generally the most useful statistics include geometric mean, geometric standard deviation, 95th percentile exposure value, and upper confidence limit for the 95th percentile exposure. There are specific references available that describe the applications of sampling statistics.[8,9]

Sometimes if the location of the contamination is unknown, or the concern is to determine the effectiveness of engineering controls, the establishment of a grid sampling pattern is useful. It can be used in bulk soil sampling as well as in air sampling situations. In setting up a grid the sampler has to take into account the size of the area to be surveyed. The best way to set up the grid is to take a diagram of the area to be surveyed and draw the grid directly on it using both horizontal and vertical lines.

Actual sampling situations can provide many challenges. The concentration of air contaminants can change rapidly over time depending on the source and the type of process. Gases, vapors, and fumes move easily with air currents and can spread throughout areas, including offices. Sometimes they are entrained in a building's ventilation system and they spread into other areas where no chemicals are in use, or they may move through doors and other openings. Outdoor air pollution can be entrained into buildings through vents, open doors and windows, and air supply systems. Thus, contaminants may spread in many directions from the source of emission. For example, sulfur dioxide emanating from the ceiling area of one room can seep under doors and through cracks into other rooms.

TABLE 18-4. Typical Velocities Found in Industrial Operations

Feet per Minute	*Location*
3	Settling velocity of heavy respirable particles
40	Random air movement in industrial settings
50	Minimum capture velocities at emission sources
100	Face velocity of a typical laboratory hood
100	Begin to feel air movement on dry skin
300	Typical eddy velocities left in wake of person walking briskly
700	Average wind velocities
1800	Maximum velocity of cooling air blowing on a person for 30 minutes
2000	Duct velocities not requiring particle transport
3000	Stack exit velocity
3500	Maximum velocity of air blowing on a person for a short time (e.g., 10 minutes at a supplied-air island)
4000	Duct velocities for transport of average industrial dusts
5000	Duct velocities for transport of heavy industrial dusts

The concentration at any given location is determined by a combination of the source of contaminants; airflow direction and velocity (Table 18-4),[10] whether it is due to wind or thermal gradients; density of the contaminant; intensity of sunlight; time of day; presence of obstructions such as trees, buildings, partitions, or machinery acting as baffles to produce turbulence; humidity; and half-life of the contaminant, and possibly other factors. Concentrations can vary significantly within a relatively short distance from the point of origin. This is especially true for particulates. Airflow

TABLE 18-5. Average Relative Humidity During Working Hours

Region	*Winter*	*Spring*	*Summer*	*Fall*
Pacific	70	56	50	59
Mountain	49	33	31	39
Central	73	68	70	72
Eastern	67	63	68	71
Gulf Coast	69	68	71	69

can be in a smooth streamline current or turbulent creating eddys and changing the dispersal pattern of the contaminant.

Temperature is another important variable. As the temperature of a process increases, the rate and amount of vapor being released also increases. In addition, some compounds are unstable and tend to decompose more rapidly at higher temperatures. If sampling a mixture, the temperature can change the composition of the vapor relative to that of the liquid, increasing exposures to the more volatile compounds. Humidity should not be ignored either as it can influence sampling results as well. The sampling efficiency of some sorbents, such as silica gel, is decreased in the presence of high humidity (see Table 18-5).[11]

Also, the sampler needs to take into account seasonal variations in exposures. In winter, doors and windows are closed and buildings may develop higher levels of contaminants as well as being under negative pressure. Variations can also occur depending on the location being sampled. For example, placement of a sampling pump in front of an open garage door or window may cause fluctuations. High production quotas can cause concentrations to increase as well. In indoor air situations, such as office buildings, air is commonly recirculated. The percent of outdoor air present will vary depending on the time of day, day of the week, and season, as well as the building temperature. In areas where power costs are high, some facilities use "off-peak" power and heavy production is emphasized on weekend, swing, and grave work shifts.

For environmental surveys samples are almost always collected outdoors. The interpretation of fenceline data can be difficult because of changes that can occur under differing meteorological conditions. The variables that may affect contaminant concentrations in these situations are greater than those typically encountered indoors: atmospheric stability, wind turbulence, wind speed and direction, solar radiation, rain, topographical situations such as valleys, mountains, and plateaus, emission rates, chemical reaction rates for formation and decomposition of pollutants, and the physical and chemical properties of the contaminant. Regular or continuous measurements of ambient temperature, relative humidity, barometric pressure, and wind speed and direction will be necessary. Continuous recording instruments are now available that simplify the measurements. Standard methods are psychrometers for humidity and wind vanes and rotating anemometers for wind.[12,13] For more information on measurement methods for temperature, pressure, and humidity, see the Appendix: Pumps and Flow Rate Calibrations.

One of the fundamental factors of concern when conducting sampling outdoors is the wind speed and direction. This must be known to effectively place air samplers to assure accurate and useful information. This is especially important when doing boundary line sampling. Wind constantly changes direction. Generally, a low wind speed with constant direction will cause the greatest downwind concentration. In this situation, a pollutant plume tends to stay together. Conversely, high wind speed and constantly changing direction will result in lower downwind concentration. However, the airborne substance will be dispersed over a larger area and may affect a greater population. Also the influence of terrain and obstacles such as buildings must be considered.

Wind direction is given as the direction from which the wind blows. Therefore, a wind blowing from the west to the east is termed a westerly wind. Direction is usually given using compass points, with north being 0°, east being 90°, and so forth. Wind direction can be estimated by observing the movement of trees, smoke plumes, dust, or other substances carried by the wind.

Wind speed is expressed as meters per second, miles per hour, or knots. Methods for measuring wind range from the simple observation of a wind sock, to sophisticated but bulky electronic equipment capable of ongoing monitoring. Wind speed sensors are available in a number of differ-

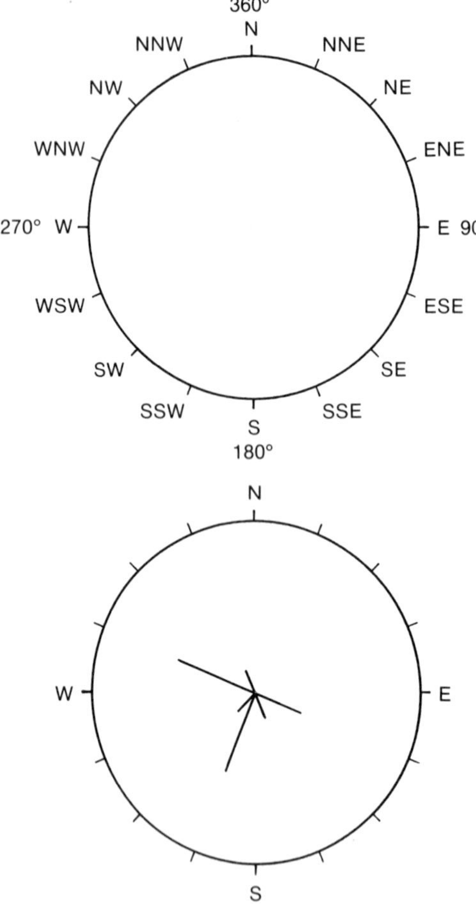

Figure 18-1. Wind rose.

ent designs with the most common being the rotational cup and propeller anemometers. Most wind direction sensors consist of a vane rotating on a fulcrum. This type of wind vane measures only the horizontal wind direction. A bidirectional wind vane is free to move through 360° and can also measure wind direction varying ±50° from the horizontal. Often an approximation of the direction is made using a wind sock and data from one or more of several government or private organizations that routinely provide such data. The convention for placement of monitoring equipment is at a height of 10 meters above ground or at a distance from obstructions such as trees of 10 times the height of these obstructions.

Wind direction and wind speed data are displayed in diagrams called wind roses (Fig. 18-1). The most common type of wind rose consists of a circle, from which 16 lines emanate. Each line is located at a compass point, and is proportional to the frequency of wind from that direction. The frequency of calm conditions is entered as a concentric circle at the center. Lines or bars are entered to show the direction of the wind. The longer the line, the more predominant that direction.

In source-oriented ambient air sampling particularly, samplers are usually located downwind (at different distances) of the source and others are placed upwind to collect background samples. Shifts in wind direction can require the samplers to be relocated or can require corrections being made for the shifts. Therefore, at a minimum, a site should be equipped with a wind sock, accurate thermometer, and barometer. Accordingly wind speed and direction should be measured during the initial survey and during subsequent activities for all sampling outdoors. The wind sock should be visible from all points, especially if there is a potential for leaks or releases, so that personnel on site can see which way to evacuate in case there is a leak of vapor or gas. Buildings or large piles of waste on the site should be identified, since there may be microenvironments in which the wind does not blow in the same direction as it does on most of a site. Table 18-6 lists wind speed and effects.[14]

FIELD WORK PREPARATION

If any aspect of sampling cannot be emphasized enough it is field work preparation. Methods must be researched, media ordered, pumps charged, and travel scheduled. If sampling is to be done in an industrial facility, the process must be studied before the survey is made.

If personnel samples are collected, notify the employer of OSHA's requirement to notify employees of the results of the sampling. Notice can be accomplished in writing, verbally, or by posting a copy of the sampling results on a bulletin board accessible to the employees.

Depending on the type of survey, the sampler may or may not be told ahead of time what contaminants the operation wishes to have sampled. One example is an annual survey to

TABLE 18-6. Wind Speed and Effects

Miles per Hour	*Condition*	*Effects*
8–12	Gentle Breeze	Wind extends light flag; leaves and small twigs in constant motion
13–18	Moderate Breeze	Loose paper and dust blown about
19–24	Fresh Breeze	Small trees sway; white caps form on fresh water
25–31	Strong Breeze	Large branches and whole trees move; wind whistles in wires; flag flutters vigorously
32–38	Moderate Gale	Flag whips wildly; loose objects may be lifted from the ground; difficult to walk against the wind

review levels of contaminants that have previously been sampled, or the type of operation may be described in enough detail to estimate the most likely exposures. Sometimes the only information available is that employees have complained or have become ill and the sampler must come prepared to sample on the first visit. In this situation, the sampler must also ascertain whether employees' symptoms can be tied to known warning signs and effects of one or more chemicals used in the process.

Often a company may have had prior air sampling done either by a consultant or by its worker's compensation carrier. These can be useful in determining what past levels and sampling methods were used as well as whether recommendations were made and if they were carried out.

When selecting chemicals and methods, be aware that many compounds are usually found in mixtures and rarely by themselves. This is sometimes a result of industrial synthesis. For example, toluene diisocyanate is a mixture of toluene-2,4-diisocyanate and toluene-2,6-diisocyanate. Polychlorinated biphenyls are a family of compounds that were sold in various mixtures usually under the trade name of Arachlor. Commercial xylene is a mixture of ortho-, meta-, and paraxylene. As combustion products, nitrogen oxide is generally always present with nitrogen dioxide.

Sometimes the sampler will have to make a decision as to whether to sample for high-volume chemicals that are often of low toxicity, or for highly toxic compounds used in very low amounts. That decision has to be based on exposure likelihood. Certain carcinogens, such as β-propiolactone, do not have allowable exposure limits; consequently personal sampling is unnecessary to determine compliance with OSHA. Any potential exposure is enough to pose a violation. Do not rely only on verbal information regarding what chemicals are present unless the source is highly reliable. Ask to see labels, review material safety data sheets (MSDS), or request the name and phone number of the supplier.

Selecting equipment and sampling media is the first step in implementing sampling. If specific chemicals have been identified, try to determine if any of them will interfere with another in the sampling methods that have been selected. Most companies purchase a certain amount of equipment based on anticipated sampling needs. However, due to the high cost of some instruments or the rarity of use of others, a necessary instrument may have to be rented or purchased. Instruments that have been stored for a period of time or used by others should be checked out to make sure they are in good working order and that batteries are fully charged. For integrated sampling, whether through consultation with a laboratory or a review of the specific method, the sampling media must be specifically identified, including type of filter and pore size, type of liquid absorbent, or type of solid sorbent and the size of the sampling tube. Be sure to allow ample planning time for identifying and ordering media since some must be specially prepared in a laboratory. For example, gauze wipes for PCBs are Soxhlet extracted and filters for certain isocyanates must be impregnated with a nitro reagent and assembled in special opaque cassettes.

Some integrated procedures require the use of chemical reagents in the field. Reagent bottles should be packed in absorbent, cushioning material to prevent bumping and leakage. Labels for reagents should be of indelible material and care should be taken to separate incompatible chemicals. In many cases, the reagent chemicals themselves represent a more serious threat of harm than the materials sampled.

It is a good rule to always pack extra equipment. Some sampling methods such as impingers take a toll on pumps. Also additional employees or areas to sample may be discovered during the survey. Temperature and humidity devices should be included.

Along with sampling equipment do not forget to pack tools such as screwdrivers of several sizes, a small exacto knife, tape measure, duct tape, and ziplock bags. Duct tape is used for sealing samples, repairing equipment, sealing leaks in sampling trains, holding a tube out of the way of an employee's movement, and securing media. Ziplock bags can protect and segregate samples, hold small parts and tools, and can be used to collect dry bulk samples and wipe samples on occasion. Having to make do with a high-flow rotameter because a flowmeter with a low flow range has been forgotten is not only unnerving but limits the quality of the sampling data. Extra batteries are also important. If a new method or type of equipment is to be used it should be calibrated and tested in the laboratory prior to use. If interferences are suspected, a new flow rate is needed.

Manufacturers of sampling equipment usually provide instructions as to procedures for setup and operation. Although not always solely reliable, these manuals should be read thoroughly before attempting to use the equipment to gain as much information as possible. A condensed set of instructions should then be created and placed in clear plastic covers to protect them from weather, chemicals, dirt, and wear. If possible, the sheets should be kept in the carrying cases with the equipment.

Always try to carry copies of the procedural method. If the sampler ends up in the field without a copy of the method, and cannot remember the procedure, guessing should not be used; instead, a call should be made to get the proper information prior to starting. Most analytical laboratories will have copies of all the sampling methods and often a technical information representative will provide advice upon request. As a rule of thumb, the following flow rates are commonly used for these types of media during integrated sampling:

Media	*Flow Rate*
Sorbent tubes	50–200 ml/min
Respirable samples (cyclones)	1.7 Lpm
Total particulate samples	2 Lpm
Impingers	1 Lpm
Long-term detector tubes	20 ml/min

Visitors and contractors are no exception to the safety and health rules which are established by a facility to protect those who work there. Another aspect of field work preparation is ascertaining what safety rules and personal protective equipment are in effect at a site. For example, the noise from industrial machinery and processes can cause headaches and can interfere with verbal communication, as well as cause temporary hearing loss. Accordingly hearing protection must be worn. The most common requirements are steel-toed shoes, safety glasses, and hard hats. To prepare for a waste site, asbestos abatement project, or other investigation requiring special operating health and safety procedures, all necessary safety and protective equipment should be gathered, and personnel should have the required OSHA training. Respirator fit-testing and medical clearance may be required.

THE WALK-THROUGH SURVEY

If possible, when first visiting an operation prior to sampling, the sampler should spend time reviewing areas where samples are to be collected. Upon arrival at a new site, ask to be taken on a tour of the process starting at the beginning of the work flow. This initial walk allows the sampling professional to learn about the operation and to make an estimation of the best areas and jobs to sample. Do not be hurried. Develop a slow, deliberate pace and be persistent, making sure to get answers to questions. Do not be afraid to ask questions, especially simple ones; it is better to get all the information than to pretend to know all the information. During this walk-through it is a good practice to make a diagram of the process. It will be helpful in understanding where various chemicals are used as well as what by-products might be generated. Getting a map of the work area can also be of assistance. Some facilities can provide very detailed diagrams including the locations of machines. This is especially useful when making real time measurements. Take note of locations where concentrated chemicals are in use. Note whether employees use compressed air, since it can alter the results of dust samples. Production schedules, where available, can be useful to review.

The age of the equipment will impact the degree of exposure and the type of chemicals in use as well as the applicability of using controls such as enclosures and local exhaust. It is not unusual to find a wide variety of ages of equipment in older plants. Depending on the nature of the operation, equipment can range from the early 1900s to present-day machinery operating side-by-side. Newer equipment often has built-in controls while older equipment is usually retrofitted to decrease emissions. Equipment maintenance and age are most critical; if the equipment has become outmoded, safety valves are aging, pressure tests may not be done regularly, and maintenance also may not be done unless something breaks down. If the primary method of controlling exposures is through work practices, and these are not enforced, exposures will vary.

Nonroutine situations are often sources of peak exposures and are the most difficult to identify. Maintenance employees are often required to do repairs on a variety of operations. Interviewing employees and supervisors may be the best technique for identification in these situations.

Even in relatively routine situations, exposures may not be what the sampler expects. For example, a resin-treated sand used in making foundry molds may prove low in silica exposure but give off phenol, formaldehyde, and/or ammonia during heating, or washing cauliflower with water containing chlorine will release residual nitrogen from fertilizer in the form of ammonia. By-products of operations can also be hard to predict, especially those that can occur when temperatures are faulty, or when inappropriate releases occur to the process sewers.

Outdoor sites are often selected based on the presence of an odor near a suspected source. This is often done when the objective of sampling is to locate sources of target compounds and allows the sampler to give particular attention to the areas that are most likely to have the highest concentrations. Ultimately all areas will be sampled to assure no sources are overlooked and that comparisons can be made between them.

FIELD WORK OBSERVATIONS

Sampling professionals, such as industrial hygienists, are sometimes hired in legal cases (such as third-party personal injury suits) to make a determination as to the likelihood of exposure. Each side in such a case has its own expectations; it is important for the expert witness to determine the merit of the case from his or her own specialized point of view. Whether supporting the manufacturer, defendant, or injured party, the work often involves doing literature research, reading, giving depositions, and attempting to recreate the alleged cause of the exposure. Sampling might be done, but frequently is not, because years have passed since the accident or last exposure at issue in the case going to trial.

It is important to collect as much information as possible whenever in the field in order to effectively interpret sample results and make recommendations for controlling exposures. Information to collect when doing monitoring includes production shut-downs, severe weather, chemical spills, and any infrequent occurrences as well as temperature, pressure, and relative humidity. If dusts, fumes, mists, or odors are present, record this information as well as their source. Also determine production volume—is it an average production day or a light day? How many manufacturing variations can occur—is the product always the same? For example, metal alloys can vary considerably in the ingredients used in their manufacture. Also determine the

SAMPLE NO.	EMPLOYEE	POSITION	TYPE OF SAMPLE	EQUIP. NO.	FLOW RATE	START	STOP	TOTAL MIN	TOTAL LITERS	NOTES

Figure 18-2. Field sampling data sheet.

frequency of manufacture of a given product, any idiosyncrasies in the process as compared to "normal" production, and machine operating speeds. Include the source of the exposure. Are there opportunities for process by-products to occur? The use of a field sampling data sheet (Fig. 18-2) will assure that critical information such as flow rates and sampling-pump on and off times are not lost or confused among samples.

Always be objective when collecting data and making observations during sampling. For each employee being sampled, time should be spent observing whether activities and work practices are typical of the jobs. For example, a forklift operator may assist during the pour in a foundry or do other part-time work in between regular tasks. Keep track of employee movements, particularly if the individual is assigned to a position that "roams," for example, helps out at a number of different tasks. Be sure to note any special idiosyncrasies regarding how employees do their job, such as the use of compressed air, licking fingers, or wearing dirty work clothing home. A significant difference between personal exposure for two workers doing the same job on the same equipment is most likely due to differences in personal habits or work practices.

Try not to leave the area(s) being sampled. Check pumps and instruments routinely to make sure they are still operating and that the collection media are secure. It is not uncommon for an employee to pick up a cassette after it has fallen off and put it on backwards. If a pump stops it may not be noticed, especially with quieter, smaller models. Tubing may become kinked or pinched, blocking airflow during employee monitoring when shoulder movements can jerk the tubing from the pump's housing. One method is to check the pump or instrument within an hour of sampling and at least four times during the sampling period. In the event of a stoppage, if the pump has a timer or the instrument has a data recorder, it can be checked to see if the sampling period was sufficient. Timers on some pumps must be checked before the switch is turned off (even if they have stopped running) because the off switch will reset the timer to zero. Record any symptoms of possible overexposure that employees mention.

Evaluate as far as possible the effectiveness of controls, such as personal protective equipment and local exhaust. The mere presence of either is no guarantee that they are functioning effectively. For example, exhaust systems flow rates may not be sufficient or inlets may be blocked, allowing a buildup of contaminant that may be noted on tanks or surfaces such as floors, rafters, and machinery. The design of the system may also be inadequate for the type of contaminant. Note the presence of open garage-type doors, area fans, and other sources of air movement.

Take photos whenever possible. They can be used later as memory refreshers. Including photos with the report and noting specific conditions can make the report look more professional and make it easier to understand. Be sure to request permission to take photos ahead of time, because some companies have specific policies regarding the use of cameras on site. If a survey is being done for legal purposes, such as an OSHA survey, document everything to the best degree possible. For example, record the times when the flow rate was checked for each pump on the field sampling data sheet.

WHEN TO REJECT SAMPLE RESULTS

The first step involved in interpreting sampling data is to review and accept or reject them based on a set of criteria. These criteria can include determining whether the methods were properly carried out, if sensitivity is adequate for comparison to the standard of interest, if instruments are correctly used, and if sampling periods are sufficient. An attempt should be made to correlate values with the activities and conditions that were present during sampling, such as production levels, work practices, and local exhaust ventilation. Results for which no correlation can be found should be reviewed carefully and may require that further sampling be done. If prior sampling has been done, a comparison should be made with these measurements. If samples were collected using different methods, this may not be possible.

Ideally measurements represent only the period(s) that were sampled; however, since gaps can occur due to problems with pumps or loss of media, professional judgment is often necessary to make some assumptions as to what the exposures were during these periods. As it is often not cost effective to repeat an entire day of sampling due to a lost sample, some gaps must occasionally be tolerated. In general, time gaps should not exceed 30% of the actual workday (less breaks and lunch). It has been recommended that for noncompliance measurements (most often performed by an employer or consultant) exposures during unsampled periods should be assumed to be the same as those actually collected. If an operation is continuous and concentrations are unlikely to vary, and the employees remain within the work area, it would also be acceptable to assume that the exposure for the unsampled time is the same as what is sampled. In these situations, a TWA would be calculated by dividing the sample results by the actual time sampled and comparing the result with the 8-hour standard. The best comparison is usually the sample collected just before or just following this missed period.

For compliance measurements unsampled periods should usually be considered zero exposures.[8] For example, if the work shift is 8 hours and sampling was conducted for 7 hours and 15 minutes, a TWA would be calculated by dividing the sample results by 8 hours rather than the actual time sampled. After carefully selecting the personnel to sample, an employee may have to leave early for an appointment only to be replaced by another; therefore, the sampling professional must decide what to do with this sample—an 8-hour TWA for the job or individual samples for each employee? In any case, good field observations will assist in making good decisions about these periods of exposure.

Exposure measurements will sometimes vary from the true exposure due to the combined effects of the problems inherent in sampling and analysis. Pump flow rates may change or may not be accurately calibrated. When determining the length of stain or accuracy of the color change of detector tubes, diffusion, migration, and trailing can obscure the stain front and make readings questionable. In the laboratory, shade changes can be introduced by interferences, for example, in the Mercali (impinger) method for isocyanates, amines can change the color of the absorbing solution as well as collecting higher concentrations than for which the method was intended.[15] Furthermore, when the sampler takes into account the other variations present, such as air movement, concentration fluctuations, and process changes, it becomes apparent that regardless of the effort that goes into making a measurement there will be a certain amount of error inherent in every sampling episode (see Table 18-7).[15]

Therefore, the sampler should strive to minimize errors during sampling, since the data are only as good as the forethought and technique used in developing them. As described, there are a number of different statistical methods that can be applied to selecting the optimum number of samples and analyzing results. The use of a random numbers table and other techniques as previously discussed are examples. Since the focus of this book is collection techniques, the reader is encouraged to consult other sources that emphasize these techniques.[8,10,16] Quality control and careful sampling techniques will improve accuracy as well. There are many other techniques for minimizing error, such as including spiked samples, blanks, and sequential samples that are discussed in the overview of Integrated Sampling chapter.

Sampling results are important, but cannot take precedence over professional judgment. Generally the results of samples that have been collected using improper techniques or the wrong medium are unusable and sampling must be repeated. In some situations, such as switching types of filters, the results may be usable; however, a notation would be made in the report as to the change and how it is likely to have affected results. If the detection level of the results exceeds the value of the standard to be used, the results must be rejected and sampling must be repeated.

PRACTICAL CONSIDERATIONS

There are many circumstances out of the control of the sampling professional that can impact sampling methods and techniques. Consider that a typical foundry survey may require sampling for silica, formaldehyde, phenols, and noise, as well as possibly some ventilation measurements on local exhaust systems. This can involve carrying several cases with pumps, chargers, and flowmeters, as well as other cases with cyclones, dosimeters, a portable survey instrument, and assorted bags, tubes, filters, refrigerator packs, and so forth. Since sampling professionals are often located in centralized urban areas, several hours of travel to get to the sampling site are not uncommon. In addition, traveling to remote locations may limit the number of pumps available as well as decrease the number of representative samples collected. A 3-day survey prevents completely discharging batteries between uses and requires recharging the pumps each night, which can result in Ni-Cad batteries developing a memory. Finally, some mechanical ability is required since pumps and real time equipment may need some field repairs.

For some individuals there is a personal cost to so much travel. Since it is not usually cost effective to send two persons to do a job that one can accomplish, the sampler spends much time alone. While in some situations clients are very hospitable, more often everyone else goes home while the sampling professional is left to find dinner and prepare for the next day at the hotel. In remote areas, or traveling for the government on per diem, the accommodations may be basic at best.

Some surveys may involve potential exposures to special hazards. Mine surveys may include surface and underground mines and may involve exposure to explosives, cave-ins, and claustrophobic conditions, as well as contaminants such as carbon monoxide, nitrogen oxide, methane, hydrogen sulfide, carbon dioxide, and dusts. Thus, all personnel required to enter a mine should receive specialized training on mine safety and health procedures prior to entering. On industrial sites trains can be a hazard, particularly if they are remotely controlled and the operator cannot see persons near the train. Similarly, over-

TABLE 18-7. Common Sampling and Analytical Errors

Element	*Sources*
Sampling	Improper pump calibration inadequate frequency use of a nontraceable secondary standard incorrect flow rate calculation calibrator not corrected for temperature, pressure Variation in pump flow rate Improper media selection and/or preparation Improper flow rate Improper sample volume (high or low) Field blanks not submitted with samples Insufficient reference information Improper sample handling
Shipping/storage	Bulk samples shipped with samples to be analyzed Improper sample preservation/protection during shipment (oxygen, light, temperature) Excessive shipment time Cross contamination during storage Excessive storage time (decomposition/degradation) Sample damage during shipment/storage Failure to follow chain-of-custody procedures
Standards preparation/ spike preparation	Unknown purity of solvents, reagents Improper calibration of volumetric equipment Improper volumetric handling techniques Improper cleaning and storage of glassware Incorrect calculation of mass levels applied
Sample preparation	Nonquantitative sample transfer Improper dilution technique Cross contamination in handling sample Inadequate quality control samples
Analytical instruments	Improper instrument parameter used Nonregulated power sources Nonlinear working range Improper instrument calibration Specific instrument idiosyncrasies
Methodology	Incorrect recoveries of quality control samples Excessively high blanks Deviations from documented procedures Improper calculations of mass concentration Inadequate extraction efficiency data over the range of analysis
Data interpretation	Improper evaluation of reported results Using results where breakthrough has occurred Erroneous calculation of TWA Using the wrong standard (exposure limit) Failure to consider additive effects Ignoring unacceptable quality control results
Report generation	Typographical errors Lack of interpretation of results Lack of review Lack of information specific to the analyses of the samples Improper data and report storage

head cranes carrying materials are a concern because the operators may not be aware of pedestrians in the working area or may not be able to see persons walking into the path of a moving load. Thermal hazards may also be encountered; heat stress from ambient exposures and work activities, hot production or process equipment, hot materials, hot samples, and high-pressure steam. Surveys may also require work on heights including scaffolds, roofs, 8-story iron see-through staircases, or tanks. There are also certain situations that pose unique dilemmas. For example, in asbestos abatement work, showers are required for exit from the work area regardless of sex, and security may be poor.

Hanging sampling pumps on workers can require them to lift up shirts and pull on waistbands because many workers do not wear belts. A solution is to have them fasten the pump themselves, or provide a belt on which the pump can be hung. A recalcitrant employee who refuses to wear a sampling pump may force the sampler to take an area sample instead. Further, the possibility of several different contaminants being present at one work station along with a limited time to sample may require only one personal sample along with area samples for the other contaminants. Most workers have difficulty wearing more than two pumps.

While employees will often be on their best behavior when strangers are around, in a situation where there are significant management-labor relations problems just the opposite might occur, giving rise to the "clipboard phenomenon." In these situations industrial hygienists and other inspectors who walk around plants and mills carrying clipboards are frequently sought out by employees who have complaints. While the complaints may or may not be valid, it is important not to take sides and just to collect data. Also, sampling professionals may represent management. They may be reporting to management or they are hired by management (unless they work for the government) and therefore sampling professionals often walk a fine line. On the other hand, this should not discourage the sampling professional from talking to employees as they are often the best source of information on the process and the types of exposures that occur in their area. Frequently the sampler only has to ask an employee what he or she thinks should be done to correct a situation, and a good solution will be provided in very specific detail. This is also a good way to find out about maintenance problems and other situations that are very specific to a type of machinery or process that only individuals very familiar with it would know about.

The "you should have been here yesterday" syndrome is inevitable in that even when going to great lengths to inquire about a process to schedule the surveys during periods when production or high exposures are most likely the sampler will arrive only to find that stainless steel was being run yesterday and today it is mild steel, or that brass is occasionally run but not today. The samplers can only do their best within the limitations of the situation. In this case, the best thing to do is to document changes in operations that can occur and use the information to determine whether on different days exposures are likely to be *higher, lower,* or *different.*

Finally, any discussion of potential sampling considerations must take note that the depth of an investigation frequently depends on one's contractual relationship to an operation, and the costs versus the benefits of each sampling strategy must be evaluated. A prime concern is balancing the need to take an adequate number of samples with the cost of analyses and the amount of time required for the sampling. Increases in the number of samples in the case of integrated measurements will increase the costs of sampling because they will require more time and more media and analyses due to the increased number of samples. A consultant can only perform the scope of work that has been agreed to by the client. Even a sampling professional dedicated to a certain operation on site will have certain budgetary and time constraints for each task. This can be frustrating because the results, while adequate, may have been better with more research, the use of new methods, or more samples. If cost is no object and accuracy is the primary consideration, then the sampler should utilize statistics and multiple sampling periods in order to maximize the accuracy of results as increased numbers of samples enhance the correlation of measured concentrations with actual concentrations.

Learning to collect samples correctly is the first step; interpreting and applying the results so that the data are meaningful to a given situation is the next. At a minimum, if samples are collected according to the method with good preparations, such as noting any interferences and pre- and postcalibration, the results can be compared to a relevant standard and preferably will provide sufficient information on whether a health or physical hazard exists. Situations where no standards exist and where methods must be altered significantly will require additional practice and research so that these results are meaningful.

REFERENCES

1. Preamble to the OSHA final rule revising workplace air contaminants. *Fed. Reg.* 34(12), January 19, 1989.
2. Corn, M. Sampling strategies for prospective surveillance: Overview and future directions. *ACGIH: Advances in Air Sampling*. Lewis Pubs., Inc., Chelsea, MI, 1988.
3. Roach, S. A. A rational basis for air sampling programs, *Ann. Occup. Hyg.* **20**:65–84, 1977.
4. Roach, S. A. A commentary on the December, 1986 workshop on strategies for measuring exposures. *AIHA J.* **48**(12):A-822–A-832, 1987.
5. First, M. W. Air sampling and analysis for contaminants in workplaces. *ACGIH: Air Sampling Instruments*, 6th ed. ACGIH, Cincinnati, OH, 1983.
6. Rappaport, S. M., S. Selvin, R. C. Spear, and C. Keil. Air sampling in the assessment of continuous exposures to acutely-toxic chemicals, Part I—Strategy. *AIHA J.* **42**(11):831–838, 1981.
7. Hinton, D. O. Community air sampling. *ACGIH: Air Sampling Instruments*, 6th ed. ACGIH, Cincinnati, OH, 1983.
8. Leidel, N. A., K. A. Busch, and J. R. Lynch. *Occupational Exposure Sampling Strategy Manual*, NIOSH, Cincinnati, OH, 1977.
9. Leidel, N. A., and K. A Busch. Statistical design and data analysis requirements. *Patty's Industrial Hygiene and Toxicology*, vol. 1. Wiley, New York, 1978.
10. Burton, D. Jeff. *Simple Rules-of-Thumb for Use in Industrial Ventilation*. Occupational Health and Safety, November, 1988.
11. U.S. Dept of Commerce, Environmental Science Service Administration, Environmental Data Service. *Climate Atlas of the U.S.*
12. ASTM D 4480-85. *Standard Method for Measuring Surface Wind by Means of Wind Vanes and Rotating Anemometers.*
13. ASTM E 337-84. *Standard Test Method for Measuring Humidity with a Psychrometer.*
14. EPA. *SOPs for Work on Hazardous Waste Sites*. EPA, Washington, DC, 1986.
15. AIHA. *Quality Assurance Manual for Industrial Hygiene Chemistry*. AIHA, Akron, OH, 1986.
16. Bar-Shalon, Y., et al. *Handbook of Statistical Tests for Evaluating Employee Exposure to Air Contaminants*. NIOSH, Cincinnati, OH, 1975.

CHAPTER 19

Specific Sampling Situations

CONFINED SPACE MEASUREMENTS

By the National Institute of Occupational Safety and Health's (NIOSH) definition, a confined space has limited openings for entry and exit and unfavorable natural ventilation that could contain or produce dangerous air contaminants; and it is not intended for continuous employee occupancy (Fig. 19-1). Confined spaces can be a concern in situations other than occupational ones, such as when residences and offices built on or near sanitary landfills develop high levels of methane gas in their basements.

Figure 19-1. A confined space.

Open-topped water and degreaser tanks, open pits, and enclosures with bottom access are examples of confined spaces. They prohibit natural ventilation, are potential sources of gas generation, and can keep gases from escaping, thus causing a potentially hazardous atmosphere. Leaking materials from storage tanks, natural gas lines, underground storage tanks, process flanges, and valves can find their way into confined spaces. A number of hazards can exist depending on the nature of the leaking gas or liquid. For example, leaking gases or vapors can displace available oxygen, and can reach explosive and toxic concentrations that can be immediately dangerous to life or health.

Decomposing organic matter, such as domestic water and plant life, can produce methane, carbon monoxide, carbon dioxide, and hydrogen sulfide, and can consume existing oxygen.

Combustion products from gasoline or diesel engines, welding, cutting and brazing, and rusting metal can create hazards as well. Oxygen is consumed by the combustion or oxidation processes, or displaced by the combustion products. Carbon monoxide is produced by incomplete combustion and other gases and fumes can be produced by welding operations. Explosive and toxic gases, such as hydrogen sulfide and carbon monoxide, and a lack of oxygen cause the majority of confined space injuries and fatalities. More than 60% of the fatalities are among would-be rescuers of initial victims.

In addition to air monitoring prior to entry, if a tank or other container held an unknown compound a bulk sample of the material should be collected and analyzed prior to entry. The potential for dermal exposure should always be evaluated whenever individuals enter a process vessel that recently contained hazardous materials, since flushing vessels is difficult and often sludge and other residual material is left behind on the bottom and on various structures in the vessel. If the purpose of entry is cleaning, sometimes wipe samples are done.

When sampling in a confined space, do not enter before taking measurements of contaminants (Fig. 19-2). Instead use either the tubing provided with pump-drawn samplers or the remote sampling cord attached to a passive sensor that allows it to be dropped inside the space. Do not take a sample near an entry way and assume that it represents the atmosphere throughout the vessel. Also sampling should be done periodically and not just in the morning, especially when work is done that disturbs the interior surface. Even after an empty tank has been purged, gases can desorb from porous walls or be liberated from sludge during cleaning. Toxic gases can be liberated from sludge or from cleaning solvents, or produced by chemical reactions with cleaning solvents and other materials.

The primary confined-space measurements are for oxygen deficiency and combustible gases. However, other gases should also be tested for if there is any possibility they may be present, such as hydrogen sulfide and carbon monoxide continuous monitoring over the entire course of the job. There are a number of instruments for this application that can continuously monitor for all of these compounds simultaneously.

Oxygen above the normal level of 21% increases the flammability range of combustible gases or material and causes them to burn violently. Be very careful when suspending passive detectors into tanks where there may be liquids or in which there is uncertainty of the contents. Submersing the detector in liquids will require repair or replacement.

The type of work to be done inside the confined space will determine what is an acceptable measurement reading. For example, if welding is to be done and there is any possibility that chlorinated materials may have been inside of a tank at any point, in addition to combustibles and oxygen deficiency, the atmosphere should also be tested for very low levels of halogenated hydrocarbons. The welding arc as well as other high heat sources can convert these compounds to phosgene, which is toxic at very low levels.

Procedure

1. All instruments should first be checked for a proper zero indication for combustible and toxic gases and for 20.9% oxygen in fresh air. Check alarm response using a test source.

2. Sample through a pick hole or open the cover slightly on the downwind side before opening the cover completely.

3. If sampling inside a tank, hook the extension tube onto the instrument. Note that the addition of the extension tube increases the response time of the instrument. Check the hose for leaks by placing a finger over the end of the hose inlet while listening to the pump. If leakage is occurring, the pump will not labor.

4. Sample at several levels and take multiple samples. In general, levels of gases will form gradients based on their densities relative to air. The lack of normal ventilation allows these gases to stratify because normal air currents that cause mixing and dilution are not present.

5. Check any areas where gases can "pocket," such as below steel gratings or between rafters.

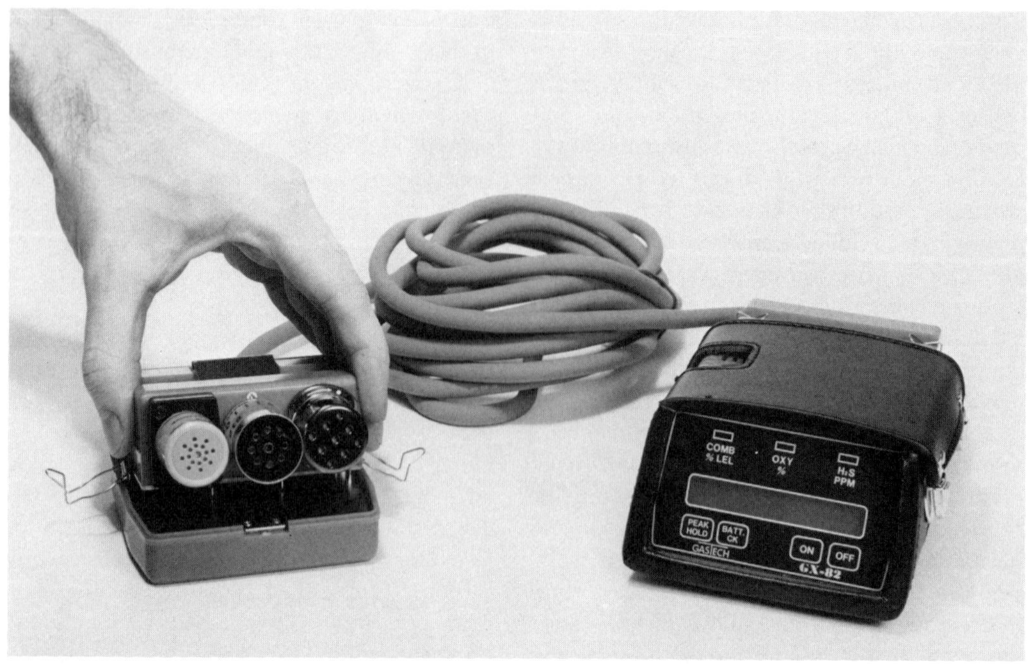

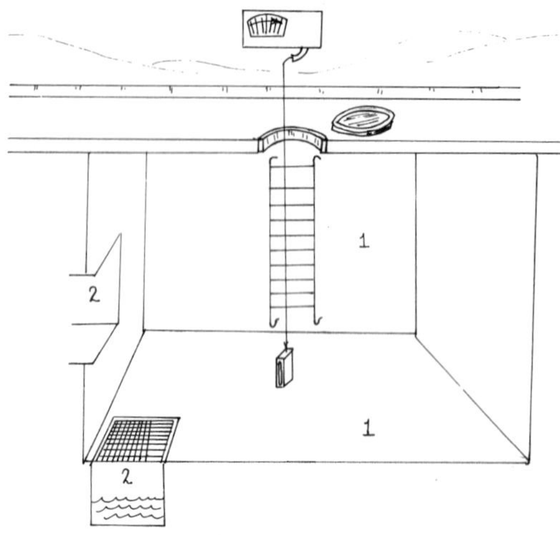

Figure 19-2. Testing apparatus (photo) and a diagram of recommended testing points (1 = test before entry; 2 = test after entry) for confined spaces. (Photo and diagram courtesy Enmet.)

6. Once work begins, sample frequently or continuously, because conditions can change. As work progresses, a once-safe atmosphere can become hazardous due to leaks, combustion, cleaning processes, or other operations such as welding. For example, scraping scale off a tank's interior can release trapped gases such as hydrogen or hydrogen sulfide.

There are many additional requirements for making a confined space entry. For this reason, it is important to be aware of all the actions needed to do a safe confined space entry even if the sampler's responsibilities are limited to making the measurements only.

INDOOR AIR SURVEYS

Indoor air quality has been dramatically affected by building designs produced in response to energy concerns. In attempts to save energy, buildings have been sealed and insulated, fresh air intakes have been reduced, and existing air has been recirculated often without concern for the fact that sources of contaminants, for example, diazo printers, are commonly incorporated into office buildings. The outcome has been "tight building syndrome." In many cases, occupants complain of a variety of health problems that disappear when they go home. Symptoms include headaches, fatigue, dizziness, skin irritation, dry skin, rash, dry mucous membranes, coughing, itching, nausea, throat irritation, and nonspecific hypersensitivity reactions.[1]

Combustion by-products and organics are two of the most commonly found indoor pollutant sources. Indoor combustion sources include tobacco smoke, gas-fired stoves, unvented gas-fired space heaters, gas-fired water heaters, and kerosene heaters. The major by-products are carbon monoxide, nitrogen oxides, formaldehyde, and respirable particulate. Outdoor sources include gasoline-fueled autos and diesel-fueled trucks. Trucks at loading docks are often left to idle. Exhaust can enter through elevator shafts and building ventilation intakes. If exhaust vents for indoor sources are improperly placed, this material can be reentrained indoors as well.

Formaldehyde is used in a variety of products, mainly in urea, phenolic, melamine, and acetal resins. These resins are present in insulation materials, particleboard, plywood, textiles including upholstery coverings and carpet, and adhesives. Formaldehyde is also given off in cigarette smoke. Since other aldehydes may also be present, sampling for total aldehydes may be more useful than using a method specific for formaldehyde.

Organic compounds may be given off by building materials, cleaning products, tobacco smoking, furnishings, consumer products including paints, building occupants, and most microbes.[2,3] Table 19-1 contains a summary of sources and types of indoor air pollutants. Pesticides may also be present, since spraying of indoor areas around the perimeter of rooms (baseboard) is often done for insect control. Often a total hydrocarbon sample is collected on a charcoal tube. Another approach is to use several different types of tubes in order to collect both volatile and semivolatile contaminants from different families such as aromatics, aliphatics, alcohols, and ketones. When comparing the results of total hydrocarbon samples to samples collected at a different time or by different sampling personnel, the compound whose response factor was used for quantitation should be identified as this can vary.[4] The following organic compounds have been detected in office buildings:[2]

Hydrocarbons
- *n*-hexane
- *n*-heptane
- *n*-octane
- *n*-nonane
- *n*-undecane
- 2-methylpentane
- 3-methylpentane
- 2,5-dimethylheptane
- methylcyclopentane
- ethylcyclohexane
- methylcyclohexane
- pentaethylheptane

Aromatics
- benzene
- toluene
- xylene

Halogenated Hydrocarbons
- trichloroethane
- trichloroethylene
- tetrachloroethylene

Miscellaneous
- hexanal
- methylethylketone

Other concerns during indoor air surveys are low relative humidity (RH) and temperature changes. RH can cause irritation and other discomfort to mucous membranes if it is too low. It is also an important factor controlling the viability of both airborne and surface microorganisms. Thermal discomfort is a stressor and may be perceived as poor air quality.[3]

Indoor air surveys are best done utilizing a team involving sampling professionals such as industrial hygienists, ventilation engineers, and physicians. Develop a questionnaire for building occupants regarding their symptoms. Map the reported symptoms on a building floor plan. Generally there will be a wide spectrum of response,

TABLE 19-1. Summary of Sources and Types of Indoor Air Pollutants

Sources	*Types of Pollutants*
Outdoor	
Stationary sources	Sulfur dioxide, carbon monoxide, nitric oxide, nitrogen dioxide, volatile hydrocarbons, particulates
Motor vehicles	Carbon monoxide, nitric oxide, nitrogen dioxide lead, particulates, sulfur dioxide
Soil	Radon
Indoor	
Building construction materials	
Concrete, stone	Radon and other radioactive elements
Particleboard	Formaldehyde
Insulation	Formaldehyde, fiberglass
Fire retardant	Asbestos
Adhesives	Organics
Paint	Organics, lead, mercury
Building contents	
Heating and cooking	Carbon monoxide, nitric oxide, nitrogen dioxide, sulfur dioxide, particulates
Copy machines	Ozone
Drinking water	Radon
Human occupants	
Metabolic activity	Hydrogen sulfide, carbon dioxide, ammonia, organics, odors
Biologic activity	Microorganisms
Human activities	
Tobacco smoke	Carbon monoxide, nitrogen dioxide, hydrogen cyanide, organics, odors, particulates
Cleaning and cooking	Organics, odors
Hobbies and crafts	Organics, metals, particulates, odors

from individuals with no symptoms to those with pronounced symptoms. Not everyone will experience the same symptoms. The areas where the highest incidence of symptoms reside can be targeted for priority inspections and monitoring.

The building inspection involves collecting information on potential sources of contaminants inside and outside of the building. Find out if any changes occurred in the building since the symptoms were noted. For example, a carpet may have been installed. A number of different glues are used to install carpets and the carpets themselves contain various chemicals. Bulk samples are often collected of materials such as carpet that are suspected of emitting contaminants. In a recent study, 4-phenylcyclohexene, dichlorobenzene, *bis*(2-ethyl-hexyl) phthalate, triethyl phosphate, (-caprolactam), and methylene-*bis* (4-isocyanatobenzene) were detected in a carpet sample.[5] Although these might be categorized as organics, it is of interest that none are currently specifically sampled for during indoor air studies, which shows that there is still more information to be learned about these situations.

Identify outdoor sources of contamination that might enter the building. If there are factories nearby or specific sources of a chemical such as sulfur dioxide or ammonia, these things should also be taken into consideration. A walk through the neighborhood should be done to identify sources of pollutants such as factories within a 1-mile radius. It is also useful in the case of a new building to research the history of the property. For example, sanitary landfills release a number of different gases including vinyl chloride and methane. Leaking underground storage tanks can contaminate neighboring properties, causing significant odors inside basements. Combustion products are especially important due to their ubiquitous presence and the fact that often they migrate through vents as a result of discharge sources that are located too near them. Building exhaust and intake vents should be identified including their proximity to each other. A misplaced flue pipe for a hot-water heater could result in nitrogen oxides being pulled inside the building. Figure 19-3 contains information that can be used to identify sources of indoor and outdoor contamination.

There are many areas indoors where growing conditions are favorable to microorganisms, such as stagnant pools of water in basements and sumps, air-conditioning reservoirs and ducts, and cooling towers. Microorganisms can also enter from outdoors. If individuals are symptomatic, air sampling should be conducted to at least rule out microorganisms as a source of disease. If a reservoir of a contaminant is suspected, a bulk sample should be collected.

Where strong odors are noted, a general survey instrument such as a PID analyzer can be useful. The area where odors are the strongest should be monitored first. If a basement or lower level has the highest levels and no intakes of outdoor air are present, then an underground source should be suspected. Since odors fluctuate, every effort should be made to perform surveys on days when odors may be present.

A good practice is to document the background levels of outdoor contaminants in addition to the levels of contaminants inside in order to determine whether these contaminants are being entrained from the outside by the ventilation system or being generated by sources within. Therefore, both indoor and outdoor samples should be collected for contaminants such as carbon monoxide, nitrogen oxides, total particulate, ozone, and volatile organic hydrocarbons. Portable GCs have also been used to take samples of indoor and outdoor air to identify whether a source was inside or outside the building. Sampling results should be entered on a floor diagram of the building. Radon and asbestos do not normally cause symptoms in building occupants; however, they are commonly sampled due to concerns for their carcinogenicity.

Initiate a discussion of the building heating, ventilating, and air-conditioning (HVAC) system with maintenance personnel to identify how often and what percent of outside air is allowed into the building. It is not uncommon for most buildings to recirculate air 100% over certain times of the day such as evenings. It should be noted that while most often indoor air quality problems are remedied by increasing the ventilation rate in the affected area, this remedy is in conflict with strategies for reducing the energy consumption by lowering ventilation rates; therefore, sources and chemical characteristics of indoor air pollutants should be identified so that energy-efficient control strategies can be implemented to eliminate them.[2]

Carbon dioxide is used as a surrogate measure for the lack of ventilation in a building and the potential that other contaminants could be building up. An increase of this by-product of human respiration over the course of a day is an indicator. Carbon dioxide is measured with an infrared spectrometer. As mentioned, temperature and humidity should also be monitored, since they can have a significant impact on the discomfort of building inhabitants. One way is through the use of an instrument that will measure all of these factors simultaneously and log the data for printout and analysis.

The AQ-501 is an example of an instrument designed to monitor the variables associated with ventilation problems in indoor air situations. The instrument has three dedicated channels for monitoring carbon dioxide, temperature, and humidity, and one extra channel. Through an auxiliary sensor, the extra channel can be utilized for monitoring additional variables, such as air velocity or toxic gases. The measurements are automatically stored in the data-logging memory of the instrument. Carbon dioxide is monitored by a nondispersive infrared (IR) sensor. The term *nondispersive* is used to describe the fact that no prisms or gratings are used in the monitor to disperse the IR energy source into component wavelengths. Instead the measurement gas itself, carbon dioxide in this case, is used in the detector to detect wavelength, and hence species specificity.[6] Figure 19-4 (page 469) shows the monitoring apparatus and an

(Text continues on page 470.)

I. General Area Description

1. Building Address:

   Provide location on sketch.

2. Describe geographic characteristics of the immediate (½-mile radius) neighborhood. Include:

   a. Percent open land ________ %
   b. Is a stream located in the area? ☐ yes ☐ no
   c. Is the area hilly? ☐ yes ☐ no
   d. Are there tall structures that affect wind flow? ☐ yes ☐ no
   e. Are there any condemned or demolished structures in the area? ☐ yes ☐ no

3. Describe each of the neighboring buildings or open areas. Indicate location of each on sketch.

Relative Location	*Type of Structure/Open Area*	*Approx. Age of Structure*	*Known or Observed Chemical Sources*

Figure 19-3. Indoor air screening data. (From Environmental Protection Agency 600/6-88009A, August 1988.)

4. Describe any potential point source for the chemicals or pollutants of interest within an area described as follows. Establish a ½-mile radius centered on the structure of interest. Extend the area an additional ½ mile upwind, creating an oval shaped area. Indicate the location of each point source on the area sketch.

Relative Location	*Distance*	*Type of Chemical*	*Comments*

5. Are there any major freeways within the area created in Section 4? If so
   a. Draw freeway on sketch
      Average traffic levels during:
      Rush hours ________ vehicles/hour
      Other times ________ vehicles/hour
   b. Indicate any major city streets
      Average traffic levels during:
      Rush hours ________ vehicles/hour
      Other times ________ vehicles/hour
   c. Average traffic flow on nearest street: ________ vehicles/hour

6. Has there been any exterior pesticide application at any neighboring buildings or areas within the last 30 days?

7. a. Describe prevailing winds or attach a wind rose.

   b. Outside temperature: high ________ °F low ______ °F

Figure 19-3. *(continued)*

c. Outside RH:

d. Wind speed and direction:

e. Barometric pressure and tendency:

II. Building Characteristics

1. Describe the site in terms of usage and surroundings.

2. Age of building. Include information on additions and major renovations.

3. a. Approximate square footage of the building, including each floor if multistory.

   b. What is the approximate ceiling height in the majority of the structure?
   ________ feet

4. Number of floors above the substructure: ________ floors

5. What are the structural materials of the exterior of the building?

6. Does the building have an attached or enclosed garage or a loading dock?

Figure 19-3. *(continued)*

7. a. What is the source of water for the structure?

   b. What is the primary source of public supply?

8. Describe the internal construction characteristics of the building:

   a. Are there:

False walls?	☐ yes	☐ no
Movable walls?	☐ yes	☐ no
Movable partitions?	☐ yes	☐ no
False ceilings?	☐ yes	☐ no
Interfloor spacing?	☐ yes	☐ no

   b. What are the surface materials of:

   Walls ______________________________

   Floors ______________________________

   Ceilings ______________________________

   Nonfixed structures ______________________________

9. a. Main type of heating:

   b. Secondary sources of heating:

Figure 19-3. *(continued)*

c. Main type of heating fuel:

d. Type of fuel used for cooking in the main kitchen:

10. What is the source of hot water for the site?

11. What type of air-conditioning system(s) are present?

12. a. Are there any fixed ventilation systems, other than the heating and air conditioning, present in the structure? ☐ yes ☐ no
Specify type, location, control mechanism, filters, and particle collectors.

b. Describe all secondary ventilation devices, including portable fans, and kitchen and bathroom exhausts.

13. Describe openings in walls and ceilings.

a. How many windows are present? ________ windows

b. What percent can be opened? ________ %

Figure 19-3. *(continued)*

c. Describe any windows of unusual type or construction.

d. How many doors and other penetrations of walls are present?

_______ doors

_______ other penetrations

e. How many penetrations through the ceilings and floors are present?

14. If known, what is the general rate of air exchange for this building?

15. What are the normal temperature and RH settings for the common areas of the building?

	Winter		*Summer*	
	Day	*Night*	*Day*	*Night*
Temperature				
RH				

16. Changes in the building in the past six months:

a. Describe any major contruction, renovations, or modifications.

b. Describe any specific weatherization, or building tightening actions taken during the same period.

Figure 19-3. *(continued)*

c. Have any changes been made in this building that include the addition of:

Furniture, drapes	☐ yes	☐ no
Synthetic carpet	☐ yes	☐ no
Wallpaper	☐ yes	☐ no
Interior paint	☐ yes	☐ no
Ceiling, floor finishes	☐ yes	☐ no
Plywood, particleboard	☐ yes	☐ no
Foam insulation	☐ yes	☐ no
Interior paneling	☐ yes	☐ no

17. a. How often is the interior of this building treated with pesticides?

b. When was the last time this building was treated?

c. By whom was the building last treated?

d. What product(s) were used and in what quantities?

e. Describe formulation, application instructions, and other details of application.

Figure 19-3. *(continued)*

18. Describe cleaning products generally used in this building. Include soaps, waxes, deodorants, disinfectants, polishes, etc.

Product	*Quantity Used in Average Month*	*Frequency of Application*	*Last Application*

19. In the location, is there:
    a. Foam insulation ☐ yes ☐ no
    b. Polyurethane products ☐ yes ☐ no
    c. Plywood subflooring ☐ yes ☐ no
    d. Wall paneling ☐ yes ☐ no
    e. Composition board ☐ yes ☐ no
    f. Particleboard ☐ yes ☐ no

20. Identify and give the age of furniture in this location.

Item	*Age*	*Construction Material*	*Approx. Surface Area*

Figure 19-3. *(continued)*

21. Describe the type, surface area, and age of any wall coverings in the location including backing material and glue/paste.

22. Describe any carpeting:
    a. Age:
    b. Backing:
    c. Glues:
    d. Surface Area:

23. Describe any pressure gradients present in this structure, especially those near potential sources of the pollutants of interest.

24. Average number of occupants in the location.
    Day:
    Night:

25. Number of smokers present during the monitoring period:

26. Were there any pets present during the monitoring period and were flea collars or flea powders in use?

27. Did anyone engage in any of the following hobbies in the location during the monitoring period?
    a. Woodworking ☐ yes ☐ no
    b. Painting ☐ yes ☐ no
    c. Ceramics/Pottery ☐ yes ☐ no
    d. Photographic developing ☐ yes ☐ no
    e. Other Type ______________________

Figure 19-3. *(continued)*

28. a. Describe the number and type of pest strips in use.

    b. Identify any other pesticides in use and time of use during monitoring.

29. Were any windows in the location open during the monitoring period?

    Percent of time open ________ %

30. a. Is there a gas cooking stove at the location?

    b. Is it vented?

    c. Does it have a pilot light?

    d. What percent of time was it in use during monitoring?

31. a. Is there a gas or kerosene space heater in the location?

    b. Is it vented?

    c. Does it have a pilot light?

    d. What percent of the time was it in use?

32. a. Was a free-standing stove or fireplace in the location in use during the monitoring period?

    b. Percent of time in use.

    c. Energy source.

    d. Is it vented?

33. a. Was a clothes dryer in use during the monitoring period?

    b. Percent of time in use.

    c. Energy source.

Figure 19-3. *(continued)*

34. a. Was a humidifier in use in the location during the monitoring period?
    b. Percent of time in use.

35. a. When was the area last vacuumed?
    b. When was the area last dusted?

36. a. What is the rate of air exchange in the kitchen?
    b. What is the rate of air exchange in the rest of the location?

37. Household products

Product	*Number of Hours Since Last Use*	*Amounts Used*	*Frequency of Use*	*Surface Area Covered*
a. Cleaning				
b. Aerosol				
Drugs/Bronchodilator				
Vaporizer				
Hair Products				
Personal Hygiene				
Deodorant				
Foot Spray				
c. Housekeeping				
Disinfectants				
Waxes				
Bathroom/Kitchen Deodorants				

Figure 19-3. *(continued)*

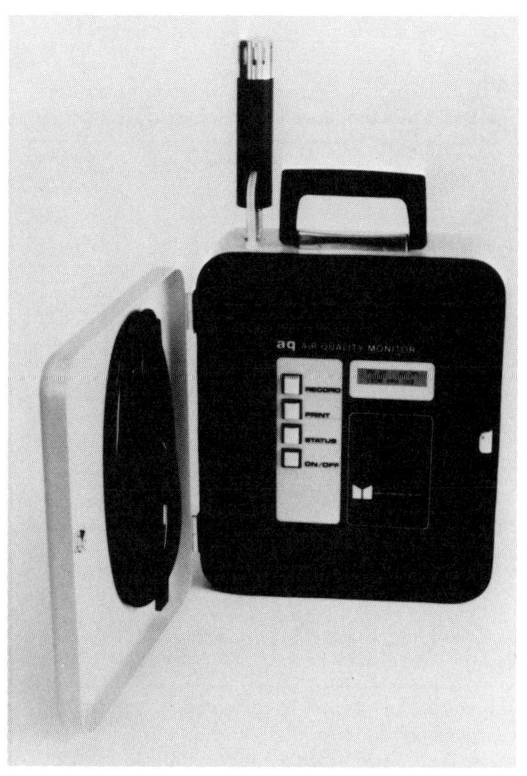

Figure 19-4. The AQ-501 air quality monitor (photo) and the report after 17 hours and 51 minutes of sampling for carbon dioxide, temperature, and RH during monitoring for tight building syndrome. (Photo and report courtesy Metrosonics, Inc.)

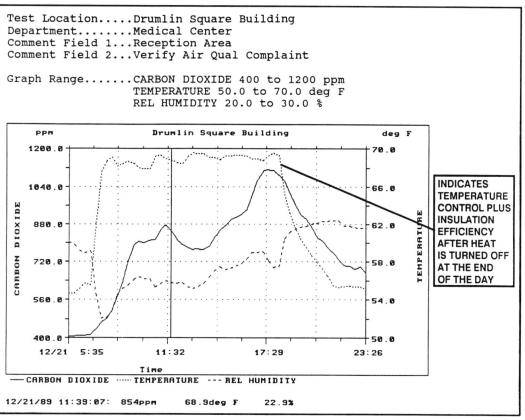

```
Test Location.....Drumlin Square Building
Department........Medical Center
Comment Field 1...Reception Area
Comment Field 2...Verify Air Qual Complaint

Graph Range.......CARBON DIOXIDE 400 to 1200 ppm
                  TEMPERATURE 50.0 to 70.0 deg F
                  REL HUMIDITY 20.0 to 30.0 %
```

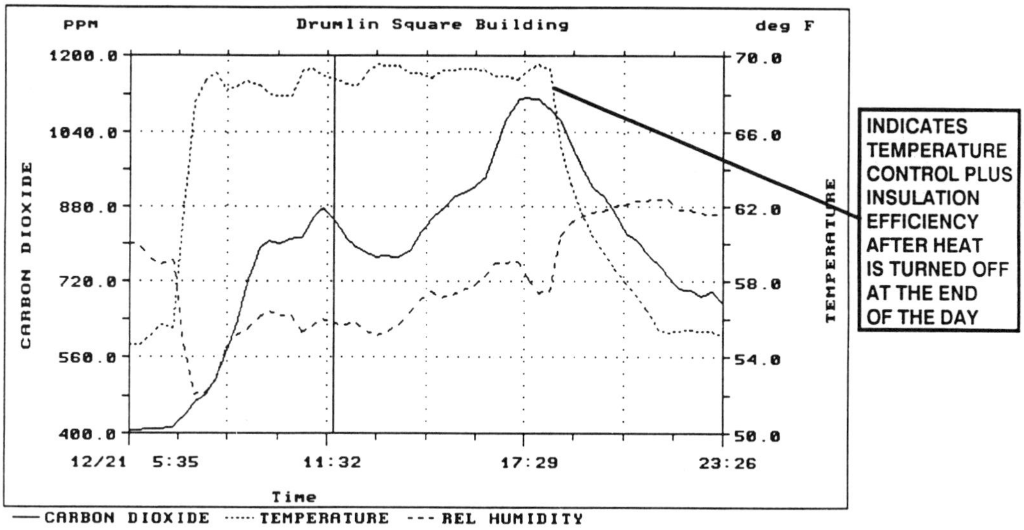

— CARBON DIOXIDE ····· TEMPERATURE --- REL HUMIDITY

12/21/89 11:39:07: 854ppm 68.9deg F 22.9%

TABLE 19-2. Formaldehyde Guidelines for Indoor Air Situations

Level (ppm)	*Interpretation*
0.01–0.05	Background (outdoor) levels
0.04–0.09	Average conventional home
0.09–0.62	Average mobile or prefab home
≤0.10	ASHRAE guideline for indoor air
≤0.40	HUD standard for prefab homes
>0.1	Mild irritation or allergic sensitization in some people
>0.5	Irritation to eyes and mucous membranes
>1.0	Possible risk of nasopharyngeal cancer
>3.0	Respiratory impairment and damage

Source: Office of Pesticides and Toxic Substances, US EPA. *Assessment of Health Risks to Garment Workers and Certain Home Residents from Exposure to Formaldehyde.* Washington, D.C., 1987.

TABLE 19-3. Comparison of Daily Outdoor Levels with Indoor Levels for Selected Chemicals

	Average Daily Concentration (ppb)	
Compound	*Indoor*	*Outdoor*
Acetone	7.955	6.927
Benzene	5.162	2.800
Carbon tetrachloride	0.40	0.168
Chloroform	0.832	0.630
1,3-Dichlorobenzene	3.988	0.889
1,4-Dichlorobenzene	0.932	0.996
1,4-Dioxane	1.029	0.107
Formaldehyde	49.4	8.293
α-Pinene	0.547	0.484
1,1,2,2-Tetrachloroethane	0.014	0.101
Tetrachloroethene	3.056	0.853
Toluene	7.615	0.775
Trichlorobenzene	0.065	0.016
1,1,1-Trichloroethane	48.90	0.911
Trichloroethene	1.347	0.495

Source: Shah, J. J., and H. B. Singh. Distribution of volatile chemicals in outdoor and indoor air. *Environ. Sci. Technol.* **22**(12):1381–1388, 1988.

example of a printout obtained during a 17-hour, 51-minute testing period in an office building.

The American Society of Heating, Refrigeration, and Air Conditioning Engineers (ASHRAE) has developed a number of guidelines for indoor air contaminants, in most cases based on state or EPA standards or guidelines. In some cases, a level can be compared to a known health effect occurring at that level. An example of a compound typically measured in residences and indoor air situations is formaldehyde. Table 19-2 describes guidelines that have been developed for formaldehyde levels in indoor air.[7] In these situations, workplace levels are unlikely to be present and therefore these standards would not be applicable. ASHRAE standards are useful for interpreting results in indoor air. For example, ASHRAE recommends that carbon dioxide indoor concentrations be maintained below 0.08%. When adapting NIOSH methods to collect samples to compare to these low-level standards, modifications such as extended sample times are required.

Outdoor samples (Table 19-3)[8] should be compared to a relevant EPA standard if one exists. In addition, whenever collecting outdoor samples, an estimation of typical levels present in the background should be made. This information is available from a number of sources, including a data base established by the EPA. The ratio of the results of indoor:outdoor sampling can be used to provide insight as to the source of contamination. For example, for volatile organic hydrocarbons such as benzene and xylene, ratios close to one suggest an outdoor source.[4]

ASBESTOS ABATEMENT SURVEYS

Asbestos abatement and its associated monitoring has evolved into a multifaceted sampling strategy involving air, bulk, and surface sampling. Air sampling is conducted on personnel inside and outside containment areas where abatement is taking place. A specialized type of sampling, termed *clearance sampling,* involves collecting large volumes of air using high-volume pumps. A similar technique to clearance methods is used when conducting risk assessments in buildings where owners have elected to leave asbestos in place. Bulk sampling has been conducted widely; however, as concerns increase regarding nonfriable asbestos-containing materials (ACM) such as roofing felts and vinyl asbestos tile it is likely that bulk samples will continue to be collected. Also it is not uncommon for new suspected sources to be identified during abatement, and additional samples will have to be collected.

Areas where more sampling is likely to be done are residences and small private buildings. Concerns developing as a result of environmental audits for real estate transfer will generate a need for bulk sampling. Surface sampling, often using adhesive-backed tape, is done when it is suspected that releases have occurred as a result of maintenance or poor abatement activities. Surface contamination has resulted in large-scale disposal of building contents including books and furnishings.

Asbestos Bulk Sampling

The first step in asbestos surveys is generally a bulk sampling survey in which a wide variety of materials are sampled for the likelihood of containing asbestos. These would include insulation, wallboard (especially transite), ceiling tile, and floor tile and its mastic (adhesive). Several problems can be encountered in the course of collecting asbestos-suspect materials. These include (1) difficulty in collecting representative samples; (2) exposure to the person collecting the sample; (3) contamination of facilities; (4) exposure to others in the area; (5) risks associated with the use of ladders, scaffolds, and lifts; (6) risks associated with the entry of confined spaces, such as crawl spaces; (7) building tenant/public relations problems associated with concerns generated when sample collection is conducted by samplers wearing "space suits."

Problems with representative samples include not collecting enough samples of similar materials, contamination of samples with materials from other samples due to not decontaminating equipment, and existence of multilayers of different materials. Exposures to sample collectors can occur even when wet methods are employed, because the wetting agent may not be able to penetrate through a covering or penetrate deeply enough to get to the hard substrate, therefore leaving the materials dry enough to release fibers during sampling. Wetting of high surfaces, such as vaulted ceilings, is a difficult task involving risk. Materials may be knocked loose during sampling, which poses a source of exposure to bystanders as well as offering the potential for fibers to be tracked throughout other areas of the building. If sampling is being conducted inside an area that is a plenum, and fibers are released, the materials can get into the HVAC system and be transported to other areas of the building.

Prior to the survey, assemble floor plans if available and develop a sampling strategy. If sampling for the Asbestos Hazard Emergency Response Act (AHERA) regulations, a specific number of samples for a given square footage is required. In addition, both friable and nonfriable sources of asbestos should be considered. In many cases, three samples of each suspect type of material may be sufficient; however the number of samples will not identify each and every pipe elbow, for example, those containing asbestos, but should be sufficient to identify trends. In most cases, if all three samples are less than 1%, unless the state regulations differ, such as in California where a content of 0.1% is sufficient to declare the sample asbestos, these samples are sufficient to show that asbestos is not present. There are several methods for collection of bulk asbestos-suspect samples currently in use (Fig. 19-5). The most popular include

1. Use of a knife and tweezers to cut and pull a sample, with collection in a container such as a zip-lock bag
2. Use of cork borers either manually or using electric drills with a "hole-saw" bit
3. Use of sharpened pieces of pipe (similar to using cork borers)
4. Use of film canisters or plastic tubing (NIOSH/EPA method)
5. Physically breaking suspect material by hand or with other tools
6. Picking up loose debris from the ground or floor.

In addition, other specialized tools are now marketed for the collection of asbestos samples (Table 19-4). An example is the PL-1 bulk sampler. This device consists of a cutting tool and a low-volume, high-velocity air curtain that collects any fibers released during sample collection. All air is collected using a vacuum cleaner equipped with

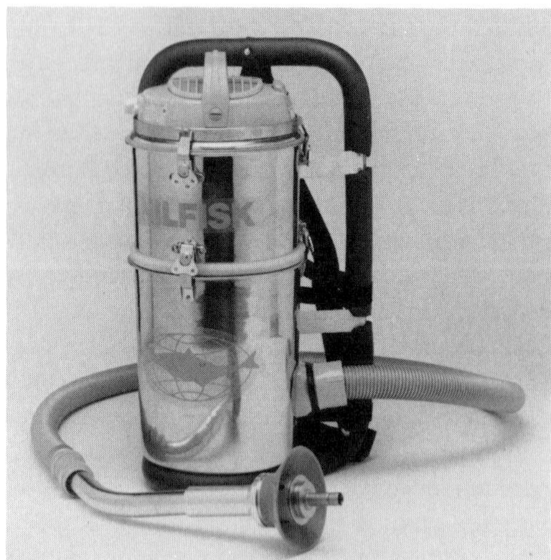

Figure 19-5. The PL-1 bulk sampler attached for use with HEPA-filter vacuum (photo). The diagram shows an exploded view. (Photo and diagram courtesy Nilfisk, Inc.)

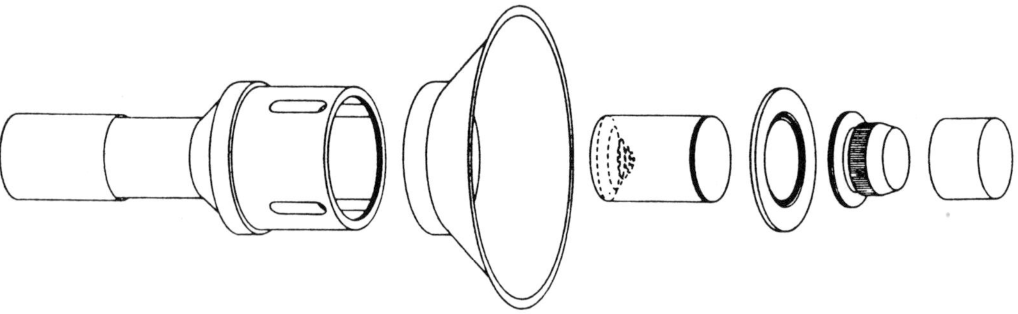

a high-efficiency particulate air (HEPA) filter. Adding extensions to the vacuum cleaner hose can enable sampling of relatively high surfaces.

Procedure

1. Assess the physical characteristics of the building prior to the survey and determine the following: date of construction; dates of renovations, repairs, and additions; function and purpose of the building.

2. Request copies of floor plans. These will be used to map sample and suspect material locations, and to ensure the building is thoroughly inspected. Depending on the detail and accuracy of the floor plans, they may also assist in determining material square footage and pipe length as appropriate.

3. Categorize suspect materials into friable and nonfriable categories. *Friable* includes any materials that will crumble with hand pressure and that have been known to contain asbestos, such as spray on insulation, finishing plaster, and pipe insulation. *Nonfriable* includes materials bonded well together, and they can be cementitious or softer, but they require some disturbance to occur, such as breaking, sanding, or other destructive activity to release fibers. Such materials include transite, floor tile, and roofing and siding shingles.

Suspect materials that do not have visual characteristics of fiberglass or other known nonasbestos substitutes, or for which some doubt as to their composition exists, are labeled *other.* They include materials sampled under previous surveys that lack documentation as to proper sampling technique and laboratory analyses; areas where discontinuity exists, such as patched or other additions

TABLE 19-4. Bulk Sampling Kit for Asbestos

Quantity	*Item*
1	Small hammer
1	1-inch or 1.5-inch chisel edge scraper
1	3-inch or 4-inch chisel edge scraper
1	20-foot to 50-foot tape measure
1	Cold steel chisel
1	Utility/carpet knife
1	Set of extra blades for carpet knife
1	Regular slotted screwdriver
1	Phillips screwdriver
1	Slip joint pliers or tweezers
1	Set of cork borers
1	Cork borer sharpener
1	Roll of cellophane tape
1	Spray can of clear sealant
1	Roll of duct tape
2–5	Disposable Chopstick sets (as sample plungers)
1	Package pH paper
50–100	Sample vials
1	Box of microscope slides
10–20	Zip-lock bags
10–20	Paper towels
1	Sprayer bottle
1	Flashlight
20–30	Sample data sheets
2–3	Felt-tip pens, large and small sizes

or visual inspection is not possible, such as inside walls, behind other materials, or in pipe chases.

4. Prior to the survey, determine whether any constraints exist, such as limited access to certain areas, need for samplers to wear personal protective equipment, and best hours for sampling. Ask whether any complaints have been received from the building's tenants. Make sure personnel assigned to collect samples are briefed about the results of this meeting. For example, it may be decided not to inform the tenants that a concern exists about asbestos in the building until after the sample results are received. Therefore, a specific text to use when questioned by occupants should be developed.

5. The first step in the survey is to conduct a visual evaluation to record locations of similar materials. The evaluation should include type of material, likelihood of friability, condition of the material, existence of water damage or abrasion, approximate size of exposed material, activity and movement in the exposed rooms, size of the exposed rooms, and presence of a ventilation plenum or other air disturbance. Determine whether any of the sampling activities, such as temporarily removing suspended ceiling tiles, may expose occupants to unknown conditions and unnecessary health risks, and therefore special precautions will be required.

6. Assemble equipment. Locate an inconspicuous, relatively inaccessible area of the material to be sampled.

7. Wear proper personal protective equipment. What to wear will depend on the type of sampling and the extent of the job, but should include disposable latex rubber gloves, a half-mask respirator with high-efficiency cartridges, and Tyvek coveralls if contamination of personal clothing is possible. A hood and booties may also be required, especially when working overhead.

8. Wet the surface with a spray of wetting agent.

9. For friable materials, use a coring device to penetrate the surface of the material. The surface may include an encapsulating coating, making penetration more difficult. If so, a knife may be needed to cut away a section of the coating. Follow with more water underneath, and then use the boring device to collect a sample. Be sure that the device goes down to the hard substrate. Use a chopstick to push the sample from the borer into a glass vial. If the vial is too small, use a zip-lock bag.

10. Record a dedicated sample number on the sample vial and also enter the sample number, place where material was collected, and type of material on the sample data sheet.

11. Cover the area with either duct tape on which the sample number is written or with bond sealant if the sampled area is a sensitive public area.

12. Rinse out the borer in between samples by holding it inside a plastic bottle and using the wetting spray; follow by wiping with the paper towel. Dispose of the contaminated water and zip-lock bags as asbestos-containing material following the survey.

13. Nonfriable materials can generally be cut with a knife. In the case of cementitious materials, if some has been broken off and is definitely derived from the material of interest, it can be used for the sample. Otherwise, the chisel must be used to break off a piece.

14. Bulk samples for asbestos analyses should be sent to a laboratory participating in the EPA asbestos bulk sample analysis program, and the laboratory should be using polarized light microscopy (PLM) protocols recommended by the EPA. In some cases, such as when there are significant amounts of vermiculite mixed in with chrysotile, it is difficult to confirm the presence of chrysotile using PLM. A good confirmation is to have any samples that a laboratory indicates as having $<1\%$ analyzed by a second laboratory. This procedure will help to compensate for variations among microscopists.

Asbestos Air Monitoring

Asbestos air monitoring requires several types of air sampling depending on the situation. Personal and area sampling using integrated methods, real time monitoring for contamination outside containment, and high-volume clearance sampling are all likely to be conducted. Personal monitoring involves the use of small portable battery-operated pumps, and clearance samples that require high-volume pumps use electric-powered current and are large, heavy, and minimally portable. A variation is sometimes done using the personal monitoring method to collect area samples during abatement work to approximate occupational exposures. General personal samples are always compared to the OSHA PEL, and clearance or high-volume samples are compared to either the EPA guidelines or the detection limit of the method used to collect them. In the case of phase contrast microscopy (PCM) the limit is often considered 0.01 fib/cm^2. For risk assessments to identify if there is any exposure to building occupants where asbestos is left in place, TEM is the preferable analytical technique due to the need for high accuracy in identifying asbestos fibers and eliminating interferences.[9] Table 19-5 describes flow rates and collection volumes for different types of asbestos air monitoring.

Personal Monitoring for Asbestos

1. Following calibration using the method for opened-face cassettes, a fresh filter is attached to the pump and the top cover is taken off. A knife or screwdriver may be required to pry it loose. A flow rate of 1 Lpm to 2.5 Lpm should be used for personal samples.

2. For personnel sampling, a 25-mm filter cassette with a 50-mm cowl is used with a 0.8-μm mixed cellulose ester filter (MCEF) inside. Position the cassette so that the opening faces downward to prevent dust from settling in it. If personnel sampling is done for compliance, in addition to getting the employee's name and job title, the social security number should be recorded.

3. For area sampling done during removal, samples are usually collected both inside and outside of the containment area. Flow rates of 1 Lpm to 2.5 Lpm can be used, and either a 25-mm or

TABLE 19-5. Flow Rates and Volume for Asbestos Sampling

	Suggested Maximum Flow Rate (Lpm)	*Suggested Volume per Cassette (L)*
Preremovals and finals	12.0	3,000 – 10,000
During removal		
Dusty	2.5	≤400
Clean	2.5	800 – 1,000

Note: Dusty situations are likely to occur during demolition, spray-on removal, and boiler removal. Clean situations are most likely to be present outside HEPA exhaust, tile removal, and outside containment.

37-mm cassette with a cowl extension and an 0.8-μm MCE filter can be used. The same PCM method, NIOSH 7400, should be used to analyze these cassettes as the personal samples, or the results will not be comparable.

Area Monitoring During Work

1. Be certain that the pumps are recharged and the rotameter used for calibration has been calibrated. The pump should have been calibrated in the laboratory using an opened-face filter and jar.

2. Assign a number to the filter cassette and record this information on the sampling data sheet along with the location being sampled.

3. Attach the cassette to the pump with tubing.

4. Transport the pump to the location being sampled.

5. Remove the plug on the face of the cassette, take off the top cover, and turn on the pump. Record the start time. Store the caps and cassette top in a clean area during the sampling period.

6. At the end of the sampling period turn off the pump, record the time, and put the top and caps into the cassette. Store the cassette in a clean container, face up. Record the stop time on the sampling data sheet. (Area monitoring can also be done using the fibrous aerosol monitor [FAM] as described in the chapter on Real Time Sampling Methods for Aerosols.

Asbestos Clearance Sampling

Following the abatement work, the determination of whether any asbestos fibers have been left suspended in the air is important. For this purpose, high-volume pumps are used and a large sample volume (at least 3,000 L) is collected (Fig. 19-6). If the sampling is to be done using aggressive methods, then fans are set up to blow on surfaces during this sampling. They assure that if any loose fibers are present, they will be detected.

Clearance sampling is associated with abatement projects. Clearance sampling is done using high-volume pumps, and can be done using aggressive blowers to stir up dust or nonaggressive methods. However, the same technique can be used to collect quarterly samples for operations and maintenance (O&M) programs. Depending on the goal of the sampling, the results will be analyzed by either TEM or PCM. The primary difference in the methods is the type of filter material used.

For clearance samples, higher flow rates, from 8 Lpm to 12 Lpm, are generally used, and if TEM analysis is desired, 0.45-μm polycarbonate filters are used. Sampling is also done open-face. A 0.8-μm MCE filter can be used for TEM sampling if no polycarbonate filters are available, although it is less desirable. A minimum volume of 3,000 liters is required for clearance samples.

Procedure

1. Use either polycarbonate filters with a pore size of 0.45 μm or MCE filters with a pore size of 0.8 μm.

2. Select a flow rate between 8 Lpm and 12 Lpm, and collect at least 3,000 liters to maintain a minimum sensitivity of 0.005 asbestos fib/cm^3 per filter.

3. Collect a minimum of 13 samples for each testing site as follows: a minimum of 5 samples per abatement area; a minimum of 5 samples per ambient area.

4. Two field blanks should be included: one for each area. Remove the caps for 30 seconds and replace. An additional sealed blank is carried with the filter cassettes but is not opened in the field. If there is any concern that loose asbestos fibers may be present, tape samples may be collected for residual dust as described in the chapter on Surface Sampling Methods for Dermal Exposure.

INDUSTRIAL SURVEYS

Virtually all industrial processes generate one or more contaminants that must be evaluated. These range from the more general type of evaluation

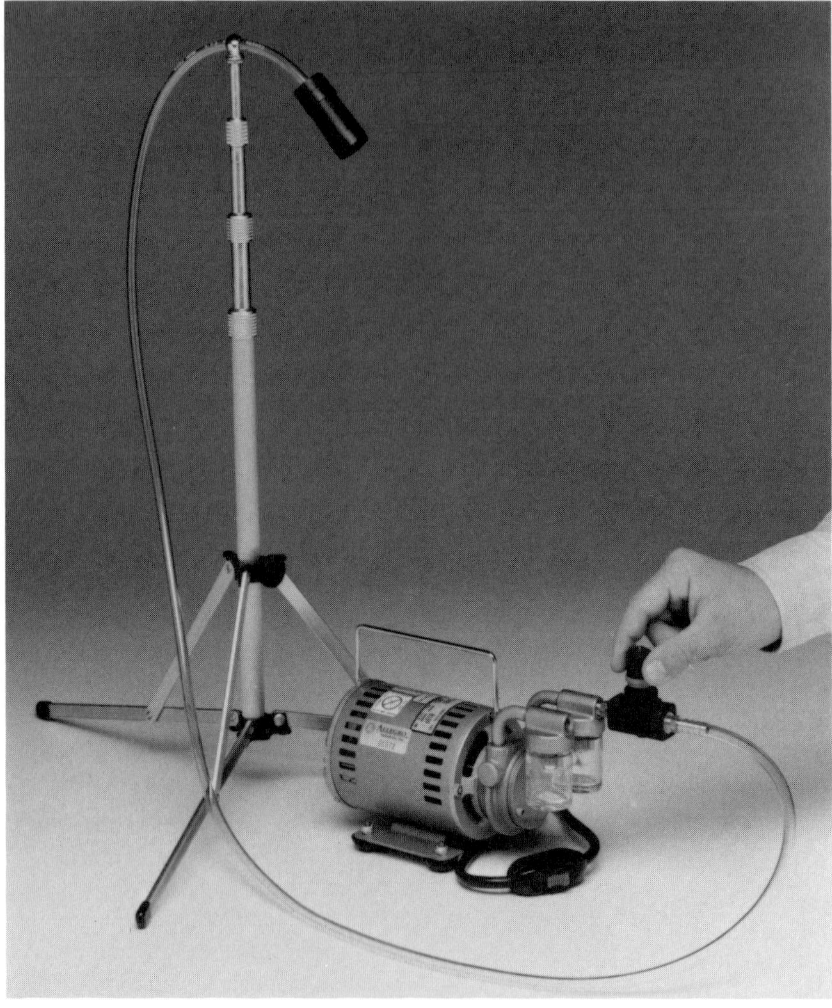

Figure 19-6. High-volume clearance air sampling for asbestos. (Courtesy Allegro Industries.)

for leaks of total hydrocarbons and volatile organic compounds (VOCs) to personal and area sampling for specific contaminants. Welding is one of the most common processes found in industry and is associated with both maintenance and manufacturing activities, yet development of the sampling strategy can be complex. Carbon monoxide is a commonly measured air contaminant due to combustion processes being so widespread, and cases of illness to employees due to overexposures to this compound are common. Ink manufacturing is an example of a process where mixing and blending are done and many different solvents are likely to be used in the same area, the result being additive exposures.

Leak Testing: Fugitive Emissions Monitoring

Fugitive emissions monitoring performed with general survey instruments such as the flame ionization detector (FID) and photoionization detector (PID). Fugitive emissions are generally defined as leaks from process components, such as valves, pumps, compressors, pressure relief devices, sampling systems, process drains, storage tanks, and open-ended lines in volatile organic compounds (VOCs) service. As defined by the Environmental Protection Agency (EPA), these are limited to emissions that do not occur as part of the normal operation of plants, but are the result of the effects of age, lack of maintenance, improper equipment specifications, or externally caused damaged.[10] Other types of emissions which are monitored include breakthrough from carbon canisters, and emissions from stacks, vents, and roof monitors. Emissions sampling can involve a determination of emission factors, the emission rate for a specific piece of equipment, or the concentration and composition of the ambient plant air. Identifying leak sources, which often are not repaired until a process can be shut down for maintenance, is also important for protecting plant personnel when working in these areas.

Historically leaks have been detected using a bubble solution that is sprayed on a potential leak source and then observing the potential leak sites to determine if any bubbles are formed. If no bubbles were observed, the source was presumed to have no detectable emissions or leaks. This method is still used, but only for sources that do not have continuously moving parts, that do not have surface temperatures greater than the boiling point or less than the freezing point of the soap solution, that do not have open areas to the atmosphere that the soap solution cannot bridge, or that do not exhibit evidence of liquid leakage. In general, the soap solution is best reserved as a gross screening technique, and sources passing this test should be followed up with a monitoring instrument.

The first portion of the sampling program involves determining the actual number of potential leak sources, such as valves in a plant. The second portion involves testing these sites. Regardless of the source being tested, it should be remembered that a regular, scheduled inspection and maintenance program is the key to reducing fugitive emissions.

There are two types of standards used to evaluate the results of fugitive emissions surveys. The first involves classifying leaks compared to a concentration defined in a particular standard. The second involves classifying leaks based on the difference between this measurement and a previous background measurement. A response factor must be determined for each compound that is to be measured.[1] The EPA has specified that a VOC analyzer, such as an organic vapor analyzer (OVA), which is an FID-based instrument, used for fugitive emissions must be calibrated at 10,000 ppm for methane or hexane, and that daily recalibration checks must be performed. Figure 19-7 shows a fugitive emissions monitor.

While the procedure below describes a method for identifying leaks, it should be remembered that the leak rate is more important. However, determining this is more difficult than just using a spot measurement. One technique that has been used is to enclose the leak source within an air bag of known volume and to measure the increase in concentration inside the bag at regular intervals.[11] The leak rate can then be calculated.

Procedure for Leaks Based on Defined Concentrations in Standards[10]

1. The sampling professional should be alert for audible indications of leaks and odors during the survey. Several problems can occur when sampling gases that have too high a concentration for an instrument. These include flame-out of FIDs and condensation on the lamp of PIDs. For these reasons, the user should monitor the hydrocarbon concentration while slowly approaching the valve stem, pump shaft seal, or other source. If the concentration exceeds the leak definition before the probe is placed close to the leak site, there is no reason to continue monitoring at that spot.

2. When monitoring sources such as valves and pumps that handle heavy liquid streams at high temperatures, relatively nonvolatile liquids can condense in the probe and the detector. This can slow down the instrument's response.

3. Place the probe inlet at the surface of the interface where leakage could occur. Move the probe along the interface periphery while observing the meter readout. If an increased reading is

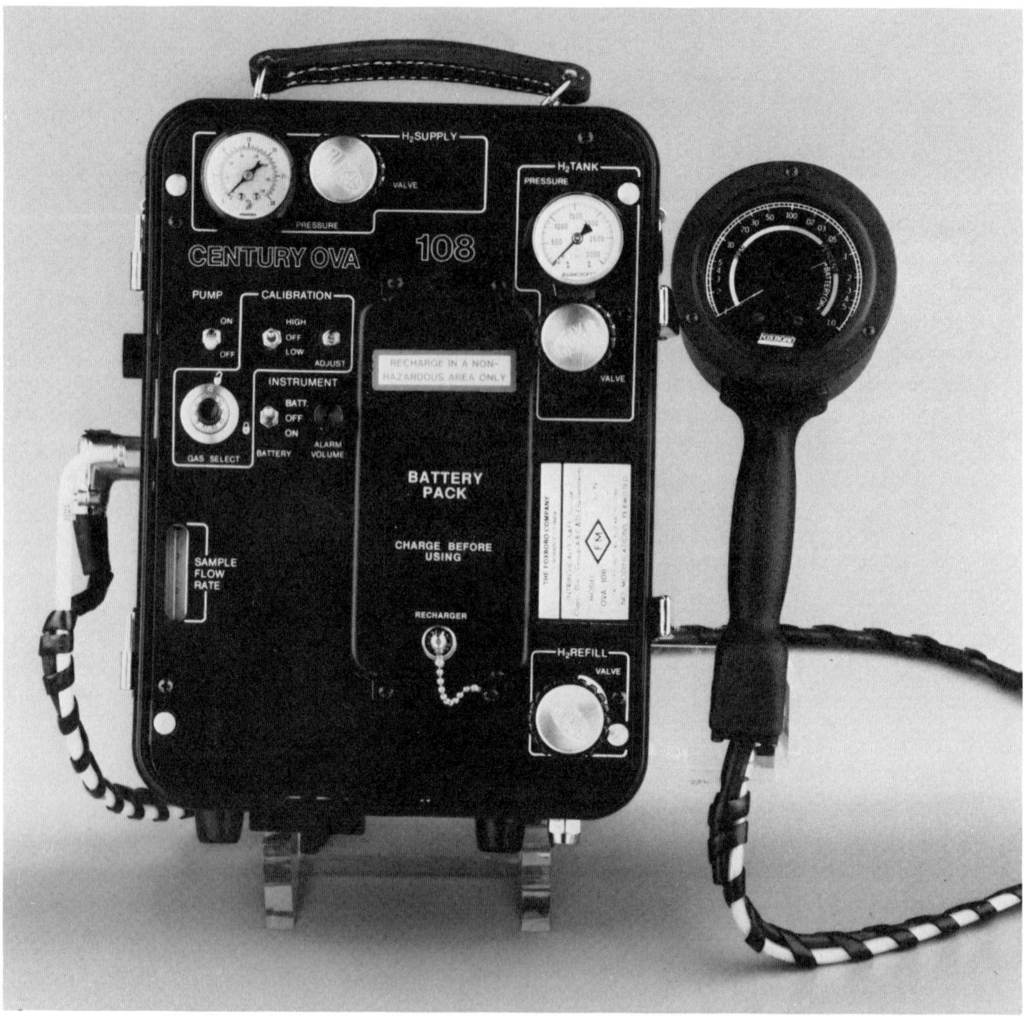

Figure 19-7. Fugitive emissions monitor. (Courtesy Foxboro Company.)

observed, slowly sample the interface where leakage is indicated until the maximum meter reading is obtained.

Valves: The most common source of leaks from valves in closed systems is at the seal between the stem and housing (valve stem packing gland). Place the probe at the point where the valve stem leaves the packing gland. The normal procedure is to circumscribe this location within 1 cm of the valve stem. Also place the probe at the interface of the packing gland take-up flange seat and sample the periphery. Valves used at the end of drains or sample lines have two sources of leakage: the valve stem and the valve seat. The probe is usually placed at the center of the discharge pipe for monitoring.

Flanges and other connections: For welding flanges, place the probe at the outer edge of the flange–gasket interface and sample the circumference of the flange.

Pumps and compressors: Fugitive emissions from pumps occur from the pump shaft seal used to isolate the process fluid from the atmosphere. The most commonly used seals are single mechanical seals, double mechanical seals, and packed seals. Conduct a circumferential traverse at the outer surface of the pump or compressor shaft and seal interface. If the source is a rotating shaft, position the probe inlet within 1 cm of the shaft–seal interface for the survey. In addition, sample all other joints on the pump or compressor housing where leaks could occur.

Pressure relief devices: The configuration of most pressure relief devices prevents sampling at the sealing–seat interface. For those devices equipped with an enclosed extension, or horn, place the probe inlet at approximately the center of the exhaust area to the atmosphere.

Process drains: For open drains, place the probe inlet at approximately the center of the area open to the atmosphere. For covered drains, place the probe at the surface of the cover interface and conduct a peripheral traverse.

Open-ended lines or valves: Place the probe inlet at approximately the center of the opening to the atmosphere.

Seal system degassing vents and accumulator vents: Place the probe inlet at approximately the center of the opening to the atmosphere.

Access door seals: Place the probe inlet at the surface of the door–seal interface and conduct a peripheral traverse.

4. Leave the probe inlet at this maximum reading location for approximately two times the instrument response time.

5. If the maximum observed meter reading is greater than the leak definition in the applicable regulation, record and report the results as specified in the appropriate regulation.

Procedure for Leaks Based on "No Detectable Emissions" or Compared to a Background Measurement[10]

1. Determine the local ambient concentration by moving the probe inlet randomly upwind and downwind at a distance of 1 meter to 2 meters from the source. Use the most sensitive scale for this reading.

2. Next move the probe inlet to the surface of the source and determine the concentration by using the procedures described in the previous section for the various sources.

3. The difference between the ambient concentration and those detected around the leak sources determines whether there are any detectable emissions.

Welding and Burning Fumes

Welding fumes cannot be classified simply. The composition as well as the amount of fume released depends on the base metal being welded and type of welding process, including the type of electrodes used. Stick welding produces high levels of fume. Gas metal shielded welding tends to produce lower levels of fume and higher levels of gases, such as ozone. Fumes often contain iron oxide, manganese, silicon dioxide, and other metals such as chromium and nickel depending on the system involved. For example, welding on stainless steel will produce chromium and nickel oxide. Some coated and flux-cored electrodes are formulated with fluorides, and the fumes associated with them may contain significant amounts of fluorides.

In most fusion welding operations, the base metal does not contribute significantly to the total welding fume. Coatings, such as primer paints or galvanizing, can contribute significantly to the fume. Therefore, the coating constituents should be analyzed in the fume when they are present on the base metal; otherwise, the composition of the constituents depends largely on the filler metal rod being used and the process conditions. Metals are predominantly present as mixed metal oxides together with some silicates and unoxidized metal. Fume from welding of stainless steels may contain about 10% chromium predominantly in the hexavalent state.[12] One way of determining the constituents of the rods, electrodes, or sticks is a review of the appropriate material safety data sheet (MSDS).

Welding fume particulate contains a wide range of airborne species from fine chains to spherical particles.[11] Particle size sampling of welding fume is complicated by the fact that the particles form agglomerates so that smaller particles decrease in number, but the number of larger-size particles increases along with a change in the overall dispersion of particles.[13] The type of welding process will also affect particle size. For example, particles from the shielded metal arc welding process are somewhat larger in size than those from other welding processes, ranging up to 2 μm in size.[14]

Therefore, it is important to measure the respirable fraction of the fume as well as to determine its individual constituents. Stick welding fume consists of chains and clusters of submicron particles and glassy coated spheres up to 10 mm in diameter.[12] Larger particles originating from electrode spatter are also present, including unoxidized metal. Spatter particles are typically above the respirable range, but as they oxidize during passage through the air outside the arc, they may contribute to the fume.

When selecting which welders to sample, a sampling priority can be developed. Welding in confined spaces presents the greatest hazard of exposure and is the first priority to sample. Persons using filler or base metals that contain highly hazardous components, such as lead or cadmium, are next. If highly visible fumes are present, they are an indicator that high exposures may be occurring. Another way of categorizing operations is based on current settings and arc times. In general, the higher the current settings and the longer the arc times, the greater priority for sampling.[15] Following is a list of observations to make when doing sampling for welding fumes.[16]

Identify the welding area and population near the work and diagram.
Determine if the work is being done in a confined space.
Measure the size of the work area or room.
Determine the number of welders in the work area.
Identify any barriers in the work space that will obstruct general dilution ventilation, such as curtains, partitions, trucks, or other machinery.
Review chemicals to determine if there are any solvents that are chlorinated or flammable in use in the area.
Identify which welding processes are in use, including the specific types of rods, size of wire, and grades of gasses.
Determine the welding conditions: amperage, voltage, and polarity.
Identify the base metal and any coatings it may have, such as paints, rust inhibitors, or plating.
Review schedules to determine if the job being sampled is typical or nonroutine.
Determine if the rate of production is light, heavy, or normal.
If the work is being done at a constant rate, identify the longest arc time.
Review the work practices of the welders to determine if their posture causes them to work in the welding plume, lean away from the plume, or if it varies with the location of the arc and the plume.
Identify if any personal protective equipment is in use, including helmets, face shields, and respirators.
Determine what type of ventilation is in use.
For local exhaust, determine the type, location with respect to the arc and distance from the arc, and the size of the hood and its flow rate.
See if the positions of local exhaust hoods are fixed or mounted on flexible ducting.
Identify any area fans and their placement with respect to the local exhaust and the welder.
For air makeup units, determine their location relative to the welder and local exhaust hoods, as well as the air movement direction relative to the welders.

The cassette should be placed outside the welding hood if the sampler is following OSHA procedures. If determining the employee's actual exposure, two pumps and filters can be mounted on the employee: one placed underneath the welding hood and one placed outside the welding hood. The results are then set up of concentration outside the hood divided by concentration inside the hood to determine the protection factor of the hood. Welding hoods have been reported to provide protection factors ranging from 3.3 to 15.[17] The extent of reduction of fume concentrations at the breathing zone due to the use of welding helmets ranged from 29% to 64% in one study, depending on the type of welding and the welders' postures.[18] The effectiveness of this type of hood should not be taken for granted.

The American Welding Society has defined the welder's breathing zone to be the area immediately adjacent to the welder's nose and mouth, inside the welder's helmet when worn, or within 9 inches of the nose and mouth when a helmet is not worn.[19]

In addition to particulate exposures, it should be remembered that many gases (Table 19-6) are generated during welding operations as well.[16] For more information on how to sample these, see the chapters on Integrated Sampling for Gases and Vapors and on Monitoring Instruments Dedicated to a Single Chemical.

Procedure

1. Use a three-stage cassette with a 0.8-μm MCEF filter inside. The pump should have been calibrated using the technique for open face filter sampling. The top should be taken off the cassette.

2. If using filters inside and outside of the welding hood, they should be taped in place.

3. Analysis will depend on whether any toxic elements are present in the welding rods, base metal, or if a metal coating exists. Sampling for fumes does not eliminate the need to evaluate the potential for hazardous gases to be present.

Total welding fume samples are analyzed gravimetrically, that is, they are weighed. This type of sample provides baseline information and can be compared to the TLV for total welding fume, but does not provide information on metals of concern such as Ni or Cr^{+6}. The filters can then be analyzed using emission spectroscopy for various elements. Results should be converted to the relevant compounds, for example, iron oxide (Fe_2O_3) is the constituent generated during welding whereas iron (Fe) is what is measured during the analysis.

TABLE 19-6. Gases Associated with Specific Arc Welding Processes

Welding Process	CO	*Fluorine*	NO_X	*Ozone*
Shielded metal arc welding	X		X	
Flux cored arc welding	X	X	X	
Argon helium shielding			X	
Carbon dioxide shielding	X		X	
Gas tungsten arc welding			X	X

Carbon Monoxide from Forklifts

A typical situation involving the use of propane-fueled forklifts is moving packed fruits into and out of cold storage. The exhaust from propane can contain significant amounts of carbon monoxide. These situations often are aggravated by the fact that doors are blocked with plastic strip curtains to minimize leakage of warm air into the buildings. The forklifts usually move in and out of various storage rooms, also with plastic curtains, as well as going outside occasionally. Often the survey is a result of employee complaints of symptoms such as headaches while on the job (Table 19-7).[20]

TABLE 19-7. Symptoms Caused by Various Amounts of Carbon Monoxide Hemoglobin in the Blood

Blood Saturation (% COHB)	*Symptoms*
0–10	None
10–20	Tightness across forehead, possible slight headache, dilation of cutaneous blood vessels
20–40	Headache and throbbing in temples, severe headache, weakness, dizziness, dimness of vision, nausea, vomiting, collapse
40–50	Same as previously noted with increased likelihood of collapse and syncope, increased respiration and pulse
50–60	Syncope, increased respiration and pulse, coma with intermittent convulsions and Cheyne-Stokes respiration
60–70	Coma with intermittent convulsions, depressed heat action and respiration and possible death
70–80	Weak pulse and slow respiration, respiratory failure and death

Area air samples are collected with a real time instrument that is specific for CO, usually electrochemical, with a strip chart recorder and/or data logger attached. The use of the strip chart allows notations to be made as to amount of traffic at a specific time and to account for specific peaks. Generally this setup is stationed inside the main traffic area and left to sample in the same place for the entire day. Typical levels that might be generated in a situation where four propane-powered forklifts were in operation for the majority of the day are as follows:

Time	*CO Level (ppm)*
8 A.M.	50
9 A.M.	125
10 A.M.	180
11 A.M.	200
Noon	200
1 P.M.	160
2 P.M.	160
3 P.M.	200
4 P.M.	200

As compared to an OSHA permissible exposure limit (PEL) allowing a 35-ppm exposure expressed as an 8-hour TWA, it is clear from these data that employees are overexposed and controls such as replacement of the propane-powered lift trucks with battery-powered and additional ventilation should be installed.

Personal samples could be collected for carbon monoxide using long-term detector tubes and low flow personal sampling pumps. The disadvantage of this method is that the accuracy at the PEL is ±25%, and this method does not provide information on changes in levels over the workday. Another option would be to use a personal real time carbon monoxide monitor with data-logging capabilities. In this case, unless more than one monitor were available, at least two days worth of sampling would have to be done to get a sample that is representative of building levels, due to differences in employee work practices and duties. In addition, if employees are complaining of symptoms such as headaches and nausea, biological sampling can be done. This can be blood samples for carboxyhemoglobin or carbon monoxide in end-exhaled air. Smokers should be identified if biological samples are collected.

Multiple Solvents in Printing Ink Manufacture

Printing ink manufacture is a typical situation where many different solvents are in use in one operation. For example, isopropanol, 1,1,1-trichloroethane, methyl ethyl ketone, and toluene might be used in various blending operations going in the same room along with other solvents.

In this type of survey, both air samples and biological monitoring may be needed. In addition, several OSHA standards would be of concern affecting the way that samples are collected. All of these compounds have 15-minute short-term exposure limits (STELs), as well as 8-hour TWA standards.

Integrated air samples would be collected using multiple-tube holders, since methyl ethyl ketone requires a different type of sorbent tube than the other solvents and isopropanol should be analyzed separately from toluene and 1,1,1-trichloroethane. A large charcoal tube would be used for the others. Tubes would be changed approximately every 1 to 2 hours, except for sampling periods designed to measure STELs, in which case tubes would be changed after 15 minutes.

Biological monitoring would be done if there was a concern that employees were overexposed. For example 1,1,1-trichloroethane can be sampled in the end-exhaled air, or its metabolite trichloroethanol can be measured in the urine. Toluene can be monitored as its metabolite, hippuric acid, in urine, in venous blood, or in end-exhaled air. While ethanol is commonly monitored in the breath of drivers accused of intoxication, the levels that can be achieved by drinking are unlikely to occur from inhalation while working around this material.

When results are received, in addition to comparisons with these standards, the potential for additive effects must be considered. It is not uncommon for individual exposures to be less than the PEL, but for the additive exposure — the ratio of the levels of each compound to its PEL added together—to exceed one and represent an overexposure.

REFERENCES

1. Jones, W. Sick building syndrome. *Appl. Occup. Environ. Hyg.* **5**(2):74–83, 1990.
2. Hollowell, C. D. *Indoor Air Quality.* Lawrence Berkeley Laboratory, Energy and Environment Division, June 1981. US DOE Contract W-7405-ENG-48.
3. Burge, H. A., and M. E. Hoyer. Focus on indoor air quality. *Appl. Occup. Environ. Hyg.* **5**(2):84–93, 1990.
4. Montgomery, D. D., and D. A. Kalman. Indoor/outdoor air quality: Reference pollutant concentrations in complaint-free residences. *Appl. Ind. Hyg.* **4**(1):17–20, 1989.
5. Pleil, J. D., and R. S. Whiton. Determination of organic emissions from new carpeting. *Appl. Occup. Environ. Hyg.* **5**(10):693–699, 1990.
6. EPA. Method IP-3A: Determination of carbon monoxide (CO) or carbon dioxide (CO_2) in indoor air using nondispersive infrared (NDIR). In *Compendium of Methods for the Determination of Air Pollutants in Indoor Air.* Research Triangle Park, NC, Sept. 1989.
7. Office of Pesticides and Toxic Substances, US EPA. *Assessment of Health Risks to Garment Workers and Certain Home Residents from Exposure to Formaldehyde.* Washington, DC, 1987.
8. Shah, J. J., and H. B. Singh. Distribution of volatile chemicals in outdoor and indoor air. *Environ. Sci. Technol.* **22**(12):1381–1388, 1988.
9. Perkins, J. L., and M. S. Cleveland. Phase contrast microscopy and transmission electron microscopy—no relationship? *AIHA J.* **51**(8):A-555–A-556, 1990.
10. EPA Reference Method 21. *Determination of Volatile Organic Compound Leaks.*
11. Jones, A. L. The measurement and importance of fugitive emissions. *Ann. Occup. Hyg.* **28**(2):211–215, 1984.
12. Hewitt, P. J., and C. N. Gray. Some difficulties in the assessment of electric arc welding fume. *AIHA J.* **44**(10):727–732, 1982.
13. Glinsmann, P., and F. S. Rosenthal. Evaluation of an aerosol photometer for monitoring welding fume levels in a shipyard. *AIHA J.* **46**(7):391–395, 1985.
14. American Welding Society. *Effects of Welding on Health.* AWS, Miami, FL, 1979.
15. American Welding Society. *Evaluating Contaminants in the Welding Environment: A Sampling Strategy Guide.* AWS F1.3-83, AWS, Miami, FL, 1982.
16. American Welding Society. *The Welding Environment.* AWS, Miami, FL, 1979.
17. Goller, J. W., and N. W. Paik. A comparison of iron oxide fume inside and outside of welding helmets. *AIHA J.* **46**(2):89–93, 1985.
18. American National Standards Institute/American Welding Society. *Method for Sampling Airborne Particulates Generated by Welding and Allied Processes.* Method No. F1.1-1978.
19. National Institute of Occupational Safety and Health. *A Guide to the Work-Relatedness of Disease.* NIOSH Pub. No. 79-116. NIOSH, Cincinnati, OH, 1979.

Appendixes

APPENDIX A

Pumps and Flow Rate Calibrations

Pumps are the primary device used for actively collecting integrated samples on media. In addition, many real time monitoring instruments utilize pumps. Other types of pumps are used for pulling samples onto detector tubes. What all these devices have in common is a need for accurate flow rates and therefore calibration.

Calibration devices are of two types: They either measure total volume or flow rate. There are a number of different devices within each category. Examples of devices that measure volumes are bubble burets, dry gas, and wet test meters. Flowmeters measure the velocity of air going through them. They include rotameters and critical orifices. Velocity meters, which are also flow rate devices, include mass flowmeters and pitot tubes.

Calibration of any pump or instrument should include enough values to cover a representative range of the flow rates generally used, and to determine the degree of variation associated with its scale. The pressure drop resulting from the media must be taken into account when a sampling system is calibrated volumetrically. Rotameters and critical orifices should be calibrated with the media in line, or a volume correction should be done to compensate for deviations from ambient conditions. In systems based on filter collection, the pressure drop across the system may steadily increase as the filter medium loads up with retained particulates and as a result many pumps are engineered to compensate for this.

Calibrators are further broken down into classifications as primary or secondary standards. Bubble burets are considered primary standards. The primary standards have the least amount of error. Secondary standards include wet test meters, dry gas meters, and rotameters. Other calibrators tend to have more sources of error than these, although it does not necessarily mean they should be ignored; more care, including a backup calibration program, is required when they are used.

The result of calibrating flow rates or volumes at several points is a calibration curve. Calibration curves are actually control charts. These charts display the test data in a form that graphically compares the variability of all test results. In most cases, the sampler is attempting to see if a straight line can be drawn through all the points on the curve, or at what points the line cannot be connected to form a curve. If the procedure is "in control," the results will fall within the established control limits.

If calibrating a secondary standard, such as a rotameter with a bubble buret, it is important that the primary standard be at the end of the calibration train and not in between the secondary standard and the pump; otherwise, it will cause a pressure drop.

VOLUME CALIBRATORS

The two basic types of commonly used devices that calibrate by measuring volume are bubble burets and dry gas meters.

Bubble Burets

Bubble burets are the most common primary standard. Burets can either be purchased specifically set up as calibration devices or can be as simple as a laboratory buret turned upside down with a piece of tubing at the tip and the stopcock left open. Airflow is measured by pulling a soap film (bubble) through the precision glass tube and recording the time it takes for the film to travel between two calibration marks.

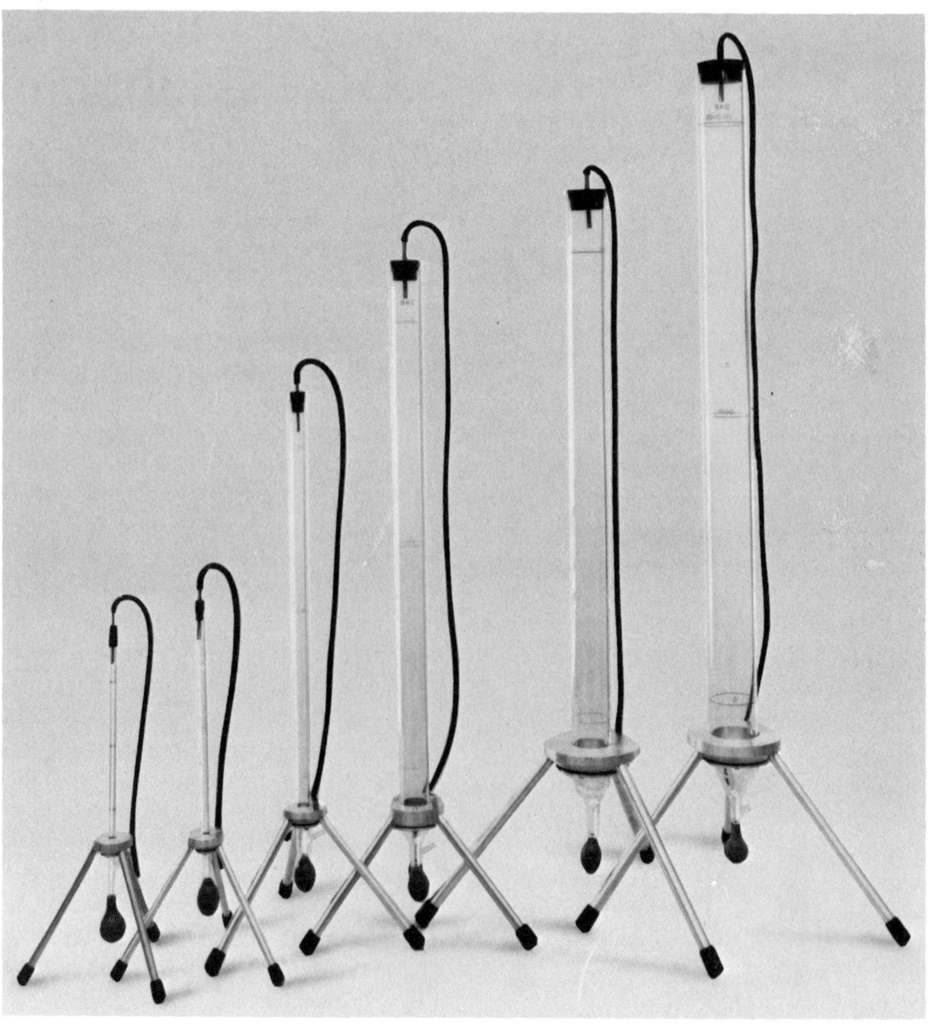

Figure A-1. Bubble burets. (Courtesy SKC, Inc.)

Bubble burets (Fig. A-1) are available in a variety of sizes. In general, all burets are in metric (millimeters) measurements. Use of laboratory-type bubble burets is limited to pumps that do not pull more than 5 Lpm. Most commonly, they are used for personal sampling pumps, detector tube pumps, and vacuum pumps on direct reading instruments. A bulb is commonly installed into the bottom to allow for squeezing bubble solution into the buret. Water-jacket burets are available to ensure temperature consistency. According to the manufacturer, these are ±2% accurate. If purchasing a laboratory model, do not use one with a glass stop-cock; use one with Teflon instead. The size selected for calibration will depend on the flow rate being measured. For example, a 1,000-ml buret is ideal for 1 Lpm and up to 5 Lpm, which is the upper limit of high flow personal air sampling pumps. A 100-ml buret can accommodate pumps with much lower flows.

Glass burets are easily broken, and plastic burets are more susceptible to interior corrosion and scratching. Models are available that will allow transport to the field. These are mounted on stands or directly into a carrying case. Accuracy is at least ±0.5%.

Electronic Bubble Meters

A recent innovation involves incorporating a microprocessor and a printer to create an electronic bubble meter (Fig. A-2). A bubble generator produces a bubble film that is carried by a pump's airflow from the bottom to the top of a cylinder. As the bubble traverses past two infrared sensors, one mounted at the bottom and the other at the top of this flow tube, each sensor transmits a signal to a microprocessing unit that stores the time and performs the necessary calculation to determine the flow rate. Timers are usually capable of detecting a soap film at 80-μsec intervals. Under steady flow conditions this speed allows an accuracy of ±0.5% of display readings. These units can provide almost instantaneous airflow readings and a cumulative averaging of multiple samples. They can accommodate flow rates up to 30 Lpm, although accuracy at higher flow rates may not be as good as that at lower flow rates.

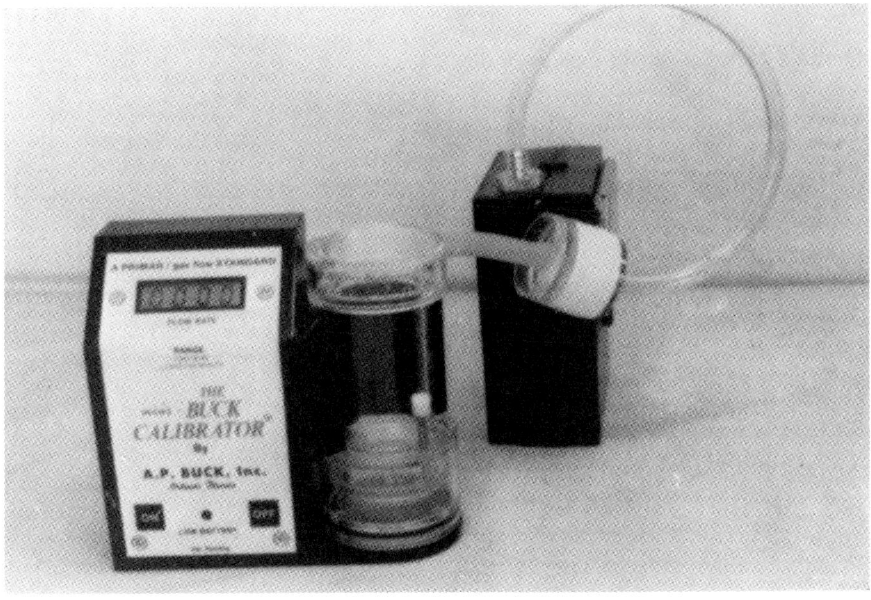

Figure A-2. An electronic bubble meter. (Courtesy A. P. Buck, Inc.)

The user can save calibration runs and average them. Electronic bubble meters are accepted by OSHA as a primary standard.

A variation on the electronic bubble meter was described previously in the chapter on Integrated Sampling for Gases and Vapors. The Buck Calilogger allows sample numbers to be input and stored along with calibration data at the end of the day. Postcalibrations are done and the data for run time, temperature, and pressure are entered for each sample and the unit prints out a chain-of-custody form with the volumes calculated.

Electronic bubble meters are factory calibrated using a standard traceable to the National Institute of Standards and Technology, formerly called the National Bureau of Standards. OSHA recommends using a 1-L glass bubble buret and conducting a test at 1 Lpm when verifying the calibration of an electronic bubble meter with a manual bubble buret. Different size cylinders are available to accommodate different ranges of flow. At higher flow rates (>8 Lpm) it is possible for bubbles to start out flat and then curve or become concave, the result being that the timing may be off.

The bubble soap provided with these calibrators is a precisly concentrated and sterilized solution formulated to provide a clean frictionless soap film bubble over the dynamic range of the unit. It is important to use a sterile soap solution in order to prevent residue buildup in the flow cell center tube, which could cause inaccurate readings. Generally it is best not to substitute for the manufacturer's soap.

Prior to use, electronic bubble meters should be cleaned. This involves removing the flow cell and gently flushing it with tap water and wiping it with a cloth. If a stubborn residue persists, the bottom plate can usually be removed from these units. A few drops of soap squirted into the slot between the base and flow cell will ease removal. Generally these burets should not be stored for more than 1 week with soap in the flow cell. Since most cylinders are made of plastic, they should never be cleaned with solvents. Acrylic flow cells can be easily scratched. The center of the cell where sensors detect the bubble should never be allowed to get dirty or scratched. The flow cylinders should not be pressurized.

When transporting electronic bubble meters it is important that one side of the seal tube that connects the inlet and outlet boss be removed for equalizing internal pressure within the calibrator. Do not transport these units with soap solution or storage tubing in place.

Problems can occur if a bubble is in front of a sensor when the unit is turned on. Very high ambient light levels may interfere with the sensor's ability to detect. If a bubble or other obstruction blocks the sensor path for too much time after normal operation, a unit may fault. Finally, some units are sensitive to having too many bubbles or imperfect bubbles go up a cylinder.

Dry Gas Meters

The interior of dry gas meters (Fig. A-3) consists of two bags interconnected by mechanical valves and a cycle counting device. The air or gas fills one bag as the other bag empties. When the cycle is complete, the valves are switched, and the second bag fills as the first one empties. The

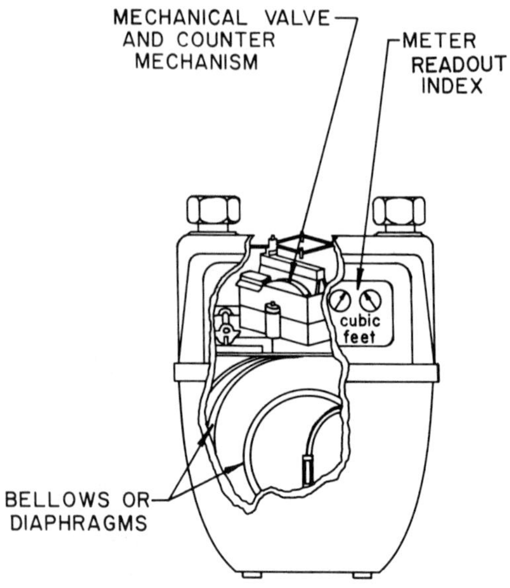

Figure A-3. Dry gas meter. (From NIOSH. *The Industrial Environment—Its Evaluation & Control.* Washington, D.C., 1973.)

maximum flow rate at which these devices are used should be limited to the volume in cubic feet equivalent to the *bagged* capacity. As an example, a meter with a bag stamped DTM-200 is a dry test meter that will safely pass a maximum of 200 dry feet per hour (dfh) at ½-inch water column differential. The inlet is marked at the tip of the meter. If condensate collects on the inlet side of the meter, suitable drips or traps should be provided for its removal to prevent inaccuracies or freezing. Gas containing dust should be filtered ahead of the meter to prevent damage or premature wear and loss of accuracy. Dry gas meters are very useful for making up calibration mixtures of gases and vapors as they can be used to deliver accurate volumes of air to Tedlar bags.

When meters, particularly on high-pressure lines, are placed in operation, the manifold or piping valves should be opened very slowly to permit the measuring compartments to become equalized with the pressure in the line. Abrupt turning on or off, or violent fluctuations in pressure or vacuum, can damage the operating mechanism.

FLOW RATE CALIBRATORS

Rotameters

Rotameters (Fig. A-4), also called variable area flow meters, are the most common devices used in the field. The precision rotameter consists of a vertically mounted internally tapered tube (the top being wider than the bottom), with a float placed inside that is free to move up and down. As air is passed through the tube, the float rises until the rate of flow is sufficient to hold the float stationary. The flow rate is adjusted using the pump's flow regulator. Rotameters are read at the widest diameter of the float, for example, a ball is read in the middle.

There are many types of rotameters, all varying in precision and materials, including glass and plastics. Rotameters are available for a wide variety of flow rates, and can be used with personal air sampling pumps and with ultra-high flow pumps (flows of 8 Lpm and up). Usually at higher flow rates they will read in cubic feet per minute or hour (cfm, cfh) rather than Lpm, so a conversion calculation may be necessary. The degree of accuracy also varies. For example, acrylic Visi-Float flowmeters are ±5% full scale for the 2-inch-long series and ±3% for the 4-inch series. Gilmont glass and Teflon calibrated and correlated flowmeters are specified to have an accuracy of ±2% or ±1% division, whichever is greater for flow rates ranging from 1 ml/min to 280 ml/min and 3 Lpm to 77 Lpm. In general, the statement "you get what you pay for" is true for rotameters.

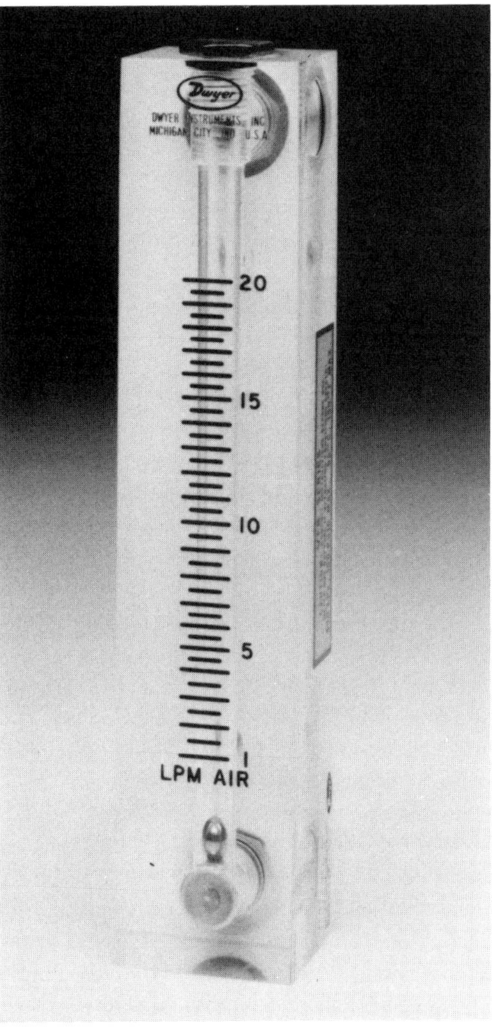

Figure A-4. Rotameter. (Courtesy Allegro Industries.)

Some rotameters are marked with the flow they are designed to measure, and others are set up with a graduated scale and the corresponding flow rate is read off a calibration curve. In general, the ones with the graduated scale are less misleading, because the others usually do not perform at the specific flow rates engraved on the tube. For example, at the 1-Lpm mark a rotameter may actually be measuring 1.1 Lpm. For these types, it is important that information on the corresponding correct flow rates be written on the side of the rotameter and that a calibration curve be constructed. The reasons a unit prestamped at the factory is not going to operate at those specified rates include variations during manufacture, for example, the diameter of the inside of the tube's bore may vary slightly from specifications, or the weight of the ball may vary.

Dual-ball rotameters are used for combination low/high flow readings. The balls are different weights with the lighter-weight ball on top of the heavier one. At low flows the light ball is read, and the heavier ball is read for the higher flow rates. A typical set of ranges is 5 ml/min to 245 ml/min on the low flow and 0.235 Lpm to 6.6 Lpm on the high flow.

Rotameters built into sampling pumps are not usually precision rotameters and therefore should not be used to calibrate the pump's flow rate. These are primarily for use as flow indicators. For example, by monitoring the rotameter during pump use the sampler can ascertain that the pump is still running and that the flow rate is approximately the same as the rate at which the pump was calibrated.

Rotameters must be inspected regularly for buildup of dust or other contamination inside of the tube or on the ball. Problems with prestamped flows have already been discussed. Fouling from particulate matter or corrosion will affect the ability of the float to move smoothly. Metal floats are attacked by corrosive gases and some reagents, such as iodine. Plastic tube models require protection from solvents that soften or cause the interior of the bore to get tacky. Fluids (water, oil, solvents) must be prevented from entering them, because a liquid film will cause the float to stick to the side of the bore.

In order to get the most reliable use of precision rotameters, they should be calibrated against a bubble buret on a regular basis. OSHA recommends that this calibration be done monthly. The date of calibration should be marked on each rotameter, and a record of each calibration should be kept in a log book. The temperature and barometric pressure should be recorded on the calibration curve. Every attempt should be made to calibrate rotameters at the altitudes where they will be used and 25°C. Alternatively, if the sampler plans to use a unit on a regular basis under different conditions, it may be worth developing a curve under these conditions.

A small chart containing the exact flow rate at each point tested should be taped onto the side of a rotameter if it is large enough, or a copy of the curve should be kept in a box with the rotameter so it is available in the field. Some precision rotameters are calibrated in the factory and come with their own curves. These should also be checked against a bubble buret for discrepancies. Note the conditions under which the rotameter was calibrated in the factory. The altitude and temperature should be the same as the location where it will be used.

For compliance work OSHA allows the precision rotameter to be used for calibrating personal sampling pumps in lieu of bubble meters provided the rotameters are calibrated with an electronic or manual bubble meter on a regular basis (at least monthly); disassembled, cleaned as necessary, and recalibrated when contaminated; not used at a substantially different temperature and/or pressure from those conditions present when the rotameter was calibrated against the primary source; and used such that pressure drop across them is minimal.

Procedure for Rotameter Calibration

1. Hook up a pump capable of the flow range of interest to the proper-size bubble buret. For example, for low flows the pump would be set to run from 50 ml/min to 200 ml/min, and the buret would be 100 ml in volume. For high flows a 500-ml or 1-liter buret is used. Use a trap to prevent bubbles from entering the pump.

2. Adjust the pump's flow rate to the maxi-

mum reading of the rotameter to be calibrated by timing several bubbles up the buret.

3. Disconnect the bubble buret and attach the rotameter. Record the reading on the rotameter for the flow rate on the pump. Always read the middle of the ball for consistency.

4. Repeat the procedure using this flow rate. Average the times.

5. Determine the average flow rates as follows:

High flow:

$$\frac{500\text{ ml}}{\text{avg. sec}} \times \frac{1\text{ L}}{10^3\text{ ml}} \times \frac{60\text{ sec}}{\text{min}} = \_\_\_\_\_\text{ Lpm}$$

Low flow:

$$\frac{100\text{ ml}}{\text{avg. sec}} \times \frac{60\text{ sec}}{\text{min}} = \_\_\_\_\_\text{ ml/min}$$

6. Determine the percent variation as follows:

$$\frac{100\,(\text{buret reading} - \text{rotameter reading})}{\text{rotameter reading}} = \_\_\_\_\_\%$$

7. If the variations exceed ±5%, clean the rotameter and repeat the calibration steps. The rotameter may be cleaned with a mild detergent solution, rinsed with water, and air dried. A rinse of isopropyl alcohol will facilitate drying. If the variations are within ±5%, continue calibration at other rotameter readings.

8. Pick at least four more flow rates spread over the range of the rotameter's capability, including one in the middle and one at the lowest rate that the pump–rotameter combination is capable of measuring, and repeat steps 2 and 3 for each flow rate. Generally, these are at intervals the sampler might select for a flow rate, such as 1 L, 1.5 L, 2 L, and 2.5 L.

9. If the variation exceeds ±5% at any value (even after cleaning), the new value should be used.

10. Plot a curve of actual flow rate versus the rotameter reading on linear graph paper. The plotted points should form a straight line. If more than one point fails to fall on the straight line graph, either the rotameter or the calibration technique is faulty. Take note of the temperature and the barometric pressure, and record these on the graph.

Critical Orifices

As used in air sampling, the critical orifice is a simple metal hole inserted downstream of a sampling device and upstream of the sampling pump. Its purpose is flow regulation. The term *critical* simply means that the air in the small nozzle throat is moving at the speed of sound. Critical orifices will pull a constant flow rate once they reach critical flow regardless of what pressure is downstream. Devices of this type have long been used for controlling the flow rate through filter holders, cascade impactors, and impingers.

When the pressure downstream from a critical orifice is about 50% less than the pressure immediately upstream, the velocity through the orifice reaches the speed of sound and no further increase in flow rate will occur (under suction conditions, that is). Once the speed of sound has been reached, creating a lower vacuum will not increase the air speed and the volume of air moving through the nozzle will be independent of further vacuum changes.

Critical orifices are sold as separate units and also incorporated into sampling devices. For example, sampling manifolds in multiple tube holders have critical orifices. Several sizes of critical orifices are available. Each size represents a different flow rate and the flow rate is often stamped on the unit by the manufacturer. However, flow rates for critical orifices must be determined by calibration, because they may differ from the manufacturer's specifications. When used with other devices or filter media, the flow rate will be different and should also be checked with a rotameter or wet test meter.

Although the critical orifice is an attractive device for maintaining a constant flow rate through a sampling instrument, certain limitations exist. The suction pump must have the ability to maintain a particular negative pressure at the desired

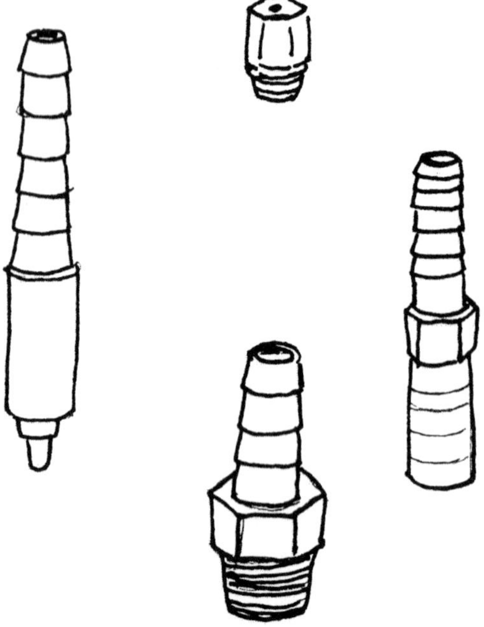

Figure A-5. Critical orifices.

flow rate. This information can be ascertained by consulting the manufacturer. If these restrictions are strictly observed, the critical orifice (Fig. A-5) is a very economical and simple technique for maintaining constant airflows.

Procedure

1. Select a critical orifice, note its number, and hook it via tubing to a vacuum source with a manometer in line. Attach the other end of the orifice to a bubble buret. The vacuum source and the manometer must be downstream from the critical orifice.

2. Turn on the vacuum gradually and increase its flow rate in increments, and measure the amount of time in seconds for the bubble to transverse the buret. This action should create a critical flow through the orifice.

3. Calculate the flow rate of each trial as related to increasing downstream pressure. Record pressure, temperature, and flow rate.

4. The flow rate of air should increase in a linear fashion with the increasing downstream pressure, up to the point of sonic velocity. At this point, the flow rate will stabilize and any degree of increased vacuum will not affect the stable flow rate.

Electronic Flowmeters

Electronic mass flow monitors (Fig. A-6) can measure flows and generate a voltage output. These meters are hot wire devices in which measurement of the flow is based on thermal conduction. The temperature of the airstream depends on the mass rate of flow and the heat input. Some meters have heated wires and some have heated thermistors and thermocouples. Mass flowmeters are incorporated into high-volume air sampling apparatus and are also available as portable instruments for field use. The readout is in actual flow units, so no extrapolation or calculations are required. If properly calibrated against a bubble buret so that their accuracy is within ±3%, these units provide ease of use in the field, including a wide range of flow rates with a single instrument; they have a wide operating temperature range; they are battery-powered; and the miniaturized versions are very portable. Some are now available with digital readouts.

Tylan mass flowmeters are used in conjunction with high-flow sampling apparatus for environmental aerosol collection. In these sensors the temperature rise of a gas is a function of the amount of heat added, the mass flow rate, and the properties of the gas being used. To measure the airflow rate, instruments generally use small stainless steel sensor tubes. Two external, heated resistance thermometers are wound around the sensor tube, one upstream and one downstream. When gas is flowing, heat is transferred along the line of flow to the downstream thermometer, thus producing a signal proportional to the gas flow. The greater the flow, the greater the differential between the thermometers.

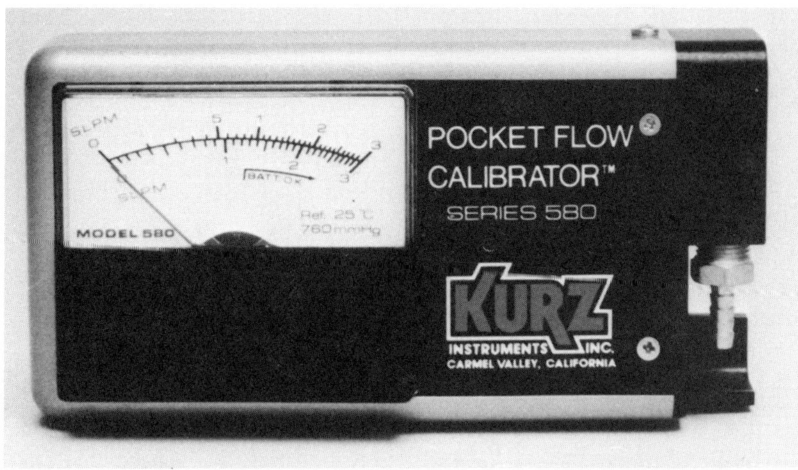

Figure A-6. Electronic mass flowmeters: the pocket flow calibrator *(top)* and a digital flow calibrator *(bottom)*. (Both photos courtesy Kurz Instruments Inc.)

If volumetric flow rate is needed, then the following calculation is necessary.

$$Q_{act} = Q_{ind}\left(\frac{d_s}{d_a}\right)$$

where

Q_{act} = actual flow
Q_{ind} = indicated flow
d_s = air density at standard conditions of 25°C and 760 mm Hg
d_a = actual air density inside the transducer

The calibration on these instruments must be checked periodically. It can be done with a bubble buret; however, if an instrument requires substantial adjustment, it is better to send it to the manufacturer who can use sources such as a wind tunnel to calibrate. Adjustment of the calibration involves opening up the unit and adjusting potentiometers used to zero and span the various range scales on the meter. For use, most of these instruments require orientation so that air always flows in a specified direction. Some units incorporate "flow straighteners" on the inlet, which are usually tubes, to eliminate turbulence in the airstream.

It is sometimes difficult to calibrate open-face filters with these flowmeters at the end of the sampling train, which is preferred, unless a calibration jar is used. Some manufacturers recommend that the flowmeter be placed in the middle of the sampling train, which is undesirable as flow rates are based on the amount of air entering the media. Some units have overcome this limitation by providing adapters that allow the unit to be used with filters (either closed-face or open-face) or sorbent tubes. Mass flow sensors should not be placed in line following bubble burets, since moisture will affect these units.

Continued use in dirty environments may necessitate periodic cleaning of the sensor. Flowmeters based on thermal conductivity are affected by dust and reactive gases, but in a different manner than rotameters and wet test meters. Corrosion will alter the electrical conductivity and leads ultimately to failure of the element, and coating by inert material will reduce heat transfer and ultimately reduce the response of the instrument. Chlorides and lead are especially destructive. These units should never be used in flammable or explosive atmospheres.

CALIBRATION PROGRAM FOR PUMPS

Traditionally for flow rates from 20 ml/min to 200 ml/min sampling professionals have used stroke volume pumps, and for flow rates from 1 Lpm to 3 Lpm another type has been used, commonly called low flow and high flow, respectively. Pumps capable of both flow ranges are now available. These high-flow pumps are not to be confused with the high-volume pumps used for area and environmental sampling designed for airflows up to 20 Lpm. Pumps used for air sampling generally fall into two categories: diaphragm and rotary vane. Personal air sampling pumps are usually constructed on the diaphragm principle and high-volume pumps are of the rotary vane type.

Most current battery-powered personal air sampling pumps use various techniques to dampen pulsing and microprocessors to monitor pump speed to improve steady flows and thus are considered to have a constant flow. The exception is stroke volume pumps whose design precludes this. Some pumps are capable of two different modes: constant flow or constant pressure. The constant pressure mode is designed to be used with manifolds that contain one or more critical orifices. More than one sorbent tube is used at a time. In the constant-flow mode, only one sampler can be used, but no manifold is required.

In order to have accurate volume calculations as well as be assured that flow rates appropriate for the type of sampling to be done are in place, flow rate calibrations are essential. Calibrations for flow rate must be done on the pumps that pull the air, on the entire sampling train as it is to be used, and on secondary calibrators such as rotameters.

Pumps that can be changed from high to low flows must have their back pressure checked; therefore, calibration kits for these pumps usually contain both a rotameter and a manometer.

There are two types of calibrations performed on pumps. A *multipoint* calibration compares the rotameter setting of the pump to a range of flow rates representing the pump's accurate range to see if the pump's performance is changing from the previous calibration. A *single point* calibration prior to and following sampling assures maintaining a specific flow rate. Single point pump calibrations in the field can be done with either a bubble buret or a precision rotameter. Multipoint calibrations should be done in the laboratory with a bubble meter.

The frequency of calibration depends on the use, care, and handling of the pump. Manufacturers furnish calibration charts, but some sources recommend the pump be recalibrated at least after each 50 hours of use, or whenever misuse of the pump might affect its performance. In addition, pumps must be recalibrated if they have been abused, and after each repair. Ordinarily, pumps should be recalibrated in the laboratory and the field, both before field use and after each field survey.

In order to determine if the desired flow rate is being maintained during sampling, one of two methods may be used. The first involves visually checking the rotameter on the pump to see if there is any deviation. The second involves checking with a precision rotameter, usually on an hourly basis. Certain OSHA standards require hourly checks with rotameters. However, it should be noted that these conventions generally predate the availability of constant flow pumps that have overcome the problems of significant decreases in flow rate over an 8-hour sampling period for the most part. In most cases, hourly checks with a rotameter are unnecessary and too time consuming to be practical. Most fluctuations in flow rate with older pumps were due to changing load because of increases in resistance from sampling media and/or battery voltage changes.

If it is determined that the flow rate has decreased, check the filter first for excessive dust loading. Do not adjust flow rates in the middle of sampling. Instead, make a note of the decrease, and if it is not more than 10% allow the sampling to continue until enough volume is collected to meet the minimum sampling volume for the method. Then replace the media and adjust the flow rate. For samples where the initial flow rate is different than the final flow rate, the rates should be averaged and the average should be used to calculate the sample volume.

When doing OSHA compliance sampling a different method is used. If the initial and final flow rates are different, the volume should be calculated using the highest flow rate. Following the results, calculations are done to determine if the employer is out of compliance. If noncompliance cannot be established using the highest flow rate, the result is recalculated using the lowest flow rate. If compliance is not established using this rate, resampling is generally necessary.

Personal Air Sampling Pumps

Constant-flow pumps (Fig. A-7) of both the low-flow and high-flow types can cope with changing back-pressure conditions such as filter loading and still maintain a constant sampling rate within certain limits by measuring and controlling the actual input flow. Changes in pressure over what is originally set are detected by flow control and compensation circuitry that result in a feedback signal to the electronics that adjusts the speed of the motor so as to maintain flow rates within ±5% or less of the initial set point as the pump is subjected to changes in load. Changes in the pump's load are sensed via changes in the motor's current demand. Most systems then compensate for these changes by applying a proportional voltage to the motor, adjusting pump speed to maintain flow.

Most constant-flow pumps with low-flow ranges use active pneumatic regulators that are set to 20 inches water gauge to control the pressure drop across variable or fixed orifices to achieve the desired low flow. The regulator consists of a sensing diaphragm with a preloaded spring that automatically opens or closes an air intake valve to recycle the air within the pump. When the regulator is in the system, approximately 20 inches of water pressure causes the regulator to open and maintain that pressure. Generally, adjustment to low flow involves opening up the valve to the pressure regulator, which is kept shut for higher (0.5 L to 5 L) flows.

Problems can occur with constant-flow pumps if there is a drift in the control circuitry, or if the pump has not been correctly calibrated before

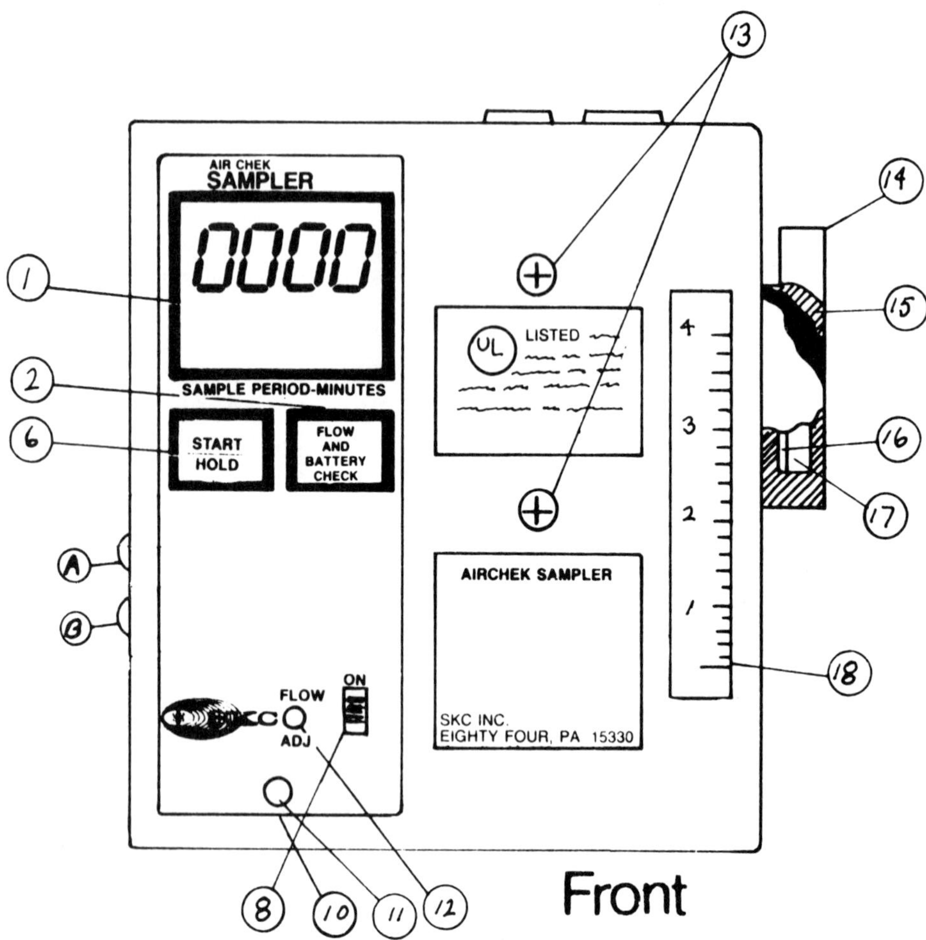

Figure A-7. Personal constant flow air sampling pump. (Courtesy SKC, Inc.)

	Description	*Function*
1	LCD display	Indicators for all sampler functions
2	Flow and battery check key	Allows setting flow rate and testing battery condition
6	Start/Hold key	Used when ready to begin the sampling cycle, pause the sampling cycle, and restart the cycle after pause
8	"ON/OFF" switch	Allows the pump to be shut down completely, clears time display
10	Antitamper cover	Protects controls from incidental contact or tampering
11	Cover screw	Fastens antitamper cover
12	Flow adjustment control	Adjusts flow from 750–5,000 ml/min
13	Accessory mounting screws (2)	Secure accessories such as impinger and trap holders
14	Filter housing (intake)	Air intake port and trap
15	Screws (4)	Secure filter housing
16	Filter O-ring	Positive leak seal for filter in housing
17	Filter (10-micron nylon)	Filters particulates before entering pump
18	Built-in flowmeter	Monitors for flow changes

taking the sample. Constant-flow pumps react to changes in mass flow rate rather than the volumetric rate at which the air is being sampled, and therefore they are sensitive to changes due to altitude differences, and as a result they should be calibrated at the same altitude at which they are used.[1]

A new concept from the Gilian Instrument Corp. that makes changes easier from low flow to high flow mode involves the use of low-flow modules. In order to convert to low flow a button on the top of the module is pushed, which is a simplification of the more extensive procedure that requires several adjustments to convert high-flow pumps to low-flow pumps. Modules screw onto the top of a pump that operates at 0.5 Lpm to 3 Lpm without the modules. One of the modules is a constant-flow regulator that allows flow rates of 5 ml/min to 500 ml/min and the other is a constant-pressure regulator that allows flow rates of 0 ml/min to 750 ml/min. The constant-flow regulator maintains a constant flow of up to 25 inches of water gauge back pressure for sampling with a single tube and the constant-pressure regulator maintains a suction pressure of about 20 inches water gauge and must be used with a manifold system for multiple tube sampling. The modules must be installed in a clean environment while the pump is off. Contamination of the small air passage between the pump and the module could shut down the system.

Many pumps can also measure time. One manufacturer of constant-flow pumps offers purchasers a choice of three options: fault shutdown function with elapsed time clock display; fault shutdown, clock display, and programmable stop time; fault shutdown, clock display, and programmable start and stop times. The fault indicator usually lights if a unit is operated out of its performance envelope for a period of 15 seconds to 30 seconds, or if the unit shuts down. The clock display can allow a sample to be recovered from a pump that has shut down as long as the display is read before the switch is turned off. The timer module provides preprogrammable turn-on and shut-off ability. The user can select 10-minute increments up to 990 minutes. At the end of the sampling period, the pump is shut off. If using this option during sampling, be sure to record the time prior to turning off the pump, because the display returns to zero. However, if the pump faults the display will still be available and usually the timing stops at this point, making it a valuable option, because it allows samples to be saved in many cases.

The timing features allow for a reasonable time estimation in order to calculate a volume when submitting the samples for analyses, although the flow rate's consistency may still be in doubt. Generally, it is best to note the time using a watch as well, since the timing mechanisms of many pumps are not always correct. Most pumps designed to be used for personal air sampling have a way of locking off the controls so that they cannot be tampered with. Another feature is a motor shutdown. Operating pumps under a heavy pneumatic load, such as the one caused by crimped hoses and heavily loaded filters, will greatly decrease the life of the motor. The point (pressure blockage) at which the pump shuts down can often be adjusted using an internal potentiometer.

A variation in collecting area samples has been made possible with the advent of pumps with timers incorporated into them. These allow the sampler to set the on and off time, and the pump will automatically turn on and collect the sample for the predetermined period. Strategies to take advantage of this feature have been proposed by some of the pump manufacturers. One strategy involves using a pump to collect a sample using an on—off strategy over the course of a workday, the result being that the total recommended sampling time, for example, 90 minutes, and sample volumes are not exceeded, but the collection of the sample takes place over the entire workday. The pumps can also be set to do a similar on—off cycle over longer periods of time, for example, to collect 24-hour averaged samples. Another proposed strategy involves using these pumps to automatically turn on and sample at undesirable times, such as on the graveyard shift.

The location of rotameters and other flow measuring devices is important. They should be at the beginning of the sampling train rather than in the middle or at the end. The sampler must account for the pressure drop that can be expected to increase as particles and moisture collect on sampling media. (Table A-1).

TABLE A-1. Sample Tube, Filter, and Impinger Initial Pressure Drop

	Pressure Drop (Inches of Water Column)		
Collector	*200 ml/min*	*500 ml/min*	*1,000 ml/min*
Midget impinger (1 mm)	—	—	3.0
Micro impinger (0.508)	—	11.0	—
Charcoal tube, std. (50/100)	2.0		16.0
large (200/400)	2.0		11.0
Silica gel tube, std. (50/100)	2.5		21.0
large (200/400)	2.5		14.5
Alumina tube, std. (50/100)	1.5		11.0
large (200/400)	2.0		8.5
Molecular sieve, std. (50/100)	1.5		12.0
large (200/400)	1.5		10.0
Nonpolar partition, std.	2.0		17.5
Polar partition, std.	2.5		20.0
Porous aromatic, std.	1.5		12.0

Source: DuPont Catalogue.

Procedure for Calibration with an Electronic Bubble Meter

1. Turn on the pump and allow it to run for 5 minutes to stabilize.

2. Assemble the bubble meter with a cell whose range is good for the flow rate to be measured.

3. Attach the media to a tubing and attach the tubing to the pump. If calibrating a filter cassette and using a cassette adapter, the same adapters used for calibrating should be used for sampling, since these adaptors can cause moderate to severe pressure drop at high-flow rates in the sampling train.

4. Connect the collection device, tubing, pump, and calibration apparatus together (Fig. A-8). Make a visual inspection of all tubing connections.

5. Wet the inside of the electronic flow cell with the soap solution by pushing on the button several times.

6. Turn on the pump and adjust the pump rotameter to the appropriate setting.

7. Press the button on the electronic bubble meter. Visually capture a single bubble and electronically time the bubble. Different flow rates can require faster or slower releases of the button. For extra low flows the button may need to remain pressed until a bubble is formed and starts up the center of the flow cell. A good technique is to watch the bubble pass properly across the sensor zone to ensure a good test has been completed. Most units beep when a test is completed. The printer attached to the electronic bubble meter will print out the readings. This step should be repeated until two readings are within 5%.

8. While the pump is still running, adjust it if necessary.

Procedure for Calibration Using a Manual Bubble Buret

1. Turn on the pump and allow it to run for 5 minutes prior to calibration. This action will allow the pump to stabilize.

2. Check the pump battery with a voltmeter to assure adequate voltage. Charge the battery if necessary.

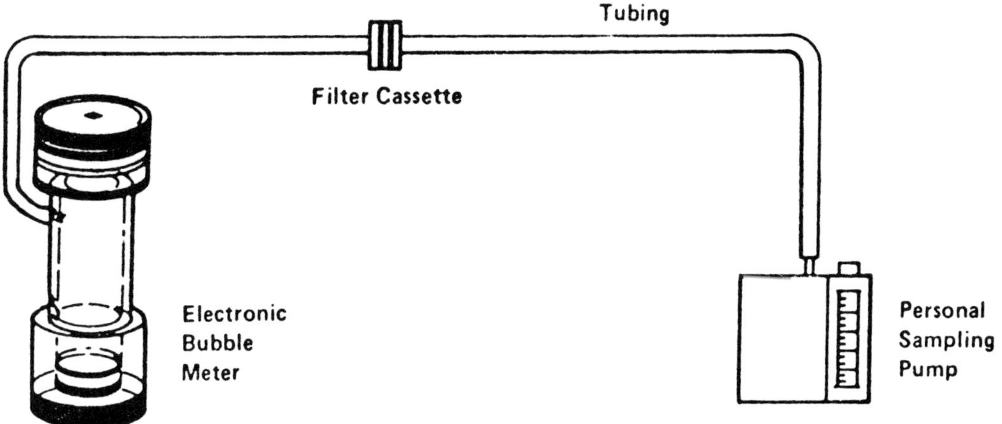

Figure A-8. Calibration setup for a personal air sampling pump using an electronic flowmeter. (From *OSHA Technical Manual.*)

3. Wet the inside of the buret with bubble solution, using a small amount and rotating the buret on its side to coat. Install it on a laboratory stand *upside down* with a flat dish or beaker underneath containing bubble solution.

4. Connect the sampling device with a piece of tubing to the top of the buret. Connect the pump with a second piece of tubing to the sampling device.

5. Have a stopwatch available.

6. Bring the soap dish momentarily in contact with the bottom end of the bubble buret to collect enough soap to form a single film. Turn on the pump. The soap bubble should be drawn up through the length of the buret. Bring up the soap dish and collect just enough film to present one bubble. It may require some practice, because the higher the flow rate, the faster the solution will be picked up, causing lots of bubbles to move up the buret at once. It will be difficult to time just one bubble. In some models, the soap solution can be raised into the buret by squeezing a bulb attached to the bottom. Wait for a few bubbles to move up the buret in order to be sure that they move smoothly and do not break. If they do not, recoat the interior of the buret with bubbles. Draw bubbles upward until they travel the entire length of the buret without breaking.

7. Adjust the pump to the approximate flow rate using the markings on the pump's rotameter.

8. Time the movement of a bubble up the buret. Divide the volume of the buret by the number of minutes it took for the bubble to reach the top. Once the desired flow rate is reached, repeat the measurement two to three times, depending on how close the measured times are, in order to achieve a timing accuracy of ± 1 second. Average the results and calculate the flow rate by dividing the calibration volume by the average time.

9. Record the following data: volume measured, elapsed time, pressure drop, air temperature, atmospheric pressure, serial number of the pump, pump model, date, and name of operator.

10. Repeat this procedure for all pumps to be used for sampling. The same cassette and filter or tube can be used for all calibrations involving the same sampling method.

Cyclones

For respirable dust sampling a cyclone is used. Correct positioning of the cassette within the cyclone is important. The cassette is placed upside down so the inlet to the cassette is in line with the inlet of the cyclone.

Procedure

1. Hook the tubing attached to the cyclone to the fitting on the inside of the lid of a 1-liter jar. The lid is then closed. A tubing is attached to the other end of this fitting and run to the pump. A second fitting in the middle of the lid is to attach a tubing that goes to a bubble buret or rotameter (Fig. A-9).

2. The pump is turned on, and the same procedures for using a bubble buret or rotameter are used. The flow rate should generally be 1.7 Lpm.

Pumps with High- and Low-Flow Ranges

When converting from high flow to low flow, must pumps require the flow rate to be set to 1 Lpm or 1.5 Lpm (Fig. A-10). It does not have to be exact, and the pump's rotameter can be used for this adjustment. Changing high-flow pumps to low flow generally involves opening up a regulator valve and adjusting it. Two rotameters are

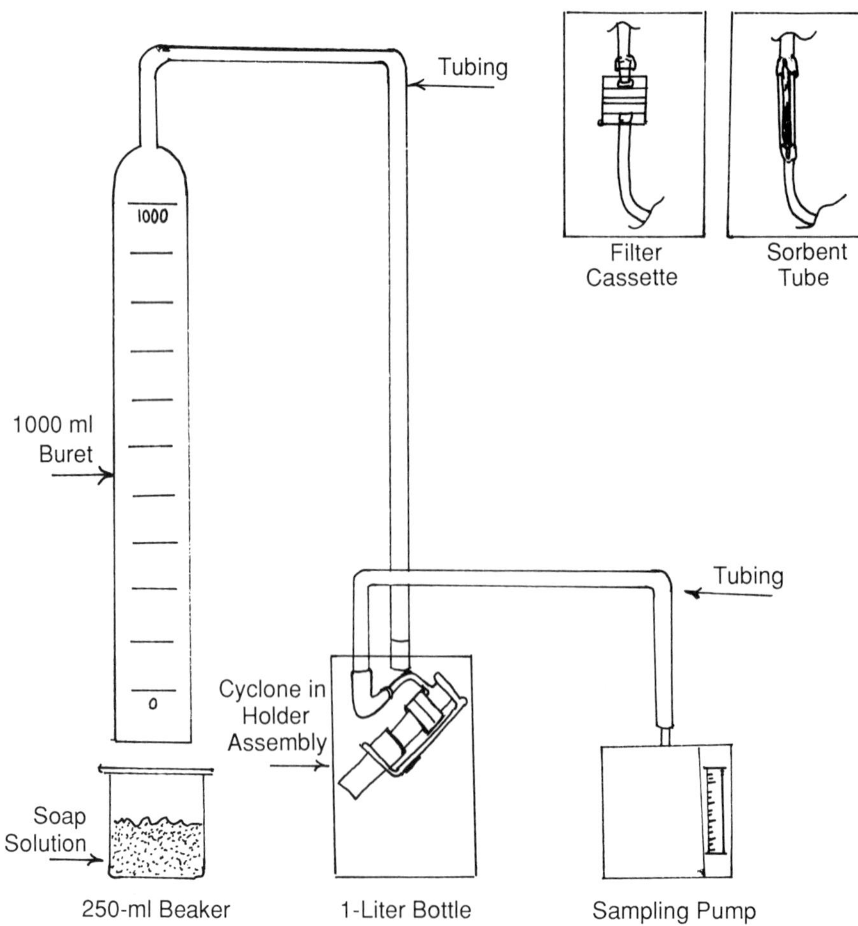

Figure A-9. Calibration setup for personal sampling pumps. (From NIOSH. *Manual of Analytical Methods,* 3d ed.)

necessary: one capable of measuring 0.5 ml/min to 3.5 ml/min, and the other capable of measuring 10 ml/min to 250 ml/min.

Procedure

1. Set the flow rate to 1 Lpm by attaching the pump with a tube to the high-flow rotameter.

2. Open up the regulator valve by turning the appropriate screw. It will be under a screw on top of the pump. Usually opening up the regulator valve results in the screw moving toward the top of the case. For example, in the case of a Gillian pump, the sampler is instructed to turn the screw clockwise five times.

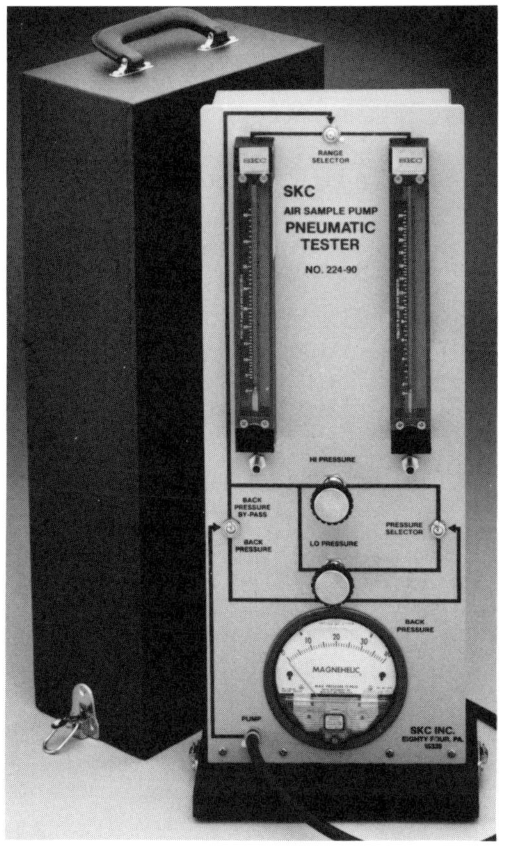

Figure A-10. Calibration apparatus for high-flow to low-flow conversions on personal air sampling pumps. (Courtesy SKC, Inc.)

3. Add the manifold tube holder designed for the type and number of tubes planned for use by hooking up the tubing to the pump. It should be of the type designated by the manufacturer for use with this model pump. Break off the ends of the tube(s) planned for sampling use and insert in the manifold. Screw the cover onto the manifold. A small fitting is usually available to screw into the cover that allows a tube to be placed onto the end of this cover. This tube is then also hooked onto the end of the low-flow rotameter. At this point, the sound of the pump should have changed, and the flow should have lowered within the range of the low-flow rotameter.

4. If the flow rate is still too high, or the pump "cuts out," another adjustment is necessary. Under a second screw on the top of the pump is the regulator adjust valve. Often it must be adjusted by turning a screw in order to get a stable low flow.

5. At this point the pump should be pulling air at less than 250 ml/min. In order to further adjust this flow rate, first use the regulator flow adjustment screw, which is usually on the front of the pump. This adjustment should get the pump to within 5 ml/min to 10 ml/min of the desired flow rate.

6. The next step is to fine-tune the flow. Usually tube manifolds that contain flow controllers have a bottom that will unscrew. Under this bottom is another screw, which is the fine flow adjustment. Turn it to adjust the flow rate to the desired flow. By this time the flow should be smooth and reliable.

Stroke Volume Pumps

Stroke volume pumps have been available for many years and also operate on the diaphragm principle. Rather than using flow rate to calculate volume with these pumps, a mechanical counter is linked to the pump drive mechanism, and it records the number of strokes of the diaphragm.

In stroke volume pumps each revolution of the motor produces one pump stroke and increases the counter reading by one. The diaphragm draws a known volume of ambient air through the sampling medium and through the pump inlet. Total volume can be determined by subtracting the initial count from the final count on the pump's digital register. However, this method only records sample volume indirectly, since the pump mechanism will continue to rotate and record an increasing count even when the input is totally blocked or when leakage is occurring due to faulty valves.

Some stroke volume pumps have a high/low-flow switch. It does not extend the flow range; instead high applies to the full scale (usually 5–200 ml/min), and low provides enhanced resolution of the flow rate adjustment by decreasing the full scale range, allowing for adjustments of smaller increments in flow rates.

In some pumps voltage varies linearly with flow rate. Therefore, the pumps can be calibrated using a voltmeter, which is known as adjusting the voltage set points. A voltmeter is hooked up to the pump. Usually this hookup requires removal of the pump's case with the positive lead on the set point voltage test point and the negative lead in the circuit common. The voltage is measured at various flow settings on the dial and compared to a bubble buret flow rate measurement.

Stroke volume pumps must be calibrated frequently, because slippage and wear in the valve system can introduce errors. Stroke volume pumps cannot be calibrated with rotameters, because the flows are not constant and the ball will go up and down. It is normal and has to do with the operating principle of the pump. So for these pumps a bubble buret must be used. The number of strokes will be recorded for a given volume that the bubble travels, and this volume will be divided by the number of strokes. After a number of trials over which the results should be within ±5, the trials are averaged to determine the stroke volume. Note the start and finish readings over the volume traveled by the soap film. Then the milliliters per stroke or count is calculated.

$$\text{ml/count} = \frac{\text{volume}}{\text{final reading} - \text{initial reading}}$$

The following calculation is used for pumps that operate on stroke volume to determine total volume.

$$\text{Volume sampled (ml)} = \text{(final stroke count} - \text{initial stroke count)} \times \text{(ml/stroke)} \times \text{(multiplier factor for orifice used*)}$$

High-Volume Pumps

High-volume pumps are used to collect large quantities of aerosol contaminants on filters for weighing and analysis. The equipment is not useful for collecting personal exposures. These pumps have been used primarily for air pollution measurements. High-volume pumps usually operate at flow rates of 15 Lpm and higher. These are generally rotary vane pumps. High-volume pumps used for indoor sampling have an inlet to which tubing with a filter cassette can be attached. The flow rate is set by a fixed or manually variable orifice. The pump sits on the floor for sampling and a tripod or other means are used to suspend the filter cassette. When used for outdoor samples, high-volume pumps are mounted inside enclosures directly below filter holders. The flow rate is monitored with an electronic mass flow controller or rotameter. If a timer is used, it is installed separately.

These pumps may or may not be oil lubricated. For some types of sampling this may be a concern. They generally require AC current and therefore have limited portability. Line voltage fluctuations can also be a problem. Flow rates have been found to decrease linearly with voltage. The result will be lower volumes than anticipated and filter efficiencies may also be affected.[2]

High-volume pumps must be calibrated when first purchased, after major maintenance, any time the flow measurement device (e.g., rotameter, mass flowmeter) has been replaced or repaired, or any time a one-point calibration check devi-

---

*Specified in pump operations manual.

ates from the calibration curve by greater than ±6%.

High-volume pumps used for indoor sampling can often be calibrated using a rotameter, usually marked in cfm. Due to their increased complexity, outdoor sampling pumps have more detailed calibration procedures.

EPA methods have traditionally required high-volume samplers for ambient aerosols to be calibrated using a certified variable resistance orifice and a continuous flow rate recorder.[3] However, where a sampler is equipped with a mass flow controller that simplifies calibration this is preferred. These units may also be calibrated by means of a standard positive displacement rotary-type meter. Although it is an accurate calibration method, it requires considerable time. A simpler and acceptable procedure uses a calibrating orifice assembly and water manometer that have been factory calibrated against the positive displacement meter.[4]

Procedure for Outdoor Sampling

1. Remove the motor/blower filter holder assembly from the shelter by lifting it up and out through the rectangular hole in the support pan.

2. Remove the filter holder from the motor/blower unit and replace it with the orifice calibrator using the resistance plate with the largest number of holes supplied with the orifice calibrator. This is the highest flow setting on the orifice.

3. Connect the pressure recorder to the pressure tap on the side of the motor/blower unit.

4. Connect the water manometer to the pressure tap of the calibration orifice.

5. Plug the sampler into a 120-V source while checking the manometer to ensure that the orifice pressure drop does not exceed the range of the manometer. Let the sampler run for 5 minutes.

6. Turn off the pump and install a clean recorder chart in the strip chart recorder. Check the recorder for proper operation and zero the pen if necessary.

7. Read the differential pressure as indicated by the manometer and record it on the data sheet. Convert the inches of water reading to CFM using the calibration curve supplied with the calibration orifice. Record the CFM figure on the data sheet.

8. Record the strip chart deflection on the data sheet.

9. Change the resistance plate in the calibration orifice to the one with the next fewer number of holes.

10. Turn on the sampler and convert the differential pressure indicated by the manometer to the correct flow rate.

11. Record the manometer pressure in inches of water, the actual flow rate from the calibration curve, and the recorder deflection on the data sheet.

12. Repeat steps 9, 10, and 11 for the remaining resistance plates.

13. Using the reading established with these procedures, plot a calibration curve representing the actual flow rate versus the recorder deflection (Fig. A-11). This new calibration curve is used as a direct reference to obtain the actual flow rate.

PUMP SELECTION

Quality care of pumps is important. Once equipment is in the field the industrial hygienist is dependent on it and repeating surveys is costly. In operations where equipment is shared it is important that pumps be inspected after each survey and repaired promptly. Batteries should often be discharged completely prior to charging. With rare exceptions pumps and chargers are systems, and replacement parts are specific for a given manufacturer. Therefore, it is usually best to be restricted to one or two manufacturers when purchasing pumps. Defunct pumps can be used for spare parts. Battery chargers will also be different for most systems. The maximum run time of pumps depends on the conditions of use and

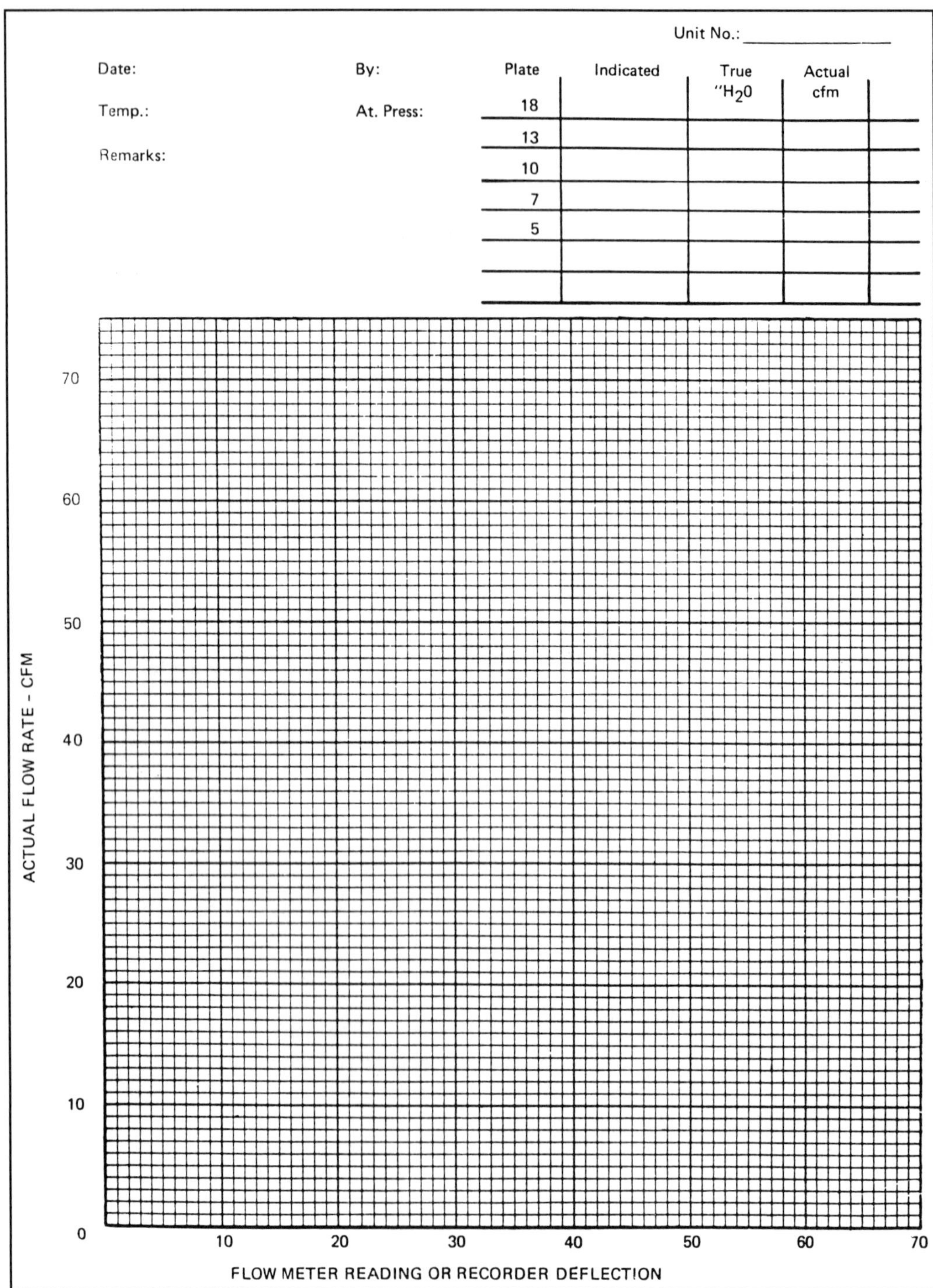

Figure A-11. Calibration data sheet for high-volume sampler. (Courtesy General Metal Works.)

the battery charge. High-pressure drop and/or high flow rates will increase battery drain, and therefore decrease run time.

Some manufacturers sell individual battery chargers designed to charge only one pump. These are useful in situations where the sampler might want to sample for long periods of time at very low flow rates, such as might be done when attempting to identify what might be present. Do not recharge a battery pack or operate a pump off the charger in areas where a potentially explosive atmosphere may be present. Five- and ten-unit chargers are also available. They are useful for recharging multiple pumps and are the most commonly used chargers. Some charging systems will provide automatic discharging for NiCad batteries.

The capability of being able to switch between high- and low-flow ranges provides increased flexibility both when in the field and when considering equipment purchases. It is not always as simple as the manufacturer directions make it sound to convert between these flow ranges, however. Problems have included

1. Manufacturer documentation is often poorly written.
2. A manometer is required to monitor back pressure to get the pump back into high flow properly, an instrument that the sampler does not typically carry into the field unless a special case is utilized.
3. The pumps do not always perform within the ranges they are specified for, especially the low flow ranges.
4. Very low flows (25 ml/min), such as those required for long-term detector tubes, can be difficult to achieve.
5. The combination pumps are heavier than the traditional low-flow pumps and will not fit into a pocket like the traditional low-flow pumps; therefore, they must be put on a worker's belt and load being transported to the site is increased.
6. The low-flow range is achieved through the use of a critical orifice inside the tube holder, so the entire system must be purchased from the same manufacturer in order to assure good performance.
7. Adjustments are more difficult than on traditional low-flow pumps, and they require more skill on the part of the sampler.
8. The need to use the tube holder means that a variety of sizes of tube holders must be purchased in order to accommodate the various sizes of tubes available.

Selection considerations for pumps include the flow range needed, cost, constant flow, ruggedness, options such as timers, and adaptability for filling bags (a more complete list follows this paragraph). Do not buy more features in pumps than are needed. For example, asbestos personal monitoring requires a 2-Lpm flow rate and usually is hard on pumps due to the water used on the projects, scaffolding, tight spaces, and daily use. Therefore, to purchase a pump with both high and low flows would result in spending more money than is necessary when a pump with a more restrictive range of flow rates could do the job.

Pump Selection Variables

Cost
Programmability
LCD display
Antitamper controls
  On/off switch
  Compensation adjustments
Interrupt "hold" feature
Timer
  Run time display
Range of flow rates
  Low flow (1 – 500 ml/min)
  High flow (1 – 3 Lpm, 1 – 5 Lpm)
  Combination flow
  High volume (8 – 20 Lpm)
Constant flow
Pulsation-free flow
Built-in rotameter
Fault features
  Low battery shutdown
  Pinched hose shutdown
  Excess back pressure shutdown
Fault shutdown with time retention
Noise level during operation

UL listed intrinsically safe
Ease of changing from high flow to low flow
Built-in fluid/particulate trap

VOLUME ADJUSTMENTS

Generally sampling is conducted at approximately the same temperature and pressure as calibration in which case no correction for temperature and pressure is required and the sample volume reported to the laboratory is the volume actually measured. Where sampling is conducted at a substantially different temperature or pressure than calibration, an adjustment to the measured air volume may be required in order to obtain the actual air volume sampled.

Portable thermometers including digital models are available that allow temperature to be measured at the sampling site. Certain instruments with multiple gas sensors have incorporated temperature sensors as well. Barometric pressure can be obtained by calling the local weather station or airport and requesting the unadjusted barometric pressure. If these sources are unavailable, a rule of thumb is to decrease barometric pressure by 1 inch Hg for every 1,000 feet of elevation.[5]

If conditions of sampling are different than those used for calibration, then the following formula can be used to correct the volume reported to the laboratory:

$$Q_A = Q_I \left(\frac{P_c}{P_a} \times \frac{T_a}{T_c} \right)$$

where

Q_A = actual or corrected flow rate
Q_I = indicated flow rate
P_c = barometric pressure at calibration site
P_a = barometric pressure at actual site
T_c = temperature at calibration site
T_a = temperature at actual site

As a rule of thumb, for every 1,000 feet of elevation, the barometric pressure decreases by 1 inch of Hg (Table A-2). The calculation for temperature adjustment (Table A-3) involves using the natural gas law.

TABLE A-2. Correction for Altitude Differences

Altitude Difference in 1000 ft	*Site Is Higher*	*Site is Lower*
0	1.000	1.000
0.5	1.009	0.991
1.0	1.019	0.982
1.5	1.028	0.973
2.0	1.038	0.964
2.5	1.047	0.955
3.0	1.057	0.955
3.5	1.067	0.946
4.0	1.077	0.937
4.5	1.087	0.920
5.0	1.090	0.912
5.5	1.107	0.903
6.0	1.117	0.895
6.5	1.128	0.887
7.0	1.138	0.879
7.5	1.148	0.870
8.0	1.160	0.862

TABLE A-3. Correction Factors for Temperature

	Site Is Hotter	*Site Is Cooler*
Temperature Difference (°F)		
0	1.000	1.000
10	1.010	0.990
20	1.020	0.981
30	1.030	0.971
40	1.040	0.962
50	1.050	0.953
60	1.060	0.944
70	1.070	0.935
80	1.080	0.926
Temperature Difference (°C)		
0	1.000	1.000
5	1.009	0.991
10	1.018	0.983
15	1.027	0.974
20	1.036	0.966
25	1.045	0.958
30	1.054	0.949
35	1.063	0.941
40	1.072	0.933

REFERENCES

1. Treatis, H. N., T. F. Tomb, and A. J. Gero. Effect of altitude on volumetric flow rate of constant flow pumps. *Appl. Ind. Hyg.* **1**(1):54–58, 1986.
2. Cohen, B. S., N. H. Harley, and M. Lippmann. Bias in air sampling techniques used to measure inhalation exposure. *AIHA J.* **45**(3):187–192, 1984.
3. EPA. *Method for Determination of Total Suspended Particulate in Ambient Air Using High-Volume Sampling.* EPA, Washington, DC.
4. *General Metal Works High-Volume Ambient Sampler Manual.* General Metal Works, Village of Cleves, OH.
5. OSHA. *OSHA Technical Manual.* Washington, D.C., Nov. 1990.

APPENDIX B

Gas and Vapor Calibrations

For calibration of direct reading and real time instruments that measure gases and vapors, a known concentration of a specific contaminant is the best way to ensure accuracy. Known mixtures of gases and vapors in air are used to set the sensitivity, range, or span of an instrument's response. They are often called span gases. They are limited to gases and vapors that will remain in the gaseous phase over the period of time needed for calibration.

One of the major problems associated with toxic gas monitoring is the availability of reliable calibration gases due to the varying physical properties of chemicals that can make storage difficult. For example, carbon monoxide is readily available in cylinders, whereas nitrogen oxides are reactive and must be generated in a laboratory using permeation tubes and special instrumentation. Depending on the specific characteristics of the compound, it could be packaged in a cylinder, made up in a bag using a syringe, mixed in a bottle using premeasured ampules, or more complex techniques may be required such as permeation tubes or dynamic gas generating systems, for example, what might be required for an atmosphere of mercury vapor. Table B-1 shows typical test atmospheres of commonly sampled chemicals.[1]

There are several methods for generating calibration gases. These include both static and dynamic systems. A static system is a mixture in a fixed volume such as a Tedlar bag or Teflon bottle. Preparation of a batch mixture in a bottle or a bag has the advantage of simplicity and convenience in some cases. Dynamic systems generate a continuous flow from a gas cylinder or from bubbling air through a liquid solvent. Flow dilution systems are capable of providing theoretically unlimited volumes at known low concentrations that can be rapidly changed if desired. The disadvantages of flow dilution systems are the skill, exactness, and complex laboratory setups required. Although these are beyond the scope of this book, information is available from other sources.[2] Permeation tubes also fall into this category, although recent advances in simplified calibration apparatus that provide sufficient accuracy for field use have made them somewhat simpler to use than in the past.

Many gases and vapors in commonly used concentrations are available from several manufacturers. Usually they are contained in compressed gas cylinders, ampules, or aerosol cans. Filling high-pressure gas cylinders is better left to the experts. When purchasing calibration or span gases, one assurance that the source is reliable is

TABLE B-1. Typical Test Atmospheres of Commonly Sampled Chemicals

Compound	*Method of Generation*
Ammonia	Permeation device or ampule
Carbon dioxide	Gas cylinder with zero air or nitrogen containing carbon dioxide in the required concentration
Carbon monoxide	Gas cylinder with zero air or nitrogen containing carbon monoxide in the required concentration
Ethane	Gas cylinder of prepurified nitrogen containing ethane as required
Hydrogen chloride	Cylinder of prepurified nitrogen containing hydrogen chloride as required
Hydrogen sulfide	Permeation device, cylinder, or ampule
Methane	Gas cylinder of zero air containing methane as required
Nitrogen dioxide	Permeation device
Oxygen	Use outdoor air at sea level
Ozone	Calibrated ozone generator
Sulfur dioxide	Permeation device
Xylene	Cylinder of prepurified nitrogen containing xylene as required or gas bag with appropriate amount of solvent to make up desired concentration

Source: 40 CFR 53 EPA.[1]

certification of the concentration and contents of the mixture from the manufacturer. The gas supplier should be asked to indicate on the certificate of analysis the analytical method used and to estimate the means used to compute the accuracy, including the calibration technique. Some instruments can accommodate a compound mixed in nitrogen; however, many instruments require oxygen to operate and therefore calibration mixtures must be made up using air.

Not every gas mixture can be pressurized. For example, the vapor pressure of toluene at 20°C is only high enough to produce a compressed gas mixture of 1,200 ppm at 300 psi; therefore, it is not possible to purchase a cylinder of toluene gas for calibration in the lower explosive level (LEL) range. For this reason, as well as others, sometimes the gas selected for calibration is not the one that will actually be monitored, but instead one for which the response of the instrument can be compared to the gas of interest. An example is the use of isobutylene for calibrating instruments to detect organic vapors in the ppm range. If a standard is to be used to calibrate a combustible gas indicator, do not prepare concentrations greater than 25% of the LEL for that compound.

Making up standards of calibration gases can be difficult unless the sampler is familiar with the problems associated with each compound and the system to be used (Table B-2). For example, isocyanates are very reactive. Other effects, such as container wall absorption, humidity, and volumetric errors, must be recognized. Appropriate precautions must be taken to avoid concentration errors derived from these seemingly extraneous sources when attempts are made to produce homogeneous, known concentrations for calibration purposes.

When calibrating any instrument, several known concentrations should be used. These should encompass the range of the instrument's measuring capability. Periodically, a calibration curve should be prepared. For field use, a calibration at one point, such as the threshold limit value (TLV), can suffice as long as the instrument is periodically calibrated at several points.

The environmental conditions, including temperature, relative humidity, and barometric pressure, of the instrument at the time of calibration should be as near as possible to what will be encountered in the field. Temperature is the most important, because changes in temperature are

TABLE B-2. Overview of Selected Calibration Methods

Method	*Typical Accuracy (%)*	*Precision*	*Range*	*Cost*	*Ease of Use*
Permeation tubes	97–99[a]	Good	ppb to low ppm	Moderate	Difficult
Gas cylinder[b]	95–99	±1% to 2%	ppm to %	Moderate	Easy
Dynamic dilution of pure gas or higher concentration	93–97[c]	±3%	% to ppm	Moderate	Moderately easy
Static dilution	90–97[d]	±3% to 10%	ppm to %	Inexpensive	Easy

Source: Becker, J. H., et al. Instrument calibration with toxic and hazardous materials. *Ind. Hyg. News*, July 1983.
[a]Depends on good temperature control, balance accuracy, and flow rate.
[b]Data from manufacturer's literature; refers to analyzed standards only.
[c]Includes inaccuracies in calibration gas plus flow rate.
[d]Depends on analyst's technique.

most often encountered in the field and can cause bias in the readings obtained. If it is not possible to calibrate at the working temperature, the user must allow sufficient time for field equilibration of temperature. Since many of these units are battery operated, it should be noted that temperature extremes can also adversely affect the performance or life of the batteries used in these devices.

Whenever possible do calibrations in a laboratory hood. Always work with the smallest quantity of material possible, including using the lowest practical concentration in gas cylinders. Use appropriate personal protection for the type of contaminant. Leak test lines prior to using. If the instrument being calibrated is a nondestructive analyzer, either use it in a hood or make sure a window or other source of air is available and "vent" the instrument in this direction. If calibrations will be performed in one given area, consider installing a local exhaust vent. For instruments like the flame ionization detector (FID), the exhaust gases should be vented as well. If the instrument's response is very rapid, large volumes of calibration mixtures are not needed.

Make sure that the calibration procedure itself does not pressurize the sensing cell. It is especially important to observe when pressurized cylinders of standard gases are used for calibration. Overpressurization of the sensor can be avoided by using a pressure regulator on the calibration gas cylinder, or by installing a "T" fitting in the line to reduce the stream to atmospheric pressure. Another method is to fill a bag with gas from the cylinder so that it can be presented to the sensor at atmospheric pressure from the bag.

It is important that appropriate accessories be used depending on the chemical in use. For example, substituting tygon tubing for Teflon tubing may cause some sample to be lost by absorption loss in the tubing walls for some compounds. Some instruments require a humidifier to compensate for the humidity effects caused by using a dry compressed gas mixed with nitrogen. Dispose of unneeded gases from bags and bottles properly. Keep bags and bottles clamped shut until ready to dispose of the gases. Decide where to release the unused contents of the calibration mixture once calibration is complete. It should be outdoors, or if the gas is soluble in water (sulfur dioxide, hydrogen chloride, chlorine) hooking up the bag to a jar of water via tubing and allowing the bag to slowly release its contents.

PREMIXED GASES AND VAPORS IN CYLINDERS

Premixed gas standards purchased commercially can be diluted, if purchased in a high enough concentration to provide a wide-range calibration curve. The major source of error is the concentration of the test gas in the cylinders. Second is working at the lower ranges of rotameters, where they are less accurate. The worst case could involve calibration values from different manufacturers determined by two different analytical techniques, such as hydrocarbons by FID or an infrared detector. In this situation, the difference between the analytical techniques could increase the error to 10% to 20%.[3]

Most scientific gas suppliers rate their mixture accuracy at 95% to 99%, depending on the type of gas and the concentration level. If using premixed gases and two different batches of gas, and the resulting measurements do not agree, a third batch of gas should be obtained and used. If it too is contradictory, change sources. For many gases, the labeled ppm value defines the analysis at the time of manufacture only. For certain gases and certain manufacturers, labeled ppm values are applicable for a year or longer. Unfortunately, not all commercially available cylinder standard gases have such stability, and some may become unreliable in 2 to 3 weeks or less.[4]

Compressed gas cylinders present a number of hazards. They are often under extremely high pressure when full, the size of the tank frequently determining how much pressure it can hold. Striking the valve at the top of the tank may shear off the valve assembly, venting the pressurized gas. In addition to the potential for fire or explosion, the velocity of the existing gas may propel the cylinder at hazardous speeds. Always check tanks for pitting and rusting. Any sign of deterioration should be reported immediately, and the tank should be removed from service. Never assume that the color of the tank indicates the contents. Color schemes are strictly the prerogative of the company that sells the cylinders. Never add adaptors or other gear to a regulator to make equipment fit. Never open the valve on a gas container without a regulator attached. As a rule of thumb, at least 200 psi of gas is required in a calibration cylinder for accurate calibration.

Often special threads and sizes are used for regulators to forewarn or prevent certain types of equipment from being used or attached to the tanks. Regulators are specific for different types of tanks and should not be switched between tanks of different manufacturers. The direction of the threads on tanks may be reversed from the normal directions used in other equipment. Never attempt to force threads or nuts. Never store tanks in direct sunlight or near excessive heat. Nonflammable gases, such as carbon dioxide, may rupture with a force equal to or greater than that of flammable gases. Do not store calibration gas tanks in rooms where individuals work. Calibration gas tank contents are under pressure; therefore, do not use oil, grease, or flammable solvents on the flow control or the calibration gas tank. Do not throw cylinders into a fire, incinerate, or puncture. It is illegal in some cases to refill certain types of tanks.

Procedure for Using Gas Cylinders

1. Attach the piece of tygon tubing connected to the bag to the beveled connector attached to the cylinder. A pressure gauge should be on top of the cylinder as well. If the gas in the cylinder is sufficiently toxic, this operation should be done inside a laboratory hood or outdoors.

2. Turn on the cylinder slowly and allow the bag to fill. When full, turn off the cylinder, clamp the bag shut, and remove the tubing.

3. Attach the tubing to the instrument. Open the clamp, and turn on the instrument, which should have been zeroed earlier in a noncontaminated area.

STATIC CALIBRATION MIXTURES

Static or batch calibration mixtures allow the users to make up their own concentrations, thus providing more flexibility and offering a wide variety of concentrations. Static calibration mixtures are prepared by introducing a known volume of contaminant into a specific volume of air inside a container. Usually the container is either a bag or a bottle (rigid chamber). Two options for rigid vessels are a 1-L Teflon bottle whose cap has been modified to accept a septum and a 1-L cylindrical glass vessel with Teflon stop-cocks in each end and a septum port projecting from the side wall (Fig. B-1). The primary disadvantage of the Teflon bottle is its thin walls, which make it

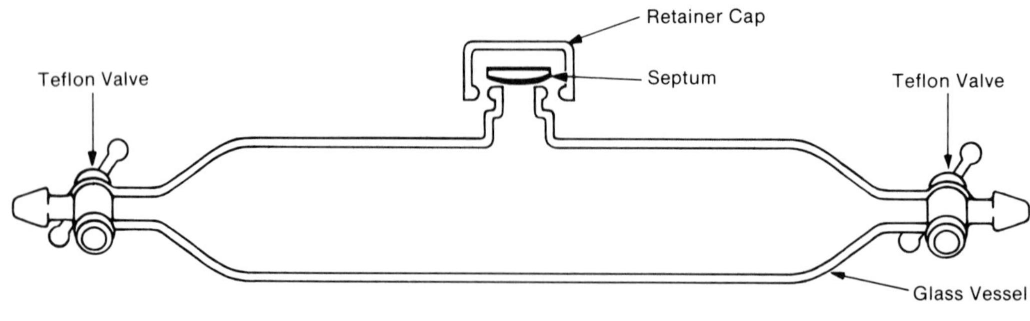

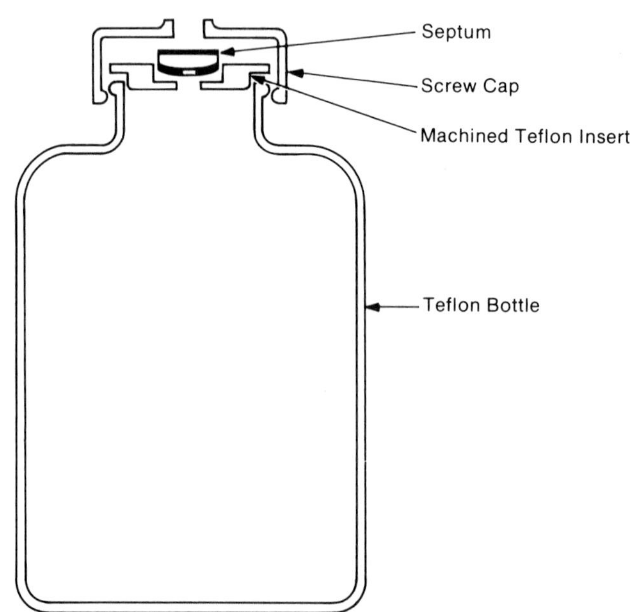

Figure B-1. Static calibration containers. (Courtesy Photovac Inc.)

susceptible to diffusion following an overnight bakeout at 100°C.[5] The glass vessel is purged by passing a stream of clean air through the two stop-cocks for 2 to 5 minutes. A way of determining if contamination is present is to inject a sample of inside air into a gas chromatograph. Bottles should also be calibrated, which can be done by filling them with water using a graduated cylinder. The most commonly used bottle is a 5-gallon narrow-neck carboy. It is large enough to allow sufficient volume for the liquid to evaporate. The sampler can utilize approximately one tenth of the volume of this container before the mixture is diluted appreciably by room air. Placing two or three carboys in series and preparing the same concentration in each reduces the dilution effects even further and allows more of the original container to be used. In general, the lower the concentration of a static standard, the more frequently it should be replaced.

The major advantage of using flexible bags is that no dilution from room air due to displacement of volume occurs as the sample is withdrawn. Bags should be tested frequently for pinhole leaks.

Testing is done by filling bags with clean air and sealing them. If no detectable flattening occurs within 24 hours, the leakage is negligible. Generally, bags of known volume can be purchased. However, it is a good idea to calibrate a bag from each lot for quality control by using a personal air sampling pump at a known flow rate to fill it.

Adsorption and reaction on the walls is not a great problem for relatively high concentrations of inert materials, such as carbon monoxide. However, low concentrations of reactive materials, such as sulfur dioxide, nitrogen dioxide, and ozone, are partly lost, even with prior conditioning of the bags. Prior conditioning is accomplished by putting a similar mixture in a bag and allowing it to sit for at least 24 hours in the bag, and then evacuating the bag just before use. Larger-size bags are preferable to minimize surface-to-volume ratio. Losses of 5% to 10% can occur during the first hour, so calibrations should be done immediately following the mixing. Precautions must be taken in order to avoid contamination of calibration gases by prior contents of bags: Bags should be dedicated to certain chemicals or new bags should be used to do calibrations.

One static calibration method involves the use of ampule kits. A kit generally contains ampules of a specific compound, a calibration bottle, and some device for breaking the ampules inside the bottle, such as ceramic balls. For calibration, the ampules are broken in the calibration bottle and the resulting gas mixture within the bottle equals the concentration specified on the label of the ampule. The calibration bottle is then placed over the head of the sensor for calibration. Ampules are available for hydrogen sulfide, ammonia, sulfur dioxide, hydrogen cyanide, and methane. Ampules are used primarily for field checks.

Static calibration mixtures using liquids are prepared by injecting either the liquid or head space vapor from a glass microsyringe, gas syringe, or pipet containing a calculated volume of the chemical of interest into a bag of known volume that contains a sufficient volume of purified air to produce the desired concentration. Generally, syringes are better for handling concentrated solutions of volatile materials than pipets. Microliter-sized auto pipets are often made of plastic. They are usually designed to deliver water-based solutions rather than compounds with a high vapor pressure (VP). With plastic pipets the VP over the liquid can displace some of the liquid, causing inaccurate deliveries. The use of syringes is discussed later in this section.

Contaminants that are gases at room temperature are usually introduced via a gas syringe. Gas-tight syringes are leak-free only when new and should be tested regularly or injections will be less than indicated, resulting in inaccurate mixtures. Syringes can absorb amounts of contaminants into the walls.

Calculating Concentrations for Static Calibrations

When making up a calibration mixture from a liquid, the amount to add to a specific volume can be calculated.

$$\text{ppm} = \frac{x_{g\ of\ cpd}\,(1\text{ mole})\,(22.4\text{ L})\,(T_2 - \text{kelvins})\,10^6}{\text{vol. of bag}\,(MW)\,(\text{mole})\,(273\text{ kelvins})}$$

where

ppm	=	desired concentration
$x_{g\ of\ cpd}$	=	grams of compound needed for desired concentration
vol. of bag	=	amount of air to be metered into bag
MW	=	molecular weight of compound
T_2	=	temperature where calibration will be performed

To convert from grams of compound to milliliters,

$$\frac{22.4\text{ L}}{\text{mole}} = \frac{22{,}400\text{ ml}}{\text{mole}}$$

$$\frac{(x_g)\,22{,}400\text{ ml/mole}}{\text{grams/mole}} = \text{ml of cpd}$$

In a 10-L bag these microliters of vapor will generate the following concentration:

$$\frac{\mu\text{l of vapor}}{10\text{ L}} = \text{ppm in bag air}$$

For compounds that are gases at room temperature,

$$\text{concentration (ppm by volume)} = \frac{(\text{vapor volume in cc})(1000)}{\text{air volume in liters}}$$

The headspace vapor technique can also be used for making up smaller amounts of calibration mixtures.[5] These concentrations can be made up using the headspace VP. The temperature of the room must be measured and referenced to the VP of the compound at that temperature. A pure gas has a vapor pressure greater than 760 mm Hg; therefore, VP is not a concern, and to make up a mixture the ratio needed is calculated using the desired final concentration.

$$\text{ppm} = \frac{\mu\text{l of compound}}{\text{volume}}$$

$$= \mu\text{l of compound}\left(\frac{760\text{ mm Hg}}{\text{VP of compound}}\right)$$

Syringes

Accurate techniques for using the syringe should be followed. A syringe twice as big as the amount of sample needed is generally required. For example, a 10-μl syringe would be used to deliver 5 μl of sample. The syringe should be flushed with the chemical several times before using. A small amount of the liquid (1 μl) is pulled into the syringe initially, and then a small amount of air. Pull the liquid–air interface in until all of the 1 μl of the liquid can be seen. Then depress the plunger until the liquid is just above the air, and pull in the amount of sample needed. Pull the sample into the barrel to make sure the amount is accurate; then push the sample to the bottom of the barrel. Inject the sample into the bag.

With a gas, a gas syringe can be used to pull air from a tygon tubing attached to a cylinder with a regulator on it. This operation is best done under a laboratory hood. The tubing should be pinched off at the end with a sturdy pinch clamp, and the cylinder should be turned on. The gas syringe can then be inserted into the tubing, and the sample can be pulled out. The sample is then injected into the bag. Gas syringes with shut-off valves are not recommended because of memory problems associated with the valves. For samples suspected of containing high concentrations of volatile compounds, disposable glass syringes with stainless steel/Teflon hub needles are used.[6] One way to increase the accuracy of measurement of the amount delivered by syringe is for the sample to be weighed out.[5]

Syringes can develop problems: Teflon plunger tips become worn and leaks can develop, seals on screw-on needles can become loose, and needles can become plugged. With large syringes a blockage can be detected because there is a resistance as the plunger is pushed in. With smaller syringes blowing air through the syringe into clean water and observing that bubbles emerge throughout the travel of the plunger down the barrel can detect blockage. Manufacturers generally provide fine wire for cleaning blocked needles or alternatively the needle can be replaced.

Preparation of Known Concentrations Using Bags

Bags can be used for compounds that are liquids or gases at room temperature. Following injection of a liquid or vapor into a bag, a sufficient amount of time to allow equilibration is needed (Fig. B-2). Recommendations range from one hour for liquids to 15 to 30 minutes for gases and headspace vapors.[7]

Procedure

1. Select a 5-L to 10-L bag. The volumes of air and gas or liquid needed should be calculated ahead of time in order to make up the desired concentration and will be dependent on the size of the bag.

2. Partially fill the bag with air that has been passed through a series of traps, one of which is

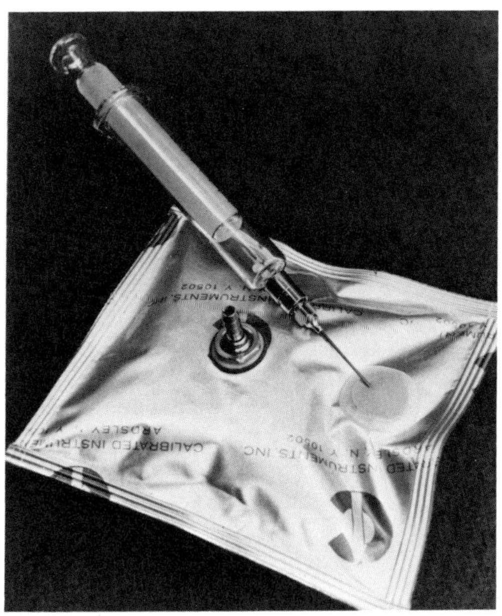

Figure B-2. Using a bag and syringe to create a calibration gas. (Courtesy Calibrated Instruments, Inc.)

filled with a solid desiccant and another that is filled with silica gel or activated charcoal, or draw clean air (or oxygen or nitrogen) from a supply cylinder into the syringe for measured transfer into the bag and then completely evacuate it using either a personal air sampling pump or a 1-L to 2-L syringe. This procedure should be repeated several times in order to condition and purge the bag.

3. Following this purging cycle, a specific volume of air should be metered into the bag using a dry test meter or personal sampling pump at a flow rate appropriate for the size of the bag. For example, 5 Lpm is appropriate for a 10-L bag. When done, close the inlet to the bag. Often it involves not only pushing the valve down but also turning it a few times to lock it.

4. Add the chemical through the bag's septum (usually located behind the valve) or through tubing attached to the valve using a microliter syringe if it is a liquid, or a gas syringe if it is a gas, using proper techniques to assure accuracy.

5. The contents may be mixed by gently kneading the bag with the hands. Following injection of a liquid, inspect the inside of the bag for signs of liquid if the bag is the see-through type. Liquid is an indication that the material has not completely volatilized. Continue to knead the bag until the material evaporates.

6. Perform calibration of the desired instrument following the manufacturer's recommendations.

Preparation of Known Concentrations Using Bottles

Generally, bottles are best used for compounds that are liquids at room temperature (Table B-3).[8]

Procedure

1. Clean and dry the bottle. Oily residue inside should be avoided, since it may absorb the gas or vapor added. The bottle should have a cap or plug in which two short glass tubes have been inserted. A short piece of tubing should be attached to one glass tube and a long one to the other. Both are then clamped off.

2. Prepare two squares of aluminum foil approximately 10 cm by 10 cm, crumple them up, and put them inside the bottle. They will be used to provide a means to agitate the mixture inside of the bottle (Fig. B-3).

3. Measure out the liquid using a microsyringe and proper technique. Unclamp one of the tubes, carefully insert the syringe, and carefully deliver the liquid into the bottle. Reclamp the tubing. Another technique is to take the top off the bottle and insert the needle, covering it with the top. Deliver the liquid, quickly remove the needle, and recap.

4. Pick up the bottle and shake it so that the foil balls rotate as they move up and down. Continue mixing until all the liquid has evaporated. The mixture is now ready to use.

TABLE B-3. Concentrations of Selected Compounds to Produce 100 ppm in a 5-gallon Bottle

Compound	*Concentration (μl)*
Acetaldehyde	4.5
Acetone	5.9
Acetonitrile	4.2
Acrolein	5.3
Acrylonitrile	5.3
Benzyl chloride	9.2
n-Butanol	7.3
Butyl acetate	10.5
Carbon disulfide	4.8
Carbon tetrachloride	7.7
Chloroform	6.4
Cumene	11.2
Decane	15.6
1,2-Dibromoethane	6.9
1,2-Dichloroethane	6.3
1,4-Dioxane	5.2
Ethanol	4.7
Ethyl acetate	7.8
Ethyl acrylate	8.7
Ethyl benzene	10.0
Ethyl bromide	6.1
Ethyl mercaptan	5.9
Freon 11	7.4
Freon 113	9.6
Heptane	11.7
Hexane	10.4
Hexanol	5.8
Isopropanol	6.1
Methanol	3.2
Methyl acrylate	7.2
Methylene chloride	5.1
Methyl ethyl ketone	7.2
Methyl isobutyl ketone	10.0
Methyl methacrylate	8.9
Methyl propyl ketone	8.5
Nitromethane	4.3
2-Nitropropane	7.2
Nonane	14.3
Pentane	9.2
n-Propanol	6.0
Propylene oxide	5.4
Pyridine	6.4
Styrene	9.2
Tetrachloroethylene	8.2
Tetrahydrofuran	6.4
Toluene	8.5
1,1,1-Trichloroethane	8.4
1,1,2-Trichloroethane	6.8
Trichloroethylene	7.2
Xylene	9.9

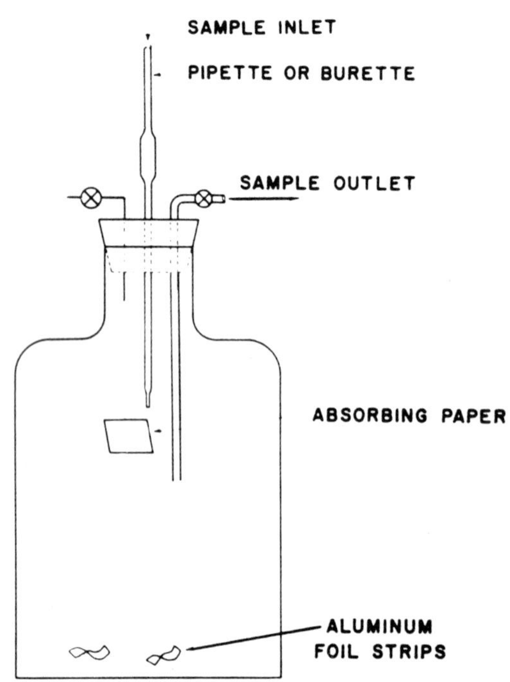

Figure B-3. Batch calibration in a 5-gallon bottle. (From NIOSH. *The Industrial Environment—Its Evaluation & Control.* Washington, D.C., 1973.)

GAS PERMEATION TUBES

Gas permeation tubes allow for very accurate calibration of some instruments at low gas concentrations. The primary value of these tubes is the ability to generate atmospheres of reactive gases, such as sulfur dioxide and nitrogen dioxide, that would be difficult otherwise. The tubes require a prolonged equilibration period prior to use, and the equipment to house them in a constant-temperature environment is bulky. The principle is that any material whose critical temperature is above 20°C to 25°C can be sealed in Teflon tubing. The material then permeates the walls of the tube and will diffuse out at a rate dependent on wall thickness and area (fixed parameters) and temperature. At constant temperature, the rate of weight loss is constant as long as there is liquid in the tube. Permeation tubes for many compounds, including sulfur dioxide, nitrogen dioxide, hydrogen sulfide, chlorine, propane, butane, and methyl mercaptan, are available. For compounds such as nitrogen oxides and formaldehyde permeation tubes may be the only choice. Dilution gas systems can be used to generate varying concentrations from individual permeation tubes.

The liquid or gas in permeation tubes maintains a constant vapor pressure in contact with the inner wall of the tube. The compound's molecules pass through the tubing wall at a very small, extremely constant, flow rate. After an initial indication period of 1 to 3 weeks, the material permeates at a uniform rate through the walls of the tubing as long as it is kept at a constant temperature, and the output rate of the tube will remain essentially constant until nearly all of the liquid has permeated through the tube. In general, permeation tubes can be used to generate concentrations between 0.1 ppm and 50 ppm.

Two levels of accuracy are used in the manufacture of permeation tubes. Certified tubes are rigorously calibrated and their rates are given a factory calibration. Batch calibrated tubes are identical in construction to certified tubes, but do not go through the same rigorous calibration method. Their calibration is based on the knowledge of the rate of certified tubes from the same batch of tubing material. Accuracy for batch tubes is generally +5%. A graph of permeation rate versus temperature is provided with permeation tubes.

Tubes can be disposable or refillable. Disposable tubes are usually short lengths of Teflon tubing with the liquid inside. The tubing used most commonly is either FEP Teflon or TFE Teflon. In general, TFE has permeation rates approximately five times higher than FEP at the same temperature.[9] The flow rate is determined gravimetrically for each tube. Refillable tubes are small stainless steel cylinders with a Teflon tubular membrane sealed inside. The cylinder serves as a large-volume reservoir for the component that is independent of the tubular membrane. The compound surrounds the membrane and permeates through the membrane wall to mix with the dilution gas flowing inside the membrane.

Refillable permeation tubes are certified by vacuum leak measurement and are supplied with graphs describing the emission rates over their entire operating temperature range. Refilling the tubes does not disturb the membrane inside, thus eliminating any need for recalibration according to manufacturers extending the tube's life. Some tubes have two layers of polymer. In this case, the liquid or gas inside the device permeates through the first polymer layer to a gaseous form that then permeates through the second polymer layer at a controlled rate allowing for much lower concentrations. Some tubes have a vial attached that serves as a reservoir, giving the device an extended life and making lower permeation rates obtainable.

The permeation rate is dependent on temperature. The sensitivity of tubes ranges from a 1% to 15% change in permeation rate per °C. Some tables give the permeation rate per cm of length of tube. In general, as the length of the tubing doubles, the permeation rate also doubles. Permeation rate can also be controlled to some extent by the wall thickness of the permeation tube and tubes having three different wall thicknesses are available. The life of the tube depends on the volume of the tube, the weight of the material inside the tube, and the permeation rate of that particular tube. Permeation tubes are calibrated by weight loss over a known time interval. Generally, a semi-micro balance is chosen to measure these weight losses. Fingerprints or additional dirt on the outside of the tube can seriously affect the accuracy of these weights. A wire loop is supplied with many diffusion tubes to help facilitate handling of the tubes. In general, more than three weighings are obtained on a tube.

If a higher concentration is desired, some systems can accommodate more than one tube. The permeation rate is then the sum of the rates of these tubes. Increasing the temperature of the calibrator if it has an oven will also increase permeation rates up to a certain point.

Some calibrators have ovens and others do not. An oven is required for tubes that are designed to operate at elevated temperatures. After selecting the desired concentration, the temperature and other data are used to calculate the flow rate of the calibrator. The calibrator is turned on and the permeation tube is placed inside the bottle

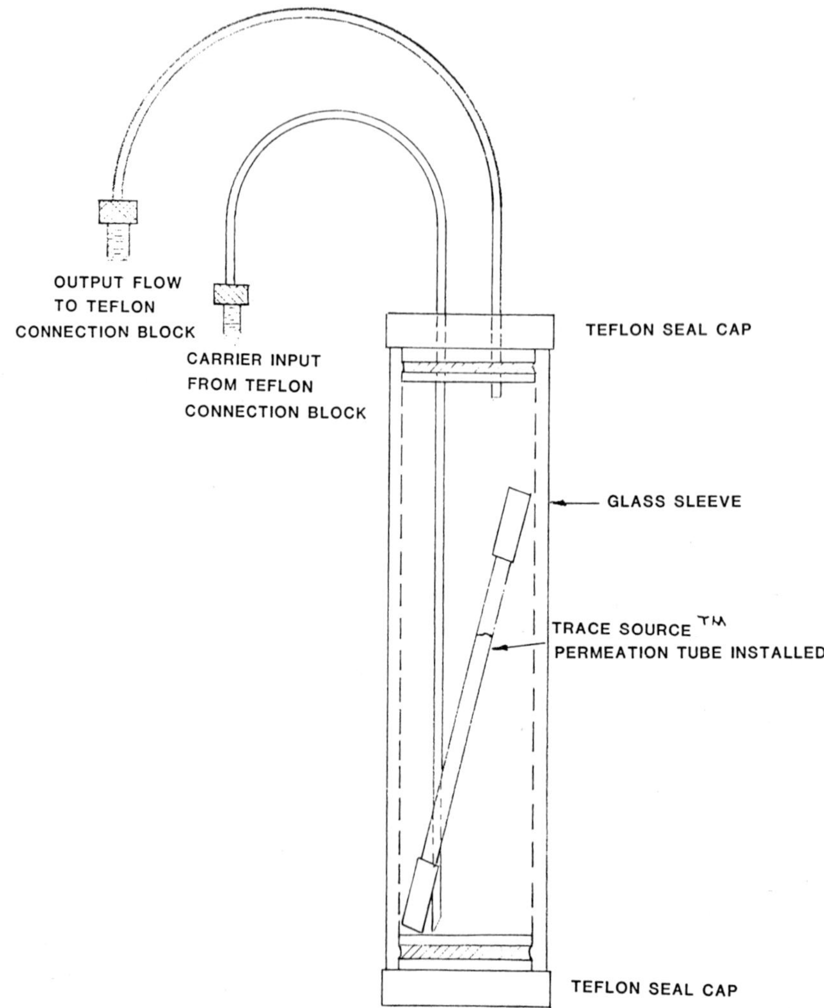

Figure B-4. Permeation tube holder for a portable permeation tube system. (Courtesy Kin-Tek Laboratories.)

or other holder and put inside of the calibrator (Figs. B-4 and B-5). The instrument is hooked up and turned on and calibration procedes. Effective use of permeation tubes requires precise control of the tube operating temperature and the flow of dilution air. In the case of an ovenless calibrator, the temperature inside is kept constant through insulation. Heated calibrators have stable ovens.

In use, a time period of several hours is required for a tube to come to thermal equilibrium, thus producing a constant permeation rate. Generally, at least 1 hour is required for warm-up and stabilization of oven units. A precision rotameter is used to measure the span gas output. Airflow to dilute the concentration to the desired flow rate is adjustable and often measured with a precision rotameter. Equilibration time for tubes inside of a calibrator ranges from 2 hours to 24 hours. Generally 2 hours are sufficient for field grade work while for high accuracy longer times are desirable.[10] Gas flow is passed across the tube. Generally the flow rate is kept fairly low to preserve the thermal stability of the permeation tube and thus its permeation rate. The low gas flow transports all of the material that has permeated

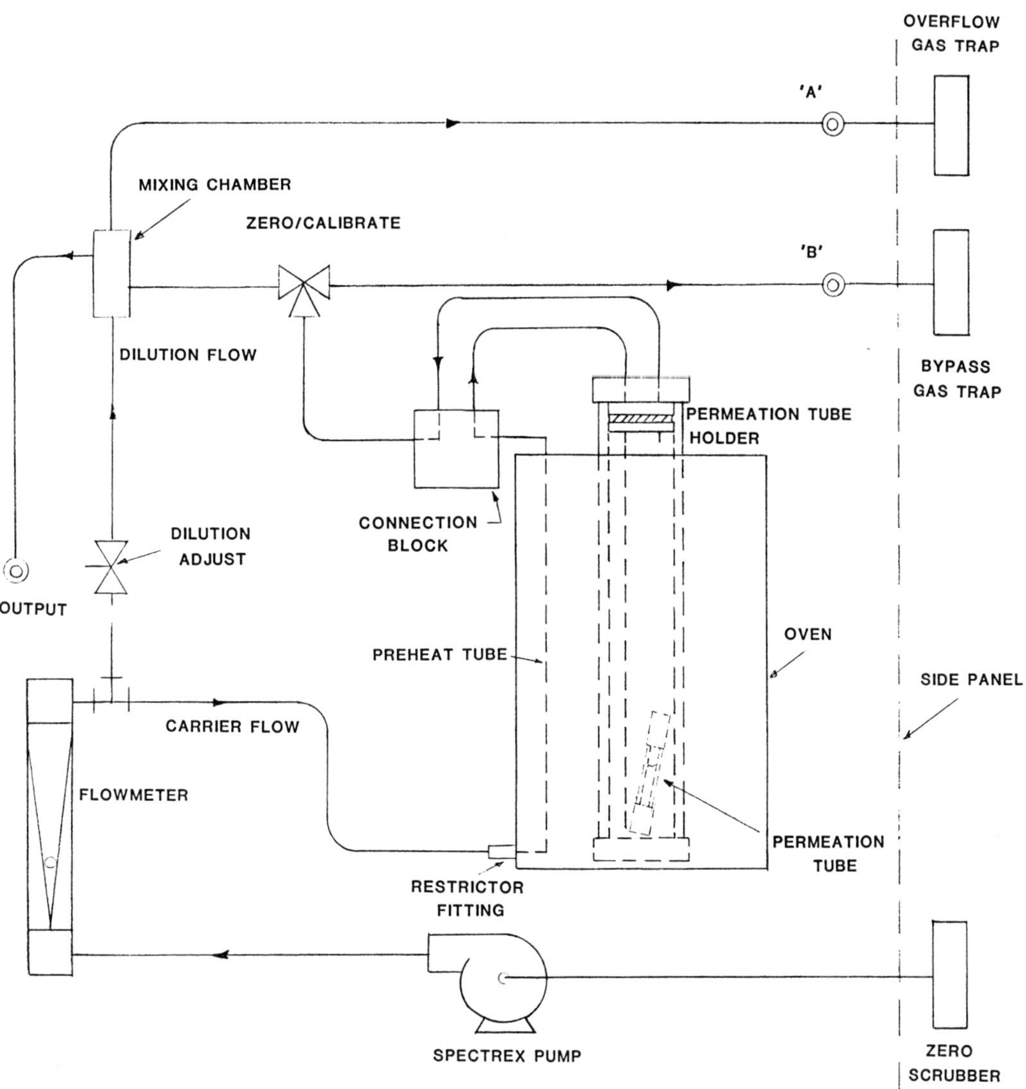

Figure B-5. Flow diagram of a calibrator for a permeation-tube setup. (Courtesy Kin-Tek Laboratories.)

from the tube and is diluted by a high airflow commonly called the diluent airflow to the desired concentration.

The Model G-CAL 2301 Calibrator is an example of an ovenless device that uses a built-in pump to draw air (carrier gas) into the unit, which then flows through a cup containing scrubbing material. The scrubber air travels through the flowmeter (controlled with a rotary metering valve) to obtain the desired flow rate. The air then passes through the "T" fitting (to which the G-CAL device has been attached) and mixes with the gas permeating from the G-CAL device. The resultant gas/air mixture is then delivered to and out of the span gas fitting on the front panel. Tubing can be attached to this fitting to deliver the span gas to gas detection or analysis equipment or to a glove box, or wherever the calibration mixture is needed. Some calibrators are battery powered; others use AC.

Tubes can be stored and reused without affecting their permeation rate. Some tubes can be stored up to a year. Storing tubes at a lower temperature than they are designed for use at will prolong their life. Generally, for tubes designed to operate at temperatures above 60°C, no special precautions are required. For tubes designed to operate at low temperatures, such as 30°C, the total tube life can be extended significantly by storing the tube at reduced temperatures. Tubes can also be stored inside a laboratory hood. The proper procedure for storing a permeation tube depends on the type of material in the tube. It is suggested that the tube be stored in a sealed container together with packets of activated charcoal to prevent contamination to the outer surface of the tube and to prevent the tube from contaminating the room by permeation. Permeation tubes continue to emit their component during storage, even at reduced temperatures. Refrigeration is appropriate; however, storage under conditions that will freeze the liquid in the tubes is not recommended.

For calculations of concentrations generated by permeation tubes and flow rates to operate calibrators, the following equation is used:

$$Co = \frac{(K)(E)(L)}{(F)} = \frac{(C_1)(F_1)}{F_1 + F_2}$$

where

Co = concentration, ppm
E = permeation tube emission rate, hg/(min − cm) = C_1F_1
L = permeation tube length, cm
K = factor for a particular gas to convert from hg/ml to ppm at 25°C
F = total dilution airflow rate, ml/min $= F_1 + F_2$
F_1 = carrier gas flow rate
F_2 = dilution gas flow rate

Some systems use only dilution gas and then the equation simplifies to *F. K* is usually provided by the manufacturer. In order to make additional concentrations, for each *Co*, a different flow rate for the dilution gas, F_2, is selected. Then a *Co* is selected; this value is plugged into the equation and a value for F_2 is selected.

Advantages of permeation tubes include no loss of material due to adsorption into the walls of the container such as happens with static calibration systems. By varying the airflow of the calibrator, multiple standards can be generated for linearity checks. Since toxic compounds are kept encapsulated inside the permeation tube, the risk of exposure to the operator is reduced.

Precautionary measures include not allowing material from a calibrator to vent continuously into a closed or poorly ventilated room. Permeation tubes should be opened outdoors or under a laboratory hood. Avoid overheating permeation tubes, since some tubes have a high vapor pressure inside and may rupture if overheated. Some less stable compounds may decompose, or polymerize violently, if overheated. Do not leave tubes inside a vehicle cab because this area is poorly ventilated. Care should be taken when removing permeation tubes from calibrators because they can be very hot.

REFERENCES

1. 40 CFR 53 EPA. *Ambient Air Monitoring Reference and Equivalent Methods.*
2. Nelson, G. O. *Controlled Test Atmospheres—Principles and Techniques.* Ann Arbor Science Pubs., Inc., Ann Arbor, MI, 1971.
3. Becker, J. H., et al. Instrument calibration with toxic and hazardous materials. *Ind. Hyg. News,* July 1983.
4. Shaw, M. *Everything You Wanted to Know about Toxic Gas Monitors and Were Not Afraid to Ask.* Interscan Corp., Chatsworth, CA.
5. *Photovac 10S50 Gas Chromatograph Operating Manual.* Photovac, Inc., Thornhill, Ontario, Canada.
6. EPA. *Compendium Method TO-14: The Determination of VOCs in Ambient Air Using Summa Passivated Canister Sampling and Gas Chromatographic Analyses.* Washington, D.C.
7. EPA. *Portable Instruments User's Manual for Monitoring VOC Sources* (EPA 340/1-86-015). Washington, D.C., 1986.
8. *Technical Data Sheet #19.* Photovac, Inc., Thornhill, Ontario, Canada.
9. *Permeation and Diffusion Devices.* Thermedics, Inc., Woburn, MA.
10. Span Chek 8700, Kin-Tek Laboratories, Texas City, TX.

Index